MANUAL OF
CULTIVATED
BROAD-LEAVED
TREES & SHRUBS

Volume II, E–PRO

Written by GERD KRÜSSMANN
Translated by MICHAEL E. EPP
Technical Editor: GILBERT S. DANIELS

B T Batsford Ltd
London

Translated and revised from the German:
Krüssmann, Handbuch der Laubgehölze
® 1977 by Verlag Paul Parey
Berlin and Hamburg
This edition © 1986 Timber Press

ISBN 0 7134 5348 6

Printed in Hong Kong
for the Publisher
B T Batsford Ltd
4 Fitzhardinge Street
London W1H 0AH

Contents of Volume II

GENERAL

MAIN TEXT

Alphabetical Reference to the Botanical Terminology

To facilitate the use of this book by readers of all language groups, the most frequently used botanical terms have been arranged in alphabetical order. The word is then followed by the scientific (generally Latin, but occasionally Greek) word and then by the equivalent terms in German, French and Dutch. Furthermore, in all cases where a plant part from the 'Guide to the Terminology' is illustrated, a reference number may be found in parenthesis after the word in the English column referring to the illustration number in volume I.

English (fig. no.)	Latin	German	French	Dutch
abscising	caducus	hinfällig	caduc	afvallend
achene	achenium	Nüsschen	achaine, akène	nootje
acrodrome 23d	acrodromus	spitzläufig (Nerven)	acrodrome	acrodroom
acuminate 20i	acuminatus	lang zugespitzt	terminé en queue	lang toegespitst
acute 20a, 21g	acutus	spitz	aigu	spits
adult leaves	folia adulta	Altersblätter	feuilles adultes	ouderdomsbladeren
albumen	albumen	Eiweiss	albumen	eiwit
alternate	alternans	wechselständig	alterne	afwisselend
alternate (of leaves) 10a	alternifolius	—(von Blättern)	à feuilles alternes	afwisselend
angiosperms	angiospermae	Bedecktsamer	angiospermes	bedektzadigen
angle of nerves		Nervenwinkel	aisselle des nerves	nerfaksel
anther	anthera	Staubbeutel	anthère	helmknop
apex, top	apex	Blattspitze	pointe	top
apiculate	apiculatus	fein zugespitzt	apiculé	puntig
applanate	applanatus	abgeflacht	aplati	afgeplat
appressed	adpressus	angedrückt	appressé	aangedrukt
arrow-headed	sagittatus	pfeilförmig	sagitté	pijlvormig
ascending 6f	ascendens	aufsteigend	ascendent	opstijgend
auriculate (eared) 21e	auriculatus	geöhrt	muni d'oreillettes	geoord
awn	arista	Granne	arête	baard (van gras)
awned 19i	aristatus	grannig	aristé	met baard (van gras)
axillary	axillaris	achselständig, seitenständig	axillaire	okselstandig
bark	cortex	Rinde, Borke	écorce	bast
bearded	barbatus	Achselbart	barbé	okselbaard
berry 34	bacca	Beere	baie	bes
biennial	biennis	zweijährig	bisannuel	tweejarig
bilabiate 30i	bilambatus	zweilippig	bilabié	tweelippig
blade 9b	lamina	Blattspreite	limbe	bladschijf
blunt 20b, 21f	obtusus	stumpf	obtusé	stomp
boat-shaped	navicularis	kahnförmig	naviculaire	bootvormig
brachidodrome 23c	brachidodromus	schlingenläufig (Nerven)	brachidodrome	brachidodroom
bract	bractea	Hochblatt	bractée	schutblad
branch	ramus	Ast	branche	tak
branchlet	ramellus	Zweiglein	petite branche	twijgje
bristle	seta	Borste	soie	borstel
bristle-pointed	setosus	borstig	seteux	borstelig
bud	gemma	Knospe	bourgeon	knop
bullate	bullatus	aufgetrieben, blasig	boursoufflé, bullé	opgeblazen
calyx	calyx	Kelch	calice	kelk
campanulate (bell-shaped) 29d	campanulatus	glockig	campanulé	klokvormig
camptodrome 22c	camptodromus	bogenläufig (Nerven)	camptodrome	camptodroom
capitula (head) 32e	capitulum	Köpfchen	capitule	hoofdje
capsule 34	capsula	Kapsel	capsule	doos
carpel 25f	carpellum	Fruchtblatt	carpelle	vruchtblad
catkin 32b	amentum	Kätzchen	chaton	katje
channeled	canaliculatus	rinnenförmig	canaliculé	gootvormig
ciliate 19b	ciliatus	gewimpert	cilié	gewimperd

English (fig. no.)	Latin	German	French	Dutch
cirrhous	cirrhus	Wickelranke	vrille	rank
clavate 29f	clavatus	keulenförmig	claviforme	knodsvormig
claw	unguis	Nagel	onglet	nagel
clawed	unguiculatus	genagelt	onguicule	genageld
climbing	scandens	kletternd, klimmend	grimpant	klimmend
clustered 10f	fasciculatus	büschelig	fasciculaire	gebundeld
coarse	grossus	grob	grossier	grof
compound	compositus	zusammengesetzt	composé	samengesteld
compressed	compressus	zusammengedrängt	comprimé	samengedrukt
concave	concavus	vertieft	concave	uitgehold
conduplicate 8h	conduplicatus	zusammengefaltet	condupliqué	samengevouwen
connate 11f	connatus	verwachsen	conné	vergroeid
connective	connectivum	Mittelband	connectif	konnektief
convex	convexus	gewölbt	convexe	gewelfd
convolute 8f	convolutus	übergerollt	convoluté	opgerold
cordate 21b	cordatus, cordiformis	herzförmig	cordé, cordiforme	hartvormig
corky 5i	suberosus	korkig	subéreux, liégeux	kurkachtig
corolla	corolla	Blumenkrone	corolle	bloemkroon
corymb 32d	corymbus	Doldentraube	corymbe	schermtros
cotyledon	cotyledon	Keimblatt	cotyledon	kiemblad
craspedrome 23b	craspedromus	randläufig (Nerven)	craspedrome	craspedroom
creeping 6a	repens, reptans	kriechend	rampant	kruipend
crenate 19h	crenatus	gekerbt	crénelé	gekarteld
crescent-shaped 16h	lunatus	mondförmig	luniforme	halvemaanvormig
crispate 19k	crispus	gekraust	crispé, ondulé	gekroesd
cross-section	sectio transversa	Querschnitt	section transversale	dwarse doorsnede
cross-wise 10d	decussatus	kreuzständig	décussé	kruisgewijs
crown	corona, cacumen	Krone	couronne	kroon
cuneate (wedge-shaped) 21d	cuneatus	keilförmig	en forme de coin	wigvormig
cupulate (cup-shaped) 29b	cupulaeformis	becherförmig	cupuliforme	bekervormig
cupule	cupula	Fruchtbecher	cupule	vruchtbeker
cuspidate 20h	cuspidatus	feinspitz	cuspidé	fijn toegespitst
cyme 33b	cyma	Scheinquirl, Trugdolde	cime, cyme	bijscherm, tuil
deciduous	deciduus	abfallend	caduc	afvallend
decumbent 6e	decumbens	liegend	décombant	liggend
decurrent 11e	decurrens	herablaufend	decurrent	aflopend
decussate (deeply cut) 19e	incisus	eingeschnitten	incisé	ingesneden
deflexed	declinatus	niedergebogen	déliné	neergebogen
delicate	gracilis	zierlich	gracieux	sierlijk
dentate 19g	dentatus	gezähnt	denté	getand
digitate	digitatus	gefingert	digité	vingervormig
dioecious	dioecus, dioicus	zweihäusig	dioique	tweehuizig
distichous (2 ranked) 10e	distichus	zweizeilig	distique	tweerijig
double	bi-, plenus (of flowers)	doppelt, gefüllt	double	dubbel-, gevuld
downy	pubescens	weichhaarig	pubescent	zachtharig
drooping	cernuus	übergebogen	penché	overgebogen
drupe	drupa	Steinfrucht	drupe	steenvrucht
dull	opacus	matt	mat	mate, dof
elliptic 13	ovalis	oval	elliptique	ovaal, elliptisch
elliptical 13	ellipticus	elliptisch	elliptique	elliptisch
emarginate 20c	emarginatus	ausgerandet	émarginé	uitgerand
enclosed/covered 7e	vestitus	bedeckt	couvert	bedekt
entire 19a	integer, -rimus	ganzrandig	entier	gaafrandig
epigynous 28c	epigynus	unterständig	épigyne	onderstandig

English (fig. no.)	Latin	German	French	Dutch
equitant 8e	amplex	umfassend	amplexe	omvattend
evergreen	sempervirens	immergrun	toujours vert	groenblijvend
exserted	exsertus	vorragend	saillant	eruit stekend
falcate 16f	falcatus	sichelförmig	en forme de faux	sikkelvormig
fascicle	fasciculus	Büschel	fascicule	bundel
fastigiate	fastigiatus	fastigiat	fastigié	fastigiaat
female flower (pistillate)	flos femineus	Stempelblüte	fleur féminin	stamperbloem
filament	filamentum	Faden	filet	helmdraad
fissured	fissura	rissig	fissurer	spleet, scheur
flagellate (whip-formed)	flagellaris	peitschenartig	flagellaire	zweepvormig
flat, plain	planus	eben, flach	plat	vlak, glad, ondiep
flexuose	flexuosus	hin und hergebogen	flexueux	zigzag gebogen
floccose	floccosus	flockig-filzig	floconneux	vlokkig, viltig
follicle	folliculus	Balgfrucht	follicule	
fragrant	odoratus	duftend	odorant	geurend
fringed	fimbriatus	gefranst	fimbrié	met franjes
fruit	fructus	Frucht	fruit	vrucht
funnelform 29c	infundibuliformis	trichterförmig	infundibulé	trechtervormig
furrowed 5f	sulcatus	gefurcht	sillóné	gegroefd
gaping	ringens	rachenblütig	fleur en gueule	mondvormig
geniculate	geniculatus	geknickt	genouillé	geknikt
genus	genus	Gattung	genre	geslacht
glabrous	glaber	kahl	glabre	kaal
gland	glans	Drüse	gland	klier
glandulary-hairy 36g	glanduloso-pubescens	drüsenhaarig	glanduleux-pubescent	klierachtig behaard
glaucous	glaucus	bereift	pruineaux	berijpt
globose	globosus	kugelig	globeux	bolvormig
glossy	lucidus, nitidus	glänzend	brillant, luisant	glanzend
gymnosperms	gymnospermae	Nacktsamer	gymnospermes	naaktzadigen
habit	habitus	Habitus, Gestalt	forme	habitus
hair-covering	indumentum	Behaarung	pubescence	beharing
hairy	pilosus	behaart	pileux	behaard
hastate 21h	hastatus	spiessförmig	hasté	spiesformig
herbaceous	herbaceus	krautartig, krautig	herbacé	kruidachtig
hermaphrodite	hermaphroditus	zwittrig	hermaphrodite	hermafrodiet
hilum	hilum	Nabel	hile	navel
hirsute	hirsutus	rauhhaarig	hirsute	ruwharig
hispid	hispidus	steifhaarig	hispide	stijfharig
husk	siliqua	Schote	silique	hauw
hypanthium	hypanthium	Blütenboden	hypanthium	bloembodem
hypogynous 28a	hypogynus	oberständig	hypogyne	bovenstandig
imbricate 8a	imbricatus	dachziegelig	imbriqué	dakpansgewijze
incised	incisus	eingeschnitten	incisé	ingesneden
inflorescence	inflorescentia	Blütenstand	inflorescence	bloeiwijze
involucre	involucrum	Hüllkelch	involucre	omwindsel
involute (rolled inward) 8b	involutus	eingerollt	enroulé	ingerold
irregular	irregularis	unregelmässig	irregulier	onregelmatig
jointed	articulatus	gegliedert	articulé	geleed
juvenile leaves	folia juvenilia	Jugendblätter	feuilles juveniles	jeugdbladeren
keeled	carinatus	gekielt	caréné	gekield
kernel, stone	nucleus	Kern	noyeau	kern
kidney-shaped 16g	reniformis	nierenförmig	en forme de rein	niervormig
lanceolate 14o	lanceatus, lanceolatus	lanzenförmig, lanzettlich	lancéolé	lancetvormig
large, broad	latus	breit	large	breed
latex	latex	Milchsaft	laiteux	melksap

English (fig. no.)	Latin	German	French	Dutch
leaf 9a	folium	Blatt	feuille	blad
leaf base	basis	Blattgrund	base de feuille	bladvoet
leaf cushion	pulvinus	Blattkissen, Blattpolster	coussinet foliaire	bladkussen
leaf margin	margo	Blattrand	contour de feuille	bladrand
leaf scar	cicatricula	Blattnarbe	cicatrice foliaire	bladmerk
leaflet	foliolum	Blättchen	foliole	blaadje
leathery	coriaceus	lederartig	coriace	leerachtig
leprous 36e	lepidisotus	schülferschuppig	lepidote	schubbig
ligulate	ligula	Zungenblüte	ligule	tongetje
limb	limbus	Saum	limbe	zoom, rand
linear 12a	linearis	linealisch	linéaire	lijnvormig
lobe, loculicidal	loba	Lappen	lobe	lob
lobed 18f	lobatus	gelappt	lobé	gelobd
locule (chamber of ovary)	loculum	Fach	loge	hok
long shoot		Langtrieb	rameau longue	langlot
male flower (staminate)	flos masculus	Staubblüte	fleur mêle	meeldraadbloem
mane-like	jubatus	mähnenartig	criniforme	manenvormig
mealy	farinosus	mehlig	farineux	melig
monoecious	monoecus	einhäusig	monoique	eenhuizig
mucronate 20g	mucronatus	stachelspitz	mucroné	gepunt
mucronulate 20f	mucronulatus	stachelspitzig	mucronulé	fijn gepunt
naked 7d	nudis	nackt	nu	naakt
narrow	angustus	schmal	étroit	smal
needle-form	acerosus	nadelförmig	acéré	naaldvormig
nodding 4d	nutans	überhängend	incliné	overhangend
not shining	opacus	glanzlos	opaque	mat, dof
nut 34	nux	Nuss	noix	noot
obcordate 16c	obcordatus	obcordat, verkehrt herzförmig	obcordate	omgekeerd hartvormig
oblanceolate 15u	oblanceolatus	oblanzettlich	oblancéolé	omgekeerd lancetvormig
oblate	oblatus	oblat	oblate	oblaat
oblique 21k	obliquus	schief	oblique	scheef
oblong 12	oblongus	länglich	oblong	langwerpig
obovate 15	obovatus	obovat, verkehrt eiförmig	obovate	omgekeerd eirond, eivormig
obvolute 8d	obvolutus	halbumfassend	obvoluté	halfomvattend
opposite (of foliage) 10b	oppositus, oppositifolius	gegenständig (bei Blättern)	opposé	tegenoverstaand
orbiculate 13m	orbicularis	kreisrund	corbiculaire	cirkelrond
outspread	patulus	abstehend	étendre	afstaan
ovary	ovarium	Fruchtknoten	ovaire	vruchtbeginsel
ovate 14	ovatus, ovulum	eiförmig, eirund	ovate	eirond, eivormig
ovule	ovulum	samenanlage	ovule	eitje
palmate 17d	palmatus	handförmig, handteilig	palmé	handdelig, handvormig
panicle 32h	panicula	Rispe	panicule	pluim
papilionaceus (butterfly-like)	papilionaceus	schmetterlingsförmig	papilionacé	vlinderbloemig
papilla	papilla	warze	papille	wratachtig
pappus	pappus	Haarkelch	aigrette	zaadpluis
pectinate	pectinatus	kammförmig	pectiné	kamvormig
pedate	pedatus	fussförmig	pedatiforme	voetvormig
pedicel	pediculus	stielchen	pedicelle	steeltje
peduncle	pedunculus	Blütenstiel	pedoncule	bloemsteel
pedunculate	pedunculatus	gestielt (Blüte)	pédonculé	gesteeld
peltate 16a	peltatus	schildförmig	pelté	schildvormig

English (fig. no.)	Latin	German	French	Dutch
pendulous (weeping) 4e	pendulus	hängend	pendant	hangend
penniform 17e	pinnatiformis	fiederförmig	penniforme	veervormig
perfoliate 11d	perfoliatus	durchwachsen	perfolié	doorgroeid
perianth	perianthemum	Blütenhülle, Perigon	périanthe	bloembekleedsel
perigynous 28b	perigynus	mittelständig	périgyne	halfonderstandig
petal 25c	petalum	Blütenblatt	pétale	kroonblad
petiole	petiolus	Blattstiel	pétiole	bladsteel
petioled 11a	petiolatus	gestielt (Blatt)	petiolé	gesteeld
phylloclades	phyllocladium			
phyllode	phyllodium	Phyllodium	foliace	phyllodium
phyllotaxis	Phyllotaxis	Blattstellung	phyllotaxis	bladstand
pinnately cleft 18b	pinnatifidus	fiederspaltig	pennatifide	veerspletig
pinnately partite 18c	pinnatipartitus	fiederteilig	pennatipartite	veerdelig
pinnatisect 18d	pinnatisectus	fiederschnittig	pinnatiséqué	veervormig ingesneden
pistil	pistillum	Stempel	pistil	stamper
pistillate (see: female flower)				
pith	medulla	Mark	moelle	merg
plaited 8g	plicatus	gefaltet	plié	gevouwen
pod	legumen	Hülse	gousse	peul
poisonous	venenatus	giftig	vénéneux	vergiftig
pollen	pollen	Blütenstaub	pollen	stuifmeel
polygamous	polygamus	vielehig	polygame	polygaam
prickle 37a	acus	Stachel	aiguillon	stekel
prickly	aculeatus	stachelig	muni d'aiguillons	stekelig
procumbent	procumbens	niederliegend	tracant	neerliggend
prostrate	prostratus	niedergestreckt	couché	neerliggend
pruinose	pruinosus	bereift	pruineux	berijpt
pubescent	pubescens	feinhaarig	pubescent	fijn behaard
pulverulent	pulverulentus	bepudert, bestäubt	pulverulent	bepoederd, bestoven
punctate (dotted) 5k	punctatus	punktiert	ponctué	gestippeld
pungent	pungens	stechend	piquant	stekend
quadrangular 5e	quadrangulatus	vierkantig	à quatre angles	vierhoekig
raceme 32c	racemus	traube	grappe	tros
racemose	racemosus	traubig	en grappe	trosvormig
radiate	radiatus	strahlig	radiaire	radiaal
receptacle	receptaculum	Blütenboden	receptacle	bloembodem
reflexed	reflexus	zurückgebogen	re1flechi	teruggeslagen
regular	regularis	regelmässig	régulaire	regelmatig
resinous	resinosus	harzig	résineux	harsachtig
reticulate	reticulatus	netznervig	reticulé	netvormig
retuse 20d	retusus	eingedrückt	émoussé	ingedrukt
revolute 8c	revolutus	zurückgerollt	revoluté	teruggerold
rhizome	rhizoma	Wurzelstock	rhizome	wortelstok
rhombic	rhombicus	rautenförmig (=rhombisch)	rhombique	ruitvormig
rooting	radicans	wurzelnd	radicant	wortelend
rotate (wheel-shaped) 29a	rotatus	radförmig	rotacé	radvormig
rough	asper	rauh	rude	ruw
rounded 21a	rotundatus	abgerundet	arrondi	afgerond
roundish 13l	suborbiculatus	rundlich	arrondi	afgerond
rugose	rugosus	runzelig	rugueux	gerimpeld
runner 6b	sobol, stolon	Ausläufer, Sprössling	drageon, stolon	uitloper, wortelspruit
salver-shaped 29g	hypocraterimorphus	stieltellerförmig	hypocratériforme	schenkbladvormig
samara 35b	samara	Flügelfrucht	samare	gevleugeld nootje
scabrous	scaber	scharf	scabre	ruw

English (fig. no.)	Latin	German	French	Dutch
scaly 10g	squama	Schuppe	squameux	schub
scattered	sparsifolius	zerstreut (Blättern)	espacé	verspreid
schizocarp 35e	schizocarpium	Spaltfrucht	schizocarpe	splitvrucht
scion	vimen, virga	Rute	verger, scion	twijg, roede
seed	semen	Samen	graine, semence	zaad
semidouble	semiplenus	halbgefüllt	demi-double	half gevuld
semi-evergreen (see: wintergreen)				
semi-terete 5b	semiteres	halbrund	demi-cylindrique	halfrond
sepals 25k	sepala	Kelchblätter	sépales	kelkbladen
serrate	serratus	gesägt	serré	gezaagd
serrulate 19d	serrulatus	feingesägt	serrulé	fijn gezaagd
sessile 7a, 10b	sessilis	sitzend	sessile	zittend
shaggy	villosus	zottig behaart	poilu	donzig
shell	epicarpium	Schale	écorce	schil van vrucht opeen gehoopt
shoot	ramulus	Trieb	pousse	scheut
short branch		Kurztrieb	rameau court	kortlot
shrub	frutex	Strauch	arbuste, arbrisseau	struik, heester
silky	sericeus	seidenhaarig	soyeux	zijdeachtig
simple	simplex	einfach	simple	enkelvoudig
sinuate 18e	sinuatus	gebuchtet	sinué	bochtig
slightly drooping 4c	cernuus	nickend	penché	knikkend
smooth	laevis	glatt	lisse	glad
solitary	solitaris	einzelstehend	solitaire	alleenstaand
spathulate 16e	spathuliformis	spatelförmig	spatulé	spatelvormig
species	species (sp)	Art	espèce	soort
spike 32a	spica	Ähre	épi	aar
spindle, rachis	rhachis, rachis	Spindel	fuseau	spil
spiral	spiralis	schraubig	spiralé	spiraalvormig
spur	crus	Sporn	éperon	spoor
stamen 25a	stamen	Staubblatt	1etamine	meeldraad
staminate (see: male flower)				
standard		Hochstamm	haute-tige	hoogstam
stellate 38f	stellatus	sternhaarig	étoilé	sterharig
stem	culmus	Halm (Gramineae)	tige	halm
stem-clasping 11c	amplexicaulis	stengelumfassend	amplexicaule	stengelomvattend
sticky	glutinosus, viscosus	klebrig	poisseux, visqueux	kleverig
stigma 34	stigma	Narbe	stigmate	stempel
stipule 9b	stipula	Nebenblatt	stipule	steunblad
straggly 4f	divaricatus	sparrig	divariqué	uitgespreid
strap-form 12b	loratus	riemenförmig	loriculé	riemvormig
strap-shaped	loratus	bandförmig	loriculé	bandvormig
striated	striatus	gestreift	strié	gestreept
strict 4b	strictus	straff	raide	opgaand
strigose 36	strigosus	striegelhaarig	à poils rudes	scherpharig
subspecies	subspecies (ssp)	Unterart	sous-espèce	ondersoort
subulate (awl-shaped)	subulatus	pfriemförmig	subulé	priemvormig
syncarp	syncarpium	Sammelfrucht	syncarpe	vruchten
tapering	attenuatus	verschmälert	attenue	versmald
terete 5a	teres	stielrund	cylindrique	rolrond
terminal	terminalis	endständig	terminal	eindstandig
terminal bud 7g		Endknospe	bourgeon terminal	eindknop
ternate 17b	ternatus	dreizählig	terné	drietallig
tessellate 22b	tessellatus	würfelnervig	tessellé	schaakbord-vormig
thorn 37b-c	spina	Dorn	épine	doorn
thorny	spinosus	dornig	épineux	gedoornd
throat	faux	Schlund	gorge	keel

English (fig. no.)	Latin	German	French	Dutch
tomentose	tomentosus	filzig	tomenteux	viltig
tooth	dens	Zahn	dent	tand
toothed 19g	dentatus	gezähnt	denté	getand
translucent	pellucidus	durchscheinend	pellucide	doorschijnend
triangular 16b	triangularis, triangulatus	dreieckig, dreikantig	à trois angles, triangulaire	driehoekig, driekantig
trichoma	trichoma	Haare	trichome	beharing
trifoliate	trifoliatus	dreiblättrig	trifoliolé	driebladig
trioecious	trioecus	triözisch	trioique	driehuizig
truncate 21e	truncatus	abgestutzt	tronqué	afgestompt
truncated	truncatus	gestutzt	tronqué	afgestompt
tube 29h	tubus	Röhre	tube	buis
tubercled	tuberculatus	höckerig	tuberculeux	bultig
twig (secondary branches)	ramulus	Zweig	branche	twijg
twining	volubilis	windend	volubile	windend
twisted 4i	tortus, tortuosus	gedreht	tortueux	gedraaid
two-edged 5d	anceps	zweischneidig	à deux faces	tweezijdig
umbel 32f	umbella	Dolde	ombelle	scherm
underside		Unterseite	face de dessous	onderzijde
undulate	undulatus	gewellt	ondulà	gegolfd
unisexual	unisexualis	eingeschlechtig	unisexuel	eenslachtig
upper side		Oberseite, oben	face de dessus	bovenzijde, boven
upright	erectus	aufrecht	dréssé	oprecht
urceolate	urceolatus	krugförmig	urcéiforme	kruikvormig
v-valved	valvatus, -valvis	klappig	-valve	klepvormig
variety	varietas (var.)	Varietät	variété	varieeteit
velvety	holosericeus	samthaarig	velute	fluweelhaarig
viscid	viscidus	schmierig	viscide	kleverig
warty	verrucosus	warzig	verruqueux	wrattig
whorled 10c	verticillatus	quirlig	verticillé	kransstandig
winged 5h	alatus	geflügelt	ailé	gevleugeld
wintergreen	semipersistens	wintergrün	semi-persistant	wintergroen
woolly 38	lanatus, lanuginosus	wollhaarig, wollig	laineux, lanugineux	wollig, 'zachtharig

Explanation of Symbols

To provide the reader with information relating to the use of the plants described in this work and their cultural requirements, the following symbols are used after many of the descriptions.

Light Requirements

○ = needs or tolerates full sun

◐ = needs or tolerates semishade

● = needs or tolerates lasting shade

Soil Requirements

m = moist soil

w = wet soil (most plants do not like standing water but some will tolerate it for varying lengths of time)

d = dry soil

S = sandy soil (light and loose)

H = humus soil (loose, dark, organic loam)

A = alkaline soil (even chalk soils)

Properties of the Plants

✧ = with ornamental flowers

✗ = fruit or other plant part edible or economically useful

∅ = especially attractive foliage or fall color

= evergreen foliage

⚭ = ornamental fruit

Ⓕ = cultivated for timber, countries where grown are indicated

Other Symbols or Abbreviations

N = North, S = South, E = East, W = West, M = Mid, Middle.

Dates refer to the year of introduction into cultivation.

✕ = hybrid, cross.

HCC (with number) = color code from the Horticultural Color Chart.

z followed by number denotes hardiness zone according to USDA Hardiness Map.

Notes on Illustrations

The abbreviation 'Fig.' always refers to illustrations within the text pages of this book. References to illustrations in other books will be found at the end of the description and are indicated by initials and numbers.

For example: DL 2: 21 = Dippel, Handbuch der Laubholzkunde, Vol. 2, Illustration #21 (or if the illustrations are not numbered, the illustration on page 21).

The list of abbreviations to the literature providing illustrations cited in this book can be found on pp. 15–17. A complete bibliography of all books used in compiling this work may be found in the last volume.

Pl. = plate (for those found in other works). Plate 31, 43 etc. refers to those found in this book.

Synonyms

Plant names in italics at the end of a description are invalid synonyms.

The cultivars are noted with single quotes '____'

Abbreviated Temperature Conversion Chart

Fahrenheit to Celcius

°F	°C
40	4.4
35	1.7
30	− 1.1
25	− 3.9
20	− 6.7
15	− 9.4
10	−12.2
5	−15
0	−17.8
− 5	−20.6
−10	−23.2
−15	−26.1
−20	−28.9
−25	−31.7
−30	−34.4
−35	−37.2
−40	−40
−45	−42.8
−50	−45.6

Celsius to Fahrenheit

°C	°F
− 5	41
0	32
− 5	23
−10	14
−15	5
−20	− 4
−25	−13
−30	−22
−35	−31
−40	−40
−45	−49

Readers should note the following points relative to hardiness:

1. The majority of the plants dealt with in this book will be found in zones 4 to 8, and occasionally zone 9.
2. Areas with the same average low temperature are described by the isotherms drawn along 5°C clines.
3. Hardiness ratings of plants is given using the U.S. Dept. of Agriculture system. It is indicated in the text as "z" followed by a number indicating hardiness zone.

Winter hardiness ratings are only guidelines. It must be realized that microclimate plays an important role so a specific location within a zone may be warmer or cooler than the zone as a whole. Protected areas in woods, southern exposures or gardens within cities as well as careful cultivation may allow one to raise plants in a zone which is normally too cold.

Hardiness Zones and the British Isles

Very few of the trees, shrubs and other garden plants cultivated in Britain are native to the British Isles. Over the centuries they have been introduced from all over the world, though especially from cool and warm temperate climates. How well they thrive in the British Isles largely depends on the climate they evolved in.

Although all plants are closely adapted to the climate of the region in which they occur wild, few have rigid requirements of heat and cold. There are other factors that decide whether a plant will thrive, e.g. soil type and amount of rainfall, but these will be mentioned later; temperature is of primary importance.

The British Isles has an equable oceanic climate which is seldom very cold, hot or dry. As a result, a wide range of the world's plants can be grown outside providing they are sited intelligently. Undoubtedly, some of these plants would prefer more summer sun or a more definite cold winter rest, but their innate adaptability is catered for in the vagaries of our climate. There is, however, a point at which a plant's tolerance ceases. Low temperature is the most important of these tolerances. If a plant cannot survive an average winter outside it is said to be tender. If a plant survives average winters but not the exceptionally hard one it is said to be half-hardy. These terms are, of course, relevant only to the area in which one lives.

Large continental land masses, e.g. North America and Central Europe, have climates that get progressively colder winters as one proceeds northwards and further inland from the sea. North America provides a familiar example, the extreme south being almost tropical, the far north arctic. In the 1930s, the United States Department of Agriculture divided the USA into 7 hardiness zones based upon an average of the absolute minimum temperatures over a period of 20 years. Later, the system was revised and refined and 10 zones recognized (zone 1 is arctic, zone 10 tropical). More recently this Hardiness Zone system has been extended to Europe, including the British Isles. Gardeners in the United States and Canada soon took advantage of the hardiness zone concept and over the years, largely by trial and error, most trees and shrubs and many other plants have been assessed and given zone ratings. Nevertheless, this system, though useful, can only be considered to give approximate hardiness ratings, especially when applied to the British Isles.

Sitting as it does on the eastern edge of the North Atlantic Ocean, the British Isles occupies a unique position. Although its total length, about 650 miles (Cornwall to Orkney), lies within latitudes 50° to 60° N, it falls into zone 8! Moved into the same latitudes in North America, it would lie entirely north of the Canadian border with the tip of Cornwall level with Winnipeg (zone 2–3). Even the eastern coastal region of Canada at these latitudes is no warmer than zones 3–4. Because of the influence of the Gulf Stream the British Isles enjoys a remarkably uniform climate. Such temperature gradients as these are run east to west rather than south to north.

It is a characteristic of temperate oceanic climates to have milder winters and cooler summers than equivalent continental ones and because of their northerly position this is even more marked in the British Isles. For this reason, a number of trees and shrubs which thrive in zone 8 in USA fail to do so well in Britain e.g., *Albizia julibrissin, Lagerstroemia indica,* etc. Such plants may live but fail to bloom, or get cut back severely by the British winters. The factor is primarily lack of summer sun rather than absolute cold.

This lack of summer warmth brings us to the several important ancillary factors which affect a plant's hardiness. Apart from lack of damaging low temperatures a plant needs the right kind of soil, adequate rainfall and humidity, plus sufficient light intensity and warmth. As with low temperature most plants have fairly wide tolerances, though there are noteworthy exceptions. Most members of the *Ericaceae,* especially *Rhododendron* and allied genera, must have an acid soil or they will die however perfect the climate. For plants near the limits of their cold tolerance, shelter is essential. Protection from freezing winds is particularly important. This can be provided by planting in the lee of hedges, fences and walls or among trees with a fairly high canopy. Individual plants can also be protected by matting or plastic sheeting or the bases can be earthed up or mounded around with peat, coarse sand or weathered boiler ash. A thick layer of snow also provides insulation against wind and radiation frost! Plenty of sunshine promotes firm, ripened growth with good food reserves, notably a high sugar content in the cell sap which then takes longer to freeze. If the summer is poor a partial remedy is to apply sulphate of potash (at 10g/per square metre) in late summer. This will boost the amount of sugars and starches in the plant. Half-hardy plants will stand having their tissues moderately frozen providing the thawing-out is gradual. For this reason it is best to grow them in a sheltered site which does not get the first rays of the morning sun. This is especially relevant for species with tender young leaves or early flowers, e.g. *Cercidiphyllum, Camellia* and *Magnolia.*

Zone 9 in the USA is warm-temperate to sub-tropical with hot summers. In the British Isles it tends to have even cooler summers than zone 8, and as a result very few truly sub-tropical plants can be grown in Britain. Most of the plants in the famous so-called, sub-tropical gardens, e.g. Tresco, Logan, Inverewe, etc., are of warm-temperate origin. For the reasons set down above, in Britain, if in doubt, it is best to consider zone 8 as zone 7 and zone 9 as zone 8 for plants of unreliable hardiness.

by Kenneth Beckett

Hardiness Zones of Europe

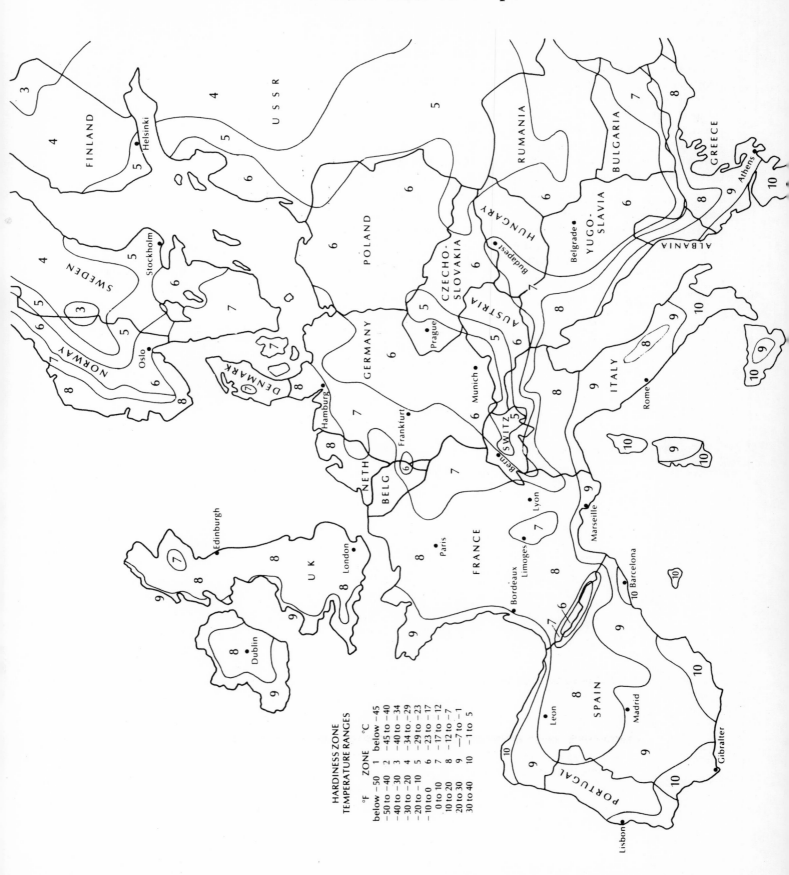

HARDINESS ZONE
TEMPERATURE RANGES

°F	ZONE	°C
below −50	1	below −45
−50 to −40	2	−45 to −40
−40 to −30	3	−40 to −34
−30 to −20	4	−34 to −29
−20 to −10	5	−29 to −23
−10 to 0	6	−23 to −17
0 to 10	7	−17 to −12
10 to 20	8	−12 to −7
20 to 30	9	−7 to −1
30 to 40	10	−1 to 5

Hardinesss Zones of North America

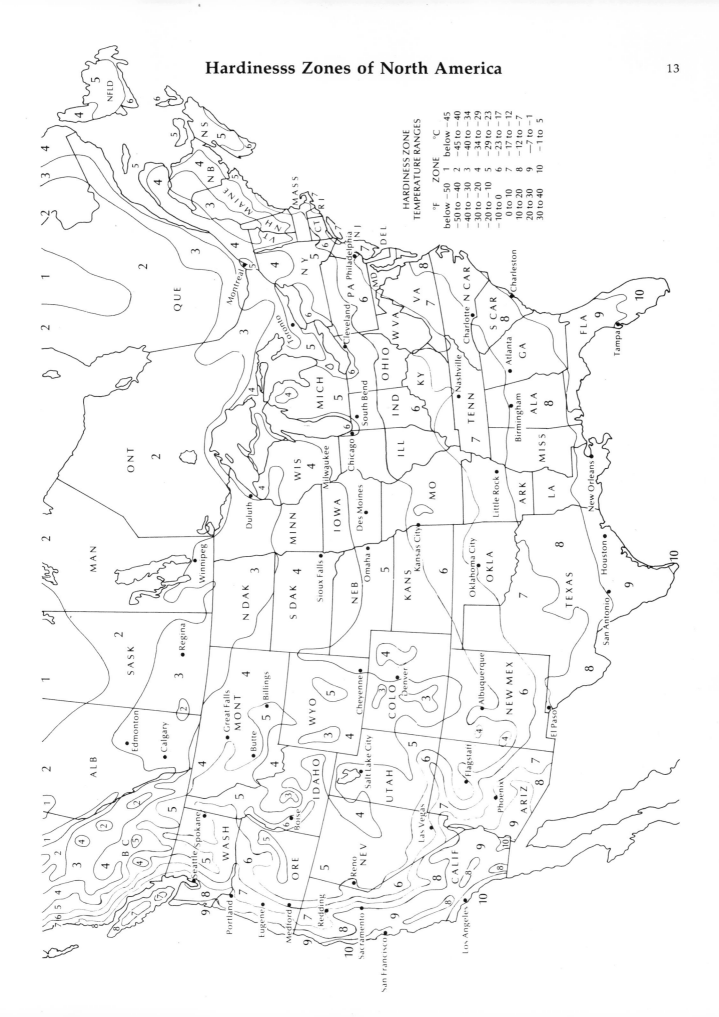

HARDINESS ZONE
TEMPERATURE RANGES

°F	ZONE	°C
below −50	1	below −45
−50 to −40	2	−45 to −40
−40 to −30	3	−40 to −34
−30 to −20	4	−34 to −29
−20 to −10	5	−29 to −23
−10 to 0	6	−23 to −17
0 to 10	7	−17 to −12
10 to 20	8	−12 to −7
20 to 30	9	−7 to −1
30 to 40	10	−1 to 5

Hardiness Zones of China

SOVIET UNION

Mongolia

Heilongjiang

Jilin

Xinjiang

Liaoning

Inner Mongolia

KOREA

Gansu

Hebei

−4° C

PEOPLE'S REPUBLIC OF CHINA

Ningxia

Shanxi

Shandong

0° C

Qinghai

Shaanxi

Henan

Jiangsu

JAPAN

Tibet

Anhui

4° C

Hubei

NEPAL

SIKKIM

Sichuan

Zhejiang

8° C

BHUTAN

Hunan

Jiangxi

INDIA

ASSAM

Fujian

12° C

BANGLADESH

Guizhou

Yunnan

Guangxi

TAIWAN

Tropic of Cancer

Guangdong

BURMA

VIETNAM

LAOS

HAINAN

THAILAND

PHILIPPINES

CAMBODIA

HARDINESS ZONE
TEMPERATURE RANGES

°F	ZONE	°C
below −50	1	below −45
−50 to −40	2	−45 to −40
−40 to −30	3	−40 to −34
−30 to −20	4	−34 to −29
−20 to −10	5	−29 to −23
−10 to 0	6	−23 to −17
0 to 10	7	−17 to −12
10 to 20	8	−12 to −7
20 to 30	9	−7 to −1
30 to 40	10	−1 to 5

N. BORNEO

INDONESIA

MALAYSIA

MALAYSIA

List of Abbreviations to Other Reference Works

The following abbreviations are found at the end of most of the plant descriptions; they refer exclusively to illustrations in the given works. Many of these abbreviations are the same as those found in American and English literature. Included in this list is the book's title and date of publication; more detailed bibliographical information may be found at the end of this work.

AAu	Audas: Native Trees of Australia. 1947
AB	Arnoldia; Bulletin of the Arnold Arboretum. 1911 →
ANZ	Allan, H. H.: Flora of New Zealand, Vols. 1–2. 1961–1970.
Bai	Baileya (periodical). 1953
BB	Britton & Brown: Illustrated Flora of the Northern USA and Canada, 3 Vols. 1896–1898
BC	Bailey, Standard Cyclopedia of Horticulture; 1950
BCi	Bolanos: Cistographia Hispanica. 1949
BD	Bulletins de la Société Dendrologique de France. 1906–1939
BFC	Clapham-Tutin-Warburg: Flora of the British Isles Illustrations, 4 Vols. 1957–1965
BFl	Barrett: Common Exotic Trees of South Florida. 1956
BIC	Bor & Raizada: Some beautiful Indian Climbers and Shrubs. 1954
BIT	Blatter & Millard: Some beautiful Indian Trees. 1954
BM	Curtis' Botanical Magazine, 1787–1947 ns (new series) 1948 →
BR	Botanical Register. 1815–1847
BRho	Bowers: Rhododendron and Azaleas. 1960
BS	Bean: Trees and Shrubs hardy in the British Isles, 4 Vols. 1970 →
CBa	Camus: Les Bambusées. 1913
CCh	Camus: Les Châtaigniers
CFQ	Carvalho & Franco: Carvalhos de Portugal. 1954
CFTa	Curtis: The endemic Flora of Tasmania, 5 Vols.
ChHe	Chapple: Heather Garden. 1961
CIS	Hu-Chun: Icones Plantarum Sinicarum, Vols. 1–5
CMi	Clements: Minnesota Trees and Shrubs. 1912
CQ	Camus: Monographie du genre *Quercus*, 6 Vols. 1936–1954
CS	Camus: Monographie des Saules de France. 1904–1905
CTa	Curtis: Student's Flora of Tasmania. 1956 to 1967
CWF	Clark: Wild Flowers of British Columbia. 1973
DB	Deutsche Baumschule. 1949 →
Dfl	Dendroflora (yearbook). 1964 →
DH	Dallimore: Holly, Box and Yew. 1908
DL	Dippel: Handbuch der Laubholzkunde, 3 Vols. 1889–1893
DRHS	Dictionary of Gardening, 4 Vols. 1951, Supplement 1969
EH	Elwes & Henry: Trees of Great Britain and Ireland, 7 Vols. 1906–1913
EKW	Encke: Die schönsten Kalt– und Warmhauspflanzen
ENP	Engler & Prantl: Die Natürlichen Pflanzenfamilien, 2nd Ed. 1924 →
EP	Engler: Das Pflanzenreich
EWA	Erickson: Flowers and Plants of Western Australia
FAu	Forest Trees of Australia. 1957
FIO	Fang: Icones Plantarum Omeiensium. 1942–1946
FRu	Focke: Species Ruborum 1–23, 1911 to 1914
FS	Flore des Serres et des Jardins de L'Europe, 23 Vols. 1845–1880
FSA	Coed/de Winter/Rycroft: Flora of South Africa
GC	Gardeners Chronicle. 1841 →
GF	Garden and Forest. 1888–1897
Gfl	Gartenflora. 1852–1938
GH	Gentes Herbarium. 1935 →
Gn	The Garden (periodical)
GPN	Gram-Jessen: Vilde Planter i Norden, 4 Vols.
Gs	Gartenschönheit. 1920–1942
GSP	Grimm: Shrubs of Pennsylvania. 1952
GTP	Grimm: Trees of Pennsylvania. 1950
Gw	Gartenwelt. 1897 →
HAl	Hara: Photo-Album of Plants of Eastern Himalaya. 1968
HBa	Houzeau de Lehaie: Le Bambou. 1906 to 1908
HF	Schlechtendahl-Langethal-Schenck: Flora von Deutschland, 5th Edition, Hrsg. von Halier. 1880–1887
HH	Hough: Handbook of the Northern States and Canada. 1950
HHD	Harlow-Harrar: Textbook of Dendrology. 1941
HHS	Harrar-Harrar: Guide to the Southern Trees. 1962
HHo	Hume: Hollies. 1953
HHy	Haworth-Booth: Hydrangeas. 1951
HI	Hooker: Icones Plantarum. 1836 →
HIv	Hibbert: The Ivy. 1872
HKS	Hong Kong Shrubs. 1971
HKT	Hong Kong Trees. 1969
HL	Hendriks: Onze Loofhoutgewassen, 2nd Edition 1957
HM	Hegi: Flora von Mitteleuropa, 13 Vols. 1908–1931
HRh	Hooker: Rhododendrons of Sikkim-Himalaya. 1849–1851
HS	Hao: Synopsis of the Chinese *Salix*. 1936
HSo	Hedlund: Monographie der Gattung *Sorbus*. 1901
HTS	Harrison: Know your Trees and Shrubs. 1965
HW	Hempel & Wilhelm: Baume und Straucher des Waldes, 3 Vols. 1889–1899
IC	Ingram: Ornamental Cherries, 1948
ICS	Icones Cormophytorum Sinicorum, 4 Vols. 1972–1975
IH	L'Illustration Horticole. 1854–1896

JA	Journal of the Arnold Arboretum. 1919 →
JAm	Jones: The American species of *Amelanchier*. 1946
JMa	Johnstone: Asiatic Magnolias in Cultivation. 1955
JRHS	Journal of the Royal Horticulture Society. 1846 →
JRi	Janczewski: Monographie des Grosseilliers. 1907
JRL	Jahrbücher Rhododendron und immergrune Laubgehölz. 1937–1942; 1952 →
KD	Koehne: Deutsche Dendrologie. 1893
KEu	Kelly: Eucalypts. 1969
KF	Kirk: Forest Flora of New Zealand. 1889
KGC	Kunkel: Flora de Gran Canaria. 1974 →
KIF	Kurata: Illustrated important Forest Trees of Japan, 4 Vols. 1971–1973
KO	Kitamura & Okamoto; Coloured illustrations of Trees and Shrubs of Japan
KRA	Koidzume: Revision Aceracearum Japonicarum. 1911
KSo	Karpati: Die *Sorbus*-Arten Ungarns. 1960
KSR	Keller: Synopsis Rosarum Spontanearum Europae Mediae. 1931
KTF	Kurz & Godfrey: The Trees of Northern Florida. 1962
LAu	Lord: Shrubs and Trees for Australian Gardens. 1948
LCl	Lavallée: Les Clématites a grandes fleurs. 1884
LF	Lee: Forest Botany of China. 1935
LLC	Lloyd, C.: *Clematis*. 1965
LNH	Labillardiére: Novae Hollandiae Plantarum Specimen. 1804
LT	Liu: Illustrations of native and introduced plants of Taiwan, 2 Vols. 1960
Lu	Lustgarden (periodical). 1920 →
LWT	Li: Woody Flora of Taiwan. 1963
MB	Mitford: The Bamboo Garden. 1896
MCea	McMinn: A systematic study of the genus *Ceanothus*. 1942
MCl	Markham: *Clematis*
MD	Newsletters of the German Dendrological Society. 1892 →
MFl	Meyer: Flieder (Lilacs). 1952
MFu	Munz: A revision of the genus *Fuchsia*. 1943
MG	Moellers Deutsche Gärtner-Zeitung. 1896 to 1936
MiB	Miyoshi: Die Japanischen Bergkirschen. 1916
MJ	Makino: Illustr. Flora of Japan. 1956
MJCl	Moore & Jackman: *Clematis*
MLi	McKelvey: The Lilac. 1928
MM	McMinn & Maino: The Pacific Coast Trees. 1935
MMa	Millais: Magnolias. 1927
MNZ	Metcalf: Cultivation of New Zealand Trees and Shrubs. 1972
Mot	Mottet: Les arbres et Arbustes d'ornement. 1925
MOT	Muller: The Oaks of Texas. 1951
MPW	McCurrach: Palms of the World. 1960
MRh	Millais: Rhododendrons, 2 Vols. 1917 and 1924

MS	McMinn: Manual of the Californian Shrubs. 1951
NBB	Gleason: The New Britton & Brown, Ill. Flora
NDJ	Yearbook of the Netherlands Dendrological Society. 1925 to 1961
NF	The New Flora and Silva. 1928–1940
NH	Nat. Hort. Magazine. 1922 → (see only those after 1955), from 1960–1974 titled 'American Horticulture Magazine'
NK	Nakai: Flora Sylvatica Koreana. 1915
NT	Nakai: Trees and Shrubs of Japan proper. 1927
NTC	Native Trees of Canada. 7th edition 1969
OFC	Ohwi: Flowering Cherries of Japan. 1973
PB	Hesmer: Das Pappel-Buch. 1951
PBl	Pareys Blumengärtnerei, 2nd edition 1958–1960
PCa	Pertchik: Flowering Trees of the Caribbean
PDR	Poln. Dendr. Rocznik (see only after 1950)
PEu	Polunin: Flowers of Europe
PFC	Pizarro: Sinopsis de la Flora Chilena. 1959
PMe	Polunin: Flowers of the Mediterranean
PPT	Palmer, E. & N. Pitman: Trees of Southern Africa, 3 Vols. 1972
PSw	Polunin: Flowers of Southwest Europe
RBa	Rivière: Les Bambous. 1878
RH	Revue Horticole. 1829 →
Rho	Rhodora. 1899 →
RLo	Rehder: Synopsis of the genus *Lonicera*. 1903
RMi	Rosendahl: Trees and Shrubs of the Upper Midwest. 1955
RPr	Rousseau: The Proteaceae of South Africa. 1970
RWF	Rickett: Wild Flowers of the United States, 13 Vols.
RYB	Rhododendron Year Book, 1–8; from Vol. 9 Rhododendron and Camellia Year Book. To 1971
SB	Satow: The cultivation of Bamboos. 1899
SC	Sweet: Cistineae
SCa	Sealy: A revision of the genus *Camelia*. 1958
SDK	Sokoloff: Bäume und Sträucher der USSR, 7 Vols.
SEl	Servettaz: Monograph. Eleagnaceae
SFP	Salmon: New Zealand Flowers and Plants in Colour. 1967
SH	Schneider: Handbuch der Laubholzkunde, 2 Vols. 1904–1912
SL	Silva Tarouca: Unsere Freiland-Laubgehölze. 1922
SM	Sargent: Manual of the Trees of North America. 1933
SME	Schwarz: Monographie der Eichen Mitteleuropas und des Mittelmeergebietes. 1936–1939
SNp	Stainton: Forests of Nepal. 1972
SPa	Sudworth: Forest Trees of the Pacific Slope. 1908
SR	Stevenson: Species of *Rhododendron*. 1947
SS	Sargent: The Silva of North America, 14 Vols. 1891–1902
ST	Sargent: Trees and Shrubs, 2 Vols. 1905 to 1913

StP	Stern: A study of the genus *Paeonia*. 1946
THe	Tobler: Die Gattung *Hedera*. 1912
TPy	Terpo: Pyri Hungariae. 1960
TY	Trelease: The Yuccaceae. 1902
UCa	Urquhart: The *Camellia*, 2 Vols. 1956 and 1960
UJD	Uehara: Japanische Dendrologie, 4 Vols. 1961
UR	Urquhart: The *Rhododendron*, 2 Vols. 1958 and 1962
VG	Viciosa: Genisteas Espanoles I–II. 1953 to 1955
VQ	Vicioso: Revision del genero *Quercus* en Espana. 1950
VSa	Vicioso: Salicaceas de Espana. 1951
VT	Vines: Trees, Shrubs and Vines of the Southwest. 1960
VU	Vicioso: Ulex. 1962
WJ	Wilson: The Cherries of Japan. 1916
WR	Willmott: The genus *Rosa*, 2 Vols. 1910 to 1914
WRu	Watson: Handbook of the Rubi of Great Britain and Ireland. 1958
WT	West-Arnold: The native Trees of Florida. 1956
YTS	Yamakai Color Guide, Flowering Garden Trees and Shrubs, Vols. 1–2. 1971
YWP	Yamakai Color Guide, Flowers of Woody Plants, 1–2. 1969.

Plant Descriptions

ECCREMOCARPUS Ruiz & Pav. — BIGNONIACEAE

Evergreen climbers; leaves opposite, doubly pinnatisect or bipinnate; flowers yellow, scarlet-red or orange; calyx campanulate, 5 parted, corolla tube elongated, limb on the described species entire and narrow, on the others more or less 2 lobed; fruit an ovate or elliptic, chambered capsule; seeds surrounded by a broad wing. — 5 species in Chile and Peru.

Eccremocarpus scaber Ruiz & Pav. Evergreen climber, with the appearance of a small-leaved *Clematis*, becoming woody and 2.5–3 m high in its habitat and in mild regions; leaves 5–7 cm long, leaflets alternate, obliquely cordate, serrate to entire; flowers scarlet-red to dark orange, 2.5 cm long, July–fall. BM 6408; BR 939. Chile. Fig. 1. z8 ✣

'Aureus'. Flowers pure gold-yellow. ✣

'Carmineus'. Flowers carmine-red. ✣

May be treated as an annual and grown from seed in colder areas; flowers in a few months but grows only 1 m high.

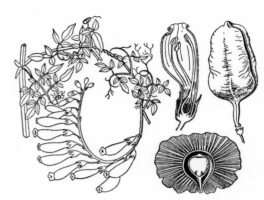

Fig. 1. *Eccremocarpus scaber*, twig, flower, fruit and seed (from Bureau, Schumann)

ECHINOCARPUS

Echinocarpus sinensis see: **Sloanea hemsleyana**

E. chinocarpus hemsleyanus see: **Sloanea hemsleyana**

E. sterculiaceus see: **Sloanea sterculiacea**

ECHINOPANAX See: **OPLOPANAX**

EDGEWORTHIA Meissn. — THYMELAEACEAE

Deciduous or evergreen shrubs, very closely related to *Daphne*; branches thick and stiff; leaves alternate, short petioled, entire, clustered at the branch tips; flowers in dense, stalked heads, axillary on the previous year's wood, appearing with or before the leaves; calyx tubular, with 5 erect lobes, petals absent; stamens 8, in 2 rings; style elongated, stigma cylindrical; fruit a dry drupe. — 3 species in E. Asia, from the Himalayas to Japan.

Edgeworthia gardneri Meissn. Very similar to the following species but evergreen; leaves tough; flowers with looser and shorter pubescence outside, hairs coarser; ovaries completely pubescent, April. BM 7180; DRHS 735; HAL 79. Nepal, Sikkim. z8 ✣

E. papyrifera Zucc. Shrub, about 1 m high and wide, very densely branched, branches densely silky pubescent at first, very tough, thick; leaves thin, membranous, narrow oblong, 7–12 cm long, entire, dark green above, gray-

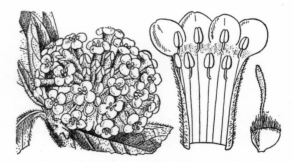

Fig. 2. *Edgeworthia papyrifera* (from Bot. Mag.)

green beneath, both sides silky pubescent at first, later glabrous above; flowers 40–50 in heads, golden-yellow, white pubescent outside; ovaries pubescent only at the tip, April. FS 289; SNp 132; HTS 193. Japan, China. 1845. Only suitable for the mildest regions. Plate 1; Fig. 2. z8 ✣

EDWARDSIA

Edwardsia macnabiana see: **Sophora microphylla**

E. tetraptera see: **Sophora tetraptera**

E. grandiflora see: **Sophora tetraptera 'Grandiflora'**

EHRETIA P. Browne — EHRETIACEAE

Deciduous or evergreen trees or shrubs; leaves alternate, glabrous or pubescent, entire or serrate; flowers usually white, small, usually in terminal panicles; calyx small, 5 toothed, tube short; stamens 5, adnate to the tube, style more or less deeply split; ovaries 4 chambered; fruit a globose drupe. — About 50 species in the warmer areas of E. Asia, No. & So. America, Africa, and Australia.

Ehretia dicksonii Hance. Deciduous tree, 6–10 m high, young branches stiff, bristly; leaves oblong-elliptic, 8–18 cm long, 5–12 cm wide, base broadly cuneate, finely serrate, glossy green above, lighter with reticulate venation beneath, tough, leathery, rough bristles on both sides; flowers white, fragrant, in terminal, 10 cm long and wide panicles, August; fruits yellowish, 1 cm thick. LF 264; LWT 328; KIF 4: 42. China, Formosa. 1900. Once planted erroneously as *E. macrophylla* Wall. A large tree grows in the Bagatelle Park, Paris, France. Plate 5. z7

E. macrophylla Wall. Deciduous tree, 5–6 m (forming a shrub in z8 at Kew Gardens, London!); leaves ovate, 10–15 cm long, 7–10 cm wide, margin unevenly serrate, base cuneate, both sides with short, rough, bristly hairs, especially on the upper side, petiole 1.5–2 cm long; flowers in terminal, short, globose panicles, corolla tube twice as long as the calyx, fruit globose, 12 mm thick. LF 265. Himalayas; W. Hupeh; Nepal. z9 # Ø

E. serrata see: **E. thyrsiflora**

E. thyrsiflora (S. & Z.) Nakai. Deciduous tree, 5–10 m high, branches gray, glabrous; leaves oblong-ovate, 8–16 cm long, short acuminate, toothed or finely crenate, glossy green above with short stiff hairs, lighter and glabrous beneath, leathery; flowers in 15 cm long, dense panicles, August; fruits orange at first, later black, globose, 4 mm thick. Bmns 440; BS 2: 15; CIS 99. China, Japan. 1880. Earlier mistaken for *E. serrata* Roxb. Fig. 3. z7 Ø

Fig. 3. *Ehretia thyrsiflora* (from Lindley)

ELAEAGANUS L. — ELAEAGNACEAE

Deciduous or evergreen trees or shrubs, branches often prickly; leaves alternate, simple, with silvery or gold lepidote scales; flowers tubular-campanulate, axillary, bisexual, with simple perianth, usually fragrant, calyx 4 lobed; stamens 4, quite short stalked; fruit a fleshy drupe with a striped pit. — About 45 species in Asia, S. Europe and N. America.

- Leaves deciduous, thin
  E. angustifolia, commutata, multiflora, umbellata
- Leaves evergreen, leathery:
  E. ebbingei (often semi-deciduous), *glabra, macrophylla, pungens, reflexa*

Elaeagnus angustifolia L. Deciduous tree or shrub, to 7 m high, branches silvery, often thorny, prickly, with

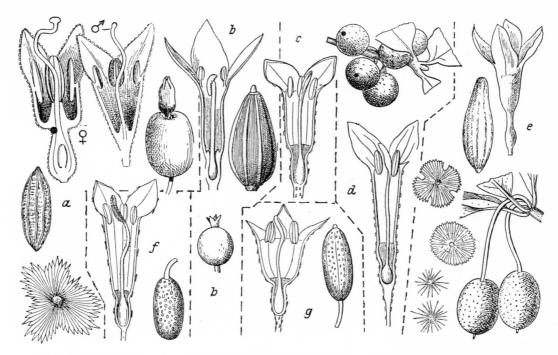

Fig. 4. **Elaeagnus** a. *E. angustifolia*; b. *E. commutata*; c. *E. pungens*; d. *E. umbellata*; e. *E. multiflora*;
f. *E. glabra*; g. *E. macrophylla* (from Baillon, Gilg, Koehne, Shirasawa, Bot. Mag., Schneider)

silvery scales when young; leaves lanceolate to oblong, 4–8 cm long, dull green above, with silvery scales beneath, petiole 5–8 mm long; flowers 1–3 axillary, on the lower branch tips, calyx tubular, 1 cm long, yellow inside, silvery outside, fragrant, calyx lobes as long as the tube, erect, June; fruits oblong, 1 cm long, yellow with silvery scales, mealy, sweet. HM 2146–47; SEl 7 (1–10). Mediterranean region, W. and Middle Asia to the Altai Mts. and Gobi Desert. In cultivation since the 16th century. S. Europe to Central Asia, the Himalayas, China; as windbreaks in the Soviet Union; being tried in Tunisia and W. Sahara. z2 Fig. 4a. ⓕ ⊗ ∅

var. **orientalis** (L.) Ktze. Branches only slightly thorny, young shoots tomentose with stellate hairs and scaly; leaves oblong-elliptic, 3–7 cm long, obtuse, base rounded, dull gray-green and scaly above, tomentose beneath with white stellate hairs; fruits to 2 cm long, yellow, edible (= *E. orientalis* L.; *E. sativa* Hort.) Eastern Mediterranean region. 1739. z4 ⊗ ✗

var. **spinosa** (L.) Ktze. Branches normally thorny; leaves wider, more elliptic, 3–7 cm long, densely scaly; fruits smaller and more rounded. BR 1156 (= *E. spinosa* L.)

E. argentea see: **E. commutata**

E. commutata Bernh. Deciduous shrub, 2–5 m high, stoloniferous, branches thornless, bowed, red-brown, reed-like; leaves ovate-oblong, short stalked, 2–10 cm long, both sides glossy silver; flowers 1–3 in clusters, funnelform, yellow inside, silvery outside, very fragrant, May–July; fruits 1 cm long, silvery, pulp dry, mealy, seed

pit slightly 8 ribbed. BM 8369; NBB 2: 575 (= *E. argentea* Pursh non Moench). N. America; along river banks and moist slopes. z2 Fig. 4b. ∅

E. crispa see: **E. umbellata**

E. × ebbingei Boom (*E. macrophylla* × *E. pungens*) Semi-deciduous shrub, evergreen in mild areas, to 3 m high and wide, first year branches brown, later becoming gray lepidote; leaves elliptic, glossy green above, silvery scaly beneath and brown, petiole 1 cm; flowers 3–6 in the leaf axils, short stalked, white, fragrant, October–November (= *E. × submacrophylla* Hort.) Developed in 1928, The Hague, Netherlands, introduced into the trade in 1939 by S. G. A. Doorenbos with 4 of 60 seedlings tolerating a frost. z6 #

In 1976 two cultivars were named by H. G. Hillier:

'Albert Doorenbos'. Large leaved form, more closely resembling *E. macrophylla*: leaves 9–12 cm long, 5–6 cm wide. Fig. 5 # ∅

'The Hague'. Narrow leaved form, more closely resembling *E. pungens*; more upright growing; leaves 7–10 cm long, 4–5 cm wide. Fig. 5. # ∅

'Gilt Edge' (Water, Sons & Crisp, 1961). Leaves narrow yellow margined. JRHS 97: 184. # ∅

E. edulis see: **E. multiflora**

E. glabra Thunb. Evergreen shrub, very similar to the more frequently planted *E. pungens*, but with thornless

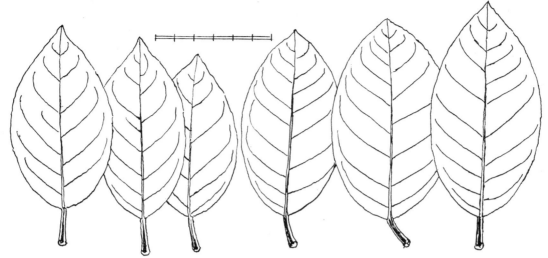

Fig. 5. Both cultivars of *Elaeagnus* × *ebbingei*: left 'The Hague', right 'Albert Doorenbos' (Original)

branches, often somewhat climbing, young branches brown, glossy; leaves longer acuminate, thinner, more metallic glossy beneath (whitish and dull on *E. pungens*); flowers white; brown scaly, fragrant, October–November. LWT 246; KO 356. Central China, Korea, Japan. 1880. z8 Plate 5; Fig. 4f. # ☉

E. longipes see: **E. multiflora**

E. macrophylla Thunb. Evergreen shrub, to 3 m high, branches erect, thornless, silvery; leaves elliptic, 6–8 cm long, dark green above, eventually smooth, densely silvery lepidote beneath; flowers usually grouped in 4's, silvery, appearing like small *Fuchsia* flowers, but only 12 mm long, very fragrant, September–November; fruits ripening the following May, red scaly, with persistent calyx tube. BM 7638; NK 17:4. Japan, Korea. 1879. z8 Fig. 4g, 6. # ∅ ☉

E. multiflora Thunb. Deciduous shrub, to 3 m high and wide, branches brown scaly; leaves broad-elliptic, 6–8 cm long, 3–4 cm wide, dark green above, eventually glabrous, silvery beneath with many glossy, brown scales; flowers 1–2 in numerous fascicles, pale yellow, May; fruits oblong, 1.5 cm long, dark red-brown, on 2–3 cm long, filamentous stalks, pulp pleasantly tangy, juicy, seed pit with 8 ribs. BM 7341; KIF 3: 54; DRHS 738 (= *E. edulis* Carr.; *E. longipes* Gray). China, Japan. 1873. z6 Fig. 4e. ∅ ☉ ⊗

var. **ovata** (Maxim.) Servettaz. Young leaves stellate pubescent above, but quickly becoming glabrous; flowers grouped 1–3; fruits brown, scaly, on 12–15 cm long stalks, pendulous. Japan. Quite variable.

E. orientalis see: **E. angustifolia** var. **orientalis**

E. pungens Thunb. Evergreen shrub, to 4 m high, much shorter in cultivation, branches thorny, brown; leaves oblong-elliptic, to 10 cm long, margins sinuate or crispate, glossy dark green above, dull silvery beneath

with a few large brown scales; flowers usually 3 in the leaf axils, silver-white, fragrant, October–November, calyx tube compressed above the ovary; fruit brown at first, then red, 15 mm long, ripening in May of the following year. SEl 11. Japan. 1830. z7 Fig. 4c, 7. #

Including a number of cultivars:

'Aurea'. Branches with few thorns; leaves oval-elliptic, obtuse at both ends, green, with a narrow, deep yellow margin. Rev. Hort. Belg. 1864: 356 (as *E. pungens foliis aurei-marginatis*). # ∅

'Dicksonii'. Very similar to 'Aurea', but with a wide golden-yellow border, many leaves with the upper third totally golden-yellow. # ∅

'Frederici'. Branches thorny; leaves elliptic, 3–4 cm long, with 1 or 2 yellow stripes in the middle, otherwise green. Before 1888. # ∅

Fig. 6. *Elaeagnus macrophylla* (from Dippel)

Fig. 7. *Elaeagnus pungens* (from Dippel)

'Golden Rim'. Mutation of *E. pungens* 'Maculata' discovered by W. J. Streng, Boskoop, Holland. Leaves with a narrower or wider golden-yellow border, otherwise like 'Maculata', but without the yellow spots in the middle of the blade. Dfl 11: 52.

'Maculata'. Leaves large, green, middle dark yellow, yellow spot quite variable, occasionally reverting back to green (= 'Aureo-variegata'). # Ø

'Simonii'. Shrub, stoutly branched, not thorny, branches gray-brown; leaves larger than those of the species, elliptic, acuminate, base rounded, 5–11 cm long, bright green above, bright silvery beneath, nearly or totally lacking brown scales (= *E. simonii* Carr.). China. 1862. # Ø

'Tricolor'. Leaves like 'Simonii' in size and form, but green with yellowish and whitish-pink coloration (= *E. pungens simonii tricolor* Hort.). # Ø

'Variegata'. Very similar to 'Aurea', but with pale yellow margin. # Ø

E. × pyramidalis Wrobl. (*E. commutata* ? × *E. multiflora*). Medium sized shrub, to 2.5 m high, compact conical

habit, short branchlets absent; leaves rather variable, rounded at the tips or slightly acute, silvery green above, silvery beneath, 3–4 cm long on small branches, apex rounded; fruits not observed.

E. × reflexa (Morr. & Decne.) (= *E. pungens* × *E. glabra*). Long branches with few thorns; leaves more ovate-lanceolate, margin not sinuate, glossy green above, dense brown scaly beneath. Gs 20: 53; HTS 196 (= *E. pungens* var. *reflexa* [Morr. & Decne.] Schneid.). From Japan. #

E. simonii see: **E. pungens** var. **simonii**

E. spinosa see: **E. angustifolia** var. **spinosa**

E. umbellata Thunb. Deciduous shrub, to 4 m high, quite wide, branches often thorny, twigs yellow-brown or also silvery; leaves elliptic-oblong, about 3 times as long as wide, 3–7 cm long, margins often crispate, with a few silvery scales above when young, silvery with brown scales beneath; flowers yellowish-white, 1–7 in the leaf axils, usually clustered on small side branches, fragrant, May–June, calyx tube much longer than the limb, gradually narrowing to the ovary; fruits globose, 6–8 mm thick, silvery-brown at first, eventually red, stalk 6 mm long, September–October. NK 17: 1; KIF 4: 33 (= *E. crispa* Thunb.; *E. crispa* var. *typica* Nakai). China, Japan. 1830. ℗ Japan; on dunes. z3 Plate 5; Fig. 4d. Ø

var. **parvifolia** (Royle) Schneid. Shrub, to 5 m high, branches silvery, erect and sparse, prickly; leaves elliptic-lanceolate, short acuminate, stellate pubescent above at first, becoming glabrous, silvery beneath, 3–7 cm long; flowers several, axillary and clustered, June, calyx tube narrow, longer than the limb, whitish inside, fragrant; fruits rounded, silvery at first, later brown, August. BR 29: 51. Himalaya, China, Japan.

For sandy to medium-heavy soils; otherwise not very particular, especially the silvery foliage types. The deciduous species thrive in the worst, dry soil. The evergreen species will survive winters to zone 6.

Further *Elaeagnus* species may be found illustrated as follows: *E. asakawana* KIF 3: 52; *E. matsunoana* KIF 3: 53; *E. montana* KIF 4: 31; *E. murakaniana* KIF 4: 32.

Lit. Servettaz, C.: Monographie des Eláeágnacées; in Beih. Bot. Centralbl. 25/II. 1-420, 1909 ● Hillier, H. G.: *Elaeagnus × ebbingei*; Groen 1976, 259.

ELAEOCARPUS L. — ELAEOCARPACEAE

Evergreen trees or shrubs; leaves alternate, occasionally opposite, entire or serrate; flowers with 4 or 5 sepals and petals, the latter incised or toothed at the apex; stamens numerous, stigma simple; fruit a hard drupe with 2–5 single seeded chambers. — About 200 species in E. Asia, Indomalaysia, Australia and the Pacific; only a few species in cultivation.

Elaeocarpus cunninghamii see: **E. dentatus**

E. cyaneus Sims. A tall tree in its habitat, but shrubby in cultivation, entire plant glabrous; leaves elliptic-oblong,

acuminate, elevated reticulate venation; flowers cream-white, in loose racemes, shorter than the leaves, petals fimbriate, stamens very numerous, July; fruit a globose, blue drupe. BM 1737; BR 657 (= *E. reticulatus* Smith). Australia. 1803. z9 # ✧

E. dentatus (J. R. & G. Forst) Vahl. A tree in its habitat, 12–15 m, bark gray, young branches silky pubescent; leaves usually clustered at the branch tips, narrow-oblong to slightly obovate, 5–10 cm long, 2–4 cm wide, tapering to the 2 cm long petiole, more or less entire or obtuse serrate, leathery tough, dark glossy green above, with whitish hairs beneath; flowers white, in numerous, 10 cm long, pendulous, axillary racemes, petals 3–5 lobed, October–February; fruit purple-gray, ovate, 15 mm long (= *E. cunninghamii* Raoul). MNZ 21. New Zealand. 1833. z9 # ⊕

E. reticulatus see: **E. cyaneus**

Greenhouse culture is recommended in cooler regions, otherwise easily grown in any good soil.

ELEUTHEROCOCCUS

Eleutherococcus henryi see: **Acanthopanex henryi**

E. leucorrhius see: **A. leucorrhius**

E. senticosus see: **A. senticosus**

E. simonii see: **A. simonii**

ELLIOTTIA Muehl. — ERICACEAE

Monotypic genus; deciduous shrub; leaves alternate, simple, entire, without stipules; flowers in terminal racemes; calyx small, 4 parted, petals 4, stamens 8, filaments short; fruit a flat-round, normally 4 valved capsule; seeds numerous, rounded or winged. — 1 species in Eastern N. America.

Elliottia racemosa Muehl. Shrub, normally 1(–3) m, occasionally a small tree, young branches thin, downy pubescent; leaves oblong to elliptic; acute on both ends, 5–10 cm long, thin, petiole about 2.5 cm long; flowers pure white, 4 cm wide, fragrant, in terminal, erect, 12–18 cm long racemes, July. BM 8413; BC 1390; NF 10: 158; NH 1956: 46. S. Carolina and Georgia; wet, sandy forest areas. 1894. In its habitat as well as in cultivation, a very rare plant, definitely one of the rarest shrubs of all. z7 to z8 Fig. 8. ⊕

Fig. 8. *Elliottia racemosa* (from Sargent)

ELSHOLTZIA Willd. — LABIATAE

Aromatic perennials or subshrubs; leaves opposite, toothed or serrate; flowers in clustered whorls, in dense, one sided false spikes grouped in larger terminal panicles; calyx tubular or campanulate, corolla nearly radiate, 4 lobed, the upper lobes emarginate, the others obtuse; stamens 4, exserted, 2 longer and 2 shorter. — About 35 species, most in China and India, south to Java; 1 species each in Europe and Ethiopia.

Elsholtzia fruticosa see: **E. polystachya**

E. polystachya Benth. Subshrub, 1 to 1.5 m high, pubescent; leaves lanceolate, 7–12 cm long, acuminate, scabrous serrate, cuneate, tapering to a short petiole, venation glandular and pubescent beneath; flowers in slender, cylindrical, 5–12 cm long, erect, terminal spikes,

white or pale yellow, June to October. Collett, Fl. Siml. 123; SNp 125 (= *E. fruticosa* Rehd.). Himalaya; W. China. z5 ⊕

E. stauntonii Benth. Subshrub, 1–1.5 m high, ascending branching habit and broadly spreading, all parts aromatic; leaves deciduous, ovate-oblong to lanceolate, 6–12 cm long, coarsely dentate, densely glandular pubescent beneath; flowers in 10–15 cm long, often panicle-like spikes, corolla 7–8 mm long, light purple, September–October. BM 8460; RH 1914: 60; MG 1910: 541. N. China. 1909. z5 ⊕

For open areas in good garden soil; looks well in mass plantings for fall flowers.

ELYTROPUS Muell. Arg. — APOCYNACEAE

Monotypic genus with the characteristics of the species.

Elytropus chilensis Muell. Arg. Evergreen, vigorous climber, 4–5 m high, branches thin, bristly pubescent;

leaves opposite, elliptic to nearly oblong, long acuminate, bristly pubescent, margin ciliate; flowers small, white with a trace of lilac, 1–2 axillary, spring; fruits green, eventually yellow. Chile, Argentina. z9 #

Protected areas in sun or semishade. Cultivated by Hillier.

EMBOTHRIUM J. R. & G. Forst. — PROTEACEAE

Evergreen trees or shrubs; leaves simple, entire, alternate, leathery; flowers very attractive, in dense panicles; corolla tube narrow cylindrical, splitting open in 4 narrow strips, the strips twisting; ovaries long stalked; fruit a leathery capsule with persistent style, unilocular, many seeded. — 8 species in the S. American Andes, Chile, E. Australia.

Embothrium coccineum Forst. Evergreen shrub, a tree to 10 m high in its habitat and milder regions, stoloniferous; leaves alternate, but occasionally somewhat opposite, oblong to elliptic, 5–12 cm long, glabrous, deep green above, glossy; flowers in terminal and axillary, dense, 7–10 cm long racemes, scarlet-red, May–July. FS 1311; PFC 170; MB 4856. Chile. 1846. z8 Plate 5. # ⊕

'Eliot Hodgkin.' Flowers yellow, somewhat like *Jasminum nudiflorum* in color. A color illustration may be found in JRHS 1977:472. Brought from Osorno, S. Chile in 1975 to England where it is cultivated by Hillier.

'Longifolium'. Leaves longer, more narrow; flowers earlier than the species, flowers all along the branches (more at the branch tips on the wider leaved type). HTS 194. z8 Plate 1. # ⊕

var. **lanceolatum** Ruiz & Pav. Leaves often deciduous, quite narrow-oblong. Andes, Chile, to 1500 m high in the mountains. z8 # ⊕

'Norquinco'. A selection of var. *lanceolatum* with especially abundant flowers. JRHS 73: Pl. 139; 87: 126. Available in the trade in England. z8 # ⊕

An *Embothrium* in flower is an unforgettable sight. Winter hardiness is quite variable due to its wide native range, reaching from 38° S. latitude (near Temuco, Chile) to 54° S. latitude (Tierra del Fuego) possibly even to Cape Horn. Closer investigation of geographical races would, therefore, be helpful in determining winter hardiness. These plants have exceptional garden merit due to a floral display that sometimes lasts for 3 months. Plants have been observed in their native range having white and yellow flowers (Bean 2: 86; 1975).

EMMENOPTERYS Oliv. — RUBIACEAE

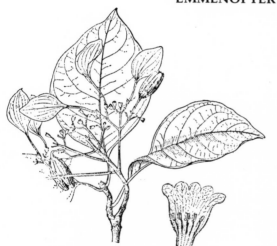

Fig. 9. *Emmenopterys henryi* (from ICS)

Deciduous trees; leaves opposite, petioled, thin leathery, simple, entire, stipules inconspicuous; flowers in terminal, many-flowered panicles, yellow or white, calyx small, 5 parted, 1 lobe occasionally leaflike, white, stalked; corolla campanulate-funnelform, limb 5 lobed; stamens 5; style filamentous; fruit a spindle-shaped, 2 chambered capsule with numerous, winged seeds. — 2 species in China, Thailand and Burma.

Emmenopterys henryi Oliv. Tree, to 10 m, totally glabrous; leaves elliptic acute, base cuneate, 10–15 cm long, midrib pubescent beneath; flowers yellow, about 2.5 cm long, in 10–15 cm long panicles, the leaflike calyx lobe about 5 cm long, white, June-July. LF 272; CIS 47; HI 1823. China; Hupeh. 1907. z8? Plate 7; Fig. 9.

Winter hardy in England, but plants have not yet flowered in cultivation; flowers very interesting. First observed in 1971 in the Villa Taranto Botanic Garden, Pallanza, Lake Maggiore, Italy (see JRHS 1971: 496–497, fig. 220).

EMPETRUM L. — Crowberry — EMPETRACEAE

Evergreen, procumbent, dwarf shrub having a heather-like appearance, with tough, needlelike leaves; flowers very small and inconspicuous, bisexual or unisexual, mono- or dioecious, axillary; fruit a berrylike drupe with 6–9 seeds. — 2 or 15–16 species (depending on interpretation of the genus) in Arctic and Subarctic parts of the temperate world. Range map Fig. 10.

Empetrum atropurpureum Fern. & Wieg. Procumbent, young branches and leaves white tomentose; leaves

linear-oblong, loosely arranged, erect, soon reflexed, 6–8 mm long on long shoots; fruits red to black-red, translucent, 5–8 mm thick. N. America, Gulf of St. Lawrence to Maine and New Hampshire. 1890. z3 # ⚭

E. eamesii Fern. & Wieg. Leaves linear-oblong, compact, mounding, later slightly outspread, to 4 mm long or less on long shoots; fruits pink or bright red, soon becoming translucent, not over 5 mm thick. Northeast N. America. z3 # ⚭

E. hermaphroditum (Lange) Hagerup. Resembling *E. nigrum*, but the whole plant is coarser, more upright, young branches green at first, then brown; leaves somewhat ovate-oblong, lighter green; flowers bisexual. GPN 670 and 671. N. Scotland, Iceland, Alps, Siberia, Canada; occurs together with *E. nigrum* but distinguished by its more slender habit. z3 # ⚥

E. nigrum L. Procumbent, mounding to 25 cm, branches glandular at first, not white tomentose; leaves linear-oblong, 4–6 mm long, erect, soon curved, margin glandular pubescent; flowers purple-pink, inconspicuous; fruit pea-sized, black, juicy, dull flavor. HM 2667. N. Europe, N. Asia, N. America. Before 1700. z2 Plate 6; Fig. 10 # ⚥

'Leucocarpum'. Fruits white. Found in the Eastern Baltic region. # ⚥

E. rubrum Vahl. Very similar to *E. eamesii*, but less densely foliate, more outspread and somewhat larger. BR 1783; PFC 84b. Antarctic America, Tristan da Cunha. z6 Fig. 27 # ⚥

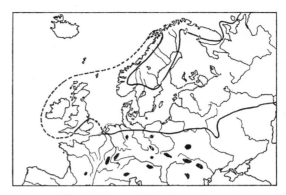

Fig. 10. Range of *Empetrum nigrum*
(from Bialobok)

Grasslike plants for very organic, peaty soils.

Lit. Good, R. D. O.: The genus *Empetrum*; in Jour. Linn. Soc. **47**, 489–523, 1927 ● Hagerup, O.: Studies on the Empetraceae; in Biol. Med. **20**, 1–49, 1946.

ENCEPHALARTOS Lehm. — ZAMIACEAE

Evergreen plants having a palm-like appearance and a distinctively tall stem, this occasionally subterranean or to 9 m high and 15–60 cm thick, usually unbranched, covered with the scales and leaf scars of the old leafs; leaves developing several at a time, pinnate, upright, the outer ones outspread, nearly in whorls, petioled, linear-oblong in outline, glabrous or shaggy, green, gray or blue-green, pinnae stiff, sessile with a broad base, thick leathery, thorny; flowers dioecious, the sporophylls grouped into a cone; the cone scales peltate at the apex and without horns. — 30 species in tropical and S. Africa; common in botanic gardens.

Encephalartos altensteinii Lehm. Stem to 5 m high and 70 cm diameter, cylindrical; leaves to 1.5 m long, pinnae very numerous, about 15 cm long, 2.5 cm wide, oblong-linear, somewhat decurrent, margin with 3–5 small erect teeth on either side, glossy green; female cones 30–35 cm long and 10 cm wide, male cones sessile, broad ovate, to 45 cm high and 25 cm wide, brown. BM 7162–63; Giddy Pl. 17. S. Africa. 1835. z9 Plate 6. # ∅ ⚥

E. horridus (Jacq.) Lehm. Stem very short or totally underground; leaves distinctively blue-green, 50–100 cm long, very stiff, reflexed at the tips, pinnae to 10 cm long, 5 cm wide, oblique ovate-lanceolate, thorny lobed; male cones cylindrical, stalked, to 30 cm long, 6 cm thick, female cones broad oval-oblong, to 35 cm long and 15–20 cm wide, rather triangular at the apex. Giddy Pl. 24. S. Africa. 1800. z9 # ∅

E. lehmannii (Eckl. & Zeyh.) Eckl. Stem to about 1.5 m; leaves glabrous, exceptionally blue, to about 1 m long, pinnae do not touch, middle pinnae 12–18 cm long, 1.5–2 cm wide, entire, with an occasional tooth on the lower margin; male and female cones blackish-red, male 25–35 cm long, 8–10 cm wide, female 45 to 50 cm long, to 25 cm wide; seeds large, bright red. Giddy Pl. 14, S. Africa. 1777. z9 Plate 6. # ⚥ ∅

E. villosus (Gaertn.) Lehm. Stem subterranean or at the tallest 30–40 cm high above ground and 30 cm thick; leaves 1.5–3 m long, under ideal conditions to 4 m, middle pinnae 15–25 cm long, 1.5–2 cm wide, with 1–3 teeth on each side, not overlapping, erect, somewhat reflexed at the apex, glossy green, pinnae becoming generally smaller toward the leaf base, eventually reduced to small thorns; male cones 60–70 cm long, 12–15 cm thick, lemon-yellow, female cones 30–50 cm long, 20–25 cm thick, dark yellow and glossy. BM 6654; Giddy Pl. 10. S. Africa. 1866. z9 # ∅ ⚥

All species require a warm climate and abundant moisture during the growing season, especially *E. villosus*, which is found in moist areas in its habitat. Little water needed during the dormant season.

Lit. Giddy, C.: Cycads of South Africa; with 30 color plates, Capetown 1974.

ENKIANTHUS Lour. — ERICACEAE

Deciduous, seldom evergreen, shrubs, branches usually whorled; leaves simple, alternate, usually clustered in whorls at the branch tips, tapering to both ends, usually finely serrate; flowers in terminal racemes or umbels; calyx with 5 small teeth, corolla campanulate to urceolate, usually 5 parted; stamens 10, enclosed, anthers with 2 appendages at the tip, opening by short slits; fruit a dry capsule (= *Tritomodon* F. Maekawa). — 10 species, from Japan to Himalaya.

Outline of the Genus

Section **Euenkianthus** Palib.
> Flowers urceolate, with 5 sack-like bulges at the base; in umbels, appearing before the leaves:
>> *E. perulatus, quinqueflorus, serrulatus*

Section **Andromedina** Palib.
> Flowers the same as in the above section, but in pendulous racemes, appearing after the leaves:
>> *E. nudipes, subsessilis*

Section **Enkiantella** Palib.
> Flowers campanulate, without the sack-like swellings, in corymbs, appearing after the leaves; corolla with 5 short lobes:
>> *E. campanulatus, chinensis, deflexus*

Section **Meisteria** (S. & Z.) Palib.
> Flowers the same as in the above section, but with the corolla limb irregularly incised:
>> *E. cernuus*

Enkianthus campanulatus (Miq.) Nichols. Deciduous shrub, narrow upright habit, to 3 m high; leaves 3–7 cm long, rhombic-elliptic, awn-like serrate, with both sides bristly pubescent, bright red in fall; flowers in pendulous corymbs, corolla campanulate, not tubercled, light yellow to light pink, with reddish venation, May; ovaries and styles glabrous. DB 1951: 149; BMns 512. Japan. 1880. z5 Plate 10; Fig. 11a.

'**Albiflorus**'. Flowers whitish. BMns 512 (= *E. pallidiflorus* Craib). Very attractive; much rarer than the species. Plate 9. ∅ ☿

'**Donardensis**' (Donard). Flowers larger, limb broader and deeper red, venation wider. Otherwise hardly differing from the species; occasionally found as a chance seedling. ☿

'**Hiraethlyn**'. Flowers cream-white with wine-red venation. Found in Bodnant Garden, Wales. 1961. ☿

var. **palibinii** Bean. Leaves narrower, rusty pubescent along the midrib; flowers red, filaments shaggy pubescent (= *E. ferrugineus* Craib; *E. rubicundus* Matsum. & Nakai). Plate 12. ☿ ∅

E. cerasiflora see: **E. chinensis**

E. cernuus (S. & Z.) Mak. Deciduous shrub to 5 m tall in its habitat; leaves elliptic to ovate-rhombic, 2–4 cm long, crenate, bright green above, venation somewhat pubescent beneath; flowers 10–12 in nodding racemes, corolla campanulate, white, 6–8 mm long, May. NT 1: 194–195. Japan. 1910. z6 Fig. 11b. ☿ ∅

var. **matsudae** (Komatsu) Makino. Leaves broad lanceolate to narrow ovate, rather coarsely serrate, with midrib usually brown pubescent beneath; calyx lobes narrow ovate to broad lanceolate, corolla broad campanulate, deep red, style somewhat exserted. Japan; Honshu. ☿

var. **rubens** (Maxim.) Mak. Leaves shorter, wider; flowers red. KO 396. Plate 10. ∅ ☿

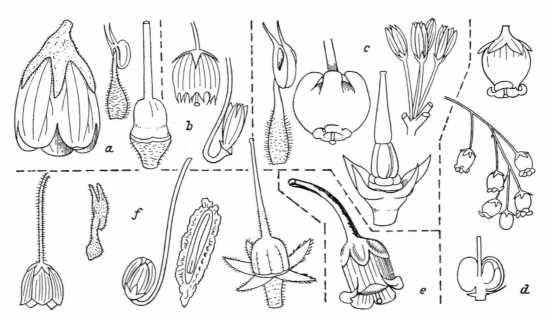

Fig. 11. **Enkianthus** a. *E. campanulatus*; b. *E. cernuus*; c. *E. perulatus*; d. *E. subsessilis*; e. *E. quinqueflorus*; f. *E. deflexus* (from Bot. Mag., Bot. Reg., Rehder)

E. chinensis Franch. Resembling *E. campanulatus,* but with leaves elliptic to oblong, more crenate, quite glabrous; inflorescence and flower stalk glabrous, flowers broad campanulate, salmon-red, corolla lobes reflexed. BM 9413; FIO 81 (= *E. cerasiflora* Lév.; *E. sinohimalaicus* Craib). Central and W. China. 1900. z7 ⊕ Ø

E. deflexus (Griff.) Schneid. Tall shrub, branches red; leaves elliptic to oblong, both sides sparsely pubescent, 3–7 cm long; flowers 1.5 cm wide, yellow-red with darker limb, with triangular limb lobes, May–June, in fascicled, often umbellate racemes; ovaries and style pubescent. BM 6460; HAL 89; FIO 81 (= *E. himalaicus* Hook. f. & Thoms.). W. China, Himalaya. 1878 and 1908. z6 Fig. 11f. Ø ⊕

E. himalaicus see: **E. deflexa**

E. japonicus see: **E. perulatus**

E. nikoensis see: **E. subsessilis**

E. nudipes (Honda) Ohwl. Very similar to *E. subsessilis,* but with flower racemes glabrous, leaves glabrous on the upper side, although somewhat pubescent when young, petiole 1–2 mm, becoming glabrous, July. Japan, mountains of Honshu. z6

E. perulatus (Miq.) Schneid. Shrub, 1–2 m high, young branches reddish and totally glabrous; leaves clustered at the branch tips, elliptic-ovate to obovate, acute, 3–5 cm long, finely serrate, yellow and bright red in fall; flowers in pendulous, terminal umbels, corolla white, rounded urceolate, with 5 sack-like bulges at the base, lobes reflexed, May. BS 1: 653; BM 5822; DRHS 746 (= *E. japonicus* Hook. f.). Japan. 1870. Hardy; seldom true in culture. z6 Plate 9; Fig. 11c Ø ⊕

E. quinqueflorus Lour. Semi-evergreen or deciduous shrub, 1–1.5 m high (in cultivation), young branches glabrous; leaves red at first, later deep green, narrow elliptic to obovate, acute, 5–10 cm long, entire, leathery; flowers in pendulous, terminal, umbellate clusters, stalks pink, corolla open campanulate, pink or pink with white, widening at the base, about 1 cm wide, May. BM 1649; HKS 34; CIS 96; JRHS 1977:67. SE. China, Hong Kong. 1810. z8 Plate 10; Fig. 11e. # ⊕

E. serrulatus (Willd.) Schneid. Deciduous, small shrub, to 6 m high in its habitat; leaves broad obovate, long acuminate, 6–8 cm long, quickly becoming totally glabrous, serrate; flowers around 5, in umbels, nodding, corolla campanulate, white, 1 cm long, filaments somewhat pubescent. FIO 81. Central and W. China. 1900. z6 Plate 10. Ø ⊕

E. sinohimalaicus see: **E. chinensis**

E. subsessilis (Miq.) Makino. Upright, strongly branched shrub, 1–3 m high; leaves obovate to elliptic, 2–3 cm long, 1–1.5 cm wide, acute, long acuminate at the base, finely toothed, midrib white pubescent above, brown pubescent beneath, petiole 1–3 mm; flowers 5–10 in pendulous racemes on a 2–3 cm long axis, axis white pubescent (!), calyx lobes ovate, 2 mm, white ciliate, corolla white, urceolate, 5 mm long (but without the bulging base as on *E. perulatus*), May–July; fruit capsules pendulous, 4 mm long, glabrous. ST 1: 25 (= *E. nikoensis* [Maxim.] Makino). Japan; Mountains of Honshu. 1892. Hardy but slow growing. z6 Plate 10; Fig. 11d.

Fast growing shrubs needing no special cultural requirements; treat them like Rhododendron and Azaleas, but allow more sun; prefer a moist humus soil.

Lit. Craib, W. G.: New Species of *Enkianthus.* — Hardy Species of *Enkianthus* under cultivation in the Royal Botanic Gardens, Edinburgh; (both) in the Not. Bot. Gard. Edinb. **11**, 155–168, 1919 ● Palibin: Rev. gen. *Enkianthus:* in Script. Hort. Bot. Univ. Petrop. **15**, 18–28, 1897 ● Wilson, E. H.: in Gard. Chron. **151**, 311 and 363, 1907.

ENSETE See: **MUSA**

ENTELEA R. Br. — TILIACEAE

Monotypic genus with the characteristics of the species. Closely related to *Sparrmannia*. The wood of this tree is exceptionally light, even lighter than cork.

Entelea arborescens R. Br. Evergreen shrub or small tree, to 6 m high; leaves alternate, large, cordate, 10–20 cm long, long acuminate, double serrate, with small triangular lobes on the upper third of both blade halves, base cordate, stellate pubescent, petiole 10–20 cm long; flowers in terminal, loose cymes, 7–12 cm wide, the individual flowers about 2.5 cm wide, white, with 4–5 sepals and petals each, stamens numerous, all distinct, anthers yellow, May; fruit a globose, bristly, loculicidal capsule, 1.5–2 cm long. BM 2480; KF 33; DRHS 747. New Zealand. 1820. z9–10 Fig. 12. # ⊕ Ø

Easily grown in frost free climates or in the greenhouse.

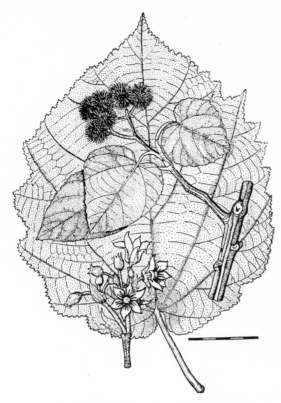

Fig. 12. *Entelea arborescens* (from Poole/Adams)

EPACRIS Cav. — Australian Heath — EPACRIDACEAE

Evergreen shrubs of heath-like appearance, often cultivated as potted plants; leaves alternate, tightly packed, small, entire, usually sharply acuminate, sessile or short stalked; flowers axillary, small, but occurring on elongated branches creating the appearance of terminal spikes or racemes; corolla tubular, limb 5 toothed, white or red to purple shades; stamens 5; ovaries 5 chambered, a capsule or fleshy fruit when ripe. — About 40 species in SE. Australia, Tasmania and New Zealand. Only hybrids usually found in cultivation.

Epacris impressa Labill. Evergreen, upright shrub, 0.3–1 m high, flowers appearing on very young plants, open branched, branches more or less pubescent; leaves sessile, linear-lanceolate to more ovate, long acuminate, apex prickly, base cordate, 15 mm long; flowers somewhat pendulous, corolla campanulate to tubular, 5–15 mm long, from pure white to pink or dark red, limb with 5 short, triangular sections, developing into a one sided "spike", winter to spring. BM 3407; LNH 58. S. Australia, Tasmania. 1825. z9 # ✿

f. **ceriflora** Grah. Flowers waxy, white, short. BM 3243 (as *E. ceraeflora*). Tasmania.

f. **nivalis** Lodd. Branches tomentose; flowers snow-white, flowers abundantly. BR 1531.

Most of the so-called *Epacris* hybrids may be included here. According to Bosse (Handb. d. Blumengärtnerei 2, 46) they number at least 40. Furthermore, around 70 cultivars are listed under *Epacris variabilis* Lodd. (*E. impressa* × *E. longiflora*). PBl II: 293 ('Apfelblüte'); Gs 1922: 57 ('Odoratissima').

E. longiflora Cav. Shrub, to 1 m high, branches usually soft pubescent; leaves ovate or more lanceolate, with prickly tips, scarcely 15 mm long, base round to cordate, markedly veined, upper leaves only about 6 mm long, very short petioled; corolla tube cylindrical, 1–2 cm long, scarlet to carmine-red, limb tips white, flowering mainly during spring. BM 922; FS 121 (as *E. miniata*); Gs 1922: 56; HTS 199. Australia; New South Wales. 1803. Commonly used species for hybridizing. z9–10 # ✿

E. miniata see: **E. longiflora**

E. microphylla R. Br. Upright shrub, branches thin, pubescent; leaves broad ovate, only 4–6 mm long, base cordate, concave, partially clasping the stem, apex short acuminate; flowers pure white, fragrant, 6 mm long, in 5–10 cm long, foliate spikes, May. BM 3658 (= *E. pulchella* Sieb.). Australia, Tasmania. Before 1840. z9–10 # ✿

E. onosmiflora see: **E. purpurascens**

E. paludosa R. Br. Shrub, to 1 m or higher in its habitat, somewhat narrow growing, branches softly pubescent; leaves linear-lanceolate, 9–15 mm long, 2 mm wide, long acuminate, margin scabrous, tapering to both ends; flowers axillary, often clustered at the branch tips, corolla tube pure white, 12 mm long, 6 mm wide at the 5 lobed limb; spring. Australia; New South Wales, Victoria. Before 1840. z9–10 # ✿

E. pulchella see: **E. microphylla**

E. pungens see: **E. purpurascens**

E. purpurascens R. Br. Upright shrub, about 1 m or more, narrow; leaves very densely arranged, ovate, 12–15 mm long, base cordate and concave, clasping the stem, long and sharp acuminate; flowers white, turning purplish or pink, apexes often lighter, corolla tube about 10 mm long, 12 mm wide at the limb, calyx teeth as long as the bowed corolla tube, March–April. BM 844; 199 (as *E. pungens*); 3168 (as *E. onosmiflora*). New South Wales. 1803. z9–10 # ✿

E. variabilis see: **E. impressa**

Cultivated somewhat like *Erica*, preferring a light, humus, acid soil; the branches should be cut back after flowering. Do not irrigate with alkaline water. Where freezing temperatures are likely, overwintering in a greenhouse is advisable.

EPIGAEA L. — Trailing Arbutus — ERICACEAE

Evergreen, creeping shrubs, branches with short pubescence; leaves alternate, entire; flowers in short, terminal racemes, 5 parted; corolla salver-shaped with 5 lobes at right angles from the tube; stamens 10, anthers opening with a long slit; ovaries 5 chambered; fruit a mealy-glutinous capsule with many seeds. — 2 species in N. America and Japan.

Epigaea asiatica Maxim. Creeping habit, only about 3–7 cm high, rooting along the branches, bristly to glandular pubescent; leaves oval-oblong, coarsely veined, abruptly short acuminate, 4–7 cm long, base cordate, deep green above with scattered bristly pubescence at first, lighter beneath with bristly midrib; flowers grouped 3–6, terminal and axillary, on pubescent stalks, corolla with 12 mm long tube, 20 mm wide at the limb, pink-red, tube pubescent inside at the base, otherwise glabrous (*E. repens* pubescent!), limb tips short and broad-ovate, calyx with ovate, 8 mm long lobes, April–May. BM 9222. Japan; Hondo and Yezo. z4 Plate 10. # ✥

E. × intertexta Mulligan (*E. asiatica* × *E. repens*). Intermediate between the parents. Flowers in February–April. z3

Including:

'Apple Blossom'. Flowers pure pink. # ✥

'Aurora' (Marchant). The type of the cross; leaves 6 cm long, to 4 cm wide, margins sinuate; flowers situated above the foliage, pure pink, limb carmine, tube white inside, pubescent, JRHS 64: 116. Introduced in 1931 by Marchant, England. # ✥

E. repens L. Creeping habit, but branches becoming 15–20 cm high; leaves long petioled, oval-rounded, 2–8 cm long, base occasionally somewhat cordate, both sides thin bristly pubescent and ciliate; flowers white to pink, tube 15 mm long, 12 mm wide at the limb, woolly pubescent inside, very fragrant, 4–6 in terminal racemes, March–April; fruit a berrylike capsule. BR 201; NBB 3: 20. Eastern N. America, in sandy-gravelly woods. 1736. Dislikes alkaline soils! z2 Plate 7, 10. # ✥

'Plena'. Flowers double. Found wild in Massachusetts, USA. Quite rare. # ✥

'Rubicunda'. Flowers especially dark pink. Rare. # ✥

Difficult to cultivate; require a humus, lime-free soil; avoid direct sunlight.

Lit. Mulligan, B. O.: *Epigaea* hybrids; in Jour. RHS 1939, 507–510; with 2 Plates.

ERCILLA A. Juss. — PHYTOLACCACEAE

Evergreen climber, climbing like *Hedera* with disc-like holdfasts; leaves alternate, petioled, simple and entire, tough and leathery; flowers small, bisexual, in axillary, tightly packed, spike form racemes; perianth 5 parted; stamens 4–8; fruits berrylike. — 2 species in Chile and Peru.

Ercilla spicata see: **E. volubilis**

E. volubilis Juss. High climbing, evergreen vine, branches round, brown, striped, with holdfasts; leaves ovate to broad-lanceolate, 3–6 cm long, short petioled, leathery tough; flowers in dense, sessile, 2–4 cm long racemes, reddish or whitish, March–April. PFC 157 (= *E. spicata* Moqu.; *Bridgesia spicata* Hook.). Chile. 1840. z9 Fig. 13. #

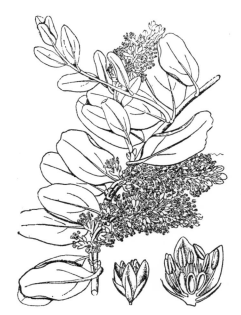

Fig. 13. *Ercilla volubilis* (from Hooker)

ERICA L. — Heath — ERICACEAE

Evergreen shrubs, tree-like in some cases, densely branched; leaves needlelike, usually in whorls, very small; flowers solitary or several to many flowered umbels, racemes or panicles, brightly colored; calyx 4 parted nearly to the base; corolla campanulate, urceolate, inflated tube form or cylindrical, usually longer than the calyx, with a 4 lobed limb, persistent; stamens 8, seldom 6–7; anthers with or without tail-like appendage; fruit a many seeded, 4 valved capsule; seed small. — About 630 species, nearly all in S. Africa, the others in Europe, Atlantic islands, Asia Minor and Syria.

A. EUROPEAN SPECIES

Key to the most important European species and hybrids

○ Leaves and sepals distinctly ciliate margined;
+ Flowers in terminal umbels or corymbs; anthers with appendages;
1. Leaves fine pubescent on both sides; flowers oblong-ovate:
E. tetralix
2. Leaves glabrous above; flowers nearly globose:
E. mackaiana
3. Leaves only sparsely ciliate, glabrous in the middle, above; flowers urceolate:
E. × williamsii
++ Flowers in terminal spikes; anthers without appendages:
E. ciliaris

○○ Leaves and sepals glabrous;
★ Anthers totally enclosed within the corolla tube;
> Branches fine pubescent; flowers in terminal umbels, racemes or panicles; corolla to 6 mm long;
1. Flowers in racemes or umbels up to 7 cm long, purple; anthers with toothed appendages:
E. cinerea
2. Flowers 4–8 in umbels, pink; anthers with entire appendages:
E. terminalis
3. Flowers in large, pyramidal panicles, white; anthers with short appendages; stigma wide, white:
E. arborea
4. Very similar to the above, but with stigma distinctly pink:
E. veitchii
>> Branches glabrous; flowers axillary, forming cylindrical spikes; corolla 2–3 mm long:
E. scoparia

★★ Anthers exserted, without appendages; flowers axillary;
× Sepals always shorter than half the tube length, usually ovate; corolla broad campanulate;
1. Leaves grouped 4–5, erect; flowers in dense spikes:
E. vagans
2. Leaves in 3's, more erect; flowers in loose spikes:
E. manipuliflora
3. Leaves grouped 5–6; flowers in dense spikes:
E. multiflora

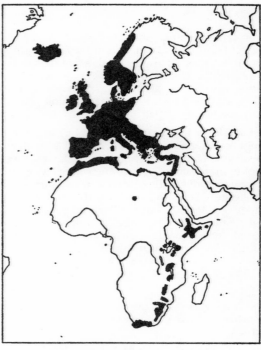

Fig. 14. Range of the genus *Erica*

×× Sepals longer than half the corolla tube length; corolla oblong-ovate;
1. Anthers nearly totally exserted; procumbent shrub:
E. carnea
2. Anthers only semi-exserted; erect shrub, to 1 m high:
E. erigena

Overview of the parentage of the described cultivars of the European *Erica*-species

'Alba'	(*cinerea*)
'Alba'	(*vagans*)
'Alba Minor'	(*cinerea*)
'Alba Mollis'	(*tetralix*)
'Apple Blossom'	(*cinerea*)
'Aragonensis'	(*australis*)
'Ardy'	(*tetralix*)
'Arthur Johnson'	(*darleyensis*)
'Atrorubens'	(*cinerea*)
'Atrorubra'	(*carnea*)
'Atrosanguinea'	(*cinerea*)
'—Reuthe's Variety'	(*cinerea*)
'—Smith's Variety'	(*cinerea*)
'Aurea'	(*carnea*)
'Brightness'	(*erigena*)
'Camla'	(*ciliaris*)
'C. D. Eason'	(*cinerea*)
'Cecilia M. Beale'	(*carnea*)
'C. G. Best'	(*cinerea*)
'C. J. Backhouse'	(*carnea*)
'Coccinea'	(*cinerea*)
'Coccinea'	(*erigena*)

'Connemara'	(*praegeri*)
'Con Underwood'	(*tetralix*)
'Corf Castle'	(*ciliaris*)
'Cornish Cream'	(*vagans*)
'Daphne Underwood'	(*tetralix*)
'Darley Dale'	(*darleyensis*)
'Darleyensis'	(*tetralix*)
'David McClintock'	(*ciliaris*)
'Dawn'	(*watsonii*)
'Diana Hornibrook'	(*vagans*)
'Domino'	(*cinerea*)
'Dr. Ronald Gray'	(*mackaiana*)
'Eden Valley'	(*cinerea*)
'Eileen Porter'	(*carnea*)
'Erecta'	(*darleyensis*)
'Foxhollow'	(*carnea*)
'Foxhollow Mahogany'	(*cinerea*)
'F. White'	(*watsonii*)
'George Rendall'	(*darleyensis*)
'Ghost Hills'	(*darleyensis*)
'Globosa'	(*ciliaris*)
'Golden Drop'	(*cinerea*)
'Golden Hue'	(*cinerea*)
'Gold Tips'	(*arborea*)
'G. Osmond'	(*cinerea*)
'Gracilis'	(*carnea*)
'Grandiflora'	(*cinerea*)
'Grandiflora'	(*vagans*)
'Gwavas'	(*williamsii*)
'Gwen'	(*watsonii*)
'Heathwood'	(*carnea*)
'Helma'	(*tetralix*)
'H. Maxwell'	(*watsonii*)
'Hookstone Pink'	(*tetralix*)
'James Blackhouse'	(*carnea*)
'Katinka'	(*cinerea*)
'Ken Underwood'	(*tetralix*)
'Kevernensis'	(*vagans*)
'King George'	(*carnea*)
'Knap Hill Pink'	(*cinerea*)
'Lawsoniana'	(*mackaiana*)
'L. E. Underwood'	(*tetralix*)
'Loughrigg'	(*carnea*)
'Lyonesse'	(*vagans*)
'Mary Grace'	(*tetralix*)
'Maweana'	(*ciliaris*)
'Mr. Robert'	(*australis*)
'Mrs. C. H. Gill'	(*ciliaris*)
'Mrs. D. F. Maxwell'	(*vagans*)
'Mrs. Dill'	(*cinerea*)
'Mrs. Sam Doncaster'	(*carnea*)
'Nana'	(*vagans*)
'Pallas'	(*cinerea*)
'Pallida'	(*carnea*)
'Pallida'	(*cinerea*)
'Pallida'	(*vagans*)
'P. D. Williams'	(*williamsii*)
'Pink Glow'	(*tetralix*)
'Pink Ice'	(*cinerea*)
'Pink Pearl'	(*carnea*)
'Pink Spangles'	(*carnea*)
'Pink Star'	(*tetralix*)
'Plena'	(*mackaiana*)
'Praecox Rubra'	(*carnea*)
'P. S. Patrick'	(*cinerea*)
'Pygmaea'	(*cinerea*)
'Pyrenees Pink'	(*vagans*)
'Queen of Spain'	(*carnea*)
'Rachel'	(*watsonii*)
'Riverslea'	(*australis*)
'Rosy Gem'	(*carnea*)
'Rosea'	(*cinerea*)
'Rosea'	(*vagans*)
'Rubra'	(*tetralix*)
'Rubra'	(*vagans*)
'Ruby Glow'	(*carnea*)
'Ruby's Variety'	(*tetralix*)
'Schizopetala'	(*cinerea*)
'Silberschmelze'	(*darleyensis*)
'Snow Queen'	(*carnea*)
'Springwood Pink'	(*carnea*)
'Springwood White'	(*carnea*)
'Stapehill'	(*ciliaris*)
'St. Keverne'	(*vagans*)
'Stoborough'	(*ciliaris*)
'Thomas Kingscote'	(*darleyensis*)
'Superba'	(*carnea*)
'Truro'	(*watsonii*)
'Urville'	(*carnea*)
'Valery Proudley'	(*vagans*)
'Vivellii'	(*carnea*)
'W. G. Notley'	(*cinerea*)
'Winifred Whitley'	(*cinerea*)
'Winter Beauty'	(*carnea*)
'Wishanger Pink'	(*australis*)
'W. T. Rackliff'	(*darleyensis*)
'Wych'	(*ciliaris*)

Breeder or developer of *Erica* and *Calluna* forms:

Georg Arends, Wuppertal-Ronsdorf, Germany
J. Backhouse, York, England
J. Brummage, Taverham, Norfolk, England
J. Drake, Aviemore, Inverness-shire, Scotland
Geoffrey Hayes Ltd., Grasmere, England
Herm. A. Hesse, Weener-Ems, Germany
J. F. Letts, Windlesham, Surrey, England.
Maxwell & Beale, Corfe Mullen, Wimborne, Dorset, England
Plantsoendienst, Driebergen, Holland
J. W. Porter, Carryduff, N. Ireland
James Smith & Son, Darley Dale, Derbyshire, England
J. W. Sparkes, Beoley near Redditch, Worcester, England
Treseder & Sons, Truro, Cornwall, England
G. Underwood & Son, Hookstone, Woking, England
Gebr. Verboom, Boskoop, Holland
P. G. Zwijnenburg, Boskoop, Holland

Erica arborea L. Tree Heath. Tall shrub or small tree, to 5 m high, young branches pubescent, hairs branched; leaves in whorls of 3, very densely arranged, 3–4 mm long, smooth, furrowed beneath; flowers on small side branches, grouped in large, panicled inflorescences 20–40 cm long, corolla rounded-campanulate, gray-white, 4 mm long, fragrant, stigma somewhat flat, white, stamens enclosed, March–April. HM 2699; KGC 37; PEu 89. S. Europe, N. Africa, Caucasus. 1658. z9 Fig. 15. # ⊕

var. **alpina** Dieck. Shrub, usually not taller than 1 m, narrow upright habit; leaves lighter green; flowers purer white. DB 1956: 173; HF 2025; JHe 33. W. Spain, mountains in Cuenca Province. 1899. z8 # ⊕

'Gold Tips' (Maxw. & Beale). Similar to the above, but new growth with gold-yellow tips. PH 31. # ⊕

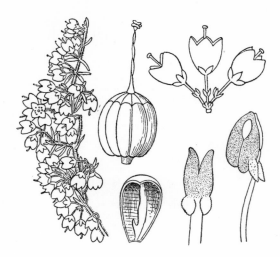

Fig. 15. *Erica arborea,* branch, 3 solitary
flowers and flowering branch (from Reichenbach)

E. australis L. Spanish Heath. A tall shrub in its habitat, to
2 m or more, a small shrub in cultivation, growth erect,
branches soft pubescent when young; leaves in whorls
of 4, bright green, linear, glandular, with prickly tips;
flowers in terminal, umbellate clusters on the previous
year's wood, pink-red, tubular, with reflexed lobes,
flowers abundantly, April–June. JRHS 66: 64; 87: 108;
BM 8045. Spain, Portugal. 1769. z9 Plate 11. # ⊕

'Aragonensis'. 1.2–1.5 m high, very similar to *E. australis,* but
with subtending leaves and sepals glabrous (pubescent on the
species); leaves finer, more densely arranged; flowers
somewhat smaller, but more numerous, pink-red, March–June.
⊕

'Mr. Robert' (R. Williams 1912). Like the species, but leaves a
lighter green. UHH 108; PH 28. Found in the wild near
Algeciras, S. Spain. More susceptible to frost damage, very
popular in England. ⊕

'Riverslea' (Prichard & Sons, Christchurch, Hampshire, Eng.
Before 1946). Narrow habit; flowers somewhat larger, 4–5 in
clusters on short side branches, fuchsia-purple, April–June.
⊕

'Wishanger Pink' (Evans, before 1957). This cultivar is very
reminiscent of *E. × darleyensis,* but with purple-pink flowers
like *E. australis.* Probably no longer cultivated. JRHS 94: 57. ⊕

E. carnea L. Snow Heath. Procumbent shrub, only rarely
over 30 cm high, young branches glabrous; leaves
needlelike, most in whorls of 4, bright green; flowers in
foliate, terminal racemes, 3–10 cm long, all turned to one
side, buds developed the previous fall, corolla oblong-
ovate, flesh-pink, anthers reaching nearly beyond the
corolla, black-brown, sepals lanceolate, nearly as long as
the corolla, December–April. HM Pl. 206; HF 2026 (= *E.
herbacea* L.; *E. mediterranea* L.). Eastern and Central Alps,
in coniferous forests, to high in the mountains, lime
tolerant. 1763. Quite hardy. z6 Fig. 16. #

Includes many cultivars–

Flowers white:
'Alba', 'Cecilia M. Beale', 'Foxhollow', 'Snow Queen',
'Springwood White'

Flowers light pink:
'Aurea', 'C. J. Backhouse', 'Gracilis', 'James Backhouse', 'Mrs.
Sam Doncaster', 'Pallida', 'Pink Beauty', 'Queen of Spain',
'Thomas Kingscote'

Flowers dark pink:
'Heathwood', 'Queen Mary', 'Rose Gem', 'Winter Beauty'

Flowers carmine:
'Atrorubra', 'Eileen Porter', 'King George', 'Loughbrigg',
'Praecox Rubra', 'Ruby Glow', 'Vivelii'

f. **alba** (Dipp.) Schelle. A white form of the species found occasionally in the wild; corolla 5–6 mm long, stalk greenish,
filaments white, anthers dark brown, reaching completely out
of the corolla, apical half of the style pink, flowers sparse, very
early. MaLR 32. ⊕

'Atrorubra'. Growth rather flat, 15–20 cm high; foliage deep
green, somewhat bluish; flowers carmine, not fading, arranged
in clusters, corolla 5–6 mm long, filaments light pink, anthers
dark brown, totally exserted, style pink, flowers later, after mid-
March. ⊕

'Aurea'. 15–20 cm high; leaves golden-yellow, especially on
new growth, later becoming lighter, somewhat reddish in
winter; flowers dark pink, abundant. PH 23. Always appears
chlorotic due to the yellow foliage color. ⊕ ∅

'Cecilia M. Beale' (Maxwell & Beale, around 1920). Low,
globose, only about 10–15 cm high, flowering branches erect;
flowers pure white, corolla large, February–March. ⊕

'C. J. Backhouse' (Backhouse). 15–20 cm; very light pink,
probably the lightest pink of all, but fading to darker pink,
loosely arranged, March–April. ⊕

'Eileen Porter' (J. W. Porter 1934). Growth open, about 15 cm
high; foliage deep green; flowers dark red, abundant and long
lasting, October–April. Very valuable form. ⊕

'Foxhollow' (J. F. Letts). Growth strong, to 20 cm, long
branched; leaves yellow-green, branch tips brownish in spring;
flowers white at first, fading to pink, calyx yellowish-pink,
blooms less abundantly, February–March. PH 24. ⊕

'Gracilis' (Backhouse). Growth very open, graceful, about 15
cm high; leaves finer than those of the species, 4–8 mm long,
deep green; light pink, holds its color well, December–March.
One of the oldest cultivars, the earliest pink. ⊕

'Heathwood' (J. Brummage 1963). 25 cm, resembles
'Loughrigg', but wider and bushier; leaves darker green;
flowers dark pink, buds yellowish, stalk dark brown, February–
April. ⊕

'James Backhouse' (Backhouse 1911). 25 cm, branches coarse,
stout; leaves light green; flowers pale pink, flowers large,
arranged completely around the flowering branch, anthers
only slightly exserted, March to April. Meritorious. ⊕

'King George' (Backhouse 1911). Completely identical to the
plant often cultivated as **'Winter Beauty'.** ⊕

'Loughrigg' (G. Hayes). 15 cm high; leaves light green, often
also somewhat bluish, branch tips bronze in winter; flowers an
intense purple, flowers abundantly, February–March. ⊕

'Mrs. Sam Doncaster' (Backhouse). Branches procumbent;

leaves rather loosely arranged, gray-green, good ground cover; light pink, January–April. ✿

'Pallida'. 15 cm; leaves small, dull green; light pink, nearly white at first, becoming darker. ✿

'Pink Pearl' (Backhouse). Growth to 15–20 cm; leaves light green; flowers light pink, quickly fading, corolla large, March–April, does not flower heavily. (= 'Pink Beauty'). ✿

'Pink Spangles' (Treseder). Growth very strong, to 30 cm, a good ground cover; dark pink-red, sepals lilac and erect, flowers profusely with large blossoms, February–April. LHb 74. ✿

'Praecox Rubra' (Backhouse). Only about 15 cm high, branches widely, a good ground cover; dark pink-red, not fading, small flowers, but lasting over a long period, December–March. LHb 53; PH 13 (= 'Rubra'). Very worthy cultivar. ✿

'Queen of Spain' (Backhouse). Growth open, 15 cm; leaf tips reflexed downward (very important distinction, observed only on this cultivar); flowers light pink, buds darker, February–April. ✿

'Rubra' see: **'Praecox Rubra'**

'Rosy Gem'. Very dense habit, about 15 cm high; bright pink, not fading, March–April, very prolific bloomer. ✿

'Ruby Glow' (Backhouse). Very similar the the well known 'Vivellii', but flowers somewhat lighter carmine-red; purer in tone, March–April; foliage likewise dark green, bronze in fall. PH 25. Very meritorious. ✿

'Snow Queen' (Verboom 1934). Branches somewhat slack; leaves light green; flowers abundant, pure white, large, well displayed against foliage, January–March. Often confused with 'Cecilia M. Beale', which is more globose with erect branches and flowers 4 weeks later. ✿

'Springwood Pink'. Growth strong, similar to 'Springwood', but not a form of it. Occurred as a seedling in a Scottish nursery, branches very long, good ground cover, 20 cm; light clear pink, very prolific bloomer, to 3 flowers in a leaf axil, especially at the branch tips, January to March. UHH 108. Not as good as 'Pink Spangles'. ✿

'Springwood White' (Mrs. R. Walker). To 25 cm high, but branches often much longer; leaves light green; pure white, buds yellowish, all along the branch, anthers light brown, January–March. GC 139: 518; HTS 226; LHb 53. Found before 1925 on Monte Correggio, Italy and named for the residence of Mrs. Walker (in Springwood, Stirling, Scotland). Excellent ground cover. Plate 11. ✿

'Thomas Kingscote' (Backhouse). Small flowers, few, pale pink, anthers dark brown, totally exserted, March to April. ✿

'Urville'. Presumably a misspelling of **'Vivellii'.**

'Vivellii' (Theoboldt 1919). 15 cm high; leaves blackish, dark green in summer, bronze to nearly red-brown in winter; flowers dark red to carmine, not fading, January–March. UHH 125. Discovered by Theoboldt, of Aulenback in Württemberg, W. Germany and named for the Vivell nursery, Olten, Switzerland. Very important, garden worthy cultivar. ✿ ∅

'Winter Beauty' (Backhouse). Very dense and compact, 15 cm; leaves small, dark green; flowers a clear, deep pink, exceptionally abundant, December–March. PH 14 (as 'King George'). See notes after 'King George'. ✿

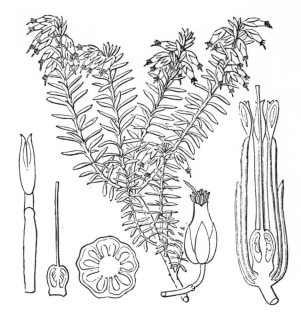

Fig. 16. *Erica carnea* (from Lauche)

E. ciliaris L. Procumbent growth habit, growing to 30 cm high, branches ascending, with short and glandular pubescence; leaves in whorls of 3, ovate, with long glandular hairs, otherwise glabrous, margin somewhat involuted, dark green above, gray-white beneath; flowers 3–4 together, grouped in terminal spikes, corolla urceolate, pink-red, 8–10 mm long, limb tips rounded, glabrous, stamens not exserted, without appendages, June–July. BM 484; DRHS 766; ChHe 82; UHH 142. S. Ireland, SW. England, S. and W. France, W. Spain, NW. Morocco; sandy areas, peat bogs. Dislikes limestone. z8 Fig. 18 # ✿

'Camla' (Ingwersen 1934). Growth wide and bushy, good ground cover; leaves green, coarse; pink-red, flowers large, very abundant, July–October. z8–9 ✿

'Corf Castle' (G. Osmond, around 1962). Wide and compact, to 30 cm; leaves green like the species, but bronze-green in winter; salmon-pink, well displayed over the foliage, July–September. PH 53. ✿

'David McClintock' (Aldenham Heather Nursery 1962). Strong grower, to 40 cm and very wide; leaves gray-green; flowers white, but limb violet-pink (bicolor), eventually becoming totally pink, July–October. PH 52. Found near Carnac, Brittany, France and named after the discoverer. ✿

'Globosa' (Maxwell & Beale, around 1930). Growth strong, to 40 cm high, branches and leaves gray-green; mallow-pink, flowers large, prolific bloomer, July–October. PH 133. The most winter hardy cultivar. ✿

'Maweana'. Grows more strongly than the species, about 30 cm, narrowly upright; flowers carmine, larger than the type, June–October. BM 8443 (as *E. ciliaris* N. E. Br. non L.). Found by George Maw in Portugal, 1872. z9 ✿

'Mrs. C. H. Gill' (Maxwell & Beale). Growth denser and less wide than the species, 25–30 cm; leaves emerald-green, soft woolly; flowers pure red, July to October. Earliest cultivar. LHb 74. ✿

'Stapehill' (C. J. Marchant). 25 cm high, long flowering branches; cream-white, turning purple, July to October. ✿

'Stoborough' (Maxwell & Beale). Very strong grower, 50–60 cm high; foliage apple-green; pure white flowers, 15–20 in heads, corolla globose, buds yellowish, July–October. ✿

'Wych' (Maxwell & Beale). 40–45 cm high; flowers cream-white, with a trace of light pink, in long inflorescences, July–October. Found in the wild in Wych Heath, Dorset, England. ✿

E. cinerea L. Gray Heath. Shrub, 30–40 cm high, branches procumbent-ascending, short haired; leaves in whorls of 3, ovate-oblong, distinctly involuted, quite fine ciliate-bristly; flowers in terminal racemes or umbels, corolla ovate-urceolate, pink, stamens enclosed, both anther halves completely separate, each with an appendage at the base, July–August. HF 2024; HM 2704. W. Europe; heaths, light forests, dry warm slopes. The most common species in England. Dislikes alkaline soil. # ✿

The most important forms of about 100 cultivars found in England are:

'Alba'. To 25 cm high; leaves a nice apple-green; white, long racemes. In cultivation since 1818. ✿

'Alba Minor'. Only 15 cm high, very dense; leaves medium green; white, flowers abundantly, June–August, then again in October. LHb 53; PH 55. ✿

'Apple Blossom' (Maxwell & Beale). To 30 cm high, upright; foliage light green; white, turning pink, large flowered, in long spikes, very prolific bloomer, June–August. JMe 65. z9 ✿

'Atrorubens' (1915). 15–20 cm high, growth broad; foliage gray-green and pubescent; ruby-red, flowers abundantly. LHb 53; HTS 225. Very attractive. ✿

'Atrosanguinea'. Resembling 'Coccinea', but somewhat larger in all respects, 15 cm high, June–August. PH 54. ✿

From 2 different sources in the trade:

'Reuthe's Variety' (Reuthe 1926). Somewhat larger flowering and taller than 'Coccinea', otherwise identical.

'Smith's Variety'. (J. Smith & Sons 1852). Resembles 'Atrorubens', but with flowers somewhat more carmine-pink, foliage dark green.

'C. D. Eason' (Maxwell & Beale 1931). Dense, bushy, to 20 cm high, branches and leaves deep green; luminous pink-red, without a trace of blue, July–August. LHb 53; ChHe 81; PH 60, 121. Highly valued. ✿

'C. G. Best' (Maxwell & Beale 1931). 30 cm high, young branches greenish-red, later gray, pubescent; salmon-pink flowers on long branches with much space between the floral whorls (!), June–August. JRHS 96: 181. ✿

'Coccinea'. (J. Smith & Sons 1852). Only 10–15 cm high, branches dark green, new growth bronze; leaves 2–3 mm long; carmine-red to carmine-scarlet, June–August. LHb 74. ✿

'Domino' (Maxwell & Beale). To 20 cm, branches light gray at first, later dark green; corolla white, calyx and floral axis ivory-white, anthers deep brown at first, eventually nearly black, June–September. One of the best white cultivars. ✿

'Eden Valley' (G. Walker 1926). Broad, spreading habit, young branches gray-green, later light green; lilac-pink at first, later tending to lilac, fading toward the base and therefore distinctly bicolored, June–September. PH 57. ✿

'Foxhollow Mahogany' (J. F. Letts). To 30 cm high, broad, open growth, branches somewhat limp; dull wine-red, abundant flowers, situated well above the dark green foliage, June–September. PH 72. ✿

'Golden Drop' (Charles Eason; Maxwell & Beale). Growth mat-like, 10–15 cm high, new growth coppery, leaves later yellow-brown, coppery-red in winter; flowers lilac-pink, but sparse, July–August. LHb 74. z9 ✿ ∅

'Golden Hue' (Maxwell & Beale). 30 cm high; leaves yellow in summer, bronze to nearly red in winter; flowers well, pink. MaLR 54; LNG 53. Easily distinguished from the previous cultivar by the light summer color and the much stronger growth habit. ✿ ∅

'G. Osmond' (Osmond; Maxwell & Beale 1931). Growth bushy, upright, to 35 cm; pale lilac, calyx and flower stalk ivory-white, in long spikes, June to September. PH 66. ✿

'Grandiflora'. 30 cm; pure purple-pink, flowers large, in large spikes, June–August. An old cultivar, but still of much value. ✿

'Katinka' (Plantsoendienst Driebergen, Holland, 1968). 30 cm;

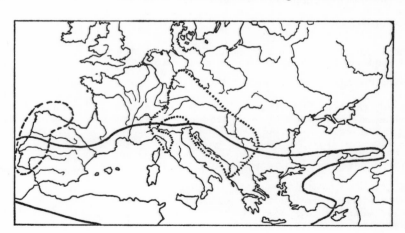

Fig. 17. Range of *Erica arborea* (———), except Canary and Madeira Islands), *E. australis* (--------) and *E. carnea* (...........) (from Underhill)

foliage dark green; blackish-purple, darkest cultivar of all, June–September. Darker and somewhat lower growing than the English cultivar 'Velvet Night'. ✤

'Knap Hill Pink' (Waterer). Branches long, to 30 cm high; leaves more olive-green; dark pink-red, July to September. PH 62. Very similar to 'Rosea' but flowers somewhat darker. ✤

'Mrs. Dill'. Only about 10 cm high, nearly hemispherical habit; leaves bright green; dark pink, very pretty color, June–August. ✤

'Pallida' (Maxwell & Beale 1927). Upright growth habit, 15–30 cm; pale lilac-pink, in long, dense spikes, June–August. ✤

'Pallas' (1970). 35 cm; pure lilac, flowers very abundantly, June–September. Good, winter hardy cultivar from the Dutch nursery trade, grown erroneously under the name 'Pallida' until 1970 and then corrected. ✤

'Pink Ice' (J. F. Letts, before 1968). 20 cm; foliage dark green, more bronze-green on new growth and in winter; pure pink, flowers large, abundant, June to September. ✤

'P. S. Patrick' (Patrick; Maxwell & Beale, around 1928). Strong grower, 40 cm; bright violet-purple, in large, slender spikes, August–September. UHH 125. ✤

'Pygmaea' (Reuthe, before 1908). Only about 10 cm high, creeping; pink-red, June–July. One of the earliest cultivars. ✤

'Rosea'. Branches long, 25 cm; light carmine-pink, June to August. ✤

'Schizopetala' (Maxwell & Beale). 35 cm; leaves green with brown limb; pale lilac, corolla split into 4 petal-like lobes, June–July (= × *Ericalluna bealeana* 'Schizopetala'). ✤

'W. G. Notley' (Maxwell & Beale). About 25 cm high; leaves like those of *E. cinerea*; dark pink, corolla split into 4 segments, June–September, style stunted (= × *Ericalluna bealeana* 'W. G. Notley'). Plate 11. ✤

'Winifred Whitley' (Waterer 1936). Differing only slightly from the above; to 30 cm high; flowers pale lilac-pink, corolla split into 4 segments, these radiating in stellate fashion, to 12 mm wide, August–September (= × *Ericalluna bealeana* 'Winifred Whitley'). Found in the wild in Cornwall, England.

E. codonodes see: **E. lusitanica**

E. crawfurdii see: **E. mackaiana 'Plena'**

E. × darleyensis Bean (= *E. carnea* × *E. erigena*). Leaves, young branches and flowers not distinguishable from the those of the parents, especially on young plants, but flowering from November to May, growth stronger than *E. carnea*, but less vigorous than *E. erigena*, about 40 cm. Developed about 1890 in the nursery of James Smith & Son, Darley Dale, Derbyshire, England. z6

Includes the following cultivars:

'Darley Dale'. The type of the cross described above; habit broad upright, rather open; flowers light lilac-pink. JHe 97; ChHe 120; LHb20; JRHS 66: 67 (= *E. darleyensis* Bean; *E. darleyensis* 'Böhlje'; *E. mediterranea hybrida* Hort.). Tolerates alkaline soil. z8 ✤

'Erecta'. Growth bushy but open upright, to 50 cm; leaves bright green; flowers purple-pink, December–May (= *E. mediterranea* 'Erecta'). z8 ✤

Fig. 18. *Erica ciliaris* (from Bot. Cab.)

'Arthur Johnson' (A. T. Johnson, before 1952) (presumably *E. erigena* 'Glauca' × *E. carnea* 'Ruby Glow'). To 60 cm high and 1 m wide; leaves light green; flowers deep pink, in nearly 20 cm long spikes, December–April. PH 18. z8 ✤

'George Rendall' (Maxwell & Beale). 50 cm; similar to 'Darley Dale', but growth denser, branch tips yellow-green in spring and early summer, later dark green; dark pink flowers, November–March. ChHe 120. z8 ✤

'Ghost Hills' (J. H. Brummage). Growth broad upright, 40 cm; dark pink-red, darker than 'Arthur Johnson', anthers and style markedly exserted, December–April. z8 ✤

'Silberschmelze' (G. Arends 1937). Foliage deep green, often somewhat reddish in winter, coarse and tough; flowers silvery-white, in very long spikes, December–April. PH 3, 36. Originated as a mutation of 'Darley Dale'. Known by a number of synonyms in England. Very winterhardy. z8 ✤

E. erigena R. Ross. Tall Spring Heath. Dense, bushy, upright shrub, 1–2.5 m high, branches glabrous; leaves in whorls of 4, linear, 4–8 mm long, deep green, coarser than *E. carnea*, with a light limbed furrow beneath; flowers ovate-cylindrical, 6 mm long, pink-red, solitary or paired in the leaf axils at the ends of the previous year's branches, buds develop during summer, in densely foliate 3–5 cm long spikes, calyx lobes narrow-oblong, slightly more than half as long as the corolla tube, anthers dark red, exserted, flower stalk 3 mm long or less, March to May. BS 2: 105; UHH 54 (= *E. mediterranea* auct. non L.; *E. mediterranea* var. *hibernica* Hook. & Arn.; *E. hibernica* [Hook. & Arn.] Syme). SW. France, Portugal, Spain, Ireland, but not in the Mediterranean region except the Balearic Islands (from I. Hansen). 1648. z9 ✤

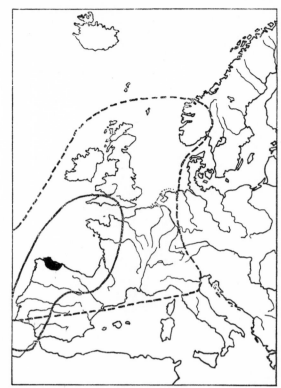

Fig. 19. Range of *Erica ciliaris* (———), *E. cinerea* (------)
and *E. mackaiana* (■■■) (from Underhill)

Includes the following cultivars:

'**Brightness**'. To 50 cm high or higher, strongly branched;
leaves dark green, tips bronze in winter, becoming smaller and
more densely packed at the branch tips; flowers purple-red
nearly like *Erica carnea* 'Vivellii'), buds bronze, March to May.
ChHe 82; GC 139: 687. From Ireland. Very worthy. ۞

'**Coccinea**'. To 90 cm; very similar to 'Brightness', but with
flowers more purple-pink, buds deep bronze, March–May.
۞

'Darleyensis' see: **E. × darleyensis** 'Darley Dale'

'**Nana**'. Growth dense and compact, 40–45 cm high; foliage dull
gray-green; flowers pale pink, March–April. PH 35. ۞

'**Superba**'. Strong grower, densely branched, to 1.8 m high;
foliage dark green, lighter in spring; flowers pink-red,
especially large, clustered in the leaf axils, with the fragrance of
honey, March to May, buds yellowish-white. PH 7. ۞

'**W. T. Rackliff**' (Maxwell & Beale). Significant improvement
over 'Alba', compact habit, to about 50 cm high; young
branches light green, older ones dark green; flowers pure
white, very numerous, on the branch tips, January–April. JHe
96; GC 139: 518; PH 29. ۞

E. herbacea L. see: **E. carnea** L.

E. hibernica see: **E. erigena**

E. hybrida is an old collective name for a number of
hybrids which today are referred to as *E. × darleyensis, E.
× stuartii, E. × watsonii* and *E. × veitchii*, which see.

E. lusitanica Rud. Portuguese Heath. Similar to *E. arborea*
in appearance, but shorter (to 3 m), flowers wide, more
campanulate, appearing earlier; young branches
pubescent, hairs unbranched; leaves needlelike, light
green, 6 mm long; flowers in terminal fascicles on short
side branches, grouped in foliate panicles, corolla white,
fading somewhat reddish, campanulate tubular, apex
acute, erect, calyx tips white, rhombic, stigmas and
anthers pink, February–May. BM 8018; JHe 81; ChHe
99; UHH 162; PEu 89; PH 23 (= *E. codonodes* Lindl.) SW.
Europe. z9 # ۞

E. mackai see: **E. mackaiana**

E. mackaiana Bab. Small shrub, 15–40 cm high, erect at
first, later broad growing, resembling *E. tetralix*, but with
leaf margins not so distinctly involuted, therefore
appearing wider; leaves in whorls of 4, usually dark
green and glabrous above, white beneath; flowers in
dense, terminal umbels (like those of *E. tetralix*), but the
corolla shorter, wider and darker pink, August to
September; seed capsule glabrous. ChHe 100 (= *E.
mackai* Hook.; *E. tetralix* 'Mackaiana'). Ireland; also found
in Spain. 1833. Earlier considered a hybrid of *E. ciliaris* ×
E. tetralix.

Most of the cultivars once included here are now placed under
E. × watsonii, some under E. × williamsii.

'**Dr. Ronald Gray**' (Maxwell & Beale 1964). Only 15 cm high;
flowers white, July–September. PH 48; JRHS 96: 184. ۞

'**Lawsoniana**'. Only 15 cm high, similar to *E. tetralix*, but with
shorter leaves, wider and dark green; flowers pale pink, July–
September. Discovered in Connemara, Ireland. ۞

'**Plena**'. About 15 cm high, loosely branched; flowers nearly
globose and densely double, in 2 pink tones, July–August.
UHH 165; PH 41 (= *E. tetralix* 'Plena'; *E. crawfurdii*). Discovered
in W. Galway, Ireland before 1914. Very pretty, but little
known. ۞

E. manipuliflora Salisb. Very closely related to *E. vagans*,
but with stiff branches, sparse, to 1 m high, gray-white;
leaves in whorls of 3, more erect, 4–6 mm long; flowers
in radially arranged whorls, pink to red, July–August.
UHH 167 (= *E. verticillata* Forsk.). Greece, the Orient. z9
۞

E. mediterranea see: **E. carnea, E. erigena**

E. multicaulis see: **E. terminalis**

E. multiflora L. Developing into a small tree in its habitat,
but only 30–50 cm in cultivation, branches ascending,
glabrous; leaves in whorls of 4–5, linear, 6–9 mm long,
somewhat pubescent at the base; flowers in erect,
terminal, 4–8 cm long, dense racemes, corolla cupulate,
pink, 5 mm long, lobes oblong, stamens exserted, sepals
long and narrow, November–February. UHH 173. S.
Europe, Algeria. Very similar to *E. vagans*, but easily
distinguished by the anthers (here both anther halves
connate but slightly emarginate at the apex, while on *E.
vagans* both halves are more or less totally distinct). z9
Fig 21. # ۞

E. × praegeri Ostenfeld. Collective name for the hybrids between *E. mackaiana* and *E. tetralix,* occurring wherever the two species are found together.

'Connemara'. Very dense, 15 cm high; flowers light pink, flowers abundantly, June–October. ChHe 100; PH 39, 40 ('Irish Lemon'). W. Ireland; Connemara, Donegal. Differing from *E. mackaiana* in the partly pubescent sepals and upper portion of the ovary. ☼

E. scoparia L. Shrub to small tree, more or less upright, to 4 m high, young branches smooth, red-brown; leaves 3–4 in whorls, linear, 6 mm long; flowers small, greenish-white, not very conspicuous, in terminal, foliate racemes, corolla rounded-urceolate, anthers enclosed, May–June. UHH 177. Western Mediterranean region. 1770. In S. France and Spain the branches are made into brooms. z9 Fig. 22. #

'Pumula'. Only about 30 cm high, growth dense; leaves a pretty glossy green (= 'Nana'). #

E. sicula see: **Pentapera sicula**

E. × stuartii Linton. About 30 cm high, closely resembles *E. tetralix;* branches dark green, leaves glossy; differing from the similar *E. × praegeri* in the distinctly bicolored corolla, light pink with dark pink tips, corolla much narrowed at the apex, June–September. PH 108 (= *E. hybrida* 'Stuartii'). Found around 1920 by Ch. Stuart in County Galway, Ireland. There is still much debate over the true parentage of this hybrid and therefore *E. mackaiana × E. mediterranea* cannot yet be substantiated. ☼

E. stricta see: **E. terminalis**

E. terminalis Salisb. Growth narrowly upright, to 80 cm high, very stiff, young branches white pubescent, very densely foliate; leaves in whorls of 4, fine pubescent; flowers 4–8 in terminal umbels, corolla tubular-urceolate, pink-red, limb tips reflexed, brownish, anthers not exserted, July–September. BM 8063; UHH 127, 179 (= *E. stricta* Donn; *E. multicaulis* Salisb.). Corsica, Sardinia, S. Spain. 1765. Thrives on alkaline soils, but not particularly attractive. z9 Fig 22. #

E. tetralix L. Moor Heath. To 40 cm high, branches and leaves glandular and gray pubescent; leaves in whorls of 4, needlelike; flowers in terminal umbels, grouped 5–12 together, the individual flowers nodding, corolla barrel-shaped, pink, stamens enclosed, calyx soft pubescent and ciliate, June–September. HM Pl. 207; HF 2023. W. and N. Europe, to Ireland; on peat bogs, moist pastures, etc. z3 Fig. 24. # ☼

Includes the following cultivars:

f. **alba** (Ait.) Braun-Blanquet. Flowers white. Commonly found in the wild. ☼

'Alba Mollis'. Vigorous grower, 20 cm; foliage silver-gray, especially during spring and summer, greener during fall and winter; flowers white, June–October. PH 49. ☼

'Ardy' (P. G. Zwijnenburg 1974). Growth broadly upright, 25 cm; foliage darker gray-green; flowers darker pink-red, small, July–August. LHG 109. The best red of the *tetralix* cultivars. ☼

'Con Underwood' (G. Underwood & Son 1938). Ornamental habit, to 35 cm high; leaves dark gray-green; flowers dull carmine-red, numerous, June–October. JHe 96; GC 139: 518. Still one of the best cultivars. ☼

'Darleyensis' (J. Smith & Sons). Open growing, to 15 cm high; leaves gray-green; flowers salmon-pink, July–August. ☼

'Daphne Underwood' (G. Underwood & Son 1953). Compact, to 25 cm; foliage gray-green; flowers abundantly, carmine-pink, June–September. ☼

'Helma' (P. G. Zwijnenburg 1967). 40 cm; flowers lilac-pink, erect on all sides (!), very prolific bloomer, July–September. LHG 73; PH 49. Found wild in Holland in 1965. ☼

'Hookstone Pink' (G. Underwood & Son 1953). Strong grower, 20–30 cm; foliage silver-gray the entire year; flowers light pink, buds light brown. ☼

'Ken Underwood' (G. Underwood & Son 1951). Broad upright habit, 30 cm; foliage dark gray-green; carmine-pink, flowers very abundantly, June–August. Quite worthy. ☼

'L. E. Underwood' (G. Underwood & Son, about 1937). Growth narrowly upright, 30 cm; leaves light gray-green; flowers yellowish-pink ("apricot colored"), buds brown-red, June–October. PH 51. Good flower display, not reliable every year. ☼

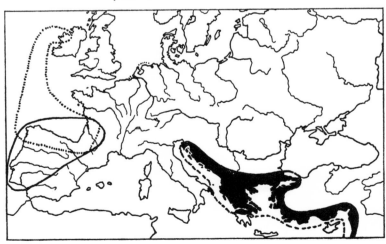

Fig. 20. Range of *Erica lusitanica* (———), *E. manipuliflora* (■■■) and *E. mediterranea* (..........) (from Underhill)

'Mackaiana' see: **E. mackaiana**

'Mary Grace' (Maxwell & Beale). 15 cm high; leaves silver-gray; pink to light red, calyx deeply incised, June–October. ⊕

'Mollis' see: **'Alba Mollis'**

'Pink Glow'. 20 cm; foliage silver-gray; pink, rather small flowers, but very numerous, July–August. ⊕

'Pink Star' (J. F. Letts). Low growing, wide, loose, 15 cm; foliage gray-green; pure lilac-pink, flowers erect all around the stem (!) and directed upward. June–October. LHG 74. ⊕

'Plena' see: **E. mackaiana 'Plena'**

'Rubra'. About 15–20 cm high; leaves medium green to olive-green; flowers a good red, July–September. # ⊕

'Ruby's Variety'. About 15 cm high; flowers mostly white, the others purple-pink, also distinctly bicolored. # ⊕

'Silver Bells'. Also 15 cm high; flowers silvery-pink, June–October. Very attractive. # ⊕

E. umbellata L. Shrub, 30–70 cm high, branches erect or twisted, young branches fine pubescent; leaves usually in whorls of 3, linear, 4 mm long; flowers 3–6 in umbels at the branch tips, May–June, corolla ovate-urceolate, pink or red, 4 mm long, limb tips wide, calyx half as long as the corolla, lobes linear-oblong, anthers exserted, incised to the middle. JHe 97; NF 4: 269; UHH 143, 186; GC 129: 74. Spain, Portugal, Morocco. z9 # ⊕

E. vagans L. Wide, vigorous grower, to 30 cm high; leaves 4–5 in whorls, dark green and glossy, to 1 cm long; flowers in dense, terminal, cylindrical racemes, corolla nearly globose, pink, long stalked, anthers exserted, split half their length, July to September. HF 2027. W. Europe, Ireland to Portugal. z7–9 Fig. 24. # ⊕

Includes the following cultivars:

'Alba'. Like the species, but with white flowers, anthers dark brown, exserted; leaves dark green. Also found in the wild. ⊕

Alba minima see: **'Nana'**

Alba minor see: **'Nana'**

'Cornish Cream' (Treseder & Sons). 50 cm; cream-white, in generally narrow, tapering spikes, August–October. Pretty, new form found in the wild. ⊕

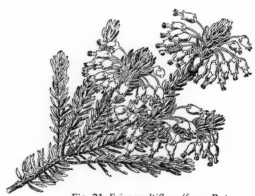

Fig. 21. *Erica multiflora* (from Bot. Cab.)

'Cream' (Smith & Son, before 1968). Strong grower, 40–50 cm; foliage dark green; flowers white, turning pink, anthers reddish-brown, general impression is a cream color, July–October. PH 110. ⊕

'Diana Hornibrook'. Dense and compact, growth 20–30 cm high; foliage deep green, lighter when young; flowers a good red, very numerous, in 5 cm long spikes. PH 134. ⊕

'Grandiflora'. Growth loose, to 70 cm high; inflorescences to 20 cm long, nearly clear pink, fading to reddish, August–September. Occasionally confused with the much lower 'Rosea', which flowers light pink. Not especially attractive. z9 ⊕

'Kevernensis' see: **'St. Keverne'**

'Lyonesse' (Maxwell & Beale 1925). Strong grower, compact, 30 cm, branches light gray-yellow; leaves deep green; flowers cream-white, very densely arranged, anthers light brown-yellow, August–September. ChHe 199. Best white cultivar. (The name "Lyonesse" refers to the legendary region beneath the sea between Cornwall and the Scilly Isles off the SW. tip of England. ⊕

'Mrs. D. F. Maxwell' (Maxwell & Beale, before 1925). To 35 cm high, young branches yellow-brown; leaves medium sized, wide, deep green; flowers dark pink, very densely compact, August–September. ChHe 102; MaLR 62: PH 117; HTS 227; LGH 106. One of the most commonly cultivated forms. Plate 8. ⊕

'Nana'. Scarcely over 20 cm high, growth very dense and compact; leaves yellowish-green in winter; flowers cream-white, August–September (= 'Alba Minima'; 'Alba Minor'). Quite winter hardy. ⊕

'Pallida' To 60 cm high, open growing; inflorescences small, flowers light pink, August–September. ⊕

'Pyrenees Pink'. (G. Underwood & Son). Compact grower, 35 cm high; very similar to 'St. Keverne' and 'Mrs. D. F. Maxwell', flowers pure salmon-pink, somewhat darker than the former, later becoming lighter, August–September. PH 115. Much valued. ⊕

'Rosea'. Strong grower, 40–50 cm high, thickly branched; flowers pink-red, but the inflorescence only of medium length, August–September. ⊕

'Rubra'. Strong, broad, bushy grower, 50 cm high; flowers dark purple-red, in long racemes, August–October. PH 124. ⊕

'St. Keverne'. Bushy and compact habit, 35 cm; leaves bright green; flowers especially campanulate, pure salmon-pink (occasional branches with lighter flowers!), very prolific bloomer, August–September. ChHe 119; MaLR 58 (= *E. vagans kevernensis* Turrill). Discovered in a moor in St. Keverne, Cornwall before 1914 by P. D. Williams of Lanarth. Quite a worthy form. ⊕

'Valerie Proudley'. (Proudley's Heather Nursery 1968). Growth broad, upright; foliage light golden-yellow; flowers white, 20 cm, August–September. PH 10. The first yellow-leaved form. ⊕

'Viridiflora'. Growth like the species, 30 cm; flowers very small, sea-green with some lilac flowers intermixed, August–September. Only of botanical interest.

E. × veitchii Bean (*E. arborea* × *E. lusitanica*). Intermediate between the parents; upright habit, hairs on the branches partly branched (like those of *E. arborea*),

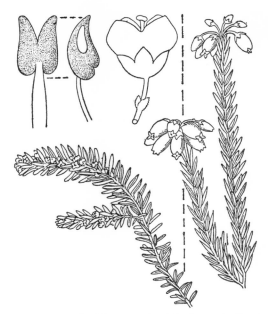

Fig. 22. Left, *Erica scoparia*; right *E. terminalis*
(from Reichenbach, Andrew)

and some unbranched (like *E. lusitanica*), young branches fine pubescent; leaves light green; flowers white, intermediate in form, stigma flattened, pink, anthers pink-red. GC 1905: 97–100. Developed about 1900 by Veitch, Exeter, England. z9 Plate 11. # ☉

'**Exeter**'. The type of the cross. LHG 72; UHH 144.

E. verticillata see: **E. manipuliflora**

E. × watsonii (Benth.) Bean. Group of hybrids *E. ciliaris* × *E. tetralix*. The type was discovered in 1839 by Bentham and described as *E. ciliaris* var. *watsonii*; it is now known as 'Truro', which see.

This group includes the following cultivars:

'**Dawn**' (Maxwell & Beale 1925). Growth broad to upright, 20–25 cm; leaves similar to *E. ciliaris*, but somewhat darker, branch tips yellow-orange in spring and early summer, glandular pubescent; flowers deep lilac-pink, in dense spikes, July–September. MaLR 71; PH 122. ☉

'**F. White**' (Maxwell & Beale). To 20 cm high, open branched, branches and leaves light green, pubescent; very similar to 'Dawn', but flowers with a trace of pink, smaller, and more prolific, June–October. ☉

'**Gwen**' (Maxwell & Beale). About 15 cm high; leaves deep green, often also coppery; flowers pale pink with a trace of lilac, similar to *E. tetralix*, June–October. ☉

'**H. Maxwell**' (Maxwell & Beale). Like *E. ciliaris* in habit and foliage, bushy, upright, 30 cm, branch tips golden-brown in spring; flowers like those of *E. tetralix*, large, dark pink, somewhat lighter than those of 'Dawn'. MaLR 69; ChHe 121; LHG 109. Pubescence not glandular. Likes moist situations. ☉

'**Rachel**'. 25 cm; flowers dark pink, well displayed over the dark green foliage, glandular pubescent, August to October. ☉

'**Truro**'. Prostrate habit, 15 cm; leaves narrow; similar to *E. tetralix* in appearance, but flowers in short racemes, corolla longer, obliquely urceolate, carmine-pink, July–October (= *E. hybrida* 'Watsonii'). Discovered in 1839 by H. C. Watson. ☉

E. × williamsii Druce. (*E. tetralix* × *E. vagans*). Shrub, low and compact, erect, to 20 cm high, branches light brown, later dark, slightly pubescent; leaves 4 together, light green, branch tips golden-yellow during winter; flowers 5–6 in umbels, 10 umbels in a terminal inflorescence, corolla ovate-campanulate, pink, stamens enclosed, anthers brown, July–September. Discovered about 1910 by P. D. Williams.

'**Gwavas**' (Miss Waterer 1924). 30 cm high; young leaves golden-yellow at first, but soon greening; flowers pink-red, July–October. ☉ ⊘

'**P. D. Williams**'. The type of this cross with the above described characteristics. PH 58. Very rare, natural hybrid. ☉

Most of the Heaths prefer a sandy-humus, acid soil, but *E. carnea*, *E. erigena* and *E. terminalis* will succeed as well in an alkaline clay soil. *E. cinerea* needs a warm, sandy soil and will not tolerate alkalinity. If the plants become leggy they should be sheared back in very early spring.

Lit. Hansen, I.: Die europäischen Arten der Gattung *Erica*; in Bot. Jahrb. **75**, 1–89, 1950 ● Chapple, F. J.: The Heather Garden; 180 pp., London 1952 ● Johnson, A. T.: Hardy Heaths and some of their nearer allies; rev. ed. 127 pp., London 1956 ● Maxwell, D. F.: The low road; hardy heathers and the heather garden; 105 pp., London 1927 ● Hondelmann, W.: Die winterharten Gartenheiden; in Deutsch Baumschule 1956, 92–102, 146–152, 161 to 173 ● Maxwell, D. F. & P. S. Patrick: The English Heather Garden; 184 pp. London 1966 ● Proudley, B. & V.: Heathers in colour; 192 pp., 141 color photos; London 1974. The best illustrated book of cultivars ● Letts, J. F.: Hardy Heaths and the Heather Garden; 127 pp., 1966 ● Underhill, T. L.: Heaths and the Heathers; *Calluna, Daboecia* and *Erica*; 256 pp., Newton Abbot 1971 ● Van de Laar, H.: Heidegärten; 160 pp., Berlin 1976.

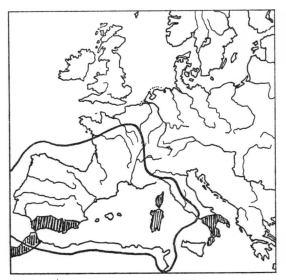

Fig. 23. Range of *Erica scoparia* (———)
and *E. terminalis* (//////)

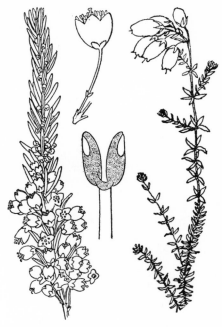

Fig. 24. Left, *Erica vagans*, branch, single flower
and stamen; right *E. tetralix* (from Reichenbach, Nose)

B. SOUTH AFRICAN SPECIES

Around 600 species are indigenous to South Africa, and a
further 10 species in the mountains of E. Africa (see range map
Fig. 14.). Although first introduced into European gardens in
the late 18th century, the large So. African *Erica* collections have
only gained attention in recent times. Johann Christoph
Wendland of Hanover-Herrenhausen, W. Germany had
amassed 150 species by 1811; Heinrich Sander of Kitzingen, W.
Germany had 300 species and Conrad Loddiges of London,
England had assembled 400 species by 1826. Generally, the
South African *Erica* species are found only in gardens of the
Mediterranean region, SW. England, and southern USA, in
botanic gardens.

It should be pointed out that the wonderful plates in the *Erica*-
works of Andrews and Wendland, as well as in Loddiges
Botanical Cabinet and the older issues of Curtis' *Botanical
Magazine* are of limited botanical accuracy. The plates depict
plants grown in the greenhouses of Kew and other botanic
gardens, where, by selection and ideal growing conditions,
they would flower much more abundantly than plants in their
native habitat. Many plants classified as "species" are actually
only diversely colored varieties and even hybrids. For an
alphabetical list of common names (from Regel 1843) with
reference to proper botanical origin, refer to Dulfer 1964, 151–
155.

Erica abietina L. Upright shrub, 0.5–1 m, branches
usually in whorls, downy pubescent, densely foliate;
leaves linear, stiff, very acute, 6–14 mm long, short
pubescent, margins glandular dentate, teeth erect;
flowers in whorls, groups in spikes at the branch tips,
corolla clavate-tubular, bowed, slack, glutinous, 20 mm
long, short pubescent, scarlet-red, lobes short, erect,

obtuse, sepals oval-lanceolate, pubescent, April–July.
AH 13 (as *E. coccinea*); WE 3: 9 (as *E. coccinea*) (= *E.
coccinea* Berg. non L.). z10 ✧

E. albens L. Upright, glabrous shrub, 30–40 cm high;
leaves in groups of 3, linear, 8–16 mm long, rather
loosely arranged, erect, very finely ciliate; flowers in
foliate, 3–9 cm long, racemes at the branch tips, corolla
oval-urceolate, 6 mm long, much narrower at the throat,
white, lobes tiny, eventually inclined inward, April–July.
BM 440; WE 6: 3; AH 2. z10 ✧

E. aristata Andr. Upright shrub, to 60 cm high, glabrous;
leaves in groups of 4, scale-like, broad-oblong, 5–6 mm
long, margins dentate, bristly ciliate; flowers grouped in
4's, corolla tubular and inflated, only slightly narrowed
at the middle, somewhat constricted at the throat, very
glutinous, 25 mm long, purple-pink with 8 dark lines,
throat deep purple, lobes 3 mm long, erect, obtuse,
August–September. BOE 44; BM 1249; NF 9: 63.
Flowers abundantly; one of the most attractive species.
z10 ✧

E. baccans L. Upright, glabrous shrub, 0.9 to 1.5 m high;
leaves in groups of 4, narrow linear, 6–8 mm long;
flowers usually in 4's at the branch tips, corolla globose, 6
mm wide, constricted under the 4 lobes, reddish-purple,
sepals lanceolate, as long as the corolla and the same
color, April–July. BM 358; AH 4; WE 6: 13; HTS 200.
1774. z10 ✧

E. bauera Andr. Upright shrub, to 90 cm, branches long
and rodlike; leaves grouped in 4's, oblong-lanceolate, 4–
5 mm long, erect to reflexed, furrowed, thick, gray;
flowers axillary in dense racemes near the branch tips,
corolla tubular, asymmetrically inflated, 16–20 mm long,
narrowed at the apex, mostly white, also pink, lobes
erect, short, anthers enclosed, August–October. BOE 10;
BMns 222; AH 252; HTS 203 (= *E. bowiana* Lodd.). 1822.
Frequently found in cultivation, but rare in the wild. z10
✧

E. blanda see: **E. doliiformis**

E. bowiana see: **E. bauera**

E. caffra L. A shrub or small tree in its habitat, to 4 m,
branches rodlike or thicker, soft pubescent; leaves in 3's,
nearly erect or outspread, 8–12 mm long, gray-green and
pubescent, open beneath; flowers in 3's or in short
stalked umbels, corolla oval-oblong to more tubular-
urceolate, white to cream-white, 5–7 mm long,
constricted at the throat, loosely coarse pubescent, occasionally downy, corollas persistent, anthers with awns to
half their length, spring. Easily confused with *E.
subdivaricata* Berg., which has been included under this
name for many years. z10 ✧

E. campanulata see: **E. pageana**

E. canaliculata Andr. Upright shrub, to nearly 2 m high,
branches ascending, gray pubescent, with many erect,
abundantly flowered branches; leaves in 3's, linear,
deeply furrowed, tomentose beneath, 4–10 mm long,
erect-outspread; flowers mostly in 3's, calyx glabrous,

lighter outside, inside red, deeply 4 lobed, middle line of the corolla distinct, corolla broad-cupulate, 3–3.5 mm long, 4 sided, glabrous, pink, lobes erect, wide, obtuse, more or less distinctly reflexed, anthers exserted, March–May. BOE 165; BMns 339; HTS 205. 1802. Well known in cultivation. z10 ⊕

E. cerinthoides L. Upright shrub, to 1.8 m high in its habitat, but usually much shorter, branches ascending, often rodlike, pubescent; leaves 4, 5 or 6 together, linear or more lanceolate, obtuse, 6–16 mm long, erect to outspread or reflexed, deeply furrowed, usually glandular-bristly and ciliate; flowers in umbels, corolla tubular, more or less bullate, 22–34 mm long, throat somewhat constricted, carmine, pink, red or white, outside shorter or longer pubescent, lobes erect or outspread, obtuse, anthers enclosed, flowers during the entire year. BOE 29; BM 220; HTS 207. 1774. One of the best known species, but quite rare in cultivation. z10 Plate 1. ⊕

E. chamissonis Kl. ex Benth. Upright shrub, to 60 cm, branches thick, ascending, fine pubescent; leaves in 3's, linear to more lanceolate, densely packed, 3–5 mm long, erect to outspread, broadly furrowed, usually bristly, the bristles with a warty base; flowers usually terminal in 3's, on short branches, racemose, corolla broad cupulate, dry, glabrous, pink, 3–5 mm long, lobes nearly circular, anthers brown, April. BOE 163; BM 6108. 1872. z10 Plate 1. ⊕

E. coarctata Wendl. Upright shrub, to 30 cm high, branches numerous, thin, fine pubescent to glabrous; leaves 3 (4) together, linear-triangular, 5–8 mm long, obtuse, slightly furrowed, imbricate, glabrous, erect; flowers usually 2 in the leaf axils, in narrow, dense spikes, very prolific, but flowers very small and nearly totally hidden in the foliage, corolla broad-campanulate, 1.5 mm long, glabrous, dry, pink to dull yellow, lobes erect to outspread, summer. BOE 118; WE 19: 37. z10 ⊕

E. coccinea L. Shrub, to 1.2 m, branches thick, with many short, foliate, pubescent branches; leaves in 3's, linear, furrowed, 4 to 8 mm long, most erect and reflexed, ciliate or glabrous; flowers 1–3, in terminal spikes, most nodding, corolla narrow ovate to tubular or tubular-inflated, dry or glutinous, 6–17 mm long, quite variable in color, yellow, green-yellow to nearly green, red, pink or orange, lobes erect, obtuse, 4–6 mm, anthers distinctly exserted, gold-brown, sepals 5–7 mm, colored, August–October. BOE 1 (= E. petiveri L.). 1774. z10 ⊕

E. conspicua see: **E. curviflora** var. **splendens**

E. curviflora L. Upright shrub, to 1.6 m high, branches thick, rodlike, pubescent; leaves in groups of 4, linear to more lanceolate, 3–7 mm long, erect, outspread or rolled inward, imbricate, ciliate; flowers usually solitary, occasionally 2–4, usually ascending, corolla clavate-tubular, 22–26(–38!) mm long, bowed, seldom straight, usually pubescent, dry, red, orange or yellow, lobes 3.5 mm long, obtuse, outspread, March–June. BOE 28; AH 16. S. Africa, moist areas, usually near rivers. z10 ⊕

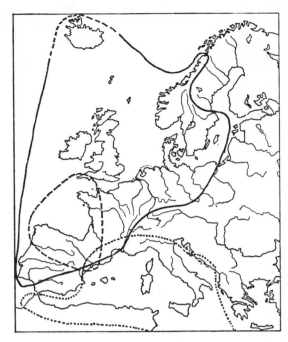

Fig. 25. Range of *Erica tetralix* (———), *E. vagans* (--------) and *E. multiflora* (...........) (from Underhill)

var. **splendens** (Wendl.) Dulfer. 30–60 cm high, branches densely covered with short branchlets; leaves 8 mm long, 3 sided, smooth; flowers in terminal clusters, outspread, corolla oblong-clavate, curved, soft pubescent, reddish-yellow or yellow, lobes reflexed, May–August. AH 14 (as *E. conspicua*); AH 222 (as *E. longiflora*) (= E. conspicua Soland.; E. splendens Wendl.). z10 ⊕

E. cylindrica Thunb. Upright, open shrub, to 1.2 m high, glabrous in all respects, branches nearly rodlike with long internodes and distinct petiole scars; leaves in groups of 4, linear, erect, furrowed, 4–5 mm long; flowers usually solitary, irregularly clustered near the branch tips, corolla cylindrical, somewhat compressed at the throat, 10–12 mm long, 1.5 mm wide, dry, pale yellow to white, sweet smelling, spring and fall. BOE 55. z10 ⊕

E. discolor Andr. upright shrub, 60–90 cm high, branches soft pubescent; leaves in groups of 3, linear, glossy, 4–6 mm long; flowers 2–3 on short side branches, corolla narrow funnelform, 18 mm long, pink-red on the lower portion, limb white with small, erect, yellowish-white or greenish lobes, November–April. AH 160; WE 5: 9. z10 Plate 8. ⊕

E. doliiformis Salisb. Stiff, upright shrub, seldom over 30 cm high, branches bristly pubescent; leaves in 6's, linear, 5–7 mm long, densely packed, outspread, furrowed, glabrous, glandular ciliate; flowers in umbels, corolla tubular-inflated, 10–14 mm long, pink, glandular pubescent, September–April; ovaries long pubescent (!). AH 152; BM 612 (as E. metulaeflora) (= E. blanda Andr. non Salisb.). z10 ⊕

E. floribunda see: **E. sparsa**

E. gracilis Salisb. Shrub, 30–45 cm high, branches thin, densely covered with short, pubescent side branchlets; leaves in 4's, linear, 3 sided, 4 mm long, light green, glabrous, erect or curved inward; flowers usually in 4's at the tips of the side branches, corolla more or less globose-urceolate, constricted at the throat, pink-red (on the wild species), 3–4 mm long, flowering in cultivation from September–November, in its habitat from October to February, anthers enclosed, globose, with short awl-shaped awns; ovaries glabrous. AH 68; NF 8: 6. 1774. z10 ⊕

Includes the cultivars:

'Glasers Rote' (Karl Glaser, Harreshausen, Hessen, 1929). Leaves more appressed; flowers luminous dark red, showing color in the early bud stage.

'Globularis' Small flowers, but very abundant, flowers usually in 8's at the branch tips, long lasting, available in several colors. From France.

E. hiemalis (also spelled *hyemalis*) Hort. Small shrub, branches upright, loose, bushy, rodlike, side branches short; leaves 5 mm long, glabrous above, ciliate, pubescent beneath; flowers 2–8 at the branch tips, forming a long, racemose inflorescence, corolla tubular-clavate, distinctly wider at the throat, 18–20 mm long, red beneath, changing to white above, anthers dark brown December–January. PB1 2: 286; GF 5: 137. Introduced and distributed from England about 1845. Origin unknown. z10 ⊕

E. hyemalis see: **E. hiemalis**

E. longiflora see: **E. curviflora** var. **splendens**

E. mammosa Andr. Upright shrub, to 1.2 m, densely branched under favorable conditions, finely pubescent, leaves in 4's or scattered, linear-lanceolate, erect-outspread, 6–10 mm long, 3 sided; flowers in dense or loose racemes, nodding, corolla tubular, obliquely inflated, 15–25 mm long, dark red or orange-red, but also pink, purple-pink, green, cream-white or white and with green tips, lobes erect, rounded, anthers enclosed, July–October. BOE 9; BMns 100; HTS 214; AH 124. z10 ⊕

E. metulaeflora see: **E. doliiformis**

E. pageana Bolus. Upright shrub, 1–1.2 m high, young branches pubescent; leaves in groups of 4, densely arranged, linear, 4–6 mm long, slightly furrowed; flowers in cylindrical spikes at the branch tips, corolla campanulate, 6–9 mm long, 6 mm wide at the throat, gold-yellow, sepals greenish, oval-lanceolate, short, anthers enclosed, March–April. BM 9133 (= *E. campanulata* Guthrie & Bolus non Andr.). Around 1920. z10 ⊕

E. patersonia Andr. Upright, glabrous shrub, to 90 cm, sparsely branched; leaves in 4's, linear, acute, densely imbricate, erect, inward curving; flowers terminal on short side branches, developing a dense spike near the branch tips, corolla tubular, constricted near the throat, 14–18 mm long, glabrous, dry, yellow with darker, short,

erect, eventually reflexed lobes, anthers enclosed, filaments bent under the anthers, May–August. BOE 21 (= *E. abietina* Benth. non L.). 1790. Frequently found in cultivation; tolerates some frost. z9–10 ⊕

E. persoluta see: **E. subdivaricata**

E. perspicua Wendl. Upright shrub, 60–150 cm high; leaves in 4's, narrow-linear, soft pubescent or glabrous, 4–6 mm long; flowers 1–3 at the ends of the short side branches, grouped in about 6 cm wide and 12–15 cm long panicles, corolla tubular, 20–25 mm long, 6 mm wide at the throat, deep pink to purple at the base, apex becoming lighter to white, May–August. WE 1: 7; AH 255; NF 9: 69. 1800. This species is probably the origin of *E. hiemalis*. z10 Plate 8. ⊕

E. peziza Lodd. Shrub, 30–60 cm high, branches soft pubescent; leaves in 4's together, linear, to 6 mm long, ciliate; flowers solitary, axillary, about 3–5 cm long, grouped in foliate racemes, corolla ovate, 5 mm wide, purple-red, April–August. HTS 216; Lodd. Bot. Cab. 265. z10 Plate 8. ⊕

E. pillansii Bolus. Upright shrub, to 1.2 m high, branches ascending, shaggy pubescent at first; leaves in 4's, linear, furrowed, 4–6 mm long, erect, curved inward, ciliate and pubescent when young; flowers solitary, terminal, grouped in long racemes, corolla tubular-clavate, 15 mm long, with tiny pubescence, dry, scarlet-red, lobes more or less erect, pubescent, anthers enclosed, April–October. BOE 27; BM 9676; NF 9: 62. 1910. Very attractive, conspicuous species. z10 ⊕

E. plukenetii L. Upright, totally glabrous shrub, to 60 cm high; leaves arranged in 3's, most 12–16 mm long, linear, densely imbricate, usually curving inward, nearly 3 sided; flowers solitary and nearly axillary, nodding, corolla more or less inflated at the base, 13–18 mm long, pink, purple, red-orange or also white with green lobes, these short, erect, obtuse, anthers linear, markedly exserted, sepals lanceolate, keeled, red, April–September. BOE 1; AH 186. z10 ⊕

E. regia Bartl. Upright shrub, to 90 cm high, branches thin and sparsely foliate, except when young; leaves in 6's, linear, 6–12 mm long, erect-outspread, more or less furrowed, glabrous to slightly pubescent; flowers in the leaf axils near the branch tips, erect or nodding, corolla broad-tubular, narrowing at the throat, more or less glutinous, 14–18 mm long, glabrous, carmine, anthers enclosed, September–October. BOE 11; HTS 218. Very well known in cultivation. z10 ⊕

Including:

var. **variegata** Bolus. Corolla white with red tips. BOE 11. ⊕

E. sparsa Lodd. 0.5–1 m high; leaves in groups of 3, about 4 mm long, erect, obtuse keeled; flowers in 3's at the tips of the short side branches, corolla campanulate, 2 mm long, light pink, lobes broad lanceolate, twice as long as the corolla tube, very small flowers, but prolific, anthers half protruding, March to May. Lodd. Bot. Cab. 1467 (= *E. floribunda* Lodd.). z10 ⊕

E. splendens see: **E. curviflora** var. **splendens**

E. speciosa Andr. Upright shrub, 0.5–1.5 m, branches limp; leaves in 3's, linear to narrow-lanceolate, 4–6 mm long, furrowed, often also ciliate; flowers 2–4 in terminal clusters, corolla tubular-cylindrical, to 25 mm long, bright red, green at the apex, floccose pubescent and glutinous, lobes small and reflexed, June–August. AH 133 and 192. 1800. z10 ✥

E. subdivaricata Berg. Densely branched shrub, 30–90 cm high, branches usually pubescent; leaves in 4's, linear 3 sided, 4–5 mm long, limp, ciliate, otherwise glabrous; flowers 4 or more together at the tips of short branches, forming racemose inflorescences, corolla white, campanulate, not narrowed at the limb, 3 mm long, spring. AH 7 (as *E. caffra*) (= *E. persoluta* L.). 1802. z10 ✥

E. ventricosa Thunb. Upright shrub, to 90 cm, branches thick, stiff, the youngest pubescent, later glabrous; leaves in 4's, linear-awl shaped, 12–16 mm long, long acuminate, furrowed, densely packed, outspread, ciliate with long, soft hairs; flowers several in dense umbels, corolla oval-urceolate, narrowed at the apex, constricted at the throat, glabrous, glossy, dry, hard to the touch, white, pink or red, lobes 3 mm long, stellate outspread, acute, October–January. BOE 50; HTS 220; BM 350. 1787. Includes many forms in cultivation. z10 ✥

E. versicolor Wendl. Upright shrub, 0.6 to 1.2 m high, nearly glabrous; leaves in 3's, linear, rigid, 5–10 mm long, dentate and glandular-ciliate at first, glabrous or short pubescent later; flowers 1–3, nearly sessile at the tips of the branches, corolla tubular, 20–25 mm long, red, yellowish-green, and glabrous at the apex, fall and spring. AH 13 and 47; WE 11: 3. Before 1840. z10 ✥

E. vestita Thunb. Upright shrub, to about 90 cm, branches rodlike, finely pubescent; leaves in 6's, linear and nearly 3 sided, 13–33 mm long, but only 0.5 mm wide, erect to outspread, densely packed, glabrous, ciliate when young; flowers axillary in spikes near the branch tips, corolla tubular-clavate, straight or bowed, more or less pubescent, dry, white, yellow, pink or carmine, 17–25 mm long, lobes short, obtuse, erect to slightly bowed, anthers enclosed, May–September. BOE 12. 1789. Very attractive, especially the red form. Common in cultivation. z10 ✥

E. ✕ **willmorei** Knowl. & Westc. Parents unknown. Shrub, to 60 cm; leaves 3 together, linear-stiff, obtuse, 6–8 mm long, very densely arranged, furrowed beneath; flowers terminal and axillary, in 15 cm long and 5 cm wide foliate racemes, corolla oblong-tubular, constricted beneath the limb, deep pink with white tips, anthers enclosed, with awn-like appendages, spring. DRHS 769; RH 1892: 202; HTS 223. Various flower colors in cultivation. z10 ✥

Lit. for the South African species: Andrews, H.: Coloured Engravings of Heaths; 4 vols., covers 288 species; 1796–1809 ● Andrews, H.: The Heathery or Monograph of the genus *Erica*; 300 species described, 6 vols., 1804–1812 ● Loddiges Botanical Cabinet, appearing annually from 1818 to 1833, with 226 descriptions ● Dulfer, H.: Revision der südafrikanischen Arten der Gattung *Erica* L.; in Ann. Naturhist. Mus. Wien, 66–68, 1963–1965 ● E. G. H. Oliver & Baker, H. A.: Ericas in Southern Africa; 167 species described and illustrated in plates; 1967 ● Regel, E.: Die Kultur und Aufzählung der in deutschen und englischen Gärten befindlichen Eriken; 189 pp., Zurich 1843 ● Vogel, F.: Azaleen, Eriken (pp. 131–186) und Kamelien; Berlin 1965 ● Waitz, C. F.: Beschreibung der Gattung und Arten der Heiden; 355 pp., Altenburg 1805.

Please refer to the notes on p. 40 concerning the reliability of the descriptions found in the literature.

ERINACEA Adans. — LEGUMINOSAE

Monotypic genus; deciduous, hemispherical habit, very densely branched thorny shrub; leaves simple to trifoliate, small, at the branch tips, quickly dehiscing, opposite or the higher leaves alternate; flowers 2–4, short stalked, in the upper leaf axils; calyx persistent, inflated, tubular, with 5 teeth; petals narrow, wing and keel adnate to the stamen tube, standard slightly auriculate at the base, ovate; pods with 4–6 seeds. — 1 species in SW. Europe, N. Africa.

Erinacea anthyllis Link. Dense cushion form, hemispherical shrub, 10–30 cm high, branches spreading widely, rigid with thorny tips, silky pubescent when young, abscising quickly; flowers violet-blue, about 2.5 cm long, very numerous, May–June, pods 2 cm long. BM 676; PEu 52; VG 38 (= *E. pungens* Boiss.). Spain, E. Pyrenees, Algeria, Tunisia. 1759. Likes limestone soil. Beautiful shrub for a sunny area in the alpine garden. z10 Plate 12. ✥ ○

ERIOBOTRYA Lindl. — Loquat — ROSACEAE

Evergreen trees or shrubs; leaves alternate, short stalked to nearly sessile, simple, coarsely dentate, venation straight, terminating in the toothed margin; flowers white, in terminal panicles; calyx tube 5 toothed, woolly; corolla with 5 petals; stamens 20–40; styles 2–5, connate at the base; fruit a pome with 1–2 very large seeds. —

About 10 species in E. Asia.

Eriobotrya deflexicalyx (Hemsl.) Nakai. Evergreen, medium-sized tree; leaves usually clustered at the branch tips, obovate-oblong to elliptic, 13–25 cm long, 4.5–5.5 cm wide, leathery, obtuse, coarsely serrate, cuneate at the base, both sides glabrous, dark green above, lighter beneath, with 13–15 vein pairs, these elevated on both sides, petiole 2.5–4.5 cm long; flowers in terminal panicles, rust-brown pubescent, 1.5 cm wide, petals white; fruit ellipsoid, 1.5–2.5 cm long, tomentose, juicy, with persistent calyx. LWT 101. China; Taiwan. Relatively recently introduced into England and distributed by Hillier. z9 ⌀

E. japonica (Thunb.) Lindl. Japanese Loquat. Evergreen shrub, 5–7 m high, branches thick and white woolly when young; leaves obovate to elliptic-oblong, acute at both ends, very tough, with distinct venation, 15–25 cm long, deep green above, brownish tomentose beneath; flowers 1–2 cm wide, white, in 10–15 cm long panicles, September; fruits pear-shaped, yellow, with 2 hazelnut-like seeds, 3–4 cm long, edible, sour. HM 998; LF 172; KIF 3: 27. China. 1787. z9 Plate 10. # ⌀ ⚭ ✂

Common in botanic gardens; cultivated in many subtropical areas for its fruits.

ERIOLUBUS

Eriolobus trilobata see: **Malus trilobata**

E. tschonoski see: **Malus tschonoski**

ERYTHEA Wats. — PALMACEAE

Palms with slender, thornless stem, 10–15 m high; leaves fan-shaped, circular, in a terminal crown; blade folded, tomentose when young, sometimes shredded at the apex; petiole rigid, with or without marginal thorns; sheaths and inflorescence thick leathery, densely pubescent, fruits rounded to ovate, without stony seed pit. — 6 species in Mexico, S. California, Guadeloupe Islands.

Erythea armata Wats. Blue Palm. Stem to 13 m high, to 40 cm thick, bark thick, cracked, corky; leaves large, circular, silver-blue on both sides, with about 30–40 segments, somewhat prickly and shredded at the tips, petiole with thorns on the margin; inflorescence to 2.5 m long, ivory colored, nodding in a wide arch, August. MM 110 (= *Brahea armata* Wats.; *B. glauca* Hort.). Baja California, Mexico. 1887. z9 Plate 7. ⌀ ☼

Not uncommon in the warmer regions of the world and very ornamental.

ERYTHRINA L. — Coral Tree — LEGUMINOSAE

Tropical trees or shrubs, deciduous or evergreen, occasionally a subshrub or with the current year's shoots partly dying back after the flowers; plants usually thorny; leaves alternate, with 3 broad leaflets; flowers usually large, brightly colored, in dense panicles; calyx usually obliquely 2 lobed, often more or less 5 toothed, occasionally incised to the base; standard wide or narrow, long clawed, wings often absent; fruit a pod with the seeds connected by a "thread". — About 30 species in the warmer, temperate zones and tropics of the world.

Erythrina americana Mill. Deciduous tree, flowers appearing with the leaves; leaves ovate, acute, terminal leaflets 7–10 cm long and nearly so wide, petioles 5–7 cm long; inflorescence open, calyx tubular, about 12 mm long, standard narrow, erect, 5–6 cm long, keel petal narrowed, keel and wings 1 cm long; pods 15–20 cm long, soft pubescent when young, somewhat constricted between the seeds, seeds red. Mexico. In cultivation often erroneously labeled as *E. corallodendron* L. z10 ☼

E. caffra Thunb. Large tree, semi-evergreen; leaves ovate, terminal leaflet 5–9 cm long, flowers scarlet-red, compact, calyx campanulate, irregularly incised on one or both sides, 12 mm long, standard wide and reflexed, 5–6 cm long and half as wide, keel petals connate, 2.5 cm long, standard similar in size and form to the keel; pods constricted between the seeds, seeds red. PPT 954; MCL 5 (= *E. insignis* Todaro; *E. constantiana* Micheli). S. Africa. Occasionally confused with *E. lysistemon*. z10 ☼

E. corallodendron L. Coral Tree. Tall deciduous shrub, woody stem, 1.5–3 m high, prickly; leaflets wide, rhombic-ovate, acute, petiole not prickly; flowers deep scarlet-red, in long racemes, appearing after the leaves, standard erect, linear-oblong, truncate, 5 cm long, May–June; pods glabrous, many seeded (= *E. speciosa* Andr.). Mexico to Brazil. 1690. z10 ☼

E. crista-galli L. Shrub or small tree, most of the flowering shoots dying after fruit development, branches with flat, stout thorns, leaf petioles and midrib also thorny; leaflets oval-oblong to oblong-lanceolate, 10–15 cm long, shorter at the branch tips, tough, entire; flowers dark scarlet-red, in large, terminal racemes, August–September; seeds brown. BM 2161; BR 313; HTS 229 (= *E. laurifolia* Jacq.; *E. pulcherrima* Todaro). Brazil. 1771. z9 Plates 12 and 13. ☼

E. humeana Spreng. Small deciduous tree, flowers appearing soon after the foliage; leaflets ovate, tapering at the tip, terminal leaflets 15–20 cm long, 15–17 cm wide; inflorescence 30–50 cm long, densely compact, calyx tubular, 1 cm long, standard rather narrow, erect, 4 cm long, 2 cm wide, scarlet-red, keel petals separate, wing and keel 12 mm long; pods constricted between seeds, seeds red. BM 2431 (as *E. caffra* Ker. non Thunb.; *E. humei* E. Mey.). z10 ✥

E. insignis see: **E. caffra**

E. laurifolia see: **E. crista-galli**

E. lysistemon Hutchins. Large semi-evergreen tree, flowers appearing with and occasionally before the foliage; inflorescence 15–20 cm long, densely compact, calyx campanulate, irregularly incised on one or both sides, standard scarlet-red, erect, both halves folded together, 5–6 cm long, 2 cm wide, keel petals connate, 6 mm long, wings somewhat longer; pods 10–15 cm long,

constricted between the seeds, seeds red. PPT 957; MCL 54. S. Africa. Often confused with *E. caffra* Thunb. z10 ✥

E. pulcherrima see: **E. crista-galli**

E. speciosa see: **E. corallodendron**

E. crista-galli is the most frequently found in cultivation. The others are less commonly found in botanic gardens and parks of the frost free climates of the world. Very impressive plants in flower.

Lit. Standley, P. C.: The Mexican and Central American species of *Erythrina*; in Contr. U. S. Nat. Herb. **20**, 175–182, 1919. ● Krukoff, B.: The American species of *Erythrina*; in Brittonia 3, 205–337, 1939 ● McClintock, E.: The cultivated species of *Erythrina* in Baileya **1**, 52–58, 1953 ● McClintock, E.: *Erythrina*; in Dict. RHS Suppl. 2nd Ed., 281–282, 1969.

ESCALLONIA Mutis — ESCALLONIACEAE

Evergreen shrubs or small trees; leaves simple, alternate, occasionally nearly opposite or whorled, margins usually dentate, teeth with a gland at the apex, without stipules; flowers bisexual, occasionally dioecious or polygamous, usually in racemes; sepals usually connate on the basal portion, seldom distinct; petals 5, distinct, but arranged in a tube; stamens 5, occasionally 4 or 6; style usually simple, stigma 2–5 lobed; fruit a dry capsule or berry, with persistent style and calyx lobes. — 50–60 species in S. America, especially in the Andes.

★WILD SPECIES

(hybrids and cultivars follow ★★)

Escallonia alba see: **E. grahamiana**

E. alpina DC. Evergreen shrub, about 1 m high, wider than tall, totally glabrous, branches angular; leaves sessile, obovate to oblanceolate, 1.5–3 cm long, glossy green on both sides; flowers in columnar panicles, to 10 cm long, corolla carmine or dark red to nearly deep brown (= *E. fonkii* Phil.; *E. glaberrima* Phil.). Chile. 1926. z9 #

E. bifida Link & Otto. Evergreen shrub, also a small tree in its habitat, very similar to *E. floribunda* and often confused with it; leaves narrow ovate, 3–7 cm long, finely dentate; flowers white, in flat-rounded panicles, calyx lobes more acute than *E. floribunda* and with tiny glandular teeth, style twice as long as the calyx lobes (*E. floribunda* calyx lobes and style equally long). BM 6404 (as *E. floribunda*) (= *E. montevidensis* [Cham. & Schlechtd.] DC.; *E. floribunda* Hort. non H. B. K.). S. Brazil. z9 Fig.26. #

E. floribunda H. B. K. Evergreen shrub, to 2.5 m high (occasionally a small tree in its habitat), branches glabrous, but rather glutinous; leaves obovate to narrow ovate, 3–10 cm long, margins finely dentate to entire, glabrous and glandular punctate beneath; flowers pure white, 12 mm wide, in terminal panicles, fragrant, July–

August. Venezuela, Columbia. Often confused with *E. bifida*. z9 #

E. fonkii see: **E. alpina**

E. glaberrima see: **E. alpina**

E. grahamiana Gill. Evergreen, upright shrub, branches more or less outspread, with sessile glands; leaves elliptic-obovate to oblong, to about 5 cm long, obtuse, tapering to the narrow petiole, finely serrate, glabrous and glossy above, finely glandular punctate beneath; flowers white, in terminal, pyramidal panicles, calyx teeth awl-shaped. Chile. This species is often found under the following names; *E. alba, E. floribunda, E. bifida* and *E. virgata*). z9 #

E. illinita Presl. Evergreen shrub, about 2 to 2.5 m high, branches with resin glands; leaves obovate to oval, 2–6 cm long, apex rounded or short acuminate, base cuneate, glabrous when young, but with resin glands, petiole 1–3 cm long; flowers white, in 7–10 cm long, conical, resinous panicles, the 5 calyx lobes linear, June–August. BR 1900. Chile. 1830. The leaves smell like a "pigsty". z8 #

E. laevis (Vell.) Sleum. Evergreen shrub, 1 to 1.5 m high, branches rigid, glabrous, angular, with resin glands; leaves narrow obovate to oval, 5–7 cm long, finely dentate, cuneate at the base, dark green above, lighter beneath; flowers in terminal, rounded, 5–7 cm long panicles, soft pink, buds darker, calyx lobes half as long as the petals. BM 4274; HI 514 (= *E. organensis* Gardner). Brazil, Organ Mts. 1844. z9 #

E. montevidensis see: **E. bifida**

E. organensis see: **E. laevis**

E. pterocladon see: **E. rosea**

E. pulverulenta (Ruiz & Pav.) Pers. Evergreen shrub, 2–3 m high, branches angular, finely pubescent, glutinous;

leaves oblong, 5–10 cm long, round, base cuneate, dentate, bristly pubescent; flowers in 10–20 cm long, 3 cm wide, dense inflorescences (important!), individual flowers very densely packed, white, July–September. Chile. z9 # ☼

E. punctata see: *E. rubra*

E. revoluta Pers. Evergreen shrub, 2–3 m high, branches angular, gray-green tomentose (important!); leaves obovate, unevenly dentate, round to acute, base cuneate, 2–5 cm long, dense gray-green pubescent on both sides (important!); flowers white, in terminal panicles, 3–7 cm long, flower stalk and calyx gray pubescent, July–September. BM 6949. Chile. 1887. Flowers sparsely. z9 #

E. rosea Griseb. Evergreen shrub, 2–2.5 m high, branches pubescent and angular; leaves narrow obovate, 1–2.5 cm long, margins dentate, dark glossy green above, cuneate at the base; flowers in 3–7 cm long, narrow, terminal panicles on short, densely foliate side branches, corolla white, 8 mm long, with narrow limb (important!), calyx glabrous, June–August. BM 4827 (= *E. pterocladon* Hook.). S. America, Patagonia. 1847. z9 Fig. 26. # ☼

E. rubra (Ruiz & Pav.) Pers. Evergreen shrub, strong grower, to 4 m high in its habitat and under favorable conditions in cultivation, branches reddish pubescent, glutinous, glandular; leaves obovate to lanceolate, 2.5–5 cm long, acuminate at both ends, with stalked glands near the base; flowers in loose, 3–7 cm long panicles, red, July–August. BM 2890; BM 6599 (as *E. punctata*) (*E. punctata* DC.). Chile. 1827. z9 Figs. 26 & 27. # ☼

var. **macrantha** (Hook. & Arn.) Reiche. Evergreen shrub, dense, round, 1.5–4 m high, very strong grower, often growing sideways, branches pubescent, glandular; leaves broad ovate to obovate, 2.5–7 cm long, doubly dentate, glossy dark green above, punctate with resin glands beneath, aromatic; flowers in terminal panicles, 5–10 cm long, flowers abundantly, corolla light carmine-red, 15 mm long and wide, calyx and flower stalk glutinous, June–September. FS 632; BM 4473; DRHS 781; HuF 1: 37 (= *E. macrantha* Hook. & Arn.). S. America, Chiloé Island. 1846. A good hedge along the seacoast; attracts bees. z9 # ☼

'Pygmaea'. Dwarf form, very dense, only 30–50 cm high; flowers very abundantly, intense red, June–August. Originated in Ireland as a witches'-broom on *E. rubra*. # ☼

E. virgata (Ruiz & Pav.) Pers. Deciduous shrub, to 1 mm high, growth dense and divaricate, branches brown, glabrous; leaves obovate, 8 to 15 mm long, finely dentate, cuneate at the base, glossy green above; flowers pure white, 1 cm wide, in axillary, foliate, racemes about 5 cm long, petals rounded-obovate, slightly clawed, June–August. BS 1: 677 (= *E. philippiana* [A. Engl.] Mast.). Chile. 1866. Hardiest species. z7–8 Figs. 26 & 27. ☼

E. viscosa Forb. Evergreen shrub, 2–2.5 m high, branches glutinous due to the numerous resin glands; leaves obovate, finely dentate, 2.5–7 cm long, round or short acuminate, cuneate at the base, glossy on both sides with glutinous glands; flowers white, in 12–15 cm long, pendulous, narrow panicles, June–August. Chile. Leaves (like *E. illinita*) smell strongly like a "pigsty". z9 #

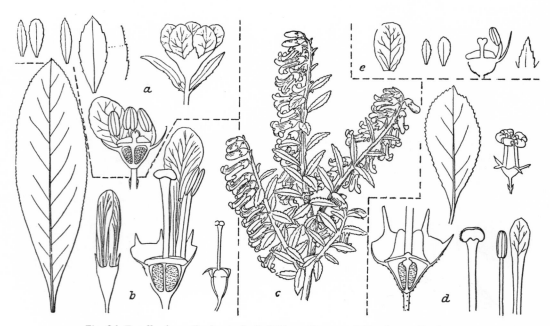

Fig. 26. **Escallonia.** a. *E. virgata*; b. *E. bifida*; c. *E. rosea*; d. *E. rubra*; e. *E. virgata*
(from Baillon, Bot. Mag., Schneider)

★★HORTICULTURAL CULTIVARS AND HYBRIDS
(Classified by Sleumer 1968)

Much of the hybridizing has been done by the Slieve Donard Nursery, Newcastle, N. Ireland. The numerous cultivars are used primarily in England and Ireland where they are more frequently planted than the species. The escallonias are very useful due to their rapid growth and tolerance of strong sea breezes.

'Alice' (*E. virgata* × *E. rubra*). To 1 m high; leaves obovate, 2.5–5 cm long, glossy above; flowers pink-red, in 7–10 cm long racemes, April–August. #

'Apple Blossom' (Donard) (*E. virgata* × *E. rubra* var. *macrantha*). To 1 m high; flowers soft pink, June–July. HTS 230. #

'C. F. Ball' (Ball 1912) (sport of *E. rubra* var. *macrantha*). Broad arching habit, to 1.5 m high; flowers tubular, large, carmine-pink, May–August. Introduced by Ball, Glasnevin, Ireland. #

'Donard Beauty' (Donard) (*E. virgata* × *E. rubra*). To 1 m high, branches thin, pendulous; flowers pink-red, very abundant, June–July. One of the more attractive forms. # ✪

'Donard Brilliance' (Donard) (*E. virgata* × *E. rubra*). To 1.5 m high, nodding habit; flowers carmine-red, May–July, anthers gold-yellow. # ✪

'Donard Gem' (Donard 1928) (*E. virgata* × *E. rubra* var. *macrantha*). To 1 m high; flowers pink, fragrant, May–June but continuing sporadically until fall, flowers abundantly. # ✪

'Donard Radiance' (Donard, before 1954) (*E. virgata* × *E. rubra* var. *macrantha*). Strong grower, 1–2 m high; leaves glossy; flowers deep pink. #

'Donard Scarlet' (Donard) (*E. virgata* × *E. rubra*). Narrow upright growth, 1–2 m high; flowers scarlet-red. # ✪

'Donard Seedling' (Donard 1919) (*E. virgata* × *E. rubra*). Strong grower, 1–2 m high, branches long arching; flowers pink at first, quickly becoming white, with the fragrance of *Crataegus*, June–July. # ✪

'Donard Star' (Donard) (*E. virgata* × *E. rubra* var. *macrantha*). Compact grower, 1–1.5 m high; leaves deep green, glossy; flowers dark pink, 2.5 cm wide, limb outspread. JRHS 93: 90. ✪

'Donard White' (Donard) (*E. virgata* × *E. rubra*; similar to *E. virgata*). Compact and round, about 1 m high; flowers white, buds pink, long flowering period. #

'Edinensis (Edinburgh 1914) (*E. virgata* × *E. rubra*). Strong grower, 1.5–2.5 m high, branches long, arching; flowers pink. Developed in the Royal Botanic Garden, Edinburgh, Scotland. #

'Exoniensis' (Veitch) (*E. rosea* × *E. rubra*). Upright, very strong grower, to 5 m high, branches glandular; leaves quite variable, evergreen, deep green and very glossy, 1.5–3 cm long; flowers white with a trace of pink; July–August. #

'Glory of Donard' (Donard) (*E. virgata* × *E. rubra*). To 1.5 m high; flowers deep carmine-red. Considered the best hybrid developed by Donard. #

'Ingramii' (1833) (sport of *E. rubra* var. *macrantha*). Similar to *E. macrantha*, habit narrowly upright, 2–2.5 m high; leaves narrow; flowers pink-red, June–August, but smaller than those of *E. rubra* var. *macrantha*. #

'Iveyi' (Veitch) (*E. bifida* × *E.* 'Exoniensis'). Strong grower; leaves deep green, glossy; flowers pure white, buds soft pink, July–August, fragrant. BS 2: Pl. 21; GC 130: 127. Discovered as a chance hybrid by Mr. Ivey, gardener at Caerhays Castle, Cornwall, England. 1893. # ✪

'Langleyensis' (Veitch 1897) (*E. virgata* × *E. rubra*). Strong grower, 2–2.5 m high, branches bowed; leaves ovate, 1.2–2.5 cm long; flowers, as distinguished from the species, on the upper side of the branches, carmine-pink, June–July. BS 1: 673. # ✪

'Pride of Donard' (Donard) (*E. virgata* × *E. rubra* var. *macrantha*). Dense, upright, 1.5–2 m high; leaves glossy, deep green; flowers bright red, June–August. JRHS 1977:56. #

Fig. 27. Left *Empetrum nigrum;* middle *Escallonia rubra;* right *Escallonia virgata* (from Dimitri)

'**Slieve Donard**' (Donard) (*E. virgata* × *E. rubra*). To 1.5 m high, branches long, arching; flowers pink-red, large, faded flowers becoming darker, May–June and fall. Quite hardy. # ☉

'**William Watson**' (Watson) (sport of *E. rubra* var. *macrantha*) Compact, to 1.5 m high; flowers bright red, June–August. #

All species prefer a humus soil in protected areas; many are intolerant of alkaline soils, especially the hardy *E. virgata*. They provide a good wind screen in mild areas.

Lit. Engler, A.: Monographische Übersicht der *Escallonia, Belangera* und *Weinmannia*; in Linnaea 36, 527–650, 1869–1870 ● Kausel, E.: Revisión del género *Escallonia* en Chile; Darwiniana 10 (2), 169–255, 1951 ● Sleumer, H.: Die Gattung *Escallonia* (Saxifragaceae); in Verh. Kon. Ned. Akad. Wet. Natuurk. 2 (58), Nr. 2; 146 pp., Amsterdam 1968.

EUCALYPTUS L'Hér. — MYRTACEAE

Tall, evergreen trees, all parts fragrant, bark on older trunks usually peeling in long strips; leaves entire, leathery tough, juvenile foliage usually opposite, alternate when mature; the juvenile foliage is also often different in form from the older leaves; flowers usually in umbels, but because of the elongated axis, often also in panicles or cymes, usually white or yellow, occasionally red; sepals absent, the 4 petals connate forming a "cap" (operculum), dehiscing when in flower; stamens numerous, either totally distinct or in 4 bundles; fruit a woody, 4 chambered, many seeded capsule with a dehiscent "lid". — Over 500 species identified to date. Nearly exclusively confined to Australia where it is a very characteristic element of the flora.

The bark of most species is an important characteristic which greatly eases identification.

The following species can be found along the West European coast, mainly in Spain, Portugal, S. France, S. England, Ireland and Scotland; Corsica, Italy, Greece, Cypress, Turkey, the Soviet Union, as well as coastal areas along the south and east Mediterranean region. They will also be found growing and often naturalized in southern California, Florida and other subtropical regions. A map of the cultivated range may be found in Penfold & Willis, Map 5.

Eucalyptus amygdalina Link. Tree, 20–40 m high, crown irregularly branched, bark with fine scales, gray; young leaves opposite, sessile to short petioled, linear to narrow lanceolate, 10–12 cm long, pale green, older leaves alternate, falcate, lanceolate, 15–25 cm long, pendulous, gray-green on both sides; flowers usually 3 in axillary, pendulous, stalked umbels, filaments white, anthers yellow. FAu 72; BM 3260; LNH 154; KEu 174 (= *E. longifolia* Link.). Australia, Tasmania. 1820. Ⓕ Australia; S. Africa. z9 #

var. **regnans** (F. v. Muell.). Tree, 50–75 m high, stem very straight, bark on the lower third of the trunk rough and in short shreds, upper portion white with long stripes; leaves glossy green; flowers 7–12 in umbels, usually several umbels together, white, January–March. LAu 106; FAu 138; KEu 160 (= *E. regnans* F. v. Muell.). SE. Australia, Tasmania. Tallest form of the genus; heights of over 97 m with 2 m trunk circumference have been recorded in the older literature but these are only estimates and not scientifically proven figures. Ⓕ Australia; New Zealand. z9 #

E. coccifera Hook. f. Small tree or shrub, bark peeling in long strips, white, smooth; young foliage broad elliptic, 3–5 cm long, mature leaves petiolate, lanceolate with finely hooked apex, 5–6 cm long, tough, both sides gray-

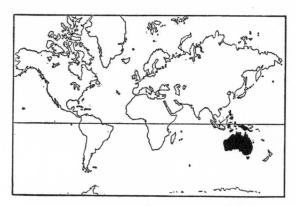

Fig. 28. Range of *Eucalyptus*

green; flowers 3 (6) in umbels, filaments yellow, in clusters. FAu 152; BM 4637; FS 736; CFTa 79; KEu 177. One of the hardier species. Ⓕ Tasmania. Plate 13. z9 #

E. cordata Labill. Small tree, to 15 m high, bark white to green-white, young branches warty; leaves opposite, sessile, circular to ovate, somewhat crenate, seldom over 7 cm long, usually white pruinose like the inflorescence; calyx broad campanulate, obtuse at the base, smooth, operculum flat hemispherical, obtuse or with sharp tips, the yellow stamen bundles 2–2.5 cm wide, November–December; fruit cupulate. BM 7835; LNH 152. Tasmania. One of the few species the foliage of which remains opposite and sessile. #

E. coriacea see: **E. pauciflora**

E. coriacea var. *alpina* see: **E. niphophilia**

E. dalrympleana Maiden. Tree, 30–35 m high in its habitat, bark smooth, peeling in large pieces, cream-white at first, then salmon-pink to light brown; young foliage opposite, sessile, broad ovate to circular, more or less cordate and occasionally stem-clasping, 4–6 cm long, green or blue-green, mature foliage petiolate, lanceolate or falcate, 10–17 cm long, 1–3.5 cm wide; flowers in umbels of 3, buds ovate, anthers white. KEu 119. Tasmania, E. Australia. #

E. delegatensis R. T. Baker. Tree, to 60 m high in its habitat, trunk straight, crown open, bark of lower trunk rough and shredded, upper portion smooth and peeling in long strips, whitish or gray-blue, young branches bluish or red; juvenile foliage only 3–4 pairs opposite, otherwise alternate, broad lanceolate, petiolate, thick, blue-green, mature foliage petiolate, lanceolate, 7–15 cm long, to 5 cm wide, often curved, dull green or somewhat blue-green, distinctly veined; flowers in umbels of 7–15, buds clavate, anthers white, fall flowering; fruits pear-shaped, 1 cm long. KEu 161. Tasmania. ⓕ Scotland, Ireland, far S. Africa and Congo. #

E. ficifolia F. v. Muell. Small tree, dense, wide crowned, often multistemmed, bark not peeling, light gray, with short shredded scales, trunk yellowish under the scales; juvenile foliage ovate to circular, petiolate, 7–10 cm long, long acuminate, thick, glossy dark green above, lighter beneath, midrib and margin yellow; flowers in terminal umbellate panicles of 3–7 flowered umbels; individual flowers 2.5 cm wide, filaments scarlet-red, anthers dark red, August. FAu 46; LAu 44; BM 7697; KEu 18; HTS 237; JRHS 90: 19. SW. Australia, in thickets along the coast. One of the most beautiful species. z9 Plate 14. # ✿

E. globulus Labill. Blue Gum. Tree, usually 45–55 m high, occasionally over 60 m, bark gray, persisting only at the trunk base, otherwise exfoliating in long strips, stem smooth beneath, bluish-gray, young branches angular or winged; juvenile foliage sessile, outspread or more appressed, ovate to broad lanceolate, blue-green, whitish pruinose, 7–15 cm long, mature foliage narrow lanceolate, somewhat falcate, petiolate, 10–30 cm long, pendulous, green; flowers solitary, occasionally in 3's, nearly sessile, in the leaf axils, white, June–November. FAu 104; AAu 310; KEu 129. SE. Tasmania. One of the most frequently planted species; very fast growing. ⓕ Entire Mediterranean region, tropical Africa and Asia, S. China, California, S. America, Australia and New Zealand; a very important forestry tree and, therefore, widely cultivated. z9 Fig. 29 and 30 #

E. gomphocephala DC. Tree, to 35 m high or less, bark persistent, rough, becoming dark; leaves thick, narrow acuminate, light green; flowers 2 cm wide, usually 3–5, sessile, operculum globose, very hard and thick, light red-brown, stamen filaments yellowish-white; fruit pear-shaped, truncate at the apex. KEu 47. W. Australia. Easily recognizable by the large, hemispherical operculum. ⓕ Spain, Morocco, Algeria, Tunisia, Cyprus, Israel, S. Africa, California, Chile. #

Fig. 29. *Eucalyptus globulus*. Left fruit capsule; right dehiscing operculum (from Kerner)

Fig. 30. *Eucalyptus globulus*. Branch with juvenile foliage in the background; foreground with mature leaves (from Hegi)

E. gunnii Hook. f. Tree, 18–24 m or higher in its habitat, branches bluish-white; juvenile foliage circular, 2–5 cm wide, often sessile, blue-green, mature foliage lanceolate, greenish, 10 cm long, acuminate, pendulous; flowers usually grouped 2–3, individual flowers 15 mm wide, yellowish-white, October–December. BS 1: 679; CFTa 66 (= *E. whittingehamensis* Hort.). Tasmania, S. Australia. Occasionally found in England as a tall tree, in Kew, Whittingeham, etc. ⓕ New Zealand, S. Africa. z9 Plate 14. # ∅

E. johnstonii Maiden. Tree, to 60 m in its habitat, with a tall, clear stem, yet develops a short, thick stem in the mountains; bark exfoliating, orange-red to brownish-green, mature trees becoming scaly at the base; juvenile foliage opposite, sessile, circular to ovate, 3–6 cm long, glossy green, margins with flat, round teeth, mature foliage petiolate, ovate to lanceolate, 5–13 cm long, dark green and glossy, leathery, margins with flat, rounded teeth; flowers in umbels of 3, sessile, conical to hemispherical, anthers soft pink, about 3 cm wide. KEu 135; CFTa 156. Tasmania, 600–1400 m. ⓕ Cultivated in England and Ireland since 1880. z8 #

E. lehmannii Preiss. Fast growing shrub to small tree; leaves lanceolate, long acuminate, stalked; flowers in globose umbels about the size of a fist, on short stalks, operculum finger-form, yellow, filaments greenish-yellow, July–September; fruit cluster globose, the individual fruits not separable. AAu 29, 40; HTS 236. S. Australia. Notable for its peculiar inflorescence. z9 Plate 14. # ☼

E. linearis Dehnhardt. Slender tree, to 15 m high, bark smooth, white, exfoliating; juvenile foliage in 5–6 opposing pairs, similar but smaller than the mature foliage, green or somewhat bluish, linear, 5–10 cm long, but only 5 mm wide; flowers 5–12 in umbels, buds clavate, stamen filament ring small, white, buds yellow-green; fruits hemispherical, very small. KEu 171. Tasmania. Very attractive foliage. Ⓕ Ireland. z8 #

E. longifolia see: **E. amygdalina**

E. nicholii Maiden & Blakely. A small tree in its habitat, scarcely over 12 m high, with a short stem, bark brown and shredded, irregularly branched, somewhat pendulous; young leaves narrow lanceolate, about 5 cm long and 6 mm wide, blue-green with a trace of purple, as are the young branches; flowers small, but numerous, cream-white, fall; fruits small, hemispherical. KEu 109. E. Australia. Ⓕ Scotland. z8 #

E. niphophila Maiden & Blakely. Snow Gum. Earlier included within *E. pauciflora*, differs, however, in the green juvenile foliage and silvery pruinose young branches, these dark red to orange-red in winter, new growth becoming bluish-white; flowers 5–10 in umbels, calyx funnelform, small, green, cap yellow-green, stamen filament ring 2.5 cm wide, white, flowers abundantly (= *E. coriacea* var. *alpina* F. v. Muell.). E. Australia, in the mountains to the timber line. Tolerates cold. Very garden worthy. Ⓕ England. z7 #

E. obliqua L'Hér. Tree, 45–60 m high in its habitat, bark exfoliating in small strips; juvenile foliage broad lanceolate, 5–7 cm long, mature foliage petiolate, very obliquely lanceolate, 7–15 cm long, rather thick, dark green on both sides; flowers 7–12 in stalked umbels, white, anthers yellow. FAu 134; AAu 330; KEu 158. Tasmania. The bark has been used by the Tasmanian natives for making roofs and boats. Ⓕ S. Africa. z9 #

E. ovata Labill. Tree, to 30 m or taller in its habitat, bark cream-white, pink or bronze, later becoming gray and peeling in long, thick bands, trunk base with persistent, furrowed bark; young leaves ovate to circular, 4–7 cm long or more, short petioled, mature foliage ovate to lanceolate, about 12 cm long or longer, acute, tapering to the base, tough leathery, usually glossy; flowers 4–7 in slender stalked umbels, buds ellipsoid, very small, stamen filament ring 1.5 cm wide, white. KEu 106; LNH 153. Tasmania, SE. Australia. Ⓕ Cultivated in Ireland, New Zealand and S. Africa. z8 #

E. parvifolia Cambage. Small tree, to 9 m, bark smooth, gray; juvenile foliage opposite, sessile to short stalked, ovate to oval-lanceolate, 3.5 cm long, dark green or somewhat blue-green, mature foliage petiolate, alternate or opposite, linear-lanceolate or more oval-lanceolate, to 6 cm long, but only 12 mm wide; flowers 4–7 in short-stalked umbels, buds ellipsoid, 5 mm long; fruits hemispherical, somewhat connected by a thread at the apex. SE. Australia. Cultivated in England (Hillier), thrives on limestone soil. z8 #

E. pauciflora Sieb. ex Spreng. Tree, 7–15 m high, stem usually distinctly twisted and crooked, bark peeling in long strips, white to dark gray, therefore spotty in appearance; juvenile foliage ovate, thick, 3–5 cm long, short petioled, mature foliage lanceolate, often curved, 6–15 cm long, acuminate, leathery, glossy green on both sides; flowers 7–12 in nearly globose umbels, white, small. FAu 23: 144; KEu 168 (= *E. coriacea* Cunn.). SW. Australia, Tasmania. Its habitat is in the coldest mountain regions. Ⓕ S. Africa, Kenya, India, Brazil. z8 Plate 14. #

E. regnans see: **E. amygdalina** var. **regnans**

E. siebereana F. v. Muell. Tree, 30–35 m high, stem straight, bark very rough and deeply grooved, not exfoliating, somewhat like *Robinia*, dark brown, hard, only exfoliating somewhat on the uppermost branches, inner trunk beneath white and smooth; juvenile foliage elliptic to broadly lanceolate, obtuse, 7–12 cm long, sessile, thick, gray-green, mature foliage lanceolate, curved falcate, both sides dull green, distinctly veined; flowers 5–12 in small, stalked, axillary umbels, white, July; fruits pear-shaped. FAu 142; AAu 317. SE. Australia, Tasmania. Ⓕ New Zealand, S. Africa. z8 #

E. simmondsii Maiden. Tree, to about 20 m high, bark on trunk persistent, dark gray and shredded, bark on branches smooth and exfoliating; juvenile foliage opposite and sessile or nearly so, ovate to broad lanceolate, acute, dark green or blue-green, margin and midrib often carmine, mature foliage petiolate, narrow elliptic, 15 cm long, 2 cm wide; flowers 7 to over 20 in dense umbels, buds clavate; fruits in thick, globose clusters and persisting for several years on the branches. W. Tasmania. Ⓕ Cultivated in Scotland. z8 Plate 13. #

E. urnigera Hook. f. Tree, 6–15 m high, young branches pendulous; juvenile foliage circular to ovate, 2.5–4 cm wide, base cordate, very blue-green, mature foliage ovate to lanceolate, 7–12 cm long, glossy dark green; flowers in 3's in small, stalked, axillary umbels on cylindrical stalks, stamen filaments yellowish-white, in a ring, February–April; fruits globose to urceolate. CFTa 119; BMns 536. Tasmania. Ⓕ England. z9 #

E. vernicosa Hook. f. Shrub or small tree, 1.5–3 m high, branches angular and somewhat warty; leaves alternate or opposite, ovate-lanceolate, thick, leathery, 2–5 cm long, gray-green on both sides, but upper surface glossier; flowers yellowish-white, 1–2 axillary, honey scented, petiole very short and thick, flat; fruits hemispherical. NF 3: 175. Tasmania. Ⓕ England. z9 #

E. viminalis Labill. Tree, 30–35 m high in its habitat, bark at the base of older trees rough and persistent, upper parts yellowish or white at first, later peeling in long strips, young branches dark red and warty; juvenile foliage opposite, ovate to lanceolate, 5–10 cm long, 1–2.5 cm wide, acuminate, sessile and occasionally stem-clasping at the base, dark green, midrib often carmine, mature foliage alternate, lanceolate or falcate, 10–18 cm long, to 2.5 cm wide; flowers 3 in nearly sessile umbels, buds ovate, small, stamen filament ring yellowish-white, 2.5 cm wide; fruits hemispherical. KEu 138; LNH 151. Tasmania, S. and E. Australia. ℗ Cultivated in Scotland, Ireland, Romania, Corsica, S. China and elsewhere. z8 #

Very fast growing in mild regions and not particular as to soil type, although this characteristic varies among species. Very difficult to transplant. Propagate easily by seed.

Gutierrez, G. de la Lama: Atlas del Eucalipto; Seville 1976–78 (4 vols.).

In this far-reaching work the winter hardiness of many species is particularly detailed. The following *Eucalyptus* species (covered here) will tolerate at least the given low temperatures:

	Celsius	Fahrenheit
E. amygdalina	−7.8	18.0
— — regnans	−7.2	19.0
— coccifera	−12.0	10.4
— cordata	−12.0	10.4
— dalrympleana	−11.7	10.9
— delegatensis	−9.0	15.8
— ficifolia	0.0	32.0
— globulus	−4.5	23.9
— gomphocephala	−2.0	28.4
— gunnii	−12.0 to −24.0	10.4 to −11.2
— johnstonii	−15.0 to −22.0	5.0 to −7.6
— lehmannii	−4.0	24.8
— linearis	−6.0	21.2
— nicholii	−10.0	14.0
— niphophila	−17.8	0.0
— obliqua	−9.4	15.1
— ovata	−9.4	15.1
— parvifolia	−14.0	6.8
— pauciflora	−15.0	5.0
— sieberi	−6.7	19.9
— simmondsii	?	?
— urnigera	−15.0	5.0
— vernicosa	−18.0	0.4
— viminalis	−15.0	5.0

Lit. Blakely, W. F.: A key to the Eucalypts, with descriptions of 500 species and 138 varieties; 2nd ed., 359 pp., Canberra 1955 ● Forest Trees of Australia; publ. by Forestry and Timber Bureau, Dept. of Interior, 230 pp., abundantly illustrated, Canberra 1957 ● Ingham, N. D.: *Eucalyptus* in California; Agr. Exp. Sta. Bull. **196**, 29–112, 1908 ● Maiden, J. H.: A critical revision of the genus *Eucalyptus*; 8 vols., Sydney 1907–1927 ● Mueller, F. v.: Eucalyptographia, a descriptive atlas of the Eucalypts of Australia and the adjoining islands; 10 vols., Melbourne 1879–1884 ● Chippendale, G. M.: Eucalypts of the Western Australian goldfields and the adjacent wheat belt; 216 pp., with nearly 300 ill., some in color, and 113 range maps of the west Australian species; Canberra 1973 ● Kelly, S.: Eucalypts; 82 pp. Text by G. M. Chippendale and R. D. Johnston, 250 color plates; Melbourne 1969 ● Penfold, A. R., & J. L. Willis: The Eucalypts; 550 pp., 61 Pl. (with a 17 pp. bibliography); London 1961 ● Pryor, L. D., & L. A. S. Johnson: A Classification of the Eucalypts; 102 pp., Canberra 1971 ● Herzog, W.: Eukalyptus, wichtigste Holzart der Welt. Erfahrung aus Südamerika; in Forstpflanzen und Forstsamen 1961, 6–58; with 21 ill. ● Baglin, D., & B. Mullins: Australian Eucalypts; 90 color photos and 32 pp. text; Sydney 1966.

EUCOMMIA Oliv. — EUCOMMIACEAE

Monotypic genus of somewhat controversial taxonomic classification; deciduous tree, leaves alternate, petiolate, serrate, without stipules; flowers dioecious, without a perianth, axillary, inconspicuous, appearing before or with the leaves; male flowers with 1 cm long, red-brown anthers; fruits winged, somewhat resembling those of *Fraxinus*, incised at the apex, single seeded.

Eucommia ulmoides Oliv. Tree, to 20 m, branches with chambered pith; leaves oval-elliptic to oval-oblong, 7 to 8 cm long, coarsely serrate, glabrous above and eventually somewhat rugose; flowers in April. MG 1912: 11 and 614; MD 1933: 2; CIS 26; BS 2: Pl. 23. Central China. 1896. Interesting for the latex content of the leaves, branches and bark, evidenced by the long filaments when a plant part is torn. ℗ Soviet Union and S. China. Plate 10; Fig. 31. z5

Lit. Harms, H.: Zur Kenntnis von *Eucommia ulmoides*; in Mitt. DDG 1933, 1–4.

EUCRYPHIA Cav. — EUCRYPHIACEAE

Evergreen or deciduous trees or shrubs; leaves opposite, simple or pinnate, with stipules; flowers white, bisexual, usually solitary, axillary, attractive; sepals 4, cap-like connate at the tips and abscising together; petals 4, stamens numerous; fruit a dry capsule, dehiscing in 5–18 chambers when ripe; seeds winged. — 5 species, 2 in Chile, 1 in Australia, and 2 in Tasmania.

Key (from Dress, simplified)

● Leaves simple;
  > Leaves crenate:
      E. cordifolia

  >> Leaves entire;
    + Leaves 3.5–5.4 cm long; with 2 small prophylls at the base of the flower stalk:
      E. lucida

    ++ Leaves usually shorter than 2.6 cm, with 3–5 small prophylls at the base of the flower stalk;
     ★ Prophylls 4(5); leaves 7 to 14 cm long:
      E. milliganii

     ★★ Prophylls usually 3; leaves 7 to 26 mm long:
      E. × *hybrida*

● ● Leaves all or in part pinnate or trifoliate;

> Leaves all compound; 5 or more leaflets;
  + Leaflets crenate:
                E. glutinosa

++ Leaflets entire;
   ★ Axillary leaflets narrow-oblong; in (1) 2–5(7)
     pairs:
                E. moorei

   ★★ Axillary leaflets oblong-elliptic, in (1) 2 (4)
     pairs:
                E. × hillieri

>> Leaves simple and trifoliate on the same plant;
  + Total margin distinctly serrate:
                E. × nymansensis

   ++ Margin only somewhat serrate on the apical half
      or nearly entire:
                E. × intermedia

Eucryphia billardieri see: **E. lucida**

E. cordifolia Cav. Tall shrub, a tree in its habitat, 15 (40)
m high; branches pubescent; leaves simple, oblong-
cordate, 4–8 cm long, scabrous dentate especially on
younger plants, dull dark green above, gray and
somewhat pubescent beneath; flowers solitary in the
leaf axils near the branch tips, milk-white, about 5 cm
wide, petals 4 (but occasionally also 5–6) fragrant,
anthers dark brown, August. BM 8209; NF 12: 106; PFC
88. S. Chile. 1851. Thrives on limestone or humus soil. z9
Fig. 32. # ✣

E. glutinosa (Poepp. & Endl.) Baill. Evergreen or often
only a deciduous shrub, seldom over 5 m high, not over
10 m in its habitat, narrowly upright, branches

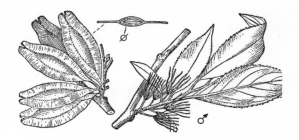

Fig. 31. *Eucommia ulmoides*. Left, fruits with
cross section, right male flower (from Engler)

pubescent at first; leaves pinnate, leaflets 3 (5), about 1–5
cm long, regularly dentate, glossy dark green above,
orange to red in fall (if deciduous); flowers 1–2 on the
branch tips, 5 cm wide, normally 4 (but very often 5–6 or
more) petals, fragrant, anthers gold-yellow, July–
August; fruit a pear-shaped, woody capsule. BS 1: 683;
Bai 4: 38; BM 7067; JRHS 70: 72 (= *E. pinnatifolia* Gay).
Chile 1859. Hardiest species, but only to z9. Fig 32. ✣

'Plena'. Loose double-flowered form often found among
seedlings. Of little merit. z9

E. × hillieri Ivens (*E. lucida* × *E. moorei*). Evergreen
shrub, 2 m high or more; leaves trifoliate or pinnate;
leaflets (3) 5 (9), terminal leaflet twice as long as the
axillary, 4 cm long, 1 cm wide, oblong-elliptic, all entire,
bluish beneath; flowers white, about 2.5 cm wide. Fig. 32.
#

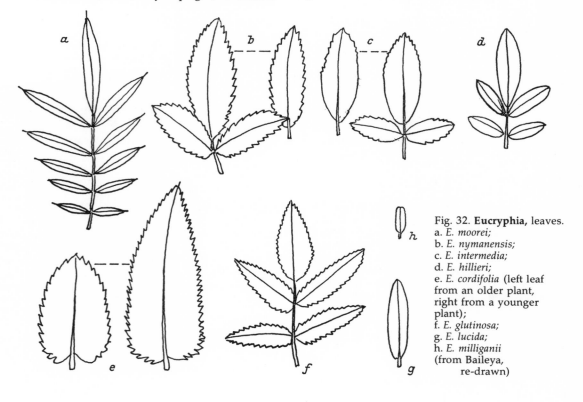

Fig. 32. **Eucryphia**, leaves.
a. *E. moorei*;
b. *E. nymanensis*;
c. *E. intermedia*;
d. *E. hillieri*;
e. *E. cordifolia* (left leaf
from an older plant,
right from a younger
plant);
f. *E. glutinosa*;
g. *E. lucida*;
h. *E. milliganii*
(from Baileya,
   re-drawn)

'Penwith' (Bolitho). A further cross; by Sir Edward Bolitho, Trengwainton, Cornwall. Before 1954. z9 # ☺

'Winton' (Hillier). The type of the cross. Developed by Hillier & Sons, Chandlersford, England. Before 1953. z9 # ☺

E. × hybrida Bausch (E. lucida × E. milliganii). Occurs in the wild among the parents, but not yet in cultivation. Tasmania. z9

E. × intermedia Bausch (E. glutinosa × E. lucida). Tall, evergreen; leaves simple or usually trifoliate, occasionally with more leaflets, the leaflets nearly entire or somewhat serrate on the apical half; flowers 1–2 in the leaf axils, white, about 4 cm wide, petals obovate, overlapping, September–October. BMns 534. Fig. 32. #

'Rostrevor'. The type of the cross. GC 1936 (Sept. 12); Bai 4: 36. Developed before 1936 at Rostrevor House, County Down, Ireland. z9 # ☺

E. lucida (Labill.) Baill. Tall shrub, 6–15 m in cultivation, to 30 m in its habitat, young branches pubescent; leaves simple, oblong, rounded at the apex, 3.5–5.4 cm long, entire, glossy green above, blue-green beneath, resinous, glabrous; flowers solitary in the leaf axils, hanging on 1.5 cm long stalks, pure white, 3 to 5 cm wide, fragrant, petals 4, stamens numerous, anthers yellow, June–July. Bai 4: 36; NF 3: 95; CFTa 33 (= E. billardieri Spach). Tasmania. z9 Fig. 32. #

E. lucida var. milliganii see: E. milliganii

E. milliganii Hook. f. Most appear simply as an alpine form of E. lucida; shrubby habit, but occasionally to 14 m high in its habitat; leaves only 8–20 mm long; flowers smaller, June–July. BM 7200; Bai 4: 37; DRHS 788; CFTa 34 (= E. lucida var. milliganii Hook. f.). Tasmania. Allegedly lime tolerant. z9 Fig. 32. #

E. moorei F. v. Muell. Tree, young branches and leaf rachis short brown pubescent; leaves pinnate, with 5–13 leaflets, these nearly sessile, narrow-oblong, entire, oblique at the base, midrib occasionally ending in a bristly tip, 1.5–7 cm long, glossy dark green above, bluish and pubescent beneath; flowers solitary in the leaf axils of the current year's wood, 2.5 cm wide, pure white, petals 4, obovate, September. BM 9411. New South Wales. 1915. Very tender. z9 Fig. 32. #

E. × nymansensis Bausch (E. cordifolia × E. glutinosa). Tree with a narrow upright habit, medium size, branches and petioles pubescent; leaves pinnate or simple, leaflets 3–5, but usually only trifoliate, terminal leaflet larger than the others, about 3–8 cm long, distinctly serrate on the entire margin, glossy dark green above, lighter and somewhat pubescent beneath; flowers solitary, axillary, pure white, to 3 cm wide, petals obovate, overlapping, anthers yellow, August. Lime tolerant. z9 Fig. 32. #

'Mount Usher' (E. cordifolia (female) × E. glutinosa). Like the type but with simple leaves predominating. GC 132: 13. z9 # ☺

'Nymansay' (Messel). The type of this cross. Flowers broadly campanulate at first, nodding, August–September. NF 12: 106; GC 130: 15; BS 2: P1. 24. Developed about 1915 by Messel in Nymans, Sussex, England. z9 # ☺

E. pinnatifolia see: E. glutinosa

Of the species and forms described E. glutinosa is by far the hardiest but still very tender in all but the mildest climates. E. × intermedia, E. × nymansensis and E. cordifolia follow in hardiness. Most species prefer a humus soil. Although some will tolerate lime, none require it.

Lit. Bausch, J.: A Revision of the Eucryphiaceae; in Kew Bull. 1938, 317–349 ● Dress, W. J.: A Review of the Genus Eucryphia; in Baileya 4, 116 to 127, 1956; with ills.

EUGENIA L. — MYRTACEAE

Evergreen trees and shrubs; leaves opposite and usually entire, finely pinnately veined, often punctate with oil glands; calyx lobes 4 or 5, very small on many species and abscising early, white or cream-yellow and erect on other species, stamen filaments numerous, yellow and very conspicuous; fruit a drupe-like berry, usually globose or pear-shaped, with 1–5 seeds. — About 1000 species in the tropics and subtropics, many with edible fruits; but hardly more than 30 species in cultivation.

Eugenia apiculata see: Myrtus luma

E. australis see: E. paniculata var. australis

E. elliptica see: E. smithii

E. paniculata Banks ex Gaertn. A tree in its habitat, to 10 m, trunk smooth, younger branches and leaves bright red; leaves oblong-lanceolate, 3–7 cm long, long acuminate, base cuneate, glossy green, short stalked; flowers white, 2 cm wide, in short, axillary and terminal panicles standing above the foliage, spring; fruits ovate, 2 cm long, pink-red. NH 38: 102. Australia. z10 # ⚭

Includes;

var. australis (Wendl. ex Link) F. M. Bailey. Shrubby; leaves on the flowering shoots acute or obtuse; flowers usually not distinctly over the foliage, in clusters of 3–15, about 2.5 cm wide across the outspread anthers, July; fruits ovate to globose, red, 12 mm wide. HTS 235; BM 2230; PBl 2: 172; JRHS 100: 15 (= E. myrtifolia Sims; E. australis Wendl. ex Link). Australia. z10 # ☺⚭

E. myrtifolia see: E. paniculata var. australis

E. pungens Berg. Medium high shrub, branches thin, brown, all parts pubescent when young; leaves elliptic-oblong, 5–7 cm long, light green, thorny acuminate, Ligustrum-like; flowers solitary in the leaf axils, white, flower stalk much shorter than the leaves; fruits flat rounded, small, soft pubescent. Brazil. z10 # ⚭

E. smithii Poir. "Lilli Pilly". A tall tree in its habitat, glabrous; leaves ovate to oval-oblong or oblong-lanceolate, tapering to both ends, 3–8 cm long, 1–2.5 cm wide, petiole short; flowers in terminal corymbs, 5–10 cm wide, filaments only 2 mm long, May–July; fruits globose, 8–12 mm thick, white, pink later or with a trace of purple. HTS 234; BM 1872 (as *E. elliptica)*; BM 5450 (as *Acmena floribunda*). Australia. 1790. z10 # ⚭

E. ugni see: **Myrtus ugni**

Can be cultivated and propagated without difficulty.

Lit. Gagnepain, F.: Classification des *Eugenia;* in Bull. Soc. Bot. France **64**, 94–103, 1917 ● Henderson, M. R.: The genus *Eugenia* (Myrtaceae) in Malaysia; The Garden's Bull. Singapore **12**, 1–239, 1949; in this work a connection with *Syzygium* is rejected.

EUODIA J. R. & G. Forst. — RUTACEAE

The earlier name *Evodia* Scop. did not meet the rules of international nomenclature and was therefore dropped in favor of the original name of 1776.

Deciduous or (tropical) evergreen trees or shrubs with aromatic foliage; easily distinguished from the very similar genus *Phellodendron* by the distinct winter buds and the leathery fruit capsule (winter buds on *Phellodendron* are hidden in the leaf petiole, fruit is a drupe, flowers 5 parted); leaves opposite, odd pinnate, leaflets largely tapering from the apex to the base, entire or finely crenate; flowers unisexual, usually 4 parted, small, but usually in large, axillary and terminal panicles or corymbs; styles 4, ovary lobed; fruits composed of 4–5 bivalved, leathery capsules with 1–2 seeds. — About 45 not easily distinguished species in the tropical and subtropical regions of Asia, Australia and Polynesia, including 20 species in China.

Euodia bodinieri Dode. Shrub, to 2.5 m high, branches soft pubescent; leaflets 5–11, elliptic, ovate to oblong, 4–10 cm long, long and obtuse acuminate, base cuneate, only the midrib pubescent above, more pubescent beneath, especially on the venation, rachis pubescent; flowers whitish, in 6–13 cm wide, loose inflorescences, August; fruits reddish, not beaked. China; W. Hupeh Province. 600–900 m. z5.

E. daniellii (Benn.) Hemsl. Shrub or small tree, 4–9 m high, bark gray, smooth, young branches densely gray pubescent; leaves 20–30 cm long, with usually 5–9 leaflets, these oblong-ovate, to 9 cm long and 4 cm wide, long acuminate, base rounded, occasionally somewhat cordate, dark green above, lighter beneath, rather glabrous, finely crenate, nearly sessile; flowers white, in 10 to 15 cm wide, fine pubescent corymbs, June; fruits 8 mm long, reddish, ellipsoid, with a short, hook-like beak. N. China, Korea. 1907. z5 Fig. 33 and 34. ⌀

E. fargesii see: **E. glauca**

E. fraxinifolia Hook. Tree, 9–12 m high, very aromatic; leaves 25–40 cm long, with 5–9 (11) leaflets, these ovate-oblong, about 9 cm long, rather entire, glossy green above, light green beneath, both sides glabrous; flowers greenish-yellow, June; fruits nearly globose, red, seeds dark brown. China, S. Yunnan Province. z6

Fig. 33. **Euodia.** Left *E. daniellii;* middle *E. ruticarpa;* right *E. glauca* (from ICS)

Fig. 34. **Euodia.** Left *E. hupehensis;* middle *E. velutina;* right *E. daniellii*
(drawn from fresh material out of Les Barres Arboretum, France)

E. glauca Miq. Tree, 6–15 m high, young branches thin and soft pubescent; leaves 15 to 25 cm long, leaflets 5–11, ovate-lanceolate to narrow-lanceolate, 8–10 cm long, finely crenate, dark green above, blue-green and glabrous beneath except for the fine brown pubescent midrib; flowers greenish, in 15–20 cm wide, rather glabrous corymbs, June; fruits 6 mm long, with tiny tips, exterior with reticulate venation, rachis red. KIF 2: 32 (= *E. fargesii* Koke). China, Japan. z6 Fig. 33.

E. henryi Dode. Tree or tall shrub, 6–9 m high, branches soft pubescent at first; leaves 16–25 cm long, leaflets 5–9, ovate to oval-lanceolate, 5–9 cm long, paper-thin, finely crenate, deep green above, lighter and somewhat bluish beneath, soon becoming glabrous on both sides; flowers whitish-pink, in loose 5–7 cm wide panicles, July; fruits woody, red-brown, with hook-like beak (= *E. daniellii* Pritz. p. p. non Hemsl.). China, Hupeh Province. 1908. z6

E. hupehensis Dode. Tree, 5–20 m high, bark reddish and very soon quite glabrous, with lenticels; leaflets (5) 7–9, stalked, narrow-ovate, 5–12 cm long, long acuminate, base oblique and rounded, usually entire, leathery, glossy green above, more blue-green beneath, venation silky pubescent; flowers whitish, in wide panicles, August; fruits red-brown, with 5 mm long beak, 2–4 seeds in each capsule. LF 189 (= *E. daniellii* Pritz. p. p. non Hemsl.). China; Hupeh Province. 1908. Ⓔ Hungary. z6 Fig. 34. ∅

E. officinalis Dode. Shrub or small tree, 6–10 m high, young branches soft pubescent; leaflets 7–15, short stalked, ovate-elliptic, 5–8 cm long, acuminate, base cuneate, glabrous above, pubescent beneath; flowers in 7–10 cm wide corymbs, June; fruits globose, reddish, warty, with glossy blue seed (= *E. ruticarpa* Pamp. non Benth.). Middle and W. China. 1907. z5

E. ruticarpa Benth. Tree, to 18 m high, branches thin, brown, densely brown pubescent; leaflets 5 to 11, oblong, long acuminate, 7–14 cm long, 3–5 cm wide, base rounded, crenate, glossy green above, densely gray tomentose beneath, thick and leathery; fruits with 4–5 capsules, small, greenish-white, in dense, terminal corymbs, seed connected by a thin filament. China; Yunnan Province, 600–3000 m. z5 Some doubt exists as to its existence in cultivation. Fig. 33.

E. velutina Rehd. & Wils. Resembles *E. henryi;* tree, 10–15 m high, young branches densely velvety; leaves to 25 cm long, leaflets 7–11, oblong-lanceolate, 5–10 cm long, long acuminate, base obliquely rounded, dull green and pubescent above, densely tomentose beneath; flowers yellowish-white, in about 15 cm wide cymes, August; fruits short-beaked, reddish-brown, pubescent, seeds black. China; W. Szechwan Province. 1908. z5 Fig. 34.

The **Euodia** have little garden merit and are therefore seldom found outside of botanic gardens. The unpleasant smell produced by rubbing the branches is quite noticeable. Prefer a fertile, clay soil, in an open area.

Lit. Dode, L. A.: Revue des espèces du continent asiatique de la section nouvelle *Evodiaceras* du genre *Evodia;* in Bull. Soc. Bot. France **55**, 701–707, 1908 (11 species described).

EUONYMUS (= Evonymus) L. — CELASTRACEAE

Deciduous or evergreen, erect or procumbent, occasionally climbing shrubs; branches usually 4 sided; leaves opposite, occasionally alternate (*E. nanus)*, simple, glabrous; flowers 4–5 toothed, bisexual or unisexual, rather inconspicuous; fruit a 3–5 chambered capsule; each chamber with 1–2 seeds enclosed within a fleshy seed coat. — Over 170 species in Asia, Europe, N. and Central America, on Madagascar and 1 species in Australia.

Key to the Series (from Blakelock)

● Subgenus **Euonymus** Beck.

Winter buds usually ovate, acute, small; stamens with filaments or nearly sessile, anthers upon opening not or incompletely merging, opening by 2 distinct slits; fruit capsules various (if broad globose, then not winged);

Section I. **Melanocarya** (Turcz.) Nakai
Fruit capsule divided to the base;

Series 1. **Alati** Blakelock
Leaves deciduous, margins finely serrate:
E. alatus, sacrosanctus, verrucosoides

Series 2. **Vyonemi** (Presl) Blakelock
Leaves evergreen, finely or coarsely serrate:
E. lucidus

Section II. **Biloculares** Rouy & Fouc.
Fruit capsules unlobed or lobed to the middle of the chambers, the lobes angular or keeled, rounded at the apex or truncate or often with a small tip, smooth or usually somewhat rough;

Series 3. **Lophocarpi** (Loes.) Blakelock
Branches smooth; leaves deciduous, opposite, the largest 9–16 cm long, lanceolate to ovate;

petiole long (the longest 6–30 mm); flowers 4 parted (5 parted on *E. occidentalis)*; petals entire or short toothed; fruit capsules obovoid to obcordate:
E. atropurpureus, bungeanus, europaeus, hamiltonianus, occidentalis, phellomanus, velutinus

Series 4. **Myrianthi** Blakelock
Branches smooth; leaves evergreen, opposite, occasionally in whorls of 3, ovate to linear-lanceolate, 5.5–14 cm long, the longest petiole 0.3–3 cm long; inflorescence usually many flowered, with 3–120 flowers, occasionally only 1–3 flowered, flower stalk usually longer than 1 cm long, to 8 cm long; flowers 4–5 parted; petals entire, short toothed or fimbriate; fruit capsule obovoid or obcordate or clavate:
E. myrianthus, tingens

Series 5. **Glomerati** (Loes.) Blakelock
Not included as confined to the tropics.

Series 6. **Pseudovyonemi** (Nakai) Blakelock
Branches warty or smooth; leaves deciduous, opposite, ovate or linear lanceolate, small, to 5 cm long (if over 5 cm long the branch becomes warty), petiole to 5 mm long; flowers 1–7; petals entire, or short toothed; fruit capsule obovoid or obcordate:
E. nanoides, oresbius, pauciflorus, semenovii, verrucosus

Series 7. **Nanevonymi** (Loes.) Blakelock
Branches smooth; leaves evergreen, alternate, very seldom opposite, linear to linear-oblong or oblanceolate, to 5 cm long, nearly sessile, petiole to 5 mm long; flowers 4–5 parted; petals entire to

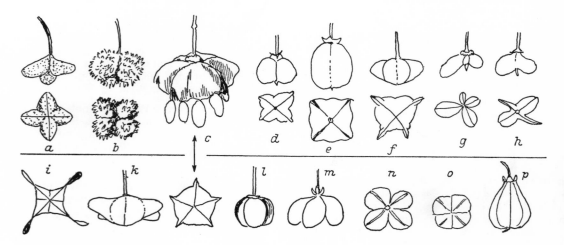

Fig. 35. **Euonymus,** fruit (the first 2 rows are viewed from the side and from beneath respectively). a. *Euonymus atropurpureus;* b. *E. americanus;* c. *E. latifolius;* d. *E. bungeanus;* e. *E. myrianthus;* f. *E. monbeigii;* g. *E. verrucosoides;* h. *E. lucidus;* i. *E. macropterus;* k. *E. sachalinensis;* l. *E. oxyphyllus;* m. *E. europaeus;* n. *E. hamiltonianus* var. *hians;* o. *E. phellomanus;* p. *E. hamiltonianus* (partly from Blakelock)

short toothed; fruit capsule obovoid to obcordate:
E. nanus

Section III. **Multiovulatus** Loes.
Fruit capsules ovate, obtuse at the apex or rounded, with a small tip, the lobes keeled, lobed to the middle of the chamber; disc large, 6–8 mm wide;

Series 8. **Grandiflori** Blakelock
Leaves evergreen or only semi-evergreen, opposite to alternate:
E. grandiflorus

Section IV. **Stenocarpus** Blakelock
Fruit capsule completely without lobes, spindle-form, acute, smooth;

Series 9. **Fusiformes** Blakelock
Leaves evergreen (not included as confined to the tropics)

Section V. **Ilicifolia** Nakai
Fruit capsule occasionally somewhat truncate at the apex, otherwise unlobed or with lobes short and round, 4 parted;

Series 10. **Japonici** (Blakelock)
Leaves evergreen:
E. aquifolium, fortunei, ilicifolius, japonicus, kiautschovicus

Section VI. **Echinococcus** Nakai
Fruit capsule very warty or bristly;

Series 11. **Tuberculati** (Loes.) Blakelock
Leaves deciduous, flowers 5 parted:
E. americanus, obovatus

Series 12. **Echinati** (Loes.) Blakelock
Leaves evergreen; flowers 4 parted:
E. echinatus, wilsonii

● ● Subgenus **Kalonymux** Beck.

Winter buds usually conical, very acute, large; stamens nearly sessile, the anthers opening by a common slit, then spreading open plate-like; fruit capsule broad globose, winged (not winged on E. oxyphyllus);

Series 13. **Macrogemmi** (Nakai) Blakelock
Leaves deciduous:
E. fimbriatus, latifolius, macropterus, monbeigii, oxyphyllus, planipes, sachalinensis, sanguineus

Series 14. **Cornuti** (Loes.) Blakelock
Leaves evergreen:
E. cornutus, frigidus

Euonymus alatus (Thunb.) Sieb. Deciduous shrub, dense, rounded, 1–2 (3) m high and wide, branches with 4 wide, thin, corky, longitudinal wings; leaves ovate-elliptic, fine and scabrous serrate, 3–5 cm long, bright deep red in fall, quite glabrous beneath; fruits reddish, normally 4 lobed, often with only 1–2 lobes, seeds brown, aril orange. BM 8823; DL 2: 234. NE. Asia to Middle China. 1860. z3 Plate 16 and 19. ∅

var. **apterus** Regel. More open growing, branches without corky wings (= E. subtriflorus [Bl.] Franch.).

'Coloratus'. Growth narrowly upright; leaves always somewhat bronze, deep red and scarlet in fall. Selected by Marchant. 1929. ∅

'Compactus' (J. W. Adams). To 1 m high and 3 m wide, very dense, flat rounded; fall color scarlet-red. Introduced to the trade before 1926 by J. W. Adams, Springfield, Mass., USA.∅

E. americanus L. Upright, deciduous shrub, 1.5 (2.5) m high, branches glabrous, winter buds to 1.5 cm long; leaves persisting well into the fall, lanceolate, crenate, 3–8 cm long, deep green, somewhat leathery; flowers reddish-green; fruits pink, 3–5 lobed, exterior with soft warty prickles, seed yellowish-white, aril bright red. BB 2365; GSP 302. N. America, New York to Florida and Texas. 1679. z5 Fig. 35.

var. **angustifolius** (Pursh) Wood. Upright, somewhat taller than the species; leaves narrow, more lanceolate, somewhat falcate, tapered at the base, green at first, dark purple in fall. N. America, Georgia.

E. aquifolium Loes. Evergreen shrub, Ilex-like in appearance, branches 4 sided; leaves nearly sessile, ovate to oblong, 4–7 cm long, wide open dentate, teeth thorny, dull green above; flowers usually solitary; fruits greenish, flat globose, more lobed than the other species of this Series, seeds purple, aril orange. W. China. 1908. Slow growing, pretty. z9 # ∅

E. atropurpureus Jacq. Deciduous shrub, 1.5–2.5 m high, narrowly upright; leaves ovate-elliptic, 4–12 cm long, finely serrate, pubescent beneath, yellow to red-brown in fall; flowers purple, 7–15 on a thin stalk, 1 cm wide, June; fruits light purple, aril scarlet, seed white. BB 2367; GSP 300 (= E. latifolius Marsh.). N. America, Ontario to Florida, west to Oklahoma. 1756. Hardy. z5 Figs. 34 & 36. ∅

E. bulgaricus see: **E. europaeus** var. **intermedius**

Fig. 36. Euonymus atropurpureus

E. bungeanus Maxim. Deciduous shrub, to 4 m high, wide, branches thin, glabrous, rodlike rounded; leaves ovate-elliptic, long acuminate, 5–10 cm long, finely serrate, light green, pale yellow in fall, petioles long (important); flowers, few in a cluster, but flowering, abundantly, yellowish, June; fruits deeply 4 lobed, 12 mm wide, yellowish with a trace of pink, seeds white to reddish, aril orange. BM 8656; DL 2: 233. Manchuria, Korea, China. 1883. z5 Plate 19; Fig. 35.

'Pendulus'. 4 m high, branches arching gracefully; petiole to 3.5 cm long; fruits less numerous.

var. **semipersistens** (Rehd.) Schneid. Foliage semi-evergreen, otherwise like the species; fruits pink, less numerous. Hardy. Plate 19. ∅

E. cornutus Hemsl. Evergreen shrub, buds very large, acute, long conical; leaves narrow lanceolate, 6–11 cm long, 8–15 mm wide, long acuminate, serrate, axillary veins at a 45° angle to the midrib; flowers 4 parted, usually grouped in 3's, each 4–6 mm wide; fruits broad globose, winged, the wings 7–13 mm long, usually somewhat reflexed when young. ICS 3097. China; Hupeh, Yunnan, Szechwan Provinces. z9 #

E. echinatus Wall. Evergreen, procumbent shrub; leaves elliptic-ovate to lanceolate, 4–7 cm long, long acuminate, serrate, cuneate at the base; flowers usually 3–7 in small cymes on 1–1.5 cm long stalks; fruits globose, 8 mm thick, densely covered with 1 mm long prickles. BM 2767. Himalaya. 1827. z9 #

E. europaeus L. Deciduous shrub, 3(7) m high, branches green, angular, with corky stripes; leaves ovate elliptic to oblong, 3–8 cm long, crenate; flowers yellowish-green, anthers yellow, May; fruits 4 lobed, pink to bright red, seeds white, aril orange. HM 1831; GPN 559–560. Europe to W. Asia. In cultivation for centuries. ⓔ In screen plantings in Czechoslovakia, Romania; as dune stabilizer in France. z3 Plate 17, 19; Figs. 35 and 37.

Includes the following cultivars:

'Albus'. Fruits white, aril yellowish-white (= f. *leucocarpus*). Before 1770.

'Aldenhamensis' (Gibbs). Leaves elliptic-ovate, 3–8 cm long, glabrous; fruits pink-red, pendulous on long stalks. 1922. ♂♀

var. **angustifolius** K. F. Schulz. Leaves narrow lanceolate, to 7 cm long; flowers 2–4 together; fruits smaller.

'Argenteovariegatus'. Leaves irregularly white variegated. Looking quite sickly. 1862.

'Atropurpureus'. Leaves narrow lanceolate, purple, fall color scarlet to violet; fruits dark red. ∅ ♂♀

'Atrorubens'. Fruits dark carmine-red. ♂♀

'Aucubifolius'. Leaves in large part with yellow or yellowish speckles, purple-red in fall. Before 1862. Not uncommon today. ∅

'Chrysophyllus'. Leaves greenish-yellow at first, later becoming yellowish-green, from Späth, Berlin. Introduced before 1921 from Hungary.

var. **intermedius** Gaud. Erect, branches more divaricate; leaves broad ovate, base rounded, larger than those of the species;

Fig. 37. *Euonymus europaeus* (from Kerner)

fruits abundantly, fruits light red (= *E. bulgaricus* Velen.; *E. europaeus* var. *ovatus* Dipp.). SE. Europe. ♂♀

var. *latifolius* see: **E. latifolius**

f. *leucocarpus* see: **'Albus'**

'Microphyllus'. Leaves elliptic, 2–3 cm long, 1 to 1.8 cm wide, base slightly keeled, rounded at the apex, venation beneath rough pubescent. Before 1921.

var. *nanus* see: var. **'Pumilis'**

var. *ovatus* see: var. **intermedius**

'Pumilus'. Dwarf form, dense, erect; leaves lanceolate, 2–3 cm long (= *E. europaeus* var. *nanus* Loud.). Before 1838.

'Redcap'. Selected by the University of Nebraska, USA; fruits abundantly, bright red, fruits especially persistent. Propagated in the USA by Interstate Nurseries, Hamburg, Iowa.

'Red Cascade'. (Jackm.). Tall growing, branches somewhat pendulus; fruits abundantly, fruits opal-pink (HCC 022), aril orange (HCC 710). Meritorious selection. ♂♀

E. fimbriatus Wall. Deciduous shrub, similar to *E. latifolius*, but leaves elliptic-oblong and to 10 cm long, abruptly short acuminate, finely double serrate to fimbriate, thin; flowers in 1.5–2.5 cm wide cymes; fruits with long, acute wings. SDK 4: 46; FS 669. Himalaya. 1920. Occasionally found in cultivation. z8 ♂♀

E. fortunei (Turcz.) Hand. Mazz. Evergreen climbing shrub, to 5 m with aerial roots or procumbent along the ground, branches fine warty, green; leaves ovate-elliptic; acute, 3–6 cm long, quite variable, base broadly cuneate, finely serrate, distinctly veined beneath, indistinct above; flowers similar to those of *E. japonicus*; fruits only about 6 mm wide (= *E. radicans* var. *acutus*

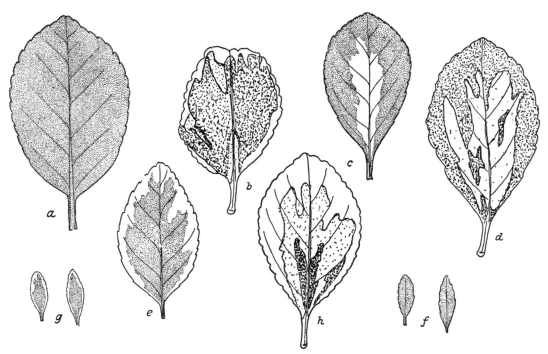

Fig. 38. Leaves of some cultivars of *Euonymus japonicus*. a. 'Macrophyllus'; b. 'Albomarginatus'; c. 'Aureus'; d. 'Duc d'Anjou'; e. 'Aureomarginatus'; f. 'Microphyllus'; g. 'Microphyllus Albomarginatus'; h. 'President Gauthier' (in part from Boom)

Rehd.). China. 1907. Hardy and very popular ground cover. z5# ∅

Includes many cultivars and varieties:

'Argenteo-marginatus' see: **'Variegatus'**

'Berryhill' (Berryhill Nursery, Springfield, Ohio, USA, before 1960). Strong grower, to 70 cm high in 5 years, erect; leaves green, 4–5 cm long. #

'Carrierei'. Bushy, low, climbing when espaliered, branches normally without tendrils or exceptionally with a few; leaves elliptic-oblong, 3–5 cm long, acute, glossy; fruits very abundantly. 1881. Possibly only the mature form of the species. # ∅ ⚭

'Coloratus'. Leaves like the species, 3–5 cm long, thin, rather coarsely serrate, deep purple above in fall, lighter red beneath. Very attractive and meritorious. Plate 19. # ∅

'Corlissii' see: **'Emerald Leader'**

'Dart's Blanket' (Darthuizener Boomkwekerijen, NL, 1969). A 'Coloratus' type, growing to about 50 cm high, branches partly procumbent, partly ascending; leaves 4–6 cm long, brown-red to light red in fall, terminal leaves remaining green. Good ground cover. # ∅

'Dart's Carpet'. A radicans type with small leaves; branches procumbent, but ascending to 35 cm; leaves small, dull dark green, coloring bronze-brown-red in winter. Introduced 1971. # ∅

'Dart's Dab'. A 'Vegetus' type with rather small, dull green leaves; branches spreading along the ground, some to 35 cm

high. 1969. Good ground cover. # ∅

'Emerald Gaiety' (US Pl. Pat. Nr. 1960). Rather robust, broad and tall growing; leaves rather large, broad-elliptic, narrow white margined, coloring somewhat bronze-brown in winter. Similar to 'Gracilis' but coarser. Good winter hardiness. # ∅

'Emerald'n Gold' (US Pl. Pat. Nr. 2231). Strong grower, but not taller than 120–150 cm and to 90 cm wide; leaves with wide bright yellow border, coloring somewhat reddish in winter. Introduced to the trade in 1967 by Corliss Bros., Gloucester, Mass., USA (this is the source of all the 'Emerald' forms). # ∅

'Emerald Leader'. Narrow upright, with a dominant leader, can become 1 m high and 50 cm wide; leaves coarse, thick, glossy, fruits abundantly (= *E. fortunei* 'Corlissii'). US Pl. Pat. 881. Plate 17. #

'Golden Prince' (US Pl. Pat. Nr. 3211). Small shrub, tips of the young branches gold-yellow. Needs full sun for best color. # ∅

'Gracilis' see: **'Variegatus'**

'Kewensis'. Procumbent, rooting along the branches, dense mat-form, the dense, thin, erect branches usually not taller than 5 cm; leaves elliptic to rounded, only about 6 mm long; sterile form. Lowest cultivar of the genus. # ∅

'Minimus'. Very similar to 'Kewensis' and usually given as a synonym, but differing in the more vigorous growth habit and the 15 mm long leaves; otherwise equally good. Plate 18. # ∅

var. **radicans** (Miq.) Rehd. Procumbent or climbing on a support by aerial roots; leaves elliptic, 2–3 cm long, crenate, dull green above, tough, venation indistinct; flowers and fruits not conspicuous. MD 1906: 219; RH 1885: 295 (= *E. radicans* Sieb.). N. and Central Japan. 1865. Very hardy and meritorious. z5 # ⌀

'Reticulatus'. Like var. *radicans*, but with the leaves distinctly white veined. # ⌀

'Sarcoxie' (Sarcoxie Nurseries, Sarcoxie, Missouri; before 1960). Strong grower, erect, to 1.2 m high; leaves green, small, only about 2.5 cm long, glossy. One of the most popular cultivars in the USA. # ⌀

'Silver Gem' (Veitch 1885). Growth less bushy, strong climber, with aerial roots, several meters high; leaves smaller than those of 'Silver Queen', white bordered and always somewhat reddish speckled. # ⌀

'Silver Queen'. Shrubby, 2.5–3 m high, branches generally not climbing; leaves to 6 cm long and 2.5–3 cm wide, border broad white variegated. Possibly the most beautiful white variegated form. # ⌀

'Variegatus'. Leaves somewhat larger than those of var. *radicans*, margin with broad, white band, gray middle, occasionally also reddish speckled (= 'Gracilis'; 'Argenteo-marginatus'). Originally introduced from Japan. # ⌀

'Vegetus'. Growth broad and bushy, but also occasionally climbing given support, branches thick, easily broken; leaves dull light green, elliptic to rounded, crenate, thick; good fruiting. MG 23: 13; MD 1927: 48 (= *E. radicans* var. *vegetus* Rehd.). Japan. One of the most attractive forms. Plate 19. # ⌀ ⚭

E. frigidus Wall. Evergreen shrub, to 4 m high, quite glabrous, young branches green, 4 sided, buds large, to 1 cm long, acute; leaves oblong to lanceolate, long acuminate, 10–12 cm long, glossy dark green, venation on both sides distinctly elevated, margin finely serrate; flowers 7–15 together in loose, long stalked cymes, petals 4, green-yellow or purple, border occasionally whitish; fruits carmine, pendulous, the 4 wings situated near the base, seeds white, surrounded by an orange-red aril. BMns 161. E. Himalaya, Upper Burma, China, Yunnan Province. 1931. Cultivated in Cornwall, England. z9 #

E. grandiflorus Wall. ex Roxb. f. **salicifolius** Stapf & F. Ballard. Semi-evergreen, glabrous shrub or small tree, to 5 m high; leaves linear-lanceolate or narrow elliptic, 5 to 10 cm long, normally acuminate, base cuneate (but some have been seen to have obovate to broad ovate leaves!), glossy dark green, finely dentate, flowers greenish to yellowish, 2 cm wide, 3–9 in cymes; fruits 4 sided, light pink, 15 mm wide, aril scarlet, seeds black. BM 9183. N. India; W. China. 1914. z9

E. hamiltonianus Wall. Deciduous shrub to small tree, very similar to *E. europaeus*, but with branches rounded, green at first, later brown-red; leaves tougher, thicker, oblong-lanceolate, short acuminate, petiole short; flowers reddish-white, anthers purple; fruits obcordate, pink. BMns 548; KIF 2: 42 (as *E. sieboldianus*) (= *E. sieboldianus* Bl.). Himalayas to Japan. z6 Plate 15; Fig. 35.

Includes the following varieties, which until 1951, were considered to be separate species (the classification of Blakelock is retained here, although in Bean, 8th ed., pp. 153 to 154, several varieties are considered to be only synonyms. It must be emphasized, however, that this difficult species requires more precise examination before final determinations can be made. It appears therefore better to retain the Blakelock nomenclature):

var. **australis** Komar. Leaves larger; flowers more numerous; fruits yellow.

var. **hians** (Koehne) Blakel. Leaves lanceolate to elliptic, 6–12 cm long, base rounded, finely serrate; flowers yellowish-white, grouped 6–7, May; fruits teardrop-shaped, carmine-pink, aril blood red, seeds blood-red. MD 1910: 110; BMns 181 (= *E. hians* Koehne). Japan. Plate 16; Fig. 35.

var. **lanceifolius** (Loes.) Blakel. Shrub or tree, to 10 m; leaves broad-lanceolate to elliptic-oblong, 8–14 cm long, acute, tough, crenate, somewhat reticulate venation beneath, petiole 4–8 mm long; flowers 1.2 to 1.5 cm wide, grouped 7–15, May–June; fruits 4 lobed, light pink, aril orange, somewhat opened, seed pink (= *E. lanceifolius* Loes.). Central and E. China. Plate 15.

var. **maackii** (Rupr.) Komar. Round shrub, to 5 m high; leaves lanceolate, 5–8 cm long, to about 3 cm wide, long acuminate, scabrous and finely serrate, gradually tapering to the base; flowers yellowish, June; fruits pink, aril orange, closed or somewhat opened, seed red. MD 1910: 106–107 (= *E. maackii* Rupr.). Ussuri, Amur, Manchuria, Japan. 1976. Ⓕ Soviet Union. Plate 15.

var. **nikoensis** (Nakai) Blakel. Tree, leaves broad-lanceolate to ovate, 9–12 cm long, acute, finely serrate, short pubescent on the venation beneath; flowers 4–7; fruits 4 sided, bright red, aril orange, seeds green (= *E. nikoensis* Nakai). Japan. 1930. Plate 15.

var. **semiexsertus** (Koehne) Blakel. Similar to var. *hians*, but leaves more oblong-lanceolate, narrower, crenate, acute; fruits whitish-pink, aril orange, wide open, the blood-red seeds half visible. MD 1910: 106 (= *E. semiexsertus* Koehne). Japan. 1895. Plate 15.

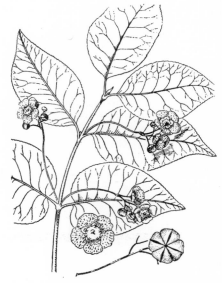

Fig. 39. *Euonymus occidentalis* (from Abrams)

var. **yedoensis** Koehne) Blakel. Shrub, narrowly upright, 3–4 m high; leaves usually obovate, 6–12 cm long, 4–6 cm wide, abruptly short acuminate, crenate, dull green, red-brown in fall; flowers many, in cymes, June; fruits pink, aril orange, closed or only slightly open. MD 1910: 106; BMns 181 (= *E. yedoensis* Koehne). Japan. 1876. Ⓕ Soviet Union. Plate 15.

E. hians see: **E. hamiltonianus var. hians**

E. ilicifolius Franch. Evergreen shrub, 1 to 2 m high, *Ilex*-like, branches angular at first, but eventually rounded, glabrous; leaves elliptic, 4–7 cm long, petioles short (!), margin coarse thorny like *Ilex aquifolium*, tough leathery; flowers small, dull olive-green, usually 3 in short stalked cymes, axillary; fruits rather globose, not lobed, gray-white, 8–12 mm thick, aril scarlet. W. China. 1930. z9 #

E. japonicus Thunb. Evergreen shrub, in cultivation scarcely over 2 m high, 5–8 m high in its habitat, branches slightly 4 sided, dark green; leaves obovate to oblong, 3–7 cm long, leathery tough, obtuse serrate, glossy dark green above, somewhat lighter beneath; flowers greenish-white, but rarely seen, 5–12 in 5 cm long cymes, June–July; fruits pink, not lobed, 8 mm long, aril orange, seeds white. BR 30: 6; HM 1831. Japan, Korea, Liukiu Island. 1804. Many decorative uses, has great garden merit in milder climates. Ⓕ Spain and Yugoslavia (gutta-percha, a rubber substance, is obtained from the roots). z8 Plate 16. #

Numerous cultivars.

Outline of the Cultivars
(from Blakelock)

● Growth broadly upright;

1. Leaves not variegated, dark green;

   Leaves obovate to narrow elliptic, 3 to 7 cm long, ∅:
   E. japonicus (type)

   Leaves elliptic, 5–7.5 cm long, ∅:
   'Macrophyllus' (= 'Latifolius') (Fig. 38)

   Leaves narrow oblong to oblong-lanceolate, 1–2.5 cm long, ∅:
   'Microphyllus' (Fig. 38)

   Leaves crispate and twisted:
   'Calamistratus'

   Leaves oblanceolate, 4.5–8 cm long, branches often pendulous:
   'Longifolius'

   Leaves oblong, seldom obovate, 4 to 6.5 cm long, venation above distinctly indented:
   'Rugosus'

2. Leaves green and white;

   Leaves ovate-elliptic (often also irregular), green, margin white, irregular, more or less white marbled:
   'Argenteovariegatus'

   Leaves large, with broad, white border:
   'Macrophyllus Albus'

   Leaves large, obtuse, dull green, with quite narrow white border, occasionally also somewhat speckled:
   'Albomarginatus' (Fig. 38)

   Leaves large, elliptic, margin broad yellowish-white, gray-green in the middle of the blade, with a very small, dark green, irregular patch in the center, ∅:
   'Président Gauthier' (Plate 17; Fig. 38)

   Leaves crispate, white bordered and marbled:
   'Crispus'

   Leaves like those of 'Microphyllus', but white variegated:
   'Microphyllus Albovariegatus' (Fig. 38)

3. Leaves yellow and green;

   Leaves yellow bordered, middle deep green, ∅:
   'Aureomarginatus' (Fig. 38)

   Leaves dark green bordered, middle irregularly speckled light yellow and light gray-green, ∅:
   'Duc d'Anjou' (= 'Viridivariegatus'), discovered by Jégu, Angers, France about 1870 (Fig. 38)

   Leaves yellow punctate:
   'Punctatus'

   Leaves dark green bordered, middle golden-yellow:
   'Ovatus Aureus' (= 'Aureovariegatus')

   Leaves dark green bordered, middle pale yellow:
   'Sulphureovariegatus'

   Leaves gold-yellow, with a very narrow, dark green marginal line:
   'Aureopictus' (= 'Aureus')

   Leaves greenish-yellow and pink:
   'Tricolor'

   Leaves like 'Microphyllus', but yellow variegated:
   'Microphyllus Aureovariegatus'

●● Habit narrowly columnar upright;

   Growth narrowly upright, compact, leaves tightly arranged, broad elliptic:
   'Pyramidatus'

   Growth habit the same, but more open, leaves narrow elliptic
   'Fastigiatus'

E. kiautschovicus Loes. Evergreen, but often only semi-evergreen, shrub, creeping or broadly upright, to 2 m high; leaves broadly ovate, 5–8 cm long, 2.5 cm wide, crenate, base cuneate; flowers greenish-white, August–September; fruits pink, not winged, aril orange, seeds brown, ripening in October (= *E. patens* Rehd.). China; Hainan Province. 1860. Quite hardy and meritorious. z5 Plate 19. # ∅

'Dupont'. American selection with smaller leaves and denser habit (Willis Nurseries, Ottawa, Kansas, USA.). #

'Newport' (Monrovia Nurs.). Another US. selection with especially dense habit and good evergreen foliage. #

E. lanceifolius see: **E. hamiltonianus var. lanceifolius**

E. latifolius (L.) Mill. Upright shrub, to 5 m high, deciduous, branches long rodlike, winter buds long, glossy, brown; leaves oblong-elliptic, occasionally obovate-elliptic, acute or short acuminate, to 12 cm long, finely crenate, deep green, petiole grooved above; fruits light carmine-red, to 2.5 cm wide, winged, the wings 2–7 mm long, rounded, aril orange, seed white. BS 2: 26; BM 2384; HM 1835 (= *E. europaeus* var. *latifolius* L.). S. Europe to Asia Minor. 1730. Ⓕ Romania z6 Fig. 35.⌀ ⌀

var. *sachalinensis* see: **E. planipes**

E. lucidus D. Don. Tall evergreen shrub or small tree, branches glabrous, pruinose; leaves narrow-ovate to lanceolate, 5–12 cm long, 2–3 cm wide, glossy red new growth, regular and deeply serrate; flowers small, greenish, in short cymes, May; fruits 12 mm wide, deeply 4 lobed, aril orange. FS 7: 70 (as *E. fimbriatus*); SDK 4: 46 (= *E. pendulus* Wall.). Himalaya, N. Assam. 1850. z9 Plate 18; Fig. 35. # ⌀

E. maackii see: **E. hamiltonianus** var. **maackii**

E. macropterus Rupr. Deciduous, glabrous shrub, broad growing, about 2.5 m high or higher; leaves obovate or elliptic, about 7 to 9 cm long, long acuminate, base cuneate, finely serrate, petiole 2–8 mm long; flowers small, greenish, 4 parted, several in stalked cymes, stalk 4–6 cm long, May; fruits winged, the 4 wings 5–10 mm long, acuminate, pink, aril deep red. KIF 2: 41; SDK 4: 45 (= *E. ussuriensis* Maxim.). NE. Asia. 1905. z5 Plate 15; Fig. 35. ⌀ ⌀

E. monbeigii W. W. Sm. Very similar to *E. sanguineus*, possibly a variety. Deciduous shrub or also a tree to 8 m; leaves ovate to oblong, densely serrate, teeth curved inward; flowers green, in large, thin-stalked cymes; fruits with triangular, 3–5 mm long wings. China, SE. Tibet. 1926. z9? Fig. 35.

E. myrianthus Hemsl. Evergreen, glabrous shrub, 2.5–3 m high; leaves elliptic-lanceolate to oblong-obovate, long acuminate, base cuneate, sparse and obtuse serrate, 5–10 cm long, dull green above, petiole 8–10 mm long; flowers in 3–4 branched cymes; fruits top-shaped, 4 sided, yellowish, several together on about 3 cm long stalks. BMns 64 (= *E. sargentianus* Loes. & Rehd.; *E. rosthornii* Loes.). W. China. 1908. z9 Fig. 35. # ⌀ ⌀

E. nanoides Loes. & Rehd. Deciduous shrub, to 1 m high, branches sharply 4 sided to narrow winged, pubescent at first; leaves linear-lanceolate, 8–20 mm long, obtuse; flowers 1–3, nearly sessile, greenish; fruits 1–2, flat globose, 1 cm wide, lobed to the middle, usually only 2 fertile chambers, aril orange, somewhat opened, seeds dark purple. W. China. 1926. Hardy. z3 Plate 19.

E. nanus Bieb. Deciduous shrub, 50 to 80 cm high, habit procumbent-ascending, branches angular, rodlike; leaves alternate or in whorls (!), linear-lanceolate, 2–4 cm long, sparsely dentate, margins involute; flowers 1–3 on a small stalk, brownish, May–June; fruits pink, stalk filamentous, aril red, seeds brown. BM 9308. E. Europe; Soviet Union (Podolia, Bessarabia); Caucasus, E. Turkestan to W. China. 1830. z2 ⌀

var. *koopmanii* see: var. **turcestanicus**

var. **turcestanicus** (Dieck) Krishtofovich. More upright growing; leaves broader, margin not involuted, otherwise like the species. SDK 4: 44 (= *E. nanaus* var. *koopmanii* Lauche ex Koehne). 1883. z2 Plate 19. ⌀

E. nikoensis see: **E. hamiltonianus** var. **nikoensis**

E. obovatus Nutt. Deciduous ground cover, branches prostrate, branches rooting or also climbing if supported; leaves obovate-elliptic, 2.5–6 cm long, light green, crenate; flowers greenish-red, grouped 1–3, 5 parted; fruits carmine-red, warty, aril red. BB 2366; GSP 304. N. America. 1820. Valuable and quite hardy ground cover. z3 Plate 19. ●

E. occidentalis Nutt. Deciduous shrub, very closely related to *E. atropurpureus*, but flowers 5 parted and leaves glabrous beneath; shrub, to 5 m high, buds to 8 mm long; leaves ovate to elliptic-lanceolate, glabrous; flowers grouped 1–5 together, purple. N. America. 1895. z5 Plate 15; Fig. 39.

E. oresbius W. W. Sm. Deciduous, glabrous shrub, 1–1.5 m high, branches 4 sided; leaves linear to oblanceolate, apex rounded, base cuneate, about 1.5–2 cm long; flowers green, small, 1–3, June; fruits 12 mm wide, nice pink-red, with 4 lobes, stalk 8–12 mm long, aril scarlet-red. China; Szechwan Province. z5 1913. ⌀

E. oxyphyllus Miq. Deciduous, glabrous, tall shrub or small tree, to 7 m high; leaves oval-oblong, acuminate, to 7 cm long, 3–4 cm wide, finely serrate, teeth with inward curving tips, dull green, red-brown in fall; flowers 5 parted, brownish-green, May; fruits globose, 5–8 mm long, 1–1.2 cm wide, with 4–5 ribs, dark red, seed scarlet-red. BM 8639; KIF 3:39. Japan, Korea, China. 1895. z5 Plate 15; Fig. 35.

var. **nipponicus** (Maxim.) Blakel. Leaves to 5.5 cm long, 1.8 cm wide; fruits globose, 1 cm long and wide. Japan.

var. **yesoensis** (Koidz.) Blakel. Leaves to 8.5 cm long, 4–5.5 cm wide; fruits 1 cm long, to 1.4 cm wide. Japan.

E. patens see: **E. kiautschovicus**

E. pauciflorus Maxim. Deciduous shrub, very similar to *E. verrucosus*, but with leaves densely pubescent beneath and flowers usually grouped 1–2; 2 m high, branches densely warty; leaves nearly sessile, elliptic to obovate, 3–6 cm long, acuminate, finely serrate, densely pubescent beneath; flowers purple; fruits 4 lobed, circular, red, lobes angular, aril red, seed black. NE. Asia. 1934. z5

E. pendulus see: **E. lucidus**

E. phellomanus Loes. Tall deciduous shrub, to 5 m high, glabrous, branches 4 sided, with broad corky wings; leaves oval-oblong to oblong-lanceolate, 6–10 cm long, crenate, base cuneate, petiole about 1 cm long; flowers normally in 7's; fruits pink, 4 sided, aril red, seeds dark brown to nearly black. DB 1950: 268. N. and W. China. 1928. Very attractive species. z5 Plates 16 and 19; Fig. 35. ⌀ ⌀

E. planipes (Koehne) Koehne. Deciduous shrub, similar to *E. latifolius* in growth habit, size and long acute buds; leaves, however, more coarsely dentate and petioles not furrowed above; fruit like that of *E. latifolius,* but more conical at the apex, the 4–5 lobes not so wing-like and thin (= *E. sachalinensis* [Fr. Schmidt] Maxim, in part, not *E. latifolius* var. sachalinensis Fr. Schmidt). Japan, Korea, NE. China. 1895. One of the most valuable species for its abundant fruit, fall foliage color, growth habit and winter hardiness (see notes at *E. sachalinensis*). z4 ⚭

E. radicans see: **E. fortunei** var. **radicans** and **'Vegetus'**

E. sachalinensis (Fr. Schmidt) Maxim. Purple or dark red flowers in small, few flowered inflorescences. In terms of current thinking (see Bean, Trees and Shrubs, 8th ed., vol. 2, p. 160) this species is presumably not in cultivation. The plant is typical of that found on Sakhalin, Island by Fr. Schmidt in 1869 and originally classified as a variety of *E. latifolius.* Then in 1881, Maximowicz gave this variety species rank and published a new description which would have included, however, plants that in reality were *E. planipes!* Therefore, plants found in cultivation under *E. sachalinensis* would be better labeled *E. planipes.* Plate 15; Fig. 35.

E. sacrosanctus Koidz. Not completely known; very closely related to *E. alatus,* possibly only a variety of it. Japan.

E. sanguineus Loes. ex Diels. Tall deciduous shrub or small tree, to 5 m high, branches nearly round (rodlike), reddish when young, glabrous; leaves oval-oblong to broad elliptic, 4–10 cm long, scabrous and densely serrate, tough, dull green, both sides with fine reticulate venation, new growth reddish, brown-red in fall and very persistent; flowers 4 parted, reddish, grouped 3–16, long stalked, May; fruits to 2.5 cm wide, somewhat lobed, wings 6–8 mm long, horizontally outspread, aril orange, seed black. China, SE. Tibet. 1900. Very attractive species. z6 Fig. 19 ∅ ⚭

E. sargentianus see: **E. myrianthus**

E. semenovii Regal & Herder. Deciduous shrub, to 2 m high, branches smooth; leaves lanceolate, 2–4.5 cm long, thickish, finely crenate; flowers purple, small, grouped 3 to several together; fruits with reddish-green seeds and orange aril, somewhat open at the apex. SDK 4: 43. Turkestan, China. 1910. Plate 19.

E. semiexertus see: **E. hamiltonianus** var. **semiexertus**

E. sieboldianus see: **E. hamiltonianus**

E. subtriflorus see: **E. alatus** var. **apterus**

E. striatus var. *apertus* see: **E. verrucosoides**

E. tingens Wall. Evergreen glabrous shrub or small, 4–5 m tall tree, branches angular; leaves narrow-ovate to lanceolate, tapering to both ends, 3–7 cm long, glossy dark green; flowers 12 mm wide, in about 3 cm wide cymes, petals white with dark red veins, May; fruits 12–15 mm wide, 4–5 sided, dark pink, aril scarlet-red. SNp 29. Himalaya, W. China. 1849. z9 # ∅

E. ussuriensis see: **E. macropterus**

E. velutinus (C. A. Mey.) Fisch. & Mey. Deciduous shrub, very similar to *E. europaeus,* but the young branches densely tomentose, as are the leaf undersides, fruits and inflorescences. Caucasus, Armenia, N. Persia, Transcaspian. z6

E. verrucosus Scop. Upright, deciduous shrub, to 2 m high, branches densely covered with fine black warts; leaves ovate-lanceolate, 3–6 cm long, crenate; flowers brownish; fruits yellowish-red, deeply 4 lobed, 6 mm wide; the black seeds not completely enclosed within the red aril. HW 3:55; HM 1834; KO ill. 302. S. Europe, Caucasus, Asia Minor, N. Persia. 1763. Fall foliage color pale lilac and yellowish. z6 Plate 19. ∅ ⚭

E. verrucosoides Loes. Deciduous shrub, very closely related to *E. alatus,* but the leaves more crenate, venation on the underside not so elevated; filaments 1.5–2 mm long, petals red; the black seeds very conspicuous within the aril (= *E. striatus* var. *apertus* Loes.). SE. Tibet, China. 1910. z6 Fig. 35.

E. wilsonii Sprague. Evergreen shrub, climbing to 6 m high; leaves lanceolate, 6–14 cm long, acuminate, shallowly dentate, base cuneate, distinctly veined beneath, petiole 6–12 mm long; flowers yellowish, in 4–8 cm wide cymes, June; fruits 4 lobed, 2 cm thick, totally covered with about 5 mm long prickles, aril yellow. GC 72: 49. W. China. 1904. z9 Plate 19. # ∅

E. yedoensis see: **E. hamiltonianus** var. **yedoensis**

All species prefer a fertile clay-humus soil. The evergreen species (except *fortunei* and its forms) need winter protection or a wooded site. While the flowers are generally inconspicuous, the brightly colored fruits and often impressive fall foliage are an embellishment in the landscape.

Lit. Blakelock, R. A.: A Synopsis of the genus *Euonymus* L.; in Kew Bulletin 1951, 210–288 ● Lawrence, G. M. H.: *Euonymus europaea, E. hamiltoniana* and relatives; in Baileya 3, 113–114, 1955 ● Sprague, T. A.: The correct spelling of certain generic names. 6. *Euonymus* or *Evonymus;* in Kew Bull. 1928, 294–296.

EUPATORIUM L. — COMPOSITAE

Perennials and shrubs or small trees (mostly the subtropical and tropical species), leaves opposite, entire or serrate, occasionally deeply incised; flowers in heads without ray florets, grouped in flat or arched cymes, occasionally in open panicles, purple, bluish or white; ovaries naked, pappus rough, bracts composing the involucre in 2 or 3 rings and overlapping, all flowers tubular with 5 incisions. — About 1200 species, most in the Americas, only a few in Europe, Asia and Africa.

Eupatorium atrorubens Nichols. Evergreen shrub, to 1.5 m high; leaves ovate, 15 to 30 cm long, serrate, short stalked, margin and venation with long reddish hairs; flower heads numerous, in hemispherical cymes, 10–15 cm wide, purple to lilac-red, February–March. EKW 134; IH 9: 310. S. Mexico. 1862. z10 # ☼

E. ligustrinum DC. Dense, evergreen shrub, to about 2 m high; leaves elliptic-lanceolate, 5–10 cm long, only slightly dentate, tapering to the base, petiole with a narrow, crispate stripe extending onto the blade, light green; flower heads small and few flowered, but numerous in large terminal cymes, arched or umbrella-shaped, to 20 cm wide, the individual flowers cream-white with a pink pappus, fragrant, September to November. NF 4: 272; GC 1905: 229 (= *E. micranthum* Lessing; *E. weinmannianum* Regel & Koern.). Mexico. 1867. z10 # ☼

E. micranthum see: **E. ligustrinum**

E. triste see: **E. vernale**

E. vernale Vatke & Kurz. Subshrub, about 1 m high, strong grower, branches pubescent; leaves oblong-ovate, acuminate, about 10 cm long, irregular and coarsely serrate, base more or less cordate, glossy above with scattered pubescence, light beneath with gray velvety pubescence; flowers white, in loose, more or less conical corymbs, February to March. GF 1873: 750 (= *E. triste* Hort. non DC.). Mexico. z10 ☼

E. weinmannianum see: **E. ligustrinum**
Unpretentious ornamentals, easily cultivated, but requiring a frost free climate; also successful grown in containers.

EUPHORBIA L. — EUPHORBIACEAE

Large, variable genus, annual and perennial plants, shrubs and trees, many species also succulents (the latter will not be dealt with here; but refer to Jacobsen, H., Handbuch der sukkulenten Pflanzen); leaves alternate, in the species covered here often clustered at the branch tips; all parts exude a milky latex if cut (this very often poisonous); flowers 5 parted and leaves glabrous beneath; inflorescence a cyathium (flowers condensed within a bracteate envelope) comprising male flowers with only 1 stamen, 1 terminal female flower with only 1 ovary; fruit a dehiscent capsule with 3 seeds, often with an appendage (caruncle). — Rather cosmopolitan with 2000 species, although most are subtropical or in the milder climatic zones.

Euphorbia acanthothamnus Heldr. & Srt. ex Boiss. Low, cushion form shrub, older branches thorny, the thorns originating in the forked rays of the corymbs; leaves elliptic-ovate, about 1 cm long, bracts of the cyathium oval, yellowish. JRHS 88: 144; PMe 89. Greece, Turkey. z10

E. balsamifera Ait. Shrub, to 2 m, branches thornless; leaves arranged in rosette form at the branch tips, oblong-spathulate, acute to obtuse, 1.5–2.5 cm long, light green to blue-green; flowers inconspicuous, flower heads solitary, nearly sessile to short-stalked, campanulate; seed capsules solitary, globose, seeds brown, rugose. BFCa 37 and 184. Canary Islands. 1779. z10 ∅

E. characias L. Subshrub, to 80 cm high, branches dull purple; leaves dark gray-green, oblong-lanceolate, to 13 cm long; numerous, densely arranged bracts, together with the cymes they form a large, cylindrical inflorescence, prophylls paired, pale green, pubescent, glands 4, truncate, deep purple, capsule densely pubescent; seeds smooth, silver-gray, caruncle light yellow. GC 1958: 436. W. Mediterranean region. z10 ∅

var. *wulfenii* see: **E. wulfenii**

E. dendroides L. Nearly hemispherical shrub, 1 to 2 m high, branches evenly forked, upper portion loosely foliate (but not in rosette form!), branches yellowish, with many, elevated leaf scars; leaves linear-lanceolate, obtuse or acute; prophylls yellowish, diamond-shaped to nearly circular, with small tips, glands truncate or half-moon shaped, capsule to 5 mm long, 6.5 mm wide, 3 lobed, glabrous, the lobes flattened on the side, seeds gray, flattened. RH 1887: 160; PMe 90. Mediterranean region, from the Balearic Islands to the coast of Turkey. z10 Plate 17; range map Fig. 40.

E. mellifera Ait. Similar to *E. dendroides*, but 2–3 m high or more, also a tree on the Canary Islands to 15 m, bark gray, smooth; leaves clustered at the branch tips, narrow lanceolate, obtuse or acute, nearly sessile, dark green; flowers in terminal panicles, prophylls abscise quickly, capsules largWilld. Poinsettia. This widely disseminated holiday plant is a shrub in the frost free regions of the world, usually 2–3 m high, many branched, often leafless at flowering time; leaves elliptic, 10–15 cm long, open lobed, acute, dark green, the apical leaves (bracts) surrounding the flower, bright red, or according to cultiver, pink, salmon or white. BM 3493; HKS 39 (= *Poinsettia pulcherrima* [Willd. ex Klotzsch] Grah.). Tropical Mexico and Central America, in moist locations. 1834. z10 ∅

E. pulcherrima Willd. Poinsetta. This widely disseminated holiday plant is a shrub in the frost free regions of the world, usually 2–3 m high, many branched, often leafless at flowering time; leaves elliptic, 10–15 cm long, open lobed, acute, dark green, the apical leaves (bracts) surrounding the flower, bright red, or according to cultiver, pink, salmon or white. BM 3493; HKS 39 (= *Poinsettia pulcherrima* [Willd. ex Klotzsch] Grah.). Tropical Mexico and Central America, in moist locations. 1834. z10 ☼

'Plenissima'. Flowers largely composed of narrow, bright red bracts, forming a nearly hemispherical inflorescence about 10 cm wide. GC 5: 17. Commonly found in the gardens of the Mediterranean region. ☼

E. regis-jubae Webb & Berth. Shrub, to 1.5 m with thick branches, or also a small tree, branches thornless; leaves narrow-linear, shorter than the flower stalks; flowers in cymes, glands with 2 short horns, prophylls small, yellow, abscising before the fruits ripen; seeds with more or less sessile appendages. BFCa 187. Canary Islands, especially on Tenerife, on dry slopes. z10

E. robbiae Turrill. Evergreen subshrub, 80 cm high, developing vigorous underground stolons, branches

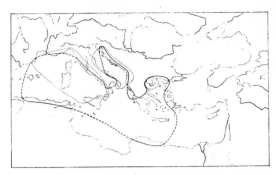

Fig. 40. *Euphorbia* range.
-------- *E. dendroides;* .... *E. spinosa*

remaining green for several years, about 1 cm thick, green, glabrous, with spiraling leaf scars; leaves oblanceolate, 6–10 cm long, 2 (4) cm wide, obtuse or rounded, leathery, dark green, lighter beneath, glabrous on both sides; inflorescences in a large racemose, compound panicle, yellowish-green. BMns 208. NW. Asia Minor. 1953. Completely winter hardy in zone 8 and a good ground cover for minimum maintenance areas. Not particular as to soil or location. # ✛

E. spinosa L. Glabrous, densely branched shrub, 10–30 cm high, not thorny (!), flower stalk and supporting shoots becoming woody after flowering, these shoots die off annually but persist as dead wood for several years, new growth generated from the base each year; leaves lanceolate, 5–15 mm long, entire; inflorescence often composed of only 1 cyathium, prophylls yellowish, capsule 3–4 mm thick, slightly furrowed, usually covered with short warty growth; seeds 2–3 mm thick, smooth, brown. PMe 88. Mediterranean region, from France to Albania. z9 Range map Fig. 40.

E. venata see: **E. wulfenii**

E. wulfenii Hoppe ex Koch. Very closely related to *E. characias,* but growing to 2 m high, although usually only about 1 m high, multistemmed, branches densely covered with linear, sessile, entire, 4–15 cm long, blue-green, soft pubescent leaves, branches 1–1.5 cm thick, basal portions leafless and with many leaf scars; inflorescence a terminal panicle to 20 cm long and 10 cm wide, flowers each with 2 nearly circular, yellow-green bracts, forming a cuplike structure around each cyathium, glands yellow-brown. BMns 482; JRHS 84: 127 (= *E. venata* Willd.; *E. characias* ssp. *wulfenii* [Hoppe ex Koch] A. R. Smith.). Yugoslavia, Greece and Turkey. z9? Plate 20. ✛

The shrubby species are not particular as to soil type, however, they do require warm locations in very mild climates.

Lit. Smith, A. R., & T. G. Tutin: *Euphorbia* L.; in *Flora Europaea* **2,** 213–226, 1968; 105 Species covered ● Smith, A. R.: *Euphorbia;* in Dict. RHS Suppl. **2,** 290–294, 1969.

EUPTELEA S. & Z. — EUPTELEACEAE

A family with only 1 genus; deciduous shrubs or also trees, quite rare in cultivation; branches with large, glossy, brown buds; leaves alternate, petiolate, pinnately veined, distinctly serrate; flowers bisexual, but without sepals and petals, appearing along the branches before the leaves, usually 6–12 in axillary clusters; stamens numerous, anthers linear-oblong; carpels 8–18, all distinct, in a whorl, each unilocular and developing into a wedge-shaped samara with 1–4 seeds. — 3 species in E. Asia.

Euptelea davidiana see: **E. pleiosperma**

E. franchetii see: **E. pleiosperma**

E. pleiosperma Hook. & Thoms. Tree, 5–10 m high; leaves broadly ovate, with a long drawn out apex, base cuneate, 5–10 cm long and 3–7 cm wide, irregularly dentate, the longest teeth around 4 mm long, green beneath, fall foliage red; fruits 2–5 cm long, with 1 to 3 seeds. LF 128; FIO 47; CIS 22 (= *E. franchetii* Van Tieghem; *E. davidiana* Baill.). W. China. z6 ∅

E. polyandra S. & Z. Tree, 5–7 m high; very similar to the previous species, but with broad ovate to circular leaves, 7–12 cm long and wide, coarse and irregularly dentate, the largest teeth 15 mm long, fall color red and yellow; fruits with only 1 seed. Central Japan, in the forests. z6 Plate 10. ∅

Generally hardy plant with peculiar foliage; only for collectors. Prefers a wooded site.

Lit. Nast & Bailey: Morphology of *Euptelea* in comparison with *Trochodendron;* in Jour. Arnold Arb. **27,** 186–192, 1946 ● Smith, A. C.: A taxonomic review of *Euptelea;* in Jour. Arnold Arb. **27,** 175–185, 1946.

EUROTIA Adanson — CHENOPODIACEAE

Deciduous subshrubs or shrubs, low growing with stellate hairs on all parts; leaves alternate, entire, narrow; flowers usually dioecious, inconspicuous, in club-shaped spikes; fruit enclosed in the tubular perianth. — 4 species in Central Asia, S. Europe, N. America.

Eurotia ceratoides (L.) C. A. Mey. Shrub, 0.7 to 1 m high, but often much wider, totally covered with gray-white stellate hairs at first; leaves lanceolate, 2–5 cm long, especially dense white stellate pubescence beneath, with 3 parallel veins; flowers in about 1–3 cm long, dense spikes, grouped in 20–50 cm long terminal panicles, July, male flowers gray and very woolly, anthers yellow, female flowers inconspicuous, axillary. Asia Minor and the Caucasus to China. 1780. Hardy. z4 d S ○

E. lanata (Pursh) Moquin-Tandon. Upright, 0.5–0.7 m high, total plant covered with gray-white stellate hairs; leaves linear to linear-lanceolate, margin somewhat involuted, similar in appearance to *Lavandula,* 1.5–5 cm long; flowers very dense in axillary, 2 cm wide heads grouped in terminal panicles, interspersed with foliate leaves, June–July. BB 1386; MS 89. NW. America, dry locations. 1894. z5 S ○

Both species are only for the collector. Require sandy-dry soil.

EURYA Thunb. — THEACEAE

Evergreen shrubs, leaves normally serrate; flowers small, unisexual, the plants dioecious (the similar genus *Cleyera* has bisexual flowers), sessile or short stalked or in axillary clusters; sepals and petals 5, the latter connate at the base (!), all imbricate, stamens usually 5, attached at the base of the petals, styles usually 3; fruit a berry. — About 130 species in E. Asia, Indomalaysia and on the Pacific Isles.

Eurya emarginata (Thunb.) Makino. Evergreen, densely branched shrub, branches dense yellow-brown pubescent; leaves thick and leathery, narrow-obovate, 2–3.5 cm long, 1–1.2 cm wide, rounded, often also emarginate, cuneate at the base, margin somewhat involuted, dark green above with indented venation, lighter beneath; flowers axillary, small, petals yellow-green, December to April; fruits globose, purple-black, 5 mm. Japan. z9 Plate 27. # ⌀

E. fortunei see: **Cleyera fortunei** Hook. f.

E. japonica Thunb. Evergreen shrub, also a small tree in its habitat, totally glabrous; leaves leathery, elliptic to oblong-lanceolate, 3–8 cm long, 1–3 cm wide, acute, base likewise acute, serrate, dark green and glossy above, light green to yellow-green beneath, short petioled; flowers unisexual, 5–6 mm wide, yellowish-green, 1–3 in the leaf axils, March–April; fruit a 5 mm thick, purple-black berry. BMns 588; LWT 231; KIF 3: 49 (= *E. pusilla* Sieb.). Japan, China, Korea, India, Malaysia. z9 Fig. 41. # ⌀

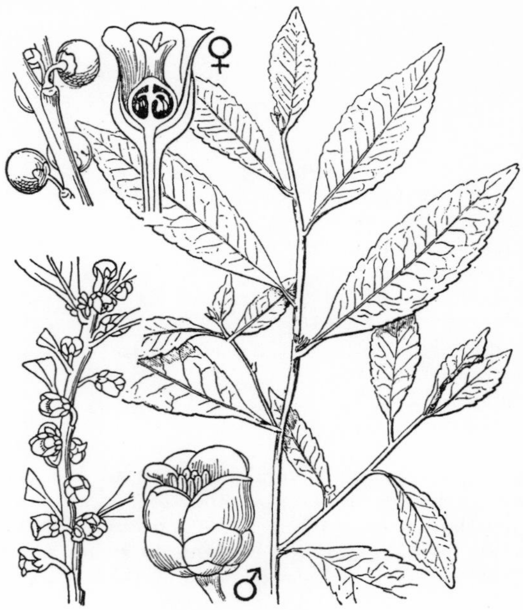

Fig. 41. *Eurya japonica* (from Thunberg and Shirasawa)

E. japonica 'Variegata' see: **Cleyera fortunei**

E. ochnacea (DC.) Szysz. see: **Cleyera japonica** Thunb.

Soil and location requirements similar to those of the more tender *Rhododendron* species. Often confused with *Cleyera* in culture.

EURYOPS Cass. — COMPOSITAE

Small, evergreen shrubs; leaves alternate, densely arranged, entire or pinnately divided; flower heads yellow, solitary, terminal or axillary, on erect, leafless stalks, attractive. —About 70 species, from S. Africa to Socotra Island and Saudi Arabia, but only the following in cultivation.

Euryops acraeus M. D. Henders. Small evergreen shrub, stoloniferous, about 80 cm high; leaves silver-gray, densely clustered at the branch tips, linear, 3 toothed at the apex, to 2.5 cm long, 1 cm wide, base stem-clasping, parallel venation, margins involuted; flowers light yellow, 2.5 cm wide, with many ray florets, in the axils of the uppermost leaves, May; fruits woolly. JRHS 89: 163. S. Africa; Drakensberg. Cultivated under the erroneous name *E. evansii.* z9 # ⌖

Need a very warm, sunny location; otherwise must be overwintered in the greenhouse.

Lit. Nordenstam, B.: The genus *Euryops;* in Opera Botanica **20,** 1968, Lund.

EUSCAPHIS S. & Z. — STAPHYLEACEAE

Monotypic genus; very much resembles *Staphylea,* but with completely different fruits. Deciduous shrub, leaves opposite, odd pinnate, stipules abscising; flowers bisexual, yellowish-white, in small, terminal panicles; calyx 5 parted, persistent; petals 5, rounded; stamens 5, widening at the base; ovaries 2–3 parted, styles connate; fruits are leathery, protruding, fleshy pods, dehiscing to a boat shape, with black seeds. — 1 species in Japan.

Euscaphis japonica (Thunb.) Kanitz. Deciduous shrub, in its habitat also a small, glabrous tree, with pithy branches; leaves 15–25 cm long, with 7–11 leaflets, these ovate to oval-lanceolate, short stalked, finely dentate; flowers greenish or yellowish, in long stalked, 7–15 cm wide panicles, May; fruits boat-shaped, dehiscent, pink-red, 1.5 cm long, with 3–5 blue-black seeds. SH 2: 121; FIO 11; KIF 2: 43. Japan. z9 Fig. 42. ⚘

EVODIA See: **EUODIA**

EXOCHORDA Lindl. — ROSACEAE

Deciduous shrubs; winter buds distinct; leaves alternate, petioled, entire or serrate; flowers large, white, in terminal, elongated, few flowered racemes; flowers polygamous-dioecious, occasionally all bisexual in cultivation; calyx tube circular, constricted in the middle; calyx 4–5 lobed, abscising; petals 5, stamens 15–30, carpels 5, sunken within and tightly adnate to the calyx tube; fruit at first a 5 ribbed or 5 furrowed capsule, later dividing into 5 bone hard, 1–2 seeded follicles. — 5 species, from Central Asia to Korea.

Key (from Rehder, completed)

- Stamens 15, 3 on each petal, the lower flowers on the raceme distinctly stalked (about 0.5 cm long); leaves on the long shoots without stipule-like lobes; fruit 1 cm high or less; petals broadly ovate, abruptly tapering into a short claw:

  E. racemosa

- ● Stamens 20–30; flowers short stalked or nearly sessile; petals generally tapering into a claw;
  - ✕ Leaves on the long shoots without stipule-like appendages;
    - ★ Leaves elliptic, oval or obovate, abruptly narrowing into a slender petiole; fruit circular, truncate or indented at the apex, 1–1.5 cm long;
      1. Leaf petiole 2.5 cm long, red or reddish; leaves nearly always entire; stamens 25 to 30:

         E. giraldii

      2. Leaf petiole 2 cm long, green, occasionally somewhat reddish; leaves on the long shoots at least with a few teeth on the apical half; stamens 20–25:

         E. giraldii var. *wilsonii*

      3. Leaf petioles 1–2 cm long; leaves always scabrous, serrate on the apical half; stamens 25:

         E. serratifolia

    - ★★ Leaves obovate to oblong, vase cuneate; leaf petiole about 1 cm long; stamens about 20:

      E. macrantha

  - ✕✕ Leaves on the long shoots, at least in part, with stipule-like appendages at the base, oblong to obovate-oblong, generally tapering to the petiole; stamens about 25; fruit usually broad oval, coming to an acute point, about 1.5 cm long:

    E. korolkowii

Exochorda albertii see: **E. korolkowii**

E. giraldii Hesse. Strong growing, broad, new growth pink; leaves nearly like *E. racemosa*, but leaf petiole and venation red, nearly always entire, only very seldom somewhat crenate; flowers 6–8 in terminal racemes, very short stalked, to 2.5 cm wide, petals often incised, wider at the apex, tapering to the base, stamens 25–30, May. BS 1: 701; MD 1909: 295; Gw 16: 450. NE. China, Shansi Province. 1897. One of the best species. z6 ✿

Fig. 42. *Euscaphis japonica* (from S. & Z.)

var. **wilsonii** Rehd. More upright habit, to 3 m high; leaves larger, narrower, on long shoots with apical blade half toothed, petioles shorter, only 2 cm long, green; flowers earlier, stamens 20–25. Central China. 1907. Plate 20.

E. grandiflora see: **E. racemosa**

E. 'Irish Pearl' (*E. racemosa* ✕ *E. giraldii* var. *wilsonii*). Strong grower, flowering shoots very long, 60–90 cm; leaves obovate, bluish-white beneath, light green above; flowers 8–10, pure white, nearly stellate form, 4.5 cm wide, May. GC 1933 (1): 24. Developed in Glasnevin, Ireland. ✿

E. korolkowii Lav. To 4 m high, more upright than *E. racemosa*, new growth commencing very early, branches slender, red-brown; leaves oval-oblong, 4–7 cm long, light green above, gray to yellowish-green, on long shoots, serrate on the apical blade half; flowers 5–8 in about 8 cm long racemes, sessile, 3–4 cm wide, stamens about 25, May. Gw 16: 451 (= *E. albertii* Regel). Turkestan. Flowers less abundantly. 1878. z6 Fig. 44.

E. ✕ macrantha (Lemoine) Schneid. (*E. korolkowii* ✕ *E. racemosa*). Growth strong, upright, flowers abundantly and in dense racemes; flowers in about 10 cm long, erect to outspread racemes, pure white, stamens about 20, May. BS 2: 27; RH 1903: 18; DRHS 802. Introduced to the trade by V. Lemoine in 1902. Plate 20. ✿

Fig. 43. *Exochorda racemosa* (from Lauche)

'The Bride' (Grootendorst, around 1938). Very compact, 1.5–2 m high eventually, branches distinctly nodding; flowers especially abundantly with large flowers. Seedling of *E. macrantha*. The only low growing form!

E. racemosa (Lindl.) Rehd. Shrub, 3–4 m high, divaricate branching habit, branches red-brown, with lenticels, glabrous; leaves oblong to obovate, 3–8 cm long, short tipped, entire, light green above, dark green beneath; flowers white, on 5–15 cm long stalks, about 4 cm wide,

Fig. 44. *Exochorda korolkowii*, fruit cluster
(from Dippel)

6–10 in erect racemes, petals rounded, stamens 15, 3 on each petal, May. BM 4795 (= *E. grandiflora* Hook.). E. China. 1849. z5 Fig. 43. ⊕

E. serratifolia Moore. Shrub, to 2 m high; leaves elliptic, 3–7 cm long, always scabrous serrate on the apical half, slightly pubescent beneath; flowers white, 4 cm wide, in loose racemes, May, petals emarginate at the apex, stamens 25. NK 4: 2. Manchuria, Korea. 1918. The rarest species in cultivation. z5

Best used in larger gardens and parks. Need a well drained, fertile soil in a sunny location. The plant form is better maintained if the long shoots are cut back after flowering. Gorgeous, early leafing and abundantly flowering shrubs.

FABIANA Ruiz & Pav. — SOLANACEAE

Small evergreen shrub, superficially resembling heather, leaves small, densely imbricate, compact; flowers solitary, terminal or axillary; corolla narrow funnelform, tubular at the base, becoming generally wider at the apex, limb short 5 lobed; stamens of unequal length, enclosed within the corolla tube; fruit a many seeded capsule, bivalved on the apex, enclosed within the calyx. — About 20 species in S. America; Bolivia, Brazil to Patagonia.

Fabiana imbricata Ruiz & Pav. Densely branched, erect shrub, 0.5–1.5 m high, resembling *Tamarix* or *Calluna*, branches densely covered with short side branches, finely pubescent; leaves dark green, densely arranged, imbricate, triangular, 1–1.5 mm long; flower solitary, terminal, about 1.5 cm long, white or light pink, the 5 small limb tips rounded and reflexed, June. BS 2: 2; BC 1471. Chile. 1838. z9 Plate 27; Fig. 45. # ⊕

'Prostrata'. Procumbent form; flowers white.

'Violacea'. Branches spreading more horizontally; leaves shorter and more appressed; flowers "slate" blue (= Comber's var.). S. Chile. 1854. Supposedly hardier than the species. # ⊕

For light, sandy soil in a protected area; well suited to seacoast plantings in mild climates.

FAGUS L. — Beech — FAGACEAE

Deciduous, usually tall trees, with smooth bark; leaves alternate, petiolate, entire, finely dentate or sinuate on the margins; male flowers in nearly globose, long stalked clusters; female flowers with 3 stigmas; fruit cup 4 parted, soft prickly exterior; fruit (beechnut) 3 sided. — 10 species in the temperate zones of the Northern Hemisphere.

Fagus americana see: **F. grandifolia**

F. crenata Bl. Tree, to 30 m high, crown rounded; leaves oval-rhombic, widest below the middle, 5–8 cm long, shallow crenate, with 7–10 veins, these silky pubescent when young; fruit cupule with long bristles, the basal

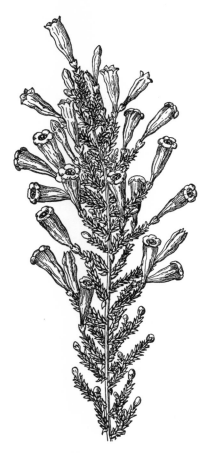

Fig. 45. *Fabiana imbricata* (from Bot. Reg.)

ones becoming spathulate bracts, stalk 1.5–2 cm long. KIF 1: 43 (= *E. sieboldii* Endl.). Japan. 1892. ⓕ Japan. z6 Plate 21, 22; Fig. 47. ✕

F. engleriana Seemen. Tree, to 20 m, usually lower, normally multistemmed from the base; leaves oval-elliptic, short acuminate, with 10–14 vein pairs, underside conspicuously blue-green and with pubescent venation; fruit cupule with linear, leaflike bracts at the base. LF 84. Central China. 1907. Plate 21, 22. ∅

F. grandifolia Ehrh. American Beech Tree, 20–30 m high, bark light gray, smooth, buds brown, glossy; leaves oval-oblong, acuminate, 6–12 cm long, coarsely serrate, teeth curved inward, usually bluish-green above, light green beneath, with 9–14 vein pairs, both sides quite glabrous, golden-yellow to leather-brown, petiole 3–8 mm long; fruit cup with thin, straight or curved bristles. BB 1225; GTP 151 (= *F. americana* Sw.; *F. latifolia* [Münchh.] Sudw.). Eastern N. America. ⓕ Eastern USA. z3 Fig. 47 ∅

This is often the plant found listed as "*Fagus silvatica castaneifolia.*"

var. **caroliniana** (Loud.) Fern. & Rehd. Leaves tougher, deeper green, ovate to obovate, not so coarsely toothed; fruit cup brown tomentose, bristles less numerous, shorter. Eastern N. America. 1836. z6

f. **pubescens** Fern. & Rehd. Leaves more or less densely and softly pubescent on the underside.

F. japonica Maxim. Tree, to 25 m high, multistemmed from the base, young branches glabrous; leaves ovate to elliptic-ovate, short acuminate, 5–8 cm long, base rounded to lightly cordate, slightly open crenate to nearly entire, bluish beneath with midrib somewhat pubescent, otherwise glabrous, with 9–14 paired veins, petiole 1 cm long; fruit cup only 6–8 mm long, with triangular nut exserted. SH 1: 91; KIF 1: 44. Japan. 1905. z6 Plate 21; Fig. 47. ✕

F. latifolia see: **F. grandifolia**

F. longipetiolata Seemen. Tree, to 25 m high, bark light gray, young branches glabrous; leaves ovate to oval-oblong, acuminate, 7–12 cm long, sparsely serrate, base broad cuneate, with 9–12 vein pairs terminating in the toothed margin, bluish and finely pubescent beneath, petiole 1–2 cm long; fruit cup 2–2.5 cm long, bristles thin, usually reflexed. LF 85–86; FIO 109; CIS 17 (= *F. sinensis* Oliv.). Central and West China. 1911. z6

F. lucida Rehd. & Wils. Tree, to 10 m, bark gray; leaves oval-elliptic or ovate, acute or acuminate, 5–8 cm long, base round to broad cuneate, both sides glossy green, the 8–12 vein pairs terminating in the small marginal teeth; fruit cup about 1 cm long, brown tomentose, densely covered with short, appressed, triangular scales, nut somewhat exserted. FIO 110; CIS 130. W. China. 1905. z6 Plate 21. ∅

F. macrophylla see: **F. orientalis**

F. moesiaca (Maly) Czeczott. Transition form between *F. orientalis* and *F. silvatica*; leaves narrower than *F. silvatica*, base cuneate, having more vein pairs; appendages on the fruit cup longer and softer. Central and Western regions of the Balkan Peninsula, especially in the mountains (Velebits, Biokovo, Orjen and Lovcen), nearly to the Adriatic coast. Generally found above 800 m in the southern areas and above 600 m in the northern regions; climbing to 2000 m. z5
The Serbian cultivar 'Zlatia' probably also belongs here, as well as the large leaved *silvatica* form 'Latifolia'. Seeds from Romania have produced young plants of this form and it is certain that whenever the parents occur together, *F. moesiaca* will be present.

F. orientalis Lipsky. Tree, to 40 m, crown somewhat conical, young branches pubescent; leaves elliptic-oblong, acute, 6–12 cm long, widest above the middle, base broadly cuneate to rounded, entire to slightly sinuate, with 7–10 vein pairs, these silky pubescent beneath, petiole 5–15 mm long; fruit cup 2 cm long, bristly, the basal bristles wide, somewhat spathulate. SH 1: 88 (= *F. macrophylla* Koidz.; *F. winkleriana* Koidz.). Asia Minor, N. Iran, Caucasus. 1904. Faster growing in youth than *F. silvatica*. ⓕ Being tried in Germany and Switzerland. z6 Plate 21; Fig. 47.

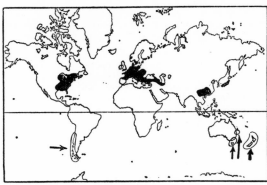

Fig. 46. Range of *Fagus*
(→ = range of *Nothofagus*)

F. sieboldii see: **F. crenata**

F. silvatica L. European Beech. A tree to 30 m high, bark gray, smooth, winter buds dull brown, somewhat silky pubescent; leaves oval-elliptic, 5–10 cm long, lightly toothed, bright green above, lighter beneath, with 5–9 paired veins, yellow to light brown fall color; fruit cup 2.5 cm long, with erect bristles. HW 2: 42, 20. Central Europe to the Caucasus Mts. Ⓕ Germany, Belgium, Austria, Czechoslovakia, Yugoslavia, Holland and Denmark. Cultivated for centuries. z5 ✗∅

Included here are many cultivars.

Outline of the Cultivars

● Differing in habit

Pendulous;

a) leaves green:
   'Bornyensis', 'Miltonensis', 'Pendula', 'Remillyensis'

b) Leaves not green:
   'Aurea Pendula' (leaves yellow), 'Purple Fountain' (leaves red), 'Purpurea Pendula'

Habit columnar;
   'Dawyck (green), 'Dawyck Gold' (yellow), 'Dawyck Purple (brown), 'Fastigiata'

c) Dwarf habit:
   'Cocleata' (upright, tight), 'Tortuosa' (looser, broad, irregular), 'Purpurea Nana' (dwarf form, brown leaved)

● ● Differing in the bark
   'Quercoides' (bark is like that of *Quercus*)

● ● ● Differing in the foliage

a) Leaves green;
   Leaves larger than normal:
   'Latifolia'

   Leaves much smaller, circular:
   'Cockleshell', 'Rotundifolia'

   Leaves developing cockscomb-like:
   'Cristata'

   Leaves elliptic or lanceolate, more or less deeply serrate:
   'Aspleniifolia', 'Cochleata', 'Comptoniifolia', 'Crispa', 'Grandidentata', 'Interrupta', 'Laciniata', 'Quercifolia'

b) Leaves not green;
   Blade monochrome, brown-red to black-red, normal shape:
   f. *purpurea*, 'Purpurea Latifolia', 'Riversii;, 'Spaethiana', 'Swat Magret'

   Blade monochrome greenish-red (somewhat yellowish on 'Rohan Gold'), form differing:
   'Ansorgei' (blade lanceolate), 'Rohanii' (blade deeply incised, red-brown, yellowish on 'Rohan Gold')

   Blade yellow:
   'Zlatia'

   Blade multicolored:
   'Albovariegata' (white bordered), 'Argenteomarmorata' (white marbled), 'Luteovariegata' (light and dark green and yellow), 'Purpurea Tricolor' (red with pink border), 'Striata' (yellowish stripes between the axillary veins), 'Tricolor' (white, green and pink), 'Viridivariegata' (light and dark green)

'Albovariegata'. Leaves often somewhat narrower, margin coarser and more irregularly sinuate, uneven yellowish-white border. PDR 13: 3; MD 52: 29. 1770.

'Ansorgei' (Ansorge). Slow growing; leaves lanceolate, entire, 1–2 cm wide, rather regular in form, dark brown-red. Developed by Ansorge in 1891 at Hamburg-Flottbek, W. Germany. Plate 21.

f. *arcuata* see: **'Tortuosa'**

'Argenteomarmorata' (Späth). Leaves on the new spring growth a normal green, secondary growth irregular, with fine white punctation or also only punctate or marbled, usually not very striking. PDR 13: 11. Introduced 1886.

'Aspleniifolia'. Fernleaf Beech. Slow growing at first, but eventually a very stately tree, crown very dense; leaves distinctly lobed and incised, some linear (to 10 cm long and 6 mm wide), also sometimes pinnatisect and variations of both. MD 52: 30; PDR 13: 4. Those grown by this name in cultivation today are rather consistent (see also, however, 'Grandidentata' and 'Laciniata'). Plate 21. ∅

'Atropunicea'. Earlier name for the grafted purple beech; this name was dropped in favor of f. *purpurea* in Holland in 1976.

'Atropunicea Macrophylla' see: **'Purpurea Latifolia'**

'Atropurpurea Pendula' see: **'Purpurea Pendula'**

f. *atropurpurea tricolor* see: **'Purpurea Tricolor'**

'Aurea Pendula' (Van der Bom 1900). Yellow-leaved Weeping Beech. Trunk erect (always?), branches outspread or weeping; leaves yellow when young, later green-yellow, some scorch evident in full sun (= 'Aureopendula'). ∅

'Aureopendula' see: **'Aurea Pendula'**

'Bornyensis' (Simon-Louis). Weeping beech with straight, erect stem, broadly columnar, branches weeping in an even arch, like a "green fountain". MD 1910: 165. Discovered about 1870 in Borny, France.

'Cochleata'. Dwarf form, compact conical habit, slow grower, scarcely over 4–5 m high and 3 m wide; leaves elliptic, 3–4 cm long, deeply toothed, margins usually undulate, blade often somewhat plaited (= f. *undulata* Jouin). Known since 1864 in Germany. Plate 21.

'Cockleshell'. Columnar with smaller more rounded leaves than 'Rotundifolia'. AB 1967: 20. Discovered in Hillier's Nursery in England in 1960.

'Comptoniifolia'. Like 'Aspleniifolia', but the portion of linear-lanceolate leaves is much greater, slower growing. In cultivation in Germany since 1864; still cultivated today in the Dortmund Botanic Garden.

f. *conglomerata* see: 'Tortuosa'

'Cristata'. Small tree, fast growing; leaves mostly arranged at the branch tips, usually cockscomb-like incised and deformed. MD 52: 29. Plate 21.

'Cuprea'. Copper Beech. A name used primarily in English nurseries to distinguish the seedlings of the purple beech from grafted trees. Leaves greenish-red, usually totally green by fall.

'Dawyck' (H. A. Hesse 1913). Growth narrowly columnar to narrowly conical, retains its form well without pruning, about 25 m high and 3 m wide eventually. MD 1912: 366. This form was discovered in the wild in 1864 in the forest near the Scottish estate Dawyck, Peeblesshire. Around 1907, the German firm Hesse in Weener received scion wood and in 1913 introduced the present form under the above name. Since the descendants are all cloned, the oldest name takes precedence (see: 'Fastigiata' also).

'Dawyck Gold' (Van Hoey Smith 1973) (Selection presumably from 'Dawyck' × 'Zlatia'). Growth narrowly columnar; leaves normal size, golden-yellow in spring, light green in summer, golden-yellow again in fall. Originated in 1968 in Rotterdam, Holland. ⌀

'Dawyck Purple' (Van Hoey Smith 1973) (Selection from 'Dawyck' × Purple Beech). Columnar habit, branch tips directed inward; leaves normal size, purple-brown. Developed in Rotterdam, Holland, 1968. ⌀

'Fastigiata'. This name is given to all the other columnar European beech trees which are encountered occasionally. One such form was propagated by Simon Louis Frères in 1873 at Metz, France; this plant is presumably no longer in cultivation.

'Grandidentata'. Leaves broad-elliptic, margins coarse and evenly dentate, teeth angular, base cuneate. MD 52: 30; PDR 13: 1. Discovered about 1810 in Germany. ⌀

f. *heterophylla* see: 'Laciniata'

f. *incisa* see: 'Laciniata'

'Interrupta' (Van Hoey Smith 1955) (Seedling of 'Rohanii'). Leaves very irregular and deformed, often part of the blade is connected only by the venation. DB 1955: 265.

'Laciniata'. Strong grower, becoming as tall as the species; leaves quite variable, usually broad-lanceolate, long acuminate, pinnately divided, among these are some normal leaves, which are also lanceolate. MD 52: 30; PDR 13: 3 (= f. *incisa* Hort.; f. *heterophylla* Hort.). Known since 1795 in Germany. Plate 21. ⌀

'Latifolia', Leaves like the species, but larger, to 15 cm long, 9–12 cm wide, much tougher. PDR 13: 1 (= f. *macrophylla* Hort.). Known since 1864 in Germany. Probably descended from the transition form *F. moesiaca*.

'Luteovariegata'. Somewhat more vigorous than 'Albovariegata'; leaves yellow variegated, margin yellow. Known in England since 1770.

f. *macrophylla* see: 'Latifolia'

'Miltonensis'. Weeping form with horizontally arranged main branches. Discovered in Milton Park, Northamptonshire, England, before 1899.

'Pendula'. Growth slow when young, main branches more or less horizontal or curved upward, side branches usually hanging vertically. Discovered in England, 1836. Plate 22.

'Purple Fountain' (Grootendorst 1975). Seedling of 'Purpurea Pendula', but differing in the narrow, upright habit with erect top branches and slightly weeping secondary branches; leaves normal size, red-brown, not so dark as 'Purpurea Pendula'. Developed in Holland. The oldest specimen thus far observed is 4 m high and 1 m wide at the base. ⌀

f. *purpurea* (Ait.) Schneid. Collective name for all the seedlings with red foliage, including those found in the wild. Leaves variable in size and color intensity, often brown in spring, becoming more brown-green during the summer. The oldest places or origin are: Buch am Irsel, near Zurich, Switzerland, before 1680; Thuringian Forest. E. Germany, before 1772; Bagarina Valley in S. Tirol before 1840. This name is used exclusively in England and Holland, based on Aiton's work in 1789. That used by Rehder (f. *atropunicea* [Weston] Domin) was based on Weston's work of 1710 which is no longer valid due to some confusion as to whether this name originally applied to *F. grandifolia* or *F. silvatica*. ⌀

Over the years a number of expecially dark red and black-red beeches have been found, named and propagated: i.e. 'Riversii', 'Purpurea Latifolia', 'Swat Magret' and 'Spaethiana'. All these forms are propagated by grafting.

'Purpurea Latifolia'. Habit normal with vigorous growth; leaves rather large, dark black-brown, but somewhat smaller

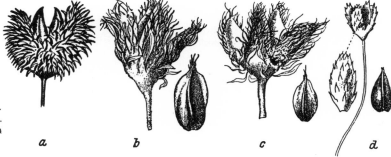

Fig. 47. **Fagus**, fruits. a. *F. grandifolia*; b. *F. orientalis*; c. *F. crenata*; d. *F. japonica* (from Sargent, Schneider)

a   *b*   *c*   *d*

than those of 'Riversii'. This plant (according to H. Grootendorst) is grown in Holland and the USA as 'Riversii', although not always true to name. The distinguishing characteristics are so few on young trees as to make positive identification nearly impossible. ⌀

'Purpurea Major' see: **'Riversii'**

'Purpurea Nana' (Van Geert). Dwarf form, oval crown, only 3 m high and 2 m wide in 40 years, densely branched; leaves rather small, brown. Cultivated in the Trompenburg Arboretum, Rotterdam, Holland. ⌀

'Purpurea Pendula'. Slow growing dwarf form, normally grafted on a standard, branches weeping in a short arch from the graft union, without a dominant leader, therefore not growing any taller than the graft union (unless the branches are trained upward); leaves normal size, glossy black-brown (= 'Purpureopendula'; 'Atropurpurea Pendula'; 'Purpurea Pendula Nova'). Developed about 1865 in Germany. A somewhat different clone is cultivated in the Dutch nurseries. ⌀

'Purpurea Pendula Nova' see: **'Purpurea Pendula'**

'Purpurea Riversii' see: **'Riversii'**

'Purpurea Roseo-marginata' see: **'Purpurea Tricolor'**

'Purpureopendula' see: **'Purpurea Pendula'**

'Purpurea Tricolor'. Tree, only occasionally to 10 m high, thinly branched especially attractive for its carmine-red foliage on new growth, best in full sun; leaves medium size or small, brown with a distinctly wide pink border. MD 52: 29 (= *F. tricolor* André in Rev. Hort. 1885: 311; *F. silvatica atropurpurea tricolor* Pynaert In Rev. Hort. Belg. 1886: 145; *F. silvatica roseo-marginata* [Cripps] Rehd. In Gard. Chron. 1888: 779; *F. silvatica purpurea roseo-marginata* Henry in Elwes & Henry, Trees of Great Britain and Ireland. 1906: 8). Until recently, nearly all authors utilized Rehder's nomenclature; however the nursery trade has retained the earlier title, 'Tricolor'. For reasons of uniformity, the above name should be utilized. Furthermore, this is the name given preference by Bean (1973) and Grootendorst (1976). Also here will be found some clones differing somewhat from one another; in particular, I have seen plants in Oregon, USA (1974) with a nearly 1 cm wide border of bright carmine-red on the new growth. ⌀

'Quercifolia'. Intermediate between 'Grandidentata' and 'Laciniata'; leaves smaller than the type, margins coarsely and deeply incised, teeth not triangular, margins rather undulate, lanceolate leaves absent. MD 52: 30; PDR 13: 2. Known in Germany since 1860. Plate 21.

'Quercoides'. Differing from the typical beech in the "Oak-like", deeply furrowed bark of the trunk and older branches. MG 43: 510; 34: Pl. 31. Discovered by Persoon in 1799 near Göttingen, W. Germany.

'Riversii (Rivers, Sawbridgeworth, England, about 1870). Larger, broad crowned tree; leaves very large, deep black-brown, glossy, coloring very little in fall. First distributed from England, but introduced by H. W. Van der Bom, Holland. Cultivated to a large degree as 'Purpurea Major'; one of the best large leaved, purple beech cultivars available today. ⌀

'Rohan Gold' (Van Hoey Smith 1973). Seedling of 'Rohanii'; leaves also incised and lobed, but yellowish. Developed in 1970 at the Trompenburg Arboretum, Holland. ⌀

'Rohanii'. (Masek). Growth strong, tree-like; leaves very similar to 'Quercifolia', but dark red, partly deformed. MG

1908: 499. Originated in 1894 as a cross between a purple beech × 'Quercifolia' by Prince Camille de Rohan of Sichrow near Turnau in Bohemia; later introduced to the trade by V. Masek also of Turnau. Plate 21. ⌀

f. *roseo-marginata* see: **'Purpurea Tricolor'**

'Rotundifolia'. Low, wide tree, although slender upright when young, branches erect; leaves rounded to circular, 1.5–3.5 cm wide, base occasionally lightly cordate. MD 52: 29. Discovered and distributed by Jackman, Woking, England, 1878.

'Spaethiana' (Späth 1920). Medium sized tree, smaller and narrower than 'Riversii'; leaves rather small, dark beneath, with distinctly lighter venation, deep black-brown (darkest of the purple beeches!), glossy above, holding its color until very late fall. New growth commences a few weeks later than most of the other beeches, leaves persist long into winter. ⌀

'Striata'. Normal habit; leaves normal, totally green at first but soon developing wide yellow-green stripes between the major veins. Discovered in Germany about 1851, but first introduced to the trade by Späth in 1892.

f. *suentelensis'* see: **'Tortuosa'**

'Swat Magret' (Frahm). A purple beech like 'Riversii', but leafing out 8–10 days earlier; also retaining its black-red color long into the fall. Introduced to the trade by Frahm, Elmshorn, W. Germany in 1895. ⌀

'Tortuosa'. Stem and branches very twisted, usually in small curls, partly prostrate on the ground; foliage normal. MD 1911: 269. First discovered in France, 1845.

> Included here are the many deformed beeches, i.e. f. *arcuata* Schwer.; f. *conglomerata* Bean; f. *suentelensis* Schelle.

f. *tricolor* see: **'Purpurea Tricolor'**

'Tricolor' (Simon-Louis). Leaves nearly white, green punctate, margin pink. Known in France before 1870. Not very viable and therefore rare; presumably no longer cultivated. ⌀

f. *undulata* see: **'Cochleata'**

'Viridivariegata' (Lombarts). Leaves dark green with light green patches. NDJ 20: 4f. Introduced to the trade by Lombarts, Zundert, Holland in 1935.

'Zlatia' (Späth). Normal growing beech, but with leaves golden-yellow at first, very glossy, later becoming yellowish-green. Discovered in the wild in 1890 near Vranja, Serbia and introduced into the trade by Späth, 1892. This form probably belongs to *F. moesiaca*, since it was found in the range of the latter. ("zlatia" is a Serbian word for "gold".) ⌀

F. sinensis see: **F. longipetiolata**

F. sylvatica castaneifolia see: **F. grandiflora**

F. winkleriana see: **F. orientalis**

The beeches need a moist, alkaline soil, except *F. grandifolia*, which prefers an acid soil. The many cultivars must be propagated by grafting.

Lit. Krüssmann, G.: Die Spielarten der Rotbuche, *Fagus silvatica*; in Mitt. DDG **52**, 111–122, 1939; with 4 plates ● Browicz, K.: Buki uprawiane w Polsce (= The Beeches cultivated in Poland); in Rocznik Dendrol. XIII, 67–93, 1959 ● Van Hoey Smith, J. R. P.: Over bruine beuken; in Groen, 30–31 ● Wyman, D.: Registration List of cultivar names of *Fagus*; in Arnoldia 1964, 1–8 (only covering *F. silvatica*) ● Grootendorst, H.: *Fagus*; examiners' report; Dendroflora 11, 3–17, 1976.

FALLUGIA Endl. — ROSACEAE

Deciduous shrub; branches straw-yellow, bark exfoliating; leaves small, pinnatisect, the lobes linear, margins involuted; flowers solitary or few grouped at the branch tips, stalked; calyx cup hemispherical, with bracts; sepals 5, petals 5, white, outspread; stamens numerous; styles numerous; fruit composed of leathery nutlets, each with a long feathery appendage when ripe. — 1 species in N. America and Mexico.

Fallugia paradoxa (D. Don) Endl. Shrub, 0.5 to 1 m high, densely branched, bark white; leaves 1–2 cm long, with 3–7 linear lobes, brown scaly beneath; flowers solitary to several, about 2.5 cm wide, white, June to August; fruits

Fig. 48. *Fallugia paradoxa*, flowers as seen from beneath

in feathery, 3 cm wide heads. MS 225, BM 6660; MG 1900: 207; BS 2: 185; MCL 176. New Mexico to Utah. 1877. ℗ SW. USA, for erosion control. z6 Fig. 48. ⊕⚭○

× FATSHEDERA (FATSIA × HEDERA)
Guillaumin — ARALIACEAE

Hybrids between *Fatsia* and *Hedera*; evergreen shrub, upright growing; leaves long petioled, 3–5 lobed, more like *Hedera*; flowers in umbels, grouped in large, terminal panicles; styles 5, very short. — Only 1 hybrid known to date.

× **Fatshedera lizei** (Cochet) Guillaumin (*Fatsia japonica* 'Moseri' (female) × *Hedera helix* var. *hibernica*). Evergreen shrub, narrow-upright, branches thick, warty, to 2 m long, ususally unbranched, rust-brown pubescent when young; leaves 10–25 cm wide, 10–15 cm long, usually 3–5 lobed, the lobes triangular, deep green, glossy, leathery, petiole as long as the blade; flowers light green, in 10–20 cm long, umbellate panicles, October–November; pollen sterile. BM 9402; RH 1924: 179. Developed by Lizé Frères, Nantes, France, 1910. z8 # ∅

'Variegata'. Leaves white bordered. ∅

Outstanding shrub for shady areas in milder climates; otherwise must be overwintered in a cool greenhouse.

Lit. Bilquez, A.-F.: Etude des causes de la stérilité de × *Fatshedera lizei*; in Jour. Agr. Trop. et Bot. Appl. 1957, 545–547.

FATSIA Desne. & Planch. — ARALIACEAE

Evergreen shrubs; leaves alternate, long petioled, palmately lobed, nearly circular in outline; flowers bisexual, petals just touching or totally distinct (not overlapping like *Aralia*), stamens 5, styles 5, distinct, disk thick, fleshy; fruit globose, fleshy, black. — 2 species in Japan and Formosa.

Fatsia horrida see: **Oplopanax horridus**

F. japonica (Thunb.) Decne. & Planch. Evergreen shrub, 2(4) m high, but usually wider, branches thick, about 2 cm in circumference; leaves palmately lobed, 15–30 cm wide, with 7–9(11) lanceolate lobes, these acute, toothed, dark green, and glossy, petiole 7–30 cm long; flowers yellowish-white, in globose, long stalked heads, these grouped into compound panicles, October–November; fruits black, BM 8638 (= *Aralia japonica* Thunb.; *A. sieboldii* Hort.). Japan. 1838. z7 # ∅

'Moseri' (Moser). More compact growing; leaves larger. Developed by Moser, Fontainbleau, France. Plate 23. # ∅

'Variegata'. Slow growing; leaves white or yellow variegated, often turning green. # ∅

F. papyrifera see: **Tetrapanax papyrifer**

Hardiness and uses like *Fatshedera*.

FEIJOA See: **ACCA**

FELICIA Cass. — COMPOSITAE

Subshrubs, occasionally annual plants; leaves alternate or opposite, entire or dentate; flowers in heads, solitary on long stalks, stellate, ray florets blue or white, disk flowers yellow, involucre hemispherical to broad campanulate, petals imbricate in several rings, margins narrow, dry membranous, pappus bristles in only 1 ring. — About 60 species in tropical and S. Africa.

Felicia amelloides (L.) Voss. Cape Aster. Subshrub or often a herbaceous perennial, 30–50 cm high; leaves opposite, oblong-ovate, 3–5 cm long, obtuse, sessile or tapering to a short petiole; flowers sky-blue, about 3.5 cm wide, disk yellow, June–October, flowering the entire year when cultivated under glass. BM 249. S. Africa. 1753. z8 ⊗

In cooler climates grown annually from seeds or cuttings, but more perennial in milder climates. Not particular as to soil. Flowers abundantly, especially in summer and fall.

FENDLERA Engelm. & Gray. — PHILADELPHACEAE

Small genus of deciduous shrubs; branches striped; leaves opposite, usually sessile, entire, 1–3 veined, without stipules; flowers 4 parted, stamens 8, ovary half inferior; flowers grouped 1–3 on short side branches, stalked; fruit a many seeded capsule, 4 valved, longer than the persistent calyx. — 4 species from Colorado to Mexico.

Fendlera rupicola Gray. Shrub, about 0.5–1 m high, branches somewhat divaricate, striped, finely pubescent and gray-yellow at first, bark later somewhat shredding; leaves lanceolate to narrow-oblong, 2–3 cm long, acute, entire, sessile, usually rough and green above; dense appressed pubescent or also glabrous beneath; flowers white, usually solitary, occasionally to 3 together on short branches, often with a reddish limb, 3 cm wide, fragrant, petals pubescent outside, May–June; fruit a brown capsule. BC 1480, SW. USA. 1879. For clay-sandy-gravelly soil in warm, dry locations. z7 ⊗

F. wrightii (Gray) Heller. Similar to the above species, but with leaves white tomentose beneath. BM 7924. Mexico, Texas. 1879. z6

FENDLERELLA A. A. Heller — PHILADELPHACEAE

Low, deciduous shrubs, densely branched with exfoliating bark; leaves small, 3 veined, entire; flowers small, with 5 sepals and petals, 10 stamens, filaments without appendages, in dense, small cymes, white. — 4 species in the southwestern USA to N. Mexico.

Fendlerella utahensis (Wats.) Heller. Erect shrub, to 1 m, many branched from a thick rootstock, entire plant more or less strigose; leaves opposite, elliptic to linear or oblanceolate, thickish, to 25 mm long and 4 mm wide, entire, petioles short; flowers unattractive, 3–7 on short, leafy shoots, terminal, 5 mm wide, white, May–August. VT 304. Texas to Mexico, dry mountain slopes. z6

FICUS L. — Fig — MORACEAE

Evergreen or deciduous trees or shrubs, or occasionally climbers with milky sap; leaves normally alternate, simple or lobed, flowers very tiny, unisexual, inside the hollow receptacle and in 3 forms: male flowers with 1–5 stamens, female flowers with ovary and style, and finally "gall flowers" (altered female flowers with disturbed egg chambers) formed by the fig wasp (*Blastophaga*). Occasionally all 3 flower types will be found in one fig (e.g. *F. religiosa*); the plants are occasionally dioecious, having plants with seed bearing figs on the one hand, and plants with male and gall flowers on the other hand. For a better understanding of the important role played by the fig wasp and the effects of "caprification", refer to the notes at *Ficus carica*. — About 800 species in the warmer regions of the entire world.

Ficus australis see: **F. rubiginosa**

F. benjamina L. In its habitat a large and very wide crowned tree, branches spreading widely and nodding, bark rather smooth, gray-green on young trees, entire plant glabrous; leaves thin and leathery, elliptic to more oval-lanceolate, 5–10 cm long, sharply acuminate, dark green and glossy above, lighter beneath, base rounded, margins somewhat sinuate, with very many, fine axillary veins; fruits in 2's, blood-red when ripe. KFi 52; BM 3305; Exo 1159; KFi 13 (= *F. nitida* Thunb.). SE. Asia, from Burma to the Philippines, but planted in the tropics as a street tree. A popular potted plant in cooler climates. z10 # ∅

The epithet "benjamina" has nothing to do with Benjamin, rather it is a latinization of the Sanskrit word for "banian" or "banyan".

F. carica L. Common Fig. Shrub or tree, deciduous, branches thick; leaves deeply 3–5 lobed, palmate, 10–20 cm long, irregularly dentate, rough pubescent above, softer pubescent beneath; flowers, refer to the genus description; fruits are the well known fig, greenish, brownish or violet. HM 87. Asia Minor. z9 ✂

Included here are a large number of cultivars, although none are suitable for cooler temperate climates.

Figs occur in 2 forms: a) the pollinator with only male and gall flowers, in which the fig wasp (*Blastophaga*) is found, and b) the "fruit" producer with only female flowers. The fig wasps when exiting the gall flowers pick up the pollen and transfer it to the stigma of the female flowers, resulting in the formation of fuller, more juicy, flavorful figs. For this reason many pollinators will be found planted in a fig plantation or branches bearing the gall flowers are hung in the crown of the fig producing trees. This process is referred to as "caprification". The latest cultivars will produce good figs without caprification, such as the Italian form 'Dottato'. The main type cultivated in the Orient is 'Sari Lop', others include 'Bardajic', 'Kassaba' and 'Cheker', the latter 3 with red fruit flesh.

The nomenclature of no other fruit species is as confused as that of the fig; many cultivars have up to 15 synonyms, and the spelling of these can vary greatly. The most complete and significant work on fig varieties is:

Condit, Ira J.: Fig Varieties. A monograph; in Hilgardia **23**, 323–538, with many ills. and 26 plates of the fruit forms; Berkeley, Calif. 1955.

An introduction to this work, by the same author, is:

Fig characteristics useful in the identification of varieties; in Hilgardia **14**, 1–69; with 1 plate; Berkeley, Calif. 1941.

F. elastica Roxb. Rubber Tree. In its habitat a tall, often multistemmed, broad crowned tree, with many aerial roots reaching from the branches to the ground, older stems with vertically flattened roots at the base and a network of thick roots over the ground, bark red-brown; leaves oblong to ellipic, 10–15 cm long, to 30 cm on young plants, rounded at both ends, apex with a short pronounced tip, secondary veins running parallel from midrib, in more than 50 pairs, very glossy above, dark green, lighter beneath, glabrous, petiole 3–5 cm long, rodlike, young leaves involuted and completely encircled by the pink-red, sheath-like connate stipules; fruits greenish-yellow, 1.2 cm long, only developing in the tropics. KFi 54; Exo 1158. Forests of E. Himalaya to Assam, Burma, Malaya and Java. 1815. The most frequently cultivated species of the entire genus, mostly as a pot or tub plant but also used in the landscape in frost free climates. z10 # ∅

'Decora' (Van Hecke). Leaves shorter, but wider and stiffer. Exo 1157. Selected as a seedling about 1938 by Gratian van Hecke of Ghent, Belgium; a very important form in culture today. # ∅

F. lyrata Warburg. Fiddleleaf Fig. Small tree, bark scaly or with many longitudinal cracks, branches without aerial roots, trunk without flattened roots at the base, young shoots about 1.5 cm thick, with large, elevated lenticels; stipules large, 5 cm long, brown, persistent; leaves broad-obovate with rounded apex, deep lyre-form, emarginate, base cordate, 50–60 cm long, 15–30 cm wide, leathery tough, glossy dark green above, venation elevated and whitish, margins sinuate, petiole thick, to 7.5 cm long, somewhat flattened above, but not grooved; fruits grouped 1–2, sessile, globose, to 3 cm thick, green with white patches. Exo 1159 (= *F. pandurata* Sander non Hance). Tropical W. Africa. 1903. z10 Plate 24. # ∅

F. macrophylla Desf. ex Pers. A large, broad-crowned tree in its habitat, with a thick trunk and very conspicuously flattened basal roots, bark dark gray to nearly black, rough and somewhat scaly, young branches green, glabrous; both stipules to 15 cm long, exterior rust-brown pubescent, leaves rather densely arranged, oval-oblong, 10–22 cm long, 7–12 cm wide, obtuse to broadly acuminate, base broad rounded, with 16–20 pairs of secondary veins, midrib light green to nearly white above, leathery tough, entire, petiole light green, 10–15 cm long, slightly flattened above; fruits usually 2, axillary, pear-shaped, 15–20 mm long, greenish to purple with yellowish patches. Exo 1174; KFi 211. Australia. 1869. Markedly less heat loving than *F. elastica*. z10 # ∅

F. microcarpa L. f. Abundantly branched, glabrous tree, usually with many aerial roots hanging from the branches and trunk; leaves oblong to obovate, leathery tough, 5–8 cm long, 3–5 cm wide, with short, blunt apex, obtuse or cuneate at the base, entire, axillary veins numerous, but not very distinct, all parallel except for the basal pair, dark green and very glossy, petiole 7–20 mm long; fruits flat globose, 8 mm thick. KFi 62 (as *F. retusa* var. *nitida*); Exo 1165. SE. Asia to India, Malaysia and Australia. 1793. z10 Plate 24. # ∅

Nearly always erroneously labeled in cultivation as *F. nitida* Thunb. or *F. retusa* L.

F. nitida see: **F. benjamina**

F. pandurata see: **F. lyrata**

F. pumila L. Evergreen, climbing species, branches climbing by tendrils with holdfasts; leaves nearly 2 ranked, small on young shoots, ovate, densely veined, dark green, 2 to 4 cm long, mature foliage and that on fruiting branches much larger, more oblong-elliptic, 5–10 cm long, leathery tough, very short petioles; fruits large, circular or pear-shaped, about 4 cm long, a turbid violet. KFi 158; Exo 1168; BM 6657 (= *F. repens* Hort.; *F. scandens* Hort.; *F. stipulata* Thunb.). China, Japan. 1721. z9 Plate 24. # ∅

F. religiosa L. The Sacred Tree of India (Peepul-tree). Evergreen tree, about 8 m high or only shrubby; leaves cordate with very long, caudate apex; fruits dark purple, in sessile pairs. KFi 67; Exo 1162. E. India. 1731. z10 Plate 24. # ∅

F. repens see: **F. pumila**

F. retusa var. *nitida* see: **F. microcarpa**

F. rubiginosa Desf. Large tree, to 12 m high and wide, crown hemispherical or broadly conical, very dense, trunk often surrounded by aerial roots on the lower portion, bark dark gray, rather smooth, with longitudinal cracks, young branches scabby-pubescent, often somewhat angular or flattened, terminal bud 2.5–5 cm long, dense rust-red pubescent; stipules to 12 cm long, of uneven length, glabrous interior, exterior pubescent; leaves broad-elliptic, apex obtuse, rounded at the base, 8–17 cm long, to 6 cm wide, with 8–12 vein pairs, tough leathery, entire, young leaves pink-red on both sides, but later more or less glabrous above, petiole about 4 cm long; fruits usually in pairs, globose to flat globose, 1–1.5 cm thick, green or brown or yellow. BM 2939; Exo 1165 (= *F. australis* Willd.). Australia. z10 # ∅

F. scandens see: **F. pumila**

F. stipulata see: **F. pumila**

F. sycomorus L. Pharaohs Fig. Generally evergreen, but often leafless for a short time, large, broad-crowned tree with a short, thick trunk, bark yellow-brown, exfoliating in large plates, aerial roots absent, branches brown, with large, raised lenticels, glabrous, except for a ring of white silky pubescence at the nodes; leaves broad-ovate, to 15 cm long, 13 cm wide, with rounded apex, base cordate, somewhat leathery, tough, with 3–5 vein pairs, deep green and glossy above, glabrous, finely pubescent and lighter beneath, margins more or less sinuate, petiole 5–6 cm long, brown, lightly silky pubescent; fruits very numerous, small, to 2.5 cm long, white pubescent, edible. Exo 1172. Originally from Ethiopia and Central Africa, but planted in Egypt, Syria and Arabia in biblical times. z10 # ∅

The "Sycamore" was one of the sacred trees of the Egyptians and Arabs and is mentioned several times in the Bible. When grown in a hot, dry climate the abundant fruit is high in sugar and is a good source of nourishment for both humans and livestock. Five cultivars are available in Israel.

Lit. Condit, I.: Fig Culture in California; Cal. Agr. Ext. Serv. Circ. **77**, 1–67, 1941 ● Condit & Enderud: A Bibliography of the Fig; in Hilgardia 1956, 1–663 ● Condit, I. J.: *Ficus*. The exotic species; 363 pp., 50 ills., 35 plates, Univ. of Calif. 1969 ● King, G.: The species of *Ficus* of the Indo-Malayan and Chinese Countries; 185 pp. and 237 plates, Calcutta 1888 ● Domke, W.: Der Gummibaum und seine Verwandten; in Gartenflora 1935, 135–138 ● Von Leick: Die Kaprifikation und ihre Deutung im Wandel der Zeiten; in Mitt. DDG 1924, 263–284 ● Solms-Laubach, H. Graf: Die herkunft, Domestication und Verbreitung des gewöhnlichen Feigenbaumes; in Abhandl. Kgl. Ges. Wiss. Göttingen **28**, 1 to 106, 1882.

FIRMIANA Mars. — STERCULIACEAE

Deciduous trees; leaves alternate, long petioled, large, palmately lobed; flowers 5 parted, unisexual, in terminal panicles; petals absent, sepals colored, carpels separated at the base, but connate at the apex and with a common style; fruits dehiscing before ripe, leaflike, leathery tough, the seeds attached to the margin. — 15 species in Asia, 1 in Africa.

Firmiana platanifolia (L. f.) Schott. & Endl. Deciduous tree, crown rounded, to 15 m high, bark gray-green, smooth; leaves alternate, palmately lobed, 15–20 cm long, base cordate, lobes acuminate, entire, glabrous above, fine tomentose beneath; flowers small, yellowish-green, in terminal, 20–30 cm long panicles; fruit a follicle, opening long before it is ripe, then leaflike outspread, 4–6 cm long, seeds pea-sized. LF 230; LWT 216; KIF 2: 62 (= *F. simplex* W. F. Wight; *Sterculia platanifolia* L. f.). China; cultivated in Japan. 1757. z9 Plate 27. ∅ ⚭

F. simplex see: **F. platanifolia**

Plate 1

Edgeworthia papyrifera
on Mainau Island, W. Germany

Embothrium coccineum var. *lanceolatum*
in the Dublin Botanic Garden, Ireland

Erica cerinthoides
in the Brummeria Park, Pretoria, S. Africa

Erica chamissonis
in the Kirstenbosch National Botanic Garden, S. Africa

Plate 2

Forsythia 'Beatrix Farrand" *Forsythia ovata*
both in the Dortmund Botanic Garden, W. Germany

Forsythia intermedia 'Lynwood'
in a private garden

Forsythia suspensa 'Atrocaulis'
in the Dortmund Botanic Garden, W. Germany

Plate 3

Cytisus decumbens
in the Royal Botanic Garden, Edinburgh, Scotland

Grevillea robusta
in a park in S. Spain

Gleditsia triacanthos 'Sunburst'
in the Hillier Arboretum, England

Halesia diptera f. *magniflora*
in the Hillier Arboretum, England

Plate 4

Hydrangea aspera ssp. *strigosa*   *Hydrangea sargentiana*
in the Dortmund Botanic Garden, W. Germany

Hebe 'Fairfieldii'
in the Edinburgh Botanic Garden, Scotland

Hymenosporum flavum
in a park in S. Spain

Plate 5

Ehretia dicksonii
in the Dortmund Botanic Garden, W. Germany

Elaeagnus umbellata
in the Lyon Botanic Garden, France

Elaeagnus glabra
in Malahide, Ireland

Embothrium coccineum before full bloom
in a park in Ireland

Plate 6

Empetrum nigrum in its native habitat in central Sweden

Encephalartos altensteinii
in Kirstenbosch, S. Africa

Encephalartos lehmannii in the Brummeria
Botanic Garden, Pretoria, S. Africa

Plate 7

Emmenopterys henryi
in Glasnevin Botanical Garden, Dublin, Ireland

Erythea armata
in Jardin Marimurtra, Blanes, Spain

Epigaea repens in the Berlin Botanic Garden, W. Germany

Photo: R. C. Jelitto

Plate 8

Erica discolor
in its native habitat in S. Africa

Erica vagans 'Mrs. D. F. Maxwell'
in the Dortmund Botanic Garden, W. Germany

Erica perspicua
in Kirstenbosch Garden, S. Africa

Erica peziza in its native habitat in S. Africa

Plate 9

Enkianthus campanulatus 'Albiflorus' in Villa Taranto Park, Pallanza, N. Italy

Enkianthus perulatus in the Royal Botanic Garden, Edinburgh, Scotland

Plate 10

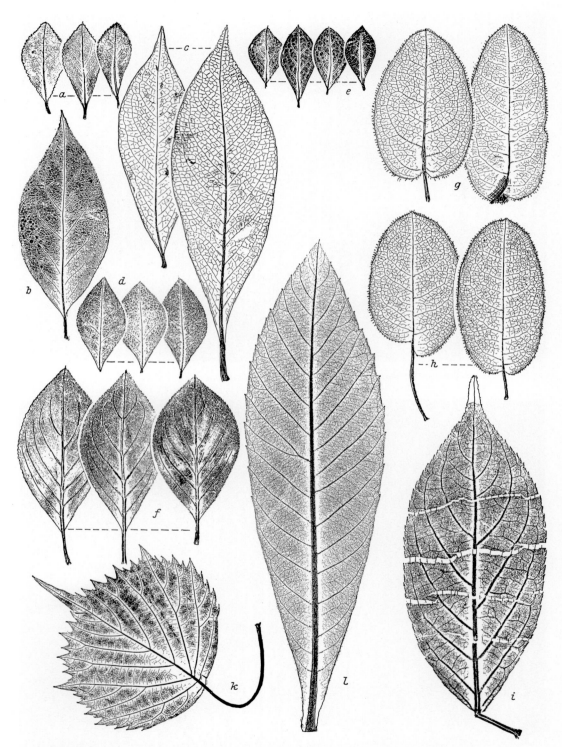

Enkianthus a. *E. cernuus* 'Rubens'; b. *E. serrulatus*; c. *E. quinqueflorus*; d. *E. subsessilis*; e. *E. perulatus*; f. *E. campanulatus*. — **Epigaea.** g. *E. asiatica*; h. *E. repens.* — i. *Eucommia ulmoides* (blade purposely torn to show the elastic filaments). — k. *Euptelea polyandra.* — l. *Eriobotrya japonica,* small leaf.

Plate 11

Erica veitchii in RHS Gardens, Wisley, England

Erica carnea 'Springwood White'
Photo: Archiv D.B.

Erica cinerea 'W. G. Notley'
Photo: Collinridge Archive, London

Erica australis
in the Fota Island Arboretum near Cork, Ireland

Plate 12

Enkianthus campanulatus var. *palibinii*
Photo: Dr. Watari, Tokyo, Japan

Erinacea anthyllis
in the Royal Botanic Garden, Edinburgh, Scotland

Erythrina crista-galli, oldest tree in Portugal, in the Coimbra Botanic Garden

Plate 13

Erythrina crista-galli in the Coimbra Botanic Garden, Portugal
(branch from the tree on Plate 12)

Eucalyptus coccifera
in the Villa Taranto Park, Pallanza, N. Italy

Eucalyptus simmondsii
in the Forestry Arboretum, Kilmun, Scotland

Plate 14

Eucalyptus pauciflora in the Australian Alps
near Cabramurra, New South Wales
Photo: Australian News and Information Bureau

Eucalyptus lehmannii, flowers and buds,
in its native habitat
Photo: Australian News and Information Bureau

Eucalyptus ficifolia, in its native habitat
Photo: Australian News and Information Bureau

Eucalyptus gunnii (E. whittingehamensis)
in Whittingeham, Scotland

Plate 15

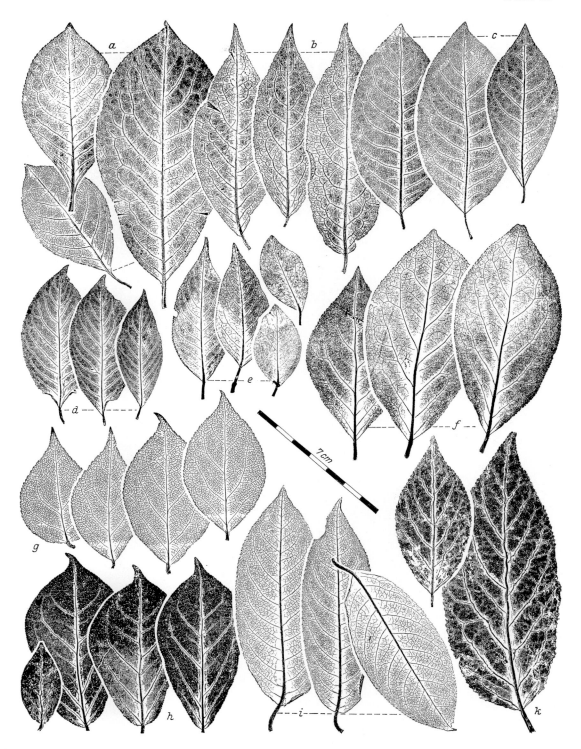

Euonymus a. *E. hamiltonianus* var. *yedoensis;* b. *E. hamiltonianus*
var. *lanceifolius;* c. *E. hamiltonianus* var. *semiexsertus;* d. *E. hamiltonianus;* e. *E. hamiltonianus* var. *maackii;* f. *E. sachalinensis;* g. *E. oxyphyllus;* h. *E. macropterus;* i. *E. hamiltonianus* var. *nikoensis;* k. *E. occidentalis* (collected from wild plants)

Plate 16

Euonymus japonicus, fruits
Photo: Dr. Watari, Tokyo, Japan

Euonymus hamiltonianus var. *hians*
in the Dortmund Botanic Garden, W. Germany

Euonymus alatus

Euonymus phellomanus

FONTANESIA Labill. — OLEACEAE

Deciduous shrubs, resembling *Ligustrum*, branches 4 sided; leaves opposite, short petioled, entire; flowers in small, axillary racemes and terminal panicles, small, with 4 distinct petals; stamens 2, longer than the petals; ovary superior, with 2 chambers; style short; fruit a small nut with encircling wing. — 2 species in E. Asia and China.

Fontanesia fortunei Carr. Shrub, about 3 m high, occasionally higher, growth narrowly upright, branches thin; leaves lanceolate, long acuminate, 3–10 cm long, entire, bright green above, glossy, persisting late into the fall; flowers small, white, May–June. DL 1: 58; DB 1955: 8. China. 1855. Fig. 49. z5 ⊘

F. phillyreoides Labill. Like the previous species, but smaller, usually not over 1–1.5 m high, much more densely branched, growth wide; leaves oval-lanceolate to elliptic-oblong, 2–7 cm long, finely serrate or rough margined, dull green; flowers greenish-white, May–June. DL 1: 59; BC 1308. Syria, Asia Minor. 1787. z6 Fig. 49. ⊘

'Nana'. Growth lower, slower, more compact.

Used like *Ligustrum*. No particular cultural preferences. Good winter hardiness, but only slightly ornamental.

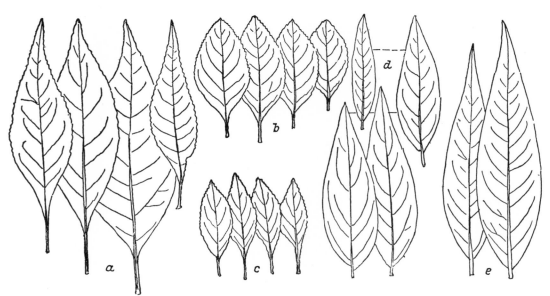

Fig. 49. **Forestiera**. a. *F. acuminata;* b. *F. ligustrina;* c. *F. neo- mexicana.*
Fontanesia. d. *F. fortunei;* e. *F. phillyreoides* (drawn from material collected in the wild)

FORESTIERA Poir. — OLEACEAE

Ligustrum-like, usually deciduous shrubs; leaves opposite, entire or finely serrate, similar to those of *Ligustrum;* flowers polygamous or dioecious, resembling *Fraxinus* flowers, not particularly attractive, greenish, without petals, in axillary clusters or racemes on the previous year's wood, some appearing before the leaves; fruit a small, usually black drupe. — About 15 species in N. America, Lesser Antilles to Brazil; the tropical species not found in cultivation.

Forestiera acuminata (Michx.) Poir. Shrub, to 1.5–2.5 m high, occasionally also a small tree, branches glabrous; leaves ovate-oblong to lanceolate, acuminate, 3–10 cm long, usually slightly serrate on the apical blade half, dull green; flowers greenish, male flowers in sessile fascicles, female flowers in small panicles, April–May; fruits

oblong, 12 mm long, purple. BB 2844; NBB 3: 51. N. America; Illinois to Indiana and Missouri, south to Georgia and Texas, in swamps and wet woodlands. 1812. z5 Fig. 49. ⊘

F. ligustrina (Michx.) Poir. Shrub, to about 2.5 m high, branches pubescent; leaves elliptic-oblong, 2–5 cm long, pubescent beneath, obtuse, appressed serrate nearly to the base; flowers in tiny inflorescences, greenish, male flowers in dense, sessile fascicles, female flowers less numerous; fruits 8 mm long, bluish, August. NBB 3: 51. Southeast USA. 1812. Occurs more commonly on sandy, gravelly soils. z6 Fig. 49.

F. neo-mexicana Gray. Shrub, to 3 m high, glabrous, branches divaricate and often prickly; leaves obtuse ovate-lanceolate, 3–5 cm long, lightly crenate, gray-green (!); flowers yellowish, unattractive, April–May; fruits short elliptic, 4 mm long, blue-black. MD 1919: 72. SW. USA, in lowlands and along river banks, but in relatively dry locations. z6 Plate 23; Fig. 49.

All 3 species usually only cultivated in botanic gardens; only slight garden merit; winter hardy. No particular cultural preference, uses similar to those of *Ligustrum*.

Lit. Johnson, M. C.: Synopsis of the United States species of *Forestiera* (Oleaceae); in Southwestern Naturalist 1957, 140–151.

FORSYTHIA Vahl — OLEACEAE

Deciduous shrubs; branches with chambered pith, hollow only on *F. suspensa;* leaves opposite, normally unlobed, occasionally 3 parted or with 3 leaflets, serrate to nearly entire; flowers 1–6, axillary, appearing before the leaves, with variable styles (heterostylous), calyx and corolla deeply 4 parted, the latter twisted in the bud; flower stalk with 2–4 small scales; stamens 2; styles long or short, with twice incised stigma; fruit a leathery or hard, somewhat inflated, bivalved, beaked capsule; seeds completely encircled by a wing. — 6–7 species in E. Asia, 1 in Europe (only in N. Albania).

Forsythia europaea Deg. & Bald. Shrub, 1.5 to 2.5 m high, rather narrowly upright, glabrous, branches with chambered pith; leaves ovate, never 3 parted, entire or with few shallow teeth, 5–8 cm long; flowers usually solitary, occasionally paired, short stalked, small, corolla

dark yellow (HCC 3/1), truncate at the apex, styles short or long, anthers very large, April; fruits smooth, tapering to the base, 17 mm long. BM 8039; MD 1930: 1. N. Albania (Mirdita) and SW. Yugoslavia (near Péc). 1899. Garden use not significant. z6 Plate 28; Figs. 50, 51.

F. fortunei see: **F. suspensa** var. **fortunei**

F. giraldiana Lingelsh. Erect shrub, 1.5–4 m high, branches dark gray, with chambered pith; leaves elliptic to lanceolate-oblong, long acuminate, 5–12 cm long, entire or also serrate, often somewhat brownish, occasionally lightly pubescent beneath; flowers usually solitary, very early, light yellow (HCC 603), campanulate, long styled, relatively few flowers, about 2 cm in diameter. BM 9662; JRHS 66; 32; EPN 238: 11. NW. China. 1910. Often flowering before *F. ovata*! Plate 28; Figs. 50, 51. ✧

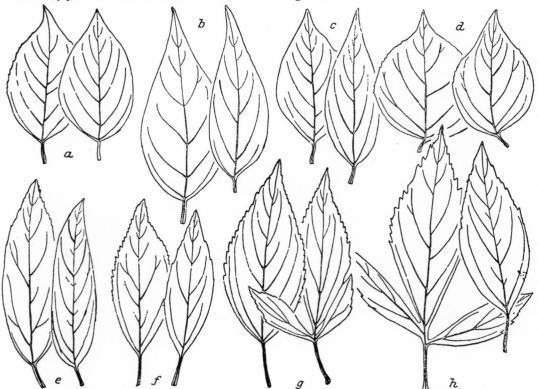

Fig. 50. **Forsythia** leaves. a. *F. europaea;* b. *F. giraldiana;* c. *F. japonica;* d. *F. ovata;* e. *F. viridissima;* f. *F. intermedia* 'Spectabilis'; g. *F. intermedia* 'Densiflora'; h. *F. intermedia* 'Primulina' (from Szuszka, altered)

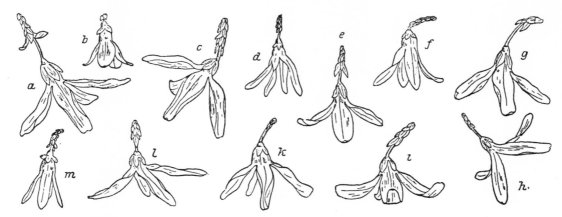

Fig. 51. **Forsythia** flowers (somewhat reduced). a. *F. suspensa* 'Variegata'; b. *F. ovata*; c. *F. intermedia* 'Vitellina'; d. *F. giraldiana*; e. *F. suspensa* var. *fortunei*; f. *F. viridissima*; g. *F. suspensa* var. *sieboldii*; h. *F. intermedia* 'Primulina'; i. *F. intermedia* 'Densiflora'; k. *F. japonica*; l. *F. intermedia* 'Spectabilis'; m. *F. europaea* (from Szuszka, altered)

F. × intermedia Zab. (*F. suspensa × F. viridissima*). Shrub, erect to broad-spreading, 2–3 m high, branches olive-yellow, with chambered pith, but usually solid at the nodes; leaves ovate-oblong to oblong-lanceolate, trifoliate on long shoots, serrate; flowers usually grouped 2–3, moderately numerous, deep yellow (HCC 4/1), long styled, April. Gfl 1185; Gn 2622. Discovered in the Göttingen Botanic Garden, W. Germany, 1878. z5–6

Includes the following cultivars:

'Arnold Giant' (Arnold Arboretum). Stiff upright habit, thick branched; flowers very large, dark yellow (HCC 4), corolla lobes over 1 cm wide, flat, but only a few flowers, late. JRHS 1951: 111. 1939. Tetraploid (developed by colchicine treatments), therefore very difficult to propagate (roots poorly). Not recommended. Plate 26.

'Beatrix Farrand' (K. Sax; Arnold Arboretum 1944) ('Arnold Giant' × 'Spectabilis'). Strong upright grower; leaves coarsely serrate; flowers usually solitary, 5–6 cm wide, the corolla lobes 1 cm wide, chrome-yellow (HCC 605), NH 1958: 112. Tetraploid, developed by crossing a colchicine treated tetraploid × *F. ovata*. Superior cultivar. Plates 2 & 26. ⊕

There are actually 2 seedling clones from the same cross bearing the name 'Beatrix Farrand', originally called No. 6 and No. 13 by the Arnold Arboretum. "No. 6" was the first released to the trade under the name 'Farrand', but was later changed to 'Beatrix Farrand'. Later "No. 13" was released to the trade and probably erroneously took on the same name. They are very similar, but "No. 6" is triploid and fruitless, while "No. 13" is tetraploid and produces much fruit. Those plants in cultivation today were originally distributed by the Gulf Stream Nurseries in Wachapreague, Virginia, USA.

'Charming'. A mutation of 'Lynwood' with intensive white variegated foliage, coloring lasts into summer. Discovered by W. Laqua, W. Germany; cultivated since 1976.

'Densiflora'. Upright grower, to 2 m, branches wide arching, young branches brownish-green, older branches more gray; flowers solitary, very large, densely compact, light yellow (HCC 603), lobes reflexed, throat narrow (5 mm), style long,

early April. Gfl 1906: 229. Introduced to the trade by Späth in 1899, but seldom cultivated today and often erroneously labeled. Plate 25; Figs. 50, 51. ⊕

'Farrand Hybrids'. A collective name published by G. P. de Wolf and R. S. Hebb in Arnoldia 31: 55 (1971) for the descendants of the crossing of 'Arnold Giant' × 'Spectabilis'; of which 'Beatrix Farrand' and 'Karl Sax' are two well-known cultivars.

'Goldzauber' (Hachmann 1974) ('Lynwood' × 'Beatrix Farrand'). Only medium-sized, thin branched; flowers resembling those of 'Lynwood', but darker yellow. Brought into the trade by H. Kordes, Bilsen, W. Germany, 1974. ⊕

'Karl Sax' (Arnold Arboretum 1944). Resembles 'Beatrix Farrand', 2–3 m high, but bushier, branches not so stiffly upright; leaves ovate, 7–10 cm long, 3.5–4.5 cm wide, scabrous serrate, deep green, lighter beneath; flowers deep yellow, more yellow than those of 'Beatrix Farrand', to 4.5 cm wide, throat with deep orange markings, short styles; fruit to 2.5 cm long and 1–1.5 cm wide. AB 20: 51; BMns 652. ⊕

'Lynwood' (Donard). Originated as a mutation on 'Spectabilis'; growth more upright, somewhat stiffer; flowers larger, varying in form, better distributed (HCC 4/1). JRHS 91: 104 (= 'Lynwood Gold' in USA). Discovered in a garden in Cookstown, Tyrone Co., Ireland. 1935. Introduced to the trade by Slieve Donald Nursery, Newcastle, England. Plates 2 and 26. ⊕

'Lynwood Gold' see: **'Lynwood'**

'Mertensiana' (Mertens). Growth compact, more creeping, short internodes; leaves ovate, acute, 3–4 cm long, 1–2 cm wide; flowers light yellow (HCC 603), lobes reflexed, flowers abundantly, styles short. DB 1950: 299. Brought into the trade by Mertens & Nussbaumer of Zurich, Switzerland, 1949. Plate 26. ⊕

'Mirabilis'. Erect habit, shoots somewhat nodding; flowers light yellow, resembling 'Spring Glory' but generally considered an improvement. Origin unknown; cultivated in Denmark.

'Parkdekor' (J. Hachmann 1976) ('Beatrix Farrand' × 'Spectabilis'). Differing from 'Beatrix Farrand' in the wide, more pendulous growth habit; flowers to 5.8 cm wide, deep yellow, very abundant. Introduced to the trade by H. Kordes, 1976. ⊕

'Primulina'. Upright grower, about 1.8 m high, branches brownish; leaves scabrous serrate on the apical blade half; flowers solitary, densely packed at the base of shorter branches, light yellow (HCC 603), lobes reflexed (like 'Densiflora', but narrower), styles short, very prolific bloomer, early April. MD 1912: 193. Plate 25; Fig. 50, 51. ⊕

'Spectabilis' (Späth). Shrub, upright, to 2.5 m high, branches brownish; leaves sometimes entire, most somewhat navicular; flowers densely arranged, dark yellow (HCC 4), lobes twisted, styles short, early April, throat wide (8 mm), flowers always solitary with 5–6 parted corolla (!). Gfl 1906: 227; JRHS 69: 129. Brought into the trade in 1906 by Späth and still widely used today. Plates 25 and 26; Figs. 50, 51. ⊕

'Spring Glory' (Wayside). Mutation of 'Primulina'; upright habit, to 1.8 m high, branches yellowish-brown; flowers usually in groups, lobes widely outspread, light yellow (HCC 603). Developed about 1930 by H. H. Horvarth, Mentor, Ohio, USA and brought into the trade by Wayside Gardens, Mentor, in 1942. An improvement on 'Primulina'. Plate 25. ⊕

'Tremonia' (Krüssmann 1963) ('Beatrix Farrand' × ?) To 2 m tall or higher, densely branched, branches thin; leaves narrow, very deeply incised, teeth nearly 1 cm long; flowers light yellow, good bloomer. Developed in the Dortmund Botanic Garden, W. Germany and distributed by the Arnold Arboretum, USA. Interesting for its foliage. ∅

'Vitellina' (Späth 1899). Very strong grower, upright habit, branches somewhat pendulous; flowers numerous, egg-yolk yellow (HCC 3), small flowers, corolla lobes outspread, twisted, throat narrow (5 mm). Not always true to name in cultivation. Fig. 51.

F. intermedia × japonica (Sax). As yet the only known hybrid:

'Arnold Dwarf' (Sax, Arnold Arboretum (*F. intermedia* × *F. japonica* var. *saxatilis*). Creeping habit and rooting along the branches; leaves small, ovate, partly three lobed, 4–6 cm long, scabrous serrate; almost never flowers, flowers greenish-yellow (HCC 64/2). Developed in the USA, 1941. To be considered only as a ground cover. Plate 26.

F. japonica Mak. Similar to *F. ovata*, but grows wider; leaves ovate to broadly-ovate, 5–12 cm long, densely serrate nearly to the base, with 4 prominent vein pairs, pubescent beneath; flowers solitary, only 1.5 cm long. NT 1: 360. Japan. z6 Fig. 50, 51.

var. **saxatilis** Nakai. Branches brownish, later more gray-yellow; leaves oval-oblong, 3–6 cm long, venation pubescent beneath when young; calyx lobes brown with white tips, corolla yellow (HCC 3/1), flowers very abundantly, but flowers small. Korea. 1924. Slight garden merit.

F. × kobendzae Seneta (*F. europaea* × *F. suspensa*). Habit narrowly upright, to 2 m high, young branches violet-purple, with chambered pith; leaves entire to scabrous serrate, trifoliate on the long shoots; flowers light yellow-green at first, then becoming light yellow, styles short or long, corolla lobes to 10 mm long, flat at first, then twisted, flowers very early, soon after *F. ovata*. Developed from seed in Warsaw, Poland, 1952; as yet no cultivar has been named.

F. ovata Nakai. Shrub, rounded, usually not over 1 m high, branches gray-yellow, twisted, (not angular!), pith chambered; leaves ovate to nearly circular, entire or somewhat serrate, 5–7 cm long, base rounded; flowers 1–2, very small, corolla lobes broadly ovate (HCC 603), March. BM 9437, NK 10: 3; DB 1957; 130. Korea. 1919. Does not flower every year, and when it does often sparsely. z5 Plate 2; Fig. 50, 51. ⊕

'French's Florence'. To 1.5 m high; flowers smaller and lighter yellow than *F. ovata*, but the flower buds more frost hardy than the type. Selection introduced in 1940 by K. W. French of West Lebanon, N. H., USA.

'Ottawa' (Ottawa Expt. Station, Canada). Growth more upright, strongly branched, flowers very abundantly. Very winter hardy. Cultivated in Boskoop, Holland (Grootendorst).

'Robusta' see: **F. viridissima 'Robusta'**

'Tetragold'. Tetraploid clone developed through colchicine treatment; low growing, bushy, about 1 m high; flowers larger than those of the type, to 3 cm wide, the corolla lobes 8 mm wide, deep yellow, flowering a few days earlier than the species. Developed in 1963 at the Proefstation V. D. Boomkwekerij in Boskoop, Holland. ⊕

F. suspensa (Thunb.) Vahl. Branches hollow, solid at the nodes. This species consists of 2 easily distinguished varieties: var. *sieboldii* Zab. (branches always nodding, leaves usually simple, flower stalk to 2 cm long, corolla tube 7 mm long, corolla lobes flat, somewhat campanulate in arrangement; with no cultivars) and var. *fortunei* (Lindl.) Rehd. (branches erect, nodding with age, leaves often trifoliate, flowers stalks 5–15 mm long, corolla tube 5 mm long, corolla lobes outspread, margin involuted; includes the cultivars 'Atrocaulis', 'Decipiens', 'Nymans', 'Pallida', 'Variegata'). z5–6

'Atrocaulis'. Very strong growing, erect, bark black-brown, new growth reddish; flowers small, light yellow (HCC 3/1), few flowers on young plants. From China. 1907. Plate 2.

'Decipiens' (Späth 1905). Erect grower; flowers solitary long stalked, moderately numerous, dark yellow (HCC 4/3), erect at first, later spreading horizontally, corolla lobes about 16 mm long, margin somewhat involuted, style long. GF1 1906: 205. ⊕

var. **fortunei** (Lindl.) Rehd. Growth strongly upright, to 3 m high, branches outspread or nodding with age, greenish-brown; leaves simple, but frequently (especially on long shoots) 3 lobed to trifoliate, ovate, to 9 cm long and 5 cm wide, scabrous serrate, base rounded; flowers usually grouped 1–2, tube 5 mm long, corolla lobes outspread, dark yellow (HCC 3), twisted, stalks 5–15 mm long, early April. Gfl 1906: 204 (= *F. fortunei* Lindl.). China. 1860. Plate 25; Fig. 51, 52. ⊕

'Nymans'. Like 'Atrocaulis', but flowers twice as large, light yellow (HCC 3/1), also flowering well on young plants. Before 1954. Presumably from Nymans, Sussex, England. Prominent cultivar. ⊕

'Pallida'. More upright habit; flowers solitary, moderately numerous, very pale yellow (HCC 603), style short, stalk 1 cm long.

var. **sieboldii** Zab. Shrub, to 2.5 m high with age, branches always nodding, creeping, fast grower with branches rooting upon contact with the ground, branches very thin; leaves usually simple, ovate to broadly ovate, 4–6 cm long, larger on long shoots, serrate; flowers moderately numerous, usually solitary, corolla campanulate at first, lobes broad and flat, somewhat erect, throat wide, distinctly red and yellow striped, early April. BM 4995; FS 1253. From Japanese horticulture. 1883. One of the oldest known *Forsythia* forms. Beautiful for its graceful, nodding habit, but few flowers. Plate 25; Fig. 51, 52. ⊕

'Variegata'. Like var. *fortunei*, but the leaves yellow variegated (ugly!); flowers dark yellow, few flowers, styles short. Fig. 51.

F. × variabilis Seneta (*F. ovata × F. suspensa* or *F. intermedia*). Growth narrowly upright, to 2 m, open branched and sparsely foliate, branches angular, violet-purple, with chambered pith; leaves oval-elliptic, resembling those of *F. ovata*, scabrous serrate, except on the apex and the base, also 3 lobed on strong shoots; flowers small, only 15–18 mm wide, with 5–6 (8) corolla lobes, the other flower parts also more numerous, flowers soon after *F. ovata*; fruits similar to *F. suspensa*, with small warts, seeds viable. Developed in 1952 from seed. Exact origin unknown.

'Volunteer' (*F. ovata × F. suspensa*). Medium tall shrub, branches dark; flowers densely arranged, deep yellow. Originated in the garden of A. Simmonds of Clandon, Surrey, England (cultivated by Hillier).

F. viridissima Lindl. Erect habit, branches green, 4 sided, pith chambered throughout, pith absent only in the internodes at the base of the stouter branches; leaves elliptic-oblong to lanceolate, usually serrate only on the apical third of the blade, otherwise nearly entire, to 8 cm long, dark green, often coloring violet-brown in fall, hardly ever trifoliate; flowers grouped 1–3, yellow with greenish traces, corolla lobes long, narrow, reflexed, late April (the latest species). FS 261; BM 4587; JRHS 69: 30. China. 1844. z5 Fig. 50, 51. ⊕

'Bronxensis' (Bronx Park). Dwarf variety, scarcely 30 cm high, very short internodes; leaves very small, ovate, 2–4 cm long, flowers abundantly (HCC 603). GC 125: 41, 43; 133: 79. Grown from a Japanese seed source at the Bronx Botanic Garden, New York, 1939. Plate 26.

var. **koreana** Rehd. Erect habit, branches outspread; leaves ovate-oblong, widest below the middle, 5–12 cm long, fall foliage violet; flowers and calyx somewhat larger than the species, light yellow (HCC 4/1). NK 10: 2. Korea; discovered near Seoul, 1917. ⊕ Ø

'Robusta' (the Dutch nurseries) Earlier cultivated as *F. ovata* 'Robusta', but differing from *F. ovata* in its more vigorous growth habit, ovate-elliptic, 12 cm long leaves and larger flowers. Placed in *F. viridissima* by Herman Grootendorst.

The Forsythia are some of the most beautiful flowering shrubs in the garden. They are easy to cultivate in any soil and full sun. The correct choice of cultivars is, however, very important. Hard pruning is not advisable; for best flowering they should be simply thinned out.

Lit. Markgraf, F.: *Forsythia europaea* und die Forsythien Asiens; in Mitt. DDG 1930, 1–12 ● Dahlgren, K. V. O.: Om odlade Forsythior och deras fruktsättning; in Lustgarden 1946, 89–100 ● Thompson, J. M.: Some features of horticultural interest in the Forsythias; in Jour. RHS 71, 166–172, 1946 ● Hyde, B.:

Fig. 52. *Forsythia suspensa* var. *fortunei* (right) and var. *sieboldii* (left) (from Dippel)

Forsythia Polyploids; in Jour. Arnold Arboretum 32, 157–158, 1951, with plates ● Duvernay, J. M.: Le genre *Forsythia* Vahl; in Revue Horticole 1953, 831–835 ● Sampson, D. R.: Studies on the progeny of triploid *Philadelphus* and *Forsythia*; in Jour. Arnold Arbor. 1955, 369–384 ● Szuszka, B.: Rodzaj *Forsythia* Vahl w Arboretum Kornickim; in Arboretum Kornickie Rocznik (Poland) I, 1955, 91–110 ● Dietrich, H.: Erfahrungen mit Forsythien-Sorten; in Deutsche Baumschule 1957, 268–275 ● Lingelsheim, A.: Oleaceae; *Forsythia*; in Engler, Pflanzenreich 72 (new printing 1957), 109–113 ● Szuszka, B.: Results hitherto obtained in breeding *Forsythia* at Kornik; in Rocznik Arboretum Kornik 1959, 205 to 225 (Polish with English summary) ● Wyman, D.: Foremost among the Forsythias; in Americ. Nurseryman, (4/15/59) ● Wyman, D.: The *Forsythia* Story; in Arnoldia 21, 1961, 35–38, ● Wyman, D.: Registration Lists of cultivar names of Forsythias; in Arnoldia 21, 39–42, 1961 ● Wyman, D.: Forsythias; in Amer. Hort. Mag. 1961, 191–197 (with the best key to all the cultivars) ● Seneta, W.: Über 2 neue *Forsythia*-Hybriden; in Rocznik Dendrol. 19, 181–192, 1965 ● De Wolf, G. P., & R. S. Hebb: The Story of *Forsythia*; in Arnoldia 31, 41–63, 1971 (very good compilation of the entire genus!).

FORTUNEARIA Rehd. & Wils. — HAMAMELIDACEAE

Monotypic genus; deciduous tree; leaves alternate, simple, serrate, short petioled, with small, quickly abscising stipules; flowers monoecious, either totally male or bisexual, but never purely female, in terminal racemes, usually with 1–3 leaves at the base; flowers 5 parted, calyx 5 lobed, petals 5, awl-shaped, somewhat shorter than the sepals; styles 2, filamentous; male flowers nearly catkin-like, developing in fall and naked

Fig. 53. Left, *Fortunearia sinensis*; middle, *Fortunella margarita*; right, *Fortunella hindsii*
(from ISC)

over winter, female flowers first appear with the foliage; fruit a woody, bivalved capsule, with 2 brown seeds. China.

Fortunearia sinensis Rehd. & Wils. Broad growing shrub, eventually also a tree, 5–10 m high, bark dark gray, branches gray-brown with somewhat soft stellate pubescence, similarly the leaf petiole and flower stalks; leaves oblong, about 10 cm long, base broadly cuneate, serrate, leathery tough, dull green and glabrous above, lighter beneath with pubescent venation; flowers reddish, in dense erect racemes, the purely male racemes 2 cm long, the female racemes to 5 cm, March; fruit ovoid, 12 mm long. LF 157; EP 18a, 171; CIS 25. China, W. Hupeh. Completely winter hardy but of little garden value. 1907. z6 Fig. 53.

FORTUNELLA Swingle — Kumquat — RUTACEAE

Small evergreen trees or shrubs with thorny branches, young branches green and glabrous, older branches more rodlike and green, gray or brown; leaves alternate, simple, oblong to elliptic, petiole narrow winged, often with a thorn at the base; flowers solitary or 3–4 axillary, fragrant, petals 5, white, thick, waxy, calyx tiny and flat lobed, stamens numerous, with very short filaments; fruit a small orange. — 4 species and numerous cultivars in E. Asia and Malaysia.

Fortunella hindsii (Cham.) Swingle. Wild Kumquat. Thorny, stiff, erect shrub, 1–2 m high, thorns thick, with sharp tips; leaves thick and leathery, elliptic, 5–7 cm long, petioles short; flowers solitary or in small axillary fascicles, white, not opening wide, fragrant, with about 15 stamens, blooming several times a year in its habitat; fruits long-rounded, about 1.3 cm thick, orange to scarlet-orange, with 2 green seeds, edible. HKS 40. S.

China, Hong Kong. z10 Fig. 53. # ⚭

F. japonica (Thunb.) Swingle. Round Kumquat. Small tree, branches thorny; leaves ovate, acute, to 10 cm long, petiole winged; flowers axillary, often solitary, white; fruits globose, golden-yellow, 2.5 cm thick. China, Hong Kong. Known only in cultivation. z10 # ⚭

F. margarita (Lour.) Swingle. Kumquat. Small shrub, to 1.5 m high, densely branched, branches thornless; leaves narrow-elliptic, 4–8 cm long, deep green, obtuse, indistinctly dentate at the apex; flowers only about 1.3 cm wide, fragrant, solitary or several at the branch tips, calyx 5 toothed, stamens 18–20; fruits ovate to ellipsoid, 2.5–4 cm long, orange, fragrant, pulp sour, fruits abundantly. HKS 41. China. z10 Fig. 53. # ⚭

Lit. Swingle, W. T.: *Fortunella*; in Bailey, Stand. Cyclop. Hort. II, 1269–1270, 1950 (with ill.).

FOTHERGILLA L. — HAMAMELIDACEAE

Low deciduous shrubs with stellate pubescence; leaves alternate, somewhat resembling *Alnus*, petioles short, obovate to oblong, usually somewhat oblique, coarsely crenate toward the apex; flowers bisexual, without petals, with 15–24 stamens, filaments thickened clavate-like, white, set off well against the anthers, in dense terminal spikes; fruit an ovate, stellate tomentose, cartilaginous capsule, bivalved at the apex and with 2 seeds. — 2 species in Eastern N. America.

(Earlier the genus was divided into 4 species by several botanists, including Rehder, but was generally not accepted by the scientific community. In Bean, 8th ed. [p. 205] only 2 species are given; that is the basis for the following species descriptions.)

- Shrub, 60–80 cm high, leaves 2.5–6 cm long; flowers appearing before the foliage:
    F. gardenii
- • Shrub, 1.5–3 m high; leaves 5–10 cm long and nearly as wide; flowers appearing with the foliage:
    F. major

Fothergilla alnifolia see: **F. gardenii**

F. alnifolia var. *major* see: **F. major**

F. carolina see: **F. gardenii**

F. gardenii Murray. Deciduous shrub, seldom taller than 60–90 cm, branches thin, crooked, often rather sparse and outspread, young branches with white stellate pubescence; leaves elliptic to obovate, 2.5–6 cm long, 2–3.5 cm wide, with a few large, unequal teeth on the apical blade half, base cordate or rounded or tapered, pubescent beneath, green or whitish, fall color frequently carmine, petiole 6 mm long, pubescent; flowers in terminal spikes, cylindrical, 2.5–3.5 cm long and 2.5 cm wide, composed of white filaments with yellow anthers, appearing before the leaves, fragrant, April–May, BM 1341 (= *F. alnifolia* L. f.; *F. carolina* Britt.). SE. USA. 1765. Because of the variable leaf forms, some cultivars have been named, however, they differ little from the species. z5 Plate 28. ⊕

F. major Lodd. Deciduous shrub, 1.5–3 m high, rather globose in habit, stems numerous, usually erect, the young branches with whitish stellate pubescence; leaves rounded to broadly ovate, 5–10 cm long, 3.5–7 cm wide, dark glossy green above, blue-green and stellate pubescent beneath, especially on the venation, petiole about 8 mm long, fall color yellow, orange-yellow or eventually totally red; flowers appear with the foliage, in numerous, terminal spikes, 2.5–5 cm long, erect on short side branches, composed of 2 cm long, white filaments with a trace of light pink and yellow anthers, fragrant, May. BS 2: 32; BM 1342 (= *F. alnifolia* var. *major* Sims; *F. monticola* Ashe). USA, in the Alleghany Mts. from Virginia to S. Carolina. 1780. z5 Plate 26, 28. ⊕

The earlier distinction of more glabrous plants as *F. monticola* was apparently unjustifiable and therefore discontinued.

F. monticola see: **F. major**

Preferably cultivated in moist, peaty, humus soil, but not in stagnant or standing water; grows well in semishade but some of the orange-yellow to glowing carmine-red fall color is then lost. Fully winter hardy.

FRANGULA See: **RHAMNUS**

This name has been used since 1768 by some authors for *Rhamnus frangula*, i.e. by Miller. This and 2 other species were separated from the genus based on a number of conspicuous characteristics (underground germination, naked winter buds, 5 parted flowers with only 1 style).

Frangula alnus see: **Rhamnus frangula**

FRANKENIA L. — FRANKENIACEAE

Subshrubs or shrubs, *Erica*-like in appearance; leaves opposite, without stipules; flowers usually bisexual, petals and sepals 5 (occasionally 4), petals with a scale-like appendage on the claw, stamens in 2 rings, the outer shorter; fruit a small, 3 sided capsule. — About 60 species, mostly in the salt marshes and salt deserts of the temperate and subtropical regions.

Frankenia laevis L. "Sea Heath". Heath-like evergreen, 15 cm high subshrublet, branches wiry, dark green, turning reddish, finely pubescent, base woody; leaves heath-like, 2–4 mm long, linear, margins involuted, glabrous above, pubescent beneath, usually densely arranged on short side branches; flowers 5 mm wide, pink, July–August. Along the coast of S. and W. England, from the English Channel to the Mediterranean region, Madeira and to Asia Minor. Without particular ornamental value. z6

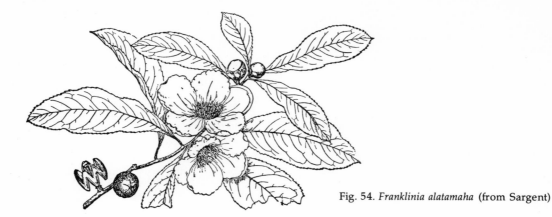

Fig. 54. *Franklinia alatamaha* (from Sargent)

FRANKLINIA Marsh. — THEACEAE

Deciduous shrub or (in its habitat) a tree; leaves alternate, serrate; flowers bisexual, solitary, axillary, nearly sessile; sepals 5, persistent, rounded, uneven, convexly arched; stamens distinct, attached at the base of the petals; styles connate, with 5 lobed stigma; fruit a rounded, 5 valved, woody capsule, dehiscing at the apex and the base when ripe, with a persistent middle column; seed flat, angular, wingless, one in each locule. — 1 species in N. America.

Franklinia alatamaha Marsh. Deciduous shrub or tree, 5–7 m high, occasionally to 10 m, branches erect, bark smooth and thin, young branches densely silky pubescent; leaves obovate-oblong, 12–15 cm long, generally tapering to a short petiole, sparsely serrate, bright glossy green above, pubescent beneath, bright red in fall; flowers cupular, 7–8 cm wide, white, petals obovate-rounded, September–October; fruits globose, to 2 cm thick. RFW 224; NF 6: 226; NH 1955: 249 to 255; GF 2: 616 (= *Gordonia alatamaha* Sarg.). N. America, mountains in Georgia. Exceptionally beautiful shrub, valuable for its late flowers and fall foliage color. Not found in the wild since 1790. z6 Plate 30; Fig. 54. ∅ ✢ ⚭

Cultivated somewhat like the large leaved *Rhododendron* species.

FRAXINUS L. — Ash — OLEACEAE

Deciduous trees (*F. uhdei* evergreen), usually tall, only rarely shrubs; winter buds with 1–2 pairs of bud scales, usually thick, black, brown or gray and lepidote; leaves opposite, pinnately compound, occasionally simple; flowers bisexual or unisexual, small, in axillary or terminal panicles or racemes, seldom attractive; calyx small, 4 parted or 4 lobed or totally absent; corolla with 2–6, usually 4 distinct petals, occasionally connate at the base or absent; fruit a single seeded nutlet with a long extended wing on the apex. — About 65 species in the Northern Hemisphere.

Classification of the Genus for the species described

Section I. **Ornus** (Neck.) DC.
Flowers in terminal and axillary panicles on leafy shoots, bisexual or polygamous.

Subsection 1. **Euornus** Koehne & Lingelsh.
Corolla present; scales of the terminal bud with entire margins:
F. bungeana, cuspidata, floribunda, griffithii, lanuginosa, ornus, paxiana, raibocarpa, retusa, sieboldiana

Subsection 2. **Ornaster** Koehne & Lingelsh.
Corolla absent; outer scales of the terminal bud leaflike:
F. chinensis, longicuspis rhynchophylla

Section II. **Fraxinaster** DC.
Flowers appear before the leaves, at axillary buds; petals usually absent.

Subsection 1. **Sciadanthus** (Coss. & Durieu) Lingelsh.
Flowers with calyx, polygamous, in dense panicles; leaves with winged rachis, leaflets 3 to 13, small, not over 4 cm long, obtuse; outer scales of the terminal buds leaflike:
F. xanthoxyloides

Subsection 2. **Petlomelea** (Nieuw.) Rehd.
Rachis not winged; petals developed:
F. dipetala

Subsection 3. **Melioides** (Endl.) Lingelsh.
Flowers with calyx, usually dioecious; leaflets large, rachis not winged; scales of the terminal bud entire:
F. americana, anomala, berlandieriana, biltmoreana, caroliniana, pensylvanica, platypoda, spaethiana, tomentosa, uhdei

Subsection 4. **Bumelioides** (Endl.) Lingelsh. Flowers without calyx and corolla (except on *F. quadrangulata* having a tiny, abscising calyx); flowers dioecious or polygamous; anthers broad-oblong to cordate; leaflets usually more than 7; outer scales of the terminal buds leaflike:

F. angustifolia, elonza, excelsior, holotricha, hookeri, mandshurica, nigra, numidica, obliqua, pallisae, potamophila, quadrangulata, rotundifolia, syriaca

Parentage of the Cultivars

'Althena'	— *excelsior*
'Atlas'	— *excelsior*
'Autumn Purple'	— *americana*
'Doorenbos'	— *excelsior*
'Emerald Ash'	— *pensylvanica*
'Eureka'	— *excelsior*
'Fallgold'	— *nigra*
'Fan-Tex'	— *velutina*
'Golden Cloud'	— *excelsior*
'Moraine'	— *holotricha*
'Patmore'	— *pensylvanica*
'Raywood'	— *angustifolia*
'Rosehill'	— *americana*
'Tomlinson'	— *uhdei*
'Veltheimii'	— *angustifolia*
'Wentworthii'	— *excelsior*
'Westhof's Glorie'	— *excelsior*
'Wollastonii'	— 'Raywood'

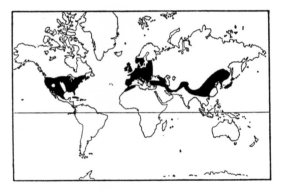

Fig. 55. Range of the genus *Fraxinus*

Fraxinus americana L. White Ash. Tree to 40 m high, with broad ovate crown, young branches glabrous, buds dark grown, new growth appears later than on *F. excelsior*; leaves 9–38 cm long, leaflets 5–9, usually 7, petiolate, ovate-lanceolate or elliptic, but quite variable in form, 6–15 cm long, 1–9 cm wide, entire to serrate or crenate, dark green above, usually obtuse, usually whitish beneath, glabrous, occasionally leathery tough, purple and yellow fall foliage; flowers dioecious, April–May; fruits narrow-oblong, 3–5 cm long. BB 2838; GTP 325; NBB 3:49; HH 456. Eastern and central USA, in moist fertile woodlands; very valuable forestry tree in its habitat. Ⓕ USA; Germany, Bulgaria, Yogoslavia; Argentina. z3 Fig. 60 ⊘

'Acuminata'. Tree, leaves entire, long acuminate, glossy dark blue-green above, nearly white beneath, purple-violet fall color. Beautiful park tree.⊘

'Ascidiata'. Leaflets often only 5, more ovate, long acuminate, base either conical or broad rounded. GC 76: 122. Around 1910. Park tree.

'Autumn Purple' (Cole). Selected for its fall color, deep purple to mahogany-brown, with violet and brown areas, persisting 2–3 weeks, depending on soil type and weather. Introduced through the Cole Nursery, Circleville, Ohio, USA. ⊘

var. **juglandifolia** (Lam.). D. J. Browne. Leaflets more or less serrate or crenate, less glossy above, less distinctly blue-white beneath, more or less pubescent. Frequently found in the northern part of its range. ⊘

ssp. *oregona* see: **F. latifolia**

'Rosehill' (Asjes 1966). Fast growing selection, branches more sparsely arranged, sturdy; leaves deep green, gray-white beneath, bronze-brown in fall; fruitless. US Pl. Pat. 2678. Introduced by the Rosehill Gardens, Kansas City, Missouri, USA. ⊘

F. angustifolia Vahl. Tree, to 25 m high, young branches glabrous, winter buds dark brown; leaflets 7–13, sessile, oblong-lanceolate, 3–7 cm long, acuminate, base cuneate, sparse and scabrous serrate, dark green above, lighter beneath, glabrous, rachis with upper surface closed, open furrowed only at the nodes; flowers in small racemes (!); fruits elliptic-oblong to oblanceolate, 3–4 cm long, base rounded. DL 1: 91; EH 245 (= *F. oxycarpa* Willd.; *F. angustifolia* ssp. *oxycarpa* [Willd.] Franco & Rocha Alphonso; *F. oxyphylla* Bieb.). S. Europe, N. Africa, Asia Minor. 1800. Ⓕ S. Africa, on a trial basis. z6 Fig. 58

Scheller includes *F. oxycarpa* in *F. angustifolia*, citing the significant variability of the former and the lack of knowledge of the range of the species. These factors render an infraspecific division unjustifiable based on the more or less pubescent leaf undersides.

var. **australis** (Gay) Schneid. Midrib of the lower leaflets and the rachis pubescent; otherwise scarcely distinguishable. S. Europe, N. Africa. Fig. 60.

var. *biltmoreana* see: **F. biltmoreana**

'Elegantissima'. Small tree, scarcely over 8 m high; leaflets usually 11, narrow-lanceolate, light green, 4 to 6 cm long, 1 cm wide, long acuminate, fine and scabrous serrate. SpB 233. This form is often erroneously included with *F. excelsior*, but should more rightly be classified here thanks to the work of Scheller. ⊘

'Lentiscifolia'. Leaves to 25 cm long, leaflets more widely spaced. 1809.

'Monophylla'. Leaves simple, often with 2 side lobes or trifoliate at the branch base, 5–12 cm long, ovate to broad-lanceolate, generally acuminate, cuneate at the base, irregularly and coarsely serrate, glabrous and bright green above, usually somewhat pubescent beneath. EH 262, fig. 3 (= *F. veltheimii* Dieck). Before 1889. Fig. 59.

'Pendula'. Branches pendulous thin. An old cultivar, but rarely seen.

'Raywood'. Selection with moderately strong growth habit, dominant central stem and open crown, but with dense, ornamental foliage; leaves dark green at first, violet-purple in fall (= 'Wollastonii'). Selected in Australia and distributed by the Notcutt Nursery in England about 1927. Frequently planted as a street tree in England and Holland. ∅

'Wollastonii' see: **'Raywood'**

F. anomala Torr. Utah Ash. Small tree, to 8 m high, branches 4 sided, slightly winged, glabrous, thin, winter buds gray-brown tomentose; leaves simple, only occasionally with 5 leaflets, ovate-rounded, 3–6 cm long, entire to lightly crenate, dark green above, lighter beneath, petiole brown-red pubescent at first; fruits obovate, 2 cm long. MM 396; SM 741; EPIv 238: 10b; DB 1951: 339. Very notable Ash from the SW. USA, inhabiting dry, hilly slopes. z6 Fig. 56. ∅

var. **lowellii** (Sarg.) Little. Tree, to 8 m, branches also 4 sided and winged; leaflets 3–7, short stalked, oval-elliptic, 5–8 cm long, sparsely shallow toothed, yellow-green, glabrous or slightly pubescent; fruits 3–4 cm long, elliptic-oblong. SM 740 (= *F. lowellii* Sarg.). Arizona. z7

F. berlandieriana DC. Tree, to 10 m high, occasionally higher, young branches glabrous; leaves 15–25 cm long, leaflets 3–5, usually 5, petiolate, lanceolate-elliptic to obovate, 7–10 cm long, long acuminate, base broadly cuneate, sparsely serrate to nearly entire, vein axils pubescent beneath or also quickly becoming glabrous; fruits oblong-ovate to spathulate, 2.5–3.5 cm long. Winged nearly to the base. SS 273; SM 751. Texas to Mexico. 1879. Ⓕ Uganda. z7–8

F. biltmoreana Beadle. Probably a natural hybrid between *F. americana* and *F. pensylvanica* ssp. *pensylvanica*. Tree, to 15 m high, branches and leaf petiole with dense short pubescence; leaflets 7–11, stalked, oval-oblong, often sickle-shaped, 8–15 cm long, entire or very shallowly dentate, dark green above, smooth, blue-green beneath, pubescent, purple-violet in fall; fruits linear-oblong, 3–4 cm long. HH 467; SM 747; SS 716 (= *F. americana* var. *biltmoreana* [Beadle] J. Wright). E. USA. z3 Fig. 57. ∅

F. bracteata see: **F. griffithii**

F. bungeana DC. Shrub, 3–5 m high, broad habit, branches pubescent, winter buds nearly black; leaflets usually 5, stalked, ovate-rounded, 2–4 cm long, serrate, glabrous, dark green; inflorescence ornamental, finely pubescent, in 5 to 7 cm long panicles, but only a few flowers, May; fruits narrow-oblong, 2.5–3 cm long, obtuse to emarginate. DL 1: 31; BC 1574; GF 7: 5 (= *F. dippeliana* Lingelsh.). N. China. 1881. z5

Fig. 56. *Fraxinus anomala* (from Sargent)

Fig. 57. *Fraxinus biltmoreana* (from Sargent)

F. caroliniana Mill. Swamp Ash. Shrub or tree, 4–15 m high, branches cylindrical, glabrous to pubescent, thin; leaves 12–30 cm long, leaflets 5–7, elliptic to oval-oblong, 5–10 cm long, acuminate, base cuneate to rounded, serrate, dark green above, lighter beneath, venation white pubescent, rachis shallowly furrowed; flowers dioecious, with calyx and corolla absent; fruits elliptic to obovate, to 5 cm long and 12–18 mm wide (!). BB 2841; HH 461; WT 190; SM 742; Fruits in Miller, *Fraxinus* (l. c.) (= *F. floridana* [Wenz.] Sarg.; *F. platycarpa* Michx.). Coastal plains of SE. USA. 1783. Notable for the very broad winged fruit. z6 ∅

F. chinensis Roxb. Tree, to 15 m high, winter buds brownish-black, crusty, young branches stout, glabrous, gray; leaves 12–20 cm long, leaflets 5–9, elliptic to ovate, acute, cuneate at the base, crenate, light green beneath (not white) and with only the major veins distinctly pubescent, the lowest pair of leaflets very small; flowers dioecious, calyx campanulate, corolla absent, in 8–10 cm long panicles, May; fruits oblanceolate, 4 cm long, 6 mm wide. LF 261; GF 6: 485; EP IV, 238: 8. China; not a forest tree, cultivated rather as a host for wax scale, a source of most of China's commercial wax (see MD 1930: 168). 1891. Probably not cultivated in the western world. Ⓕ N. China. z6 ∅

var. **acuminata** Lingelsh. Leaves more lanceolate, more slender acuminate at the apex, serrate. A large specimen grows in the Royal Botanic Garden, Edinburgh, Scotland.

var. *rhynchophylla* see: **F. rhynchophylla**

F. ciliata see: **F. dipetala**

F. coriacea see: **F. velutina**

F. cuspidata Torr. Shrub, seldom tree-like, branches thin, winter buds viscid, dark red-brown; leaflets usually 7, lanceolate to more oblong, 3–6 cm long, acuminate, base cuneate, coarsely serrate, glabrous, petiole occasionally somewhat winged, thin; flowers in 6–10 cm long panicles, fragrant, corolla 1.5 cm long. SM 738; SS 260; EP IV, 243: 7. Arizona to Mexico. 1914. z8 ⊕

F. dipetala Hook. & Arn. Shrub or small tree, 2–6 m high, branches 4 sided to cylindrical, young branches reddish at first, later gray, glabrous; leaves 5–12 cm long, leaflets 3–7(9), quite variable in form, from broad elliptic to

Fig. 58. **Fraxinus.** a. *F. numidica;* b. *F. elonza;* c. *F. syriaca;* d. *F. potamophila;* e. *F. xanthoxyloides* var. *dimorpha;* f. *F. angustifolia;* g. *F. longicuspis;* h. *F. floribunda* (from Dippel)

ovate or nearly triangular, especially the terminal leaflet, 1.5–6 cm long, nearly entire to serrate or crenate, obtuse, light green above, with reticulate venation beneath; flowers bisexual, appearing with the foliage, cream-white, with 2 petals, in 5–10 cm long panicles, May; fruits oblanceolate, 2.5 cm long. SPa 204; MM 398 (= *F. ciliata* Dipp.). California. One of the most attractive flowering ash species. z9 Plate 30. ⊘ ⊕

F. dimorpha see: **F. xanthoxyloides** var. **dimorpha**

F. dippeliana see: **F. bungeana**

F. elonza Kirchn. Small tree, young branches gray-green, later yellowish, winter buds brown; leaves 12–15 cm long, leaflets 9–11, ovate to lanceolate, very short stalked to sessile, acute, base usually rounded or cuneate, 2 to 5 cm long, bright green above, glabrous, lighter beneath, venation white tomentose, otherwise loose pubescent; fruits in loose panicles, narrow-oblong, base rounded, apex obliquely blunt-rounded to emarginate, 3 cm long. DH 1: 46. Origin unknown, probably Italy. Before 1864. z6 Fig. 58

F. excelsior L. Common European Ash. Tree to 40 m high, branches gray-green, glabrous, winter buds black; leaves 25–30 cm long, leaflets 7–11, ovate-oblong, sessile, 5 to 10 cm long, serrate, dark green above, lighter beneath, glabrous except for the midrib, new growth either green or violet-brown; fruits oblong, 3–4 cm long. HM 213. Europe, N. Asia, usually in moist to wet swampy, deep topsoil, especially in meadowlands. Ⓕ Germany, Romania, Austria, Holland, Soviet Union. z3

Includes many cultivars.

● Leaves normally pinnate

  Differing in habit
    Habit strictly pyramidal:
      'Spectabilis'
    Habit low, globose:
      'Nana'
    Branches pendulous, always green:
      'Pendula', 'Pendula Wentworthii'
    Branches pendulous, young shoots brown:
      'Pendulifolia Purpurea'
    Branches pendulous, bark yellow:
      'Aurea Pendula'

  Differing in the foliage
  × Leaves green, especially narrow:
      'Angustifolia', 'Asplenifolia'
    as above, but margin uneven:
      'Erosa'
    Leaves in whorls:
      'Verticillata'
    Leaves black-green, crispate:
      'Concavifolia', 'Crispa'
  ×× Leaves variegated
    Leaves yellow variegated:
      'Aureovariegata'
    Leaves white variegated:
      'Argenteovariegata'
    Leaves punctate:
      'Punctata'

    Leaves yellow; strong grower:
      'Jaspidaea' (common)
    Leaves yellow; slow grower:
      'Aurea' (rare)

  Differing in the bark
    Bark on older branches split:
      'Verrucosa'

● ● Leaves simple to trifoliate
  Habit normally upright:
    'Diversifolia', 'Hessei'
  Habit pendulous:
    'Heterophylla Pendula'

'Angustifolia'. Winter buds black, otherwise very similar to *F. angustifolia*; leaflets narrow-ovate, terminal leaflet long stalked. Germany. 1903.

'Argenteovariegata'. Leaflets white variegated. England. 1770.

'Althena'. Crown medium wide, conical (= 'Monarch'; "Nr. 17"). Selected in 1943 by the Dutch Forestry Department from a street tree planting between Sleeuwijk and Nieuwendijk. Cultivated in the nursery trade in Holland.

'Asplenifolia'. Leaflets quite variable, from normal to narrow-linear. Germany. 1864.

'Atlas' (van t'Westeinde 1942). Crown slender conical, branches ascending at acute angles, new growth appears late, from the terminal buds; leaves deep green; male, therefore no fruit develops. Strongly recommended as a street tree in Holland.

'Aurea'. Weak grower, winter buds very densely distributed, bark yellow, never striped; leaves yellow-green at first, later yellow. First recognized in Holland, 1807. Much rarer than 'Jaspidea' and often mistaken for it.

'Aurea Pendula'. Branches pendulous and yellow; leaves green. England. 1838.

'Aureovariegata'. Leaflets yellow variegated. England. 1770.

'Concavifolia'. Weak grower, narrow crowned, narrowly upright; leaves small, 10–12 cm long, leaflets cupped. A small specimen is cultivated in Gisselfeld Park, Denmark, and in the Chenault Garden, Orléans, France. Plate 30.

'Crispa'. Rather slow growing, but becoming a large tree (e.g. in the Copenhagen Botanic Garden, Denmark); leaflets usually 7–11 on a very short rachis, very densely arranged, black-green, long internodes (= f. *cucullata* Carr.; *atrovirens*). England. 1788.

f. *cucullata* see: **'Crispa'**

'Diversifolia'. Narrow crowned tree, open branched; leaves simple to trifoliate, usually deeply incised-serrate (= f. *heterophylla, monophylla; F. veltheimii* Hort.). England. 1789. Fig. 59.

'Doorenbos' (Doorenbos 1943). Not a park tree, rather, a selection with especially good lumber qualities, young branches olive-green, smooth, with dark, flat lenticels, first year grafts quickly develop many side branches, new growth light brown, medium late; leaves with 9–11 leaflets, petiole base without a distinct border (= 'Doorenbos Nr. 5'). Selected by S. G. A. Doorenbos, Den Haag, Holland. Widely planted after 1945, but particular as to soils and surpassed by 'Westhof's Glorie'.

'Erosa'. Leaflets 9–11, some normal, some deformed or margin scalloped (= f. *scolopendrifolia*). Germany. 1806 (Hillier.)

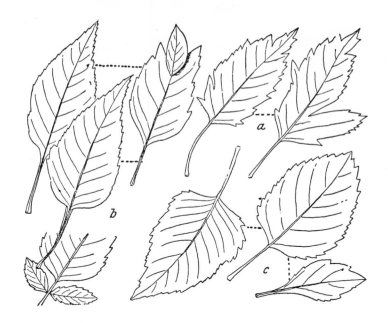

Fig. 59. **Fraxinus** forms.
a. *F. angustifolia* 'Monophylla'
b. *F. excelsior* 'Diversifolia'
c. *F. excelsior* 'Hessei'

'Eureka' (van der Have 1947). Stem grows straight upward, crown open, broadly conical, branches ascending at acute angles, young branches gray-green, with high, brown lenticels, new growth light brown, rather late; leaflets usually 11–13, petiole base with 2 small, erect borders. Selected as a street tree in Holland.

f. *globosa* see: **'Napa'**

'Glomerata'. From Simon Louis Frères, Metz, France. Presumably identical to 'Crispa'.

'Gold Cloud' (Spring Hill Nursery, USA 1963). Branches golden-yellow; leaves green. US Pl. Pat. 2286. Possibly identical with 'Jaspidea'. ∅

'Hessei' (Hesse). Selection of 'Diversifolia', but with wider crown, branches more olive-green; leaves rather large, simple to trifoliate (often 3–5 parted on 'Diversifolia'), remaining a good green until late in the fall. Introduced into the trade in 1937 by Hesse, Weener, W. Germany. Fig. 59. ∅

f. *heterophylla* see: **'Diversifolia'.**

'Heterophylla Pendula' (Späth). Branches hanging in a wide arch like 'Pendula', but leaves simple like 'Diversifolia'. Introduced into the trade in 1898 by L. Späth, Berlin, W. Germany.

'Jaspidea'. Tree, fast grower, to 15 m high or more, broadly conical crown, winter buds not densely arranged (like 'Aurea'), bark on young shoots yellow, often with green stripes, especially on the long shoots; leaves green at first, but becoming yellow during the course of the summer. France. 1802. Much more common than 'Aurea', also grows better.

'Monarch' see: **'Althena'**

f. *monophylla* see: **'Diversifolia'**

'Monstrosa'. Young branches often banded; leaves frequently alternate.

'Nana'. Dwarf form, growth more or less densely globose, 1.5–2.5 m high and wide; leaves smaller than the species, usually not over 15–20 cm long, with 9–11 leaflets (= f. *globosa*; *polemonifolia*). France. 1805. Nearly always seen grafted on a standard.

'Pendula'. Branches and twigs nodding in a wide arch; leaves normal. England. 1725. Plate 29.

'Pendulifolia Purpurea'. Like 'Pendula', but young shoots brown. France. 1864.

'Pendula Wentworthii'. Stem and terminal branch always erect, side branches however, distinctly pendulous. Cultivated in Kew Gardens (1969).

'Punctata'. Leaves yellowish punctate, the spots pink at first on young branches. Known since 1864 and occasionally seen today.

f. *scolopendrifolia* see: **'Erosa'**

'Spectabilis'. Strictly pyramidal, erect habit.

'Verrucosa'. Older branches with rough, split bark.

'Verticillata'. Leaves occasionally in whorls of 3, but also frequently alternate.

'Westhof's Glorie' (van t'Westeinde 1947). Tall tree, wide crowned, always well branched, branches broadly outspread, young branches grass-green, smooth, lenticels dark, flat, one year grafts narrowly upright, without lateral branching, new growth dark brown, very late; leaflets 11–13, leaf base with 2 distinctly protruding marginal lobes. A very common street tree in Europe.

F. floribunda Wall. A tree in its habitat, to 40 m, young branches red-blue, compressed, glabrous, white punctate, leaflets 7–9, short petioles, oval-oblong, 10–15 cm long, scabrous serrate, long acuminate, base cuneate, glabrous above, venation beneath elevated and pubescent, rachis broad furrowed and somewhat winged; flowers in 20–30 cm long panicles, white, with

Fig. 60. **Fraxinus.** a. *F. mandshurica;* b. *F. angustifolia* var. *australis;* c. *F. pensylvanica;* d. *F. nigra;* e. *F. americana;* f. *F. latifolia;* g. *F. rotundifolia* (from Dippel)

calyx and corolla, June; fruits linear spathulate, 3 cm long, emarginate. EP IV, 243: 5. Himalaya, in the mountains between 1500–3000 m. Only for the mildest climates. 1872. Ⓕ Himalaya. Fig. 58. z9 Ø ⊕

F. floridana see: **F. caroliniana**

F. greggii Gray. Small tree or shrub, 1–10 m high, bark thin, gray to light brown, peeling in large, paper-thin pieces, young branches light brown, finely pubescent at first, rough due to thick, round lenticels; leaves 2–5 cm long, usually with 3–7 leaflets, occasionally simple, leaflets 0.7–3 cm long, sessile or nearly so, linear to narrow-oblanceolate, entire or sparsely serrate on the apical half, leathery, deep green above, lighter beneath, rachis winged; fruits 1–2 cm long, lanceolate to oblong-obovate. SM 739. SW. Texas. Dry areas. z7 Fig. 61.

F. griffithii Clarke. Tree, about 6 m high, bark gray, young branches brown pubescent; leaflets 3 to 7, rachis flattened above, leaflets about 8 cm long, elliptic-oblong, short acuminate, base narrow cuneate, entire, light glossy green above, venation silvery pubescent beneath, leathery tough; flowers white, in large, terminal panicles, stalk dense silvery pubescent, panicles with small bracts throughout (= *F. bracteata* Hemsl.). China to the Phillipines. 1900. z9 ⊕

F. holotricha Koehne. Small tree, to 10 m high, branches and petioles densely pubescent, winter buds brown; leaflets 9–13, nearly sessile, lanceolate, 4–7 cm long, scabrous serrate, acuminate, base oblique, both sides gray pubescent at first; flowers bisexual, ovary pubescent. MD 1925: 282; 1910: 114. E. Balkan. 1870. z6

'Moraine' (Siebenthaler 1958). Selection with especially symmetrical, ellipsoid crown, fast growing, bark light gray; leaves persist for some time in fall. Good street tree. Ø

F. hookeri Wenz. Tree, winter buds black, bud scale margins brown pubescent, very similar to *F. excelsior*, but leaflets 5 (7), oblong-elliptic, 6–12 cm long, long acuminate, entire on the basal third, otherwise shallowly serrate, dark green above, lighter beneath, major veins pubescent; flowers like those of *F. excelsior*; fruits lanceolate, 4 cm long, furrowed in the middle. NW. Himalaya, river valleys to 3000 m high. 1920. z5

F. lanceolata see: **F. pensylvanica**

F. lanuginosa Koidz. This species is very closely related to *F. sieboldiana* but distinguished either by the glabrous surface or the short, white, nearly bristly pubescence on the young twigs, inflorescence and petioles; buds glabrous or with stiff, single celled (occasionally 2 celled) pubescence; leaflets also larger and usually distinctly serrate (= *F. sieboldiana* var. *pubescens* Koidz.). Japan, in the mountains. z6

F. latifolia Benth. Oregon Ash. Tall tree, to 20 m or more, young branches red-brown, more or less dense short pubescent and rough with tiny "warts"; leaves 15–30 cm long, with 5, 7 or 9 leaflets, these ovate to oblong, 8–15 cm long, acute, entire or indistinctly and sparsely dentate, base cuneate to rounded, dark green above and

Fig. 61. *Fraxinus greggii* (from Sargent)

thinly pubescent at first, light beneath and more pubescent, also the rachis, petiolule of the terminal leaflet to 2.5 cm, the axillary leaflets short stalked or sessile; panicles glabrous, flowers without petals, on the previous year's wood; fruits 3–5 cm, wings elliptic to oblanceolate, running nearly to the base of the seed. HHD 496; MM 379; SM 754 (= *F. oregona* Nutt.; *F. americana* ssp. *oregona* Wesm.; *F. pensylvanica* ssp. *oregona* [Wesm] G. N. Mill.). USA; mountains from Washington to California. z6–7 Fig. 60.

F. longicuspis S. & Z. Tree, 6 (15) m high, young branches slightly angular, glabrous, winter buds red-brown pubescent; leaflets usually 5(7), elliptic to ovate-lanceolate, 5–10 cm long with about 1 cm long, protruding tips, base cuneate, crenate margin, glossy above, fall color red; flowers white, in 6–12 cm long panicles, axillary and terminal on the current year's wood, June; fruits narrow-elliptic to oblanceolate, 2.5–3.5 cm long, 3 mm wide. DL 1: 29; EP IV, 238: 6; KIF 2: 76. Japan, Korea. See also *F. pubinervis*. z5 Fig. 58. Ø ⊕

F. lowellii see: **F. anomala** var. **lowellii**

F. mandshurica Rupr. Slow growing tree, to 30 m tall in its habitat, branches thick, greenish-brown, obtuse 4 sided, glabrous, winter buds black-green to black-brown, scales loosely arranged; leaves 20–35 cm long, leaflets 9–11, sessile, oblong-ovate to more lanceolate, 7–12 cm long, scabrous serrate, dull green above, often somewhat rough pubescent, venation indented, lighter and pubescent beneath, rust-brown pubescent vein axils; rachis somewhat winged; dioecious, without a corolla and calyx; fruits oblong-lanceolate, 2.5–3.5 cm long, in globose clusters. DL 1: 35; NK 19: 8; KIF 1: 97. NE. Asia. Always susceptible to frost damage due to the early new growth. Ⓕ NE. China, Japan, z6 Fig. 60. Ø

F. mariesii Hook. f. As confirmed by H. Scheller, the type of this species is identical to **F. sieboldiana** Bl.

F. michauxii see: **F. pensylvanica** and **F. tomentosa**

F. nigra Marsh. Black Ash. Tree, to 25 m high, glabrous, winter buds dark brown, somewhat crusty, branches stout, round glabrous, dull, growing steeply upward; leaves about 3 cm long, leaflets 7–11, sessile, oblong-lanceolate, 7 to 12 cm long, serrate, teeth curved inward, deep green above, lighter beneath with brown tomentose venation; flowers dioecious, without calyx and corolla; fruits oblong, 3–4 cm long, rounded at the apex. BB 2843; GTP 330; HH 458; SM 756 (= *F. sambucifolia* Lam.). USA, in swamps and cold frost pockets. Crushed leaves smell like *Sambucus nigra*. z7 Fig. 60.

'Fallgold' (D. W. G. Ronald). Strong growing form with wide forked branches; leaves deep green, golden-yellow in fall and very persistent; male and seedless. Selected at the Agriculture Research Station, Morden, Manitoba, Canada. ⊘

F. numidica Dipp. Shrubby, branches dark brown; leaflets 7–9, rounded-elliptic, acute, base cuneate, scabrous serrate, gray–green and somewhat rough pubescent above, lighter beneath, pubescent, rachis short pubescent. DL 1: 52. N. Africa. 1890. z7 Fig. 58.

F. obovata see: **F. rhynchophylla**

F. obliqua Tausch. Small tree, quite glabrous, branches olive-green, somewhat compressed, warty, winter buds dark brown; leaflets 9–11, nearly sessile, elliptic to oval-lanceolate, 4–8 cm long, terminal leaflet to 12 cm long, coarsely serrate, rachis furrowed; fruits oblong-obovate, 2.5–3 cm long. SH 2: 242c (= *F. willdenowiana* Koehne). Asia Minor. 1843. z6

F. oregona see: **F. latifolia**

F. ornus L. Manna Ash, Flowering Ash. Tree, 6–8 m high, occasionally higher, round crown, branches gray, smooth, winter buds gray-brown; leaves 15–20 cm long, leaflets usually 7, ovate-oblong, stalked, 3–7 cm long, terminal leaflet obovate, irregularly serrate, dark green above, lighter beneath, midrib-pubescent only at the base; flowers in dense, terminal panicles, 10 cm long and wide, white, fragrant, May; fruits narrow-oblong, 2.5 cm long. HM 1923; DB 1952: 128; 1956: 191. S. Europe, Asia Minor, on dry sunny slopes. 1710. In S. Italy a food "Manna" is derived by cutting the bark and allowing the sap to flow and harden in the air. ⓕ Italy, Yugoslavia, Romania, Hungary. z6 Plate 29. ⊕✗

var. **juglandifolia** Ten. Leaflets ovate-oblong, 5–10 cm long, 2.5–5 cm wide. DL 1: 33 (= var. *latifolia* Ait.)

var. **rotundifolia** Ten. Leaflets broad elliptic to circular-obovate. SLH 2: 513 (= *F. rotundifolia* Lam. non Mill.).

F. oxycarpa Willd. and *F. oxyphylla* Bieb. see: **F. angustifolia** Vahl.

F. pallisae Willmott. Small tree, resembling *F. holotricha*, but with only 5–9 (occasionally to 11) leaflets, narrow-lanceolate, sessile, sparsely serrate, light green, rachis pubescent on both sides, midrib pubescent beneath; fruits oblong, obtuse, 4–5 cm long. MD 1925: 281. Balkan. 1840. Attractive foliage, very ornamental Ash. z6 Plate 31. ⊘

F. parvifolia see: **F. rotundifolia**

F. paxiana Lingelsh. A shrub in cultivation, but a tree to 20 m high in its habitat, then broad spreading, branches thick, glabrous, winter buds rust-brown, tomentose; leaflets 7 to 9, ovate to oblong-lanceolate, 8–18 cm long,

sessile, crenate, glabrous, long acuminate; flowers in large panicles on young foliate shoots, large, white, fragrant, with large calyx, petals narrow, early June. BM 9024; MG 51: 381. China, Himalaya. 1901. Has the largest and cleanest white inflorescences of the genus, but the foliage is not attractive, especially when mature. Very hardy. z5 Plate 30. ⊕

F. pensylvanica Marsh. Red Ash, Green Ash. Tall tree, 10–18 m high or more, bark brown, slightly furrowed, young branches cylindrical, light brown to olive-green, more or less densely pubescent; leaves to 25 cm long, usually with 7–9 leaflets, these oblong, lanceolate to narrow-elliptic, 7–15 cm long, 2.5 to 5 cm wide, long acuminate, entire or indistinctly dentate, base broadly cuneate and usually entire, dull green on both sides, usually quite glabrous above; slightly pubescent beneath, leaflets stalked; flowers dioecious, calyx small, campanulate, corolla absent; fruits rather variable, spathulate to lanceolate, wing running to the middle of the nutlet. BB 2840; GTP 326; NBB 3: 49; HH 471 (= *F. pubescens* Lam.; *F. michauxii* Britt.; *F. lanceolata* Borkh.; *F. pennsylvanica* var. *subintegerrima* [Vahl] Fern.). USA, primarily east of the Rocky Mts., often on wet lowlands and swamps. The non-hairy type is called "Green Ash", the other is "Red Ash". The two are not distinguished by most American botanists today. ⓕ USA and USSR. z3 Fig. 60.

'Aucubifolia'. Leaves yellow speckled. (Hillier.) ⊘

'Emerald' (Marshall 1948; US Pl. Pat. 3088). Tree, medium height, about 12 m, bark corky rough, crown ellipsoid and symmetrical; foliage deep green; fruitless (= 'Marshall's Seedless Green'). Selected by Marshall Nurseries in Arlington, Nebraska, USA. ⊘

'Marshall's Seedless Green' see: **'Emerald'**

ssp. *oregona* see: **F. latifolia**

'Patmore'. Selection of ssp. *pensylvanica*; strong grower, stem straight, crown symmetrical-elliptic; foliage very attractive, glossy, disease free, leaves very persistent in fall. Discovered in Vegreville, Alta, Canada. In the trade in Canada. ⊘

'Variegata'. Leaves more silver-gray, cream-white bordered and speckled. (Hillier.) ⊘

F. pistaciaefolia see: **F. velutina**

F. platycarpa see: **F. caroliniana**

F. platypoda Oliv. Tree, 10–20 m high, bark gray, lightly fissured, young branches thick, stiff, gray, glabrous; leaves 15–25 cm long, leaflets 7–11, oblong-lanceolate, 8–14 cm long, 3–5 cm wide, acuminate, base cuneate, serrate, deep green above, bluish beneath with brownish pubescent midrib, sessile, rachis gray pubescent, rachis base conspicuously widened; fruits oblong, 2.5 cm long, acuminate, wings reddish. LF 262; EP IV, 243; 11c. China; Hupeh, Szechwan, Kansu, Yunnan Provinces. 1909. z5⊘ ⊗

F. potamophila Herd. Tree, to 10 m, branches olive-brown, glabrous, winter buds dark brown; leaflets 9–11, stalked, broad elliptic, 2.5 cm long, base tapering to the petiole, serrate, glabrous; fruits elliptic-oblong, 3–5 cm long. DL 1: 53 (as *F. regelii*) (= *F. regelii* Dipp.). Turkestan. 1891. Very attractive, ornamental tree. Fig. 58. ∅

F. profunda see: **F. tomentosa**

F. pubescens see: **F. pensylvanica**

F. pubinervis Bl. According to Scheller, this plant is not a separate species and should be included in **F. longicuspis.**

F. pubescens see: **F. penslvanica**

F. quadrangulata Michx. Blue Ash. Tree, 25–30 m high, young branches sharply 4 sided, glabrous, winter buds gray-white tomentose; leaves 20–35 cm long, leaflets 7–11, short stalked, ovate, 6–12 cm long, scabrous serrate, teeth curved inward, yellowish-green above, glabrous beneath except for the pubescent midrib, fall color light yellow; flowers without a corolla, but with a 0.5 mm long calyx; fruits 3–5 cm long, oblong, crenate at the apex, winged to the base. BB 2842; HH 463; SM 755. N. America; likes alkaline soils. 1823. When the inner bark is laid in water, it will turn blue. z3

F. raibocarpa Reg. Closely related to *F. bungeana;* small tree; leaflets 3–5, elliptic to obovate, occasionally oblong-obovate, somewhat leathery, 2–6 cm long, obtuse to acute or slightly emarginate, glabrous and dark green above, lighter beneath with short stellate pubescence at first, later glabrous; fruits 2.5 cm long, sickle-shaped (!). DL 1: 50; EP IV, 243: 4a–b. Turkestan, E. Bucharei. 1884. z5

F. regelii see: **F. potamophila**

F. retusa Champ. Small tree or a low shrub, young branches gray, punctate; leaflets 3–7, rounded to oblong-ovate, glabrous, 4–6 cm long, short acuminate, base round to broadly cuneate, serrate, glossy green above, lighter beneath, venation raised; flowers white, in dense panicles, calyx large, short stalked, rachis thin, glabrous; fruits oblong, 2.5 cm long, 3 mm wide. China, Kwangtung Province. Existence in cultivation unknown. z6 ✤

var. **henryana** Oliv. 3–7 m high; leaflets with 4 to 15 mm petioles, oblong, to 8 cm long, more leathery, margin regularly serrate; flowers in 10 to 15 cm long, dense panicles; fruit 2–2.5 cm long, emarginate. China; Hupeh Province. 1900 z7–9

F. rhynchophylla Hance. Tall tree, to 25 m in its habitat, young branches glabrous, yellowish; leaves 15–30 cm long, with 5 leaflets, these oblong or ovate to obovate, short acuminate, the terminal leaflet 6–15 cm long, the others shorter, quite short stalked, coarsely crenate, dark green and glabrous above, midrib somewhat pubescent beneath with brown hair fascicles at the juncture of the leaflets; flowers in 7–15 cm long panicles on the tips of foliate shoots, calyx present, petals absent, June; fruits oblanceolate, 3 cm long, 5 mm wide. BS 2: 213 (= *F. chinensis* var. *rhynchophylla* [Hance] Hemsl.; *F. obovata* Schneid.). Korea; China. 1892. It is not yet absolutely clear whether or not this species can be included in *F. chinensis* (Scheller). z6

Fig. 62. *Fraxinus sieboldiana* (from Hooker)

F. rotundifolia Mill. Shrub or small tree, to 5 m high, divaricate branching habit, branches red-brown, glabrous, winter buds black-brown; leaflets 7–13, sessile, oval-rounded, 1–3 cm long, rounded or acute at the apex, serrate, dark green, pubescent along the midrib beneath; fruits obovate, about 3 cm long, 5–7 mm wide. DL 1: 51 (= *F. parvifolia* Lam.) S. Europe, Asia Minor. 1750. Questionable species! z6 Fig. 60.

'Pendula' Branches more or less pendulous (= *F. parvifolia* var. *pendula* Dipp.).

F. sambucifolia see: **F. nigra**

F. sieboldiana Bl. Small tree, 6–9 m high, occasionally taller in its habitat, young branches thin, gray-brown with erect pubescence when young, buds tight, scales gray pubescent to nearly glabrous on the exterior; leaves 5–7 parted, 10–15(20) cm long, 1.5–3.5 cm wide, long acuminate, glabrous or with erect white hairs along the midrib beneath, toothed to nearly entire, nearly sessile, rachis pubescent when young; flowers white, in terminal and axillary panicles, 9–15 cm long, calyx tiny toothed, petals 4, about 5–7 mm long, distinct, linear-oblanceolate, June; fruit wings oblanceolate, 2.5–3 cm long. KIF 2: 77 (= *F. mariesii* Hook. f.). Japan; Central China. z6 Fig. 62. ✤

var. **pubescens** see: **F. lanuginosa**

F. sogdiana see: **F. syriaca**

F. spaethiana Lingelsh. Tree, about 10 m high, branches gray, glossy, glabrous, stout, winter buds black, surrounded by the leaf petiole base in summer; leaves to 45 cm long, leaflets 7–9, sessile, elliptic to narrow-ovate, acuminate, 8–16 cm long, crenate, dark green above, lighter beneath, glabrous except for the vein bases, leaf petioles furrowed, base reddish-brown and distinctly

Fig. 63. *Fraxinus tomentosa* (from Sargent)

swollen (!); flowers in terminal panicles, white, but inconspicuous; fruits oblanceolate, 3–3.5 cm long, blunt tipped. DL 1: 27; NT 1: 400; KIF 1: 98 (= *F. stenocarpa* Koidz.). Japan. 1873. z6 ⌀

F. stenocarpa see: **F. spaethiana**

F. syriaca Boiss. Small tree, branches erect, glabrous, often compressed (with short internodes and whorled leaves), winter buds brown; leaflets 3–5, occasionally to 7, lanceolate, 3–10 cm long, scabrous serrate, both sides light green and glabrous, open slits on both sides; flowers in short panicles; fruits obovate-oblong, 3–4 cm long, obtuse. DL 1: 46 (as *F. sogdiana*) (= *F. turkestanica* Carr.; *F. sogdiana* Bge.). Asia Minor to Central Asia. 1880. Ⓕ A street tree in the Middle East. z5 Fig. 58.

F. texensis Sarg. Tree, to 15 m; leaflets usually 5, small, 3–8 cm long, often leathery tough, acute or obtuse, usually crenate on the apical blade half, reticulate venation beneath; fruits 1.5–3 cm long. SS 270; SM 746. Central Texas. Limited to dry alkaline soils. z6

F. tomentosa Michx. f. Tree, 15–40 m high in its habitat, trunk very thick at the base, young branches thick, cylindrical, densely tomentose in the first year, second year glabrous and gray, winter buds large, deep brown, scabby; leaves 20–45 cm long, leaflets 7–9, lanceolate to more elliptic, usually about 11 cm long, 4.5 cm wide, acute, base rounded to cuneate, entire or indistinctly serrate, tough, dark yellowish-green above, lighter and soft pubescent beneath; flowers dioecious, with calyx, without a corolla, in dense panicles; fruits narrow-oblong, 5–7 cm long, about 1.2 cm wide, rounded or somewhat emarginate, nutlet round in cross section, winged nearly to the base. SS 714–715; SM 748 (= *F. profunda* [Bush] Bush; *F. michauxii* Britt.). Eastern USA, in swamps. 1913. z6 Fig. 63. ⌀

F. toumeyi see: **F. velutina**

F. turkestanica see: **F. syriaca**

F. uhdei (Wenzig) Lingelsh. Small evergreen tree, winter buds red-brown; leaves 12–20 cm long, leaflets 5–7, rachis white pubescent, oblong-lanceolate, 5–8 cm long, 2 to 3 cm wide, quite glabrous above, deep green, venation white tomentose beneath or glabrous, margin scabrous serrate; flowers in dense, 17 cm long panicles; fruit elliptic, 2.5–3 cm long, 0.5 cm wide, acute or obtuse. Central America; in dry areas. z9 #

'Tomlinson' (Tomlinson 1965). Selection; small tree, in 10 years about 4 m high and 2 m wide, narrow crowned, branches ascending; leaves leathery tough, distinctly serrate, nearly *Ilex*-like. Introduced by Tomlinson's Select Nursery, Whittier, California, USA. z10 # ⌀

F. veltheimii see: **F. angustifolia** var. **'Monophylla'**

F. velutina Torr. Arizona Ash. Shrub, or small tree; leaflets 3–7, elliptic to lanceolate-elliptic, 2–7 cm long; the apical third of the blade scabrous serrate, usually short stalked gray-green above, more or less densely pubescent beneath; fruits to 3 cm long, wings about as long as the nutlet. SPa 203; 201; MM 402; SM 752–753 (= *F. velutina* Torr.; *F. toumeyi* Britt.; *F. coriacea* S. Wats.; *F. pistaciaefolia* Torr.). SW. USA, from W. Texas to S. California; in dry areas and desert canyons. z6

'Fan-Tex' (E. Fanick 1964). Fast growing selection, crown especially symmetrical; leaves larger, dark green, regularly formed; seedless. Introduced by the Ildridge Nursery, Van Ormy, Texas, USA ⌀

F. willdenowiana see: **F. obliqua**

F. xanthoxyloides (G. Don) DC. Small tree or shrub, scarcely over 6 m high, branches divaricate, young branches pubescent at first; leaves 4 to 6 cm long, rachis narrow-winged, somewhat pubescent, leaflets 5–9, nearly sessile, the basal pinnae short stalked, oval, 2–4 cm long, obtuse, crenate, midrib somewhat pubescent beneath; fruits narrow-oblong to spathulate, 3–4 cm long, obtuse to emarginate, nutlet longitudinally furrowed. DL 1: 34; SH 2: 518k. Himalayas to Afghanistan. 1870. Ⓕ W. Pakistan. z8 Fig. 64. ⌀

var. **dimorpha** (Coss. & Durieu) Wenz. Young branches and leaves glabrous or nearly so; leaflets nearly sessile, ovate to elliptic, 1–2.5 cm long, obtuse, on sterile branches with 7–11 leaflets, shorter and wider; fruits 4–5 cm long, obtuse. DL 1: 35; EP IV 243, 9e (= *F. dimorpha* Coss. & Durieu). N. Africa. Fig. 58.

var. **dumosa** (Carr.) Lingelsh. Dense, rounded, dwarf form of var. *dimorpha*, always shrubby; leaflets rounded to ovate, 0.5–1.5 cm long. 1865.

All prefer open sunny areas in deep, fertile and moist soil. The so-called Flowering Ashes are beautiful park trees while the others are primarily forestry trees.

Lit. Lingelsheim, A.: *Fraxinus*; in Engler, Pflanzenreich, Heft **72**, 9–65, 1957 ● Miller, G. N.: The genus *Fraxinus* in North America; Cornell Agr. Expt. Sta. Memoir **335**, 1–64, 1955 (with 8 pp. of reference literature) ● Scheller, H.: Kritische Studien über die kultivierten *Fraxinus*-Arten; in Mitt. DDG 69, 49 to 163, 1977.

Fig. 64. *Fraxinus xanthoxyloides*
(from Sargent, actual size)

FREMONTIA See: **FREMONTODENDRON**

FREMONTODENDRON Coville — STERCULIACEAE

Evergreen shrubs or occasionally a small tree; leaves alternate, simple, usually leathery thick with stellate pubescence; flowers bisexual, regular, large, attractive, solitary and axillary on small shoots; sepals 5, petal-like, petals absent; stamens 5, connate on the basal half, occasionally connate nearly to the anthers, style protruding far past the stamens; fruit an ovate, dense bristly, 4–5 valved capsule, with 2–3 brown or black seeds in each chamber. — 2 species in California.

Fremontodendron californicum (Torr.) Cov. Shrub, 1.5–4 m high, young branches with dense brown stellate pubescence; leaves entire to slightly lobed, distinctly lobed only on long shoots, 5–10 cm long, with 3–7 lobes, dull green above with scattered stellate pubescence, gray-white and densely pubescent beneath; flowers wide cuplike, 2.5–3 cm wide, lemon-yellow, May to June; fruit capsule more globose; seeds dull brown. MS 405; SPa 180; MCL 182; JRHS 12: 262; FS 2349 (= *Fremontia californica* Torr.). California. 1851. z9 Fig. 65. # ⊕

ssp. *napense* see: **F. napense**

F. (*californicum* × *mexicanum*) **'California Glory'**. Very strong grower and good bloomer; leaves resemble *F. californicum*, but 5 veined from the base; flowers flat shell-like, 4–6 cm wide, lemon-yellow, exterior reddish on spent flowers, nectaries pubescent (like *F. mexicanum*, but also with some long hairs like *F. californicum*); seeds black. JRHS 92: 228. Originated in the Rancho Santa Ana Botanic Garden, California, USA in 1952. z9 # ⊕

F. napense (Eastw.) R. Lloyd. Lower habit than the previous species, branches thin; leaves entire or 3–5 lobed, 1–2.5 cm long, dull above and nearly glabrous, white pubescent beneath, later brown; flowers 3–4 cm wide, yellow, occasionally somewhat pink toned (= *Fremontia napensis* Eastw.; *F. californicum* ssp. *napense* [Eastw.] Munz.). California, Napa and Lake Counties, but not generally cultivated. z9 # ⊕

F. mexicanum Davidson. Similar to *F. californicum*,

Fig. 65. *Fremontodendron californicum*, branch, single leaf lower right; leaf from *F. mexicanum* upper right (branch from Bot. Mag.)

equally tall habit; leaves distinctly lobed with palmate venation, distinct stellate pubescence; flowers 3–6 cm wide, golden-yellow with orange specks on the ventral side of the sepals, long flowering period but mostly from May–July; fruit capsule conical, seeds black, glossy. MS 409; BM 9269; NH 1955: 204; BS 2: 235; JRHS 92: 262 (= *Fremontia mexicana* [Davidson] MacBride). Southern California. 1926. Somewhat more touchy than *F. californicum*, but a more prolific bloomer, even on young plants. Fig. 65. z10 # ⊕

All species only for the mildest regions; require a poor soil or the plants will not bloom. Difficult to transplant and short lived. Like dry conditions.

Lit. Harvey M.: A revision of the genus *Fremontia*; in Madroño 7, 100–110, 1943 ● Eastwood, A.: New species of *Fremontia*; Leaf. W. Bot. I, **12**, 139 to 141, 1934 ● Summary with short decriptions of 3 new species in Jour. RHS 1935, 379.

FREYLINIA Colla — SCROPHULARIACEAE

Glabrous, evergreen shrubs; leaves opposite to alternate, entire; flowers in terminal panicles or racemes, compounded from axillary cymes; calyx 5 lobed, corolla tubular with outspread limb; 4 fertile stamens, enclosed; fruit a capsule, seeds disc-like. — 5 species in the tropics and S. Africa.

Freylinia cestroides see: **F. lanceolata**

F. lanceolata (L. f.) D. Don. Evergreen, dense shrub, to 3 m or more, young branches angular, lightly pubescent, very densely foliate; leaves opposite, linear, acuminate at both ends, 5–12 cm long, 3–10 mm wide, dark green on both sides, quite glabrous, midrib beneath raised; flowers in 5–10 cm long axillary panicles, grouped in a 20–25 cm long terminal panicle, corolla tubular, 12 mm long, 3–6 mm wide, with 5 small limb lobes, light yellow or cream colored exterior, interior a brighter yellow, throat pubescent, fragrant, November. RCA 119 (= *F. cestroides* Colla). S. Africa. 1774. z10 Plate 27. # ⊕ Easily cultivated in a warm climate.

FUCHSIA L. — ONAGRACEAE

Deciduous shrubs or small trees, also evergreen in very mild climates; leaves opposite, occasionally whorled, seldom alternate, petiolate, dentate, with small abscising stipules; flowers axillary or also in panicles or racemes, usually very attractive, pendulous, red or reddish, some flower parts also white or blue; calyx tube campanulate to tubular, elongated over the ovary; sepals 4, usually outspread, petals 4, seldom absent; stamens 8, usually unequal, often exserted, style long exserted, stigma large, simple or 4 lobed; fruit a soft 4 chambered berry. — About 100 species, most in tropical America, some in New Zealand and Tahiti; only a few winter hardy.

Fuchsia alba see: **F. magellanica** var. **molinae**

F. coccinea Soland. Shrub, about 0.7 m high, branches thin, soft pubescent; leaves 1.5–5 cm long, ovate, acute, dentate, soft pubescent beneath, petiole short, pubescent; flowers ornamental, tube red, 6 mm long, sepals scarlet-red, 18 mm long, petals violet, obovate, 8 mm long, July–October. BM 5740; MFu 4. S. Brazil. Often confused with *F. magellanica* from Chile, but the latter having a longer tube and glabrous branches. z9 Fig. 66. ✥

F. discolor see: **F. magellanica**

F. excorticata (Forst.) L. f. Shrub, also a small tree in its habitat, bark paper-like exfoliating, branches very brittle; leaves alternate (!), oblong-ovate to oval-lanceolate, 2.5–10 cm long, acuminate, nearly entire, thin, lighter beneath; flowers solitary, pendulous, 2–4 cm long, green-yellow at first, then pink-red, pedicel long and thin, sepals 12 mm long, petals dark purple, 3 mm long, spring. BR 857; MF 60; KF 36–36a. New Zealand. z9

'Purpurascens'. Leaves purple, silvery-white beneath.

F. magellanica Lam. Shrub, to 3 m high, young branches thin, glabrous (!); leaves opposite or in whorls of 3, lanceolate-ovate, 2.5–5 cm long, acute, dentate, petiole short; flowers axillary, nodding, stalk 2.5–3 cm long, tube deep red, 8 mm long, sepals deep red, 18 mm long, petals purple, obovate, 10 mm long, stamens markedly exserted, July–October. BR 1805 (as *F. discolor*); MFu 3; PFC 150. S. Chile, Argentina. 1823. z9

A quite variable species, from which most of the so-called winter hardy fuchsias originate. The most prominent forms are described here:

var. **conica** (Lindl.) Bailey. To 2 m high; leaves grouped 2 or 3 together, wider than var. *macrostemma*, soft pubescent, dentate; flowers not very numerous, long stalked, flower buds rounder and thicker than those of the species, sepals carmine, petals dark violet, bordered, tube scarlet-red, conical (wider at the base than at the apex), filaments pink, anthers light yellow, August–October. BR 1062. Chile. 1824. ✥

var. **discolor** (Lindl.) Bailey. Compact habit, dense, branches purple; leaves small, whorled, oval-lanceolate, glabrous, glossy, margins sinuate; sepals red, ovate, directed forward,

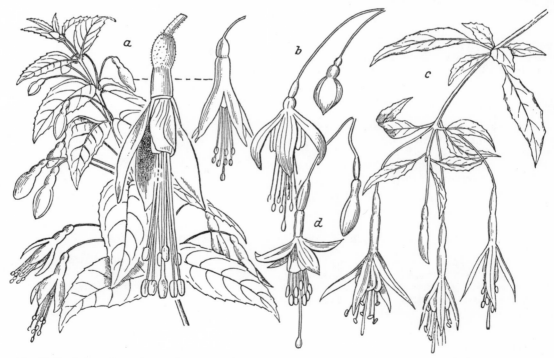

Fig. 66. **Fuchsia**. a. *F. coccinea*, enlarged at left, flower cut away; b. *F. magellanica* var. *globosa*; c. *F. magellanica* 'Gracilis'; d. *F. magellanica* var. *discolor* (from Bot. Mag. and Bot. Reg.)

petals violet, shorter than the sepals, June–October. BR 1805; BC 1300. Falkland Islands. Before 1834. Fig. 66. ⊕

var. **globosa** (Lindl.) Bailey. Shrub, 1.5–2 m high, branches glabrous, slightly outspread, mostly bowed downward, brittle; leaves ovate, acute, dentate, glabrous, venation red; flowers pendulous, stalks very thin, buds inflated like a balloon, sepals scarlet-red, outspread, apex curved inward, petals violet-blue, half as long as the sepals, June–October. BM 3364; BR 1556. Peru, Chile. Hybrid developed in England by Wood, 1832 (?). Fig. 66. ⊕

'Gracilis'. Shrub, 1.5–3 m high, branches thin, very soft pubescent, reddish; leaves usually opposite, 2.5–5 cm long, 1–2 cm wide, lanceolate to more oval, acute, dentate, petiole red; flowers very long and ornamental, tube carmine, 2–3 times as long as wide, sepals red, petals purple, stigma nearly spindle-shaped to conical, short 4 lobed, purple-red, July–fall. BM 2507; JRHS 67: 6. Chile. This plant retains its cultivar name but is actually a form of var. *macrostemma.* Fig. 66. ⊕

var. **macrostemma** (Ruiz & Pav.) Munz. Like 'Gracilis' but this is the less uniform wild plant, not found in cultivation. BR 847; BM 97 (as *F. coccinea*); BM 3521 (as *F. recurvata*). S. Chile. ⊕

var. **molinae** Espinosa. Strong grower, shrubby; leaves light green; flower stalks and ovaries green, tube, sepals, stamens and styles pink, petals lilac, flowers abundantly (but only in full sun) (= f. *alba*). S. Chile, Prov. Llanquihue. 1929. According to Munz, not a cultivar. Very winter hardy. ⊕

'Riccartonii'. Shrub, 1.5–3 m high and wide, branches thin, pendulous; leaves very small, irregular and short toothed, petioles over 0.4 cm long; flowers ornamental, hanging on long stalks, tube bright red, shorter than half sepal length, petals purple-violet, similar to var. *globosa* when open, but the buds are not inflated. BC 2: 1300. Developed in Riccarton, Scotland, near Edinburgh from seed from the Falkland Islands, 1830. This fuchsia is found as a 3 m high hedge in S. England, Ireland and W. Scotland. ⊕

F. recurvata see: **F. magellanica** var. **macrostemma**

The fuchsias mentioned here are some of the most valuable long lasting summer flowers of the milder regions; a fuchsia hedge in bloom is a wonderful sight.

Lit. Munz, P. A.: A revision of the genus *Fuchsia;* in Proc. Calif. Acad. Sci. **25**, 1–138, 1943 ● Wood, W. P.: A *Fuchsia* survey: London 1950 (with very detailed decriptions of many cultivars and a long list of references to color plates) ● Thorne, T.: *Fuchsias* for all purposes; London 1959 (most important work with 94 p. index to cultivars) ● Saunders, E.: Wagtails Book of *Fuchsias;* Vol. 1, 96 plates; Vol. 2, 104 plates; Godalming, England, 1971 and 1972.

FUMANA (Dunal) Spach — CISTACEAE

Subshrub, similar to *Helianthemum,* but with outer stamens sterile, ovules anatropous (= inverted); leaves opposite or alternate, linear or lanceolate; flowers in terminal, raceme-like florescences or solitary; sepals 5, very uneven, petals 5, yellow; stamens numerous, only the innermost fertile; fruit a capsule, with 3–12 large seeds. — 15 species in the Mediterranean region to Asia Minor.

Fumana ericoides (Cav.) Pau. Shrub, 10–35 cm high, stoutly branched, but not procumbent, nearly totally short glandular pubescent and rough to the touch; leaves alternate, linear, about 1 cm long, becoming smaller toward the branch tips; flowers solitary, seemingly axillary, some also clustered, about 1.5 cm wide, yellow, June–August; seeds black-brown, glossy. HM 2044. Mediterranean region, in dense thickets on sandy soil. z8 **d** ○

F. nudifolia (Lam.) Janchen. Heath-like subshrub, 10–20 cm high, branches procumbent to ascending, gray pubescent on the upper portion; leaves very narrow-linear, thick, all rather equally long, 8–20 mm long, alternate, somewhat appressed; flowers usually solitary or in terminal inflorescences, yellow, 2 cm wide, June–August. HM Pl. 184; 2045; GPN 583; EP IV, 193: 20d to k. Mediterranean region, on sunny dry slopes in limestone (chalk) soils. Fig. 67. **d** ○

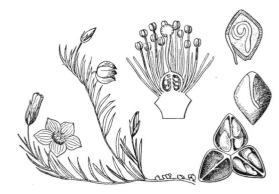

Fig. 67. *Fumana nudifolia* (from Grosser)

GARDENIA Ellis — RUBIACEAE

Evergreen shrubs or trees; leaves opposite, but on many species seemingly in whorls of 3, through the close spacing of 2 leaf pairs with the 4th leaf reduced to a small scale; flowers usually solitary, axillary or terminal, corolla funnelform with a broad limb, the limb lobes twisted together in bud, later outspread, white or yellow, occasionally violet, calyx tube ovate or an inverted cone. — About 250 species in tropical Asia and Africa, including S. Africa.

Gardenia florida see: **G. jasminoides**

G. jasminoides Ellis. Evergreen, thornless, small shrub, 0.5–1.5 m high, wide spreading, densely branched; leaves elliptic-lanceolate, tapering to both ends, about 7 cm long, 3–4 cm wide, dark green, glossy, tough; flowers solitary, nearly terminal, salver-shaped on the wild species, with 6 outspread limb lobes, white, very fragrant, June to September. BR 73 (wild species) (= *G. florida* L.; *G. radicans* Thunb.) China, Japan. 1754. Only cultivated in frost free climates. The wild form is seldom found in cultivation, rather the following cultivars with double flowers. z10 #

'Belmont'. Flowers on long shoots, very large, densely double, good as a cut flower. Cultivated especially in S. Africa in the landscape. ✥

var. *florida* see: **'Plena'**

'Fortuneana'. Leaves to 12 cm long and 5–6 cm wide; flowers regularly developed and arranged, larger than those of the species. FS 177 (= *G. jasminoides* var. *ovalifolia* Hara; *G.* 'Veitchii' Hort.; *G.* 'Fortunei' Hort.). ✥

f. grandiflora (Lour.) Makino. Flowers simple, larger than the species, limb wider. Japan, occurring in the wild. ✥

'Mystery'. Leaves larger than the other cultivars, very glossy; flowers pure white, to 10 cm wide, spirally twisted in the middle, limb camellia-like outspread and regular. Mainly cultivated in California and one of the most beautiful cultivars. ✥

var. *ovalifolia* see: **'Fortuneana'**

'Plena'. Collective name for the double flowered forms. BM 2627 (= *G. jasminoides* var. *florida* [L.] Ellis.). ✥

G. radicans see: **G. jasminoides**

Planted as a shrub in frost free climates but in the temperate zone used only as a greenhouse plant. Young plants bloom better than older shrubs.

GARRYA Dougl. — Tassel Tree — GARRYACEAE

Evergreen shrubs; branches 4 sided at first, pubescent; leaves opposite, simple, entire; inflorescences catkin-like, axillary, pendulous, often branched, more or less densely pubescent; flowers inconspicuous, dioecious, solitary or in 3's in the axils of opposing bracts, male with simple perianth, female without a perianth; fruit a globose, leathery, rather dry berry with 2 seeds. — About 18 species in western N. America.

Garrya elliptica Lindl. Shrub to 4 m or also a small tree (in its habitat), branches densely pubescent at first; leaves oblong-elliptic, leathery tough, dark green above with sinuate margin, densely woolly beneath at first, eventually glabrous; male flower catkins 10–20 cm long in clustered racemes at the branch tips, flowers greenish and brownish, female inflorescences only 10 cm long, January–March; fruits white tomentose, with reddish interior and 1–2 seeds. EP IV, 56a: 3; SPa 199; BMns 220; MS 428. USA, Oregon to California. 1828. For milder climates and very ornamental in the winter months. z9 Fig. 68. # ✥

'James Roof' (Roof). Only male, flower catkins 15 to 20 cm long, in dense clusters. Seedling, selected by Director Roof in the Regional Parks Botanical Garden, Berkeley, California, before 1950. In the trade in USA. (Hillier) Plate 32. # ✥

G. faydenii Hook. Shrub, 3–5 cm high; leaves ellipitic to oblong, with sharp tips, 3–8 cm long, glossy above, more or less pubescent beneath, leathery tough; male catkins 2 to 3 cm long, branched, female catkins not branched, to 5 cm long, both densely tomentose, spring. Jamaica, Cuba. z9 #

Fig. 68. *Garrya elliptica*, one branch each of flowers and fruits respectively (from Sudworth)

G. flavescens S. Wats. Shrub, 1.5–2.5 m high, young branches dense gray silky pubescent, soon becoming glabrous and brown; leaves elliptic, 3.5–5 cm long, somewhat arched or thickened, leathery tough, somewhat appressed pubescent above, dense silky pubescent beneath; male catkins 2 cm long, female catkins compact, 3 cm long; fruits ovate, 5 mm long, silky pubescent. SW. USA, Nevada to New Mexico. z9 #

G. fremontii Torr. Shrub, to about 3 m high, generally appearing yellow-green, branches soon glabrous; leaves oblong-elliptic, 2–6 cm long, to 3 cm wide, glabrous and glossy above, older leaves yellow-green, lighter beneath and eventually totally glabrous, entire and very flat, young leaves often dense gray pubescent; male catkins yellowish, in 7–20 cm long, fascicled simple racemes, female catkins only 3–5 cm long, to 8 cm long in fruit, January–April; fruits 6 mm thick, brownish to reddish or black, stalked, nearly totally glabrous. GC 35: 44 (1904); MS 427, 429. USA, Oregon to California. 1842. z9 #

G. laurifolia Benth. Is not to be found in cultivation, rather its var. **macrophylla** (Benth.) Wanger. Strong growing evergreen shrub, to 3 m (a small tree on Guernsey Island in the English Channel), young branches gray tomentose; leaves oblong, 5–15 cm long, rounded, cuneate at the base or also rounded, deep green and glossy above, gray-white tomentose beneath, petiole 1–2.5 cm long; male catkins 2.5–7 cm long, axillary, usually branched, May–June, female catkins in the axils of leaflike bracts on the side branches, 7–12 cm long. DL 1: 141. Mexico. 1848. *G. faydenii* (which is generally not winter hardy) is often found under this name in cultivation. z9 #

var. **macrophylla** see above

G. × thuretii Carr. (*G. faydenii* × *G. elliptica*). Shrub, fast growing, to 5 m high, branches stiff, pubescent; leaves narrow-oblong, 6–10 cm long, tapering to both ends, with small tip at the apex, eventually quite glabrous and glossy above, gray pubescent beneath; catkins more or less erect, gray, terminal and axillary, 3–7 cm long, June. Developed in 1862 by G. Thuret, Antibes, S. France. As winter hardy as *G. elliptica*. z9 #

G. wrightii Torr. Shrub, 0.5–2 m high, young branches silky pubescent, but quickly becoming glabrous; leaves elliptic-oblong, 2–5 cm long, bright green, leathery tough, with very fine tips, very quickly becoming glabrous, venation reticulate beneath; flower catkins 3–7 cm long, the individual flowers sparsely arranged; fruits glabrous, nearly sessile, July–August. EP IV, 56a: 4. SW. USA. Hardiest species, but not nearly as attractive as *G. elliptica*. z8 #

For milder climates in the landscape. Not particular as to soil, tolerates heat well; prune immediately after flowering, never in summer. Best held in containers since it is difficult to transplant.

Lit. Wangerin, W.: Garryaceae; in Engler, Pflanzenreich IV, 56a, 1–17, 1910.

× GAULNETTYA J. W. Marchant — ERICACEAE

Intergenic hybrid between *Gaultheria* and *Pernettya;* differing from *Pernettya* mainly in the fleshy calyx, which in only half as long as the fruit; flowers axillary, solitary or in short racemes. — Hybrids between several species occur in New Zealand, one hybrid has been found in Mexico, one hybrid has also originated in cultivation (= × *gaulthetya* Camp).

Gaulnettya oaxacana Camp (= *Gaultheria conzaltii* × *Pernettya mexicana*). Dwarf evergreen shrub, pink flowers. Found in the wild in Mexico. Before 1939. Not generally found in gardens. z9 #

G. wisleyensis Marchant (= *Gaultheria shallon* × *Pernettya mucronata*). Dense evergreen shrub, to 1 m high; leaves oblong-elliptic, 4–6 cm long, shallowly dentate, base rounded or cuneate; flowers 6–15 in short, glandular pubescent racemes, calyx fleshy, corolla white, June; fruits red-brown to wine red. JRHS 64 (1939): 14–16. z6 Plate 26. #

'Ruby'. Dense evergreen shrub, to 1 m, young branches thick, ruby-red; leaves elliptic-lanceolate, 2–2.5 cm long; flowers white, late May–early June; fruits numerous, small, ruby-red. Originally developed in RHS Gardens, Wisley, but propagated and named by Hillier. # ⌂

'Wisley Pearl'. The type of the cross; found in RHS Gardens, Wisley, 1929. Good fruiting character; seedlings are quite variable. Vegetative propagation has been unsuccessful thus far.

Cultivated like *Pernettya;* in a clay-humus soil.

Lit. Camp, in Bull. Torrey Bot. Club **66**, 9–28, 1939.

GAULTHERIA Kalm ex L. — ERICACEAE

Evergreen shrubs; leaves alternate, short stalked, serrate; flowers solitary or in racemes; calyx 5 parted (occasionally only 4 parted) or with 5 incisions, spent flowers becoming larger and fleshy, surrounding or enclosing the fruit capsule (which is then berrylike in appearance); corolla urceolate or campanulate, with 5 erect or outspread lobes; stamens 10, attached at the base of the corolla, with 4 awns each; fruit a 5 chambered, loculicidal capsule, surrounded by the fleshy calyx. — About 200 species in N. and S. America, West Indies, and from Japan to Tasmania, Australia and New Zealand.

Key to the Asiatic Selections and Series (from Airy-Shaw)

● Flowers solitary; low shrubs, usually with small leaves;

  ✕ Bracteoles 2, opposite, immediately beneath the calyx:

   Section **Eugaultheria**

   + Leaves over 1 cm wide; filaments and corolla (interior) pubescent:

   Series **Procumbentes**
   (*G. procumbens*)

   ++ Leaves less than 1 cm wide; filaments and corolla (interior) not pubescent:

   Series **Trichophyllae**
   (*G. sinensis, thymifolia, tricho-phylla*)

  ✕✕ Bracteoles several, alternate, not situated immediately beneath the calyx;

   * Corolla pubescent inside; filaments pubescent on the margins; anthers 4 awned; fruit calyx blue:

   Section **Brossaeopsis**
   Series **Nummularioideae**
   (*G. nummularioides*)

   ** Corolla glabrous inside; filaments glabrous; anthers without awns; fruit calyx red:
   Section **Amblyandra**
   (*G. adenothrix, humifusa, ovati-folia*)

● ● Flowers in racemes or panicles; bracteoles 2; shrubs usually with wide leaves;

  ✕ Inflorescences lacking involucral scales; bracteoles opposite, immediately beneath the calyx:

   Section **Gymnobotrys**
   (*G. cumingiana*)

  ✕✕ Inflorescences with involucral scales;

   * Bracteoles alternate to nearly opposite, in about the middle of the flower stalk (occasionally closer to the base or apex); flower stalks much shorter than the rachis; corolla urceolate (very seldom campanulate); leaves usually obovate or oblanceolate:

   Section **Leucothoides**
   (*G. caudata, cuneata, fragrantis-sima, hookeri, itoana, pyroloides, semi-infera, shallon, stapfiana, wardii*)

   ** Bracteoles nearly opposite, situated immediately at the base of the flower stalk; flower stalks as long or nearly as long as the rachis;

   > Corolla campanulate, glabrous inside; filaments glabrous:
   Section **Brossaeopsis**
   Series **Dumicolae**
   (*G. codonantha*)

   >> Corolla urceolate, pubescent inside; filaments long pubescent:
   Section **Brossaeopsis**
   Series **Atjehenses**
   (not covered in this work)

Gaultheria adenothrix (Miq.) Maxim. Shrub, dwarf, procumbent, branches wavy; leaves ovate to elliptic, acute or obtuse, 1–3 cm long, somewhat serrate toward the apex, limb somewhat bristly, leathery, with reticulate venation above; flowers solitary, pendulous, in the upper leaf axils, corolla broadly urceolate, white, 1 cm long, 6 mm wide; fruits pea-sized, bright red, pubescent, 6 mm thick. NF 10: 204; TAP 110. Japan. 1915. z9 # ◖

G. antarctica Hook. f. Low, wide shrub, scarcely 10 cm high, branches pubescent; leaves ovate to oblong, acute, 6 mm long, indistinctly serrate, leathery, more or less pubescent; flowers solitary, in the axils of the upper leaves, white, campanulate, 4–7 mm long, spring; fruits pear-shaped to flat globose, white to pink, 6–8 mm thick, calyx fleshy (= *G. microphylla* Hort. non Hook. f.). Patagonia; Falkland Islands. Known for many years but very rare in cultivation. z9 #

G. antipoda Forst. Shrub, variable in height (procumbent and less than 10 cm high in its mountain habitat, erect to 0.5–1 m high in valley locations), young branches pubescent and bristly; leaves rounded obovate to oblong-lanceolate, 8–15 mm long, thick leathery, glabrous, distinctly veined, deep green, petiole pubescent; flowers solitary in the upper leaf axils, oblong campanulate, 3 mm long, white, calyx lobes acute, oval-oblong, becoming fleshy at fruiting, June–July, fruits usually fleshy, white or red, to 1.2 cm thick, but occasionally dry on the same branch. NF 11: 212. New Zealand, both islands. 1820 z9 Plate 32. # ⚭

var. *depressa* see: **G. depressa**

G. caudata Stapf. Closely related to *G. forrestii*, but leaves with both sides nearly equally green, less densely dentate, finer acuminate; bracteoles situated on the flower stalk further from the calyx than on *G. forrestii*, branches reddish, flowers in axillary racemes. BM 9228. China, Yunnan Province. Cultivated by Hillier, 2.5 m wide, 0.8 m high. z6 Plate 33.

G. codonantha Airy-Shaw. Shrub, to 2 m high, branches gracefully arching, bristly pubescent when young; leaves 2-ranked, ovate to lanceolate, 5–15 cm long, 2–12 cm wide, acute, irregularly dentate, eventually glabrous above, underside with 2 major veins running to the apex, with reticulate venation and bristly pubescence between; flowers in axillary corymbs of 4–7, corolla cupulate, greenish-white, often reddish toned outside, nearly 2 cm wide, November; fruits flat globose, nearly 2 cm wide, purple-black. BM 9456; GC 1933: 182. Assam. 1928. Only for the mildest areas. Exceptionally attractive, large flowering species. # ⚭ ☉

G. cumingiana Vidal. Small shrub, branches thin, with long erect pubescence; leaves thick and leathery, short petioled, ovate, 5–8 cm long, 1.5–2.5 cm wide, long acuminate, base rounded to slightly cordate, serrate, glabrous, green above, somewhat lighter beneath, with 3 vein pairs; inflorescences axillary, glabrous, to 5 cm long racemes of 3–5 flowers, corolla campanulate, 7 mm long, limb 5 lobed, white; fruits globose, purple-black, 7 mm thick. LWT 285 (= *G. leucocarpa* [Blume] f. *cumingiana* [Vidal] Sleum.). S. China to the Phillipines. z10 # ∅

G. cuneata (Rehd.& Wils.) Bean. Shrub, 20–30 cm high, bushy, young branches densely pubescent; leaves ovate-oblong, tapering to both ends, 1–3 cm long, glandular serrate, acute; flowers in short racemes in the upper leaf axils, corolla wide urceolate, white, 6 mm long, ovaries silky pubescent (!!), June; fruit distinctly pea-sized, white, globose. BM 8829; NF 11: 216; BS 2: 36 (= *G. pyroloides* var. *cuneata* Rehd. & Wils.). W. China. 1909. Often confused with *G. rupestris* and *G. miqueliana*. z6 Plate 34. # ⚭

G. depressa Hoof. f. Low cushion-like habit, to 30 cm high, branches procumbent, rooting, bristly when young; leaves oval-round to broad-obovate, obtuse to acute, 4–8 mm long, leathery tough, bristly serrate when young, bronze-brown in fall; flowers solitary in the axils of the upper leaves, 3 mm long, white to pink, campanulate, May; fruits 8 to 15 mm thick, globose, scarlet-red or also white. PRP 57 (= *G. antipoda* var. *depressa* [Hook. f.] Hook. f.). New Zealand. z9 # ⚭

G. eriophylla (Pers.) Sleum. Shrub, to 0.9 m, young branches brownish-pink and densely covered with brown, frizzy hairs; leaves oval-oblong to broad elliptic, to 6 cm long, 3 cm wide, but generally becoming smaller toward the branch tips, with a short apical tip, both sides woolly pubescent at first, eventually becoming more or less glabrous above; flowers in axillary racemes, occasionally clustered at the branch tips and grouped into terminal panicles, axis, pedicels and flowers pink-red, with brown hairs, flower stalk with 2 large bracts, corolla urceolate, 6 mm long, white pubescent inside; fruits black. BMns 254 (= *G. willisiana* C. R. Davie). SE. Brazil, Serra dos Orgaos 1949. z10 # ∅

G. forrestii Diels. Rounded bush, about 1 m high in milder climates, young branches bristly pubescent; leaves lanceolate, 5–9 cm long, bristly serrate, dark green above, lighter beneath with scattered bristly pubescence which leaves brown dots after abscising; rachis, flower stalks and calyx waxy white and very conspicuous, flowers small, urceolate, milk-white, fragrant, in axillary racemes, along the entire length of the previous year's shoots, May–June; fruits pea-sized, rounded to ovate, blue, but not numerous. GC 82 (1927): 285; JRHS 67:

Fig. 69. *Gaultheria humifusa* (from Hooker)

116. China, Yunnan Province. 1908. Very attractive and hardy, but fruits sparsely. z6 Plate 34. # ⚭

G. fragrantissima Wall. Large shrub, also a small tree in its habitat, young branches glabrous and angular; leaves leathery tough, ovate to narrow elliptic, acute, tapering to the base, 3–10 cm long, 2–5 cm wide, dentate, bright green above, abundantly brown punctate beneath (from the fallen bristles), petioles 2–4 mm; flowers in 3–7 cm long, axillary racemes on the previous year's wood, densely arranged, pendulous, with small subtending leaves, corolla ovate-campanulate, 6 mm long, greenish-yellow to white, very fragrant, April; fruits dark blue, occasionally light blue, 8 mm. BM 5984 (= *G. ovalifolia* Wall.). India, in the mountains. 1850. Cultivated in SW. England; otherwise z9. # ☉

G. hirtiflora Benth. Evergreen shrub, differing only slightly from *G. odorata* Willd.; leaves usually triangular-ovate, 3–9 cm long, obtuse or acute, base often deeply cordate, margins finely serrate, finely pubescent or glabrous; flowers in 3 to 6 cm long racemes, corolla pink to red, 6 to 7 cm long; fruit capsule 4–5 mm wide. Mexico to Guatemala. (Cultivated in the Dublin Botanic Garden, Ireland.) z9 Plate 32.

G. hispida R. Br. Shrub, 0.5 m (occasionally also taller), branches, midrib and flower stalks bristly and pubescent; leaves oblong to narrow oval-lanceolate, 2.5–5 cm long, finely dentate, tapering to both ends, with short tips, scattered bristles above, these more noticeable beneath with reticulate venation; flowers in short, compact, terminal racemes, 3–7 cm long, corolla broadly campanulate, 4 mm long, white, calyx fleshy when in fruit; fruits pure white, globose, 1–1.2 cm thick. LAu 361; NF 2: 164; 11: 213: CFTa 109. Australia, Tasmania. 1810. z9 Plate 33. #

G. hookeri Clarke. Shrub, 0.7–1.5 m high, broadly upright, young branches bristly; leaves 5 to 10 cm long, ovate to elliptic or obovate, shallow and bristly dentate, base round, with small tips, smooth above, venation somewhat bristly beneath, leathery tough, reticulate venation, quite short petioled; flowers terminal and axillary, softly pubescent, dense racemes, 2–5 cm long (much like *Convallaria*), corolla ovate-urceolate, pink, 4–6 mm long, March–April; fruits globose, 5 mm thick, violet, pruinose. GC 1945, 118: 93; SNp 92; BM 9174 (= *G. veitchiana* Craib). E. Himalaya, Sikkim. 1907. z6 Plate 34. #

G. humifusa (Graham) Rydb. Small, bushy, creeping shrublet, 7–10 cm high, branches wavy; leaves elliptic to rounded, usually obtuse, 1–2 cm long, bristly serrate, glossy above; flowers solitary, axillary, campanulate, white to pink, 5 mm long, July; fruits red, flat-globose, 6 mm thick (= *G. myrsinites* Hook.). Western N. America, moist areas in wooded foothills. 1830. Hardy, but difficult to cultivate. Plate 34; Fig. 69. #

G. itoana Hayata. Small shrublet, branches glabrous or pubescent; leaves thin and tough, short petioled, oblong, 1–1.5 cm long, 3 to 7 mm wide, acute at both ends, serrate, margins slightly involuted, glabrous, venation indented above, 2–3 veined on both sides, petiole 1 mm long; flowers in 2 cm long, nearly terminal racemes, corolla ovate-tubular, 4–5 mm long, white, with tiny limb lobes, ovaries pubescent; fruits globose, white, 6 mm thick (= *G. merrilliana* Hort.). China; Taiwan, in the mountains at 2250–3600 m, in open areas. 1935. Resembles *G. cuneata*, but lower and more ornamental. z6 Plate 34. # ⚘

G. laxiflora see: **G. yunnanensis**

G. leucocarpa f. *cumingiana* see: **G. cumingiana**

G. merrilliana see: **G. itoana**

G. microphylla see: **G. antarctica**

G. miqueliana Takeda. Shrub, to 30 cm high, young branches pubescent at first, very quickly becoming glabrous; leaves ovate to obovate, 2–4 cm long, with scattered brown glands beneath, otherwise glabrous, glandular serrate; flowers in small racemes near the branch tips in the upper leaf axils, corolla campanulate-urceolate, white, calyx white, fleshy when in fruit, May–June; ovaries glabrous, fruits globose, white to light pink, 1 cm thick. BM 9629; TAP 111 (= *G. pyroloides* Miq. p. p.). Japan. z6 Plate 31, 34. #

The names *G. pyrolifolia* and *G. pyroloides* have been the source of great confusion. In 1863, Miquel named the above described white fruited, Japanese species *G. pyroloides*. But at the same time, he included (quite erroneously) a blue fruited Himalayan species encountered in Kew Gardens, London. The latter was collected by Hooker in the wild in Himalaya but not named. In 1882, C. B. Clarke named this plant *G. pyrolifolia*, a name which remained in use for some years. Takeda cleared up the confusion in 1918 by declaring the blue fruited species with 2 awned anthers *G. pyroloides* Miq. and the white fruited Japanese species, with 4 awned anthers, *G. miqueliana*.

G. myrsinites see: **G. humifusa**

G. nummarioides D. Don. Shrub, growing flat along the ground, branches long, thin, densely foliate, bristly; leaves 2 ranked, rounded, 6–15 mm long, dull green and

rugose above, lighter beneath, bristly, bristly serrate; flowers solitary, hidden under the foliage, corolla urceolate, pendulous, white to pink or also brownish, July–August; fruits blue-black, 8 mm long, ovate. DL 1: 230; GC 135: 15. Himalaya; Khasi Hills. 1850. Occasionally found in cultivation (Savill Gardens, England). z9 Plate 34, 32. #

G. oppositifolia Hook. f. Shrub, broadly upright, about 1 m high (to 2.5 m in its habitat), young branches usually quite glabrous or with single bristles; leaves opposite (!), sessile, ovate, oblong to oval-lanceolate, acute, base cordate, bristly serrate, 2.5–6 cm long, glossy green and glabrous above, underside somewhat bristly and with reticulate venation; flowers numerous, in terminal panicles, 10 cm long and wide, corolla campanulate, white, 4 mm long, May–June; calyx lobes do not become fleshy, therefore the fruit is a dry, 4 mm thick capsule. New Zealand, N. Island. z9 #

G. ovalifolia see: **G. fragrantissima**

G. ovatifolia Gray. Wide spreading or procumbent shrub, 15–30 cm high, branches partly erect, long pubescent; leaves broadly ovate to nearly circular, acute, 1.5–2.5 cm long, base cordate, thin, bristly serrate, very shallow toothed, dark green above, glossy, rugose; flowers solitary in the leaf axils, pedicel 2 mm long, corolla campanulate, 5 mm long, white to pink, calyx fleshy, June; fruits scarlet-red, flat-globose, 6 mm thick. In western N. America forests. 1890. Somewhat difficult to cultivate. Needs a moist location in semishade. z6 Plate 33. #

G. perplexa T. Kirk see: **Pernettya macrostigma**

G. phillyreifolia (Pers.) Sleum. Evergreen shrub, 0.5–2 m high in its habitat, strong branches, new shoots finely pubescent and loose bristly; leaves lanceolate, 1–2 cm long, scabrous toothed, tips prickly, glabrous, short petioled; flowers 3–10 in axillary racemes, corolla urceolate, 3–4 mm long, white, lobes reflexed; fruits red-brown. Argentina, in the mountains. z6 Fig. 70.

Fig. 70. *Gaultheria phillyreifolia* (from Dimitri)

G. procumbens L. Creeping shrublet, stoloniferous, erect shoots to 15 cm high, developing into a dense groundcover in time, glabrous or pubescent; leaves elliptic, clustered at the branch tips, 1–3cm long (usually larger in cultivation and to 5 cm long), margin crenate, glossy dark green above, glabrous beneath; flowers solitary, nodding, occasionally in small racemes, corolla white to light pink, conical to urceolate, June–August; fruits globose, 8–15 mm thick, red, very aromatic when rubbed. BB 2775; NBB 3: 20; GSP 411. Eastern N. America; in forests and clearings on sterile, sandy soil. 1762. A good ground cover for temperate climates. z3 Plate 34, 35. # ⚹

The attractive fruits persist from October to late spring. The leaves were once the source of wintergreen oil, which is produced today synthetically or from the bark of *Betula lenta*. Wintergreen oil is used in the production of candies and chewing gum.

G. pyrolifolia see: **G. pyroloides**

G. pyroloides Miq. emend. Takeda. Ground cover, spreading by stolons, a 10–15 cm high shrublet, very similar to *G. procumbens*, branches slightly pubescent; leaves clustered at the branch tips, obovate to rounded, 1.5–3.5 cm long, dentate, more entire toward the base, glabrous above, scattered pubescent beneath; flowers 2–5 in short, pubescent racemes, corolla ovate, 5 mm long, white-pink, May to June; fruits oval-rounded, glabrous, blue-black, 8 mm thick. SNp 97 (= *G. pyrolifolia* C. B. Clarke). The Himalayas. z6 Plate 34. #

var. *cuneata* see: **G. cuneata**

G. rupestris (Forst.) G. Don. Quite variable in habit, plants in cultivation erect to procumbent, 20–30 cm high, reaching to 1 m high in its habitat; leaves oblong-lanceolate, bristly serrate, 1.5–2.5 cm long; flowers in terminal or axillary racemes at the branch tips, small, corolla campanulate, white, June–July; fruit lacks a fleshy, thickened calyx, dry (!), brown, 4 mm wide capsule. NF 8; 27; PRP 59. New Zealand; only recently introduced into cultivation. (*G. rupestris* in most gardens is actually *G. cuneata*.) z7 #

G. semi-infera Airy-Shaw. Very closely related to *G. tetramera*. Low shrub, to 2 m in its habitat, young branches with red-brown or blackish, appressed bristles, later glabrous; leaves elliptic, 6–7 cm long, about 2 cm wide, acuminate, cuneate at the base, finely crenate, especially on the apical half, glabrous above, loose bristly beneath, eventually more punctate; flowers in axillary racemes, 2.5–4 cm long, corolla white, urceolate, 4 mm long, stamens 5 (very important, otherwise always 10!), May; fruits 3, somewhat pear-shaped, 8 mm long, fleshy, blue (HCC 745/1). BMns 197. Sikkim to China, NW. Yunnan Province. Fruits abundantly! z9? # ⚹

G. shallon Pursh. Salal. Dense, vigorous, stoloniferous shrub, to 1 m high or more in its native habitat, often only to 60 cm in culture, developing into thickets, branches erect, very glandular pubescent; leaves oval-rounded, 5–10 cm long, bristly serrate, both sides glabrous or loose bristly pubescent when young; flowers several in unidirectional, pendulous racemes, corolla broadly urceolate, reddish-white, May–June; fruits 1 cm thick, rounded, black-red, somewhat glandular pubescent. BM 2843; BC 1372; BR 1411. Western N. America, Alaska to California. 1862. Excellent ground cover in moist, shady areas, but difficult to transplant, new growth often lacking for up to 2 years after being moved. z6 Plate 34. # ⚹ ∅ ⌖

'Acutifolia'. Graceful habit; leaves more acuminate.

G. sinensis Anthony. Compact to procumbent habit, 10–15 cm high, young branches with small, appressed, brown bristles; leaves obovate-oblong, 1–1.5 cm long, obtuse; finely bristly serrate, glabrous, only the midrib slightly bristly beneath; flowers solitary in the upper leaf axils, corolla white, campanulate, 6 mm long, 6 mm wide, April–May; fruits usually blue, occasionally also white or pink, fleshy, 1 cm thick. Upper Burma, Yunnan. z9? Plate 33. # ⚹

G. stapfiana Airy Shaw. Stiffer, rounded shrub, 0.6–1(2) m, young branches light green, somewhat angular, loosely covered with appressed to erect bristles, these leaving dark spots after abscising; leaves short petioled, narrow to broad elliptic, obtuse to acute, 6–10 cm long, 2–3 cm wide, cuneate at the base, leathery tough, obtuse and flat serrate, smooth and bright green above, whitish-green beneath with many dark bristles or their dotted scars, venation raised, petiole 4–7 mm long; flowers in 2–4 cm long racemes in the axils of the upper leaves, corolla urceolate, 5 mm long, soft pink, flowers abundantly, May–June; fruits flat globose, 5–7 mm wide, blue-lilac. BMns 651. China, Yunnan Province; Upper Burma, N. Assam, Sikkim (?). Cultivated in England. z9 # ⌖ ⚹

G. tetramera W. W. Sm. Shrub, erect, 30–50 cm high, occasionally more broad to procumbent; leaves broadly elliptic to lanceolate or obovate, 3–7 cm long, bristly serrate, leathery tough, acute to obtuse, cuneate at the base, smooth above, bristly punctate beneath; flowers in axillary racemes, about 2.5 cm long, usually 4 parted, corolla ovate-urceolate, greenish-white, 5 mm long, stamens 8–10 (see *G. semi-infera*), May; fruits fleshy, blue (HCC 043/1), rounded, 6 mm thick. BM 9618; GC 135: 211. Tibet, W. China. z9 # ⚹

G. thymifolia Stapf. Shrub, only 10–15 cm high, branches very thin, nearly filamentous, fine bristly; leaves oblanceolate, short petioles, 0.5–1 cm long, acute, bristly serrate on the apical half, glossy dark green, lighter beneath, glabrous; flowers solitary in the axils of the upper leaves, corolla campanulate, white to reddish, very small, June; fruits fleshy, 1 cm long, pale blue. Burma, 4000 m. Rare in cultivation and not always true to name. Needs full sun! z6 Plate 153. # ⚹

G. trichophylla Royle. Low grower, nearly carpet-like, branches very thin, bristly; leaves narrow-oblong, clustered, 5–10 mm long, acute, bristly serrate, otherwise both sides glabrous; flowers solitary in the leaf axils, small, campanulate, red, pink or white, May; fruits broadly ovate, light blue (like bird eggs), about 1 cm long, 6 mm thick. BM 7635; JRHS 63: 41. W. China,

Himalaya. 1897. Quite hardy; prefers a shaded area. z8b # ⚥

G. veitchiana Craib see: **G. hookeri**

The distinguishing properties of this species have not proven to be consistent, so Airy-Shaw has included it with *G. hookeri*.

G. wardii Marquand & Shaw. Low shrub, 0.7–1 m high, branches widespreading, bristly; leaves oblong to lanceolate, acute, 3–8 cm long, base rounded to nearly cordate, bristly serrate, both sides coarse bristly, especially beneath, venation distinctly indented above; flowers 6–10 in about 2 cm long, axillary racemes, corolla white, urceolate, 5 mm long, May–June; fruits 5 mm thick, blue, pruinose, not fleshy (!). BM 9516; NF 12: 108. SE. Tibet. 1925. z8b # ⚥

G. willisiana see: **G. eriophylla**

G. yunnanensis (Franch.) Rehd. Shrub, about 1 m high, open branched, branches nodding, glabrous; leaves oblong, elliptic to lanceolate, 5–10 cm long, acuminate, base cordate, leathery, finely serrate and with reticulate venation, glabrous to some glandular bristles; flowers 12–20 in terminal and axillary, 5–7 cm long racemes, corolla greenish-white, often with brown markings, broadly campanulate, 4 mm long, May–June; fruits flat globose, 6 mm thick, black, August. DRHS 865; JRHS 65: 96–97; FIO 37 (= *G. laxiflora* Diels). China, Yunnan Province. z6 Plate 34. #

All species prefer a sandy peat soil and a semishady location (except *G. thymifolia*); many species need a moist area, like *G. nummularioides*. The best situation is in an open pine-oak forest. The tender species require winter protection. The fruits are very ornamental.

Lit. Airy-Shaw: Classification of the Asiatic species of *Gaultheria*; in Kew Bulletin 1940, 306–330 ● Besant, J. W.: *Gaultherias* (notes on culture, etc.); in New Flora and Silva 1939, 211–218 ● Burtt, B. L., & A. W. Hill: The genera *Gaultheria* and *Pernettya* in New Zealand, Tasmania and Australia; in Jour. Linn. Soc. Bot. **49**, 611–644, 1935 ● Johnson, A. J.: Notes on Gaultherias; in New Flora and Silva 1930, 161–166 ● Johnson, A. T.: Gaultherias; in Jour. RHS **63**, 201–209, 1938 (with 7 plates; contains primarily notes on cultivation of the various species) ● Sleumer, H.: The genus *Gaultheria* in Malaysia; in Reinwardtia 4, 163–188 (only the tropical species).

GAYLUSSACIA Kunth — ERICACEAE

Deciduous and evergreen shrubs, similar to *Vaccinium*; leaves alternate, entire or serrate; flowers few in axillary racemes; corolla tubular campanulate or urceolate, usually narrowed at the apical end; stamens 10, anthers drawn out into a long tube and dehiscing on the tip with a single slit; ovaries 10 chambered, each chamber within a single seed, style filamentous, with the stamens enclosed within the corolla; calyx persistent, with 5 lobes or teeth; fruit a berrylike drupe with 10 seeds. — About 9 species in N. America and 40 in S. America.

Key to the species described

- Leaves evergreen:
  - *G. brachycera*
- ● Leaves deciduous;
  - + Leaves blue beneath:
    - *G. frondosa*
  - ++ Leaves green beneath;
    - * Corolla urceolate:
      - *G. baccata*
    - ** Corolla campanulate;
      - + Hairs glandular:
        - *G. dumosa*
      - ++ Hairs not glandular:
        - *G. ursina*

Gaylussacia baccata (Wangh.) K. Koch. Deciduous, erect shrub, about 75–100 cm high; leaves elliptic-oblong, 3–5 cm long, light green above, underside (like the young branches) often yellow glandular punctate and glutinous; flowers urceolate, dull red, nodding, in dense, pendulous racemes, May–June; fruits black, 6–8 mm thick, sweet, edible. BB 2780; NBB 3: 23; GSP 415 (*G. resinosa* Torr. & Gray). Eastern N. America; on dry, acid soils, also in swamps. 1772. Beautiful scarlet and scarlet-red fall color. z6 Fig. 71. ∅ ✗ ⚥

G. brachycera (Michx.) Torr. & Gray. Evergreen shrub, broad mat-like growth habit, scarcely over 30 cm high, branches three sided, often slightly winged, pubescent when young; leaves oval, 1.5–2.5 cm long, thick, leathery, nearly like *Buxus*, shallow glandular dentate, margins somewhat involuted, glossy dark green above, lighter beneath, densely arranged along the branches; flowers few in dense, axillary clusters, corolla tubular-urceolate, 5 mm long, white, reddish striped, June; fruits black, flavorless. NBB 3: 23; BB 2782; GSP 415; Eastern North America. In dry woods, on sterile-sandy soil. z6 Plate 42; Fig. 71. #

This species very possibly contains some of the oldest plants in the entire world. It was first discovered around 1796 near Bloomfield, Perry County, New Jersey, USA in a stand that covered an area of 4 hectares (9 acres). This stand is protected as a natural preserve for its unique monoculture. Because of the fact that the plant seldom sets fruit and only rarely produces viable seed, it must be increased by stolons. It can therefore be concluded that every stand of *G. brachycera* has very likely originated from and remains genetically identical to a single plant. American botanists have estimated the age of the older colonies to be over 10,000 years, older than the Giant Sequoias.

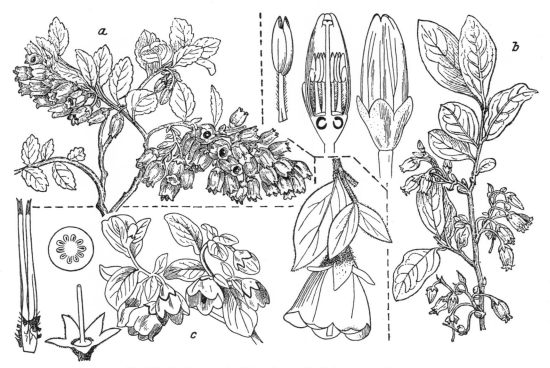

Fig. 71. **Gaylussacia.** a. *G. brachycera;* b. *G. baccata;* c. *G. dumosa*
(from Bot. Mag., Hort Then., Koehne)

G. dumosa (Andr.) Torr. & Gray. Deciduous, creeping shrub, usually not over 30 cm high, branches glandular pubescent; leaves nearly sessile, obovate-oblong, 3–4 cm long, with small tips, glossy above, lighter and glandular pubescent beneath; flowers in short, axillary, loose, pendulous racemes, corolla campanulate, white or reddish, May–June; fruits black, tasteless. BB 2781; NBB 3: 23; GSP 418. Eastern N. America; primarily on wet, but also on dry soils. 1774. z6 Fig. 71.

G. frondosa (L.) Torr. & Gray. Deciduous, to about 1 m high, occasionally taller, shrub, branches white pruinose, outspread, glabrous; leaves elliptic-oblong, 3–6 cm long, bright green above, blue-green and pubescent beneath; flowers broadly campanulate, greenish-red, May–June; fruits globose, 1 cm thick, blue pruinose, sweet, juicy, edible. BB 2779; NBB 3: 23; GSP 418. Eastern N. America; dry to moist woodlands, on sandy, acid soil. 1761. z6 ✂ ⚭

G. resinosa see: **G. baccata**

G. ursina (M. A. Curtis) Torr. & Gray. Deciduous, open shrub, 0.5–1.5 m high, young branches somewhat pubescent; leaves elliptic-oblong to obovate, acute, 3–10 cm long, slightly pubescent on both sides; flowers in short, loose, axillary racemes, corolla campanulate, white to reddish, calyx glutinous, with wide, acute lobes, May–June; fruits globose, black, sweet, to 12 mm thick.

USA, N. Carolina and N. Georgia, in dense mountain forests. 1891. z6 ✂ ⚭

Prefer soil similar to that for *Vaccinium;* acid, well drained, sandy-humus. Rarely found in cultivation outside of botanic gardens.

Lit. Camp, W. H.: The genus *Gaylussacia* in North America north of Mexico; in Bull. Torrey Bot. Club 62, 129–132, 1934 ● Camp, W. H.: A review of the North American Gaylussacieae; in Bull. Torrey Bot. Club **68,** 531–551, 1941.

GELSEMIUM Juss. — LOGANIACEAE

Evergreen, twining, glabrous shrubs; leaves opposite, simple, entire; stipules small; flowers attractive, solitary or several, axillary, yellow, fragrant; calyx short, deeply 5 parted; corolla protruding far beyond the calyx, funnelform, limb lobes rounded; stamens 5, attached at the base of the corolla, anthers oblong sagittate, inclined inward; styles thin, stigma with 4 incisions; fruit an elliptic, flat capsule, with 2 chambers; seeds small, winged. — One species each in N. America and SE. Asia.

Gelsemium sempervirens Ait. Shrub, high twining, shoots to 5 m long; leaves oblong to elliptic-lanceolate, 4–7 cm long, entire, glabrous on both sides, bright green above, glossy, somewhat lighter beneath, petioles short; flowers solitary or grouped axillary, stalk with many small scales, corolla 2.5 to 4 cm long, glossy, yellow, very fragrant, April–June; styles variable. RFW 295; MCL 109; BM 7851; HV 12. S. USA, in lowlands and valleys. 1840 z7–9 Fig 72. # ⊕

An attractive vine for milder regions; prefers a fertile, clay-humus soil.

Fig. 72. *Gelsemium sempervirens*. Branch, fruit, sectioned corolla and stigma (from Poiret and Solereder)

GENISTA L. — Broom — LEGUMINOSAE

Deciduous, thorny or thornless shrubs, occasionally subshrubs; leaves absent or simple, occasionally also trifoliate; flowers yellow, sometimes white (*G. monosperma*), either racemose, fascicled or capitate at the branch tips or solitary and axillary; calyx 2 lobed, upper lobe deeply split, wings and keel petals adnate to the stamen tube at their base, standard ovate; pods linear-oblong; seeds without hilum bulge.

Outline of the Sections
(partly from "Flora Europaea")

Section **Genista**
    Thornless shrubs with simple leaves; corolla glabrous, also usually the calyx and leaves, standard broad ovate, as long as the wings and the keel; pods narrow oblong, with 3–10 seeds:
        G. januensis, lydia, ovata, tinctoria

Section **Spartoides** Spach.
    Thornless shrubs with simple leaves; standard broad ovate, as long as the wings and the keel, usually silky pubescent, keel and pods silky also; pods narrow oblong with semi-erect hairs, with 2 or more seeds:
        G. cinerea, pilosa, sericea, tenera, villarsii

Section **Erinacoides** Spach.
    Branches thorny; leaves usually simple; standard broad ovate, as long as the keel, standard and keel usually silky pubescent; pods narrow oblong, silky, with 1 to many seeds:
        G. aspalathoides, lobelii, pumila

Section **Scorpioides** Spach.
    Shrubs with axillary thorns and alternate branching; leaves simple or trifoliate; corolla usually glabrous, standard as long as the wings and the keel; pods oblong, with 2–8 seeds:
        G. scorpius

Section **Phyllospartium** Willk.
    Shrubs with axillary thorns and simple leaves; standard usually glabrous, ovate and acute, usually shorter than the keel; pods sickle-shaped, inflated, with 4–12 seeds:
        G. anglica, berberidea, falcata

Section **Voglera** (P. Gaertner, B. Mayer & Scherb.) Spach.
    Leaves simple or trifoliate; standard usually triangular or ovate with tips, usually shorter than the keel; pods acute-ovate, with 1–2 seeds:
        G. germanica, hirsuta, hispanica, lucida, silvestris

Section **Asterospartum** Spach.
    Thornless shrubs, leaves simple or trifoliate, branches and leaves usually opposite or nearly so; standard broad or angular ovate, as long or shorter than the keel; pods long-ovate or sickle-shaped, with 1–2 seeds:
        G. aetnensis, ephedroides, nyssana, radiata, sessilifolia

Section **Acanthospartum** Spach.
    Shrubs with opposite branches, branches terminating in a thorny tip; leaves trifoliate, opposite or alternate, standard diamond-shaped, shorter or longer than the keel; pods ovate, long acuminate, with 1–2 seeds:
        G. acanthoclada

Section **Echinospartum** Spach.
    Small shrubs with opposite, thorny branches; leaves trifoliate, short petioled to sessile; calyx inflated, campanulate, the 5 teeth longer than the tube; pods ovate and long acuminate, with 1–3 seeds:
        G. horrida

Section **Genistella** Spach.

Thornless, dwarf shrubs with distinctly winged and flattened branches; leaves simple or absent; standard broadly ovate, as long as the wings and the keel, wings narrow oblong; 2–5 seeds, with or without hilum bulge:

G. delphinensis, sagittalis

Section **Retama** (Boiss.) Schneid.

Thornless shrubs, leaves simple, abscising very quickly; flowers in racemes, calyx 2 lobed, corolla white or yellow; pods ovate to globose, with 1(2) seeds:

G. monosperma, sphaerocarpa

Parentage of the Cultivars

'Goldilocks'	—*pilosa*
'Golden Showers'	—*tenera*
'Royal Gold'	—*tinctoria*

(Occasionally *Cytisus* cultivars will be listed erroneously as *Genista*, for these cultivars please refer to the former.)

Genista acanthoclada DC. Low, erect shrub, 40–50 cm, branches opposite and terminating in a thorny tip, older branches with distinct leaf scars; leaves trifoliate, opposite or alternate, leaflets narrow-oblanceolate, 5–10 mm long, 1–3 mm wide; flowers solitary in the axils of the subtending leaves, nearly opposite, near the branch tips, calyx 2 to 5 mm, loose silky, standard rectangular, yellow, 6–10 mm, silky, shorter than the keel, pods long ovate, with 1–2 seeds. PMe 60. Greece and the Aegean region. z9 ✧

G. aetnensis (Bivona) DC. Tall shrub, branches reed-like, thin, nodding when young; leaves usually absent, simple, linear, 12 mm long, silky pubescent; flowers golden-yellow, in loose, terminal, foliate racemes, calyx green, campanulate, corolla somewhat silky pubescent, July; pods drawn out to a scabrous, curved tip, with 2–3 seeds. BM 2674; NF 9: 200; GC 138: 112; BS 2: 40. Sardinia, Sicily. One of the most popular and attractive species in England. z8 Fig. 73. ✧

G. anglica L. shrub, to 80 cm high, but usually lower or procumbent, branches glabrous, thorny; leaves alternate, simple, elliptic to linear-oblong, 6–8 mm long, acute, blue-green; flowers yellow, in terminal, short, few flowered, foliate racemes, calyx glabrous, standard ovate, glabrous, June to July; pods inflated, 12 mm long. HM 1347; HF 2303 (= *G. minor* Lam.). Central and W. Europe on moist peat bogs and in forests, especially in England. z6 Fig. 73, 75.

G. anxantica see: **G. tinctoria** var. **anxantica**

G. aspalathoides Lam. Shrub, densely bushy, 10–30 cm high, stiff, branches thorny, knotty, silky pubescent when young, striped; leaves alternate, usually trifoliate, linear, tough, longitudinally plaited, finely pubescent, 6 mm long, abscising early; flowers 1–4 at the nodes of the previous year's branches, light yellow, calyx pubescent. SW. Europe, Sicily, N. Africa. 1786. z8

G. berberidea Lange. Very closely related to *G. falcata* and *G. anglica*, but the young branches with dense, erect pubescence, the calyx also, flower stalks with 1 mm long prophylls, stipules somewhat thorny; flowers yellow, standard 8 to 10 mm long, calyx 5–6 mm, lobes twice as

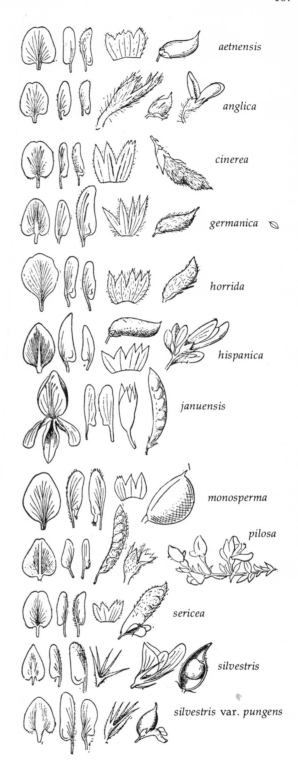

Fig. 73. *Genista*. Flower parts and fruits (from Schneider, altered)

aetnensis
anglica
cinerea
germanica
horrida
hispanica
januensis
monosperma
pilosa
sericea
silvestris
silvestris var. *pungens*

long as the calyx tube; pods with erect pubescence along the seam. NW. Spain, N. Portugal, moist meadows and swamps. z9

G. canariensis see: **Cytisus canariensis**

G. candicans see: **Cytisus monspessulanus**

G. cinerea (Vill.) DC. Dense, gray, thornless, erect shrub, to 2.5 m high, branches outspread, often somewhat nodding, silky pubescent when young; leaves alternate or fascicled, simple sessile, longitudinally plaited, oblong to lanceolate, 1.2 cm long, silky pubescent; flowers in loose, terminal racemes, yellow, standard emarginate, calyx pubescent, June–July; pods linear, 1.5 cm long. BM 8086; PMe 57. Italy, Spain, N. Africa. Flowers abundantly, one of the most beautiful species. z9 Fig. 73. ✿

G. dalmatica see: **G. silvestris**

G. delphinensis Vill. Like a form of *G. sagittalis,* only 5–8 cm high, procumbent, branches curved and wavy, winged, the wings silky beneath; the few leaves oval-elliptic, 6 mm long, silky; flowers terminal and axillary on the tips of the young branches, grouped 1–3 or in fascicles, golden-yellow, calyx silky, June–August; pods 12–18 mm long, abruptly acuminate, with 3–6 seeds. JRHS 67: 29; JRHS 67: 69; NF 1936: 55 (= *G. sagittalis* var. *delphinensis* Vill.; *G. sagittalis* var. *minor* DC.). SE. France, E. Pyrenees, on gravelly, chalk soil. z6 ✿

G. elatior see: **G. tinctoria** var. **virgata**

G. ephedroides DC. Shrub, erect, 0.3–1 m high, abundantly branched, branches somewhat pendulous, thin, striped, pubescent; leaves few, alternate, trifoliate at the base of the shoots, simple toward the tips, leaflets linear, small, pubescent; flowers rather large, very numerous, in loose, terminal racemes, light to dark yellow, very fragrant, calyx, standard and keel petals pubescent, May–June; pods elliptic, acute, pubescent. BM 2674. Sardinia, Corsica, Sicily. 1832. z9 ✿

G. falcata Brot. Very similar to *G. anglica,* nearly *Ulex*-like in appearance, young branches very stiff, furrowed, with loose silky pubescence, thorns 1.5–5 cm long, usually with 3 tips or branched, pubescent; leaves ovate to narrow obovate, 4–12 mm long, 2–6 mm wide, ciliate, silky beneath; flowers yellow, 8–10 mm long, solitary or several in the axils of the thorns, but flowers very abundantly, occasionally in 30 cm long, very narrow, dense panicles, bracts and prophylls tiny or totally absent, late April–early May; pods sickle-shaped, inflated, 12 mm long, glabrous. W. Spain, Portugal, on acid soil. z9 ✿

G. fragrans see: **Cytisus canariensis**

G. germanica L. German Broom. Shrub, erect, 30–50 cm high, very similar to *G. anglica,* but with branches shaggy pubescent, older branches covered with compound thorns, short branches thornless; leaves simple, elliptic-oblong, 1–2 cm long, short petioles, pubescent, dark green; flowers in small, 2–5 cm long terminal racemes on the young shoots, corolla yellow, pubescent, standard

Fig. 74. *Genista germanica* (from Lauche)

shorter than the keel, calyx pubescent, June–July; pods oblong, pubescent, 1 cm long. HF 3202; HM Pl. 158; 1349; PSw 17 (= *G. villosa* Lam.). Central and W. Europe, occasionally among heather in sunny dry areas. z6 Fig. 73, 74, 75.

G. grandiflora see: **Cytisus grandiflorus**

G. humilis see: **G. tinctoria** var. **humilis**

G. hirsuta Vahl. Erect shrub, branches with axillary, thick, usually unbranched thorns; leaves 6–15 mm long, 3–5 mm wide, simple, lanceolate, sparsely covered with long, erect hairs beneath and on the margins, glabrous above; flowers in dense, terminal racemes, yellow, bracts leaflike, immediately under the prophylls, the latter 3–5 mm long, calyx 9–12 mm long, lower lip longer than the upper, standard acute-ovate, glabrous or silky pubescent. PMe 59; PSw 17. W. Spain, S. Portugal, Balearic Islands. z7 ✿

G. hispanica L. Spanish Broom. Densely branched shrub, 30–70 cm high, rounded, branches ascending, thorny, dense and long pubescent; leaves simple, ovate-oblong, 1 cm long, obtuse, occurring only on the flowering branches; flowers 2–12 in terminal, erect heads, corolla golden-yellow, banner glabrous, keel pubescent, likewise the calyx, June–July; pods acute-oblong, more or less pubescent. BM 8528; BS 2: 273; PEu 52. Spain to N. Italy. 1759. One of the most beautiful species for the rock garden, needs sandy, deep, dry soil in full sun. Winter protection advisable. z7–8 Plate 36; Fig. 73, 77. ✿

G. horrida (Vahl) DC. Cushion-like habit, very dense and stiff shrub, 30(60) cm high, young branches glabrous, later stiff and prickly; leaves trifoliate, opposite, leaflets linear, 8 mm long, plaited, pubescent; flowers 1–3 in terminal heads, yellow, standard glabrous, keel silky pubescent, calyx campanulate,

Plate 17

Euonymus europaeus, fruits
Archivbild

Euphorbia dendroides
in its native habitat on Mallorca, Spain

Euonymus fortunei 'Emerald Leader'
Photo: Corliss Bros., USA

Euonymus japonicus 'Président Gauthier'
(young plant) in a nursery in Orléans, France

Plate 18

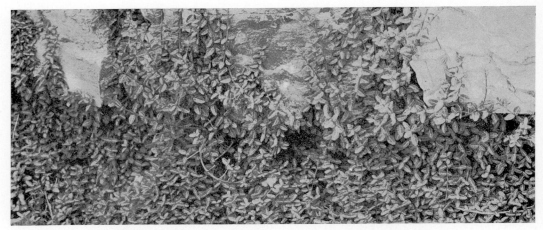

Euonymus fortunei 'Minimus' in the L. Späth Nursery, Berlin, W. Germany

Euonymus lucidus in the Bell Park, Fota Island, Ireland

Plate 19

Euonymus a. *E. bungeanus;* b. *E. bungeanus* var. *semipersistens;* c. *E. nanus* var. *turkestanicus;* d. *E. nanoides;* c. *E. verrucosus;* f. *E. semenovii;* g. *E. sanguineus;* h. *E. obovatus;* i. *E. alatus;* k. *E. kiautschovicus;* l. *E. fortunei* 'Vegetus'; m. *E. fortunei* 'Coloratus'; n. *E. wilsonii;* o. *E. phellomanus;* p. *E. europaeus* (all leaves collected from wild plants)

Plate 20

Exochorda giraldii var. *wilsonii*
in the Munich Botanic Garden, W. Germany

Exochorda macrantha
in the Edinburgh Botanic Garden, Scotland

Euphorbia mellifera in Malahide Castle Gardens, Ireland

Euphorbia wulfenii in Malahide, Ireland

Plate 21

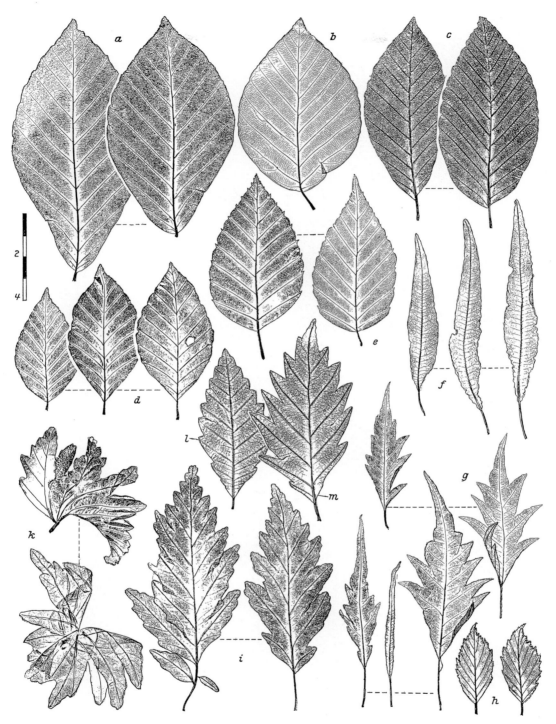

Fagus. a. *F. orientalis;* b. *F. japonica;* c. *F. engleriana;* d. *F. crenata;* e. *F. lucida;* —Forms of *Fagus sylvatica:* f. 'Ansorgei'; g. 'Asplenifolia'; h. 'Cochleata'; i. 'Rohanii'; k. 'Cristata'; l. 'Quercifolia'; m. 'Laciniata' (a–e mostly collected from wild specimens)

Plate 22

Fagus crenata, branch with fruits
Photo: Dr. Watari, Tokyo

Fagus silvatica 'Pendula' in the Dortmund Botanic Garden, W. Germany

Plate 23

Forestiera neomexicana
in the National Arboretum, Les Barres,
France

Fagus engleriana with typically branched stem,
in a garden in Holland

Fatsia japonica 'Moseri', an old plant in the park of Villa Soldati, Luganop, Switzerland

Plate 24

Ficus pumila, mature form,
in the Jardin Exotique, Monte Carlo, France

Ficus religiosa
in the Jardin Ultramar, Lisbon, Portugal

Ficus microcarpa on Ibiza Island, Balearics, Spain

Ficus lyrata as a 15 m street tree in Malaga, Spain

Plate 25

Forsythia intermedia 'Densiflora'

Forsythia intermedia 'Spectabilis'

Forsythia intermedia 'Primulina'

Forsythia intermedia 'Spring Glory'

Forsythia suspensa var. *fortunei*

Forsythia suspensa var. *sieboldii*

All photos: Dieterich, Plön

Plate 26

Fothergilla major
Archivbild

Gaulnettya wisleyensis
in the Glasnevin Botanic Garden, Dublin, Ireland

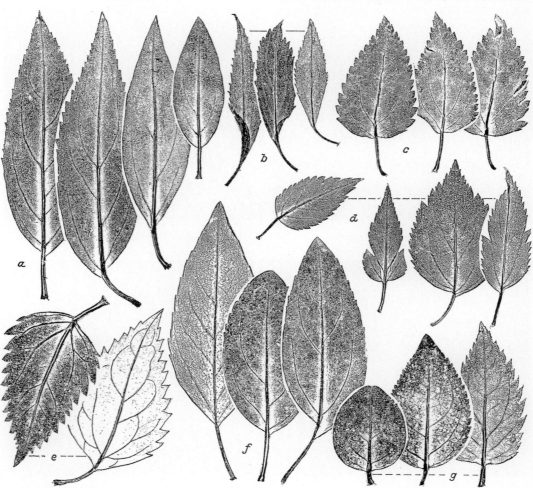

Forsythia. Leaves a. *F. intermedia* 'Spectabilis'; b. *F. viridissima* 'Bronxensis'; c. *F. intermedia* 'Mertensiana'; d. *F.* 'Arnold Dwarf'; e. *F.* 'Beatrix Farrand'; f. *F. intermedia* 'Lynwood'; g. *F.* 'Arnold Giant'

Plate 27

Firmiana plantanifolia
in the Coimbra Botanic Garden, Portugal

Freylinia lanceolata
in the Suchumi Botanic Garden, USSR

Fabiana imbricata
in the Edinburgh Botanic Garden, Scotland

Eurya emarginata
in the Kyoto Botanic Garden, Japan

Plate 28

Forsythia europaea in its native habitat
at Péc, near the Albanian border

Forsythia giraldiana
in the Dortmund Botanic Garden, W. Germany

Fothergilla gardenii
in the Dortmund Botanic Garden, W. Germany

Fothergilla major
in Royal Botanic Gardens, Kew, England

Plate 29

Fraxinus excelsior 'Pendula' in the Dortmund Botanic Garden, W. Germany

Fraxinus ornus, flowering branch
Archivbild

Plate 30

Fraxinus paxiana
Photo: Berndt

Fraxinus dipetala
in Royal Botanic Gardens, Kew, England

Fraxinus excelsior 'Concavifolia'
in the Chenault Garden, Orléans, France

Franklinia alatamaha in spring,
Arnold Arboretum, USA

Plate 31

Fraxinus angustifolia and *F. pallisae* in its native habitat
on the Ropotamo River in Eastern Bulgaria
Photo: K. Browicz

Gaultheria miqueliana with white fruits, in the Dortlund Botanic Garden, W. Germany

Plate 32

Garrya elliptica 'James Roof'
Photo: Maunsell van Rensselaer

Gaultheria antipoda
at Malahide, Ireland

Gaultheria hirtiflora
at Malahide, Ireland

Gaultheria nummarioides
in the Caerhays Park, S. England

yellow pubescent, July–September; pod oblong, 2.5 cm long, silky tomentose. GC 53: 140. S. France, Spain. 1821. Only for very dry, sunny areas. z9 Fig. 73. ⊕

G. januensis Viv. Procumbent shrub, to 30 cm high, branches erect or ascending, 3 sided, somewhat winged, the wings membranous and translucent; leaves simple, usually clustered at the base of the flowering shoots, oblong-lanceolate, 12–25 mm long, lightly ciliate margin, but otherwise glabrous; flowers in foliate racemes on the tips of erect short shoots, corolla bright yellow, glabrous like the calyx, as are the pods, 2.5 cm long, May–June. HM 1340; BM 9547; NF 5: 120; JRHS 67: 31 (= *G. scariosa* Viv.). SE. Europe. 1850. z8b Fig. 73, 75. ⊕

G. lobelii DC. Dwarf or prostrate shrub, older plants 15–25 cm high, young branches pubescent and furrowed, branch tips later becoming thorned; leaves alternate, elliptic to obovate, 2–5 mm long, 1–2 mm wide, silky beneath, abscising quickly without thorny stipules; flowers usually 1–2 per branch, light yellow, 1 cm long, calyx pubescent, lobes narrow-triangular, flower stalks thin, 4–9 mm long, June; fruits 1.5 cm long, acute, pubescent. S. Spain, SE. France. Likes alkaline soils. z9 ⊕

G. lucida Camb. Erect shrub, with dense axillary thorns; leaves narrow-elliptic, 3–8 mm long, 3 mm wide, simple, silky pubescent beneath, stipules thorny; flowers in terminal racemes, yellow, bracts leaflike, prophylls tiny, in the middle of the flower stalk, calyx 5 mm, silky, standard ovate, 10 mm, somewhat acute, glabrous, shorter than the keel. Mallorca. z9 ⊕

G. lydia Boiss. Shrub, to about 50 cm high, prostrate, branches glabrous, gray-green, ascending, thornless, but with somewhat prickly tips; leaves simple, linear-elliptic, acute, about 1 cm long, glabrous; flowers few on short, thorn tipped side shoots, blooms very richly, corolla gold-yellow, glabrous, calyx tubular, glabrous, May–June; pods flat, about 2.5 cm long, glabrous. BMns 292; BS 2: 42 (= *G. spathulata* Spach.). Balkan Peninsula, Syria. Very attractive, flowers abundantly, good winter hardy species. z5–6 ⊕

G. mantica see: **G. tinctoria** var. **humilior**

G. minor see: **G. anglica**

G. monosperma Lam. Erect shrub, 0.5–1 m high, with bowed, quadrangular branches, gracefully nodding, with fine gray appressed pubescence, only a few simple leaves, these linear-lanceolate to obovate, 6–15 mm long, abscising early; flowers white (!), few in axillary racemes, very fragrant, February–March; pods ovate, somewhat rugose, brownish-yellow, with 1–2 seeds. BM 683; BR 1918; Gw 15: 412 (= *Retama monosperma* Boiss.). S. Spain, Portugal, NW. Africa. 1690. Ⓕ SW. Spain; Morocco, for stabilizing dunes. z9 Fig. 73.

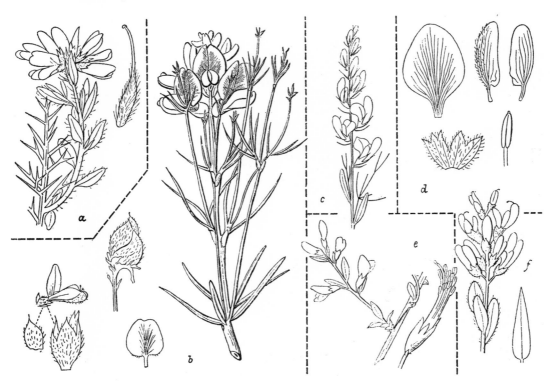

Fig. 75. **Genista**. a. *G. anglica*; b. *G. radiata*; c. *G. silvestris*; d. *G. villarsii*; e. *G. januensis*; f. *G. germanica*

G. nyssana Petrovic (also spelled *nissana*). Erect shrub, to 60 cm high, branches slender, sparsely branched, foliate, all parts with erect silky, shaggy pubescence (branches, leaves, flowers and pods); leaves trifoliate, leaflets linear, acute, 2–2.5 cm long; flowers in terminal, foliate racemes, 15 cm long, June–July; pods broad-oblong, about 1 cm long, acute. Serbia, Albania, Macedonia. 1889. Named for Nissa, a village in Serbia. Very attractive species. z6 ⌾

G. ovata Waldst. & Kit. Erect shrub, 20 to 60 cm high, very similar to *G. tinctoria*, but with erect shaggy pubescence; leaves oblong-lanceolate to ovate, 2–3 cm long, shaggy pubescent; flowers numerous, in 3–6 cm long racemes, June–July; pods linear, slightly bowed, with erect and densely rough pubescence. HF 2301 (= *G. tinctoria* var. *ovata* Schultz). S. France to Romania. 1819. Usually not correctly labeled in cultivation; often confused with *G. tinctoria* or one of its forms, but easily distinguished by the much wider leaves and the conspicuous pubescence.

G. pilosa L. Procumbent habit, rooting along the branches, branches ascending, angular, quite short, plants 10–30 cm high, seldom higher, branches occasionally growing erratically through one another, finely pubescent; leaves all simple, solitary at the branch tips, more clustered on the base, oblong, 6–15 mm long, deep green above, glabrous, pubescent beneath, often longitudinally plaited; flowers in terminal racemes on short side branches, composed of 1–3 axillary flowers, corolla rather small, golden-yellow, standard and keel fine silky pubescent on the exterior, May–July; pods to 2.5 cm long, linear-oblong, pubescent. HM 1344; HF 2296; GPN 496. Europe, from S. Sweden to the Mediterranean; in dry sunny areas. 1790. z6 Fig. 73. ⌾

'Goldilocks' (W. J. van der Laan 1970). Strong grower, broad and bushy, 40–60 cm high, but much wider, flat globose; flowers very numerous, golden-yellow, May–June. Selected in Boskoop, Holland. ⌾

G. pulchella see: **G. villarsii**

G. pumila (Debeaux & Reverchon ex Hervier) Vierh. Very similar to *G. lobelii*, but the branches thick and stout, the flowers usually in short racemes, stalks 3–4 mm long. PSw 17. S. and SE. Spain, in the mountains on chalk soil. z9 ⌾

G. radiata (L.) Scop. Erect shrub, to about 80 cm high, but much wider in time, long-lived (old plants can cover 3–4 m²), branches opposite, pubescent to glabrous, thornless; leaves trifoliate, linear-lanceolate, 1 cm long, glabrous above, pubescent beneath, abscising quickly; flowers 3–20 in terminal heads, yellow, May–June; pods ovate, curved, 0.5 cm long, with 1–2 seeds. HM Pl. 159; HF 2318; BM 2260. SE. Europe, Central Asia, SE. France. Around 1750. z6 Fig. 75, 76, 77. ⌾

G. sagittalis L. Shrub, procumbent, branches erect, evergreen, with 2 broad wings, 10–15 cm high, with only a few, simple, lanceolate leaves; flowers golden-yellow, numerous, in short racemes, May. HM Pl. 159; HF 2320; BMns 332; PSw 18 (= *Genistella sagittalis* [L.] Gams). S.

Fig. 76. *Genista radiata* (from Guimpel)

and Central Europe, Balkan Peninsula. 1588. Easily distinguished from all the other species. Very attractive for sterile sandy soil, in full sun. z6 Plate 36; Fig. 77. ⌾

var. *delphinensis* see: **G. delphinensis**

var. *minor* see: **G. delphinensis**

G. scariosa see: **G. januensis**

G. scorpius (L.) DC. Erect, occasionally wide shrub, erratically branched, shoots gray, angular, with axillary, thick, simple or branched thorns; leaves linear, 3–11 mm long, 2 mm wide, somewhat pubescent beneath, dehiscing as early as June; flowers in axillary clusters on the apical portion of the branches and on the thorns, yellow, small, standard 7–12 mm long, glabrous, calyx 3–5 mm long, glabrous, June; pods 2–4 cm long, with 3–7 seeds. PSw 17. Spain; S. France; in dry areas. z9 ⌾

G. sericea Wulf. Bushy habit, scarcely over 30 cm high, usually only half as high, outspread, stoutly branched, branches ascending, appressed pubescent; leaves simple, lanceolate, 1–2 cm long, bright green above, margin and underside silky pubescent, whitish; flowers 2–5 in racemes, golden-yellow, banner and keel distinctly pubescent, May–June; pods 1 cm long, silky pubescent. HF 2297; HM 1344. Southern Alps, on chalk soils. Sensitive, especially to wetness, but very attractive. z8–9 Plate 36; Fig. 73. ⌾

G. sessilifolia DC. Erect shrub with few, nearly opposite, bowed branches arising from the base; leaves trifoliate, sessile, leaflets linear, 5–25 mm long, 1 to 2 mm wide, often involuted, glabrous above, silky beneath; flowers in long, loose racemes, alternate to nearly opposite, yellow, bracts leaflike, trifoliate, calyx 4–5 mm, silky, standard triangular, 7–10 mm, yellow. Yugoslavia, Bulgaria, on chalk. z7–8 ⌾

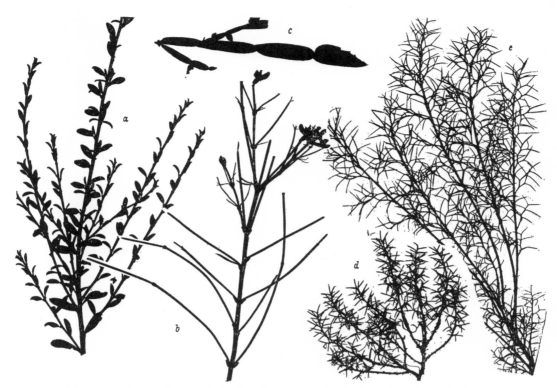

Fig. 77. **Genista.** a. *G. tinctoria;* b. *G. radiata;* c. *G. sagittalis;* d. *G. silvestris;* e. *G. hispanica*
(Original)

G. silvestris Scop. To 20 m high, densely branched (but more open than the varietal form), erect, branches bowed, ascending, more or less thorny; leaves simple, usually linear-lanceolate, quickly becoming glabrous, rigid, 6–12 mm long; flowers in narrow, loose racemes, bright to golden-yellow, calyx silky pubescent, May to July; pods short, with a short upward curving beak, single seeded. HM 1350; HF 2304 (= *G. dalmatica* Bartl.). Adriatic Coast; on dry flatlands and mountain meadows. 1818. z7 Fig. 73, 74, 77. ⊕

var. **pungens** (Visiani) Rehd. Like the species, but more densely branched, with stiff and erect, rough pubescence, thorns very stout; leaflets very narrow; tips of the standard and keel downy pubescent. BM 8075. Dalmatia, Bosnia, NW. Balkan Peninsula. Plate 36; Fig. 73. ⊕

G. sphaerocarpa (L.) Lam. Erect, very densely branched shrub, to 2 m, branches erect or ascending, thin, striped, pale green, usually glabrous; leaves simple, only on young plants, abscising quickly, linear-lanceolate, silky; flowers yellow, only 3–4 mm long, but exceptionally abundant and densely arranged, 1.5–3 cm long, axillary racemes, standard circular, glabrous, May–June; pods globose, 9 mm wide, single seeded (= *Lygos sphaerocarpa* [L.] Heywood; *Retama sphaerocarpa* [L.] Boiss.). E. Portugal, Spain, N. Africa; in dry areas, especially on sandy soil. 1780. z9 Plate 42. ⊕

G. spathulata see: **G. lydia**

G. tenera (Jacq.) O. Kuntze. Deciduous shrub, in its habitat to 2 m high and wide, young branches striped; leaves simple, gray-green, sessile or nearly so, 12 mm long, 3 mm wide, margin somewhat revoluted, silky beneath; flowers 12 mm long, in 3–5 cm long, terminal racemes, flowers abundantly, standard round, 12 mm wide, golden-yellow, calyx silky, June–July, flowering intermittently through October; pods 2.5 cm long, densely silky, with 3–5 seeds. BM 2265 (= *Cytisus tenera* Jacq.; *G. cinerea* Hort. non DC.; *G. virgata* [Ait.] Link. non Lam.). Madeira, Teneriffe; much planted in England. 1777. Very closely related to *G. cinerea.* z9 ⊕

'Golden Shower' (Hillier, after 1920). Selection with especially abundant flowers, golden-yellow. Long in the trade but only recently named. ⊕

G. tinctoria L. Erect, 1 m high shrub, branches furrowed, pubescent at first, soon totally glabrous; leaves simple, elliptic, 1–2.5 cm long, nearly glabrous, ciliate, bright green; flowers many, in branched, 6 cm long racemes, often arranged in long panicles, golden-yellow; pods 2 cm long, with 6–10 seeds, glabrous, June–August. HF 2299; HM Pl. 159. Europe to Asia Minor. z6 Fig. 77. ⊕

Includes a large number of forms and cultivars.

var. **alpestris** Bertol. Procumbent habit, branches woody, thin, only 10–20 cm high; leaves narrower (= *G. humilis* Ten.). S. Tirol.

f. **angustata** (Schur.) Rehd. Leaves elliptic narrow-oblong, 1–2 cm long, 2–3.5 mm wide; flowers in somewhat short racemes.

var. **anxantica** (Ten.) Fiori. Dwarf form, branches procumbent, thin, totally glabrous; flowers larger than those of the species. HF 2298 (= *G. anxantica* Ten.). Found near Naples, Italy. 1818. ⊕

var. **hirsuta** DC. More strongly branched than the species, young branches and leaves pubescent; ovary pubescent; pods normally glabrous (= var. *pubescens* Lang).

var. **humilior** (Willd.) Koch. Similar to var. *alpestris,* also procumbent, to 50 cm high, branches more distinctly pubescent, often nearly bristly; pods silky pubescent, flowering earlier, branches with only 1 raceme (= *G. mantica* Polloni). Tessin, N. Italy.

f. **latifolia** (DC.) Rehd. Leaves broad-elliptic, 3 to 3.5 cm long, dark green.

'Royal Gold' (J. van der Smit 1960). Low growing selection, only 50–80 cm high; leaves like the species; flowers golden-yellow, very numerous, July. ⊕

'Plena'. Procumbent grower; flowers double, more orange-yellow. Before 1835. Attractive in the rock garden. Plate 36. ⊕

var. *pubescens* see: var. **hirsuta**

var. **virgata** (Willd.) Koch. More vigorous growing, more densely branched, totally glabrous, to 2 m high; otherwise like the species; pods with 3–6 seeds. BS 2: 66 (= *G. virgata* Willd.; *G. elatior* Koch).

G. villarsii Clementi. Procumbent, dense shrub, erratic branching, branches rooting, ascending or outspread at first and very pubescent; leaves simple, nearly sessile, usually clustered at the branch base, the apical leaves alternate, linear-lanceolate, 5–7 mm long; flowers in small, terminal racemes, often with only one flower, corolla yellow, silky pubescent, June, calyx and the 12 mm long, flat pods very pubescent. JRHS 67: 27 (= *G. pulchella* Vis.). Dalmatia, Istria (Yugoslavia), S. France. Very abundantly flowering species, needs a dry sunny area. z6 Plate 36; Fig. 75.⊕

G. villosa see: **G. germanica**

G. virgata see: **G. tenera** and **G. tinctoria** var. **virgata**

The low growing species are fine rock garden plants; all species perform best in a light, gravelly, sterile soil. Development and flowering is poorer in a fertile soil. Dryness in winter is advisable, especially for the more sensitive species.

Lit. Like that of *Cytisus* ● Also: Gibbs, P. E.: *Genista;* in Flora Europaea **2**, 94–100, 1968 ● Gibbs, P. E.: A revision of the genus *Genista;* Notes Roy. Bot. Gard. Edinb. **27**, 11–99, 1966 ● Hegi: Flora von Mitteleuropa IV, **3**, 1198–1212.

GENISTELLA

Genistella sagittalis see: **Genista sagittalis**

GESNOUINIA Gaudich. — URTICACEAE

Evergreen shrub or small tree; leaves alternate, oblong-elliptic, long acuminate, base round to cuneate, 10–12 cm long, 3.5–6 cm wide, 3 veined, pubescent, without stipules; flowers monoecious, not very conspicuous, in slender axillary and terminal racemes on the branch tips, male flowers with a 4 part calyx, with 4 stamens, female flowers with a 6 part calyx and enclosed ovary, style short, fruit a small nutlet, dry, surrounded by the calyx. — Only 1 species.

Gesnouinia arborea (L. f.) Gaudich. Tree-like shrub with the features of the genus, 4–6 m high, densely branched from the base upward, stem to 20 cm thick; flowers pink-red, star-shaped, racemes 5–10 cm long, nodding, February–May. KGC 9; BFCa 4 (= *Urtica arborea* L. f.). Canary Islands; Teneriffe, 650 m. Seldom seen in culture outside of the Canaries. z10 # ⊕

GEVUINA Molina — PROTEACEAE

Evergreen trees, leaves alternate, pinnately compound; flowers bisexual, in racemes, sepals 4, abscising, stamens 4, petals very narrow and reflexed, ovaries nearly sessile, fruit a nut (= *Guevina* Juss.). — 3 species in New Guinea, Australia and Chile.

Gevuina avellana Molina. Chilean Hazelnut. Small evergreen tree or shrub, to 12 m high in its habitat, young branches dense brown tomentose; leaves simple or bipinnate, 15 to 40 cm long, 10–25 cm wide, with 1–15 pinnae, each with 1–5 leaflets of varying size, largest on young plants, then 16 cm long and 7 cm wide, the smallest only 3 cm long, ovate, acute, scabrous dentate, leathery, glossy bright green, petioles and rachis brown tomentose; flowers in about 10 cm long, axillary racemes, the flowers in about 20–25 pairs, small, ivory-white, petals very narrow, August; fruit a nut, red at first, then purple, eventually black, the size of a hazelnut, edible. BM 9161; PFC 167; BS 2: 37. Chile. 1826. z9 Plate 41. # ∅ ✗

Very interesting plant but only for very mild climates.

GLEDITSIA L. — Honeylocust — LEGUMINOSAE

Tall deciduous trees, limbs and branches usually armed with stout compound thorns; leaves alternate, pinnate and bipinnate occurring on the same tree; leaflets sometimes slightly crenate, stipules inconspicuous; flowers polygamous, small, white or greenish, in axillary racemes, occasionally in panicles; calyx lobes and petals 3–5, stamens 6–10, longer than the petals; style short, with large stigma; fruit a large, flat, leathery pod with few to many, flattened rounded seeds. — About 12 species in N. America, Central and East Asia, tropical Africa and S. America.

The identification of the species by foliage alone is very unreliable, the fruits and young branches with thorns are central to identification.

Key (from Rehder, supplemented)

● Pods many seeded, 8–40 cm long;

  + Thorns flattened at least at the base; fruit pod walls papery or leathery; pods flat, usually spirally twisted, not punctate:

  1. Leaflets usually acute, to 3.5 cm long, usually more than 20 on once pinnate leaves; ovaries pubescent:
G. triacanthos

  2. Leaflets obtuse or emarginate, 2–6 cm long, usually less than 20 on once pinnate leaves; ovaries glabrous or pubescent only on the margin:

  × Young branches always distinctly pubescent:
G. delavayi

  ×× Young branches totally glabrous;

  * Young branches green; leaflets 12–20; pods bowed, to 20 cm long:
G. caspica

  ** Young branches orange-gray; leaflets 16–30; pods bowed, to 14 cm long:
G. ferox

  *** Young branches purple; leaflets 16–20; pods inflated to spiral, 25–30 cm long:
G japonica

  ++ Thorns cylindrical; pods straight or sickle-shaped, not twisted, convex, walls thick and woody, finely pitted-punctate; leaves very seldom bipinnate;

  1. Leaflets 6–12; flower stalks 1 cm long:
G. macrantha
  2. Leaflets 8–16; flower stalks 2–4 mm long:
G. sinensis

● ● Pods 1–3 seeded, 2.5–5 cm long;
Leaflets entire, pubescent beneath; pods with 2–3 seeds:
G. heterophylla

Leaflets finely crenate, glabrous beneath; pods with 1–2 seeds:
G. aquatica

Gleditsia aquatica Marsh. Tree, about 15 m high, slow growing, thorns to 10 cm long, compound, young branches with lenticels; leaves singly to double pinnate, to 20 cm long, pinnate leaves with 12–18 leaflets, these oval-oblong, 2–3 cm long, round or emarginate, usually finely crenate only on the apical half, bipinnate leaves with 6–8 pinnae, glabrous; flowers in 7–10 cm long racemes, greenish, June–July; pods oblique rectangular, about 4 cm long, with 1(2) seeds. BB 2042; SS 127 to 128 (= *G. monosperma* Walt.; *G. inermis* Mill. non L.). Southern USA. 1723 z6

G. caspica Desf. Tree, 10–12 m high, very densely covered with stout, compound, flattened, 15 cm long thorns, young branches green; leaves usually pinnate, then 15 to 25 cm long and with 12–20 oval-elliptic, 2–5 cm long, finely crenate leaflets, rachis pubescent, apex round to emarginate, bipinnate leaves with 6–8 pinnae, midrib always pubescent on both sides, the rachis likewise; flowers nearly sessile, greenish, densely arranged in 5–10 cm long, pubescent racemes, June; pods bowed sword-like, to 20 cm long, thin. SH 2: 61–o, 7d; SL 185. N. Persia, Transcaucasus. 1822. Often confused with *G. sinensis*. ℗ USSR. z6 Fig. 79.

G. delavayi Franch. Tree, 5–10 m high, young branches finely soft pubescent (!), thorns to 25 cm long, compound, scabrous, flattened at the base; leaves pinnate, leaflets often more numerous and smaller on young plants, to bipinnate, leaflets 8–16, oblong, about 5 cm long and to 2.5 cm wide, the basal leaflets much smaller than the apical, lightly crenate to nearly entire, dark glossy green above, with reticulate venation and fine pubescence beneath, rather leathery tough, sessile; flowers in fascicled racemes, pubescent, ovaries glabrous; pods 20–50 cm long, 5–7 cm wide, thin, leathery, spirally twisted. SH 2: 6g to h, 7b. SE. China. 1900. z8 Fig. 78. ⚮

G. ferox Desf. Tree, to 10 m high, or a tall shrub, thorns very thick, scabrous, compound, young branches orange-gray; leaves pinnate and bipinnate on the same tree, leaflets 16–30, oval-oblong to oval-lanceolate, 2–5 cm long, rather symmetrical, crenate, glossy green above, lighter beneath; flowers in long racemes, corolla with 5 distinct petals, dense gray silky; pods to 14 cm long, curved, thin, leathery. SH 2: 6d–f. SW. China. 1800. Often confused with *G. japonica* in cultivation. z6

G. fontanesii see: **G. macrantha**

G. heterophylla Bge. Shrub, usually not over 3 m high, densely branched, branches thin, light gray, thorns thin, simple or 3 tipped; leaves pinnate, leaflets about 20–28, asymmetrical, small, oblique-oblong, 1–3 cm long, obtuse, entire, gray-green and pubescent beneath, or bipinnate with 3–4 pinnae, each with 12–20 leaflets from 0.5–1 cm long, rachis finely pubescent; flowers in dense, pubescent, usually tightly clustered spikes; pods

Fig. 78. **Gleditsia.** Left *G. sinensis;* middle *G. delavayi;* right *G. heterophylla*
(from ICS)

obliquely elliptic, 3–5 cm long, thin, glabrous, with 2–3 seeds. SH 2: 7g. NE. China. 1907. z6 Fig. 78.

G. horrida see: **G. japonica**

G. inermis see: **G. aquatica**

G. japonica Miq. Tree, to about 20 m, trunk and limbs very thorny, thorns 5–8 cm long, compound, somewhat flattened, young branches dark red-brown, glabrous, glossy; leaves to 30 cm long, pinnate with 16–20 leaflets, oblong to lanceolate, usually 2 to 4 cm long, entire to sparsely crenate, glossy above, midrib finely pubescent or glabrous beneath, bipinnate leaves with 2–12 pinnae, leaflets smaller, only 1–2.5 cm long, rachis pubescent on the channel edges; pods 25 to 30 cm long, blistery and twisted, seeds near the middle. GF 6: 27; SH 2:6i–k; BC 1653 (= *G. horrida* Mak. non Willd.). Japan; China. Around 1800. z6 Plate 35. ⚭

G. macracantha Desf. Tree, to 15 m high, branches thick, glabrous, bark gray, thorns very large, compound, cylindrical; leaves pinnate, leaflets 6–12, oval-oblong, 5–7 cm long, finely crenate, glabrous except for the pubescent midrib, with distinctly reticulate venation, rachis with pubescent channel edges; flowers in simple racemes; pods not twisted, slightly bowed, finely punctate, 15–30 cm long, acute elliptic in cross section. LF 176 (= *G. fontanesii* Spach). Central China. 1800. z6 Fig. 79. ⊘

G. monosperma see: **G. aquatica**

G. sinensis Lam. Tall, thick branched tree, 10–12 m high, thorns thick conical, compound; leaves pinnate, with 8–14 (18) oval-oblong leaflets, 3–8 cm long, obtuse, finely crenate, dull yellow-green above, with reticulate venation beneath, midrib pubescent on both sides, rachis pubescent on the channel edges, likewise the petiolules;

flowers in slender, pubescent racemes; pods straight, 15–25 cm long, 2–3 cm wide, with solid interior. BS 2: Pl. 41 (= *G. horrida* Willd. non Mak.). China, Mongolia. 1774. z5 Plate 42; Fig. 78.

G. × texana Sarg. (*G. triacanthos × G. aquatica*). Tree, to 30 m high or more, bark smooth, gray, young branches thornless, glabrous; leaves pinnate or bipinnate, 10–20 cm long, with 6–7 paired pinnae, each with 6 to 16 pairs of leaflets, these oval-oblong, 1–2.5 cm long, rounded, shallowly dentate, nearly sessile, deep green and glossy, rachis pubescent at first; flower racemes glabrous, 7–10 cm long, male flowers dark orange; pods 10–12 cm long, 2.5–3 cm wide, dark red-brown, dry. Texas. Found in the wild near Brazoria, 1892. Foliage like *G. triacanthos,* but with thornless branches; differing from *G. aquatica* in the many seeded fruit pod. z6

G. triacanthos L. Tree, to 20 m high, with ornamental habit, open growing with a long trunk and attractively arranged branches, densely covered with simple or compound flattened thorns; leaves pinnate (with 20–30 leaflets) or bipinnate, leaflets acute oval-oblong, 2–3 cm long, bright green, sparsely finely crenate, golden-yellow in fall; flowers greenish, very short stalked, in 5–7 cm long racemes, June–July; pods flat, sickle-shaped and twisted, 30–40 cm long, glossy dark brown, not punctate. BB 2041; GTP 255; SS 125–126. N. America. Around 1700. Ⓕ USSR, Romania, Yugoslavia. z3 One of the more easily cultivated, winter hardy, attractive park and shade trees. ⚭ ⊘

'Bujotii'. Branches thin, pendulous; leaflets smaller, quite narrow (= 'Pendula'). Before 1845.

'Columnaris' see: **'Elegantissima'**

'Elegantissima'. Dwarf form, about 2 m high, habit nearly columnar, but later more open, then more ovate, very densely branched, stem and branches thornless; leaves persist well into fall (= 'Columnaris').

f. **inermis** Willd. Like the species, but completely thornless. Occasionally occurring from seed. Plate 35.

Included here are a number of selections from American nurseries. All of these selections are protected by plant patents and must be propagated by grafting. Since they are all thornless, it is presumed that they all belong to f. *inermis*. They are usually described as male (fruitless); but since *Gleditsia* are polygamous (having male, female & bisexual flowers on the same plant), some fruit development is a possibility.

'Imperial' (D. B. Cole 1957). 9–10 m high, open crowned, broad-globose, with a dominant central leader, branches widespreading, thornless, good foliage; leaves bright green, US Pl. Pat 1605. ⊘

'Majestic' (D. B. Cole 1956). 18–20 m high, growth straight, upright, oval, open crowned, symmetrical, branches horizontally arranged, upper branches ascending, thornless; foliage deep green. US Pl. Pat. 1534. ⊘

'Moraine' (Siebenthaler 1949). Tall tree with broad crown, middle and upper branches ascend slightly, lower ones horizontal to pendulous, thornless; very densely foliate; fruitless, allegedly having only male flowers. Selected in 1937. US Pl. Pat 836. One of the most widely disseminated forms to date. ⊘

'Pendula' see: **'Bujotii'**

'Rubylace' (William Flemer III, 1961). Young foliage dark red, later becoming bronze-green. US Pl. Pat. 2038. ⊘

'Shademaster' (William Flemer III). Tall tree, erect habit, but very broadly crowned, branch tips somewhat nodding, branches thornless, one year grafts growing to 180 cm high; leaves deep green, the foliage persists very late in fall. US Pl. Pat. 1515. ⊘

'Skyline' (D. B. Cole 1957). 12–15 m high, crown broadly conical, very symmetrical, upper branches ascending, the middle and lower ones flatter to nearly horizontal, thornless; leaves dark green, golden-yellow in fall. US Pl. Pat. 1619. ⊘

'Sunburst' (D. B. Cole 1954). Fast growing, 9–12 m high, broadly conical habit, thornless, branches spread horizontally or ascend slightly, the apical 20–25 cm of the young shoots and leaves golden-yellow, later generally greening; fruitless (= *G. triacanthos inermis aurea* Cole). US Pl. Pat. 1313. The most popular selection in culture today. Plate 3. ⊘

All *Gleditsia* grow best in a good soil, but will tolerate sandy soils. They are best planted in protected areas since the branches are somewhat susceptible to wind damage. Transplanting is sometimes difficult. Very valuable park and shade tree with no disease problems.

G. triacanthos is sometimes erroneously referred to as "Christ's Thorn". Christ's crown of thorns must have been woven from the branches of *Zizyphus spina-christi* or *Paliurus spina-christi*, both thorny shrubs growing in Palestine.

Lit. Wagenknecht, B. L.: Registration Lists of Cultivar names in *Gleditsia*. Arnoldia **24**, 31–34, 1961.

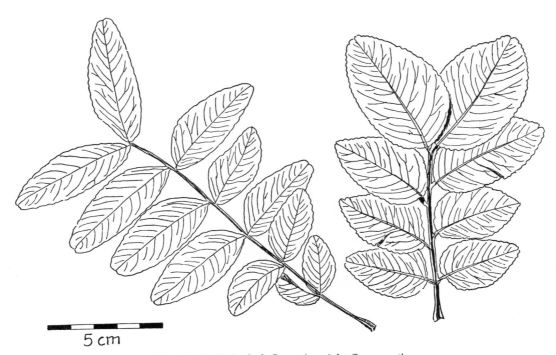

5 cm

Fig. 79. **Gleditsia.** Left *G. caspica*; right *G. macrantha*

GLOBULARIA L. — Globe Daisy — GLOBULARIACEAE

Low shrubs, subshrubs or perennials, with creeping rootstocks, often woody; leaves alternate, the basal ones often in rosettes, flower stalk often with small leaves, flowers in terminal heads, blue, base of the inflorescence with a multibracteate involucre; the individual flowers with a tubular, appressed pubescent calyx and tubular, 2 lobed corolla, stamens 4, markedly exserted; fruit a nutlet enclosed within the persistent calyx. — 28 species in S. Europe, on the Canary and Cape Verde Islands, also in Asia Minor.

Globularia alypum L. Shrub Globe Daisy. Evergreen shrub, 30–60 (100) cm high, branches woody, brittle; leaves lanceolate or oblong, stiff, leathery, with 1–3 thorned teeth at the apex, base tapering to a small petiole; flowers numerous, light blue, heads 1.5–2 cm wide, sweet smelling, involucral scales elliptic with membranous margin, flowers from winter to spring. PEu 129. S. Europe. 1840. Poisonous!! z10

G. cordifolia L. Mat-like habit, dwarf evergreen shrub, only 2–5 cm high, stoutly branched; leaves in rosettes, spathulate, 2–2.5 cm long, emarginate at the apex, deep green, leathery; flowering shoots to 8 cm high, stalk with 1–2 small, lanceolate leaves or these absent, flower heads 1–1.5 cm wide, globose, light blue-lilac, corolla 6–8 mm long, upper lobe with 3 incisions, May–July. HF 18: 77; HM Pl. 245; PEu 129; BS 2: 43. Pyrenees, Alps, Carpathians, Balkans, in the mountains on chalk soil. 1633. Popular rock garden plant. z5 Fig. 80.

ssp. *bellidifolia* see: **G. meridionalis**

ssp. *meridionalis* see: **G. meridionalis**

G. meridionalis (Podp.) O. Schwarz. Very similar to *G. cordifolia*, but larger; leaves lanceolate to oblanceolate, 20–90 mm long, 2–5 mm wide, acute, occasionally

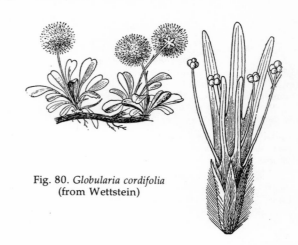

Fig. 80. *Globularia cordifolia*
(from Wettstein)

emarginate, flat; flower heads dark blue, on a short stalk, involucral scales ovate with short tips, calyx teeth linear-lanceolate (= *G. cordifolia* ssp. *bellidifolia* [Ten.] Hayek and ssp. *meridionalis* Podp.). SE. Alps, Central and S. Apennines, mountains of the Balkan Peninsula. z6

G. nana see: **G. repens**

G. repens Lam. Plants prostrate, very similar to *G. cordifolia* in appearance, but with smaller leaves, 1–2 cm long, plaited, usually acute, not emarginate, thickish; flower heads only about 1 cm wide, bluish, very short stalked. PSw 48 (= *G. nana* Lam.). Mountains in SW. Europe. 1824. z6 Plate 41.

The latter 3 species easily grown in the rock garden in normal soil and full sun.

Lit. Schwarz, O.: in Bot. Jahrb. **69**, 318–373, 1938.

GLOCHIDION R. J. & G. Forst. — EUPHORBIACEAE

Leaves alternate, entire, with stipules; flowers mono- or dioecious, in axillary clusters, sepals 3–6; male flowers with 3 stamens, without ovaries; style connate or distinct, short and thick, occasionally incised; fruit a dehiscent capsule. — About 300 species in the tropics and Polynesia; only 1 semi-hardy in a temperate climate (cultivated in Les Barres, France).

Glochidion bodinieri see: **G. puberum**

G. puberum (L.) Hutchins. Deciduous shrub, 0.5–1.5 m high and wide, densely branched, branches short pubescent, leaves obovate to oblanceolate, 3–8 cm long, acute, base broadly cuneate, dull green above, bluish with pubescent venation beneath, petiole 1–2 mm long; flowers in small fascicles, inconspicuous, greenish-white, June–September (= *G. bodinieri* Lév.). E. China. 1922. Fall foliage color scarlet-red. Only for the collector. z9

GLYCINE

Glycine sinensis see: **Wisteria sinensis**

GOMPHOCARPUS R. Br. — ASCLEPIADACEAE

Perennials and shrubs, new shoots ascending and well branched; leaves petiolate, opposite, alternate or in whorls; flowers in stalked umbels, corolla deeply lobed, the lobes compressed on the dorsal side or sheath-like, with or without furrowed ventral side; anthers with a membranous appendage on the apex; pollen masses somewhat aglutinated and individually pendulous; fruits solitary or paired follicles. — 50 species in tropical and S. Africa, rarely found in cultivation.

Gomphocarpus fruticosus (L.) R. Br. Milkbush. Shrub, 1–1.5 m high, branches rodlike, pubescent; leaves opposite, linear-lanceolate, sharply acuminate, 5–10 cm long, glabrous, but with midrib often pubescent; flowers axillary in the upper nodes, corolla lobes 7 mm long, oval-elliptic, cream-yellow, margin ciliate, rather square, sides compressed, furrowed on the ventral side, June; follicles ovate, drawn out to a beak, finely pubescent and with soft bristles (= *Asclepias fruticosa* L.). S. Africa. In cultivation at the Trsteno Arboretum, Yugoslavia. z10 Plate 41. ⚭

GORDONIA ELLIS. — THEACEAE

Evergreen trees or shrubs; leaves alternate, entire or dentate, lanceolate to elliptic; flower solitary, axillary, long stalked, sepals and petals normally 5 each; stamens numerous; in groups of 5 inside the calyx, their bases adnate to a fleshy ring; fruit a 3–6 chambered, woody, dehiscent capsule with a central column; seeds distinctly winged on the apical end. — About 40 species, mostly in tropical and subtropical Asia.

Gordonia alatamaha see: **Franklinia alatamaha**

G. anomala see: **G. axillaris**

Fig. 81. *Gordonia lasianthus*
(reduced; from Michaux)

G. axillaris (Roxb.) D. Dietr. Evergreen shrub, also a small tree in its habitat (7–10 m), branches glabrous; leaves oblanceolate to oblong, 6–15 cm long, 2–5 cm wide, glabrous, dark green and glossy, serrate to entire; flowers solitary, very short stalked, in the leaf axils at the branch tips, 7–12 cm wide, yellowish-white, November–May, stamens very numerous, orange. BS 2: 44; BM 4019; LWT 232; NF 1: 172 (= *G. anomala* Don). China; Formosa. 1818. z9 # ✿

G. chrysandra Cowan. Evergreen shrub, 3–10 m high in its habitat, young branches carmine, finely pubescent at first; leaves elliptic to obovate-elliptic, rounded, long cuneate, tapered at the base, 6–11 cm long, leathery, dark glossy green above, light green beneath, margins sparsely crenate on the apical half; flowers usually solitary in the axils of the upper leaves, short stalked, corolla 5–6 cm wide, with 6–7 petals, white with a trace of yellow, January–February, stamens numerous, short, yellow. BMns 285. China; Yunnan Province. 1931. z9 # ✿

G. lasianthus Ellis. Evergreen shrub, also a tree in its habitat, to 15 m; leaves obovate-lanceolate, acute, cuneate, tapered at the base, 10–15 cm long, lightly serrate, deep green above, glossy, light green beneath, somewhat glossy; flowers solitary, axillary at the branch tips, 5–7 cm wide, white, petals concave, limb somewhat crenate, silky pubescent outside, stamens numerous, short, yellow, July–August; fruit woody, 3 cm long. BM 668; SS 21. Southeast USA. 1768. z9 Fig. 81. # ✿

G. sinensis Hemsl. & Wils. Shrub, also a tree in its habitat, to 10 m high, young branches gray-brown, glabrous; leaves elliptic-ovate, about 11 cm long, 5 cm wide, acute, cuneate at the base, crenate, leathery tough, glossy green above, light green beneath, petiole 2 cm long; flowers solitary, stalk 2 cm long, white, calyx with 5–7 sepals, margin with white ciliate hairs; fruit a 2 cm long, woody capsule. China; W. Szechwan Province. z9 # ✿

Uses and culture somewhat like that of *Camellia*.

GREVILLEA R. Br. — PROTEACEAE

Evergreen shrubs or trees; leaves alternate, simple or lobed or pinnatisect; flowers stalked, usually in racemes, fascicles or umbels, attractively colored and conspicuous; flower bracts abscising; perianth straight or reflexed, style usually long, thickened at the ends, ovaries with a disc or fleshy base, fruit a 1 to 2 seeded follicle, without partitions between the wingless, or narrowly membranous winged, seeds. — 199 species in E. Malaysia, New Hebrides, New Caledonia and Australia.

Grevillea acanthifolia A. Cunn. Evergreen shrub, 1–1.5 m high, young branches angular, somewhat white pubescent at first, soon glabrous; leaves doubly pinnatisect, 5–7 cm long, 2.5–3 cm wide, the lobes divided into 3 triangular, thorn tipped lobes, dark dull green above, soon glabrous; flowers densely arranged in terminal and axillary, 4–7 cm long, one sided, erect, shaggy pubescent racemes, pink and greenish, May. BM 2807. Australia; New South Wales. 1820. z10 # ∅

G. alpestris see: **G. alpina**

G. alpina Lindl. Low shrub, densely branched, 0.5–1 m high, branches with erect tomentum, leaves densely packed, narrow oblong to elliptic or linear, 1–2.5 cm long, dark green above, silky pubescent beneath, obtuse, margin somewhat revolute, sessile; flowers few in short, terminal clusters, corolla 12 mm long, red and swollen at the base, more yellowish at the apex, style exserted to 6 mm, April–May. BM 5007; LAu 347; FS 1449 (= *G. alpestris* Meissn.). Australia. z9 # ✧

G. glabrata (Lindl.) Meissn. Erect shrub, 1–2 m, totally glabrous, branches thin; leaves broadly cuneate, short 3 lobed, 2.5–4 cm long, the lobes coarse and sharply dentate; flowers small, white, in axillary, 2.5–5 cm long racemes, these occasionally grouped into terminal panicles, 10–15 cm long, stigma pink, May (= *G. manglesii* Hort.). W. Australia. Around 1845. z10 # ✧

G. manglesii see: **G. glabrata**

G. juniperina see: **G. sulphurea**

G. ornithopoda Meissn. Shrub, to about 1 m high, branches thin, nodding, totally glabrous; leaves 7–10 cm long, deeply 3 lobed on the apex, the lobes 2.5–3 cm long and 3–4 mm wide, linear, acute, bowed outward; flowers white, in pendulous, 4–5 cm long, 2 cm wide racemes, style short, yellowish, April. BM 7739. W. Australia. 1850. z10 # ✧

G. preissii see: **G. thelemanniana**

G. robusta A. Cunn. ex R. Br. Australian Silk Oak. Tall tree, to 50 m in its habitat, branches silvery pubescent; leaves pinnately compound, 15–20 cm long, the pinnae usually further bipinnate, lanceolate, margin involuted, dark green above, silky pubescent beneath; flowers a gorgeous golden-yellow in one-sided, 7–10 cm long racemes, but flowering only as a tree in a subtropical climate. BM 3184; HTS 276; MCL 19. Australia; Queensland, New South Wales. Around 1835. A valuable shade tree in the tropics and subtropics; a popular potted plant in cooler climates. Ⓕ S. America, Central America, E. Africa, Israel. z9 Plate 3. # ✧

G. rosmarinifolia A. Cunn. Shrub, 1.5–2 m high, branches thin, densely pubescent; leaves linear-lanceolate, 3–5 cm long, acute, sessile, dark green, resembling rosemary, silky pubescent beneath; flowers in dense, terminal clusters, 2.5 to 3 cm wide, red, summer. BM 5971; LAu 210; HTS 277; PBl I: 532. Australia; New South Wales. Around 1830. z9 # ✧ ∅

G. × semperflorens F. E. Briggs ex Mulligan (*G. sulphurea* × *G. thelemanniana*). Erect shrub, 1.5–1.8 m high, young branches densely pubescent; leaves linear, simple or forked near the apex, 3–4.5 cm long, with dense silky pubescence beneath; flowers many, in loose, terminal and axillary racemes, 4 cm long, orange-yellow at the base, limb pink toned, apex green. BMns 535. Developed in England in 1927. z9 # ✧

G. sulphurea A. Cunn. Erect shrub, to 1.8 m high, young branches softly pubescent; leaves densely arranged, linear to nearly needlelike, 1.2–2.5 cm long, apex prickly, margin involuted, silky pubescent beneath; flowers 12 or more together in terminal racemes, tube sulfur-yellow, silky pubescent, style 2.5 cm long, May–June. BS 2: 75 (= *G. juniperina* var. *sulphurea* [A. Cunn.] Benth.). Australia; New South Wales. z9 # ✧

G. thelemanniana Huegel. Erect shrub, to 1.5 m high, branches softly pubescent; leaves pinnate to bipinnate, 2.5–5 cm long, pinnae linear to filamentous, grooved beneath, bluish, not prickly, nearly *Artemisia*-like in appearance, silky pubescent; flowers in about 5 cm long and wide, terminal racemes, tube pink, the reflexed lobes yellow-green, style red. BM 5837 (as *G. preissii*). W. Australia. 1838. z10 # ✧

Cultivation limited to seasonal tub plants outdoors in all but the mildest, frost free regions. All dislike alkaline soils. *G. robusta* can become a very large tree.

GREWIA L. — TILIACEAE

Fig. 82. *Grewia biloba* var. *parviflora* (from Gfl.)

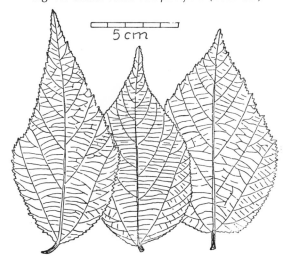

Shrubs or trees with stellate tomentose branches; leaves simple, alternate, entire or serrate, stellate tomentose, stipules persistent; flowers 5 parted, rather small, solitary or in small, terminal or axillary cymes; stamens numerous, distinct; ovaries 5 chambered; fruit a small drupe, fleshy or fibrous with 1 to several seed pits. — 150 species in the tropical and subtropical regions of Asia, Africa and Australia, of which only a few are present in cultivation.

Grewia biloba D. Don. Deciduous shrub, to 2.5 m high, young branches stellate pubescent; leaves very thin, ovate to rhombic-lanceolate, 5–12 cm long, acuminate, base cuneate to round, 3 veined, unevenly simple to doubly serrate, dark green above, nearly glabrous, usually with dense stellate pubescence beneath; flowers about 1–1.5 cm wide, grouped 5–10 in pubescent cymes, bright yellow, July–August; fruits orange to red, usually 2 lobed. E. China; Formosa. 1890. z8 Fig. 83.

var. **parviflora** (Bge.) Hand. Mazz. Shrub, 1.5–2 m high, branches tomentose; leaves oval-rhombic, 4–9 cm long, base round to broadly cuneate, scabrous above, densely stellate pubescent beneath. NK 12: 16 (= *G. parviflora* Bge.). N. China, Korea. 1883. Hardy! z6 Fig. 82, 83.

G. oppositifolia Roxb. A small deciduous tree in its habitat, branches stellate pubescent; leaves alternate, 2 ranked, ovate, 6–10 cm long, base rounded, crenate, stellate tomentose beneath, petiole 4–8 mm long; flowers 2.5–3 cm wide, 8–20 in stalked fascicles opposite the leaves, yellowish, August–September; fruits pea-sized, blackish. NW. Himalaya. 1914. Identification often confused in cultivation. z9 Fig. 83.

G. parvifolia see: **G. biloba** var. **parvifolia**

Without particular ornamental merit, only of botanical interest.

Fig. 83. **Grewia.** Left *G. biloba;* middle *G. oppositifolia;* right *G. biloba* var. *parviflora* (drawn from material collected in the wild)

GREYIA Hook. & Harv. — GREYIACEAE

Shrubs or small trees with sparsely foliate branches; leaves alternate, simple, somewhat umbrella-shaped (resembling *Pelargonium*), crenate lobed and dentate, white tomentose or glabrous with tiny glands when young; flowers in terminal, short, racemes; the individual flowers small, with 5 petals and 5 sepals, scarlet-red, fruit a 5 valved, many seeded capsule. — 3 species in S. Africa.

Greyia sutherlandii Hook. & Harv. Small tree or shrub, open branched; leaves only on the branch tips, broadly ovate, 5–7 cm long, coarsely dentate, base deeply cordate; flowers scarlet-red, grouped many together in 7–10 cm long and about 5 cm wide racemes, stamen filaments red, markedly exserted, attractive, flowering in its habitat August–September. BM 6040; BMns 374; FS 1739; JRHS 88: 137; MCL 69. S. Africa; Natal Province. z10 ⊕

GRINDELIA Willd. — COMPOSITAE

Subshrubs, perennials or biennials, somewhat resembling *Chrysanthemum maximum,* but larger and with large yellow flowers; leaves alternate, often rigid, dentate or ciliate dentate, sessile to stem-clasping; flower heads yellow, 2.5–5 cm wide, solitary at the branch tips. — 60 species in N. and S. America, outside the tropical areas of C. and S. America.

Grindelia chiloensis (Cornelissen) Cabrera. Evergreen subshrub, branches glabrous, dull green, glutinous, to 60 cm high; leaves from narrow-lanceolate to obovate, to 12 cm long, the upper leaves smaller, more or less stem-clasping; flower heads golden-yellow, to 7.5 cm wide, with about 50 ray florets, buds with a thick, white, glutinous coating (!), disk florets orange, July–August; perianth scales in 4 rings, the outer ones drawn out to a fine tip. BM 9471 (= *G. speciosa* Hook. & Arn.). Argentina (not Chile!); in dry locations. Cultivated in England since 1850, but only recently more widely distributed. z9 ⌖

G. speciosa see: **G. chiloensis**

Very easy to cultivate; propagated by cuttings.

GRISELINIA Forst. — CORNACEAE

Evergreen shrubs, also a small tree in milder regions, branches cylindrical to angular; leaves alternate, base often oblique, very leathery, petiole flared to a small sheath, venation indistinct; flowers small, dioecious, in small axillary panicles; fruit a single seeded berry. — About 6 species, 4 in Chile and 2 in New Zealand.

Griselinia littoralis Raoul. Shrub, about 5 m high, branches often wavy; leaves ovate to elliptic, rounded at both ends, 3–10 cm long, yellowish-green, thick leathery, both sides glossy and glabrous; flowers in 5–7 cm long panicles, but not particularly ornamental, October–November; fruits 8 mm long. KF 42. New Zealand. z9 # ∅

'Variegata'. Leaves white variegated. Plate 41. # ∅

G. lucida Forst. Shrub, stiffly erect, about 2.5 m high; leaves very obliquely ovate or obovate-oblong, 7–15 cm long, rounded at the apex, base asymmetrical, light yellow-green, glossy, very leathery thick; female flowers without petals; fruits 8 mm long, purple. KF 41. New Zealand z7 # ∅

Well suited for coastal planting in milder regions. Also an attractive potted plant.

GUEVINA See: **GEVUINA**

GYMNOCLADUS Lam. — LEGUMINOSAE

Tall, deciduous, thornless trees with thick, knotty branches; leaves very large, bipinnate, leaflets ovate, thin; flowers polygamous, with 5 narrow calyx lobes and 4–5 oblong petals; stamens 10, shorter than the petals; fruit a thick, fleshy pod with large, flat very hard seeds. — 2 species, in N. America and China, respectively.

Gymnocladus canadensis see: **G. dioica**

G. chinensis Baill. Tree, to 10 m high; leaflets smaller than the following species, pubescent on both sides; flowers lilac, appearing before the leaves; fruits 7–10 cm long, very thick. China. 1888. The fruit pulp is used by the Chinese in washing for its saponin content. z9

G. dioica (L.) K. Koch. Tree, to about 25 m high, branches rigid, knotty, the younger branches bluish-gray pruinose, with large, recessed leaf scars, leafing out very late; leaves 30–80 cm long, young leaves pink, with 3–7 pinnae, leaflets oval-elliptic, 5–8 cm long, dark green, pubescent beneath at first, fall color golden-yellow; flowers whitish; pods 8–15 cm long, thick, brown, pruinose. BB 2043; SS 123; SM 459 (= *G. canadensis* Lam.). N. America. 1748. Hardy. z5 Plate 43. ∅

Likes a moist fertile soil, slow growing.

HABROTHAMNUS See: **CESTRUM**

HAKEA Schrad. — PROTEACEAE

Evergreen trees and shrubs; leaves alternate, exceptionally variable in form, sometimes needlelike and cylindrical to flat and oblong, entire or deeply lobed and dentate or pinnate, usually with a few parallel longitudinal veins; flowers bisexual, paired, these pairs usually packed in dense racemes or globose clusters, usually sessile in the leaf axils, perianth tubular, the 4 lobes concave and clinging together long after the tube opens, the 4 anthers sessile at the base of the lobes, styles long or short, but always widened at the apex; fruit a hard, woody capsule, opening with 2 valves, containing compressed, winged seeds. — About 100 species, confined exclusively to Australia and Tasmania, but frequently found in cultivation in Mediterranean climates and most botanic gardens.

Hakea acicularis see: **H. sericea**

H. baxteri R. Br. Shrub, about 2 m high; leaves stiff, broadly fan-shaped (somewhat resembling small *Gingko* leaves), thorny dentate at the apex, 5–8 cm wide, drawn out to a rigid petiole; flowers sessile, in axillary clusters, perianth reddish tomentose, tube 6–7 mm long; fruit about 35 mm long, 25 mm wide, rugose, short beaked. HI 437. W. Australia. z10 # ∅

H. eucalyptoides see: **H. laurina**

H. laurina R. Br. Tall shrub, or a tree in Australia; leaves elliptic to lanceolate, 12–15 cm long, 1.2–2.5 cm wide, tapering to the petiole, with 3–7 nearly parallel veins; flowers carmine, in axillary, sessile, globose heads, surrounded by an involucre, 4–5 cm wide, the golden-yellow style exserted about 2 cm, flowering in winter; fruit capsule ovate, 3 cm long, 2 cm wide, short beaked. BM 7127; LAu 143; DRHS 948; HTS 284 (*H. eucalyptoides* Meissn.). W. Australia. 1830. Ⓕ Australia. The most beautiful species of the entire genus. z10 # ☉

H. lissosperma R. Br. Shrub or small tree, 2–6 m high, with many thick branches, inflorescences, young branches and leaves silky pubescent, otherwise glabrous, leaves twisted, stiff, prickly, 3–15 cm long, outspread or nearly erect or curving upward; flowers 8 or more in nearly sessile, axillary clusters, white, perianth lobes thick, narrow-lanceolate; fruit 2–3 cm long, both valves arched and very thick, deep purple-brown, more or less warty. SW. Australia, Tasmania, in the mountains around 1300 m. z10 #

H. microcarpa R. Br. Low shrub, 0.6 to 2 m, glabrous or with young branches finely pubescent; leaves usually twisted, rigid, with a long, prickly apex, young leaves (and occasionally also some mature leaves) flat, linear-lanceolate, with thick midrib and margins, 2–8 cm long, directed upward; flowers in nearly sessile, axillary clusters, white to yellowish; fruits rather flat, oblique, obovate, 8–12 mm long, valves leathery to thin and woody, each with a small spur on the tip. BS 2: 304; BR 475. SW. Australia. z9 #

H. oleifolia R. Br. Shrub to small, round crowned tree, 4–6 m high, branches softly pubescent; leaves oblong-lanceolate, 2.5–6 cm long, rounded and with a small tip, base cuneate, short petioled; flowers white, in thick, axillary clusters, stalk not very long, rachis woolly pubescent. EKW 317. W. Australia. 1794. z10 #

H. pectinata see: **H. suaveolens**

H. sericea Schrad. & J. Wendl. Shrub, 2–4 m high, sparsely branched; leaves cylindrical, often thin, prickly, rigid, 3–6 cm long, base rather broad at a wide angle to the branch; flowers 2–6 in nearly sessile, axillary clusters, corolla lobes narrow linear, white or also a light pink tone, stalk silky pubescent, 3–4 mm long, May–June; fruit 2.5 to 3 cm long, valves very thick, deep purple-brown, smooth to rugose, short beaked and often with 2 small thorns on the apex. BMns 229 (as *H. tenuifolia*); HTS 287 (= *H. acicularis* [Sm. ex Vent.] Knight; *H. tenuifolia* [Salisb.] J. Britt. non Dum.-Cours.). Australia, Tasmania. Before 1840. Widely planted as a hedge. z10 Plate 43. #

H. suaveolens R. Br. Globose shrub, 2–5 m high; leaves cylindrical, 5–10 cm long, with stiff thorny tips, usually simple, often also compound, in 1–5 stiff, cylindrical lobes of varying length; perianth glabrous; flowers white, fragrant, in small axillary racemes near the branch tips, summer, rachis 1–2 cm long; fruits ovate, 2.5 cm long, tapering to the apex and drawn out to a small horn (= *H. pectinata* Colla). W. Australia. 1803. z10 #

H. tenuifolia see: **H. sericea**

In their habitat all species found in hot, dry areas; but will tolerate a light frost. Young seedlings are often transition forms with leaves varying from simple to multipinnate.

HALESIA Ellis. — Silverbell; Snowdrop Tree — STYRACACEAE

Deciduous trees or tall shrubs, with scattered stellate pubescence; leaves alternate, simple, petiolate, serrate, very thin; flowers in axillary clusters on the previous year's wood, pendulous; corolla 4 parted or 4 lobed, white, campanulate, ovaries subinferior to nearly inferior; fruit a large, 3–5 cm long, 2 or 4 winged drupe, with 1–3 seeds. — 3 species in eastern N. America, 1 species in E. China.

Halesia carolina L. Tall shrub or tree to 6 m high, branches widely spread, stellate pubescent when young, flowers appear before the foliage; leaves ovate-elliptic, 5–10 cm long, finely serrate, tomentose above at first, but soon glabrous, pubescent beneath; flowers in fasicles of 2–5, stalk 1–2 cm long, corolla 1–1.5 cm long (!), pendulous, white, stamens 12–16, style glabrous, April–May; fruit 2.5–3.5 cm long, with 4(!) wings. BM 910; SS

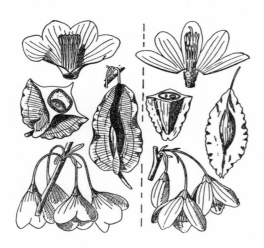

Fig. 84. **Halesia.** Flowers, corolla cross section, fruit and fruit cross section. Left *H. carolina;* right *H. diptera* (from Perkins, Sargent)

257; NBB 3: 47 (= *H. tetraptera* Ellis). N. America, in moist, fertile woodlands. 1756. Hardy. z5 Plate 50; Fig. 84. ☼

f. **dialypetala** (Rehd.) Schneid. Corolla split nearly to the base (= *H. tetraptera* f. *dialypetala* Rehd.).

'**Meehanii**' (Meehan). Leaves smaller, tougher, deeper green, serrate, pubescent beneath; flowers smaller, more deeply incised. GF 5: 535; BC 1429. 1890.

var. **mollis** (Lange) Perk. Leaves broader, short acuminate, pubescent above, tomentose beneath; stamens 8 to 12, corolla larger, 4 parted to the middle.

H. diptera Ellis. More shrubby in habit, to 5 m high, branches widely spread, young shoots pubescent, later glabrous; leaves elliptic to obovate, 6–12 cm long, abruptly acuminate, sparsely finely dentate, tomentose above at first, soon becoming glabrous, eventually only the venation pubescent beneath; flowers white, about 2.5 cm (!) long, 4 parted nearly to the base, stamens usually 8, style densely pubescent to just short of the stigma, June; fruit to 5 cm long, flat 2 winged (!). DL 1: 201. SW. USA. Quite hardy but flowers less abundantly. z6 Plate 50; Fig. 84. ☼

var. **magniflora** Godfrey. Flowers larger, 2–3 cm long. KTF 168. N. Florida, in hilly woodlands (the species, on the other hand, is found in the forested plains!). More attractive and flowers more abundantly than the species. Plate 3.

H. monticola (Rehd.) Sarg. A tall tree in its habitat, to 25 m high, flaking bark, young branches quickly become quite glabrous; leaves elliptic to oval-oblong, scarcely distinguishable from those of *H. carolina* in size and form; flowers grouped 2–5, white, about 2.5 cm (!) long, campanulate, May; fruit to 5 cm long, 4 winged. BS 2: Pl. 9. Northeastern USA. 1930. Beautiful species, but only used to any degree in recent years. z6 Plate 50. ☼

'**Rosea**'. Flowers pale pink. Before 1921.

var. **vestita** Sarg. Leaves white woolly beneath at first and persisting until fully expanded, base more rounded, to 12 cm long; flowers as large as those of the species, occasionally slightly reddish, 3–5 in the leaf axils, pendulous, late May. JRHS 83: 121. USA; N. Carolina, Arkansas. 1912. ☼

H. tetraptera see: **H. carolina**

All species prefer a cool, moist, humus soil, with the crown open to the sun, but the plant somewhat protected.

Lit. Perkins, J.: Styracaceae; in Engler, Pflanzenreich **30**, 94–99, 1907 ● Reveal, J. L., & M. J. Seldin: On the identity of *Halesia carolina* L. (Styracaceae); in Taxon **25**, 123–140, 1976.

× HALIMIOCISTUS Janchen — CISTACEAE

Hybrids between *Halimium* and *Cistus;* characteristics intermediate between the parents.

× **Halimiocistus 'Ingwersenii'** (= *Halimium umbellatum* × *Cistus hirsutus*). Evergreen shrub, about 30–50 cm high, branches white woolly; leaves linear, obtuse, 2–3 cm long, pubescent; flowers white, 2–2.5 cm wide, stalk 7–10 cm long, in conspicuous panicles, 15–30 cm wide (= × *H. ingwersenii* E. F. Warb.). 1929. Discovered as a natural hybrid in Portugal. z9 ☼

× **H. revolii** (Coste & Soulié) Dansereau (*Halimium alyssoides* × *Cistus salviifolius*). Low shrub, 30–50 cm high, densely branched, branches covered with white, erect pubescence; leaves elliptic to broad-elliptic, 6–18 mm long, 3–6 mm wide, obtuse to rounded, base cuneate, lower leaves green above and short petioled, upper leaves sessile and gray pubescent above, with raised pinnate venation beneath; flowers in terminal cymes, sepals usually 5, with silky exterior, petals white with yellow or light yellow base, June–July (= *Cistus revolii* Coste & Soulié). Discovered in the Cevennes Mountains, S. France in 1914. z9 # ☼

× **H. sahucii** (Coste & Soulié) Warburg (= *Cistus salvifolius* × *Halimium umbellatum*). Evergreen shrub, to 1 m high; leaves linear to more lanceolate, 1.5–2.5 cm long, pubescent; flowers white, 3 cm wide, grouped 2–5 in clusters, June, sepals 4 or 5, occasionally only 3, rounded at the base. BS 2: Pl. 46. Discovered before 1911 in S. France. z9 # ○ ☼

Cultivated like *Cistus.*

HALIMIUM Spach — CISTACEAE

Evergreen shrubs, subshrubs or perennials; differing from *Cistus* in the fruit capsule and the often yellow flowers; differing from *Helianthemum* in the very short and always straight style and often only 3 sepals; leaves opposite on the base of the branch, alternate toward the apex; flowers yellow or white, in racemes or umbellate cymes; sepals 3 or 5, petals 5; styles short, straight (!), with capitate or 3 lobed stigma; fruit a 3 valved capsule. — About 10 species in the Mediterranean region and Asia Minor.

Halimium alyssoides (Lam.) K. Koch. Small evergreen shrub, to 50 cm high, branches outspread, thin, gray pubescent, partly stellate pubescent; leaves narrow-obovate to oval-lanceolate, round or obtuse at the apex; 0.8–3 cm long, 3–12 mm wide, base cuneate, dense gray pubescent; flowers terminal and axillary, golden-yellow, not speckled, sepals 3, ovate, dense and short pubescent, May–June. DRHS 949; PSw 29 (= *Cistus alyssoides* Lam.). SW. Europe. z9 # ⊕

H. atriplicifolium (Lam.) Spach. Shrub, to about 1 m high, broad growing, branches softly pubescent with short hairs and white scales; leaves broadly ovate, obtuse, 2.5–5 cm long, both sides silvery and lepidote, 3 or 5 veined; flowers 2–8, golden-yellow with brown basal spot, to 4 cm wide, sepals 3, oval-lanceolate, 8 mm long, short white pubescent with long, purple hairs, stalks glutinous, June. PMe 108 (= *Cistus atriplicifolium* Lam.). S. Spain, Morocco. Probably introduced in 1768. z9 # ⊕

H. formosum see: **H. lasianthum**

H. halimifolium (L.) Willk. & Lange. Shrub, 0.7–1 m high, branches scaly and pubescent; leaves narrow obovate to oblong, 2–5 cm long, obtuse, cuneate at the base, gray, stellate pubescent; flowers golden-yellow, 3 cm wide, petals with small black spot near the base, in long stalked, erect cymes, with few flowers, sepals 3–5, stellate pubescent, not silky pubescent, June. EP IV, 193: 11c–H. Mediterranean region. 1650. z9 Fig. 85. # ○ ⊕

H. lasianthum (Lam.) Spach. Small, erect shrub, 30–70 cm high, branches short, gray tomentose and lightly stellate pubescent; leave elliptic to oval-oblong, 1.8–2.5 cm long, obtuse, 3 veined, margins often involuted, pubescent above, dense stellate tomentose beneath; flowers golden-yellow, petals with a large dark red spot on the base, May–June. EP IV, 193: 10c–H; BS 2: 313; PSw 29; BM 264 (as *Cistus formosus*) (= *H. formosum* Willk.). Portugal, S. Spain. 1780. z9 # ○ ⊕

H. ocymoides (Lam.) Willk. & Lange. Shrub, 0.5–0.7 m high, branches slender, white haired; leaves narrow obovate to oblong, 1.5–2.5 cm long, obtuse, cuneate at the base, 3 veined, white pubescent at first; flowers golden-yellow, erect, long stalked panicles with few flowers, but grouped into 10–30 cm wide inflorescences, petals triangular, with a large purple spot on the base, sepals 3, June. BM 627 (= *Cistus algarvensis* Sims). Portugal, Spain. 1880. z9 Plate 42. # ○ ⊕

H. umbellatum (L.) Spach. Erect shrub, about 40 cm high, branches pubescent and glutinous; leaves sessile, linear, 1–3 cm long, light green above, white tomentose beneath; flowers white, 2 cm wide, in erect, terminal clusters, petals obcordate, yellow at the base, sepals 3, June. BS 2: 86; BM 9141. Mediterranean region. 1731. z9 # ○

Cultural requirements like those of *Cistus*, which see.

Lit.: refer to *Cistus*.

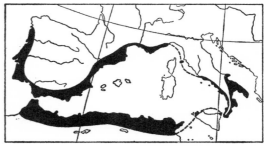

Fig. 85. Range of *Halimium halimifolium*
(from Fenaroli)

HALIMODENDRON Fisch. — Salt Tree — LEGUMINOSAE

Thorny, deciduous shrub, resembling *Caragana*, but flowers lilac, in short racemes, pods inflated. — Only 1 species; Siberia.

Halimodendron halodendron (L.) Voss. Deciduous, erect, thin branched shrub, to 2 m high, branches light gray, pubescent at first, buds white tomentose; leaves alternate, even pinnate with 2–4 sessile leaflets, these oblanceolate, 2–3.5 cm long, both sides gray-green, rachis later prickly; flowers light violet, grouped 2–4 in small racemes on older wood, June–July, calyx, corolla,

stamens and style otherwise like *Caragana*, ovaries stalked; pods 2 cm long, yellow-brown, inflated. BM 1016; HM 1455–56. Siberia. 1779. Ⓕ Canada (Saskatchewan), in windbreaks. z2 ○⊕

'Purpureum' (Späth). Flowers darker, bright purple-pink and white. Disseminated in 1893 from Späth, Berlin, but illustrated in RH in 1876 as "*H. speciosum*" Carr.

For light, well drained sand or sandy clay soil; thrives in salty coastal soils. Should never be pruned.

HALOXYLON Bge. — CHENOPODIACEAE

Nearly leafless shrub or small tree with sectioned branches; leaves reduced to scales; flowers perfect, axillary, with 2 wide bracts; sepals 5, distinct, stamens 2–5, stigmas 2–5; fruit circular, somewhat fleshy. — 10 species in the Mediterranean region and Asia Minor, in the steppes and deserts.

Haloxylon ammodendron (C. A. Mey.) Bge. Small tree, with hard wood, 4(6) m high, branches pale green, leafless, cylindrical, sectioned, pendulous; scale leaves short triangular; fruit calyx with 4 large, rounded wings. SH 2: 568. Ural and Atlas Mts. to Persia and Turkestan, in the steppes and deserts (used for camel fodder). Quite winter hardy, but only for collectors. ⓕ NW. China, Inner Mongolia, for reforestation on sandy soils. z5 Plate 47. ○

HAMAMELIS L. — Witch Hazel — HAMAMELIDACEAE

Deciduous shrub to small tree, somewhat resembling *Corylus* in appearance, stellate pubescent; leaves alternate, short petioled, unequal sided, open dentate; flowers normally bisexual, few in axillary, short stalked clusters, after leaf fall or before the new spring growth; calyx 4 parted, petals 4, linear, crumpled in the bud stage; stamens 4, filaments short, anthers nearly globose, 4 ligulate staminodes; ovaries 2 chambered, styles 2; fruit a woody capsule dehiscing from the apex, with 2 seeds. — 6 species in N. America and E. Asia.

Key to the species

- Leaves densely white pubescent, base cordate:
  H. mollis

- • Leaves totally glabrous or quickly becoming so, base indistinctly cordate or truncate;

  + Flowers in fall; leaves green beneath, glossy:
    H. virginiana

  ++ Flowers in winter or spring;
  . > Leaves obovate, with 5 vein pairs, often gray-green beneath; petals 1–1.5 cm long:
    H. vernalis
  >> Leaves broadly ovate, with 7 vein pairs; petals about 2 cm long:
    H. japonica

Outline of the cultivar parentage

"Adonis"	→ 'Ruby Glow'
'Allgold'	→ intermedia
'Arborea'	→ japonica
'Arnold Promise'	→ intermedia
'Brevipetala'	→ mollis
'Carmine Red'	→ intermedia
'Carnea'	→ vernalis
'Combe Wood'	→ mollis
"Copper Beauty"	→ 'Jelena'
'Diane'	→ intermedia
'Feuerzauber'	→ intermedia
"Fire Charm"	→ 'Feuerzauber'
'Goldcrest'	→ mollis
'Hiltingbury'	→ intermedia
'Jelena'	→ intermedia
'Lombart's Weeping'	→ vernalis
"Magic Fire"	→ intermedia
'Nina'	→ intermedia
"Orange"	→ 'Orange Beauty'
'Orange Beauty'	→ intermedia
'Pallida'	→ mollis
'Primavera'	→ intermedia
'Red Imp'	→ vernalis
'Ruby Glow'	→ intermedia
'Sandra'	→ vernalis
'Squib'	→ vernalis
'Sulphurea'	→ japonica
'Winter Beauty'	→ intermedia
'Zuccariniana'	→ japonica

Hamamelis × intermedia Rehd. (*H. japonica × H. mollis*). Strong growing shrub, about 4 m high, erect and wide, terminal buds to 10 mm long, axillary buds slightly shorter, densely brown stellate pubescent; leaves obovate, 10–15 cm long, resembling *H. mollis*, but narrower, more or less pubescent beneath; flowers deep yellow, petals not so campanulate as those of *H. japonica*, usually curved only on the apex. Dansk Dendrol. Arsskr. 1953:141–142 (= *H. japollis* Lange). z5

Includes the following named cultivars:

'Adonis' see: **'Ruby Glow'**

'Allgold'. Shrub, medium-sized, ascending habit; leaves broadly elliptic to oval, 11–18 cm long, yellow in fall; flowers butter-yellow, medium-sized, petals twisted, the same color as *H. mollis* 'Brevipetala', but long and narrow, calyx purple-red inside, fragrant. (Hillier) ✧

'Arnold Promise' (Arnold Arboretum 1963). Shrub, medium-sized, to 3 m high and 4 m wide; flowers medium-sized, in dense clusters, petals 15 mm long, deep sulfur-yellow, calyx interior reddish-green. AB 23 Pl. 6 + 7. Developed in the Arnold Arboretum, 1928. Cultivated in USA, England and Holland. ✧

'Carmine Red' (Hillier). Medium-sized, broad shrub; leaves oval to circular, 13 × 13 cm in size, yellow in fall; flowers large, petals 2 cm long, narrow, plaited and twisted, pale bronze, more coppery at the apex, base red, carmine in bud, calyx interior wine-red, late flowering, March. Poor bloomer!

'Copper Beauty' see: **'Jelena'**

'Diane' (de Belder). Shrub, very vigorous; leaves an intensive yellow and scarlet in fall; flowers medium-sized to large, in dense clusters, petals to 17 mm long and 2.3 mm wide, rather straight, carmine-red, calyx violet inside. One of the best red flowering forms, better than 'Ruby Glow'. ✧

'**Feuerzauber**' (Herm. A. Hesse 1958). Medium-sized shrub, strong grower, erect; leaves oval to broad-elliptic, base oblique cordate, about 13–17 cm long, 8–10 cm wide, yellow fall color; flowers large, petals coppery-orange, overlaid with red, similar to 'Ruby Glow', but larger, 16–18 mm long, twisted. Known also as 'Fire Charm' or 'Magic Fire' in the trade. ⊕

'Fire Charm' see: '**Feuerzauber**'

'**Hiltingbury**' (Hillier). Large shrub, broad growing or ascending; leaves circular to more obovate, 8–12 cm long, 7–10 cm wide, yellow in fall, orange, coppery and red; flowers medium to large in size, 16–18 mm long, pale copper-red, with overtones of red, calyx interior purple. Flowers of little merit but fall color and foliage gorgeous. Fig. 86. ∅

'**Jelena**' (Korr/de Belder 1955). Large shrub, ascending habit; leaves ovate, 16–17 cm long, 12–15 cm wide, base cordate, fall color orange, bronze, scarlet and red; flowers in dense clusters, large, yellow with coppery traces, appearing orange from a distance, petals to 2 cm long, 2 mm wide, twisted and wavy, calyx interior dark wine red. JRHS 90: 17; 99: 5 (= 'Copper Beauty'). Selected by A. Kort in the Kalmthout Arboretum, 1937, but named in 1955 by the late owner, de Belder. One of the most beautiful cultivars. ⊕

'Magic Fire' see: '**Feuerzauber**'

'**Moonlight**'. Shrub, medium-sized to large, branches ascending; leaves rather circular, 12–18 cm long, 9–12 cm wide, obliquely cordate, fall color yellow; flowers in dense clusters, medium-sized to large, petals to 18 mm long, plaited and twisted, pale sulfur-yellow, base wine-red, as is the calyx interior, very fragrant (larger, but lighter and less densely arranged than those of *H. mollis* 'Pallida').

'**Nina**' (Lange, Denmark 1953). Flowers very large, petals to 2.7 cm long (!), deep yellow, in many flowered fascicles. Dansk Dendr. Arsskr. 1953: P1. 140. Discovered in the Charlottenlund Arboretum, Copenhagen and named by J. Lange, but apparently not yet introduced to the trade.

'**Orange Beauty**' (Heinr. Bruns, before 1955). Strong upright grower; leaves like those of *H. mollis*, but somewhat more acute, yellow in fall; flowers medium-sized, petals 15 mm long, deep golden-yellow to orange-yellow, calyx interior greenish-brown at first, later brown-red. Originally named 'Orange' by Bruns, then renamed 'Orange Beauty' by Vuyk van Nes ('Orange Beauty' is also sometimes erroneously used for 'Jelena'). ⊕

'**Primavera**' (de Belder). Flowers medium-sized, in dense clusters, petals 17 mm long, sickle-shaped, primrose-yellow, calyx interior violet, early February. Not widely distributed. ⊕

'**Ruby Glow**'. Shrub, medium-sized, branches ascending to erect; leaves oval to circular, 12–16 cm long, 8–12 cm wide, orange, bronze and scarlet in fall; flowers medium-sized to large, petals to 18 mm long, coppery, overlaid with red, calyx interior wine-red, early (= 'Adonis'; *H. japonica* 'Flavopurpurascens Superba'). Developed in 1935 by Kort, Kalmthout Arboretum, Holland. Introduced 1946, by the Moreheim Nursery, Dedemsvaart, Holland. Flower color somewhat better than 'Hiltingbury', but fall color less intense; surpassed by 'Feuerzauber'. See 'Diane'. Fig. 86. ⊕

'**Winter Beauty**' (Wada, Japan 1962). Vigorous grower; flowers similar to those of 'Orange Beauty', but somewhat larger and brown-red at the petal base, otherwise deep golden-yellow to orange-yellow, 15–20 mm long. Cultivated in Holland. ⊕

H. japollis see: **H. × intermedia**

H. japonica S. & Z. Shrub, about 2.5 m high, habit wider, branches outspread, ash-gray, stellate pubescent when young; leaves broadly ovate, 5–10 cm long, with 6–9 veins (usually 7), light green beneath, glabrous or somewhat pubescent, tough, base asymmetrical; flowers in small, axillary heads, sepals usually reddish to brown-red, reflexed, petals bright yellow, to 2 cm long, January–

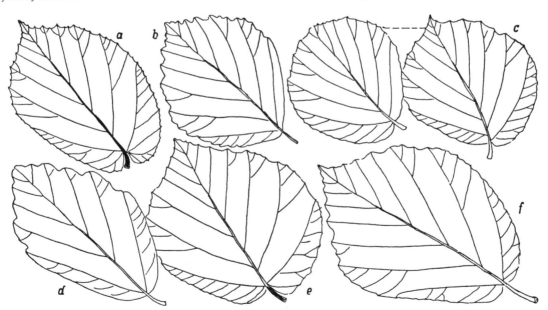

Fig. 86. **Hamamelis.** a. *H. mollis* (small leaf); b. *H. japonica*; c. *H. intermedia* 'Ruby Glow'; d. *H. vernalis* 'Lombarts Weeping'; e. *H. intermedia* 'Hiltingbury'; f. *H. mollis* 'Brevipetala'

March. Gs 1927: 64; JRHS 99: 3. Japan, mountain forests. 1862. z6 Fig. 86. ⊕

'Arborea'. Shrub, vigorous grower, to 5 m, branches widespreading to horizontal, even when young; leaves obovate to broadly obovate, short acuminate, 8–14 cm long, 4–9 cm wide, crenate, dull green, yellow in fall; flowers small, petals 12–13 mm long, crispate, yellow, calyx brown, March. Introduced by Siebold from Japan in 1862 and propagated vegetatively since that time. Of only slight garden merit. Plate 51.

var. **flavo-purpurascens** (Mak.) Rehd. Large shrub, similar to 'Arborea' in habit, but somewhat lower; leaves obovate to more rounded, to 10 cm long and 7 cm wide, open dentate on the apical half, fall color yellow; flowers small, petals 11–12 mm long, crispate, sulfur-yellow with a trace of red on the basal half, calyx interior red. Occurs in the wild in Japan (Oshima, Mutsu and Musashi Provinces). 1919. Of only slight garden merit; improved upon by the hybrids.

'Flavopurpurascens Superba' see: **H. × intermedia 'Ruby Glow'**

'Sulphurea' (Russell 1958). Large shrub, branches ascending at first, later outspread; leaves broadly obovate to nearly circular, to 11 cm long and 8 cm wide, crenate, yellow in fall; flowers small to medium-sized, 13 mm long, very crispate, sulphur-yellow, calyx interior purple-red, flowers abundantly. ⊕

'Zuccariniana'. Large shrub, erect at first, later outspread; leaves obovate to more rounded, base oblique, to 10 cm long and 7 cm wide, crenate, yellow in fall; flowers small, petals 11 to 12 mm long, sulfur-yellow, twisted and crispate, calyx interior green to green-yellow, flowers in late March. BMns 420; JRHS 99: 4. Introduced from Japan in 1891. Plate 51. ⊕

H. macrophylla Pursh. Very closely related to *H. virginiana*, often included within this species; leaves usually obtuse or rounded, lobes less numerous, less distinct, more rounded, base less uneven, more or less pubescent at first and eventually somewhat rough from the persistent bases of the fascicled pubescence; flowers smaller, yellow, 8 mm long, calyx 5 mm wide, interior yellow, with red stripes, December–February. SM 330. USA, Georgia to Texas. This plant is presumably the "late blooming *H. virginiana*" of garden culture (from Harms). z6

H. mollis Oliv. Shrub, to 5 m high, also higher in its habitat, young branches densely white pubescent; leaves oval-rounded, 8–16 cm long, to 11 cm wide, finely open dentate, somewhat metallic glossy above, densely fascicled pubescent, tomentose beneath, base usually cordate; flowers golden-yellow, petals rather wide, fragrant, straight (not curly), calyx with brown tomentose exterior, purple interior, January–March. BM 7884; DB 6: 46; 3: 35; JRHS 99: 1. China; Hupeh, Kiangsi Provinces. Widely distributed. 1879. Prettiest species for the garden. Only selected clones found in cultivation; plants imported from Japan are occasionally darker yellow. z6 Plate 51; Fig. 86. ⊕

'Aurantiaca' see: **'Brevipetala'**

'Brevipetala' (Chenault 1935). Larger, more upright shrub; leaves like the species, but undersides more distinctly bluish-green, persisting (dry) longer in the winter; flowers small, but in dense clusters of 3–12 together, petals only 10 mm long, 2

mm wide, not sinuate, orange-yellow, calyx interior yellow-green to light brownish, flowers early. JRHS 99: 2. Introduced in 1935 by the Grandes Roseraies du Val de Loire (then Chenault) of Orléans, France, first named 'Aurantiaca'. Fig. 86. ⊕

'Coombe Wood' (Veitch, around 1887). Large shrub, broad growing; leaves more or less oval, to 16 cm long and 11 cm wide, flat crenate on the apical half, yellow in fall; flowers medium-sized, petals to 19 mm long and 2 mm wide, flat, apex rolled inward, otherwise straight, golden-yellow, base somewhat reddish, calyx interior purple-red, very fragrant, January–February. Poor bloomer.

'Goldcrest' (Lord Aberconway 1961). Large shrub, branches ascending; leaves like the species, yellow in fall; petals to 20 mm long, 2 mm wide, somewhat bowed, apex crispate, golden-yellow, going to wine-red at the base, calyx purple-red inside, very fragrant. Developed in Bodnant, Wales, from seed introduced by H. E. Wilson from Japan. ⊕

'Pallida' (Wisley Gardens 1932). Medium-sized shrub, branches ascending; leaves like the species, pubescent beneath (not tomentose!), yellow in fall; flowers medium-sized to large, densely covering the branches, petals to 18 mm long, 2 mm wide, crispate and sinuate, sulfur-yellow, interior wine-red, very fragrant, January–February. JRHS 83: 43; 90: 18; 99: 6. A very beautiful form, widely planted in England. ⊕

H. vernalis Sarg. To 2 m high, stoloniferous shrub, bark on older stems very distinctly checked, branches more persistently pubescent; leaves obovate-oblong, 8–12 cm long, coarsely open dentate on the apical half, dark green above, usually distinctly blue-green beneath, often somewhat tomentose, usually with 5 vein pairs, fall foliage orange; flowers smaller than all the other species, bright yellow, slightly fragrant, petals often reddish on the base, calyx interior dark red, January–March. BM 8573; RH 1913: 131; DB 3: 64; NF 9: 221. USA, along river banks from Missouri to Louisiana. 1909. z5 ⊕ ∅

'Carnea'. Petals and calyx pale flesh pink inside. Before 1928.

'Lombarts Weeping' (Lombarts). Branches very pendulous, pubescence very persistent; leaves exceptionally blue-green; flowers not particularly attractive, reddish. Selection of P. Lombarts, Zundert, Holland; introduced to the trade in 1954. Fig. 86. ∅

'Red Imp' (Hillier 1966). Petals wine-red on the base, calyx likewise, becoming lighter toward the apex.

'Sandra' (Hillier 1962). Young leaves violet-purple, later green, but with the reddish coloration beneath persisting, fall color an intense orange, scarlet and red. JRHS 1976: 103. Selected by Hillier.

'Squib' (Hillier 1966). Petals yellow, calyx interior green.

f. **tomentella** Rehd. Leaves especially tomentose on the venation beneath; petals pure yellow. Mot 102.

H. virginiana L. Shrub to 5 m high, habit open rounded; leaves obovate, 8–15 cm long, coarsely crenate, uneven sided, light green, a beautiful yellow in fall, eventually quite glabrous; flowers light yellow, appearing just before or with leaf fall, very fragrant, calyx dull yellow-brown inside; fruits ripen the following year, then burst and throw the seeds up to 4 m. BB 1879; GSP 148; SS 198. Eastern N. America; in thickets at the edge of forests in deep fertile soil. 1736. The leaves were once used by the Indians as a medicine; the bark is still used as a source of witch hazel extract. z3 Plate 51. ∅

var. **angustifolia** Nieuwland. Branches long; leaves 6–12 cm long, 3–6 cm wide, petioles 1.5–2 cm long; petals 1.5 cm long, very narrow. USA; Indiana, Hudson Lake, Tamarack White Pine Bog. 1913

var. **orbiculata** Nieuwland. Branches glabrous, smooth, gray; leaves small, nearly circular, 1.5–5 cm long. USA; Indiana, dune region of Lake Michigan, Tamarack.

'**Rubescens**'. Petals reddish to the base, calyx interior yellowish to brownish green. Before 1922.

All species succeed best in a deep, moist, fertile soil; best used in large gardens and parks since most species are long lived and become very large. Because of their winter flowering character they are most effective when planted near a walkway for easier access.

Lit. Harms, H.: Unsere Freiland-Hamamelidaceae; in Mitt. DDG 1932, 7–11 ● Krüssmann, G.: Die *Hamamelis*-arten und-formen unserer Gärten; in Dtsh. Baumschule 1954, 44–47 (with ills.) ● Grootendorst, H. J.: *Hamamelis*; in Dendroflora **2**, 11 to 17, 1965 ● Lancaster, R.: Complete Guide to *Hamamelis* — the witch hazels; in Gard. Chronicle **167**, Nr. 21–24, May–June 1970.

HAMILTONIA

Hamiltonia oblonga see: **Leptodermis oblonga**

HAPLOPAPPUS Cass corr. Endl. — COMPOSITAE

Annuals, perennials and shrubs, usually with resinous or glandular foliage; leaves simple, alternate, occasionally the basal ones opposite, entire or distinctly lobed, flower heads small to large, either solitary or in fascicles, with disc and ray florets; involucre hemispherical to obconical; otherwise as described below. — 150 species in California, Mexico and Chile, usually in dry sandy soil.

Haplopappus ericoides (Lessing) Hook. & Arn. Evergreen shrub, compact habit, 0.3–1 m high, with numerous, thin, densely foliate branches with resinous stripes and yellow flowers; leaves filamentous, very small, 6–12 mm long, straight or bowed, usually furrowed on the dorsal side, resinous punctate, clustered at the nodes; flower heads in cymes on long shoots, forming large inflorescences, the individual heads with 2 to 6 ray florets and 8–12 disc florets, August–September. MS 680 (= *Diplopappus ericoides* Lessing). California, on sandy soil and dunes along the coast. z9 # ۞

HARRIMANELLA Coville See: CASSIOPE

Cassiope stelleriana and *C. hypnoides* are given their own genus according to Coville on the basis of the leaf arrangement and flower characteristics which differ from the other *Cassiope* species. Here, however, the previous classification will be retained.

HEBE Comm. ex Juss. — Shrub Veronica — SCROPHULARIACEAE

Evergreen shrubs, occasionally small trees in their native habitat; leaves opposite, very densely arranged, either conspicuously scale-like or more or less lanceolate to rounded or ovate, leaving distinct leaf scars when abscising; flowers in axillary racemes or small heads, white or pink, hybrids red to violet; corolla with a short tube, flared limb, 4 lobed; stamens 2, protruding; fruit a dry, septicidal capsule, compressed, thick walled. — Between 100–140 species, of which 90 are from New Zealand; the others from New Guinea, Australia, temperate S. America and the Falkland Islands. 15 species were considered to be another genus (see, *Parahebe*) by W. R. B. Oliver in 1944.

Outline of the Species described
● (from Cheeseman)

Leaves entire (occasionally with tiny incisions as on *H. salicifolia, diosmifolia, colensoi,* etc.);

1. Large shrubs to small trees, erect or procumbent; leaves 2.5–15 cm long, broad or narrow, loose, outspread, never imbricate; racemes simple, usually longer than the leaves, many flowered;
   a) Leaves obovate to oblong-lanceolate, usually wider than 8 mm:
      H. salicifolia, speciosa, and cultivars
   b) Leaves narrow, linear-lanceolate to linear-oblong, never wider than 8 mm:
      H. parviflora

2. Large shrub to small tree, erect or procumbent; leaves 4–30 mm long, usually very densely arranged, often imbricate, flat or concave or keeled; racemes or spikes usually short, simple or branched, often clustered at the branch tips;
   a) Flowers racemose; racemes more or less branched in umbellate panicles, occasionally simple:
      H. colensoi, diosmifolia
   b) Flowers racemose; racemes simple, occasionally branched:
      H. balfouriana, brachysiphon, canterburiensis, darwiniana, elliptica, glaucophylla, rakaiensis, traversii, vernicosa
   c) Flowers in spikes (often also in racemes as *H. decumbens*); spikes usually simple; leaves densely compact, imbricate, concave, round or keeled beneath:
      H. albicans, anomala, buchananii, buxifolia, carnosula, decumbens, pimeleoides, pinguifolia

3. Small shrubs, erect or procumbent; leaves dimorphic (on older plants small, short and thick, densely 4 ranked imbricate, occasionally in further spaced pairs; on young plants larger, more outspread, entire or irregularly finely lobed or pinnatisect); flowers clustered at the branch tips, usually in 2–4 flowered, short spikes;
   a) Procumbent habit, occasionally somewhat erect, flowers 2–4 in short spikes near the branch tips:
      H. tetrasticha
   b) Upright habit or broad or occasionally procumbent; flowers 3–8 near the branch tips, in small, terminal heads;
      + Leaves densely imbricate in opposite, connate and densely appressed pairs, completely surrounding the branch:
         H. hectori, lycopodioides, ochraceae
      ++ Leaves in widely spaced pairs, otherwise similarly tiny and appressed:
         H. cupressoides

4. Small, procumbent or prostrate shrubs; branches short, ascending; leaves small, 6–16 mm long; flowers in ovate, terminal heads; corolla tube long and narrow, with a small limb:
   H. epacridea

•• Leaves crenate or serrate;
   × Flowers in racemes:
      H. macrantha
   ×× Flowers sessile, in large panicles:
      H. hulkeana

The positive identification of the *Hebe* species is often only possible when the fruit capsules are present; the form and arrangement of the foliage should also be examined with a hand lens, especially those species with scale-like leaves.

Hebe albicans (Petrie) Cock. Low shrub, 30–50 cm high, but much wider, branches dense, spreading, stout; leaves erect to imbricate, broadly ovate to oblong, thick, blue-green, 1.5–3 cm long, 8–15 mm wide; flowers white. New Zealand; cultivated and fully winter hardy in England (Hillier). z9 #

H. amplexicaulis (Armstr.) Cock. & Allan. Small, usually procumbent shrub, 30–50 cm high, branches thickish, stiff, with circular leaf scars; leaves decussate, loosely imbricate, 12–15 × 8–12 mm large, broad oblong, slightly concave, base usually nearly stem-clasping and cordate; flowers white, in dense, oblong spikes. New Zealand. Plate 45.

H. × andersonii see: cv. 'Andersonii' p. 133.

H. angustifolia see: **H. parviflora** var. **angustifolia**

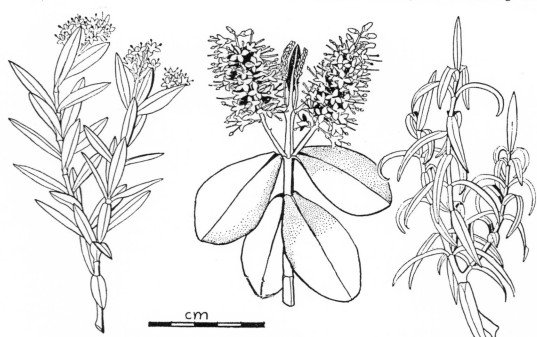

Fig. 87. **Hebe.** Left *H. subalpina;* middle *H. speciosa;* right *H. parviflora* (from Poole/Adams)

H. anomala (Armstr.) Cock. Shrub, erect, narrow, 0.7–1.5 m high, branches glabrous; leaves elliptic-lanceolate to narrow oblong, 8 to 20 mm long, glossy dark green above, acute, base tapering to a short petiole, somewhat cuneate; flowers white to light pink, in clustered spikes, of 5–10 flowers each and each 3 cm long, anthers blue, June to July. BS 3: 478; BM 7360. New Zealand. 1883. One of the hardiest species. z6 #

H. armstrongii Hort. Plants cultivated under this name are actually **H. ochracea,** which see. (The true *H. armstrongii* is illustrated in Fig. 89b).

H. balfouriana (Hook. f.) Cock. & Allan. Shrub, about 0.5–0.7 m high, erect, branches reddish, finely pubescent over the leaf axils; leaves elliptic, 6–18 mm long, margins reddish, nearly sessile, otherwise totally glabrous; flowers light violet, in 5–7 cm long, usually opposite paired racemes, corolla about 1 cm wide, anthers brownish, June–July. BM 7566. New Zealand. 1895. z7 #

H. brachysiphon Summerhayes. Rounded, 1.5 m high shrub in milder climates, much wider than high; leaves narrow ovate to oblong or lightly obovate, 12–25 mm long, 4–6 mm wide, short petioled; flowers white, in 5 cm long racemes, sepals as long as the corolla tube (!), at most twice as long, July. BM 6390 (as *Veronica traversii*) (= *H. traversii* Hort. non Hook. f.). New Zealand. 1868. z7 # ☼

H. buchananii (Hook. f.) Cock. & Allan. Shrub, low growing, 15–25 cm high, dense, compact, branches stiff, erect, rodlike; leaves very densely arranged, oblong to rounded, thick, leathery, concave, 3–6 mm long, blue-green; flowers white, usually 2–4 in clustered spikes near the branch tips, each spike about 1.5–2 cm long, stalk densely pubescent, June–July. New Zealand. z7 Fig. 88. # ∅

H. buxifolia (Benth.) Cock. & Allan. Erect shrub, 0.5–1 m high, young branches light green, glabrous or nearly so; leaves 4 ranked (decussate), oblong-obovate, 8–12 mm long, acute, base obtuse, glossy dark green above, lighter and finely punctate beneath; flowers white, in dense, 2.5 cm long spikes in the axils of the upper leaves, June–July. New Zealand. The cultivated plants differ somewhat from the wild species, according to Bean. z7 # ∅

H. canterburiensis (J. B. Armstr.) L. B. Moore. Shrub, 30–90 cm high, broad growing, branches very finely pubescent at first; leaves densely packed on the

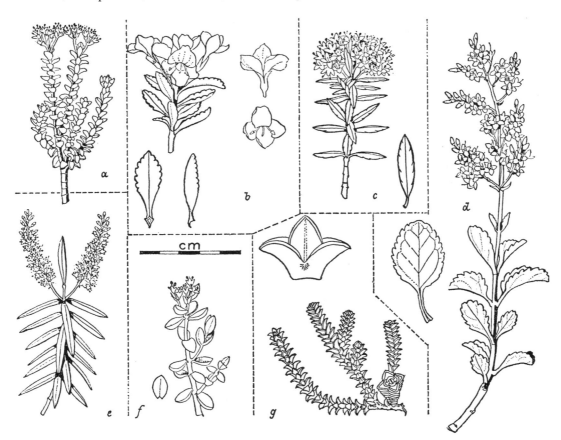

Fig. 88. **Hebe.** a. *H. buchananii;* b. *H. macrantha;* c. *H. diosmifolia;* d. *H. hulkeana;* e. *H. traversii;* f. *H. pinguifolia;* g. *H. epacridea* (from Poole/Adams)

branches, elliptic to obovate, acute, tapering to a short petiole at the base, 6–12 mm long, 3–4 mm wide, glossy dark green above; flowers white, 6–8 mm wide, in paired racemes near the branch tips, with 4–8 flowers, about 2–2.5 cm long, June–July. BMns 136. New Zealand. 1910. z9 # ☉

H. carnosula (Hook. f.) Cock. & Allan. More or less procumbent, occasionally upright shrub, scarcely over 0.3 m high in cultivation; leaves densely imbricate, broad obovate to rounded, 8–20 mm long, leathery, bluish-green, not keeled, 4 ranked; flowers white, sessile, about 6 mm wide, in numerous compact heads, grouped into dense, terminal spikes, sepals erect, about as long as the corolla tube, finely ciliate, July–August. New Zealand. Very hardy. z6 # ∅

H. colensoi (Hook. f.) Wall. Shrub, 0.3 to 0.5 m high, branches densely foliate; leaves obovate, abruptly acuminate, slightly petiolate, 2–4 cm long, 4–12 mm wide, rather blue-green when young, but later becoming dark green, totally glabrous, normally entire, but occasionally sparsely finely dentate; flowers white, in dense, about 2.5 cm long racemes in the upper leaf axils, corolla tube shorter than the calyx, July–August. New Zealand, Ruahine Mts. (Illustration in BM 7296 by Cheeseman, not typical!) z6 Fig. 90. #

H. cupressoides (Hook. f.) Cock. & Allan. Erect, usually nearly globose shrub of cypress-like appearance, about 0.5 m high, very fine and densely branched, branches often finely pubescent; leaves scale-like and densely appressed on mature plants, about 1.5 mm long, acute, not completely covering the stem, leaves linear on young plants and to 6 mm long; flowers pale blue-lilac, usually 3–8 in small heads at the branch tips, anthers red-brown, June–July. BM 7348; LNz 150. New Zealand. Attractive, popular, hardy species. z6 Plate 45; Fig. 89. # ∅

H. darwiniana see: **H. glaucophylla**

H. decumbens (Armstr.) Cock. & Allan. Low growing shrub, procumbent stem with erect branches, 30–70 cm high, blackish-red (!); leaves very densely arranged, ovate to oblong or obovate, 8–20 mm long, thick, fleshy, dark green, red bordered (!); flowers white, in 1–2 raceme pairs near the branch tips, each raceme 12–20 mm long, July–August. New Zealand. 1888. Very hardy. z6 Plate 45. # ∅

H. diosmifolia (Cunn.) Cock. & Allan. Small shrub, well branched, erect, 0.5–1 m high (occasionally to 5 m high in its habitat), young branches glabrous to densely floccose, covered with leaf scars; leaves densely arranged, outspread to nearly erect, linear to narrow lanceolate, acute, 12 to 25 mm long, dark green above, lighter beneath, with 2–4 short teeth on either side; flowers light blue to white, in 2.5 cm long, cymose racemes, calyx usually 3 parted, occasionally 4 parted, July. BM 7539 (as var. *trisepala*). New Zealand. 1835. z7 Fig. 88. #

H. elliptica (Forst. f.) Pennell. Bushy shrub, about 1 m (to 5 m in its habitat) high, branches rodlike, gray pubescent all around or only in 2 stripes; leaves densely arranged, erect, linear-oblong, 8–18 mm long, flat, leathery, not glossy, base truncate, midrib distinct; flowers few in short racemes, these grouped into loose corymbs at the branch tips, white to fleshy pink, July–August. BM 242 (as *Veronica decussata*); LNz 150. New Zealand, Chile, Tierra del Fuego. 1776. Notable for the nearly 2 cm wide corollas. z7 # ☉

H. epacridea (Hook. f.) Cock. & Allan. Shrub, low or prostrate, densely branched, completely covered with overlapping 4 ranked leaves; leaves stiff, ovate, V-shaped cuneate, acute, 5 mm long, reflexed, opposite leaf pairs connate at the base, dull dark green, limb lighter; flowers white, in ovate, dense, terminal, 1.5–3 cm long heads, corolla 3 mm wide, July. LNz 150; PRP 48. New Zealand. 1860. z7 Fig. 88. #

H. fairfieldii Hook. f. Probably a hybrid of *H. hulkeana* × *H. lavaudiana*; low growing and stiffer than the former, but the flowers similar, lilac, but in large racemes; leaves only 12–25 mm long. BM 7323. z7 Plate 4.

H. glaucophylla (Cock.) Cock. Erect shrub, about 0.7 m high, very similar to *H. brachysiphon* (but with blue-green foliage and pubescent corolla throat); leaves 4 ranked, oval-lanceolate, acute, 15–20 mm long, somewhat concave, blue-green on both sides, scarcely petiolate; flowers white, 2–4 in racemes near the branch tips, each about 2–3 cm long, flower stalks pubescent, July–August (= *H. darwiniana* sensu Cheesem. p.p.). New Zealand. z7 #

H. hectori (Hook. f.) Cock. & Allan. Shrub, erect, 0.2(0.5) m high, branches very numerous, erect, densely arranged, cylindrical; leaves scaly, broadly ovate, 2–3 mm long, densely imbricate appressed, glossy, floccose on the margin; flowers light pink to white, in small, ovate, terminal heads, July. BM 7415; LNz 150; PBl II: 534; PRP 50. New Zealand. 1895. z7 Plate 45; Fig. 89. # ∅

H. hulkeana (F. Muell.) Cock. & Allan. Loose habit, wide and divaricate, usually not over 30 cm high (1.5 m or more in its habitat); leaves broadly ovate, 3–5 cm long, coarsely dentate, rounded at the base, in sparsely arranged pairs, dark glossy green above; flowers soft lilac, in 15–45 cm long panicles, half as wide as long, May–June. BM 5484; JRHS 78: 125; HTS 289. New Zealand. 1860. Without a doubt the prettiest flowering of the shrubby types, however very tender. z9 Fig. 88. # ☉

H. loganioides (Armstr.) Cock. Shrub, usually procumbent, conifer-like appearance, 15–35 cm high, branches erect, gray-white floccose; leaves only 3–4 mm long, dark green; flowers white, in small heads, June–July. BM 7404. New Zealand. z6–7 # ∅

H. lycopodioides (Hook. f.) Cock. & Allan. Shrub, rigid and narrowly upright, 30–70 cm high, resembling *H. hectori*, but branches always 4 sided (!); leaves appressed,

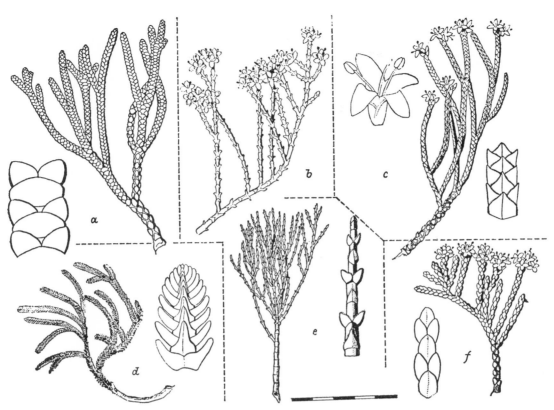

Fig. 89. **Hebe.** a. *H. hectori;* b. *H. armstrongii* (true!); c. *H. lycopodioides;* d. *H. tetrasticha;*
e. *H. cupressoides;* f. *H. ochracea* (from Poole/Adams)

scale-like, cuneate, in small part connate at the base; flowers white, in small, terminal heads, anthers blue, June. BM 7338; LNz 151, 155. New Zealand. z7 Fig. 89. # Ø

H. macrantha (Hook. f.) Cock. & Allan. Erect shrub, to over 2 m high in its habitat, 0.5 m in cultivation, branches glabrous; leaves densely arranged, obovate, often narrow, 1–2.5 cm long, distinctly dentate on the apical half, thick fleshy, especially thick on the limb, glabrous, light green; flowers pure white, corolla 2 cm (!) wide, grouped 3–8 in the upper leaf axils. BMns 177; LNz 156; BS 2: 338; PRP 49. New Zealand. Seldom seen in cultivation but quite hardy. Very attractive and large flowered. z6 Fig. 88. # ✣

H. ochracea B. M. Ashwin. Low shrub, 40–60 cm high, with thick, blackish, outspread main branches, axillary branches bowed, often also horizontal, young shoots only on the upper side of the branches; leaves only 2 mm long, triangular, tapering to a cuneate, obtuse apex, olive-brown (!) colored, somewhat more outspread in young plants, otherwise tightly appressed; flowers white, 4–8 in terminal clusters, not very attractive, anthers purple, but only a few flowers, July–August. LNz 150 (= *H. armstrongii* Hort. p.p.). New Zealand. 1899 (?). Very winter hardy. This is the plant generally

cultivated as *"Hebe armstrongii";* the true *H. armstrongii* (J. B. Armstr.) Cock. & Allan is as yet not cultivated. z6 Plate 44; Fig. 89.

H. parviflora (Vahl) Cock. & Allan. Found in cultivation as the following 2 varieties:

var. **angustifolia** (Hook. f.) L. B. Moore. Shrub, 0.7–1.5 m high, loose, thin, branches slender, erect, glabrous, glossy, eventually dark brown; leaves linear, drawn out to an apical point, 3–7 cm long, 3–6 mm wide, totally glabrous; flowers numerous, white with a trace of lilac, in 5–12 cm long, axillary racemes, near the branch tips, July to September. BM 5965 (= *H. angustifolia* [A. Rich.] Cock. & Allan). New Zealand. Around 1868. z7 Plate 44. #

var. **arborea** (Buchan.) L. B. Moore. Loose branched shrub, to 1.8 m high or more (5–7 m in its habitat), young branches thin, the leaf pairs 8–12 mm apart; leaves linear to more lanceolate, acute, sessile, 2.5–6 cm long, 4–6 mm wide, finely pubescent on the limb, otherwise glabrous; flowers white, with a trace of lilac, densely arranged in slender, 7 cm long racemes near the branch tips, numerous. New Zealand. The tallest form; largest measured to date is 8.4 m high with a 60 cm diameter trunk. z7 Fig. 87. #

H. pimeleoides (Hook. f.) Cock. & Allan. Shrub, procumbent or also sometimes erect, to 40 cm high, branches pubescent; leaves elliptic, ovate to obovate, 6 mm long, light blue-green on both sides, somewhat cuneate, outspread or directed upward; flowers pale lilac, in 2 to 4 cm long, simple or branched spikes. June–August. BM 8967 (= *Veronica glaucoerulea* Hort.). New Zealand. z7 #

H. pinguifolia (Hook. f.) Cock. & Allan. Thick branched, procumbent or more erect shrub, branches 0.3–0.7 m high, whole leaf scars, young branches soft pubescent; leaves obovate to rounded, blue-green, 6–18 mm long, thick, leathery, rather fleshy, usually red margined, base rounded (not cuneate); flowers white, in small, compact spikes at the branch tips, anthers blue, June–August. BM 6587 (as *H. carnosula*); BM 6147; PRP 47. New Zealand. 1868. Good winter hardiness. z6 Fig. 88. #

H. propinqua (Cheesem.) Cock. & Allan. Shrub, well branched, 30–90 cm high, procumbent to erect, branches outspread, rather twisted, thin, cylindrical, about 1 mm thick including the leaves, these triangular, appressed, leathery thick; flowers 12 together in spikes near the branch tips, white. New Zealand. Usually more cushion-like in cultivation, 30–50 cm high. z6 Plate 44. #

H. rakaiensis (J. M. Armstr.) Cock. Dense shrub, about 50 cm high, but wider than tall, branches with 2 pubescent stripes on each internode; leaves outspread, elliptic to more oblong or slightly oblanceolate, 12–18 mm long, 5–6 mm wide, soft, somewhat glossy green above, dull beneath, glabrous except for the ciliate margins; flowers in simple racemes, 5 cm long, white, June–July. New Zealand. Good ground cover but usually cultivated under the name of *H. subalpina*. z6 Fig. 90.

H. salicifolia (Forest. f.) Pennel. Shrub, 3–4.5 m high in its habitat, branches green, glabrous; leaves lanceolate or more oblong, 5–15 cm long, 1.2–2.5 cm wide, apex long drawn out, base tapering to a short, wide petiole, rather thin, glabrous except for the lightly pubescent midrib; flowers in slender, compact, cylindrical racemes, 10–15(25) cm long, 2 cm wide, corolla small, white, lilac toned, corolla lobes narrow, not outspread, June–August. BR 32: 2. New Zealand. z7

H. speciosa (R. Cunn.). Cock. & Allan. Erect, vigorous, 1–2 m tall shrub, branches thick, angular; leaves obovate to oblong, 5–10 cm long, rounded at the apex, leathery, midrib softly pubescent above, dark green; flowers to 8 mm wide, dark purple to blue-purple, several in 5–7 cm long, terminal racemes, 2–2.5 cm wide, July–September. BM 4057; NF 4: 176; FS 17; 2317. New Zealand. 1835. z7 Fig. 87. # ✦

For large leaved *Hebe* cultivars with inflorescences longer than the leaves, see Cultivars p. 133.

H. subalpina (Cock.) Cock. & Allan Fig. 87 — Plants under this name in cultivation are actually **H. rakaiensis.**

H. tetrasticha (Hook. f.) Andrs. Small, low growing shrub, only 10–30 cm wide, branches very numerous, procumbent-ascending, young branches 4 sided; leaves densely 4 ranked imbricate, ovate-triangular, apex

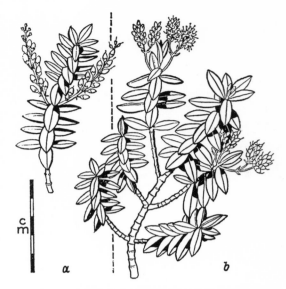

Fig. 90. **Hebe.** a. *H. rakaiensis;* b. *H. colensoi* (from Poole/Adams)

outspread from the stem, 2–2.5 mm long, with short apex, dorsal side not keeled, margins ciliate; flowers white, 2–4 in small spikes at the branch tips, May–June. LNz 150, 151; PRP 52. New Zealand. Very hardy. z6 Plate 45; Fig. 89. # ⊘

H. traversii (Hook. f.) Cock. & Allan. Shrub, broad growing, to 2 m high in its habitat, branches floccose at first; leaves densely arranged, elliptic-oblong to obovate-oblong, about 2.5 cm long, acute, cuneate at the base; flowers white, in 3–6 cm long racemes, July. LNz 153. New Zealand. Differing from the very similar *H. brachysiphon* in the corolla tube which is 2–5 times as long as the sepals. z7 Fig. 88. # ✦

H. vernicosa (Hook. f.) Cock. & Allan. Shrub, erect or procumbent, 0.3–0.7 m high, branches outspread, finely soft pubescent when young, later totally glabrous; leaves very numerous, densely arranged, outspread from the stem, 8–15 mm long, obovate-oblong, obtuse or with small tips, glossy deep green above (polished!); flowers white, in 2.5–4 cm long racemes at the branch tips, June–August. BS 3: 491. New Zealand. z7 Plate 45. # ✦

CULTIVARS

Breeding. The particulars are sparse and scattered in the literature. However, it doubtlessly began in Scotland where Isaac Anderson-Henry developed 'Andersonii' in 1849 and 'Andersonii Variegata' in 1856. In 1867, 'Imperialis' appeared from Bourschalat in Lyon, France (FS 2317). Around the turn of the century Victor Lemoine of Nancy, France followed with 'Bolide', 'Enchantresse', 'La Séduisante' and 'Simon Deleaux'. Also around 1900 Th. Smith began his work at Daisy Hill Nursery, in Newry, N. Ireland, while Veitch started in Exeter, England, in 1911. In 1930 a few more English hybrids arrived on the scene ('Alicia Amherst', 'Hielan Lassie', 'Marjorie', etc.).

Then breeding work apparently subsided until 1956, when Keessen (Holland) began his work on *Hebe*. These cultivars were, however, primarily for greenhouse culture and are listed separately following cultivars.

The following cultivars are mainly to be found in the gardens of England and Ireland in the milder coastal regions.

'Alice Amherst' (Hillier, Cat. 1950). Shrub, to about 1 m high; leaves elliptic to more ovate, much reduced at the apex and drawn out to an obtuse tip, 7–8 cm long racemes, corolla tube wide, scarcely longer than the calyx, flowering after August. FS 658. Closely related to *H. speciosa*. ⬧

'Andersonii' (Isaac Anderson-Henry, about 1849). Presumably *H. salicifolia* × *H. speciosa*. Broad growing, to 1.8 m high; leaves elliptic, 7–11 × 2.5–3.5 cm, large, rather obtuse, slightly acuminate, dark glossy green; flowers in slender, horizontal to ascending racemes, 10–14 cm long and about 3 cm wide, corolla violet, spent flowers fading to nearly white, stamens markedly exserted, flowers from August on. Plate 44. ⬧

'Andersonii Variegata' (before 1856). Mutation of 'Andersonii. Leaves ivory-white bordered, gray-green and normal green in the center. Very susceptible to frost damage. Several Dutch clones derive from this cultivar. See following section. ⬧

'Autumn Glory' (Smith Nursery, Newry, N. Ireland, about 1900.) Seedling of 'Tobarcorranensis'. Erect habit, to 50 cm, open branched, branches dark; leaves broad elliptic to more obovate, 2.5–3 cm long, fine red margined when young, then glossy green with a trace of blue-green; flowers in wide, dense compound racemes, 3.5–4.5 cm long, deep purple-blue, corolla tube white, flowering from July on. BS 2: 50. ⬧

'Bowles Hybrid' (Hillier, Cat. 1950). Origin unknown, perhaps *H. diosmifolia* × *H. parviflora*. Shrub, scarcely 50 cm high; leaves narrow-elliptic to more oblong, pale green, somewhat glossy, to 2.5 cm long; inflorescences usually composed of 2–5 racemes, 7–10 cm long, the individual flowers loosely arranged, pale purple-lavender. ⬧

'Carl Teschner' (*H. elliptica* × *H. pimeleoides*). Low shrub, procumbent, to 30 cm high and 2–3 times as wide, the older stems nearly black; leaves deep green, elliptic, 1 cm long, acute; flowers in 3–4 cm long racemes, violet, July. Valerie Finis, England, named this variety. However, it seems the plant is listed in the New Zealand literature as *Hebe* **'Youngii'**, named in 1928 by James Young, curator of the Christchurch Botanic Gardens, where the plant still grows. ⬧

'Hielan Lassie' (Hillier, Cat. 1950). Compact habit; leaves narrow; flowers in 5–7 cm long racemes, corolla tube white, limb tips intense violet-blue, fading to lighter color, filaments white. ⬧

'La Séduisante' (V. Lemoine). Close to *H. speciosa*. Leaves elliptic, 5–7 cm long, bronze-green, purple beneath when young; flowers in about 10 cm long racemes, dark carmine or purple. MD 1933 Pl. 3. Before 1897. ⬧

'Marjorie' (Hillier, Cat. 1964). Low, wide shrub, 50 cm high or more, but wider than tall; leaves like the 'Andersonii' types, but shorter and more finely dentate when young; flowers in short racemes, pure light blue. Supposedly perfectly hardy. ⬧

'Midsummer Beauty' (J. Cheal & Sons, before 1959). Hybrid of *H. salicifolia*. About 1 m high and wide; leaves oblong-elliptic, to 10 cm long, acuminate, very reddish beneath when young; flowers in rather dense, 12 cm long racemes, purple-lilac, not fading, very prolific bloomer from July to fall. ⬧

'Mrs. Winder' (Hillier, Cat. 1964). Shrub, 0.7 to 1 m, rather dense growing, older branches red-brown, glabrous; leaves narrow elliptic-oblong, 2.5–3 cm long, 6 mm wide, crosswise rugose at first with a red limb, midrib raised beneath and red toned on the base; flowers in 6–7.5 cm long racemes, white. ⬧

'Simon Deleaux' (V. Lemoine). Hybrid of *H. speciosa*. Leaves ovate, 5 cm long, purple margined; flowers in short racemes, lobes carmine, concave, tube and anthers purple. ⬧

'White Gem' (Hillier, Cat. 1964). Not from *H. brachysiphon*, but more probably *H. brachysiphon* × *H. pinguifolia*. Low shrub; flowers very abundantly, white, from early June. ⬧

'Youngii' see: **'Carl Teschner'**

THE DUTCH *Hebe*-HYBRIDS

The origins of *Hebe* culture in Aalsmeer, Holland are not known. However, before 1900, Keessen was growing 'Jacob Keessen' and 'Catherine' as well as the white variegated 'Andersonii Variegata'. Around 1920 all documented cultivation and hybridization practically ceased, doubtlessly a result of the war years. Around 1955 with *Hebe* being cultivated for export, work on the genus resumed, and with good results. At that time Keessen received a few plants of the 3 above mentioned cultivars and commenced his work. All these forms bloom from July–October and are predominantly used as potted plants.

Green leaved types

'Albert Keessen'. 35–50 cm high; flowers red. Discovered in Cornwall and introduced to the trade by Keessen in 1975. ⬧

'Catherine'. 35 cm; flowers pink. Very old cultivar. ⬧

'Jacob Keessen'. To 50 cm high; flowers crimson-red, in 12 cm long racemes. Old cultivar. ⬧

'Mathilde'. 35 cm high; flowers blue. Long in the trade. ⬧

Variegated types

All sports of *Hebe* 'Andersonii Variegata', developed by Keessen since 1965 and introduced to the trade in 1969:

'Dineke'. 35 cm; leaves green in the center, margin white; flowers an attractive marine-blue. ∅

'Herfstzon'. 30–35 cm; leaves yellow variegated, normal size; flowers few, indistinctly light blue. Meritorious only for the attractive yellow foliage. ∅

'Mickey'. 20–25 cm high; small leaved, white variegated; flowers insignificant, purple-red. Poor grower. ∅

Not difficult to cultivate; for sunny areas and a light sandy soil. Winter hardiness is often better than expected. The following species, if given a light cover, will tolerate temperatures to −15°C (5°F) or lower: *H. anomala, buxifolia, carnosula, cupressoides, hectori, pinguifolia, tetrasticha, traversii*. The dwarf species from the mountains of New Zealand are doubtlessly hardy, but they have yet to be proven in cultivation. (Except for gardens in England and Scotland where they seem to be perfectly hardy.) The *H.* 'Andersonii' hybrids are very suitable flowering shrubs in milder climates.

Note: Every new hybrid of *Hebe*, before it is introduced to the trade, should be reported to the Office of International Registration (Royal New Zealand Institute of Horticulture, P.O. Box 450, Wellington, N.Z.).

Lit. Cheeseman, T. F.: Manual of the New Zealand Flora; 490–547, 1906 (84 species fully described). ● Cockayne, L., & H. H. Allan: The present taxonomic status of the New Zealand species of *Hebe*; Transact. N. Z. Inst. **57**, 11–47, 1926. ● Allan, H. H.: Flora of New Zealand 1, 885–952. 1961. ● Souster, J.: Notes on some cultivated Veronicas (IV); Anderson's *Veronica* and some similar cultivars; in Jour. RHS **87**, 34–38, 1962.

HEDERA L. — Ivy — ARALIACEAE

Evergreen shrubs, climbing with aerial roots (some cultivars not climbing); leaves alternate, simple, those on the long shoots coarsely dentate or lobed, those on the fruiting branches entire, leathery tough, smooth, usually long petioled; flowers greenish-yellow, small, in umbellate racemes, 5 parted; styles 5, connate; fruit a 3 or 5 seeded, black or yellow berry. — 5 species in Europe, N. Africa and Asia.

For absolute identification of a species or cultivar, the properties of the stellate pubescence must be examined; otherwise one must look at many branches in various stages of growth.

Of the many cultivars developed as potted plants, such as 'Pittsburgh', 'Emerald Gem', 'Chicago', 'Lee's Silver', etc., only a few will be covered here since they are seldom found in gardens. Not much has been reported on their hardiness, although many will survive unharmed outdoors in a temperate climate.

Key to the species (from Lawrence)

A. Characteristics of mature branches
(branches radial, without holdfasts or aerial roots)

  ● Stellate hairs with 4–10 rays (occasionally to 14 rays):

        H. helix

  ● ● Hairs usually scale-like, with 8–30 rays;

    + Branches and petioles usually wine-red; leaf blade relatively thin; pubescence stellate and scale-like:

        H. canariensis

    ++ Branches and petioles green; leaf blade distinctly leathery; pubescence only scale-like;

      * Leaf blade ovate, base broadly cuneate to cordate, crushed leaves have a strong scent of celery or the Umbelliferae; stellate hairs usually with 25–30 rays:

        H. colchica

      ** Leaf blade usually elliptic-lanceolate, base cuneate, crushed leaves not or only slightly scented; stellate hairs usually with 15–20 rays;

        1. Fruits orange; ovaries at flowering time only reaching slightly beyond the base of the calyx lobes; leaves usually 2–3 cm long, much longer than wide:

          H. nepalensis

        2. Fruits black; ovaries hemispherical and considerably surpassing the base of the calyx lobes; leaves usually 1.5–3 cm long, slightly longer than wide:

          H. rhombea

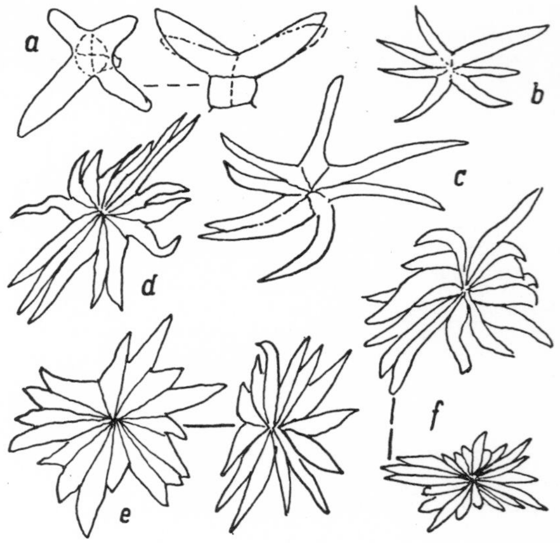

Fig. 91. **Hedera.** Stellate hairs (enlarged approx. 100 ×). a. *H. helix* (from the side and from above); b. *H. helix poetica*; c. *H. helix taurica*; d. *H. canariensis*; e. *H. colchica*; f. *H. rhombea* (from Tobler)

B. Characteristics of juvenile (sterile) branches
(branches dorsiventral, with holdfasts or aerial roots, these usually at or near the nodes)

  ● Hairs always stellate, 4–10 rays (occasionally to 14 rays):

        H. helix

  ● ● Hairs all scale-like or partly scale-like and stellate;
    + Leaf blade triangular-ovate to triangular-lanceolate, unlobed or with 1–10 penniform lobes on either side:

        H. nepalensis

    ++ Leaf blade usually broadly ovate, entire or with few, palmate lobes;
      * Branches and petioles usually wine-red; blade thin and usually glossy; leaves slightly scented when crushed; hairs predominantly with 15 rays (13–22 rays), partly scale-like partly stellate:

        H. canariensis

      ** Branches and petioles green; leaf blade thick and leathery, often dull (not glossy); hairs always scale-like;
        1. Leaves entire or slightly lobed, with strong celery or Umbelliferae scent when crushed, usually 7–25 cm long:

          H. colchica

        2. Leaves 3–5 lobed, not or only slightly scented; only 2.5–10 cm long:

          H. rhombea

Hedera acute see: **H. colchica 'Amurensis'**

H. amurensis see: **H. colchica amurensis**

H. canariensis Willd. Tall growing vine; leaves 3–5 lobed, ovate-triangular, 12–20 cm long, 5–12 cm wide, base cordate, pubescence stellate and scale-like, usually with 15 rays, leaves on mature branches smaller, ovate, usually unlobed, petioles and branches dull wine-red; fruits black, 6–10 mm thick. BFCa 49; GH 6: 77a; HIv 1; THe 19. Canary Islands, Azores, Madeira, NW. Africa. z7 Plate 46; Fig. 91. #

'**Azorica**'. Especially strong growing, large leaved, leaves 5 or 7 lobed, 7–15 cm wide, lobes ovate, obtuse, bright green, young leaves and branches with a thick brownish tomentum. Introduced from St. Michael, Azores, by Osborn, Fulham, England.

'Gloire de Marengo' see: '**Souvenir de Marengo**'

'**Margino-maculata**'. Similar to 'Variegata', but limb lighter yellow, blade more speckled, more green marbled, the green center patch more dull yellowish, less color contrast, entire to 3 lobed. GH 6: 77d. #

'**Souvenir de Marengo**'. Leaves large, white variegated, margins broad yellowish-white, center gray-green to blue-green, petiole and branches dark wine-red. GH 6: 77b (= 'Variegata'; 'Gloire de Marengo'). Discovered in Algeria in Villa Marengo. Plate 48. #

'**Striata**'. Leaves 3 lobed, large, margins wide black-green, center pale green or yellowish-white. GH 6: 77C. #

'Variegata' see: '**Souvenir de Marengo**'

H. colchica K. Koch. Caucasus Ivy. High climbing vine, branches very stiff, pea-green, dense scaly, scales usually yellow with 20 rays; leaves broadly ovate to elliptic, 10–20 cm long, usually entire, occasionally somewhat lobed, very leathery, tough, base cordate or round, dull green above, yellow scaly beneath; flowers with a distinct calyx; fruits larger than those of *H. helix*, 6–8 mm thick, eventually blue-black. HM 2285; THe 21–30; GH 6: 76 (= *H. helix colchica* K. Koch). SE. Europe, Asia Minor, Caucasus, Transcaucasus to N. Iran. z6–7 Plate 46; Fig. 91. # ∅

'**Amurensis**'. Leaves to 30 cm long, usually entire with a few small, acute teeth, occasionally also lobed (= *H. amurensis* Hibb.; *H. acuta* Hibb.). Origin unknown, but not from the Amur River region as the name would imply. Introduced by Bull's Nursery, Chelsea London in 1887. Poor climber, must therefore be tied up. #

'**Dentata**'. Like the species, but with leaves thinner, lighter, less glossy, margins with small, sharp, sparsely arranged teeth. GH 6: 76b (= *H. helix dentata* Hibb.; *H. colchica* "amurensis" in many nurseries). Frequently occurs with the species, common in cultivation and hardier. z6 Plate 48. #

'Dentata Aurea Striata' see: '**Sulphur Heart**'

'**Dentata Variegata**'. Like 'Dentata', but margins cream-white with an adjacent gray-green patch, center green. GH 6: 76c. Beautiful form, cultivated at least since 1907. #

'Paddy's Pride' see: '**Sulphur Heart**'

'**Sulphur Heart**'. Leaves large, deep green, with one or several, small to very large yellow to yellow-green patches in the center, leaves of the juvenile branches slightly dentate (= *H. colchica* 'Dentata Aurea Striata'; 'Paddy's Pride', 1970). Attractive old form, renamed in Holland in 1968. #

H. helix L. Common English Ivy. Vine, creeping on the ground or climbing to 30 m on trees and walls, with aerial roots, young branches gray stellate pubescent,

hairs with 4–6 rays; leaves on the vegetative branches 3–5 lobed, 4–10 cm long, dark green above, often with whitish venation, yellowish dark green beneath, leaves of the flowering branches oval-rhombic, entire; flowers in globose umbels, these grouped into racemes, greenish-yellow, September–October; fruits globose, black, ripening in the following year. HM 2286–2294; GH 6: 78. Europe to the Caucasus (Scotland to Spain). Cultivated for centuries. z5 Fig. 91. # ∅

Of the many varieties and cultivars, the following are the most important:

Outline of the forms discussed
(from Lawrence & Schulze, simplified),

● Branches without aerial roots; plants shrubby; Leaves usually unlobed:
'Arborescens'(fruits black); var. *poetica* (fruits yellow)

●● Branches usually with aerial roots, at least at the nodes; plants climbing or shrubby; leaves usually lobed;
  * Leaves green, at least in summer;
    + Habit stiff, shrubby, dense:
'Congesta', 'Conglomerata', 'Conglomerata Erecta', 'Cristata', 'Donerailensis', 'Ivalace', 'Maegheri', 'Ovata'
  ++ Habit always climbing or with long shoots;

    1. Leaves normally without side lobes; basal lobes present or absent:
'Deltoidea', 'Glymii', 'Ovata', 'Scutifolia'

    2. Leaves usually lobed, at least on the side:
'Atropurpurea', 'Baltica', 'Crenata', 'Digitata', 'Green Ripple', 'Helvetica', 'Palmata', 'Pedata', 'Sagittifolia', 'Scutifolia', 'Shamrock', 'Walthamensis'

      a) Leaves especially large, 8–14 cm long:
'Hibernica'

      b) Leaves small, usually 5–8 cm long;
· Stellate hairs often with 8 rays:
'Baltica', 'Gracilis'

·· Stellate hairs seldom with more than 6 rays:
helix (type), 'Scutifolia'

  ** Leaves white or yellow variegated or totally yellow;
    + Only the margins colored:
'Cavendishii', 'Cullisii'

    ++ Colored portion in the center of the blade or irregularly distributed, no variegated margins;

      1. Blade punctate or sprinkled with white or yellowish:
'Glacier', 'Maculata', 'Marmorata Minor', 'Tricolor'

      2. Blade with one or more variegated patches, either punctate or striped:
'Aureovariegata', 'Goldheart'

      3. Blade totally yellow:
'Buttercup', 'Emerald Gem'

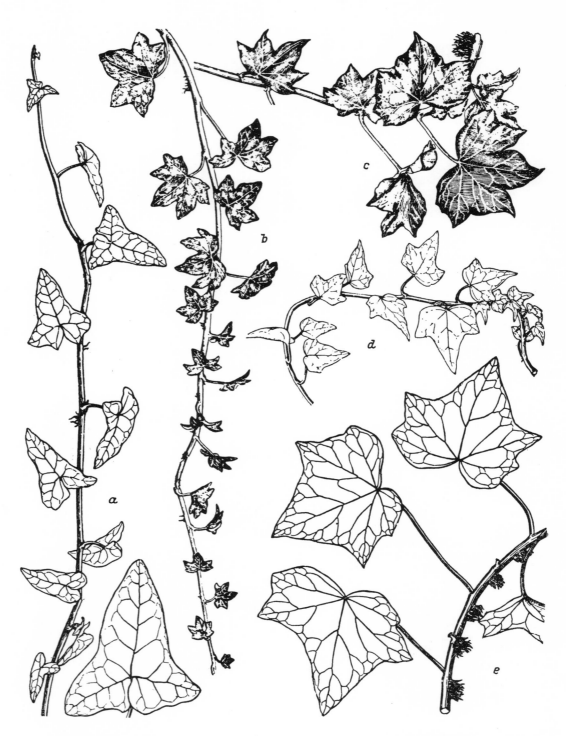

Fig. 92a. *Hedera helix* (juvenile forms). a. 'Sagittifolia'; b. 'Digitata'; c. 'Crenata'; d. 'Gracilis'; e. var. *hibernica*
(from Lawrence & Schulze)

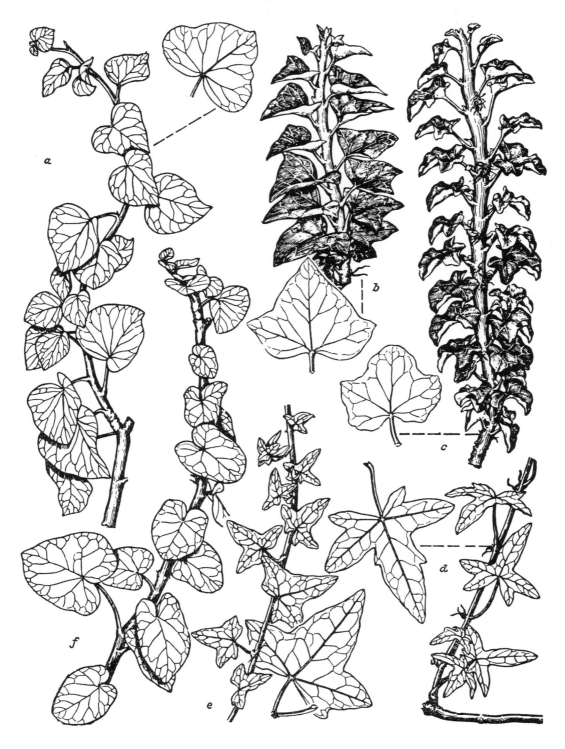

Fig. 92b. *Hedera helix* (juvenile forms). a. 'Glymii'; b. 'Conglomerata Erecta'; c. 'Conglomerata'; d. 'Pedata';
e. 'Donerailensis'; f. 'Deltoidea' (from Lawrence & Schulze)

'**Alt Heidelberg**' (Brother Ingobert Heieck of Neuberg, near Heidelberg, W. Germany). Very densely foliate, leaves very small, rhombic, dull above and beneath, the middle of each blade half with a rather indistinct, hemispherical lobe, 2 cm long, 1–1.5 cm wide, petiole 2–4 mm long. One of the more delicately textured forms.

f. *arborea* see: '**Arborescens**'

'**Arborescens**'. A mature form, held constant by vegetative propagation of the flowering twigs; habit rounded, short branched, not climbing; leaves smaller, unlobed (= *H. helix* f. *arborea* Hort.). #

'**Atropurpurea**'. Leaves with a long middle lobe and 3 short side lobes, many nearly unlobed and ovate, blackish-green, becoming darker in winter and then often bronze with green venation. #

'**Aureovariegata**'. Leaves partly totally yellow or light green, some yellow and green variegated on the same plant. HIv 1 (Pl.) (= *H. helix chrysophylla* Hibb.). # Ø

'**Baltica**' (Rehder). Scarcely differing from the species, except for the much better winter hardiness; small leaved, stellate hairs predominantly with 8 rays (the type 4–6 rays). AB 6: 1. Discovered by Rehder in 1907 near Riga, Latvia. z5 #

'**Buttercup**'. Leaves obtuse 3 or 5 lobed, base deeply cordate, venation elevated, golden-yellow in the summer. #

ssp. *caenwoodiana* see: '**Pedata**'

'**California Fan**'. Very compact habit; leaves broader than long, 3–5 cm wide, with 5–9, but usually 7, short, abruptly acuminate lobes, the sinuses crispate, apple-green, somewhat darker in summer, petiole 2–3 cm long. Fig. 93.

'**Cavendishii**' (Paul 1863). Remaining low, branches and petioles greenish; leaves usually 3 lobed, usually 2.5–3.5 cm long, middle lobe usually an equilateral triangle, base cordate, margins broad, variegated white to cream-white, turning somewhat pink in fall and winter. GH 6: 80b; MD 1927: Pl. 10d (= *H. helix marginata minor* Hibb.). #

var. *chryocarpa* see: var. **poetica**

ssp. *colchica* see: **H. colchica**

'**Congesta**'. Similar to '**Conglomerata**', but margins with lobes and not undulate; slow growing, low, dense, branches stiffly erect; leaves small, triangular, 3 lobed, base cordate. #

'**Conglomerata**' (Haage & Schmidt). Dwarf form, stem and branches stiff, erect at first, soon twisted and collapsing, then procumbent, green with reddish traces, densely gray stellate pubescent, stellate hairs with 4–6 rays, aerial roots numerous, internodes 4–8 mm long; leaves often 2 ranked, triangular to ovate, 1.2–3.5 cm long, dull green, channeled, lobes 3–5, side lobes obtuse or round, often absent, base deeply cordate, margins distinctly and stiffly undulate, venation green, petiole 6–15 mm long. GH 6: 85B; DB 1951: 63; JRHS 97: 29. Introduced to the trade by Haage & Schmidt, Erfurt, E. Germany in 1875; but known previously in England. Fig. 92b. #

'**Conglomerata Erecta**'. Similar to '**Conglomerata**', but branches tightly upright, not twisted, somewhat stocky, 30–50 cm high; leaves distinctly 2 ranked, channeled, acute triangular, side lobes small, acute, base flat cordate. GH 6: 85a; DB 1943: 158; JRHS 97: 23. Fig. 92b. #

'**Crenata**'. Resembling '**Digitata**', but branches stiffer, aerial roots less numerous; leaves somewhat glossier, margins very undulate (not "crenate"), 2 to 6 cm long, base truncate to cordate, petioles 2.5 to 5 cm long, also longer on long shoots. GH 6: 84B. Fig. 92a. #

'**Crista Curlilocks**' see: '**Curly Locks**'

'**Cristata**'. Very vigorous; leaves medium-sized, roundish, margin undulate and crispate lobed, bright green (= 'Crispa'; 'Curlilocks'). Plate 47. #

ssp. *chrysocarpa* see: **H. nepalensis**

ssp. *chrysophylla* see: '**Aureovariegata**'

'**Cullisii**'. Very similar to '**Cavendishii**', but margin becoming reddish in fall, petiole always red. HIv Pl. 3 (as *H. helix marginata rubra*). #

'**Curlilocks**' see: '**Cristata**' and '**Curly Locks**'

'**Curly Locks**'. Leaves very small, 5 lobed, middle lobe much longer than the others, irregularly crenate and very undulate-crispate, mid-vein often distinctly white, petiole 2–3 cm (= 'Cristata Curlilocks', 'Curlilocks'). American cultivar. Fig. 93.

'**Deltoidea**'. Branches light reddish to green, gray stellate pubescent, stellate hairs with 4–6 rays, internodes 1 to 2.5 cm long; leaves ovate in outline to obtuse triangular, 1.5–7.5 cm long, dull dark green, side and basal lobes coming together and hardly distinguishable, entire, base deeply cordate, basal lobes overlapping (!), venation gray-green, slightly raised. GH 6: 86A; MD 1927: 13b–h (= *H. helix palmata* Bosse). Discovered before 1836 in Ireland. Fig. 92b. #

ssp. *dentata* see: **H. colchica** '**Dentata**'

'**Digitata**'. Branches green, densely gray stellate pubescent, stellate hairs with 4–6 rays, aerial roots numerous, internodes 3–5 cm long; leaves 5–7 lobed, 2–4.5 cm long, mostly wider than long, blade concave, deep dull green, venation gray-white, middle and side lobes nearly equal pointed, petiole green. 1.5–4 cm long. GH 6: 84A; MD 1927: 13b–h (= *H. helix palmata* Bosse). Found in Ireland before 1836. Fig. 92a. #

ssp. *discolor* see: '**Marmorata Minor**

'**Donerailensis**'. Branches green to reddish, bowed, gray stellate pubescent, hairs with 4–6 rays, internodes 2 to 5 cm long; leaves small, middle lobes narrow triangular, also the side lobes, margins undulate, deep green, coloring purple-bronze in winter. GH 6: 82A; HIv 76; MD 1927: 12d–f (= *H. helix minima* Hibb.). Possibly originated in Doneraile, County Cork, Ireland. Cultivated at least since 1854. The characteristics are those for plants in pot culture; when planted in the landscape, the leaves will be distinctly larger and then more similar to '**Pedata**'. Fig. 92b. #

'**Elegantissima**' see: '**Tricolor**'

'**Emerald Gem**'. From var. *poetica*; leaves medium-sized, a good golden-yellow. Popular pot plant. #

'**Erecta**' see: '**Conglomerata Erecta**'

'**Eugen Hahn**'. Sport of '**Pennsylvanian**'. Leaves paper thin, nearly translucent, triangular-ovate, usually completely unlobed or with some lobes on one side of the base, 4–5 cm long, 3 cm wide at the base, both sides rounded and cordate, quite densely finely white punctate and marbled, nearly effectively gray, petiole 10 to 15 mm long. Introduced in 1977 by the Stauss Brothers, Möglingen, W. Germany. Fig. 93.

'**Gavotte**' (Brokamp, Ramsdorf, Krs. Borken, Westf.). Leaves rather small, usually unlobed, occasionally with 1–2 small lateral lobes, acute, dark green, 5–7 cm long, apexes usually

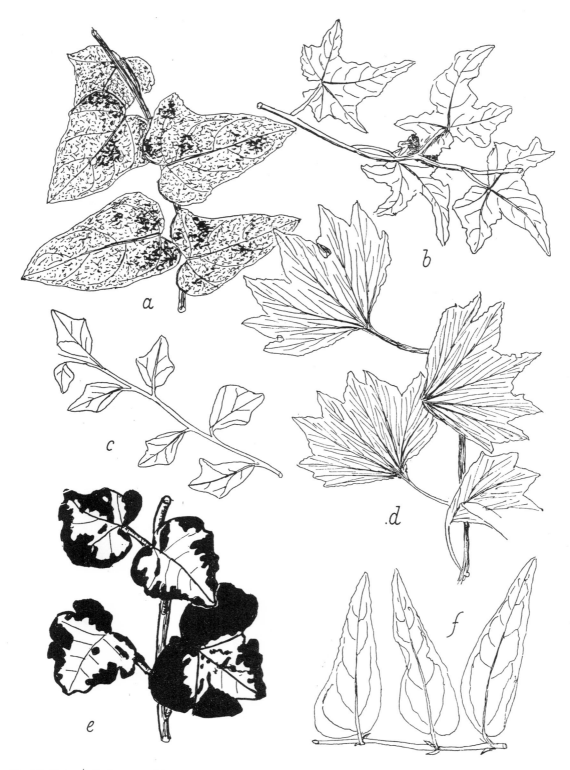

Fig. 93. *Hedera helix* forms. a. 'Eugen Hahn'; b. 'Curly Locks'; c. 'Alt-Heidelberg'; d. 'California Fan'; e. 'Stift Neuberg'; f. 'Gavotte' (Original)

rounded, blades very thin, petiole 5–10 mm long. Found as a mutation on 'Star' in 1953, introduced in 1956. Fig. 93.

'Glacier'. Leaves 3 lobed, unequal in size and shallow, base cordate to truncate, margins variegated, unevenly white, middle silver-gray and dull green, light marbled, very consistent. JRHS 97: 28. Pretty. #

'Glymii' (Paul). Branches green, stiff, climbing, stellate hairs usually with 4–6 rays, aerial roots numerous, internodes 6–15 mm long; leaves apple-green, glossy, usually wider than long, triangular to ovate in outline, 1.5–4 cm long, unlobed except for the basal lobes which are normally not (!) overlapping. GH 6: 86B (= H. helix tortuosa Hibb.). Origin unknown; although before 1867 in England. Fig. 93. #

'Golden Jubilee' see: **'Goldheart'**

'Goldheart'. Young branches reddish, later golden-yellow; leaves unequal in size, small to medium, nearly triangular, acute, margin deep green, light yellow to golden-yellow center (= 'Goldherz'; 'Jubiläum Goldherz'; 'Golden Jubilee'; 'Oro di Bogliasco'). Origin unknown (perhaps Italy?). Very vigorous with long shoots in the landscape, with better color than with pot culture. #

'Goldherz' see: **'Goldheart'**

'Gracilis'. Branches thin, dull red, young branches distinctly gray stellate pubescent, stellate hairs often with 8 rays, internodes 3–4 cm long, aerial roots at the nodes, petioles 2–5 cm long, reddish; leaves 2–4(5) cm long, thin, leathery, middle lobe the largest, margins finely undulate, lobes obtuse, base broad and flat cordate, venation gray-white. GH 6: 78D; HIv 67. Fig. 92a. #

'Green Feather' see: **'Meagheri'**

'Green Ripple' (Hahn, Pittsburgh). Leaves rather small, cuneate, dark dull green, middle lobes drawn out long and narrow, lateral lobes likewise, inclined forward, basal lobes shorter and more blunt, base usually rounded or cuneate, light green veined (= 'Hahn's Green Ripple'). Plants cultivated by this name in USA, Germany, England and Switzerland are not totally alike! The true American form has crispate, lightly undulate leaves. #

'Hahn's Green Ripple' see: **'Green Ripple'**

'Helvetica'. Branches and petioles green or reddish; leaves ovate-hastate in outline, 2.5–7 cm long, dark green, venation white to greenish, 3 lobed, basal lobes hardly recognizable, margin convex, base deeply cordate, lobes slightly overlapping. GH 6: 83A. Discovered in Switzerland, brought to the USA before 1932. #

'Hibernica'. Fast growing, branches greenish-brown, darker red near the apex, densely gray stellate pubescent, hairs on the twigs usually with 6–8 rays, the leaf hairs however only 4 rayed, internodes 3 to 5 cm long; leaves 5 lobed, 5–12 cm wide, occasionally to 15 cm long on long shoots, dull dark green, venation lighter green (not white!), middle lobe largest, as long as wide, base deeply cordate, petiole 5–20 cm long. GH 6: 79A. Ireland. Common in cultivation since 1838. z7 Fig. 92a. # ∅

'Hibernica Maculata'. Like 'Hibernica', but leaves white to yellowish speckled or margined. HIv 98: Pl. 2. #

'Ivalace'. Clinging vine, very densely foliate; leaves 5 lobed, about 2.5 cm long and wide, deep green, very glossy, leathery, margin bowed upward creating a nearly pointed effect, middle lobes more conspicuous than the lateral ones, sinuses deep. American cultivar.

'Kolibri' (Brokamp, Ramsdorf). Apparently a mutation of 'Sagittifolia', but with smaller leaves, 5 lobed, the middle lobes much larger than the laterals and both basal lobes, about 3 cm long and wide, very white variegated, with white, light green, dark green and gray-green parts, internodes 3–4 cm long, petiole 1.5 to 2 cm long.

ssp. *lucida* see: **'Scutifolia'**

ssp. *marginata minor* see: **'Cavendishii'**

ssp. *marginata rubra* see: **'Cullisii'**

'Marmorata Minor'. Small leaved form like the species, but with white speckled or punctate blade, the spots often grouped into larger patches, many leaves also white with green spots. GH 6: 80a; HIv 78 (Pl.) (= H. helix discolor Hibb.). #

'Meagheri' (Danker). Shrubby habit, procumbent, branches twiggy, to 1 m long, green to reddish, stellate pubescent, hairs with 4–6 rays, internodes 3 to 20 mm long; leaves usually 3–5 lobed, 2–3 cm long, green with lighter midrib, middle lobe largest, elongated, side lobes triangular, base truncate to cordate, petiole 1–2 cm long. GH 6: 81A (= H. helix 'Green Feather'). Introduced to the trade in 1940 by F. Danker, Albany, N.Y., USA. (Minier.) #

'Minima' see: **'Donerailensis'**

'Oro di Bogliasco' see: **'Goldheart'**

'Ovata'. Similar to 'Deltoidea', whose base has 2 deeply incised, overlapping lobes; leaves elliptic-oval, base acute rounded, unlobed or with quite flat lobes on the long shoots, bright green. #

ssp. *palmata* see: **'Deltoidea'**

'Palmata'. Similar to 'Digitata'; leaves distinctly 5 lobed, the lobes triangular, acute, base usually truncate, venation raised beneath. #

'Pedata'. Branches green, occasionally reddish, slender and flexible, stellate pubescent, internodes 2.5 cm long, aerial roots at the nodes; leaves leathery, 2.5–5 cm long, usually 5 lobed, apple-green, glossy, later much darker, venation white, raised, middle lobes lanceolate, usually 3–6 times longer than wide, often with 1–2 teeth in the margin, petiole 1.5–3.5 cm long. HIv 76; SH 2: 287i–k (= H. helix caenwoodiana Nichols.). Often atypical in the trade. Fig. 92b. #

var. *poetica* West. Branches green or reddish; leaves broadly ovate, usually with 3–5 shallow lobes, glossy green, occasionally yellowish, acute, base cordate, margin undulate, mature foliage not lobed; fruits yellow(!). THe 11–14 (as H. poetarum) (= H. helix var. chrysocarpa Ten.). SE. Europe, especially in S. Italy and Greece, Crete, Turkey. Fig. 91. #

'Sagittifolia' (Paul). Branches reddish; stellate hairs with 4–12 rays, internodes 1–5 cm long; leaves pubescent, dull green, usually indistinctly 5 lobed, middle lobes long triangular, side lobes distinctly hastate, basal lobes small, 2–5 cm long, margins occasionally undulate, mature foliage similar to the typical H. helix, but with 8–12 rayed stellate hairs; fruits allegedly yellow (?). GH 6: 83B; MD 1927: 12p–r; HIv Pl. 1; JRHS 97: 24 (= H. helix var. taurica [Hibb.] Rehd.). Before 1864. Plate 48; Fig. 92a. #

'Sagittifolia Alba'. Resembling 'Sagittifolia Variegata', but with the white variegated parts distinctly separate, the white variegation sometimes making up half the leaf blade, or only on the margin or the lobe apexes ('Sagittifolia Variegata' is indistinct on the variegated parts, often only densely punctate).

Plate 33

Gaultheria ovatifolia
in the Glasnevin Botanic Garden, Dublin, Ireland

Gaultheria caudata
in the Glasnevin Botanic Garden, Dublin, Ireland

Gaultheria sinensis
in Gibson Park, Glenarn, Scotland

Gaultheria hispida
at Malahide, Ireland

Plate 34

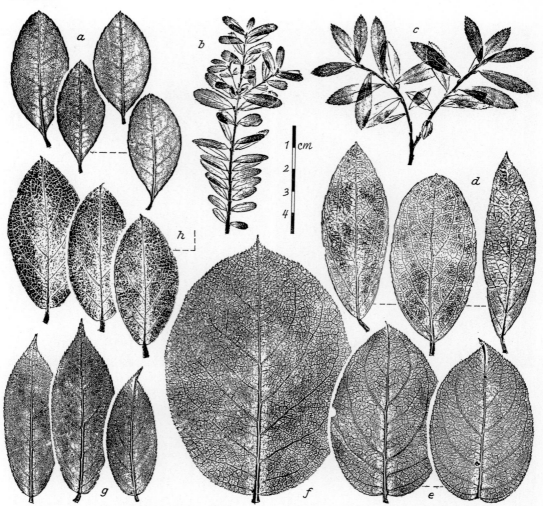

Gaultheria. a. *G. procumbens* (from cultivation); [b., see *Pachistima!*]; *c. G. cuneata* (from cultivation); d. *G. forrestii* (wild); e. *G. shallon* (from Canada); f. *G. shallon* (from USA); g. *G. yunnanensis* (from cultivation); h. *G. hookeri* (from China)

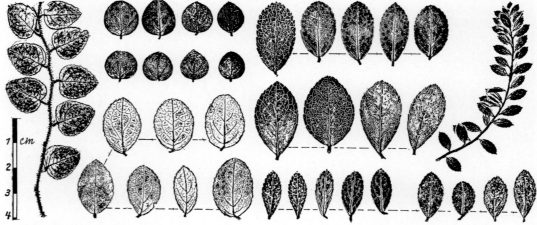

Gaultheria. Left, branch from *G. nummularioides;* next right, 8 leaves of *G. humifusa* from USA (above) and 7 leaves from *G. pyroloides* from the Himalayas (beneath); right, branch from *G. itoana* (from cultivation); next left, 5 leaves from *G. miqueliana* (wild) from Japan (upper row), from cultivated plants (middle row) and 9 leaves from 2 different origins of *G. cuneata* of W. China (lower row)

Plate 35

Gaultheria procumbens in a hoarfrost

Gleditsia japonica
in the Dortmund Botanic Garden, W. Germany

Gleditsia tricanthos var. *inermis*
in the Dortmund Botanic Garden, W. Germany

Plate 36

Genista hispanica in Royal Botanic Gardens, Kew, England   *Genista sericea* in Edinburgh, Scotland

Genista silvestris var.*pungens*
in the Dortmund Botanic Garden, W. Germany

Genista sagittalis
in the Dortmund Botanic Garden

Genista tinctoria 'Plena'
in the Dortmund Botanic Garden

Genista villarsii
in the Lausanne Botanic Garden, Switzerland

Plate 45

Hebe hectori
Photo: C. A. Purpus

Hebe cupressoides in Logan, Scotland

Hebe decumbens in Edinburgh, Scotland

Hebe tetrasticha in Edinburgh, Scotland

Hebe vernicosa
in Glasnevin, Dublin, Ireland

Hebe amplexicaulis
in the Lausanne Botanic Garden, Switzerland

Plate 46

Hedera. a.*H. canariensis* (3 leaves); b. *H. nepalensis* (right, 2 leaves of the juvenile form; left, 2 of the mature form); c.*H. colchica* (2 leaves of the juvenile form); d.*H. rhombea* (below, 2 leaves of the juvenile form; above, 2 of the mature form); e.*H. nepalensis* var. *sinensis* (3 leaves of the mature form) (samples collected in the plants' native habitat)

Plate 47

Haloxylon ammodendron in the Tashkent Botanic Garden, USSR
Photo: K. Browicz, Poznan

Hedera helix 'Shamrock'
in the Linz Botanic Garden, Austria

Hedera helix 'Cristata'
in Malahide, Ireland

Plate 48

Hedera colchica var. *dentata* (transition to the mature form) in the Dortmund Botanic Garden, W. Germany

Hedera canariensis 'Gloire de Marengo'
in the Jardin Marimurtra, Blanes, Spain

Hedera helix 'Sagittifolia'
in the Jardin Pinya de Rosa, Santa Cristina, Spain

'Scutifolia'. Branches and petioles green to reddish; leaves ovate, usually 3 lobed, deep green, very glossy, venation lighter, base distinctly cordate. GH 6: 79C (= *H. helix lucida* Hibb.). #

'Shamrock'. Leaves small, middle lobes nearly elliptic, both side lobes more or less distinct, very small, light green, internodes short. Fully winter hardy and an excellent ground cover; very popular form. Plate 47. #

var. **taurica** (Tobler) Rehd. Leaves on vegetative shoots at most sagittate and glossy light green; fruits a normal dark green (not yellow, as suggested by Tobler) (= *H. taurica sensu* Pojark.; *H. poetica* var. *taurica* Tobler). Krim. Fig. 91. #

ssp. *tortuosa* see: 'Glymii'

'Stift Neuburg' (Brother Ingobert Heieck, Stift Neuberg, near Heidelberg, W. Germany). Mutation of 'Glacier', but very different; leaves irregularly circular, flat round lobed, partly also triangular, 2–4 cm long and wide, center of blade pure white with a few small, deep green spots or marbling, otherwise deep green, very distinct against the white center, base deeply cordate, petiole 2–3 cm. Gorgeous form. Fig. 93.

'Stuttgart' (Stauss Bros. 1972). Somewhat stronger growing sport of 'Ivalace', densely foliate, leaves glossy deep green, 3.5 cm wide and long, margins finely crenate and undulate, usually distinctly 5 lobed.

'Tricolor'. Leaves small, scarcely lobed, more or less truncate at the base, gray-green, with a broad silvery margin, margin turning pink in winter (= 'Elegantissima'; 'Marginata Elegantissima'). Cultivated since 1867. #

'Walthamensis'. Branches and leaf petiole base light reddish-brown, stellate pubescent; leaves 3–5 lobed, usually only 1.5–3 cm long, occasionally 4 cm, equally wide, middle lobe as long as wide, sinuses usually acute, lobes round or acute, venation raised, whitish. GH 6: 78C. England. Before 1897. Little known. #

H. himalaica see **H. nepalensis**

H. japonica see: **H. rhombea**

H. nepalensis K. Koch. High climbing, branches with 12–15 rayed scale hairs, thin, reddish when young, aerial roots at every node; leaves on long shoots ovate-triangular, with 2–5 obtuse teeth on either side or entire, thin leathery, 5–12 cm long, base truncate, mature form with entire, oblong-ovate to more lanceolate leaves, petiole scaly, leaf blade slightly so; flowers 8–20 in umbels, these grouped into racemes, stalk very scaly; fruits orange. THe 31–38; GH 6: 75; HAL 68 (= *H.*

himalaica Tobl.; *H. helix chrysocarpa* Hibb.). Afghanistan to Kashmir, Himalaya. 1880. z8–9 Plate 46. #

var. **sinensis** (Tobl.) Rehd. Leaves of the juvenile form entire to 3 lobed, the mature form elliptic-ovate to elliptic-lanceolate; fruits yellow, to 1 cm thick. THe 39–42 (= *H. sinensis* Tobl.; *H. himalaica* Tobl.). China. Plate 46. #

H. poetarum see: **H. helix** var. **poetica**

H. rhombea (Miq.) Bean. High climbing, scaly hairs with 7–14 rays; leaves glossy green, juvenile form shallow and obtuse 3–5 lobed, 2–7 cm wide, broadly ovate to nearly kidney-shaped, the middle lobes much larger than the side ones, tips usually short acuminate, leaves of the mature form unlobed, usually obtuse-rhombic in outline, 3–7 cm long, base cuneate; flowers 12–30 in umbels, these grouped into panicles; fruits black, globose. THe 43–44a (= *H. japonica* Tobl.; *H. tobleri* Nakai). Japan (except Hokkaido), S. Korea. z7–8 Plate 46,49; Fig. 91. #

'Variegata'. Margin white variegated, irregularly and obtuse toothed. #

H. sinensis see: **H. nepalensis** var. **sinensis**

H. taurica see: **H. helix** var. **taurica**

H. tobleri see: **H. rhombea**

Valuable ground cover in the landscape, especially for humus soil in wooded areas with sufficient moisture, although it tolerates some dryness; roots easily. The variegated forms are best cultivated as pot plants.

The American Ivy Society, at the Cox Arboretum, 6733 Springboro Pike, Dayton, Ohio, 45449, has further information regarding dates and other facts on the genus *Hedera*. See also × Fatshedera.

Lit. Hibberd, S.: The Ivy, a monograph; London 1872 (completely outdated today, but well illustrated) ● Tobler, F.:Die Gattung *Hedera*; 151 pp., 57 ills., Jena 1912 ● Tobler, F.:Die Gartenformen der Gattung *Hedera*; in Mitt. DDG 1927, 1–33 (with 6 ills. and 11 plates). ● Lawrence, G. H. M., & A. E. Schulze: The cultivated Hederas; in Gent.Herb. 6, 107 to 173, 1942 (12 Pls.; most important work) ● Lawrence, G. H. M.: The cultivated Ivies; in Morris Arb. Bull. 7, 19–31, 1956 ● Jenny, M.:Araliaceae. *Hedera*; in Jahr. Verein ehem. Oeschberger, 1965, 93–102, Zurich (with 46 photos) ● Pierot, S.: The Ivy Book ● Schaepman, H.: Preliminary Checklist of cultivated *Hedera* (with 400 names).

HEDYSARUM L. — LEGUMINOSAE

Shrubs, subshrubs or perennials; leaves odd pinnate on wavy branches; leaflets opposite, without stipules, entire, often transparent punctate; flowers in axillary spikes or racemes, supported by dry membranous subtending leaves and prophylls; calyx campanulate, with 5 uneven teeth; standard obovate, slightly clawed, wings very short, keel obliquely truncate; upper stamens distinct, the others connate; fruit a flat sectioned pod with circular or square sections. — About 70 species in the temperate zone of the Northern Hemisphere, nearly all perennials.

Hedysarum multijugum Maxim. Deciduous shrub, broad growing, to 1.5 m high, branches thin, gray-yellow, wavy; leaves 10–15 cm long, with 15–27 leaflets, these oval-oblong, 6–15 mm long, gray-green, dense silky pubescent beneath; flowers purple, in erect, 25 cm long racemes, June–August; pods 3 cm long. Mongolia. Not in cultivation!
var. **apiculatum** Sprague. Like the species, but the leaflets acuminate (the species obtuse or emarginate at the apex). BM 8091; BS 2: 93; HM 1518 (= *H. multijugum* Hort. non Maxim.). Mongolia. Only this variety known in cultivation. z5 ⊕ Sandy, dry soil.

HEIMIA Link. & Otto — LYTHRACEAE

Subshrubs or perennials; branches 4 sided, leaves opposite or in whorls or alternate, entire; flowers axillary or in 3 forked panicles, yellow, purple or blue; calyx campanulate, with hornlike appendages (epicalyx) between the lobes, petals 5–7, stamens 10–18; fruit a 4 chambered capsule. — 3 species, from southern N. America to Argentina.

Heimia myrtifolia Cham. & Schlecht. Deciduous shrub, about 1 m high, branches erect, very twiggy, glabrous, densely foliate; leaves linear, willow-like, opposite on the basal part of the branch, alternate toward the apex, 3–5 cm long, 3–6 mm wide, glabrous; flowers quite short stalked, solitary in the leaf axils, yellow, 8–12 mm wide, petals 5–7, July–September. Brazil, Uruguay. 1821. (Kew Gardens). z9

H. salicifolia (H. B. K.) Link & Otto. Deciduous shrub, to 3 m high in its habitat, but usually much lower; leaves usually opposite, sessile to short petioled, linear-oblanceolate, alternate on the apical half, to 5 cm long, 1 cm wide; flowers solitary and short stalked in the leaf axils, calyx campanulate, 5–9 mm long, petals 5–7, orange-yellow, ovate, 12–17 mm long, July–September. HI 554 (= *Nesaea salicifolia* H. B. K.). Texas, northern Mexico to Argentina. 1821. Cultivated in the Cambridge Botanic Garden, England. z9

No special soil requirements, but a warm location is important. The branches will often freeze back to the ground even in warm climates, but the new growth is always strong in the spring.

HELIANTHEMUM Mill. — Sunrose — CISTACEAE

Evergreen or semi-evergreen, dwarf subshrubs; leaves small, opposite, the upper ones also alternate, pinnate venation, with or without stipules; flowers in raceme-like cincinnus; sepals 5, very uneven, the 3 inner ones much larger than the 2 outer; petals 5, yellow, orange, pink, red or white, usually yellow or orange on the base; stamens numerous, all fertile; carpels 3; style always bent at the base or bowed in an S-curve (!); fruit a 3 valved capsule. — About 80 species in Europe, the Mediterranean region and Asia Minor.

Key to the species described

- Style longer than the stamens; leaves with stipules;
  - \+ Stipules all linear-awl shaped, the lower ones and middle ones as long as the petiole; flowers white;
    - × Calyx small, 3–5 mm long:
      - *H. pilosum*
    - ×× Calyx larger, 5–8 mm long:
      - *H. apenninum*
  - \++ Stipules lanceolate to narrow lanceolate, longer than the petiole; flowers usually yellow;
    - × Leaves stellate pubescent on both sides; flowers yellow;
      - \> Inner sepals elliptic, stellate tomentose:
        - *H. sulphureum*
      - \>> Inner sepals finely tomentose, ovate:
        - *H. glaucum*
    - ×× Leaves glabrous or with only the underside pubescent, occasionally somewhat tomentose;
      - * Leaves gray tomentose beneath:
        - *H. nummularium*
      - ** Leaves green beneath, scattered stellate pubescent:
        - *H. nummularium* ssp. *grandiflorum*

- •• Style shorter than the stamens; leaves without stipules;
  - \+ Leaves gray pubescent beneath; flowers 8–12 mm wide:
    - *H. canum*

- \++ Leaves green on both sides;
  - \> Flowers 1.5–2 cm wide; fruit capsule ovate:
    - *H. oelandicum* ssp. *alpestris*
  - \>> Flowers 6–8 mm wide; fruit capsule globose:
    - *H. oelandicum*
  - \>>> Flowers solitary long stalked, yellow with an orange spot at the base of each petal:
    - *H. lunulatum*

Helianthemum apenninum (L.) Mill. Subshrub, to 40 cm high, branches and leaves gray tomentose to whitish; leaves petiolate, elliptic-oblong to linear, 1–3 cm long, margins usually involute, either both sides gray tomentose or sometimes green above; flowers white, about 3 cm wide, 3–10 in simple racemes, petals 5, May–August. PEu 78; HM 2030; HF 1261 (= *H. polifolium* Pers.; *H. pulverulentum* DC.). W. and S. Europe to Asia Minor. Before 1768. z5 Fig. 94. ⊕

var. **roseum** (Jacq.) Schneid. Leaves green, glabrous above; flowers pink (= *H. apenninum* var. *rhodanthum* [Dun.] Bean; *H. rhodanthum* Dun.). NW. Italy. Not to be confused with the cultivar 'Rhodanthe Carneum'.

H. canum (L.) Baumgart. Subshrub, 10–20 cm high, very well branched at the base, flower stalks erect, pubescent; leaves linear to oval-lanceolate or obovate, to 3 cm long, 6 mm wide, green to gray tomentose above, always white tomentose beneath, long haired; flowers 3–15 in simple racemes, dark yellow, about 1.2 cm wide, May–June. HM Pl. 184; 2040. Central and S. Europe. Quite variable. z6 Fig. 94. ⊕

H. chamaecistus see: **H. nummularium**

H. croceum (Desf.) Pers. Habit grasslike, dense, branches procumbent or erect; leaves petiolate, the basal ones rounded, the middle and upper leaves ovate-oblong to oblanceolate, 1 to 2 cm long, both sides stellate pubescent, blue-green beneath; flowers 3–15 in simple racemes, yellow or white, about 2 cm wide, May–July (= *H. glaucum* Pers.). S. Europe, N. Africa. 1815. z7 Plate 43.

H. glaucum see: **H. croceum**

H. lunulatum (All.) DC. Grasslike habit; leaves elliptic-oblong, usually obtuse, glabrous or sparsely pubescent; flowers usually solitary, on 2.5 cm long stalks, yellow, with an orange spot at the base of each petal, 1.5 cm wide, outer sepals linear, inner ones oval, acute. EP IV, 193: 15A. S. Europe. z7

H. nummularium (L.) Mill. Subshrub, to 35 cm high, branches ascending, green pubescent; leaves petiolate, elliptic to lanceolate, 2 to 5 cm long, margins flat or only slightly involuted, both sides green or gray tomentose beneath; flowers many in loose racemes, yellow, about 2.5 cm wide, sepals 5, style 2–3 times as long as the ovary, May–July. HM Pl. 184; HF 1260; GPN 581; BS 2: 365 (= *H. tomentosum* Smith; *H. vulgare* Gaertn.; *H. chamaecistus* Mill.). Nearly all of Europe and Asia Minor. z5 Fig. 94. ⊕

Includes a very large number of cultivars, usually crosses between species; please refer to the end of this section.

ssp. **grandiflorum** (Scop.) Schinz & Thell. Shrub, procumbent or erect, to 40 cm high; leaves petiolate, broadly ovate to oblong, 1–3 cm long, pubescent above, with scattered fascicle hairs beneath; flowers grouped many together, yellow, 2.5–3 cm wide, May to July. HM 2035. Europe, Asia Minor. 1860. ⊕

H. oelandicum (L.) Swartz. Subshrub, branches without stipules; leaves oblong-lanceolate to lanceolate, 6–10 mm long, fleshy, glabrous or pubescent on the margin and midrib beneath; flowers grouped 3–4, very small, only 6–8 mm wide, petals and sepals about equal length, style as long as the ovary. HF 1258; GPN 582. Spitsbergen, Norway; Oeland, Sweden. z5

ssp. **alpestre** (Jacq.) Breitst. Habit dense, grasslike, 3 to 12 cm high, loosely tomentose, often glandular on the upper parts; leaves lanceolate to oblanceolate, 6–18 mm long, 2–6 mm wide, margins involuted or flat, both sides green; flowers grouped 2–6, yellow, about 1.5 cm wide, June–August. HM 2041. Alps and Mts. of S. Europe to the Balkan Peninsula, high in the mountains on limestone soil. z5 Fig. 94. ⊕

H. pilosum (L.) Benth. Low subshrub; leaves linear to oblong, 2 cm long, pubescent, margins often involuted; flowers white with yellow center, about 2.5 cm wide, sepals 3–5 mm long, the outer ones linear, the inner ones oval. Mediterranean region. z9 ⊕

H. polifolium see: **H. apenninum**

H. pulverulentum see: **H. apenninum**

H. rhodanthum see: **H. apenninum** var. **roseum**

H. × sulphureum Willk. (= *H. apenninum* × *H. nummularium*). Shrub, low; leaves petiolate, linear-lanceolate, 1–2 cm long, green and stellate pubescent above, gray stellate tomentose beneath, stipules linear awl-shaped, longer than the petiole; flowers few in racemes, light yellow, sepals stellate tomentose. ⊕

H. tomentosum see: **H. nummularium**

H. vulgare see: **H. nummularium**

CULTIVARS # **d** ⊕ ○

It would be superfluous to include the botanical parentage of the numerous cultivars; those of concern are primarily hybrids between *H. nummularium*, *H. apenninum* and *H. grandiflorum*. Most of them originate from England, Holland and a few from Germany.

The most prominent hybridizers are: John Nicoll, Monifieth, Scotland; E. M. Christy, Emsworth, England; H. den Ouden, Boskoop, Holland; B. Ruys, Dedemvaart, Holland.

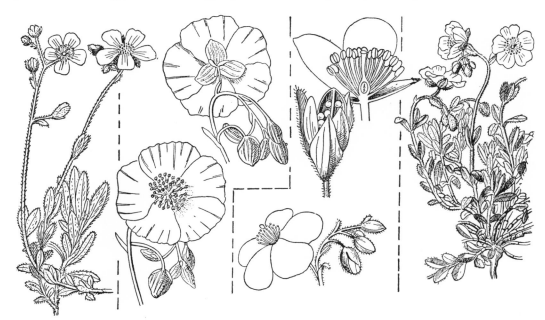

Fig. 94. **Helianthemum.** From left to right: *H. canum; H. apenninum; H. nummularium; H. oelandicum* ssp. *alpestre* (from Riechenbach, Jordan, Thomé)

'**Amabile Plenum'**. Dark red, double.

'**Amy Baring'**. Procumbent; glowing orange, single.

'**Attraction'**. Orange, center darker.

'**Avalanche'** (Lemoine). White, single.

'**Ben Alder'** (Nicoll). Brownish, buds carmine, single.

'**Ben Attow'** (Nicoll). Light yellow, center darker, single.

'**Ben Avon'** (Nicoll). Foliage light green; flowers pink and orange.

'**Ben Dearg'** (Nicoll). Bright red, center orange, single.

'**Ben Fhada'** (Nicoll). Gold–yellow, center orange, single, large flowers.

'**Ben Heckla'** (Nicoll). Only 10 cm high; flowers orange-yellow with brown basal ring, single.

'**Ben Hope'** (Nicoll). Leaves gray; flowers carmine (HCC 025/1), center dark orange, single.

'**Ben Lawers'** (Nicoll). Light orange, center darker, single.

'**Ben Ledi'** (Nicoll). Leaves dark green; scarlet, center carmine, single.

'**Ben Macdui'** (Nicoll). Salmon pink, center orange, single.

'**Ben More'** (Nicoll). Orange-red, center carmine, single.

'**Ben Nada'** (Nicoll). Yellow, single.

'**Ben Nevis'** (Nicoll). Dark lemon-yellow (HCC 5/1), center light brown, single.

'**Ben Vane'** (Nicoll). Light brown-yellow, center darker, single.

'**Ben Venue'** (Nicoll). Foliage dark green; flowers orange-red, center darker.

'**Ben Vorlick'** (Nicoll). Terracotta, center orange, single.

'**Blutströpfchen'**. Dark red, semidouble.

'**Butter and Eggs'**. Soft orange, double.

'**Chocolate Queen'**. Brown-red, double, flowers pendulous.

'**Croceum'**. Bright canary-yellow.

'**Fireball'**. Vermilion-red, single.

'**Fire Dragon'**. Leaves gray; flowers a brilliant scarlet, single.

'**Fireflame'**. Dark red, center orange, single.

'**Fireking'**. Dark red, semidouble.

'**Garibaldi'**. Brown-pink, large flowers, single.

'**Gloriosa'**. Clear pink, single.

'**Golddollar'**. Orange-yellow.

'**Golden Ball'**. Golden-yellow, double.

'**Golden Queen'**. Yellow, single.

'**Goldie'** (Christy). Foliage dark green; flowers golden-yellow, double.

'**Jock Scott'** (Christy). Foliage dark green; flowers dark cherry-red, center orange, single.

'**Jubilee'**. Straw-yellow (HCC 2/2), somewhat reddish, double.

'**Lachskönigin'**. Salmon-pink, large flowers, single.

'**Lawrenson's Pink'** (T. A. Lawrenson, Newcastle). Pink, orange center, single.

'**Magnificient'** (Christy). Foliage gray; flowers coppery-pink, single.

'**Mrs. C. W. Earle'**. Dark red (HCC 20), center orange, double.

'**Mrs. Mould'**. Foliage gray; flowers salmon-pink, single.

'**Orange'** (Christy). Foliage gray; flowers orange-yellow, center brown, single.

'**Orange Double'**. Orange-brown, double, flowers pendulous.

'**Praecox'**. Leaves gray; flowers lemon-yellow, single.

'**Red Dragon'**. Dark red, single.

'**Rhodanthe Carneum'**. Foliage gray; flowers carmine-pink, center orange.

'**Rosabella'** (den Ouden). Pink-red, single.

'**Rose Perfection'**. (Ruys). Pure carmine, single.

'**Rose Queen'**. Light pink, single.

'**Rubens'**. Flesh pink, center orange-yellow, single.

'**Rubin'**. Dark red, double.

'**Salmonea'**. Dark flesh-pink, double.

'**Snowball'**. White, double.

'**Snow Queen'**. Leaves gray; white, center yellow, single.

'**Starlight'** (Nicoll.) Foliage dark green; flowers primrose-yellow, center darker.

'**Sudbury Gem'**. Carmine-pink, single.

'**Sulphureum Plenum'**. Light yellow, double.

'**Sunbeam'**. Dark orange, center darker, single.

'**Supreme'**. Deep red, single.

'**The Bride'**. Foliage gray; flowers white, center yellow, single.

'**Watergate Orange'** (Christy). Dark orange.

'**Watergate Rose'** (Christy). Quite low; leaves gray; flowers wine-red.

'**Wisley Primrose'**. Pale lemon-yellow, single.

'**White Queen'** (den Ouden). White, center yellow, single.

'Yellow Queen' = '**Golden Queen'**.

All species and cultivars need full sun; not particular as to soil types, but not too fertile as the plants will grow too vigorously.

Lit. Grosser, W.: Cistaceae (*Helianthemum* pp. 61 to 123); in Engler, Pflanzenreich **14**, IV, 193; Leipzig 1903 ● Janchen, E.: Die Cistaceen Österreich-Ungarns (*Helianthemum* pp. 27–29); in Mitt. Naturwiss. Ver. a. d. Univ. Vienna **7**, 1909 ● Venema, H. J., & J. Koopman: Het sortiment van Helianthemums in het Arboretum 1949; in Boomk. **5**, 196–197, 204 to 205, 1950.

HELICHRYSUM L. — Strawflower — COMPOSITAE

Perennials, annuals or shrubs, usually more or less tomentose, leaves alternate to nearly opposite, simple; flowers in stalked heads, solitary or in nearly sessile umbellate racemes, surrounded by an often very attractive involucre, whose bracts are hard, dry and very persistent. — About 350 species in Europe, Asia, Africa, Australia and New Zealand.

Helichrysum angustifolium see: **H. italicum**

H. antenaria see: **Ozothamnus antennaria**

H. bellidioides (Forst. f.) Willd. Procumbent dwarf shrub, evergreen, main branches lying on the ground and rooting, to 60 cm long, very thin, branched, loose woolly pubescent; leaves broad rounded to spathulate, 5–6 × 3–4 mm in size, occasionally to 1 × 1.5 cm, eventually glabrous above, appressed white woolly beneath; flowers usually solitary, terminal, on a woolly shaft to 10 cm high, heads 2–3 cm wide, with numerous, attractive, white bracts, very abundant flowering, May (= *Gnaphalium bellidioides* Hook. f.). New Zealand. z9 Plate 51.# ☉

H. coralloides (Hook. f.) Benth. & Hook.f. New Zealand Coral Shrub. Evergreen, compact shrub, to 60 cm high and wide in its habitat, main branches thick, side branches also thick, cylindrical, very densely arranged and much branched, erect, with the leaves 7–10 mm thick; leaves scale-like, imbricate, obtuse oblong, about 5 × 2.5 mm large, leathery on the apical half, exterior green, glabrous, interior loose white woolly, margin reflexed, leaves larger on young plants; flowers appear infrequently, heads solitary, terminal, sessile, 6 mm wide, yellowish-white, but without ornamental value (= *Ozothamnus coralloides* Hook. f.). New Zealand, in the mountains. More similar in appearance to a thick branched conifer than a strawflower. z9 Fig. 95. #

H. italicum (Roth) G. Don. Subshrub, 10–50 cm high, entire plant aromatic, branches outspread, gray-brown, flowering shoots appressed tomentose, abundantly foliate, gray; leaves narrow linear, leathery, 3–6 cm long, margins distinctly involute, slightly tomentose above, dense, gray tomentose beneath; flowers many in terminal umbellate racemes, the individual heads campanulate, 4 mm long, 1 mm wide, yellowish, with numerous involucral scales, June–July. HM 6: 239 (= *H. angustifolium* DC.). Mediterranean region, from Spain to Cyprus. Leaves with a strong curry scent. z8–9 # Ø ☉

H. ledifolium see: **Ozothamnus ledifolius**

H. purpurascens see: **Ozothamnus purpurascens**

H. rosmarinifolium see: **Ozothamnus rosmarinifolius**

H. selago (Hook. f.) Benth. & Hook. f. Strongly branched, evergreen shrub, to 30 cm high, with short, thick, stem, branches compact, twigs tomentose, erect, stiff, very dense, total plant very similar to *Cupressus;* leaves scale-like, ovate-triangular, 3–4 mm long, leathery, completely

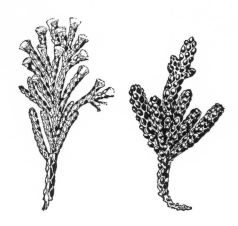

Fig. 95. **Helichrysum.** Left, *H. selago;* right, *H. coralloides* (from Poole/Adams)

covering the stem, exterior glossy green, glabrous, interior and margins woolly; flower heads solitary, terminal, sessile, 6–7 mm wide, dull white, but with only a few flowers (= *Ozothamnus selago* Hook. f.). New Zealand. z9 Fig. 95.#

H. splendidum (Thunb.) Less. Evergreen shrub, to 1.8 m high in its habitat, shoots divaricately spreading, not angular, densely foliate; leaves sessile, linear-oblong, to 3 cm long, 6 mm wide, white woolly on both sides, margin involute, with 3 parallel longitudinal veins; flowers in nearly hemispherical cymes, 3 cm wide, the solitary heads 6 mm wide, golden-yellow, very persistent. Africa, in the mountains from Ethiopia to the Cape and in the Drakensberg Mts. z9#

HELWINGIA Willd. — CORNACEAE

Deciduous plants; interesting for the unique situation of the flowers on the upper end of the leaf midrib; flowers small, greenish, dioecious, with 3–5 petals and 3–4 chambered ovary; female flowers solitary or in small groups, male 10–12 in small umbels. — 3 species in E. Asia.

Helwingia chinensis Batal. Shrub, to 2 m high; leaves ovate-lanceolate to more linear, to 12 cm long, acute, base cuneate, the margin teeth inclined somewhat forward, entire toward the base; stalk of the male flowers 7–25 mm long. EP IV, 229: 36; LWT 284. W. China. 1910. z9 Fig. 96.

H. japonica (Thunb.) F. G. Dietr. Deciduous shrub, about 1–1.5 m high, with stolons, densely branched, glabrous; leaves membranous thin, acute ovate, petiolate, serrate, 3–7 cm long, glabrous and light green on both sides; flowers tiny, greenish or reddish, in the middle of the upper leaf surface; fruits globose, 6 mm thick, black. DL

3: 139; SNp 73 (= *H. rusciflora* Willd.). Japan, China, in the mountains. Around 1830. The young leaves are eaten as a vegetable in Japan. z8 Plate 49; Fig. 96.

H. rusciflora see: **H. japonica**

For moist soil in semishade. Very interesting due to the arrangement of the flowers and fruits.

Fig. 97. *Hemiptelea davidii* (natureprint)

Hemiptelea davidii (Hance) Planch. Deciduous shrub to small tree, scarcely over 3 m high, very densely branched, with divaricate branching, dense with stout thorns expecially on young plants; leaves 2 ranked, short petioled, elliptic-oblong, 2 to 5 cm long, simple and coarsely serrate, with 8–12 vein pairs, eventually somewhat sparsely pubescent beneath; flowers inconspicuous, appearing with the leaves, April–May. LF 112; NK 19: 14 (= *Planera davidii* Hance; *Zelkova davidii* Bean). N. China, Korea, Manchuria. 1899. z5 Fig. 97.

Quite winter hardy but without ornamental merit; easily cultivated. Planted in its native habitat as an impenetrable thorny hedge.

Lit. Czerepanov, D.: Revisio specierum generum *Zelkova* Spach et *Hemiptelea* Planchon; in Botanitscheskie Materialy **18**, 70–72, 1957.

Fig. 96. **Helsingia.** Above, *H. chinensis;* below, *H. japonica* (from ICS)

HEMIPTELEA Planch. — ULMACEAE

Monotypic genus, similar to *Zelkova* and *Ulmus;* leaves small, with pinnate venation; branches with stout, 2–10 cm long, thorny short twigs especially on young plants; flowers insignificant, axillary, the pistillate flowers axillary on the young shoots; stamens 4–5; fruits like those of *Ulmus,* small, with one-sided oblique wing on the apical side.

HERTIA Less. — COMPOSITAE

Procumbent shrubs or subshrubs with woody branches; leaves alternate, 2 ranked (distichous) when young, sessile, somewhat fleshy, occasionally somewhat dentate; flowers capitate, stalked, solitary or in foliate panicles, flower receptacle flat or nearly so, naked; ray flowers in a single row, the female flowers fertile and either ligulate or truncate-tubular; disk flowers bisexual, tubular, sterile; fruits of the ray flowers indistinctly 5–10 ribbed, with white pappus.

Hertia cheirifolia (L.) O. Ktze. Evergreen, erect to procumbent, somewhat succulent shrub, to 50 cm high, branches erect at the tips; leaves sessile, lanceolate, widest near the apex, to 10 cm long, 2 cm wide, wide and round at the apex, 3 veined, margins membranous, generally tapering to the base, blue-green and pruinose; flower heads at the branch tips, terminal, 3.5 cm wide, golden-yellow, June. BMns 392 (as *Othonnopsis cheirifolia*) (= *Othonna crassifolia* L.; *Othonnopsis cheirifolia* [L.] Benth. & Hook. f.) N. Africa; Algeria, Tunisia. 1752. z10 # ∅ ✧

Needs a very dry, sunny location; easily propagated.

HESPEROYUCCA

Hesperoyucca whipplei see: **Yucca whipplei**

HIBBERTIA Andr. — DILLENIACEAE

Evergreen shrubs, many heath-like in appearance or climbing; leaves alternate; flowers usually terminal, solitary, with 5 sepals and 5 yellow or white petals, stamens numerous, style filamentous, carpels surrounded by the calyx and opening inward. — About 100 species in Madagascar, New Guinea, Australia and Polynesia. Only the following species in cultivation.

Hibbertia bracteata (R. Br.) Benth. Erect, densely branched small shrub, young branches soft pubescent; leaves narrow oblong to oblanceolate, 12–20 mm long, 1–2 mm wide, short acuminate, base cuneate; flowers yellow, 2 cm wide, terminal on side branches, petals incised at the apex, sepals densely silky, May–June. New South Wales. z9 #

H. dentata R. Br. Shrub or only a subshrub, twining or with the branches lying on the ground, pubescent when young; leaves oval-oblong, to 5 cm long, with 3–4 teeth on either side, base rounded; flowers dark yellow, to 5 cm wide, petals obovate and mucronate, otherwise entire, spring and summer. BM 2338; DRHS 994. New South Wales, Victoria, Australia. z9 #

H. scandens (Willd.) Dryand. ex Hoogl. Procumbent or twining shrub to 1 m high, young branches silky pubescent; leaves obovate to lanceolate, 4–10 cm long, 1.5–3 cm wide, tapering to a stem-clasping petiole, glabrous above, silky pubescent beneath; flowers golden-yellow, to 5 cm wide, unpleasant smelling, petals obovate, entire, sepals to 2.5 cm long, ovate, long acuminate, flowering in summer. BM 499 (as *Dillenia speciosa*); LAu 294. Australia; New South Wales, Queensland. 1790. z9 #

Outside of the very mildest regions, can be used outdoors only in summer and must be overwintered in a greenhouse.

HIBISCUS L. — Shrub Althea, Rose Mallow, Rose of Sharon — MALVACEAE

Deciduous or (tropical) evergreen shrubs or perennials; leaves alternate, palmately veined and lobed; flowers usually solitary and axillary, large, corolla broad campanulate, with 5 petals, adnate to the base; calyx 5 toothed or 5 parted, surrounded by connate bracts; stamens fused into a column, with numerous anthers; stigma 5 branched; fruit a 5 valved, dehiscent, many seeded capsule. — About 300 species, most in the tropics and subtropics.

Hibiscus hamabo S. & Z. Bushy shrub, 1–2 m high; branches, leaf undersides, exterior of the stipules, bracts and calyx gray-yellow stellate tomentose; leaves deciduous, rather thick, broad rounded to obovate or obliquely rectangular, 3–6 cm long, 3–7 cm wide, finely obtuse serrate; flowers solitary axillary, 5 cm wide, pale yellow with a dark red center, petals 4 to 5 cm long, obovate, July–August. KIF 2: 61; HKS 45. Japan, Korea. z10 ✧

H. mutabilis L. Shrub, or tree in its habitat; leaves broad ovate to rounded in outline, 3–5 lobed, 10–20 cm wide, cordate at the base, lobes triangular; flowers axillary in dense clusters, corolla widely outspread, 7 to 10 cm wide, white at first, turning pink during the day and becoming a deep red by nightfall. BIC 234; YWP 1: 89. China. 1960. z9 Plate 52. ✧

H. paramutabilis Shrub, 4–5 m high in its habitat; leaves rounded, 3 lobed, lobes acute, base truncate; flowers broad cupulate, 8–10 cm wide when fully open, petals obovate, white with a red basal spot, June, flowers abundantly. GH 1: 50. China, Kuling Province. Winter hardiness unknown, but probably hardy. (Available in the USA through Pioneer Seed Company, Dimondale, Michigan; and F. W. Schumacher, Sandwich, Massachusetts). Cited as a good garden plant for future use. Fig. 98. ✧

H. rosa-sinensis L. Shrub, 2–3 m high (a small tree in the subtropics), open branched, erect glabrous; leaves rather large, ovate to more elliptic or oval, base broadly cuneate to rounded or somewhat cordate, 6–10 cm long, more or less coarse and obtuse or scabrous toothed on the apical half, thin, glossy green, petiole 2–4 cm; flowers solitary and axillary in the upper leaf axils on young shoots, stalked, 10–15 cm wide, outspread, with a markedly exserted stamen column, the type of the species pink and simple, cultivars available in various colors, double and semidouble, involucre composed of 7 linear, narrow, distinct leaflets, these as long as the calyx. DRHS 996; HKS 46. Probably originating from China, but widely planted today in the tropics and subtropics. Cultivated since 1730 or earlier. z10

Of the many cultivars available only a few will be mentioned here. Nearly all of the over 250 forms in cultivation are described in Palmer, K. & M.: *Hibiscus* unlimited, and how to know them; Florida 1954.

'Anneli' see: **'Mulle'**

'Anita Buis' (Buis, Aalsmeer). Orange-yellow, very large flowers, single.

'Carl'. Bright red, single (practically identical to the type of the species).

'Chérie' (Nielsen, Odense, Denmark). Apricot colored, center darker, double.

'Dan' see: **'Mulle'**

'Hamburg' see: **'Mulle'**

'Lagos' (Spanay, Wageningen, Holland). Deep orange, light pink striped, center deep red, margin carmine-pink, 14–16 cm wide, single.

'Lateritia'. Leaves deeply incised; flowers orange-yellow, with a deep red center, single.

'Miami' (Spanay, Wageningen). Pure yellow, with red markings and blood-red center, 14–16 cm wide, single.

'Miss Betty'. Golden-yellow, single.

'Mulle'. Carmine-red, double. Giving rise to many sports including **'Anneli'**, brick-red, **'Hamburg'**, carmine-red and densely double, and **'Dan'**, double pink; these sports are more often cultivated in the commercial nursery.

'Sumatra' (Frederiksoord Experiment Station, Holland). Orange, black-red center.

'Yellow Queen' (Spanay, Wageningen). Lemon-yellow to orange, center orange, 10 cm wide, double.

H. sinosyriacus Bailey. Tall, erect shrub, similar to *H. syriacus,* but with leaves much larger, acutely dentate, young branches pubescent; leaves cuneate-ovate, 3 lobed, 7–9 cm long, coarsely and acutely dentate, base 3 veined, petiole 2–4 cm long; flowers large, open, 8–9 cm wide, lilac, petals nearly circular and widely overlapping, bracts lanceolate-oblong, acuminate, 2.5–3 cm long, about 0.5 cm wide, leaflike, extending beyond the calyx. HG 1: 50; JRHS 86: 31. China; Kuling Province. z8–9 Fig. 98. ⊕

Includes a few cultivars (Hillier):

'Autumn Surprise'. White with cherry-red markings. ⊕

'Lilac Queen'. White, with a trace of lilac, base garnet-red. ⊕

'Ruby Glow'. White, base cherry-red. ⊕

H. syriacus L. Deciduous, to 3 m high, erect shrub, young branches soft pubescent, later glabrous; leaves oval-rhombic, 5–10 cm long, 3 lobed, coarsely dentate; flowers solitary and axillary, violet on the type, broadly campanulate, to 6 cm wide, involucre nearly as long as the calyx, composed of 6–7 broad lanceolate leaves, August–September. HM 1969; BB 2438. China, India (not Syria). z5–6 Fig. 98.

Includes numerous cultivars; many (according to Grootendorst) under several names in cultivation

● Cultivars with single flowers
  White:
    'Albus', 'Diana', 'Dorothy Crane', 'Snowdrift', 'Totus Albus', 'William R. Smith'
  White with a red spot:
    'Monstrosus', 'Purpureus Variegatus', 'Red Heart'
  Pink and red:
    'Hamabo', 'Pink Giant', 'Rubis', 'Woodbridge'
  Blue
    'Blue Bird', 'Coelestis', 'Meehanii', 'Russian Violet'

●● Cultivars with double flowers (semi to densely double)
  White:
    'Admiral Dewey', 'Jeanne d'Arc'
  White with a spot (or pink-white):
    'Carneus Plenus', 'Comte d'Hainaut', 'Lady Stanley', 'Leopoldi', 'Speciosus'
  Pink to red:
    'Boule de Feu', 'Duc de Brabant', 'Puniceus Plenus', 'Roseus Plenus', 'Ruber Semiplenus'
  Violet-blue:
    'Amplissimus', 'Ardens', 'Coeruleus Plenus', 'Purpureus Variegatus', 'Souvenir de Charles Breton', 'Violet Clair Double'

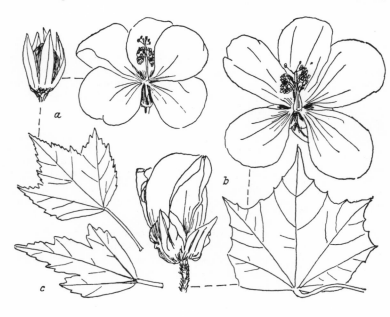

Fig. 98. **Hibiscus.**
a. *H. sinosyriacus;*
b. *H. paramutabilis;*
c. leaf of *H. syriacus*
(a–b from Bailey)

'**Admiral Dewey**'. Pure white, double, good bloomer, 5 cm wide. Developed before 1900 in the USA. ✣

'**Albus**'. White, single.

'Albus Plenus' see: '**Jeanne d'Arc**'

'Amarantus' see: '**Roseus Plenus**'

'**Amplissimus**'. Dark violet-pink, later more purple-violet, double (= 'Paeoniflorus').

'**Ardens**'. Broad growing; flowers blue-violet, densely double, good bloomer, very early. Dfl 5: 27. Before 1873. ✣

'Bicolor' see: '**Lady Stanley**'

'**Blue Bird**' (Croux Fils, Chatenay-Malabry, before 1958). Blue-lilac with a small red heart, single, outspread, about 12 cm wide (= 'Oiseau Bleu'). Developed in France. US Pl. Pat. No. 1739. Much better and more vigorous than the similar, but smaller flowered, 'Coelestis'. The flowers close in rainy weather. ✣

'**Boule de Feu**'. Densely double, deep red, without white stripes. AB 18: 47. Before 1856. ✣

'**Carneus Plenus**'. A good double, pale pink with a small, dark red center. Similar to 'Lady Stanley', but smaller.

'Celestial' see: '**Coelestis**'

'**Coelestis**'. Narrow upright habit, short branched; flowers violet-blue, single. Dfl 5: 56 (= 'Celestial'; 'Celestial Blue'). A good cultivar, but surpassed now by the larger flowered 'Blue Bird'. Plate 52. ✣

'**Coeruleus Plenus**'. Rather good double, violet-blue. Surpassed by 'Violet Clair Double'.

'Comte de Flandre' see: '**Boule de Feu**'

'**Comte d'Hainaut**'. Strong grower; double, white, pink shaded, center darker, buds globose. Poor bloomer.

'**Diana**' (Egolf; US Nat. Arboretum 1963). Erect habit, 2 m; flowers pure white, to 12 cm wide, margins undulate, flowers abundant and long lasting, triploid (sterile). (Minier.) ✣

'**Dorothy Crane**' (Notcutt, before 1935). White, red center, large flowers.

'**Duc de Brabant**'. Densely double, dark red, petals with a single white stripe, buds brown, poor bloomer, late. Introduced before 1872.

'Duchesse de Brabant' see: '**Duc de Brabant**'

'Elegantissimus' see: '**Lady Stanley**

'Grandiflorus Superbus' see: '**Lady Stanley**' (?)

'**Hamabo**'. Very vigorous; single, pale pink with reddish stripes and markings, especially on the basal half. Dfl 5: 26. Not to be confused with the Japanese species *H. hamabo* S. & Z.! ✣

'**Jeanne de Arc**'. Semidouble, pure white, buds yellowish. 1894. Flowers later than 'Admiral Dewey'.

'**Lady Stanley**' (? Barron's Nurseries, Elvaston, near Derby England, before 1875). Weak growing as a young plant, narrow upright; flowers semidouble, white, pink shaded, center deep red, early. Dfl 5: 57. ✣

'La Veuve' see: '**Speciosus**'

'**Leopoldi**'. Densely double, white with shades of pink, center deep red. Similar to 'Lady Stanley', but leaves deeply incised, and stronger growing as a young plant.

'Luteus Plenus' see: '**Jeanne d'Arc**'

'**Meehanii**'. Low growing; leaves yellow margined; flowers single, lavender-blue. Interesting only as a variegated leaved form.

'**Monstrosus**'. Single, white with pink shades, center dark red, large flowers. 1873. ✣

'Oiseau Bleu' see: '**Blue Bird**'

'Paeoniiflorus' see: '**Puniceus Plenus**' and '**Amplissimus**'

'**Pink Giant**' ('Woodbridge' × 'Red Heart'). Differing from 'Woodbridge' in the more abundant, earlier, larger flowers, pure pink, with a darker red basal spot, single. (Minier.) ✣

'Pulcherrimus' see: '**Lady Stanley**'

'**Puniceus Plenus**'. Double, red, large flowers, early, good bloomer, like a semidouble 'Rubis'. Dfl 5: 28. ✣

'**Purpureus Variegatus**'. Strong grower; leaves white variegated; densely double, lilac-red, poor bloomer. Interesting only for the variegated foliage.

'Ranunculiflorus Plenus' see: '**Jeanne d'Arc**'

'**Red Heart**'. Pure white with a very red center, single, very large flowers. Dfl 5: 2. Prettier than 'Monstrosus'. ✣

'**Roseus Plenus**'. Short branched; densely double, dark violet-pink, not fading, but often blooming poorly. Dfl 5: 28. ✣

'Ruber Semiplenus' see: '**Puniceus Plenus**'

'**Rubis**' (Froebel 1899; originally introduced to the trade as 'Rubin'). Short branched; single, red, rather small flowers. ✣

'**Russian Violet**' ('Blue Bird' × 'Red Heart'). Growth stronger than 'Blue Bird'; flowers more numerous, earlier, dark mauve-lilac. (Minier.) ✣

'**Snowdrift**'. Single, pure white, very similar to 'Totus Albus' in flower, but leaves different. 1911.

'**Souvenir de Charles Breton**'. Strong grower; semidouble, violet, large flowers, somewhat lighter than 'Violet Clair Double'. Before 1886. Often incorrectly labeled in cultivation.

'**Speciosus**'. A good double, pure white with a dark red spot in the middle, good bloomer. Dfl 5: 26 (= 'Speciosus Plenus'; 'La Veuve'). ✣

'Spectabilis Plenus' see: '**Lady Stanley**'(?)

'**Totus Albus**'. Single, pure white, funnelform, but rather small flowers. AB 18: 47. Before 1855. Surpassed by 'William R. Smith'.

'Variegatus' see: '**Purpureus Variegatus**' and '**Meehanii**'

'**Violet Clair Double**'. Growth upright; semi to fully double, violet-blue with only a few visible red center spots, large flowers. Dfl 5: 27. Plate 52. ✣

'**William R. Smith**'. Flower single, pure white, very large, wide-open. Dfl 5: 17, 24. Named for the Director of the Botanic Gardens in Washington D.C. at the time (1916), therefore not "Rev. W. Smith". One of the best white cultivars. ✣

'**Woodbridge**' (Notcutt, before 1935). Descendent of 'Rubis', but the flowers larger and somewhat lighter; single, deep red, very large flowers, darker center. Dfl 5: 16. ✣

The true origin of the cultivars, many cultivated for over 200 years, is yet to be established.

All species prefer a moist, fertile soil in a sunny, warm location; those cultivars with dense double flowers bloom poorly in cold weather. In all areas, some winter protection is advisable, especially for young plants.

Lit. Grootendorst, H.: *Hibiscus syriacus;* in Dendroflora **5**, 23–28, 1968 (with ills.) ● Wyman, D.: The hardy shrub Altheas (*Hibiscus syriacus*); in Louisiana Soc. Hort. Res. **2**, 183–188, 1964 ● Blackwell, C., W. D. Kimbrough & R. H. Hanchey: *Hibiscus* breeding at the Louisiana Agricultural Experiment Station; in Louisiana Soc. Hort. Res. **2**, 189–205, 1964 ● Bates, D. M.: Notes on the cultivated Malvaceae. 1. *Hibiscus;* in Baileya **13**, 56–128, 1965 ● Wyman, D.: The Shrub Altheas; in Arnoldia **18**, 45–51, 1958 ● Palmer, K. & H.: *Hibiscus* unlimited and how to know them; St. Petersburg, Florida, 1954, 120 pp. ● Dickey, R. D.: *Hibiscus* in Florida; Univ. Fla. Agr. Exp. Sta. Bull. **467**, 1–32, 1952 (both of the latter works deal only with cultivars of *H. rosa-sinensis*).

HIPPOPHAE L. — Sea Buckthorn — ELAEAGNACEAE

Deciduous trees or shrubs; branches with thorny short shoots, young branches with stellate or lepidote pubescence; leaves alternate, narrow oblong to linear, short petioles; flowers dioecious, with single perianth, in the leaf axils of the previous year's shoots. Male flowers: calyx with nearly absent tube, 2 parted nearly to the base, stamens 4, without a rudimentary ovary. Female flowers: calyx with oblong, indistinctly 2 parted tube, short, constricted just above the ovary. Flowers in short spikes or racemes, the female stalked with its axis growing to a thorny short shoot; fruits drupe-like, with hard, ovate seeds. — 3 species in Europe and Asia.

Hippophae rhamnoides L. Shrub or tree, to 6 m high, with divaricate branching, branches thorny, plants often stoloniferous; leaves linear-lanceolate, 5–7 cm long, also longer on young plants, glossy silvery scaly on both sides; flowers not particularly attractive, appearing before the leaves, March to April; fruits oval-rounded, 6–8 mm long, juicy, orange-yellow. BM 8016; HM 2148–55; GPN 598 to 599. Europe, W. Asia, the Caucasus to E. Asia; on moist to dry, gravel and sand, on rocky ledges and screes from the plains into the mountains. Cultivated for centuries; more frequently since 1940 for the vitamin rich fruits. Ⓕ Holland, for dune stabilization. z3 Plate 52, 54. ⚬ ∅

var. **procera** Rehd. Strong grower, branches shaggy pubescent at least when young; flowers lanceolate and stellate pubescent above. W. China. 1923. Ⓕ China, reforestation of sterile (loess) slopes. z8–9

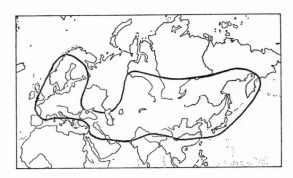

Fig. 99. Range of *Hippophae rhamnoides* (from Servettaz)

H. salicifolia D. Don. Shrub or tree to 15 m tall, branches pendulous, not thorny, young branches scaly and brown shaggy pubescent; leaves lanceolate, dull green above (not silvery!), to 1 cm wide, with brown midrib and stellate tomentum beneath; fruits yellow. MD 1918: 27. Himalaya. 1822. Plate 54. ∅ ⚬

Not particular as to soil type; sandy, well drained with sufficient moisture preferred; also well suited to slopes.

Lit. Darmer, G.: Der sanddorn als Wild- and Kulturpflanze; Leipzig 1952, 90 pp. ● Eichholz, W.: Die Nutzung der Sanddornbeeren; in Die Nahrung **2**, 156–168, 1958.

HOHERIA A. Cunn. — MALVACEAE

Deciduous or evergreen shrubs to small trees; leaves alternate, tough, dentate; flowers white, in axillary clusters; calyx 5 toothed; petals 5; stamens numerous, in 5 bundles; carpels 5, dehiscing from the central column when ripe. — 5 species in New Zealand.

Key to the Species (from H. H. Allan)

1. Carpels 5–8, broad winged, marginal
   teeth scabrous 2
   Carpels 10–15, not or indistinctly winged,
   teeth crenate 4

2. Leaves narrow oblong, obtuse or acute, seldom longer than 3 cm, teeth fine thorny:
   H. angustifolia
   Leaves ovate, usually long acuminate, longer than 5 cm teeth not thorny 3

3. Inflorescences usually with 5–10 flowers; flowers 25 mm wide or more; carpels 5(6):
   H. populnea
   Inflorescences usually with 2–5 flowers; flowers not larger than 20 mm wide; carpels usually (5) 6–7 (8):
   H. sexstylosa

4. Leaf pubescence usually dense; stigma surface decurrent:
   H. lyallii
   Leaf pubescence usually loose; stigma capitate:
   H. glabrata

Hoheria angustifolia Raoul. Evergreen, slender shrub or tree, to 9 m; leaflets linear-oblong, obtuse or acute, 2–3 cm long, with large, sparsely arranged thorny teeth; flowers usually solitary, in the leaf axils, occasionally in

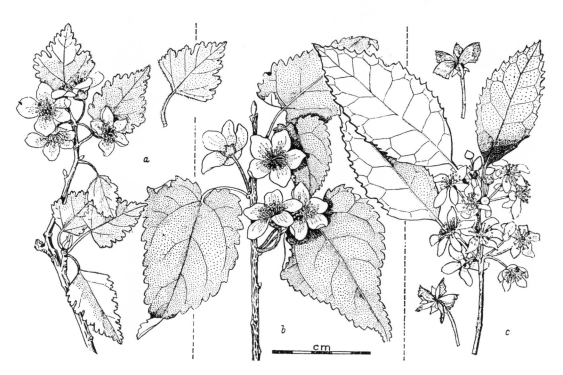

Fig. 100. **Hoheria.** a. *H. lyallii;* b. *H. glabrata;* c. *H. populnea* (from Poole/Adams)

clusters of 2–4, white, 12 mm wide, December–March. MNZ 39; NF 6: 247 (= *H. populnea* var. *angustifolia* [Raoul] Hook.). New Zealand. z9 Fig. 101. # ✧

H. glabrata Sprague & Summerhays. Deciduous shrub to small tree, gracefully arching young branches, leaves and inflorescences stellate pubescent; leaves ovate, acute, 5–10 cm long, usually doubly crenate, base cordate; flowers 2–5 in axillary clusters, white, corolla 2–2.5 cm wide, stamens yellow, June–July, flowers very abundantly. BS 2: Pl. 39; KF 134; NF 11: 9–10 (= *H. lyallii* var. *glabrata* E. H. M. Cox; *Plagianthus lyallii* sensu Hook. 1871 non 1867). New Zealand. 1871. One of the prettiest flowering plants of the forest edge in its habitat. z9 Fig. 100. ✧

H. lyallii Hook. f. Deciduous shrub to small tree, to 6 m high, branches densely stellate pubescent; leaves ovate on older plants, base cordate to truncate, 5–10 cm long, 3 to 5 cm wide, acuminate, coarsely and usually doubly dentate, both sides soft gray pubescent, but denser beneath, petiole half as long as the blade, leaves on young seedlings broad ovate, deeply lobed, lobes finely crenate; flowers pure white, 2.5–3 cm wide, grouped 2–5 together in the leaf axils of the new shoots, flowers abundantly, July–August. BM 5935 (as *Plagianthus lyallii*); BS 2: 376 (= *Plagianthus lyallii* [Hook. f.] Hook. f.). New Zealand; South Island, in the mountains. 1871. Fig. 100. z9

var. *glabrata* see: **H. glabrata**

H. populnea A. Cunn. A small tree in its habitat, 3–9 m high, only the young shoots, petiole and calyx pubescent; leaves quite variable, ovate to lanceolate and otherwise linear (especially on young plants), 9–12 cm

Fig. 101. **Hoheria.** a. *H. sexstylosa;* b. *H. angustifolia.* juv.=juvenile branches (from Poole/Adams)

long, scabrous and coarsely serrate, evergreen; flowers snow-white, 5–10 in axillary clusters, September. KF 54–55. New Zealand. z9 Fig. 100. # ⊕

var. *angustifolia* see: **H. angustifolia**

var. *lanceolata* see: **H. sexstylosa**

Includes 11 cultivars from New Zealand; some cultivated elsewhere:

'**Osbornei**' (Duncan & Davies 1926). Leaves purple toned beneath; flowers with blue stamens. Discovered in 1910 (Garnish Island, Ireland).

'**Purpurea**' (Duncan & Davies 1935). Leaf undersides and venation brownish-purple.

'**Variegata**' (Duncan & Davies 1926). Leaves yellow-green, margin dark green.

H. sexstylosa Col. Evergreen shrub or small tree; leaves oval-lanceolate, 5–10 cm long, scabrous and narrowly dentate, acuminate, tough, leathery, glossy green above; flowers pure white, 2–2.5 cm wide, 2–5 in axillary clusters, July–August, flowers very abundantly, very fragrant; fruits winged. MNZ 27; BS 2: Pl. 49; NF 6: 247; BM 8843 (as *H. populnea lanceolata*) (= *H. populnea* var. *lanceolata* Hook. f.). New Zealand. z9 Fig. 101. # ⊕

Easily cultivated where it is winter hardy. Prefers a good fertile, humusy, well drained soil. Gorgeous flowering shrubs for mild climates.

HOLBOELLIA Wall. — LARDIZABALACEAE

Evergreen, twining plants, resembling *Stauntonia*, but flowers in corymbs, with petal-like sepals and distinct stamens; leaves 3–9 parted; flowers axillary, monoecious, in corymbs; perianth 6 parted, petaloid, outermost valvate, 6 small nectaries; male flowers with 6 distinct stamens; female flowers with 3 ovaries and 6 sterile, rudimentary stamens; fruit a many seeded berry. — About 10 species from N. India to China.

Holboellia coriacea Diels. Evergreen shrub, twining to 5 m high, vigorous grower, young shoots reddish; leaves 3 parted (!), middle leaflet oval to obovate, 5–15 cm long, widest in the middle, lateral leaflets ovate, dark green, glossy, tough, distinctly veined; male flowers reddish, 12 mm long, in terminal panicles, female flowers greenish-white, somewhat reddish, 3–4 in the axils of the lower leaves, April–May; fruits 5 cm long, sausage-shaped, purple. BS 2: 380; BMns 447. W. China. 1907. z9 # ∅ ⧉

H. cuneata see: **Sargentodoxa cuneata**

H. fargesii Réaubourg. Leaflets 5–9, oblong-lanceolate to oblanceolate, 5–12 cm long, blue-green beneath (!) and indistinctly veined; flowers nearly 2 cm long, otherwise like *H. coriacea*. Central China. z9 # ∅ ⧉

H. latifolia Wall. non Gagnep. High twining, branches light gray-blue; leaves 3–7 parted, leaflets oval-oblong, 5–12 cm long, green with reticulate venation beneath; male flowers greenish-white, fragrant, 1.5 cm long, in short corymbs, female flowers purple, March; fruits sausage-shaped, purple, 5–8 cm long, edible. BR 32: 49; RH 1890: 348. Himalaya. 1840. z9 # ∅ ⧉

The hardiest species is *H. coriacea*, which is good for open areas in mild climates; otherwise only for southern exposures in walled gardens.

HOLODISCUS Maxim. — ROSACEAE

Deciduous shrubs, tall to tree-like; leaves alternate, simple, pinnate venation, pinnately lobed, dentate, tomentose, without stipules; flowers small, very numerous, grouped into large, generally pendulous panicles; petals 5, stamens numerous, about 20, of which the outer 15 are connate at the base; carpels 5, long pubescent, with 2 ovules; fruit a single, occasionally 2 seeded, indehiscent achene. — 8 closely related species in western N. America.

Holodiscus discolor (Pursh) Maxim. Shrub, 1–5 m high, strongly branched, erect, brown or gray-brown, branches thin, often nodding; leaves petiolate, broadly ovate or elliptic-oblong, 1.5–7 cm long, obtuse, lobes incised, the lobes crenate on the tips, leaf base entire, glabrous above and somewhat rugose, white tomentose beneath, pinnately veined; flowers yellowish-white, in large, nodding panicles, July–August. MS 234; BS 3: 352 (= *Spiraea discolor* Pursh). Western N. America. 1827. z5 Plate 54.

var. **ariaefolius** (Sm.) Aschers. & Graebn. This is the more common form in cultivation, leaves more deeply lobed, gray-green pubescent beneath; otherwise hardly different. GF 4: 617; Gs 1928: 247; DB 1593: 19. ⊕

'**Carneus**' (Späth). Like var. *ariaefolius*, but the flowers soft pink. Developed as a seedling in the Späth Nursery, Berlin, W. Germany; introduced to the trade in 1911. ⊕

var. **dumosus** (Nutt.) Maxim. Low growing, 0.3–0.7 m high, often procumbent, branches erect; leaves about 2–4 cm long, nearly sessile, obovate, soft pubescent above, white tomentose beneath; flowers yellowish-white, in simple, narrow, 1–15 cm long, erect panicles, flower stalks with subtending leaves and prophylls, June–August. RH 1859: 108 (= *Spiraea discolor* Hook.). Utah to N. Mexico, in rocky mountainous terrain at 2000–3500 m high. 1879. z4

No particular cultural preferences. Highly recommended for the large inflorescences in summer, especially for planting in open wooded areas. For moist, fertile soil.

Lit. Ley, A.: A taxonomic revision of the genus *Holodiscus* (Rosaceae). Bull. Torrey Bot. Club **70**, 275–288, 1943.

HOMALOCLADIUM (F. v. Muell.) L. H. Bailey — POLYGONACEAE

Evergreen shrub, closely related to *Muehlenbeckia*, also once included in that genus, but differing in the erect, distinctly flattened branches and bisexual flowers.

Homalocladium platycladum (F. v. Muell.) L. H. Bailey. Evergreen, to about 1 m high, erect shrub, all branches and twigs flattened (ribbonlike) and 1–2 cm wide; leaves scattered along the twigs, but abscising quickly, lanceolate, 1.5–6 cm long, tapering to both ends; flowers white, small, in fascicles, sessile on the edge of the branches, the fleshy perianth later develops into a purple-red fruit husk around a small nut. BM 5382; PBl 1: 544 (= *Muehlenbeckia platyclada* F. v. Muell.). Solomon Islands. 1863. z10 Plate 53. #

Cultivated only in frost free climates; otherwise nonpretentious but grown in most botanic gardens.

HOMOCELTIS See: **APHANANTHE**

HOVENIA Thunb. — RHAMNACEAE

Deciduous trees or shrubs; leaves alternate, somewhat uneven at the base, 3 veined, serrate or (occasionally) entire, stipules small, abscising quickly; flowers in terminal, many flowered, usually twice forked cymes; calyx persistent, with 5 lobes; petals 5, white, very small, concave, enclosing the 5 stamens; fruits 3 chambered, berrylike, indehiscent, on the thick, fleshy flower stalks. — 5 species in E. and S. Asia.

Hovenia dulcis Thunb. Raisin Tree. Deciduous, tall shrub, or tree in its habitat, branches pubescent; leaves broadly ovate, 10–15 cm long, simple, coarsely serrate, petiole 3–5 cm long, pubescent; flowers inconspicuous, greenish, in 4–6 cm wide cymes, June–August; fruit stalk eventually fleshy, thickened and reddish (eaten in India, Japan and China; bland flavor), fruits pea-sized, light gray-brown, often not ripening in cultivation. LF 225; HM 1885; BM 2360. China, Himalayas; planted in Japan. 1812. Hardy. Ⓕ N. China, Inner Mongolia, Argentina, for reforestation of sandy soil. z6 Plate 54, 61; Fig. 102. ⌀ ⚭

var. **glabra** Mak. 15–20 m high, stem to 80 cm thick; leaves ovate, 8–15 cm long, 5–8 cm wide. Japan, N. China.

var. **koreana** Nakai. Small tree or tall shrub, 3–5 m high, stem to 20 cm thick. Korea.

Fig. 102. *Hovenia dulcis*. Flowers, fruit seeds; right, fruits with fleshy, thickened stalk (from Weberbauer)

var. **latifolia** Nakai. 15–20 m high, stem to 80 cm thick; leaves elliptic-ovate or cordate, always large, 11–14 cm long, 8.5–11 cm wide. Japan.

Easily grown into a tall shrub or occasionally a small tree; thriving on any good garden soil, in a sunny location. Branches susceptible to frost damage but rejuvenate well in spring.

Lit. Kimura, Y.: Species and varieties of *Hovenia* in Bot. Mag. Tokyo **53**, 471–479, 1939.

HUDSONIA L. — CISTACEAE

Small, evergreen shrubs or subshrubs, densely branched, heather-like; leaves scale to needle -like, densely imbricate, without stipules; flowers solitary, terminal or axillary, clustered at the branch tips; calyx 3 parted, corolla 5 parted; stamens numerous; fruit a unilocular, 3 sided, dehiscent capsule with 3 valves and 1–3 seeds. — 3 species in eastern N. America.

Hudsonia ericoides L. Evergreen dwarf shrub, 15–20 cm high, densely branched; leaves linear, 2–7 mm long, erect or ascending, not imbricate, dense gray-green pubescent; flowers golden-yellow, 1 cm wide, on 3–10 mm long stalks, May–July. SC 36; BB 2473; NBB 2: 551. Newfoundland to Virginia, along the beaches and on the dunes. 1805. z4 Fig. 103. #

H. tomentosa Nutt. Habit dense and bushy, 20–30 cm high, densely woolly pubescent; leaves oval-lanceolate, 1–4 mm long, tightly appressed and imbricate; flowers golden-yellow, 6–10 mm wide, stalks 1–5 mm long, May–July. SC 36; BB 2474; NBB 2: 549. Coast of eastern

N. America and around the Great Lakes; on beaches and dunes. 1826. z4 #

Although winter hardy and attractive, somewhat difficult to cultivate and short lived; thrives on nearly sterile sandy soil.

Fig. 103. *Hudsonia ericoides* (from Dippel)

HULTHEMOSA

Hulthemosa hardii see: **Rosa × hardii**

H. kopetdaghensis see: **Rosa × kopetdaghensis**

HUODENDRON Rehd. — STYRACACEAE

Deciduous trees or shrubs, very closely related to *Styrax*, but flowers very small, in many flowered, terminal and axillary panicles; stamens with broad, flat filaments, tip of the anthers drawn out to 2 small teeth; petals 5, distinct at the base and reflexed; ovaries sub-inferior, 3 chambered; fruit a small capsule with many, very small seeds, irregularly winged on the apex, appendaged at the base. — 6 species in China, Thailand, Indochina.

Huodendron biaristatum (W. W. Sm.) Rehd. Shrub or (in its habitat) tree, 5 to 10 m high; leaves thin, broadly lanceolate, long acuminate, 8–20 cm long, entire; flowers white, in terminal panicles, corolla 8 to 10 mm long. China; W. and S. Yunnan to Kweichow; Burma, Indochina; in the mountains at 3000–3500 m. z6 Fig. 104. ✧ ⌀

H. tibeticum (Anthony) Rehd. Tree or shrub, 5–20 m high, with small leaves and flowers, from SE. Tibet, NW. Yunnan. Still very little known. Of only slight garden merit. z6

Neither species is generally found in Western cultivation, although the former is a heavily flowering, garden worthy species.

Fig. 104. *Huodendron biaristatum* (from Hu)

Lit. Hu, H. H.: On some interesting new genera and species of Styracaceae in China; in New Flora and Silva **12**, 146–160, 1942.

HYDRANGEA L. — HYDRANGEACEAE

Deciduous or evergreen shrubs, occasionally trees or climbers; leaves opposite, simple, entire or lobed; flowers in terminal corymbs, green, white, blue or red; all flowers either fertile (bisexual) or many to nearly all of the outer flowers of a corymb or panicle infertile; the calyx of infertile flowers enlarged and colored petal-like; normal petals much reduced to totally absent, anthers without fertile pollen; calyx of fertile flowers adnate to the ovary, 4–5 lobed, petals 4–5, valvate in bud, stamens 8 or 10, ovary distinctly or incompletely 2–4 chambered, styles 2–4, occasionally 5, distinct or connate only at the base, capsule dehiscing between the styles, many seeded. — 80 species in E. and SE. Asia, N. America to the Andes of S. America.

Outline of the genus (from McClintock)

* Section I. **Hydrangea** McClintock

  Deciduous, erect or climbing shrubs; inflorescences not enclosed by wide bracts before opening (except on *H. involucrata*);

  × Ovaries inferior at flowering and fruiting; capsules truncate on the apex; styles distinct to the base;

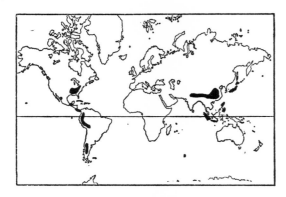

Fig. 105. Range of the genus *Hydrangea*

Subsection **Calyptranthe** (Maxim.) McClintock

Seeds with encircling wing; petals abscise together as a "cap"; plants climb by means of aerial roots:
H. anomala, petiolaris

Subsection **Asperae** (Rehd.) McClintock

Seeds without encircling wing; petals abscise individually; plants erect; seeds with tail-like appendage:
H. aspera, involucrata, longipes, robusta, sargentiana

Subsection **Americanae** (Maxim.) McClintock

As above, but seeds without tail-like appendage:
H. arborescens, quercifolia

XX Ovaries sub-inferior at flowering and fruiting; capsules conical on the apex;

Subsection **Heteromallae** (Rehd.) McClintock
Seeds with tail-like appendage; petals truncate at the base; styles at fruiting time shorter than the conical disc of the fruit capsule:
H. heteromalla, paniculata

Subsection **Petalanthe** (Maxim.) McClintock
Seeds without tail-like appendage; styles at fruiting longer than the conical capsule; petals clawed at the base:
H. hirta, scandens

XXX Ovaries inferior at flowering time, sub-inferior at fruiting; capsule conical at the apex;

Subsection **Macrophyllae** McClintock

Other characteristics like the following section:
H. macrophylla, serrata

** Section II. **Cornidia** (Ruiz & Pav.) Engl. & Prantl

Evergreen shrubs, erect or climbing with aerial roots; leaves leathery; inflorescences enclosed by wide bracts before opening;

Subsection **Monosegia** Briq.

Inflorescences composed of a singular, terminal, flat or round tipped cluster:
(plants from Mexico, not covered in this work)

Subsection **Polysegia** Briq.

Inflorescences composed of a number of round tipped clusters; only 1 species covered:
H. serratifolia; plants from S. America

Outline of the cultivar parentage
(G. Hort. = "Garden Hortensia" in the list, Lcp. = "Lacecaps" in the list)

'Admiration'	→ G. Hort.
'Alpenglühen'	→ G. Hort.
'Annabelle'	→ *arborescens*
'Bachtel'	→ G. Hort.
'Belzonii'	→ Lcp.
'Blauer Ball'	→ G. Hort.
'Blue Bird'	→ *serrata*
'Blue Wave'	→ Lcp.
'Bretschneideri'	→ *heteromalla*
'Coerulea'	→ Lcp.
'Diadem'	→ *serrata*
'Eldorado'	→ G. Hort.
'Enziandom'	→ G. Hort.
'General Patton'	→ G. Hort.
'Grayswood'	→ *serrata*
'Holstein'	→ G. Hort.
'Hortensis'	→ *involucrata*
'Hörnli'	→ G. Hort.
'Impératrice Eugenie'	→ *serrata*
'Intermedia'	→ *serrata*
'Lanarth White'	→ Lcp.
'Lemmenhof'	→ G. Hort.
'Maculata'	→ Lcp.
'Mariesii'	→ Lcp.
'Mme E. Moullière'	→ G. Hort.
'Morgenrot'	→ G. Hort.
'Otaksa'	→ *macrophylla*
'Rosalba'	→ *serrata*
'Preziosa'	→ *serrata*
'Prolifera'	→ *serrata*
'Säntis'	→ G. Hort.
'Sea Foam'	→ *macrophylla*
'Sir Joseph Banks'	→ *macrophylla*
'Soeur Thérèse'	→ G. Hort.
'Tricolor'	→ Lcp.
'Veitchii'	→ Lcp.
'Whitewave'	→ Lcp.

Hydrangea acuminata see: **H. serrata**

H. anomala D. Don. Climbing shrub, branches and inflorescences pubescent; leaves ovate, base cuneate or truncate or cordate, 6–15 cm long, acute to acuminate, 1.5–2 times longer than wide, glabrous to a few scattered hair clusters along the midrib beneath; inflorescences flat, much branched, petals 5, attached at the apex and abscising together as a cap, white, stamens 9–20, about 3–4 times longer than the style, ovaries inferior, styles 1–3 mm long; seeds to 2 mm long, with encircling wing; sterile margin flowers present, white, July. CIS 142. Japan, Korea, in forests to 2000 m; E. Himalaya, Central China. z5 Plate 56.

ssp. *petiolaris* see: **H. petiolaris**

H. arborescens L. Erect, well branched shrub, 1–3 m high, young branches and inflorescences strigose; leaves ovate, elliptic or broad ovate, acute to acuminate, 6 to 16 cm long, serrate, base rounded or somewhat cordate or cuneate, glabrous, pubescent or tomentose beneath; flowers white, in 5–10 cm wide, rounded, branched inflorescences, numerous fertile flowers, white, petals 5, abscising individually, stamens 10, with 5 twice as long as the others, styles 2, occasionally 3, July to August; capsule slightly conical at the apex. Eastern USA. z3

Currently includes 3 subspecies:

ssp. **arborescens** (L.) McClintock. Leaves green beneath, glabrous or nearly so (in the latter case pubescent only along the major veins); sterile flowers usually not present. BM 437; NF 4: 79; 9: 34; NBB 2: 275. Eastern USA. 1736.

ssp. **discolor** (Ser.) McClintock. Leaves gray beneath, velvety pubescent or tomentose, but the hairs usually not found in close spaces, hairs not rough warty when enlarged 50 times; individual marginal flowers present or absent. BB 2: 231 (= *H. cinerea* Small; *H. arborescens* var. *deamii* St. John). Eastern USA. 1906. Plate 56, 57. ✛

'**Grandiflora**'. The double form of ssp. *arborescens,* with 12–18 cm wide, greenish-white flower balls. Discovered wild in Ohio before 1900. Plate 56. · ✛

Included here also is the selection '**Annabelle**', with up to 20 cm wide inflorescences. Introduced by the Gulf Stream Nursery, Watchapreague, Virginia USA. ✛

ssp. **radiata** (Walt.) McClintock. Leaf undersides densely white tomentose, occasionally gray tomentose, the hairs showing no wartiness when enlarged 50 times or at least only sporadically warty; flowers in large cymes, white at first, later reddish, the sterile margin flowers always present, white, June–July (= *H. radiata* Walt.). Carolina, Tennessee. Plate 57. ✛

'**Sterilis**'. The double form of ssp. *discolor;* all flowers sterile, to 1.5 cm wide, grouped 15–20 in wide, hemispherical inflorescences, white, similar to the garden hortensias. Gs 4: 27 (= *H. cinerea* f. *sterilis* Rehd.). 1910. ✛

H. aspera D. Don. Shrub, occasionally a small tree, 1–4 m high, branches and inflorescences pubescent with stiff, appressed or loose, erect pubescence; leaves quite variable in form, lanceolate to ovate, broad ovate to nearly triangular, 5–35 cm long, base rounded to cuneate or cordate, serrate to bi-serrate, petiole 4–20 cm long, similarly pubescent as the branch; flowers in 10–30 cm wide, flat, many flowered cymes, fertile flowers white or blue, numerous, sterile marginate flowers always present, white, 4 lobed, the lobes 1–3.5 cm long, entire or ciliate, July–August. Himalaya, forests at 1500–2300 m; China, Taiwan, Sumatra, Java. z7

ssp. **aspera**. Leaves lanceolate to narrow ovate, 9–24(30) cm long, 2–9 cm wide, base rounded, petiole 1–5 cm long, leaf undersides with velvety, erect, somewhat loose pubescence, also on the petiole and young branches. NF 7: 177 (as *H. villosa*); CIS 28 (= *H. villosa* Rehd.; *H. kawakamii* Hayata; *H. longipes* var. *lanceolata* Hemsl.). China; Yunnan and surrounding provinces; Burma, Tibet, Taiwan. Plate 54. ✛ Ø

var. *macrophylla* see: ssp. **strigosa**

'**Mauvette**'. Similar to 'Macrophylla', but with a more erect habit; leaves smaller; fertile flowers and margin flowers light violet, August. Origin unknown, earlier listed erroneously as "*H. villosa*" by M. F. Van Der Smit, Reeuwijk, Holland; the above name given in Boskoop, Holland in 1973.

ssp. *robusta* see: **H. longipes** and **H. robusta**

ssp. *sargentiana* see: **H. sargentiana**

ssp. **strigosa** (Rehd.) McClintock. Leaves lanceolate to narrow ovate, about the same form as ssp. *aspera*, but larger, 5–35 cm long, 2–13 cm wide, with strigose pubescence beneath, densely appressed (erect on ssp. *aspera*), hairs straight, base rounded, petiole 1–5 cm long; marginal flowers white to pink. BM 9324; NF 9: 33 (as *H. strigosa* var. *macrophylla*) (= *H. strigosa* Rehd.; *H. aspera* var. *macrophylla* Hemsl.). China; Hupeh, Szechwan, Yunnan and other Provinces; Burma, Taiwan. 1907. Plate 4, 53, 57. ✛ Ø

H. bretschneideri see: **H. heteromalla 'Bretschneideri'**

H. chinensis see: **H. scandens** ssp. **chinensis**

H. chungii see. **H. macrophylla** ssp. **chungii**

H. cinerea see: **H. arborescens** ssp. **discolor**

H. davidii see: **H. scandens** ssp. **chinensis**

H. dumicola see: **H. heteromalla**

H. heteromalla D. Don. Twiggy shrub or tree-like, 0.5–7 m high, bark of older branches dark brown, but not peeling, young shoots and inflorescences pubescent, with lenticels; leaves ovate to broadly ovate, 8–20 cm long, 3–14 cm wide, distinctly densely pubescent beneath or nearly glabrous except for the venation, scattered pubescent or glabrous above; inflorescences flat-rounded, 8–30 cm wide (most important distinction from *H. paniculata*), fertile flowers numerous, white, stamens 10–12, filaments widening at the base, infertile flowers always present, white, sepals 4, about 1–3 cm wide, June–July. CIS 137 (as *H. xanthoneura*) (= *H. xanthoneura* Diels; *H. vestita* Wall.; *H. dumicola* W. W. Sm.; *H. hypoglauca* Rehd.). Himalaya; N. India, Nepal, Bhutan, SE. Tibet, NE. China, in the mountains. Plants in cultivation of this widely distributed species show a few, insignificant differences. Not, however, according to McClintock, enough to justify further dividing the species. z6 Plate 54. ✛

'**Bretschneideri**'. Shrub, 2.5–3 m high, stiffly upright, branches chestnut-brown in the second year and peeling (!!); leaves oblong to ovate, 7–12 cm long, 2.5–6 cm wide, with scattered pubescence beneath; inflorescences flat, 10–15cm wide. NF 1: 81; CIS 27 (= *H. bretschneideri* Dipp.; *H. pekinensis* Hort.). Presumably cloned from a seedling of seed collected by Bretschneider in the mountains near Peking. ✛

H. hirta (Thunb.) Sieb. Shrub, 0.5–2 m high, branches and inflorescences appressed bristly pubescent; leaves ovate, 3–10 cm long, completely serrate to 3–8 mm deep (!); flowers all fertile, bluish, in 2–10 cm wide, flat corymbs. KO ill. 199. Japan; Honshu, Okinawa. z7 Nettle-like in appearance, completely lacks sterile marginal flowers.

H. hortensis see: **H. macrophylla** var. **normalis**

H. hypoglauca see: **H. heteromalla**

H. integerrima see: **H. serratifolia**

H. involucrata Sieb. Shrub, 1–2 m, but often only 0.5 m high in cultivation, branches coarsely and densely pubescent; leaves ovate-lanceolate, 10–25 cm long, with appressed pubescence on either side; inflorescences to 15 cm wide, fertile flowers whitish, infertile margin flowers pink-lilac, inflorescences globose before opening, 2–3 cm thick, totally enveloped by 2–3 large, rounded bracts, July–September. YWP 2: 45 (= *H. longifolia* Hayata). Japan; only on Honshu. 1840. z7 Plate 55. m ○ ✧

'Hortensis'. Stronger growing; inflorescences pendulous, all flowers double, in loose cymes, silvery-pink. RS 187; GC 130: 40. Other colors are also grown in Japan (FS). Plate 59. ✧

H. kawakamii see: **H. aspera** ssp. **aspera**

H. liukiuensis see: **H. scandens** ssp. **liukiuensis**

H. longifolia see: **H. involucrata**

H. longipes Franch. Broad growing shrub, 1.5–2.5 m high, young branches more or less loosely pubescent at first; leaves oval, 7–17 cm long, 4–8 cm wide, rough to the touch, abruptly acuminate, scabrous serrate, with short bristly pubescence on both sides, especially beneath, petiole 3–8 cm long, slender, bristly when young; flowers in 10–15 cm wide, umbrella-shaped corymbs, marginal flowers white, 2–4 cm wide, June–July. CIS 140 (= *H. aspera* ssp. *robusta* [Hook. f. & Thoms.] McClintock p. p.). Central and W. China. 1901. Rare in cultivation.

var. *lanceolata* see: **H. aspera** ssp. **aspera**

H. luteovenosa see: **H. scandens** ssp. **liukiuensis**

H. macrophylla (Thunb.) Ser. Shrub, 1–3 m, young branches and inflorescences glabrous or with erect pubescence; leaves ovate, broadly ovate to somewhat obovate, thin to fleshy, 5.5–19 cm long, 3–13 cm wide, coarsely serrate, petiole 1–4 cm long; inflorescence a flat umbellate panicle, 10–20 cm wide, fertile flowers white, ovaries inferior at flowering time, sub-inferior at fruiting (!!), sterile flowers white or blue, sepals petaloid, 1–3 cm wide, entire or serrate, June–July. HKS 50; YWP 2: 43. Himalayas, S. China, Japan. z5–6 Plate 54.

Includes the following subspecies. For the most important cultivars of the pot grown hortensias, refer to the next section.

ssp. **chungii** (Rehd.) McClintock. Shrub, 1–2 m high; stem, young shoots and leaves rough with erect, 1.5–3 mm long pubescence, intermixed with short, curly hairs; leaves 10–19 cm long (= *H. chungii* Rehd.). China; Fukien. Not yet found in cultivation.

var. **normalis** Wils. Young branches, inflorescences and leaves quite glabrous; leaves broad ovate to more or less obovate, 7.5–19 cm long, 4–13 cm wide, fleshy, thick, coarsely serrate, petiole to 4 cm long; petals 3–4 mm long, style at flowering time 2 mm long; fruit capsules 6–8 mm long (!) (= *H. hortensis* Sm.; *H. maritima* Haw.-Booth). Japan, on the coast of the Chiba Peninsula, SE. of Tokyo and on the near southward islands.

'Otaksa'. Low growing, young branches green, black punctate; leaves rounded obovate, somewhat rugose; flowers all sterile, ball-shaped, pink or blue, sepals all of equal size, entire. FS 1732. Originally from China, but cultivated in Japan and brought from there by Thunberg in 1860. z8

Of the cultivars, **'Sea Foam'** (a sport of 'Sir Joseph Banks') is the most similar to the wild type; leaves bright green until frost; margin flowers bluish-pink, 2 cm wide, fertile flowers blue.

ssp. *serrata* see: **H. serrata**

'Sir Joseph Banks'. Tall growing; flowers very abundant, inflorescences wide, hemispherical, all flower sterile, sepals pink, often with a thin blue stripe. HHy 5. Introduced from China by Banks in 1789. z8 ✧

GARDEN HORTENSIAS

This list of plants, according to E. McClintock, were all developed from *H. macrophylla* ssp. *macrophylla* and not hybrids with other species. Of the nearly 200 forms available, only those most prevalent in cultivation today are covered.

The more prominent Hortensia breeders are:

Germany: Brugger, Tettnang; Fischer, Wiesbaden; Matthes, Ottendorf-Okrilla near Dresden; Schadendorff, Wedel; Steiniger, Vorst near Krefeld; Wintergalen, Münster; (Horticultural Research Institute), Friesdorf.

France: Cayeux, Le Havre; Gaigne; Lemoine, Nancy; Mouillère, Vendome.

Belgium: Draps-Dom, Strombeek near Brussels.

Holland: Baardse, Aalsmeer.

Switzerland: Moll, Zurich; Eidg. Vers. Anstalt (EVA), Wädenswil.

'Admiration' (Draps-Dom 1950). Compact habit; flowers pure red, brightly colored, inflorescence flat-hemispherical, 17 cm wide, individual flowers 4.5 cm wide, lenticels black-brown. ✧

'Alpenglühen' (Brugger 1950). Red, inflorescences flat-globose, 18 cm wide, individual flowers 4.5 cm wide, lenticels black-brown. ✧

'Bachtel' (EVA Wädenswil 1948). A good salmon-pink, inflorescence round-globose, 20 cm wide, individual flowers 7.5 cm wide, lenticels black-brown. Resembles 'Holstein'. ✧

'Blauer Ball' (Steiniger 1950). Blue with a silvery cast, inflorescence flat-rounded, 24 cm wide, individual flowers 6 cm wide, lenticels red-brown, very prolific bloomer. ✧

'Eldorado' (Draps-Dom 1950). Clear pink, inflorescence flat hemispherical, 20 cm wide, individual flowers 4 cm wide, lenticels black-brown. ✧

'Enziandom' (Steiniger 1950). Leaves medium-sized; flowers pure gentian-blue, inflorescences flat-globose, 20 cm wide, individual flowers 5.5 cm wide, lenticels dark brown. ✧

'General Patton' (Draps-Dom 1950). Low growing, compact; leaves small; flowers clear pink, inflorescences hemispherical, 21 cm wide, individual flowers 4 cm wide, lenticels dark brown. ✧

'Hörnli' (EVA Wädenswil). Low growing, graceful; small flowers bright red, inflorescence flat, 12 cm wide, individual flowers 2.5 cm wide, lenticels red-brown. ✧

'Holstein' (Schadendorff 1928). Pure salmon-pink, with some light blue, inflorescence 4–8 cm wide, lenticels light violet. ✧

'Lemmenhof' (Steiniger 1947). Blue, inflorescence flat-globose, 22 cm wide, individual flowers 5 cm wide, lenticels red-brown. Best blue form. ✧

'Mme. E. Mouillère' (Mouillère 1909). Sparsely branched; white, inflorescences flat globose, 22 cm wide, individual flowers 5.5 cm wide, lenticels white. An old standard in the trade. ⟡

'Morgenrot' (Steiniger 1950). Stoutly branched; light red, inflorescences globose, 17 cm wide, individual flowers 4 cm wide, lenticels whitish-green. ⟡

'Säntis' (EVA Wädenswil 1948). Bright red, inflorescences globose, 19 cm wide, individual flowers 5 cm wide, lenticels dark red-brown. ⟡

'Soeur Thérèse' (Gaigne 1954). Leaves large; flowers white, inflorescences flat, 22 cm wide, individual flowers 6 cm wide, lenticels whitish-green. ⟡

For further reference, Haworth-Booth briefly describe 340 cultivars, and Möhring details 115 cultivars with a photo of each.

"LACECAPS"

This group of cultivars (classified by the British) is more similar to the wild forms; the inflorescences are flat with a wreath of sterile marginal flowers surrounding the small fertile flowers. This group is especially prized in English gardens

'Belzonii'. Leaves usually in whorls of 3 (!); flowers in flat "umbrellas" with 5–6 pink, lilac or blue margin flowers, sepals 4. Siebold, Fl. Jap. Pl. 55. ⟡

'Blue Wave' (Haworth-Booth). Shrub, about 1.5 m high, broad bushy, branches stiff; leaves acute elliptic; inflorescences umbrella-shaped, the fertile flowers blue, the marginal flowers pink, lilac to gentian-blue. HHy 3 (= *H. mariesii perfecta* Lemoine 1904; renamed by Haworth-Booth). One of the most popular forms in England. ⟡

'Coerulea'. Japanese cultivar, with normal and sterile flowers, inflorescence umbrella-form, all flowers dark blue. BM 4253. ⟡

'Lanarth White'. Resembles 'Coerulea'; shrub, low growing, about 0.5–0.7 m high, dense; leaves usually yellowish-green; flowers in flat "umbrellas", fertile flowers blue, marginal flowers pure white, ring-like around the inflorescence, sepals usually 4, acute, entire. ⟡

'Maculata'. Conspicuously narrow upright habit; leaves yellowish-white margined and speckled; inflorescences with only 3–4 milky-white flowers, all sterile. HHy 22. Collector's form from Japan. ∅

'Mariesii'. About 1 m high; leaves narrow, evenly acute (!); inflorescences hemispherical-arched, the fertile flowers surrounded by a double ring of pink-red to light blue, entire or somewhat toothed marginal flowers, a sterile flower will also often be found in the center of the inflorescence. HHy 10. Brought from Japan in 1879 by Maries. Often not true in cultivation. ⟡

'Tricolor' (Rovere). Leaves dark green, light green and pale yellow, tricolored; flowers about like those of 'Mariesii', somewhat lighter. FS 696. Discovered by Rovere, Pallanza, Italy before 1860. ∅

'Veitchii'. Shrub to 1.8 m high; leaves bright green, often somewhat spoon-shaped arched; inflorescences small, flat, marginal flowers very large, white, light pink later. Gfl 1533; Gw 7: 582; HHY 9. Japan. 1880. Plate 58. ◐ ⟡

'Whitewave' (Haworth-Booth). Shrub, to 1.2 m high; leaves thick, acute-elliptic, spoon-shaped arched; inflorescence flat, fertile flowers blue, marginal flowers about 8, very large, pure white, the 4 sepals attractively sinuate toothed. HHy 1 (= *H. mariesii grandiflora* Lemoine). 1902. ● ⟡

H. mariesii see: **'Blue Wave'** and **'Whitewave'**

H. maritima see: **H. macrophylla** var. **normalis**

H. paniculata Sieb. Twiggy shrub, to 2 m high, or also a tree in its habitat, to 7 m high, young shoots and inflorescences with appressed pubescence; leaves opposite or in 3's, petiolate, ovate, 7–15 cm long, acute, serrate, venation pubescent on both sides; flowers in a l; leaves opposite or in 3's, petiolate, ovate, 7–15 cm long, acute, serrate, venation pubescent on both sides; flowers in a large, foliate, terminal, 7–25 cm long, conical panicle, petals 5, pistils 3, sterile flowers within the panicle, white, later pink, August. Gfl 530; NBB 2: 275; BMns 301. China, Japan, Sachalin. 1861. z3 ◐ ⟡

'Floribunda'. Like the species, but with every branch terminating in an inflorescence, panicles overweighted with fertile flowers, sterile flowers cream-white with a small, red "eye". HHy 20. Quite a popular form. Plate 58. ○ ⟡

'Grandiflora'. Panicles 15–30 cm long, composed almost entirely of sterile, white, later pink to red flowers. FS 1665; Gfl 530; HHy 13. Introduced from Japan, 1862. Quite hardy and much more prevalent in cultivation than the species. Plate 58. ○ ⟡

'Praecox'. Like the species, but flowering in early July, panicles shorter, smaller, sepals of the marginal flowers longer and more narrow. ○ ⟡

'Tardiva'. Very similar to 'Praecox', but flowering in the middle of September, form of the inflorescence the same; leaves longer, narrower, margins finely serrate (seen in Whitnall Park, Chicago, USA in 1974). Plate 58. ⟡

H. pekinensis see: **H. heteromalla 'Bretschneideri'**

H. petiolaris S. & Z. Climbing Hydrangea. Deciduous shrub, climbing 10–20 m, with aerial roots, even on smooth walls, young branches glabrous or pubescent, older branches with exfoliating bark; leaves oval, 4–11 cm long, 2–8 cm wide, with a short apical tip, margins regularly finely and scabrous serrate, base truncate to cordate, glossy green and glabrous above, with small hair fascicles in the vein axils beneath, petiole 1–10 cm long; flowers in flat corymbs, 15–25 cm wide, white, with a circle of marginal flowers 2.5–3.5 cm wide, June–July. BM 6788 (= *H. anomala* ssp. *petiolaris* [S. & Z.] McClintock; *H. tiliifolia* Lévl.). Japan, Taiwan, Korea. 1878. z5 Plate 57. ⟡

H. quercifolia Bartr. Shrub, 1–2 m high, lush grower and stoloniferous, young branches, leaf petioles and inflorescences densely tomentose or pubescent; leaves oval-rounded in outline, 8–25 cm long, usually deeply 5 lobed, the lobes coarsely and unevenly serrate, a good crimson-red in fall, leaf undersides pubescent or tomentose; flowers in large panicled bouquets, 15–25 cm long, with numerous, white, fertile flowers, usually one sterile flower on each panicle branch, these white, later becoming reddish, July–August. BM 975; NF 4: 80; NBB 2: 275. SE. USA. Quite hardy and attractive. z5 Plate 55, 57. **m** ◐ ⟡ ∅

H. radiata see: **H. arborescens** ssp. **radiata**

H. robusta Hook. f. & Thoms. Differing from the similar *H. longipes* in the more ovate to broadly ovate, 9–22 cm long, 7 to 13 cm wide leaves, tougher and thicker, undersides with more densely strigose pubescence, also the petiole and branches, the hairs straight and appressed. CIS 141 (= *H. aspera* ssp. *robusta* [Hook. f. & Thoms.] McClintock; *H. rosthornii* Diels). Himalayas to W. China. Existence in cultivation unknown. z6 Plate 57.

H. rosthornii see: **H. robusta**

H. sargentiana Rehd. Shrub, 2(3) m high, stem very thick, branches finger thick, branches and leaf petioles densely covered with fleshy, 2–5 mm long, erect, pink-red (at first) shaggy trichomes, the trichomes split at the apex; leaves broadly ovate, 15–35 cm long, velvety pubescent beneath, with scattered trichomes as on the branches; marginal flowers white. BM 8447; NF 12: 39; 9: 32; CIS 138 (= *H. aspera*, ssp. *sargentiana* [Rehd.] McClintock). China; only in Hupeh Province, 1908. Often erroneously labeled in cultivation, although easily recognizable. z7 Plate 4, 57, 59. ✶∅

H. scandens (L. f.) Ser. non Poepp. Shrub, 1(4) m high, thinly branched, more or less densely pubescent; leaves elliptic, 2–18 cm long, 1–8 cm wide, glabrous or pubescent on the venation, petiole 1–3 cm long, pubescent; inflorescence umbrella-shaped, 2–18 cm wide, terminal or axillary, fertile flowers few to numerous, yellow or blue, sterile flowers present, sepals yellow or white or blue, later fading to red or yellow, 1–3 cm long and wide, margin serrate, W. to SW. China, Taiwan, Philippines, Japan. Not generally found in cultivation. z9 Plate 54. ◗

ssp. **chinensis** (Maxim.) McClintock. Shrub to small tree, 1–4(6) m high, branches soft and curly pubescent to glabrous; leaves elliptic, 5–18 cm long; inflorescences normally terminal, fertile flowers usually numerous. CIS 132 (as *H. davidii*) (= *H. chinensis* Maxim.; *H. davidii* Franch.). W. China.

ssp. **liukiuensis** (Nakai) McClintock. Leaves 2–5 cm long, 1–2 cm wide, yellow striped along the midrib and the major veins, glossy above, usually bluish beneath (= *H. liukiuensis* Nakai; *H. luteovenosa* Koidz.). Japan. z8

ssp. **scandens**. Shrub, usually not over 2 m high; leaves 3.5–9 cm long, 1–3 cm wide, dull green above; inflorescences usually axillary, fertile flowers usually 10–15 (= *H. virens* Thunb.). Japan. z9 Plate 54. ◗

H. serrata (Thunb.) Ser. Small, thin branched shrub, about 1 m high, young branches, leaves and inflorescences glabrous or with appressed pubescence; leaves ovate, acute or acuminate, 5.5–15 cm long, 3–6 cm wide, with somewhat appressed pubescence on both sides, finely to coarsely dentate, dull green, not fleshy, petioles 1–3 cm long; flowers in flat or arched, 4–8 cm wide umbellate panicles, blue or white, with few, rather small, white, pink or blue marginal flowers, fertile flowers with 2–3 mm long petals, styles at flowering time 1–2 mm long, July–August, capsules 2.5–6 mm long. YWP 2: 43 (= *H. macrophylla* ssp. *serrata* [Thunb.] Mak.; *H. acuminata*

S. & Z.; *H. yesoensis* Koidz.). Japan, S. Korea, in mountain forests; from 70–1500 m. z6 ✶

'SERRATA' CULTIVARS

These are similar to the garden Hortensias, but the inflorescences are smaller, more umbrella-shaped, sepals entire or serrate; leaves coarsely or finely serrate.

f. *acuminata* see: **'Intermedia'**

'Bluebird'. Small, vigorous, thickly branched shrub; leaves abruptly long acuminate; inflorescences arched, fertile flowers blue, the large marginal flowers reddish purple, light blue on acid soil. JRHS 74: 191; HHy Pl. 7. Probably only a selection of *H. serrata*. ✶

'Diadem'. Compact habit, 70 cm high; leaves very reddish when in full sun; flowers on the axillary branches, marginal flowers scabrous serrate, pink or pure blue. In the trade before 1963. ✶

'Grayswood'. Shrub, 1.5 m high, erect at first, later spreading outward, leaves dull yellow-green, red-brown bordered, elliptic, acuminate; inflorescences umbrella-shaped, fertile flowers pink or blue, marginal flowers about 9, sepals 4, of which the outer ones (lower ones) are larger than the others, more acuminate, somewhat sinuate toothed, white at first, often somewhat reddish, but soon carmine-red, often holding its color until frost. HHy 21; JRHS 86: 33. ✶

'Impératrice Eugenie' see: **'Rosalba'**

'Intermedia' (Haworth-Booth). Shrub, to 1.5 m high, branches reddish, thin; leaves convex, usually reddish, with red venation; inflorescences often with only 3–4 marginal flowers, these white at first, later red. JRHS 65: Pl. 104 (= *H. serrata* f. *acuminata* Hort.). ● ✶

'Preziosa' (G. Arends 1961). Shrub, about 1 m high, branches red-brown; leaves turning red; inflorescences flat ball-shaped, nearly totally composed of sterile flowers (as on the Hortensias, but smaller), deep pink, later more purple-red, flowers abundantly, very winter hardy. Developed as a cross between *H. serrata* × a Hortensia by G. Arends of Wuppertal, W. Germany. ✶

'Prolifera'. Marginal flowers double, sepals narrow and acuminate. HHy 23; FS 1890 (= *H. serrata* var. *stellata* Wils.). Japan. 1864. ✶

'Rosalba' (van Houtte). Shrub, 0.7–1 m high; leaves dull yellowish-green; inflorescences umbrella-shaped, fertile flowers pink or blue, marginal flowers 6–7, white at first, soon becoming partly carmine-pink, sepals usually somewhat toothed. FS 1649–50; 1865–1867; BR 61. Included here is also the hybrid 'Impératrice Eugenie'. ● ✶

var. *stellata* see: **'Prolifera'**

H. serratifolia (Hook. & Arn.) Phil. Evergreen shrub or climbing to 15 m high with aerial roots; leaves elliptic, 7–14 cm long, 3–5 cm wide, entire or sparsely serrate, glabrous on both sides; flowers white, all fertile, small, but in numerous, dense, terminal and axillary cymes, filaments very long, sterile flowers occurring only sporadically. BMns 153 (= *H. integerrima* [Hook. & Arn.] Engl.). Argentina, Chile; Chiloe Island. Totally different in appearance from all the other species. z9 Plate 54, 59. # ◗ ✶∅

H. strigosa see: **H. aspera** ssp. **strigosa**

H. tiliifolia see: **H. petiolaris**

H. vestita see: **H. heteromalla**

H. villosa see: **H. aspera** ssp. **aspera**

H. virens see: **H. scandens** ssp. **scandens**

H. xanthoneura see: **H. heteromalla**

H. yesoensis see: **H. serrata**

All the hydrangeas prefer a fertile, sufficiently moist, humus soil; normally in semishade or shady locations. However, many species will also thrive in full sun given sufficient moisture. Most species are quite winter hardy unless otherwise noted.

Each new cultivar of *Hydrangea* should be recorded, before its introduction, with the Société Nationale d'Horticulture de France, at 84 Rue de Grenelle, Paris, France.

Lit. Haworth-Booth, M.: The Hydrangeas, 3rd ed., London, 1959 (185 pp., 23 pls.) ● McClintock, E.: The cultivated Hydrangeas; in Baileya **4**, 165 to 175, 1956 ● McClintock, E.: Hydrangeas; in Nat. Hort. Magazine **36**, 270–279, 1957 ● McClintock, E.: A monograph of the genus *Hydrangea*; in Proc. Cal. Acad. Sci. **29**, 147–256, 1957 ● Möhring, H. K., H. Kuhlen & G. Bosse: Die Hortensien; geschichtliche Entwicklung, Systematik, Sortenentwicklung; Aachen 1956 (238 pp., 147 ills.).

HYMENANTHERA R. Br. — VIOLACEAE

Evergreen or semi-evergreen, divaricately branched shrubs; leaves alternate, stipules often clustered, small, entire or dentate; flowers small, inconspicuous, axillary, solitary or in groups; sepals and petals 5; anthers nearly sessile, tubular connate; fruit a small berry with 1–2 globose seeds. — 7 species in E. Australia, New Zealand and on Norfolk Island.

Hymenanthera angustifolia R. Br. Semi-evergreen thorny shrub, erect, divaricate, about 1 m high, young branches pubescent, later glabrous and warty; leaves linear-spathulate, 0.6–2.5 cm long, deep green, later reddish; flowers yellowish, 1 cm wide, grouped 1–2 in the leaf axils, May; fruits globose, 3 mm thick, reddish (= *H. dentata* var. *angustifolia* Benth.). Tasmania. z7 Plate 62. # ⚘

H. crassifolia Hook. f. Semi-evergreen shrub, divaricate, 1(2) m high, branches thick, bark gray-white, pubescent when young; leaves clustered, obovate, 1–2.5 cm long, thickish, entire, rounded or emarginate at the apex, short petioles; flowers yellowish-white, 4 mm wide, solitary, axillary; fruits oblong, 6 mm long, white to gray-lilac, attractive. BM 9426. New Zealand. z7 Plate 62; Fig. 106. # ⚘

H. dentata R. Br. Shrub, to 1.5 m high; leaves oblong-elliptic, obtuse or acute, tough-leathery, sessile or drawn

Fig. 106. *Hymenanthera crassifolia* (from Hooker)

out to a small, narrow petiole; flowers small, yellow, April. BM 3163. New S. Wales. 1824. z9 # ⚘

var. *angustifolia* see: **H. angustifolia**

Only for collectors. The ornamental fruits sessile on the branch underside and scarcely visible from above. *H. crassifolia* is the best known species; it grows in nearly sterile sand.

HYMENOSPORUM R. Br. ex F. Muell. — PITTOSPORACEAE

Evergreen tree, very closely related to *Pittosporum*, but differing in the large yellow flowers and the flat winged seeds (*Pittosporum* has thick, unwinged seeds); otherwise with the characteristics of the species, as follows:

Hymenosporum flavum F. v. Muell. Evergreen, narrow crowned tree, to 6 m high, or only a tall shrub; branches glabrous; leaves usually alternate, obovate to oval-oblong, 7–15 cm long, base cuneate, slender acuminate, somewhat pubescent beneath; flowers in terminal corymbs, 10–20 cm wide, the individual flowers cream-white, later golden-yellow, 2.5 cm wide, with 5 oval-lanceolate sepals, corolla tubular, 3 cm long, limb with 5 obovate, widely spread lobes, silky pubescent on the underside, stamens 5, pubescent, April, fragrant. MCL 21; BM 4799 (as *Pittosporum flavum*). Australia. z9 Plate 4.

Cultivated like *Pittosporum*.

HYPERICUM L. — HYPERICACEAE

Deciduous or evergreen shrubs, subshrubs or perennials; leaves opposite or occasionally in whorls, short petioled or sessile, usually entire and translucent or black punctate; flowers in terminal cymes or panicles; calyx and corolla 5 parted; sepals often connate; petals yellow; stamens numerous, all distinct or connate in 3 or 5 bundles; ovaries with 3 or 5 distinct or (occasionally) connate styles; fruit a loculicidal capsule or berrylike. — About 400 species (nearly all perennials) in the temperate to subtropical zone of the Northern Hemisphere; only a few species are woody shrubs.

The nomenclature of this genus, and especially the section *Ascyreia*, was so confused as to require a number of corrections after the work of N. Robson. Regrettably many of the most important garden species must be renamed. To make this transition easier, an alphabetical list of synonyms is included at the end of the outline to the genus.

Outline of the genus

Section **Euhypericum** Boiss.
  Sepals fused at the base, even or uneven, entire, glandular toothed or ciliate; petals usually abscising; stamens numerous, connate in 3 bundles at the base; styles 3; capsule 3 valved:

  > *H. canariense, coris, floribundum*

Section **Ascyreia** Choisy
  Sepals even or uneven, petals more or less asymmetrical, stamens in 5 bundles, covering the petals and dehiscing with them after flowering; styles 5, occasionally only 4, more or less connate; capsule 5 chambered.

  _____

  (Because of the significant garden presence of this section and the revised nomenclature of 1970, the following classification represents the new grouping by N. K. B. Robson)

Leaf undersides with dense and distinct reticulate venation; pistil 1.5–5 times as large as the ovary = Group 1:
  > *H. calycinum, monogynum, oblongifolium*

Leaf undersides without or with only light reticulate venation; styles 0.2–1.5 times as large as the ovary (or occasionally 4 times as large in Group 4);
  Branches procumbent or ascending, rooting on 2nd year branches = Group 9:
  > *H. reptans*

Branches erect or more or less widely spreading, not rooting;
  Style 0.2–0.5 times as long as the ovary; stamens 0.25–0.35 times as large as the petals, 60–80 in a bundle; petals usually deep yellow, curved inward = Group 2:
  > *H. hookeranum, leschenaultii, 'Rowallane'*

Style 0.5–1.5 times as large as the ovary, stamens 0.35–0.85 times as large the petals (or, if the style and/or stamens are relatively shorter, like *H. bellum*,

then the stamens are bundled 20–40); petals varying;
  Leaves sessile, the upper ones stem-clasping; flowers 4–6 cm wide; sepals broad oblong to broadly elliptic, obtuse to rounded = Group 5:
  > *H. augustinii*

Leaves with short petioles; flowers and sepals otherwise;
  Petals acute with a prominent tip; flower buds about twice as long as wide, usually acute or acuminate; sepals acute or acuminate = Group 3:
  > *H. dyeri, kouytchense, stellatum, wilsonii*

  Petals acute, the tip more or less rounded; flower buds 1–1.2 times as long as wide, obtuse to acute or round; sepals varying;
    Sepals long acuminate to obtuse, not round and without a mucro; branches usually 4 sided or with 4 ridges, but occasionally with only 2 ridges or cylindrical, like *H. beanii* = Group 4:
    > *H. acmosepalum, beanii, pseudo-henryi*

    Sepals rounded or occasionally with a short mucro; branches varying;
      Sepals erect in bud, usually with a short mucro; petals with margin entire or finely toothed = Group 6:
      > *H. lobbii*
      Sepals erect or ascending in bud, with or without a mucro; petals with entire or finely toothed margin;
        Branches cylindrical, without ridges; leaf tip with or without a tiny mucro = Group 7:
        > *H. bellum, forrestii, 'Hidcote'*
        Branches with 2–4 ridges or 4 sided; leaves with or without short prominent tips = Group 8:
        > *H. moseranum, patulum, uralum*

  _____

Section **Myriandra** Spach
  Sepals uneven, petals abscising; stamens loosely 5 bundled, abscising; styles 3, connate at the base, appressed to each other at the apex; capsule 1–3 chambered:
  > *H. desiflorum, kalmianum, nothum, prolificum*

Section **Brathydium** Spach
  Sepals uneven, petals usually abscising; stamens very numerous, slightly connate in several bundles; styles 3, connate nearly to the apex; capsule unilocular:
  > *H. frondosum*

Section **Androsaemum** (All.) Dipp.
  Sepals uneven, entire, reflexed on spent flowers; petals and the stamens (in 5 bundles) abscising quickly; styles 3, distinct; capsule unilocular, berrylike, remaining closed or incompletely 3 chambered and dehiscing at the apex:
  > *H. androsaemum, grandifolium, hircinum, inodorum*

Summary of the Synonyms of the most important garden species

To outline the nomenclatural changes by N. K. B. Robson (1970), the discarded synonyms are listed alphabetically below (sensu Hort.= garden origin, p.p.= pro parte, in part).

Hypericum
— *aureum* Bartr. non Lour. → **frondosum**
— *chinense* L. → **monogynum** L.
— *dyeri* sensu Hort. → **stellatum**
— *elatum* Ait. → **inodorum**
— *henryi* Lév. & Vaniot → **patulum**
— *henryi* sensu Hort. → **acmosepalum**
— *hookeranum* sensu Hort. p.p. → **lobbii**
— *hookeranum* 'Buttercup' → **uralum**
— *hookeranum* 'Gold Cup' → **beanii**
— *hookeranum* 'Hidcote' → **'Hidcote'**
— *hookeranum* 'Rowallane' → **'Rowallane'**
— *kouytchense* sensu Hort. p.p. → **acmosepalum**
— *kouytchense* sensu Milne-Rehd. (in BM 9345) → **wilsonii**
— *lysimachioides* sensu Hort. → **stellatum**
— *lysimachioides* Wall., non Boiss. & Noé → **dyeri**
— *nepalense* K. Koch → **uralum**
— *oblongifolium* sensu Hook. (in BM 4949) → **lobbii**
— *oblongifolium* sensu Hort. p.p. → **acmosepalum**
— *oblongifolium* sensu Wall. p.p. → **hookeranum**
— *patulum* var. *forrestii* Chitt. → **forrestii**
— *patulum* 'Goldcup' → **beanii**
— *patulum* var. *grandiflorum* Hort. → **kouytchense** Lév.
— *patulum* var. *henryi* sensu Hort. p.p → **acmosepalum**
— *patulum* var. *henryi* sensu Hort. p.p. → **pseudohenryi**
— *patulum* var. *henryi* Veitch ex Bean → **beanii**
— *patulum* 'Hidcote' → **'Hidcote'**
— *patulum* 'Hidcote Gold' → **'Hidcote'**
— *patulum* var. *oblongifolium* sensu Hort. p.p. → **acmosepalum**
— *patulum* var. *oblongifolium* Hort. p.p. → **hookeranum**
— *patulum* 'Sungold' → **kouytchense** Lév.
— *patulum* var. *uralum* (Buch. Ham. ex D. Don) Koehne → **uralum**
— *penduliflorum* Hort. → **kouytchense** Lév.
— *persistens* F. Schneid. → **inodorum**
— *ramosissimum* K. Koch, non Ledeb. → **uralum**
— *rogersii* Hort. → **hookeranum**
— *uralense* Lavallé → **uralum**

Hypericum acmosepalum N. Robson. Semi-evergreen shrub, about 1 m high, branches drooping, young shoots reddish, flattened, 4 sided under the inflorescence; leaves with quite short petioles, elliptic to oblong or oblanceolate, acute or rounded, 2–6 cm long, 6–20 mm wide, base cuneate, dark green above, more bluish beneath, venation (except for the margin) scarcely visible; flowers solitary or grouped 2–3, 3.5–4.5 cm wide, buds acute ovate, sepals triangular-ovate to lanceolate or linear-lanceolate, long and drawn out, petals golden-yellow, outspread, stamens about ¾ as long as the petals, in 5 bundles, styles distinct to the base, June–October; fruit a red capsule. JRHS 95: 238, 240 (= *H. patulum* var. *henryi* Hort. p. p. non Bean; *H. kouytchense* Hort. p. p. non Lévl.). China; Yunnan, Kweichow. Long in cultivation but erroneously labeled under the synonyms. z7 Fig. 113. ⊕

H. androsaemum L. Shrub, semi-evergreen, to 1 m high and wide, young branches 2 edged; leaves oval-oblong, 5–10 cm long, base cordate, whitish beneath, aromatic when crushed, sessile, finely translucent punctate; flowers 1–9, golden-yellow, terminal, 2–2.5 cm wide, styles 3, as long as the ovary, June–September; fruit a dry berry, red-brown at first, then glossy black. HM 1999. W. and S. Europe, Asia Minor. z7 Plate 60; Fig. 107, 112. ⚭ ⊕

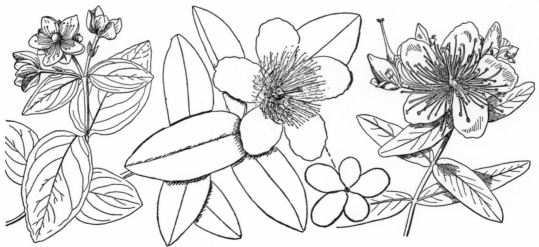

Fig. 107. **Hypericum.** From left to right: *H. androsaemum; H. calycinum* (calyx beneath); *H. monogynum* (from Bot. Mag., Reichenbach and Original)

H. × arnoldianum Rehd. (*H. galioides* × *H. lobocarpum*). Shrub, erect, to 1 m high, hemispherical, deciduous; leaves linear-oblong, 3–6 cm long, 4–8 mm wide, dark green and somewhat glossy above, whitish beneath, ornamental; flowers in terminal umbellate panicles, flowers abundantly, petals 6 mm long, narrow oblong, golden-yellow, sepals narrow oblong, uneven, shorter than the petals, stamens also shorter than the petals, as long as the 3–5 styles; capsule oblong, 6 mm long. Developed before 1910 in the Arnold Arboretum USA. z6

H. ascyron L. Subshrub, annual growth 0.5–1 m; leaves narrow oblong, 3–12 cm long, rounded at the apex, stem-clasping; flowers about 5 cm wide, golden-yellow, 3–12 in cymes, sepals uneven, ovate-oblong, petals wide at the apex, tapering to a claw at the base, July–August. BB 1444; MJ 957. N. America, N. Asia. Of only slight garden merit. z5

'Vilmorinii'. Flowers about 7 cm wide or more; otherwise like the type, but quite a valuable garden plant. BM 8557. Korea. Developed from Korean seed by Vilmorin.

H. augustinii N. Robson. Small shrub, under 1 m high, branches bowed outward, apical internodes of the branches distinctly 4 ridged or slightly 4 sided; leaves ovate to broad ovate, 4–6.5 cm long, 2.5–5 cm wide, obtuse, base broadly cuneate to rounded, sessile, the leaves stem-clasping immediately beneath the inflorescence (!!) and slightly cordate; flowers in groups up to 19 in terminal corymbs, much branched, pure yellow, the individual flowers to 5.5 cm wide, petals obovate, 2.5–3.5 cm long, rounded, stamens in 5 bundles, about as long as the petals or somewhat shorter, styles 5, about 6–7 mm long, sepals rather unevenly long and wide, oblong to elliptic, erect or somewhat outspread (= *H. leschenaultii* Hort. p. p. non Choisy). China; S. Yunnan. 1898. z7 ⊕

H. balearicum L. Shrub, about 60 cm high (in its habitat to 1.2 m), branches ascending, with 4 ridges when young and more or less warty (!); leaves opposite, ovate to oblong, 8–10 mm long, leathery, rounded, margins very undulate, very warty beneath, upper surface depressed over the warts; flowers solitary, terminal, yellow, 2–4 cm wide, fragrant, petals narrow, sepals circular, reflexed when in fruit, June–September. BM 137; PSw 28. Balearic Islands, in dry forests and on rock outcroppings. 1714. z10 # ⊕

H. beanii N. Robson. Semi-evergreen shrub, to about 1 m high, branches somewhat compressed, with 4 wings toward the apex, especially beneath the flowers; leaves not distinctly 2 ranked, ovate, obtuse, without a mucro; flowers grouped several together, buds broadly ovate, petals rounded, overlapping, dark yellow, sepals ovate-elliptic, acute to acuminate (!), to twice as long as wide, stamen filaments about half as long as the petals, style ⅔ as long as the ovary. JRHS 95: 235 (= *H. patulum* var. *henryi* Veitch). China; S. Yunnan. Introduced by A. Henry in 1898 and distributed by Veitch. z7 ⊕

'Gold Cup'. To about 1 m high; leaves appearing 2 ranked, lanceolate, slightly reddish in fall; flower buds narrow ovate, twice as long as wide, flowers cupulate, bright yellow, but somewhat lighter on the petal margin, pistil and stamens shorter than on the species. JRHS 95: 236. ⊕

H. bellum Li. Small, attractive, evergreen shrub, to about 50 cm high, young branches reddish, somewhat compressed under the flowers, but not angular; leaves broad oval to nearly circular, 1.5–3 cm long and somewhat narrow, obtuse, base rounded to lightly cordate, underside somewhat lighter than above, becoming reddish in fall; flowers grouped 1–3, terminal, golden-yellow, about 3.5 cm wide, slightly cupulate, stamen filaments about ⅓ as long as the petals, in 5 bundles, sepals ovate to elliptic, obtuse or occasionally

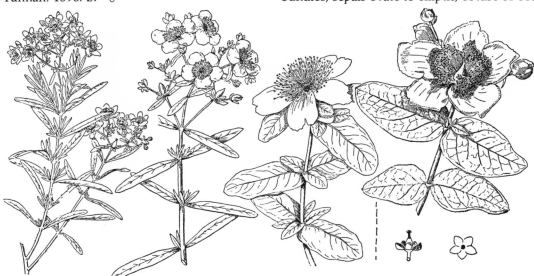

Fig. 108. **Hypericum.** From left to right: *H. densiflorum; H. kalmianum; H. frondosum; H. moserianum,* with calyx and ovary, and calyx from beneath (from Garden & Forest, Dippel and Original)

with a small apical tip, styles distinct, somewhat shorter than the ovary. BS 2: 409; JRHS 95: 486. Himalayas; Assam, SE. Tibet to SW. China. 1911 and 1914. Similar to *H. forrestii*, but differing in the smaller flowers and circular leaves. z6 # ☉

H. calycinum L. Evergreen, very strongly stoloniferous subshrub, 20–30 cm high, branches with 4 ridges; leaves oblong, 5–10 cm long, short petioles, deep green above, leathery-tough, obtuse, bluish beneath; flowers usually solitary (or 2–3), large, to 7 cm wide, terminal, golden-yellow, sepals large, obovate, erect, stamens in 5 bundles, long and attractive, with reddish anthers, styles 5, shorter than the stamens, outspread, July–September. BM 146; HM 1993. SE. Europe to Asia Minor. z5–6 Plate 60; Fig. 107, 113. ☉

H. canariense L. "Granadillo". Evergreen, glabrous shrub, 1–2.5(4) m, without glands on the leaf margin and sepals; leaves linear-lanceolate to narrow elliptic, 4–8 cm long, 3 cm wide, obtuse to acute; flowers many in large, terminal, dense panicles, golden-yellow, 2 cm wide, petals elliptic, 1 cm long, styles 3, about 3 times as long as the ovary, widely outspread, stamens ¾ as long as the petals, March–June; fruit a somewhat fleshy capsule, dry and hard when ripe. KGC I: 33; BFCa 198. Canary Islands, in dry thickets and on the forest edge, very well distributed from 100–800 m. z10 # ☉

var. *floribundum* see: **H. floribundum**

H. coris L. Dwarf evergreen shrub, seldom over 30 cm high, usually lower, prostrate to ascending, branches thin, angular, reddish; leaves 3–5 in whorls, linear, 1–2 cm long, 1–3 mm wide; flowers axillary in the upper whorls, grouped panicle-like, corolla about 1.5 cm wide, golden-yellow, July–August. BM 6563; DRHS 1035. S. Europe. Pretty plant for the rock garden. z7 # ○

H. densiflorum Pursh. Shrub, evergreen, erect, about 1 m high, to 2 m in the wild, branches 2 sided; leaves linear-oblong, 1–5 cm long, acute, margins involuted; flowers in dense, many flowered panicles, corolla 1–1.5 cm wide, golden-yellow, sepals unevenly elliptic-

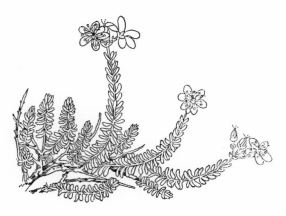

Fig. 109. *Hypericum empetrifolium*
(from Sibthorpe)

oblong, July–September. GF 3: 527; BB 2447; GSP 329 (= *H. prolificum* var. *densiflorum* Gray). Southeast USA; in swamps and marshes on acid soil. z6 Fig. 108. #

H. dyeri Rehd. Shrub, evergreen to semi-evergreen, glabrous, 0.7–1 m high, branches thin, angular, outspread; leaves nearly sessile, oval-oblong, 2–3 cm long, acute, whitish beneath; flowers golden-yellow, 3 cm wide, in loose foliate terminal cymes, petals lanceolate, 8 mm long, stamens in 5 bundles, August–September (= *H. lysimachioides* Wall.). W. Himalaya, Sikkim. 1904. Presumably not cultivated to date; plants cultivated by this name (*H. dyeri* Hort.) are really *H. stellatum!* z8–9 # ☉

H. empetrifolium Willd. Dwarf evergreen shrub, 25–30 cm high, branches erect, thin, angular, or also procumbent, a ground cover rooting along the branches; leaves usually in whorls of 3, linear, 2–12 mm long, margins involuted, sessile; flowers either in long panicles or cymes or also solitary, each flower about 12–18 mm wide, light golden-yellow, petals and stamens abscising, sepals small, oblong, with black glands on the margin, July–September; fruit 3 chambered, 8 mm long, with the erect sepals on the base. BM 6764. Greece and the Aegean Region; in rocky areas. 1788. z10 Fig. 109. ☉

H. floribundum Ait. Semi-evergreen shrub in its habitat, 1–3 m high, densely branched, branches cylindrical; leaves tough, nearly sessile, oblong-lanceolate, 2–4 cm long, 5–15 mm wide, distinctly 3 veined, obtuse; flowers 5–20 in terminal and axillary cymes, flowers very abundantly, the individual flowers about 4 cm wide, petals oval-lanceolate, obtuse, golden-yellow, 2 cm long, 1 cm wide, styles 3, erect at first, later outspread at the apex, sepals ovate, obtuse to acute, June–August; fruit a dry, hard capsule. KGC I: 32; MCL 71 (= *H. canariense* var. *floribundum* [Ait.] Bornm.). Canary Islands and Madeira. z10 ☉

H. forrestii (Chitt.) N. Robson. Deciduous shrub, about 1(1.5) m high, densely branched, main branches erect, side branches horizontal or drooping, somewhat flat compressed only under the flowers, otherwise without angles or ridges; leaves lanceolate to ovate, 2.5–4.5 cm long, 1.2–2 cm wide, obtuse with a mucro, base rounded to cuneate, often attractively red in fall; flowers solitary or 3–5 in cymes, golden-yellow, 5–7 cm wide, petals flattened and plate-like, flowers abundantly, stamens in 5 bundles, half as long as the petals, sepals broadly ovate to broad-elliptic, usually entire and round or broadly truncate, styles distinct, about half as long as the ovaries, July–September. JRHS 1923: 26; 1970: 242; DRHS 1037 (= *H. patulum* var. *forrestii* Chitt.; *H. patulum* f. *forrestii* [Chitt.] Rehd.; *H. patulum* var. *henryi* Hort. p. p. non Bean). China; Yunnan, W. Szechwan; Himalayas, Assam and Upper Burma. 1906. Quite winter hardy. 1906. z5 Fig. 111, 113. ☉

H. frondosum Michx. Deciduous, erect, occasionally only single stemmed shrub, 0.7–1 m high, branches reddish, 2 ridged, later exfoliating, glabrous; leaves

Fig. 110. *Hypericum floribundum* (from Webb)

oblong, 3–6 cm long, blue-green, bluish beneath, densely translucent punctate; flowers grouped 1–3, in nearly sessile cymes, orange-yellow, 3–5 cm wide, sepals leaflike, uneven, distinctly translucent punctate, styles 3, connate, July–August; fruit a large, red acute capsule, 12 mm long. BB 2885; BM 8498 (as *H. aureum*); NH 38: 127 (= *H. aureum* Bartr. non Lour.). SE. USA. Most beautiful of the American species. z6 Fig. 108. ☼

H. grandifolium Choisy. Malfurada. Shrub, to 1 m high, glabrous, without glands on the leaf margin and sepals; leaves broadly ovate, 4–7 × 5–5 cm large, obtuse, nearly sessile; flowers grouped 2–4 in loose inflorescences, the individual flowers about 4.5 cm wide, golden-yellow, petals lanceolate, 2 cm long, styles 3, about as long as the ovary, stamen filaments about ¾ as long as the petals, sepals acute ovate, flowers in spring; fruit a hard, dark brown capsule when ripe. BFCa 43. Canary Islands, especially in dry laurel forests and in the conifer zone, also on Lanzarote. z10 ☼

H. 'Hidcote' (possibly a garden hybrid of *H. forrestii* × *H. calycinum* ?). Semi-evergreen shrub, erect habit, to 1.5 m high and wide, branches cylindrical, brown; leaves oval-oblong or lanceolate, 4–5 cm long, widest under the middle, dark green above, light bluish-green with reticulate venation beneath; flowers few in terminal cymes, flowering continuously from June to October, corolla plate-like, 5–7 cm wide, golden-yellow, petals overlapping only on the basal ⅓, margin often

somewhat open incised (!), stamens only ⅓ of the petal length, anthers orange, styles 5, distinct, about as long as the ovary; fruits rarely developed. JRHS 95: 244; 80: Pl. 129 (= *H. patulum* 'Hidcote', in USA 'Hidcote Gold'). Origin not certain; in England between 1920 and 1930. Widely distributed in cultivation and very attractive. z7 Fig. 112. # ☼

H. hircinum L. Semi-evergreen, round habit, 0.5–1 m high and wide, branches brownish, glabrous, slightly ridged; leaves oval-oblong, thin, sessile, 3–6 cm long, goat-like odor when crushed; flowers 1–3 in large terminal or axillary inflorescences, corolla 3 cm wide, sepals small, uneven, lanceolate, abscising before fruit ripens, styles 3, stamens in 5 bundles, July–September. Northern Mediterranean region. 1640. Hardy and attractive. z7 Fig. 112. ☼

H. hookeranum Wight & Arn. Semi-evergreen, erect shrub, about 1 m high (or higher in the wild), branches reddish, round (!), glabrous; leaves oval-oblong, rounded acuminate, leathery, 4–7 cm long, dark and bluish-green above, slightly glossy, lighter beneath, with reddish mid-vein; flowers 6 or more in terminal cymes, corolla plate-like, 5 cm wide, bright yellow, petals tough, broadly obovate, sepals obtuse-obovate, usually outspread, often finely toothed, stamens only ⅓ as long as the petals, in 5 bundles, styles 5, connate nearly to the base, August–October. BM 4949 (as *H. oblongifolium*) (= *H. oblongifolium* sensu Hook. non Chois.). S. India, Sikkim, W. Szechwan, Yunnan, Assam. Before 1853. z7–8 Plate 54; Fig. 114. ☼

'Rogersii'. More compact habit, erect, 0.5 to 0.7 m high; flowers about 5 cm wide, cupulate, deep golden-yellow, flowers abundantly, but surpassed today in all respects by 'Rowallane'. Originated in the garden of Sir John Ross of Bladensburg in Rostrevor House, Co. Down, Ireland, around 1921. ☼

H. inodorum see: **H. xylosteifolium**

H. × inodorum Mill. non Willd. (= *H. androsaemum* × *H. hircinum*). Semi-evergreen shrub, to 1 m high, nearly hemispherical, many branches, reddish, somewhat 2 sided; leaves oblong-ovate, 3–8 cm long, obtuse or rounded, sessile, somewhat bluish beneath, aromatic (more or less) when crushed; flowers, many in terminal cymes or in the axils of the upper leaves, 2.5 cm wide, golden-yellow, sepals narrow to broadly ovate, shorter than the petals and persisting until the fruit ripens, then reflexed, stamen filaments longer than the petals, styles 3, about twice as long as the ovary, July–October; fruits dark brown or reddish, less fleshy than *H. androsaemum*, longer acuminate (not globose!). NH 3: 129 (= *H. elatum* Ait.; *H. persistens* F. Schneid.; *H. multiflorum* Hort.). Naturalized in England and France from cultivation; appearing in the wild on Madeira. z8 Fig. 113, 114. ☼ ⚘

'Elstead' (W. Ladhams 1933). More compact habit; fruits nearly scarlet-red, very ornamental; more winter hardy. BMns 376. Developed in the Ladhams Nursery, Surrey, England. More frequently found in cultivation than the species. ☼ ⚘

H. kalmianum L. Evergreen shrub, about 0.5–0.7 m high, branches 4 sided, side branches often only 2 sided;

Fig. 111. **Hypericum.** Left, *H. kouytchense*; middle, *H. forrestii*; right, *H.* 'Rowallane'
(from G. S. Thomas, in Gard. Chron.)

leaves oblong-linear, 3–5 cm long, bluish-green above, light blue-green beneath; flowers in 3's in terminal and axillary cymes, corolla golden-yellow, 1.5–2.5 cm wide, sepals leaflike, oblong, styles 3, connate at the base, August; capsule ovate, 5 furrowed. BB 2445; BM 8491; GF 3: 24. Eastern USA, Canada. z5 Fig. 108, 112. # ☉

H. kouytchense Lév. Semi-evergreen shrub, often deciduous, 0.5–1 m high, globose, glabrous, loosely branched, branches red-brown, rodlike to slightly compressed; leaves ovate, 4 to 6 cm long, 1.5–2.5 cm wide, acute, bluish beneath, short petioles, translucent punctate; flowers to 6 cm wide, light golden-yellow, petals outspread or reflexed, curved hook-like at the apex, stamens in 5 bundles, about ¾ as long as the petals, sepals lanceolate to oval-lanceolate, slender acuminate, scarcely outspread, styles distinct, about as long as the ovary, flowers abundantly, from June–October; fruit capsule turning bright red after the petals abscise. JRHS 95: 240, 241 (= *H. patulum* var. *grandiflorum* Hort.; *H. patulum* var. *forrestii* Hort. non p. p. non Chitt.; *H. patulum* 'Sungold'; *H. penduliflorum* Hort.). China; Kweichow Province. About 1900. z6 Fig. 111, 113. ☉

H. leschenaultii Choisy. Evergreen shrub, 1–2 m in the mildest areas, to 3 m or more in its habitat, open branched, branches reddish-brown, thin; leaves oval-oblong, obtuse, 4–6.5 cm long, 2–3 cm wide, dark bluish-green above, blue-green and densely translucent punctate beneath; flowers solitary or in 3's terminal, 6–7 cm wide, petals rounded-obovate, golden-yellow, 2.5

cm wide, concave, somewhat overlapping, stamens scarcely half as long as the petals, very numerous, in 5 bundles, sepals 1–1.5 cm long, narrow oblong to oblanceolate, acute, often rather uneven, styles 5, half as long as the ovary, reflexed; fruits rather large, conical. BM 9160. Java, Sumatra, Lombok and SE. Celebes. 1853. Prettiest of all the *Hypericum* species, but not very hardy. z10 # ☉

H. lobbii N. Robson. Evergreen or semi-evergreen shrub, erect, 1–1.5 m high, branches cylindrical; leaves ovate to nearly triangular, round to acute, 2.5–5 cm long, 1.5 to 3.5 cm wide, truncate at the base, blue-green beneath, short petioles; flowers to about 15 in terminal cymes, the individual flowers to 5 cm wide, cupulate, petals wide and overlapping, inner margin very toothed, stamens in 5 bundles, these ½–¾ as long as the petals, styles somewhat longer than the ovary, sepals erect to somewhat outspread, oval-oblong to broad elliptic, round, often irregularly toothed on the apex, August–October. BM 4949 (= *H. oblongifolium* sensu Hook. f.; *H. hookeranum* Hort. p. p.). Assam. Around 1853. z9 # ☉

H. monogynum L. Evergreen or semi-evergreen shrub, scarcely over 50 cm high, sparsely branched, branches cylindrical, without ridges; leaves sessile, oblong, obtuse, 3–7 cm long, 1.5 to 2.5 cm wide; flowers solitary, terminal, or grouped 3–7, golden-yellow, 3.5–6 cm wide, stamens in 5 bundles, some as long as the petals, styles 5, fused into a 1.5–2cm long column, but with 5 short erect stigmas at the apex, sepals oval-oblong, acute, black

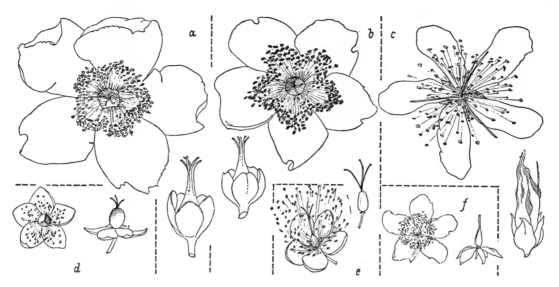

Fig. 112. **Hypericum.** a. 'Hidcote'; b. *H. moseranum;* c. *H. olympicum;* d. *H. androsaemum;*
e. *H. hircinum;* f. *H. kalmianum* (from F. Schneider, names altered)

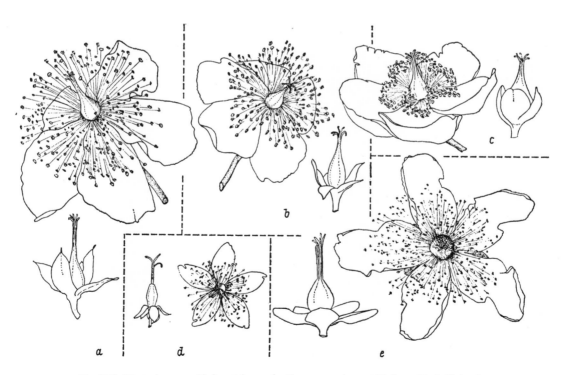

Fig. 113. **Hypericum.** a. *H. kouytchense;* b. *H. acmosepalum;* c. *H. forrestii;* d. *H. inodorum;*
e. *H. calycinum* (from F. Schneider, names altered)

Fig. 114. **Hypericum.** Left, *H. hookeranum;* middle, *H. patulum;* right, *H. inodorum* (from Bot. Mag., Jaub & Spach)

punctate, July–fall; fruit capsule ovate, acute, brown. BM 334; BS 2: 419; HKS 51 (= *H. chinense* L.). China, W. Hupeh, Kiangsi, Szechwan Provinces; Formosa. 1753. z10 Plate 54; Fig. 107. # ☉

H. × moseranum Luquet ex André (*H. calycinum × H. patulum*). Small semi-evergreen shrub, about 50 cm high, erect, branch tips somewhat nodding; leaves ovate, obtuse with small tips, 4–5 cm long, dull green above, lighter beneath; flowers 1–3, golden-yellow, corolla 5–6 cm wide, petals widely overlapping, sepals uneven, leaflike, ovate to oblong, anthers reddish, July–October; capsule acute. HM 1993; Gn 1898: 490; RH 1889: Pl. 116+117. Developed by Moser, Versailles, France, 1887. Very valuable garden plant. z8–9 Fig. 108, 112. ☉

'Tricolor' (Moser). Habit sparse and unattractive; leaves white variegated, pink margined; flowers smaller. Originated in 1891 as a sport in the Moser Nursery. z9 ∅ ●

H. multiflorum see: **H. × inodorum**

H. × nothum Rehd. (*H. densiflorum × H. kalmianum*). Shrub, low, somewhat outspread, to 0.5 m high, bark brown, exfoliating; leaves narrow oblong, obtuse, 3–4 cm long, 5–8 mm wide, glossy dark green above, whitish beneath; flowers golden-yellow, in large corymbs, petals 8 mm long, oblong, sepals linear-oblong, uneven, 4 mm long, styles 3–5; fruit conical, 6 mm long. Developed in the Arnold Arboretum before 1910.

H. oblongifolium Choisy (= *H. cernuum* D.Don). From the Himalayas (W. Pakistan to Nepal) is presumably not in cultivation. z9

H. patulum Thunb. Evergreen or semi-evergreen shrub, to about 1 m high, densely branched, branches brown to red, glabrous, very compressed at the tips with 2–4 ridges, 4 sided at the base, mature branches however cylindrical; leaves often 2 ranked, ovate to oblong-lanceolate, 2.5–6 cm long, obtuse and with a small mucro, bright green and somewhat glossy above, light bluish to whitish green beneath, short petioled; flowers solitary or in few flowered terminal cymes, golden-yellow, about 5 cm wide, petals uneven sided, obovate, concave, later outspread, sepals erect, broadly ovate to nearly circular, finely dentate, stamens half as long as the petals, in 5 bundles, styles 5, distinct at the apex, as long as the ovary, July–October; capsule ovate, more or less furrowed. BM 5693; MJ 975. China, W. Szechwan Province; Himalayas; Japan. 1862. The true species is hardy only to z10 and seldom found in cultivation. See the synonyms, varieties and cultivars. Fig. 114. ☉

'Sungold' see: **H. kouytchense**

H. penduliflorum see: **H. kouytchense**

H. prolificum L. Shrub, evergreen, erect, densely bushy, about 1 m high, branches 2 sided; leaves thin, narrow oblong to oblanceolate, acute, glossy, 3–7 cm long, often somewhat clustered in the leaf axils, margins somewhat involuted; flowers many, grouped in foliate panicles, terminal and axillary, corolla about 2.5 cm wide, petals obovate, sepals obovate, styles 3, connate at the base, July–September. BB 2446. Southeast USA. Around 1850. Similar to *H. densiflorum*, but larger in all respects. z5 # ☉

var. *densiflorum* see: **H. densiflorum**

H. pseudohenryi N. Robson. Small hemispherical shrub, branches drooping, young branches very compressed, with 4 ridges at first, later with only 2, eventually nearly cylindrical; flowers 3–5 cm wide, golden-yellow, petals outspread, similar to *H. kouytchense*, but not hooked at the apex, stamen filaments about ¾ as long as the petals, flowers abundantly, July–August (= *H. patulum* var. *henryi* sensu Rehd. p. p. non Bean). W. China. 1908. Very rare in cultivation; *H. forrestii* occasionally found under this name. z6 ☉

H. reptans Hook. f. & Thoms. Procumbent, dwarf shrub, rooting along the branches, scarcely 10 cm high, branches very thin, nearly filamentous, glabrous, 2 edged; leaves oval, obtuse, densely arranged, 8–12 mm long; flowers solitary, terminal, about 4 cm wide, golden-yellow, petals obovate, sepals elliptic, 12 mm long, obtuse, June–September; fruit capsule globose, 12 mm thick. Himalayas. 1881. Attractive for the rock garden. z9 ✣

H. 'Rowallane' (*H. hookeranum* 'Rogersii' × *H. leschenaultii*). To 1.8 m high, erect, branches cylindrical; leaves oval, 5–7 cm long, glossy green above, more bluish beneath; flowers few together, corolla flat and exceptionally plate-like, petals nearly circular, in large part overlapping, sepals broad and obtuse, 1 cm long, styles distinctly spread at the tips. BS 2: Pl. 51. Discovered about 1932 by L. S. Slinger in the garden of Armytage Moore, Rowallane, Staintfield, Co. Down, Ireland. One of the best cultivars. z9 Plate 63; Fig. 111. # ✣

H. stellatum H. Robson. Evergreen or only semi-evergreen, about 1 m tall, widely arching shrub, branches 2 or 4 sided at the apex; leaves ovate to oblong-lanceolate, 2–4 cm long, acute to obtuse, with a mucro, pale green beneath; flowers in corymbose terminal clusters, the individual flowers 2.5–4 cm wide, golden-yellow, petals obovate, twice as long as wide, with a small apical tip on one side, stamens in 5 bundles, about ⅔ as long as the petals, style about as long as the ovary to 1.5 times longer, sepals linear-lanceolate, already outspread in the bud stage, also after flowering, stellate, attractively reddish; fruit capsule about 1 cm long. JRHS 95: 237 (= *H. lysimachioides* Hort. non Wall. ex Dyer; *H. dyeri* Hort. non Rehd.). China, Yunnan, W. Szechwan Provinces; SE. Tibet. 1893. Earlier erroneously cultivated as *H. dyeri*. z6 # ✣

H. uralum D. Don. Semi-evergreen shrub, 0.6 to 1 m high, branches bowed outward; leaves 2 ranked, oblong, short petioles, only 2–3 cm long, aromatic when crushed, bluish beneath; flowers 3–15 in terminal cymes, corolla cupulate, only 2.5 cm wide, golden-yellow, styles 5, sepals broad oval, August–September. BM 2375 (= *H. patulum* var *uralum* [D. Don] Koehne). Central and E. Himalayas, Indochina and Sumatra. 1820. The name is derived from "urala swa" in the Nepalese language, not from the Ural Mts. z9 Plate 54. ✣

H. wilsonii N. Robson. Small, deciduous shrub, branches outspread, compressed and with 4 wings under the inflorescences, the middle internodes with only 2 ridges, the basal ones cylindrical; leaves usually narrow ovate, obtuse, 2.5–3 cm long, gray-green beneath, short petioled; flower buds slender, conical, flowers solitary or in 3's, terminal, 4–5 cm wide, golden-yellow, petals outspread, with hook-like tips on the apex (like *H. kouytchense*), stamens in 5 bundles, about half as long as the petals, styles 5, distinct, about as long as the ovary, sepals oval-oblong or more lanceolate, long acuminate. BM 9345 (= *H. kouytchense* sensu Rehd. non Lév.). W. China; Hupeh, Szechwan Provinces. 1907. (Wisley.) z6 ✣

H. xylosteifolium (Spach.) N. Robson. Evergreen shrub, with creeping rootstock, to about 1 m high, vigorous grower, shoots arching gracefully, usually unbranched, very densely foliate, compressed or 2 edged toward the tip; leaves oval-oblong, 2.5 to 5 cm long, rounded, dull dark green, not scented when crushed; flowers grouped only on strong shoots, otherwise solitary, terminal, only 2 to 2.5 cm wide, petals narrow, very brittle, stamens somewhat longer than the petals, sepals linear, July–September. DB 1: 260 (= *H. inodorum* Willd. non Mill.). SW. Caucasus; Georgian; NE. Turkey. 1870. Good for stabilizing dry slopes due to its vigorous stoloniferous habit. Quite winter hardy. z5 # ✣

Many species are very very popular summer flowering shrubs for sunny areas in good soil, but not for dry regions. Winter protection advisable for the tender ones, although most will resprout well after winter frost. *H. calycinum* and *H. xylosteifolium* are especially well suited for planting on slopes.

Lit. Engler, H.: Guttiferae; in Engler & Prantl, Nat. Pfl. Fam. ed. 2, **21**, 154–237, 1925 ● Juliano, V.: Guttiferae; in de Candolle, Monogr. Phanerog. **8**, 1–669, 1893 ● Lott, H. J.: Nomenclatural Notes on *Hypericum*; in Jour. Arnold Arb. **19**, 149–195, 1938; **19**, 279–290, 1938 ● Svenson, H. K.: Woody species of *Hypericum*; in Rhodora **42**, 8–19, 1940 ● Plaisted, R. L., & R. W. Lighty: The ornamental Hypericums; in Nat. Hort. Mag. **38**, 122–131, 1959 ● Thomas, G. S.: Notes on some shrubby Hypericums; in Gard. Chron. **147**, 226–227, 254–255, 1960 ● Robson, N. K. B.: Shrubby Asiatic *Hypericum* species; in Jour. Roy. Hort. Soc. **95**, 482–497, 1970 ● Robson, N. K. B.: *Hypericum*; in Flora Europaea **2**, 261–269, 1968.

IBERIS L. — Candytuft — CRUCIFERAE

Dwarf evergreen shrubs, annuals or biennials; branches procumbent, erect; leaves evergreen (on those species described here), alternate, entire or crenate, somewhat fleshy; flowers in terminal, often densely compact racemes; outer petals longer than the inner ones; fruit pod normally winged, often notched on the apex. — About 40 species in S. Europe to Asia Minor.

● Flowering shoots branched at the apex:
      I. gibraltarica

●● Flowering shoots not branched at the apex;
   * Leaves obovate, to 1 cm wide:
      I. semperflorens

   ** Leaves linear oblanceolate, to 5 mm wide;
   + Flower racemes elongated; glabrous:
      I. sempervirens

   ++ Flower racemes umbellate; finely pubescent;
   1. Flowers white; leaves acute-linear:
      I. saxatilis

   2. Flowers white; leaves spathulate:
      I. corifolia

   3. Flowers often pink, leaves obtuse linear:
      I. tenoreana

Iberis corifolia (Sims) Sweet. Dwarf shrub, 5 to 10 cm high; leaves evergreen, linear spathulate, 1.5–2 cm long, not ciliate on the margins; flowers white. S. Europe. Existence in cultivation is doubtful. z7 #

I. 'Correifolia' (*I. sempervirens* × ? *I. tenoreana*). Very strong grower, to about 30 cm high and 1.5 m wide; leaves spathulate, glossy dark green, to 4 cm long; flowers in about 7 cm long, hemispherical corymbs, late May. Developed before 1857 by H. Turner, in the Bury St. Edmunds Botanic Garden, England. Often confused with *I. corifolia*. # ⊕

I. gibraltarica L. Shrub, 20–30 cm high, somewhat divaricate; flowering shoots branched at the tip (!); leaves oblanceolate, to 5 cm long, entire, but somewhat toothed at the tip (!), thick, fleshy, dark green; flowers white, often also somewhat lilac toned, May–June. BM 124; BC 1944; DRHS 1041. S. Spain; Gibraltar; Morocco. 1732. Sensitive in cultivation, not very ornamental and often mistakenly labeled. z9 #

I. saxatilis L. About 10 cm high, evergreen, procumbent, stoutly branched, branches and inflorescences finely pubescent; leaves acute linear, somewhat half-round in cross section, not pubescent, but finely ciliate, 1–2 cm long; flowers in 2 to 2.5 cm wide cymes, white, anthers white, May (3 weeks earlier than *I. sempervirens*, often flowering again in fall) (= *I. vermiculata* Willd.). S. Europe; from Spain to the Crimea. z6 # ⊕

I. semperflorens L. Shrub, to 0.5 m high, bushy, glabrous; leaves obovate-oblong, to 1 cm wide, obtuse, fleshy, to 5 cm long; flowers white, occasionally somewhat reddish, fragrant, in about 3 cm wide inflorescences, more umbellate (at first), later elongated racemose, winter bloomer (October–March). S. Italy, Sicily. Very attractive, but not long lasting. z9 # ⊕

I. sempervirens L. Evergreen shrub, to 30 cm high, branches green, glabrous, often procumbent; leaves linear-oblong, obtuse, 1.5–3 cm long, entire; inflorescences 2–3 cm wide, gradually changing from a flat form to a longer racemose form, not branched, flowers white, anthers often violet on the tip, May–June; fruit pod oval, deeply emarginate. S. Europe, W. Asia. z5 # ⊕

var. **garrexiana** (All.) Cesati. Lower than the species; leaves narrow, acute; flowers smaller, umbellate at first, later racemose. Gn 62: 393 (= *I. garrexiana* All.). S. Europe; in the Alps and the Pyrenees. 1968. # ⊕

Includes the following cultivars:

'Climax' (J. Grieve, before 1875). About 20–30 cm high, very vigorous; leaves more spathulate, to 6 cm long, deep green; flowers large, pure white, April–June. Bk 7: 2. ⊕

'Elfenreigen' (Lindner, Eisenach). Similar to 'Schneeflocke', but fewer flowers, larger, pure white; leaves remaining a bright green longer into fall. Before 1930.

'Little Gem' see: **'Weiser Zwerg'**

'Nana'. Only 15 cm high, erect; leaves narrow, lanceolate, dark green; flowers pure white, very abundant, April–May. Good cultivar, better than 'Weiser Zwerg' ['Little Gem']. ⊕

'Perfection'. Vigorous grower, rounded, compact; flowers abundant and long lasting. ⊕

'Schneeflocke' (T. Smith, Ireland, around 1925). About 20 to 25 cm high, and often 60–70 cm wide; leaves deep green, short and wide; flowers pure white, April to June. Bk 7: 2 (= 'Snow Flake'; 'Snow Queen'). ⊕

'Snow Flake' see: **'Schneeflocke'**

'Snow Queen' see: **Schneeflocke'**

'Weiser Zwerg' (Arends 1894). Only 10–15 cm high; leaves shorter, more linear; flowers smaller, pure white. Bk 7: 2 (= 'Little Gem').

'Zwergschneeflocke' (Lindner). Only 10–15 cm high, but otherwise like 'Schneeflocke'.

I. tenoreana DC. To 35 cm high, branched and woody at the base; lower leaves rosette-like, spathulate-ovate, upper leaves smaller, more oblong-linear, all somewhat fleshy, 1.5 to 3 cm long, entire to dentate, ciliate; flowers in cymes, 4–5 cm wide, white, outer ones often pink-red, July–August. BM 2783. S. Europe. z6 #

I. vermiculata see: **I. saxatilis**

Easily grown subshrubs for sunny areas in the rock garden; valuable as a ground cover and as a white spring bloomer.

Lit. Koopmans, L.: De benaming van het winterharde *Iberis*-sortiment; in Boomkwekerij 7, 2–4, 1951.

IDESIA Maxim. — FLACOURTIACEAE

Deciduous tree; leaves alternate, long petioles, simple; flowers dioecious or polygamous; sepals usually 5, petals absent; stamens numerous; flowers with many staminodes; ovaries unilocular; fruits berrylike, in pendulous racemes. — 1 species in Japan and China.

Idesia polycarpa Maxim. Medium-sized tree, habit uniform, to 15 m high, branches widespreading, bark gray-white; leaves cordate to ovate, resembling large poplar leaves, acuminate, sparsely crenate, deep green above, blue-green beneath and glabrous, 10–15 cm long, petiole red, 10–18 cm long; flowers greenish-yellow, fragrant, in 10–20 cm long, pendulous panicles, May–June; fruits very numerous, pea-sized, green at first, eventually orange-brown, especially ornamental after leaf drop. BS 2: 428, BMns 649; FIO 77; LF 237. S. Japan, Central and W. China. Around 1865. z5 ∅ ⚭

var. **vestita** Diels. Leaf undersides densely tomentose, at least when young; fruits brick-red. FIO 77. China; W. Szechwan Province, in the mountains above 2400 m. 1908.

Prefers a moist, well drained soil in semishade.

ILEX L. — Holly — AQUIFOLIACEAE

Evergreen or deciduous trees or shrubs; leaves alternate, petioled, simple, entire or serrate, often with thorny teeth; flowers usually dioecious, solitary or grouped in small clusters; petals connate at the base, oblong to obovate; sepals persistent; fruit a berrylike drupe.— About 400 species in the temperate and tropical zones of both hemispheres, except W. North America and Australia.

Outline of the species described
Subgenus **Aquifolium** (Gray) Rehd.

Leaves evergreen, usually leathery; inflorescences solitary and axillary or solitary at the base of young shoots, or clustered in the leaf axils, single flowered or branched, occasionally in mock umbels or developed into a raceme or panicle; flowers 4–5 parted; ovules only 1 per locule;

Section **Aquifolium** Maxim.
Flowers or inflorescences predominantly in the leaf axils, rarely solitary, on the previous year's wood. Includes:
I. altaclarensis, aquifolium, beanii, bioritsensis, canariensis, centrochinensis, ciliospinosa, corallina, cornuta, dipyrena, fargesii, franchetiana, georgei, hookeri, integra, intricata, kingiana, koehneana, latifolia, melanotricha, perado, pernyi, rugosa, vomitoria

Section **Lioprinus** Loes.
Flowers or inflorescences always solitary, in the leaf axils or at the base of the young shoots. Includes:
I. atenuata, cassine, chinensis, coriacea, crenata, glabra, myrtifolia, opaca, pedunculosa, sugerokii, yunnanensis

Subgenus **Prinus** Loes.
Leaves deciduous, paper thin or membranous; flowers or inflorescences solitary in the leaf axils or clustered;

Section **Euprinus** Loes.
Leaves and flowers never clustered; fruits 4 to 8 mm thick; fruit pod smooth on the dorsal side. Includes:
I. geniculâta, laevigata, serrata, verticillata

Section **Prinoides** (DC.) Gray.
Flowers and leaves often clustered on short shoots; fruits 8–12 mm thick, fruit pod ribbed or striped on the dorsal side. Includes:
I. ambigua, amelanchier, decidua, macrocarpa, macropoda, montana

Ilex × altaclarensis (Loud.) Dallim. Hybrid between *I. aquifolium* and *I. perado* (var. *platyphylla*). Tall evergreen shrub, resembling *I. aquifolium*, but leaves larger, 6 to 10 cm long, leaf blade with smooth margin, partly more or less entire or small and regularly dentate; flowers and fruits larger.

Includes the following cultivars:

'Atkinsonii'. Branches dark green; leaves rather rough, deep green, to 10 cm long, 5 cm wide, glossy above, margin regularly dentate, somewhat larger than the otherwise similar 'Mundyi'; female. (Kew; Hillier.) z8 Plate 66. # ∅

'Balearica'. Clone of *I. aquifolium* var. *balearica* (which see) or a hybrid of it. Upright habit, somewhat conical when young; leaves oval-elliptic, flat, entire or occasionally with a thorn on the apex, glossy green above; fruits abundantly. (Hillier.) z9 # ∅ ⚭

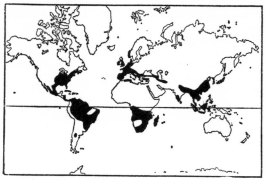

Fig. 115. Range of the genus *Ilex*

'Belgica'. Vigorous grower, young branches green; leaves bright green, ovate to more oblong, about 8 cm long, margin very distinct and stoutly toothed. DH 138; NH 36: 21 (= *I. perado* Dallim.; *I. "perado"*). z8 # ∅

'Belgica Aurea' (M, Koster & Zn. 1908). Habit like that of 'Belgica', but remaining lower, 3 m; leaves also like 'Belgica', but usually somewhat smaller and margin less thorny, irregularly broad golden-yellow, center dark green and gray-green, 7–10 cm long, 3–4.5 cm wide (= *I. "perado aurea"*). Very attractive. (Boskoop.) z7–8 # ∅

'Camelliifolia'. Habit regularly conical, young branches, petiole and midrib reddish; leaves oblong, deep green, new growth somewhat brownish, to 13 cm long and 5 cm wide, usually entire, occasionally also some leaves with 1–8 marginal thorns; fruits abundantly, fruits large. DH 24, 78; NH 36: 21, 28. z7–8 Plate 66. # ∅ ⚭

'Camelliifolia Variegata'. Leaves very glossy, deep green with light green marbling and golden-yellow margin, also occasionally half or totally golden-yellow leaves. (Hillier.) z8 # ∅

'Golden King'. Sport of 'Hendersonii'. Leaves 5 to 7 cm long, otherwise somewhat like the species, but with a broad golden-yellow margin, irregular, with few, quite small marginal thorns; fruits well (= *I. aquifolium* 'Golden King' and 'Aurea Rex'; *I. aquifolium hodginsii aurea*). Developed shortly before 1876 in the Lawson Nursery, Edinburgh Scotland. z8–9 ∅ ⚭

'Hendersonii'. Rather compact habit, bark dark green; leaves oblong-elliptic, base rather truncate, usually entire or with a few small marginal teeth, dull green, 6–8 cm long, 4–5 cm wide, venation indented above; fruits large, but not very abundant. DH 142; NH 36: 22; HHo 93 (= *I. aquifolium hodginsii*; not to be confused with the following cultivar!). Developed in 1800 by Thomas Hodgins in Dunganstown, Ireland. z8 # ∅

'Hodginsii'. Vigorous growing shrub, bark on the young shoots purple (!); leaves broadly ovate or more elliptic, 7–10 cm long, 5–6 cm wide, broadly cuneate or truncate and rounded at the base, dark green, margins partly with a few large teeth or nearly entire, especially on older plants; male (!). DH 142. Developed by Th. Hodgins and distributed, as was *I. aquifolium shepherdii*, as *I. aquifolium hodginsii*. Highly prized and widely planted in England. Very good pollution tolerance. # ∅

'Lawsoniana'. Actually nothing more than a variegated form of 'Hendersonii'. Young branches red-brown; leaves ovate to obtuse-elliptic, 6–8 cm long, margins sparsely dentate, but teeth regular and margins quite evenly flat, glossy green, center bright yellow bordered or speckled, form of the specks quite variable, limb bicolored green, usually with 10 thorns on either side. DH 142; NH 36: 22; HHo 87 (= *I. lawsoniana* Dallim.). One of the most beautiful yellow varegated forms. # ⌀

'Maderensis'. Medium-sized tree or tall shrub, branches dark; leaves ovate to more oblong, short acuminate, quite flat, 7–8 cm long, 4–4.5 cm wide, margins regularly dentate, teeth directed forward, all teeth lying flat; male. Origin of the clone is unclear, probably a seedling or hybrid from *I. perado*. (Hillier.) z9 # ⌀

'Maderensis Variegata'. Branches dark red-green; leaves dark green with golden-yellow center spot; male (Hillier.) z9 # ⌀

'Marnockii'. Leaves like 'Camelliifolia' in size, form and margin, but the apical half distinctly twisted; female. z8 # ⌀

'Mundyi'. Strong grower, bark green; very attractive form with oval, 10 cm long and 6 cm wide leaves, dull green above, finely rugose above, margins regularly thorned and undulate on one side; male. DH 142 (= *I. mundyi* Dallim.). z8–9 # ⌀

'Nobilis'. DH 138; NH 36: 22. By today's standards, identical with the species; the same is true of *I. urquhartii* Dallim. z8–9 #

'Purple Shaft'. Sport of 'Balearica'. Strong grower, bark dark purple, as are the young shoots; fruits very abundantly. (Hillier.) z9 # ⌀ ⚭

'Shepherdii'. Very strong grower; leaves rigid and tough, 5–7 cm long or more, broadly ovate, short acuminate, rarely entire, usually totally toothed or occasionally only sporadically toothed, margins flat, occasionally somewhat wavy because of the stout teeth, bright green (therefore distinguishable from the very similar, but dark green 'Hodginsii'). DH 136; HHo 90 (= *I. aquifolium shepherdii* Dallim.). z9 # ⌀

'Silver Sentinel'. Probably a sport from 'Balearica'. Strong upright habit; leaves 8.5 to 10 cm long, flat, tough, margins with only a few teeth, deep green, with light green and gray markings, with regular but distinctly cream-white or cream-yellow margins (= *I. perado* 'Variegata' Hort.). One of the best variegated forms. z9 # ⌀

'Wilsonii'. Strong grower, young branches green, somewhat furrowed, reddish in the sun; leaves to 12 cm long and 6 cm wide, ovate, evenly acuminate, venation distinct, base always round, margins evenly toothed with very sharp, widely spaced thorns, only very rarely entire, slightly glossy above, light green beneath, leaf petiole 5–7 mm long and very thick; female. HHo 90, 92. z8 Plate 63. # ⌀

'W. J. Bean' (Handsworth Nurseries, before 1929). Dense habit, compact; leaves resembling *I. aquifolium*, but somewhat larger, very undulate and with large marginal thorns; fruits bright red. HHo 90. (Hillier.) z8–9 # ⌀ ⚭

I. ambigua (Michx.) Torr. Carolina Holly. Deciduous shrub, but also a small tree in its habitat, branches thin, glabrous, dark brown, with raised lenticels; leaves elliptic to obovate, usually short acuminate, entire to flat crenate, 2.5–3 cm long, base obtuse to acute, glabrous or sparsely pubescent on both sides; flowers small, 4–5 parted, arranged in the leaf axils at the base of the new shoots; fruits red, ellipsoid or occasionally nearly globose, 6 mm thick. KTF 130; VT 653. Florida. z9

I. amelanchier M. A. Curtis. Deciduous shrub, to 2 m high, with a few short branches, young branches finely pubescent, later glabrous; leaves broadly ovate to elliptic oblong, 4 to 12 cm long, entire or indistinctly and quite finely, densely serrate, with dense reticulate venation beneath, short gray shaggy pubescence on the venation; inflorescences solitary, occasionally appearing with the leaves, male flowers grouped 6–9 together, female flowers solitary; fruits globose, 8–10 mm thick, red, petiole 7–10 mm long, October–November. GF 2: 88; NH 36: 2; NBB 2: 500; VT 657 (= *I. dubia* [G. Don] Britt. & al.). USA; Virginia to the Carolinas and Louisiana, mainly in the swamps. Should be winter hardy. z7

I. aquifolium L. English Holly. Evergreen shrub, occasionally also a tree (observed to 16 m high), branches dark or light green, glabrous; leaves evergreen, leathery-tough, dark green and glossy above, light green beneath, ovate to oblong-lanceolate, 3–5–8 cm long, margins usually undulate, coarse thorny toothed, teeth varying by cultivar and age, nearly absent on very old specimens, leaves then nearly entire and flat; flowers dioecious, arranged in the leaf axils, usually 3 in small cymes on male plants, solitary on the female, calyx normally 4 parted, seldom 5 parted, usually finely pubescent, corolla white or somewhat reddish, May to June; fruit a globose to slightly ovate drupe with coral-red pod, ripening in September, but persisting to March. HHo 23; HM 1825. Habitat W., Central and S. Europe, N. Africa, Asia Minor to N. Persia (see map). z7 # ⌀

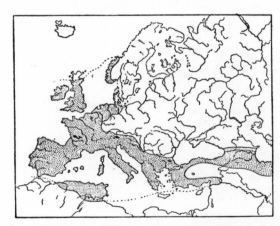

Fig. 116. Range of *Ilex aquifolium* (from Oltmanns)

Outline of the *Ilex aquifolium* cultivars described

(Descriptions of 115 cultivars may be found in NH 36: 69–85)

1. Habit unusual:
   'Argentea Pendula' (weeping)
   'J. C. van Tol' (very divaricate)
   'Pendula' (weeping)
   'Pyramidalis' (strictly conical)

2. Leaves green;
   + Normal form, margins with or without thorns:
   'Alaska' and 'Atlas' (both especially sharp thorned)
   var *balearica* (thornless)

Plate 49

Hedera rhombea, mature form in flower, taken in its native habitat
Photo: Dr. Watari, Tokyo

Hedwingia japonica in its native habit
Photo: Dr. Watari, Tokyo

Plate 50

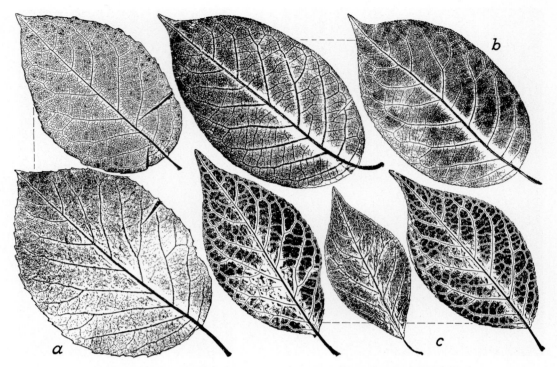

Halesia. a. *H. diptera;* b. *H. monticola;* c. *H. carolina* (collected in its native habitat)

Halesia diptera
on the Bella Island, Lake Maggiore, Italy

Halesia monticola
in the Dortmund Botanic Garden, W. Germany

Plate 51

Helichrysum bellidioides
in Royal Botanic Garden, Kew, England

Hamamelis virginiana, flowering in fall
Photo: F. Kammeyer

Hamamelis. Left, *H. japonica* 'Zuccariniana'; middle, *H. japonica* 'Arborea'; right, *H. mollis*
Photo: Otto, Wolbeck

Plate 52

Hibiscus syriacus, cultivars. Left, 'Coelestis'; right, 'Violet Clair Double'
Photo: Archivbilder D. B.

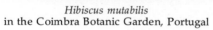

Hibiscus mutabilis
in the Coimbra Botanic Garden, Portugal

Hippophae rhamnoides
with male flowers

Plate 53

Hydrangea aspera ssp. *strigosa*

Homalocladium platycladum
in the Freiburg Botanic Garden, W. Germany

Hydrangea aspera ssp. *strigosa* in the Munich Botanic Garden, W. Germany
Photo: W. Schacht

Plate 54

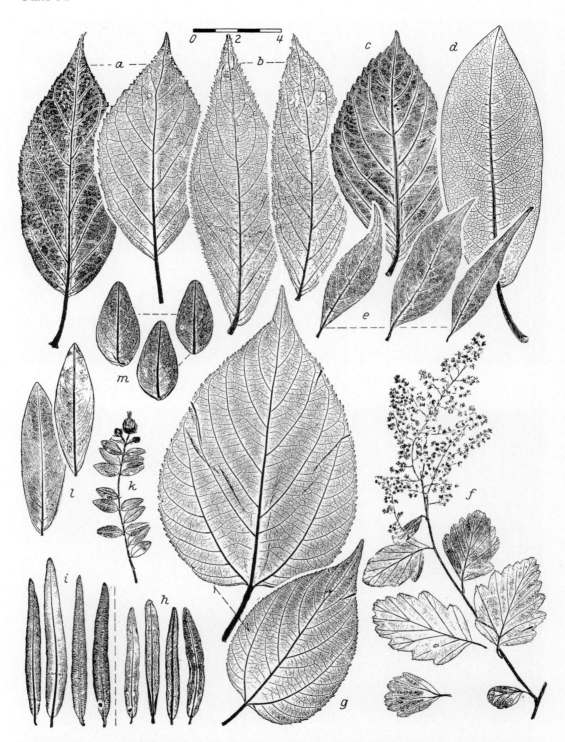

Hydrangea. a. *H. heteromalla*; b. *H. aspera* ssp. *aspera*; c. *H. macrophylla*; d. *H. serratifolia*; e. *H. scandens.* — f. *Holodiscus discolor.* — g. *Hovenia dulcis.* — **Hippophae.** h. *H. salicifolia*; i. *H. rhamnoides.* — **Hypericum.** k. *H. uralum*; l. *H. monogynum*; m. *H. hookeranum* (samples collected in the plants' native habitat)

Plate 55

Hydrangea quercifolia in the Berlin Botanic Garden, W. Germany

Hydrangea involucrata
Photo: Dr. Watari, Tokyo

Plate 56

Hydrangea arborescens 'Grandiflora'
in the Späth Arboretum, Berlin, W. Germany

Hydrangea petiolaris
in the Späth Arboretum, Berlin

Hydrangea arborescens ssp. *discolor* in the Dortmund Botanic Garden, W. Germany

Plate 57

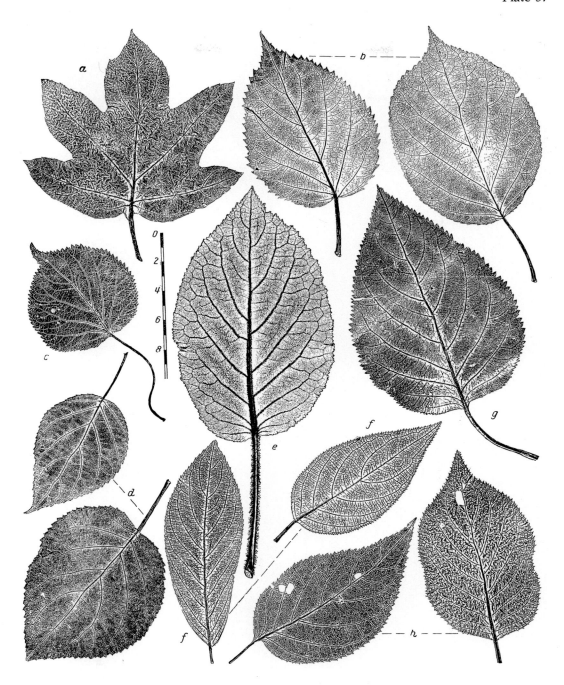

Hydrangea. a. *H. quercifolia;* b. *H. arborescens* ssp. *discolor* (2 leaves); c. *H. petiolaris;* d. *H. arborescens* ssp. *radiata* (2 leaves); e. *H. sargentiana;* f. *H. aspera* ssp. *strigosa;* g. *H. anomala;* h. *H. robusta* (2 leaves) (all leaves taken from wild material)

Plate 58

Hydrangea paniculata 'Grandiflora'
in the Oslo Botanic Garden, Norway

Hydrangea paniculata 'Tardiva'
in Whitnall Park, Chicago, USA

Hydrangea paniculata 'Floribunda'
Photo: M. Haworth-Booth, England

Hydrangea macrophylla 'Veitchii'
in the Wageningen Arboretum, Holland

Plate 59

Hydrangea sargentiana
in the Dortmund Botanic Garden, W. Germany

Hydrangea serratifolia
in Royal Botanic Gardens, Kew, England

Hydrangea involucrata 'Hortensis'

Photo: M. Haworth-Booth

Plate 60

Hypericum calycinum
Photo: Archivbild D. B.

Hypericum androsaemum with fruits in the Dortmund Botanic Garden, W. Germany

Plate 61

Hovenia dulcis
in the Geisenheim Arboretum, Rheingau, W. Germany

Jacaranda mimosifolia
in a park in Malaga, Spain

Ilex crenata 'Stokes'

Plate 62

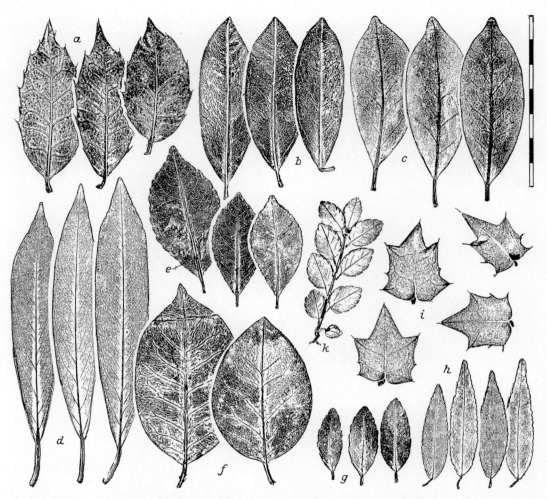

Ilex. a. *I. dipyrena;* b. *I. cassine;* c. *I. integra;* d. *I. fargesii;* e. *I. longipes;* f. *I. rotunda;* g. *I. vomitoria;* h. *I. rugosa;* i. *I. pernyi;* k. *I. yunnanensis* (collected from wild specimens)

Hymenanthera angustifolia
in the Glasnevin Botanic Gardens, Ireland

Hymenanthera crassifolia
in the Glasnevin Botanic Gardens, Ireland

Plate 63

Hypericum 'Rowallane'
in Malahide, Ireland

Iochroma coccineum
in the Lisbon Botanic Garden, Portugal

Ilex altaclarensis 'Wilsonii' in RHS Gardens, Wisley, England

Plate 64

Ilex aquifolium 'Scotica'
in the De Belder Arboretum, Belgium

Ilex aquifolium 'Crispa'
in the De Belder Arboretum, Belgium

Ilex aquifolium 'Crassifolia'
in the Nantes Botanic Garden, France

Ilex pedunculosa var. *continentalis*
in the Dortmund Botanic Garden, W. Germany

'Latispina' (very large thorns)
'Laurifolia' (partly thornless)
  ++ Form unusual, lanceolate to ovate or otherwise differing:
    'Angustifolia', 'Beetii' (circular), 'Ciliata', 'Crassifolia', 'Donningtonensis', 'Foxii', 'Handsworthiensis', 'Hastata' (boat-shaped), f. *heterophylla*, 'Myrtifolia', 'Nigricans', 'Ovata', 'Serrata'
  +++ Blade twisted or thorny on both sides:
    'Crispa', 'Ferox', 'Fisheri', 'Monstrosa', 'Recurva', 'Scotica'

3. Leaves variegated;
  + Green variegated:
    'Lichtenthalii' (bicolored green)
  ++ White variegated;
    a) Center green, margins white variegated:
    'Argentea Longifolia', 'Argenteomarginata', 'Elegantissima', 'Ferox Argentea', 'Grandis', 'Handsworth New Silver', 'Silver Queen'
    b) Center white variegated, margin green:
    'Argentea Medio-picta'
  +++ Yellow variegated;
    a) Center green, margins yellow variegated:
    'Aureomarginata', 'Aureomarginata Ovata', 'Golden van Tol', 'Laurifolia Aurea', 'Mme Briot', 'Myrtifolia Aurea', 'Ovata Aurea', 'Pyramidalis Aureomarginata', 'Rubricaulis Aurea', 'Watereriana'
    b) Center yellow variegated, margins green:
    'Angustifolia Aureomaculata', 'Ferox Aurea', 'Flavescens' (totally yellow-green), 'Golden Queen', 'Myrtifolia Aureomaculata'

4. Unusual fruit color:
    'Amber' (amber-yellow), f. *bacciflava* (yellow), 'Leucocarpa' (white), 'Pyramidalis Fructuluteo' (yellow)

'Alaska' (J. Nissen, around 1960). Habit rather narrowly upright; leaves 3–6 cm long, 3–4 cm wide, resembling the species, but more undulate and coarsely toothed, very glossy dark green; male. Very winter hardy. # ⌀

'Albomarginata' see: **'Argenteomarginata'**

'Amber' (Hillier, before 1955). Fruits amber-yellow, large. Very attractive form. # ⚭

'Angustifolia'. Habit narrowly conical, slow growing, short branches, bark green to reddish; leaves narrow or broadly ovate to lanceolate, 3.5–6 cm long, 1–2(3) cm wide, with 5 to 7 small thorns on either side, always directed forward on an even plane; male. DH 74; DB 2: 110; HHo 94 (= 'Bromeliifolia'; 'Pernettyifolia'; 'Serrata'). Known since 1838. Occasionally found as a sport inside an older plant and quite variable. All these forms should be classified under one name. Fig. 118. # ⌀

'Angustifolia Aureomaculata' (H. van Nes, before 1878). Conical habit, slow growing; leaves like those of 'Angustifolia', usually with an irregular green-yellow spot in the center, leaf form quite variable; male. (Boskoop Experiment Station.) # ⌀

'Argentea Longifolia'. Very similar to 'Argenteomarginata', but young shoots purple; leaves elliptic to more ovate, 6–7.5 cm long, stoutly toothed, margins irregular and usually only narrow, cream-white. (Kew; Boskoop Experiment Station.) # ⌀

'Argentea Marginata Elegantissima' see: **'Elegantissima'**

'Argentea Medio-picta'. Bark green on young shoots; leaves ovate, base cuneate, to 5 cm long, margins distinctly undulate and coarsely toothed, deep green with a cream-white spot in the center, often more toward the basal end. NH 1970: 165 (= 'Silver Milkmaid'). # ⌀

'Argentea Pendula'. Habit low, slow growing, branches nodding, young shoots purple; leaves 6–7.5 cm long, dark green in the center with gray-green marbling, margins broad cream-white; good fruiting (= 'Perry's Weeping'). (Hillier.) # ⚭ ⌀

'Argentea Regina' see: **'Silver Queen'**

'Argenteomarginata'. The usual, white variegated *Ilex*. Young branches and leaves always quite green, never purple; leaves otherwise like the species, to 7 cm long and 5 cm wide, broadly ovate, margins white, center deep green; fruits well. HHo 89, 94 (= 'Albomarginata'; 'Argenteovariegata'; 'Silver Beauty'; 'Silver Princess'). # ⚭ ⌀

'Atlas' (Boskoop Experiment Station 1961). Bushy upright, young branches green; leaves like those of the species, sharply toothed; male. Selected from a number of male plants. # ⌀

'Aurea Regina' see: **'Golden Queen'**

'Aureomarginata'. A collective name for a number of very similar yellow variegated forms with sharply toothed leaves and undulate margins. DH 106; HHo 88. # ⌀

'Aureomarginata Ovata'. Vigorous habit, young branches dark purple; leaves oval, green with gray markings, margins irregularly yellow, with large, regular teeth. (Hillier.) # ⌀

'Aureo Rex' see: I. × *altaclarensis* **'Golden King'**

f. **bacciflava** (West.) Rehd. Fruits lighter or dark yellow (= f. *xanthocarpa* Loes.; f. *flava* Sweet). Occasionally found in the wild. # ⌀ ⚭

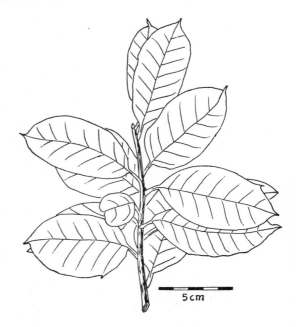

Fig. 117. *Ilex aquifolium* var. *balearica* (Original)
(Collected from the wild on Mallorca, Spain)

var. **balearica** (Desf.) Loes. Shrub, leaves elliptic, broadly ovate to obovate, 6–8 cm long, 3.5–5 cm wide, short acuminate, entire or with a few small marginal thorns, tough, deep green (= *I. balearica* Desf.). Balearic Islands. z9 Fig. 117. # ⌀

'Beetii'. Leaves short and wide, nearly circular, very thorny, about 4 cm long and wide, with long, outspread thorns. DH 86. Fig. 118.

'Bromeliifolia' see: **'Angustifolia'**

f. *calamistrata* see: **'Crispa'**

f. *carnosa* see: **'Donningtonensis'**

var. *chinensis* see: **I. centrochinensis**

'Ciliata'. Resembles 'Angustifolia', but leaves wider; growth conical, low, bark reddish; leaves ovate to lanceolate, 3–4 cm long, 1.5–2 cm wide, marginal thorns directed forward. DH 76: 5. Fig. 118.

'Crassifolia'. Leaves linear-lanceolate, to 7 cm long and about 2 cm wide, leathery-fleshy, very thick, with open sinuses, and many thorny teeth. DH 92. Not attractive, but unusual. Plate 64; Fig. 118.

'Crispa'. Leaves ovate, thorny toothed, very sinuate, more or less distinctly rolled back to the underside, partly entire. DH 80 (= f. *calamistrata* Goepp.; f. *tortuosa* Moore). Of only slight ornamental merit. Plate 64; Fig. 118.

'Donningtonensis'. Leaves oval-lanceolate, acute, thorny toothed or obtuse and nearly entire, occasionally also stunted, midrib often somewhat reflexed. DH 80, 82; NH 36:22 (= f. *carnosa* Goepp.). Fig. 118.

f. *echinata* see: **'Ferox'**

'Elegantissima'. Young branches green; leaves oblong to elliptic, 5–6 cm long, margins undulate and stoutly toothed, cream-white, dark green with gray-green markings in the center; male (= 'Argentea Marginata Elegantissima'). (Hillier.) # ⌀

'Ferox'. Bark purple on young shoots; leaves rather small, about 3–4 cm long, thick, leathery, also covered over the entire upper surface with clustered or fused thorns; male. DH 98, 102 (=f. *echinata* [Mill.] DC.). Fig. 118.

'Ferox Argentea'. Like 'Ferox', but somewhat slower growing, branches purple; leaf margins and teeth white variegated; male. BS 2: 55. # ⌀

'Ferox Aurea'. Like 'Ferox', but leaves with a green margin and teeth, center yellow-green to dark yellow; male. BS 2: Pl. 55; HHo 91. # ⌀

'Fisheri'. Large leaved form; leaves partly entire, partly stout thorny, also partly curling like 'Crispa', 10–15 cm long and 3–5 cm wide. DH 84; NH 36: 22.

f. *flava* see: f. **bacciflava**

'Flavescens'. Leaves like the species, but an outstanding golden-green especially when young, later yellow-green and gradually turning green; female. DH 102 (= 'Moonlight Holly'). ⌀

'Foxii'. Leaves ovate to broad elliptic, tough, 4 to 6 cm long, 2–3 cm wide, margins thorny toothed, usually cuneate at the base, obtuse or more acute, thorns going from a broad triangular base to a fine ciliate tip, straight, oriented toward the leaf apex; male. DH 90; NH 36: 22. Also known in a variegated form. Fig. 118.

'Golden King' see: **I. × altaclarensis 'Golden King'**

'Golden Queen'. Bark green on young shoots; leaves broadly elliptic, 6–8 cm long, very spotted, with a wide, intermittent, golden-yellow margin, occasionally with one side of the blade totally yellow from the margins to the midrib, marginal thorns erect on all sides; male. NH 36: 23; HHo 89 (= 'Aurea Regina'). Very attractive, but sensitive form. z8 ⌀

'Golden van Tol' (W. Ravestein & Zn., Boskoop 1969). Sport from 'J. C. van Tol'. Habit and branching the same, but leaves dull glazed, small, somewhat rounded, with few thorns, regularly golden-yellow margins. Very valuable new form. # ⌀

'Grandis'. Attractive form with black-purple young shoots; leaves deep green, 7 cm long, to 5 cm wide, rather flat, but margins tough thorned, dark green in the center with gray-green markings, margins broad cream-white; male. (Hillier.) #

'Handsworthiensis'. Leaves lanceolate, undulate, about 4.5 cm long, margins densely covered with long thorns, these usually inclined forward, but also in all directions. DH 76; NH 36: 22.

'Handsworth New Silver'. Bark purple on the young shoots; leaves oblong, about 6–9 cm long, marginal thorns all very regular, all white, blade green with gray-green spots and distinct, rather regular, white margin. DH 98; HHo 89, 93; NH 36: 22. Plate 37. ⌀

'Hastata'. Bark purple; leaves clustered at the branch tips, about 2–3 cm long, to 1.5 cm wide, oblong, with 1–3 thorns on either side of the basal half, apical blade half entire, elliptic, often somewhat boat-shaped. DH 86 (= f. *kewensis* Loes.). Notable, but without great ornamental merit. Fig. 118.

f. **heterophylla** (Ait.) Loes. Collective name for wild forms with distinctive leaf shapes. Well known, quite variable form, often a small tree; leaves elliptic to lanceolate, 7–14 cm long, weak or stout thorned or also entire on the same branch, more or less tough, leathery, often rather long petioled. DH 82; NH 36: 22 (= f. *laurifolia* [Kern.] Loud.). Fig. 118. ⌀

ssp. *hodginsii* see: **I. × altaclarensis 'Golden King'** and **'Hendersonii'**

'J. C. van Tol' (van Tol 1904). Shrub, broad growing, branches outspread, young branches purple; leaves usually ovate, 5–7 cm long, glossy above, always slightly raised between the major veins (!), light green underside, base round, margins usually with only a few teeth on the apical half; fruits abundantly. NH 36: 27 (= f. *laevigata polycarpa* Hort. Holl.; f. *polycarpa* Hendr.). Developed in Boskoop about 1895. Fig. 118. ⚭ ⌀

f. *kewensis* see: **'Hastata'**

f. *laevigata polycarpa* see: **'J. C. van Tol'**

'Latifolia Aureomarginata' see: **'Rubricaulus Aurea'**

'Latispina'. Branches with purple bark; leaves nearly rectangular, 5–7 cm long, apex very attenuate and bent downward, somewhat stunted, thick, with only 2–3 margin thorns, often somewhat twisted. DH 88; NH 36: 22. Very conspicuous form. Fig. 118.

f. *laurifolia* see: f. **heterophylla**

'Laurifolia'. Tall growing, tree-like, conical, not totally uniform, partly with green, partly purple branches; leaves larger or smaller, more or less thorny, 5–7.5 cm long, 2.5–3 cm wide; male. Very good pollinator. # ⌀

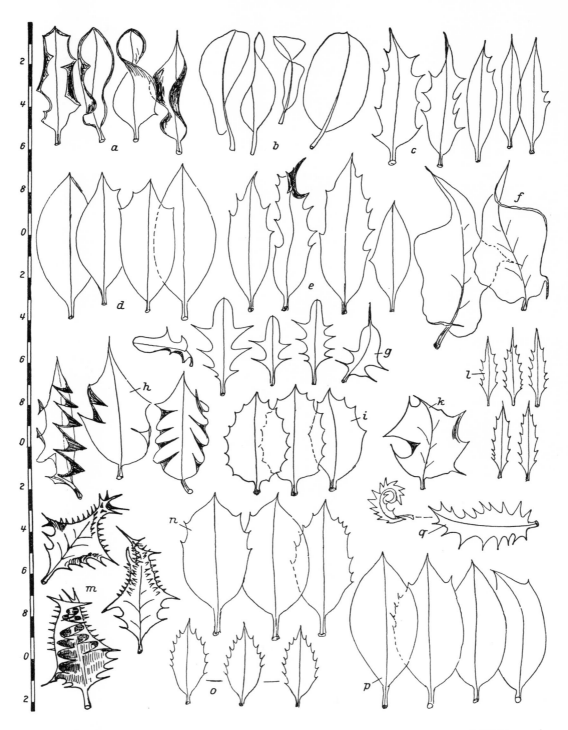

Fig. 118. Forms of *Ilex aquifolium*. a. 'Crispa'; b. 'Scotica'; c. 'Donningtonensis'; d. 'Heterophylla'; e. 'Foxii'; f. 'Latispina'; g. 'Hastata'; h. 'Monstrosa'; i. 'Ovata'; k. 'Beetii'; l. 'Angustifolia'; m. 'Ferox'; n. 'J. C. van Tol'; o. 'Ciliata'; p. 'Pyramidalis'; q. 'Crassifolia' (Original. f. k. and q. from Bean)

'Laurifolia Aurea'. Like 'Laurifolia', but branches with purple bark; leaves deep green with narrow yellow margins; male. Known in England before 1883 and still quite popular. (Hillier.) # ∅

'Leucocarpa'. Fruits whitish, occasionally also somewhat pink toned. Very rare in cultivation.

'Lichtenthalii'. Leaves narrow-oblong, about 3 times longer than wide, deep green above and glossy, but light green along the midrib and margin. # ∅

'Mme Briot'. Resembles 'Golden Queen', but branches with purple bark and orange-red fruit; leaves very distinctive, broad golden-yellow margined, occasionally nearly totally golden-yellow (!), marginal thorns stout and erect, upper leaves much more intensely colored than the lower ones. Very attractive and winter hardy. ∅

'Monstrosa'. Resembles 'Latispina' in the leaf tips and thorns, but thorns more numerous, margins more or less angular in outline, thorns alternately directed up and down, triangular. DH 88 (= f. trapeziformis K. Koch). Fig. 118.

'Moonlight' see: 'Flavescens'

'Myrtifolia'. Small, very dense shrub; leaves lanceolate, long acuminate, 2–3 cm long, occasionally longer, 0.5–1 cm wide, margins quite fine thorny, glossy green above; male. DH 76; NH 36: 22, 27.

'Myrtifolia Aurea'. Young bark purple; leaves like 'Myrtifolia', but dark green and spotted in the middle, margins narrow dark yellow; male. (Hillier.) # ∅

'Myrtifolia Aureomarginata'. Young bark purple; leaves small, regularly thorned, dark and light green, with an irregular golden-yellow spot in the center. (Hillier.) # ∅

'Nigricans'. Branches and leaf petioles dark brown-violet; leaves ovate or lanceolate, rarely elliptic, dark green, leathery, thick to nearly fleshy, thorny toothed, rarely nearly entire, 5–6 cm long, about 2 cm wide. Distributed by L. van Houtte.

'Ovata'. Shrub, scarcely taller than 2 m, branch bark nearly black; leaves ovate, truncate at the base, flat, short petioles 4–5 cm long, 2.5–3 cm wide, rather dense and short thorny toothed, teeth small, erect, deep green above, tough, leathery; male. DH 90; NH 36: 22. Pretty form. Fig. 118. ∅

'Ovata Aurea'. Like 'Ovata', but branches with dark purple bark; leaves like 'Ovata', with short, tough marginal teeth and narrow golden-yellow limb. Very attractive. (Hillier et al.) # ∅

'Pendula'. Branches erect at first, then nodding, developing into a very dense crown; leaves like those of the species, but also available in a variegated form (see, 'Argentea Pendula'). HHo 87.

'Pernettyifolia' see: 'Angustifolia'

'Perry's Weeping' see: 'Argentea Pendula'

f. polycarpa see: 'J. C. van Tol'

'Pyramidalis'. Habit narrowly conical upright, young branches with a yellow-green bark; leaves elliptic, acuminate at both ends, usually however open toothed in the middle, with small thorns on the apex, very glossy and navicular above, light green beneath; fruits very abundantly. One of the most popular cultivars. Fig. 118. ⚭ ∅

'Pyramidalis Aureomarginata' (van der Kraats 1910). Habit and leaf form like that of 'Pyramidalis', with a narrow yellow limb; fruits abundantly, therefore worthy of cultivation, although less attractive than other yellow variegated forms. # ∅

'Pyramidalis Fructuluteo'. Like 'Pyramidalis' but the fruits are yellow. (Hillier.) # ∅ ⚭

'Recurva'. Slow growing, low, branches purple; leaves very densely arranged, 3–4.5 cm long, 1–2 cm wide, curved and twisted, acuminate, with 6–8 marginal thorns on either side; male. (Hillier.) #

'Rubricaulis Aurea'. Broad conical shrub to a small tree, branches very dark violet-brown; leaves rounded, rather large, dull dark green, with a quite narrow yellowish limb; very abundant fruiting, berries orange-red (= 'Latifolia Aureomarginata'). Not particularly attractive but very winter hardy. (Boskoop.) # ∅ ⚭

'Scotica' (Germany 1872). Shrub, large and wide, branches purple, leaves ovate, nearly crispate-twisted, margins thornless, but fleshy thick, occasionally with a single thorn on the apex or margin, 3–5 cm long, glossy dark green above, light green beneath. DH 86; NH 36: 22. Also includes a form with yellow spotted leaves. Plate 64; Fig. 118.

'Serrata' see: 'Angustifolia'

'Serratifolia'. Very similar to 'Angustifolia', but leaves stiffer, margin thorns directed outward (not forward!). DH 76: 2.

ssp. shepherdii see: I. × altaclarensis 'Shepherdii'

'Silver Milkmaid' see: 'Argentea Medio-picta'

'Silver Queen'. Conical habit, upright, slow grower, bark always purple; leaves rather small, dark green with gray-green markings and a broad, white limb; male. NH 36: 21, 23; HHo 94 (= 'Argentea Regina'; 'Silver King' in USA). One of the best white variegated forms, but somewhat sensitive. z8 # ∅

f. tortuosa see: 'Crispa'

f. trapeziformis see: 'Monstrosa'

'Watereriana'. Dense, compact, slow growing shrub, branches with green, yellow striped bark; leaves oblong to obovate or elliptic, 4–6.5 cm long, 1.5–3 cm wide, with dark green center marbled by gray-green and yellowish markings, margins irregularly broad yellow, usually totally thornless or with a very small thorn on the leaf apex; male. HHo 88, 92. (Hillier.) # ∅

f. xanthocarpa see: f. bacciflava

I. × aquipernyi [1]) (I. aquifolium 'Pyramidalis' × I. pernyi). Attractive hybrids between the above species; leaves resemble I. pernyi, but twice as large. Developed by J. B. Gable, Stewartstown, Pennsylvania, USA. 1933. #

'Aquipern'. The type of the cross. Small, erect shrub; leaves 3–5 cm long, 1.5–2.5 cm wide, very undulate, with sharp marginal thorns, but not as strong as those of I. pernyi, dark green; fruits red, 7 mm thick. NH 36, 42. #

'Brilliant' (W. B. Clarke) (I. aquifolium 'Golden Beauty' × I. pernyi). Leaves dull green, 3.5–4.5 cm long, 1.3–2 cm wide, tough, sharp thorned; fruits large, bright red. NH 36: 55. Developed by W. B. Clarke, San Francisco, Calif. USA, 1935. #

---

[1]) Not yet scientifically described.

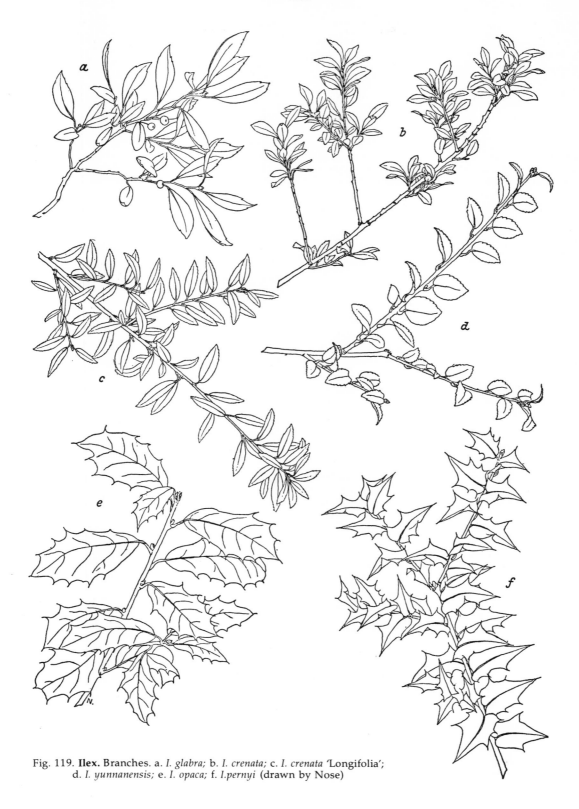

Fig. 119. **Ilex**. Branches. a. *I. glabra;* b. *I. crenata;* c. *I. crenata* 'Longifolia';
d. *I. yunnanensis;* e. *I. opaca;* f. *I.pernyi* (drawn by Nose)

I. attenuata Ashe. (*I. cassine* × *I. opaca*). Large shrub to small tree, conical; leaves elliptic to obovate-oblong, 3–8 cm long, long thorny acuminate, with 1–4 erect thorns on either side, lighter green beneath with somewhat pubescent midrib; inflorescences pubescent; fruits red, 6 mm thick, solitary or to 3 in stalked clusters. USA; Florida. Natural hybrid, occurring with the parents. z9 Plate 66. #

I. balearica see: **I. aquifolium** var. **balearica**

I. × beanii Rehd. (*I. aquifolium* × *I. dipyrena*). Branches, leaves, buds and fruit similar to *I. dipyrena* in appearance, but flowering in May (between the parents), therefore differing from *I. aquifolium,* but occasionally included as such; leaves also obtuse, not glossy green. DH 4; BS 2: 443 (= *I. dipyrena elliptica* Dallim.). 1900. # ∅

I. bioritsensis Hayata. Very similar to *I. pernyi* in appearance, but leaves larger, 4–5 cm long, with 4–5 erect thorns on either side, leaf apex not so distinctly bowed; fruits red, usually with 3 or 2, occasionally with 4 seeds (like *I. pernyi*). NH 36: 4; ICS 3031; HHo 134 (= *I. pernyi* var. *veitchii* [Veitch] Bean). China; Taiwan. 1914. z7 # ∅

I. canariensis Poir. Acibiño. Tall evergreen shrub or small tree, 5–10 m high in its habitat, strongly branched; leaves elliptic-ovate, acute, tough-leathery, glabrous, entire or with an occasional thorn, 4–7 cm long, 3–4.5 cm wide, dark green and glossy, base cuneate to rounded, petiole 1 cm; flowers 3–9 in the axils of the upper leaves, in about 3–5 cm long clusters, white, 1 cm wide, April–June; fruits globose, dark red, 1 cm thick. KGC 28. Canary Islands and Madeira. z10 #

I. cassine L. non Walt. Dahoon. Evergreen shrub, occasionally a tree in its habitat, to 12 m, young branches glabrous or finely pubescent; leaves obovate-oblong to lanceolate, entire or finely dentate near the apex, 4–10 cm long, white-gray and dense short pubescent beneath; flowers 4 parted, in small, stalked cymes; fruits red, 6–8 mm thick. BB 2357; SS 46; DH 4; NH 36: 2; HHo 135 (= *I. dahoon* Walt.). USA, along the coast from Virginia to Florida and Texas, in pine forests and marshes. 1726. z7 Plate 62. # ∅

I. centrochinensis S. Y. Hu. Large shrub or small tree, 5–7 m high, densely conical with short erect branches, young branches finely pubescent; leaves oval-oblong to oval-lanceolate, 7–10 cm long, thorny toothed, with 6–8 stout thorns on either side; flowers white, fragrant, short stalked; fruits globose, 4–8 mm thick, red, with 4 seed pits. NH 36: 14 (= *I. aquifolium* var. *chinensis* Loes.). Central China, growing together with *Metasequoia glyptostroboides.* 1901. z6 #

I. chinensis Sims. Evergreen tree, to 13 m tall in its habitat, glabrous, branches ash-gray; leaves elliptic-lanceolate, seldom ovate, 5–11 cm long, 2–4 cm wide, long acuminate, margins crenate, base obtuse to cuneate, with 6–9 vein pairs, thin, petiole 12 mm long; flowers on the young shoots, grouped 7–15 together, lilac or red(!); fruit ellipsoid, 10–12 mm long, glossy red. ICS 3019; KIF

2: 57; HHo 100; BM 2043 (= *I. purpurea* Hassk.; *I. oldhamii* Miq.). China. 1810. Very decorative tree; fruiting branches very popular New Year's decoration in China. z9 # ⚬

I. ciliospinosa Loes. To 4 m high , evergreen shrub, similar to *I. dipyrena,* but young branches cylindrical, dense and short fine shaggy pubescent; leaves smaller, about 2.5–4 cm long, elliptic, base round or cuneate, margins with small, nearly ciliate-like teeth, directed forward, dull green above, light green beneath, glabrous; fruits ellipsoid, in clusters, often grouped in 2's, in the leaf axils, usually with 2 seed pits. MD 1919: Pl. 2a–b; NH 36: 14; FIO 161. Central China; mountain forests in W. Szechwan. 1908. Frequently cultivated and quite hardy. z6 #

I. corallina Franch. Evergreen shrub, somewhat resembling *I. centrochinensis,* but leaves larger and quite variably dentate 5–12 cm long, 2–3.8 cm wide, elliptic to oval-lanceolate, thin to tough-leathery, margins stout and densely thorny, thorns directed backward and outward, usually short, very glossy above; fruits only 3–4 mm thick, in small clusters, red. DH 130; NH 36: 14; FIO 168; ICS 3035. Central China, mountain forests in Hupeh, Szechwan, and Yunnan Provinces. 1900. z7 #

I. cornuta Lindl. & Paxt. Evergreen shrub, 2–3 m high (also a small tree in its habitat), rounded, bushy, young branches glabrous, pale green, somewhat angular at first; leaves leathery tough, quite variable in form, usually somewhat rectangular, with a large thorn on each corner, about 4 to 10 cm long, another thorn on the apex; fruits globose, to ellipsoid, bright red, to 10 mm thick. HHo 13, 25, 82; BM 5059; NH 36: 20. Central and Eastern China. 1846. z7 Plate 65. # ∅

'Burfordii' (Howell). Leaves entire, occasionally with 2 quite short thorns on the apex. NH 36: 20. Discovered in the Westview Cemetery in Atlanta, Georgia, USA; introduced in 1939 by J. R. Howell, of Knoxville, Tenn. z9

'Rotunda' (McIlhenny). Wide growing, scarcely over 90 cm high, but often equally wide, very densely branched; leaves like the species, thorns longer; sterile. Originated from seed in the E. A. McIlhenny Nursery, Avery Island, Louisiana, USA, before 1930. z9

I. crenata Thunb. Evergreen shrub, upright, 2–3 m high, but often lower, very divaricate and densely branched, young branches somewhat angular and quite finely pubescent at first, dark brown; leaves clustered, short petioles, elliptic to oblong-lanceolate, 2–3 cm long, crenate, dark green above, glossy, glabrous, tough; flowers 4 parted, male flowers grouped 3–7, female solitary, May–June; fruits black, 6 mm thick, globose. HHo 98; NH 36: 11; DH 118; MJ 1106. Japan. 1864. Good hardiness. z6 Fig. 119. # ∅

Includes many forms:

'Aureovariegata'. Leaves 2–4 cm long, yellow spotted or marbled. Discovered in England. 1887.

f. *bullata* see: **'Convexa'**

f. *convexa* see: **'Convexa'**

'Convexa'. To 2 m high, habit more broad or also conical; leaves densely arranged, stiff, elliptic, arched upward, leathery tough, glossy, 1–2 cm long. HHo 200; NH 36: 25; AB 20; 11 (= f. *bullata* Rehd.; f. *convexa* Mak.). Japan, 1919. Very valuable. Hardy. ∅

f. **fukasawana** Makino. Habit dense and upright, branches stout, angular; leaves thin, lanceolate to narrow elliptic, 2.5–4 cm long, obtuse toothed, bright green to yellowish-green when young. Japan. (Hillier.) #

'Golden Gem' (L. Konijn, Reeuwijk). Low growing and widespreading, dwarfed; leaves somewhat smaller than those of the type, golden-yellow in large part, later somewhat greening. Attractive. # ∅

'Helleri' (Heller 1936). Low growing, wide; leaves elliptic, 1–2.5 cm long, acuminate at both ends, dark green above, lighter beneath, with 2–4 teeth on either side. NH 36: 15; HHo 97, 200. Introduced to the trade by J. Heller, Newport, Rhode Island, USA.

'Hetzii' (Hetz' Nurseries, USA, 1943). Wide growing, similar to 'Convexa'; leaves broad ovate, to about 2 cm long and 1–1.3 cm wide, flat. Good winter hardiness, but defoliates easily. # ∅

f. **latifolia** (Goldring) Rehd. Shrub, to 1.5m high; leaves very densely arranged, thin, leathery, flat, elliptic to oblong, 2–3.8 cm long, rounded at both ends. NH 36: 11; AB 20: 11 (= *I. fortunei* Miq.). 1887.

f. **longifolia** (Goldring) Rehd. Branches dark red-brown; leaves linear-lanceolate, about 3.5 cm long, to 9 mm wide, acute at both ends. NH 36: 11. Fig. 119.

'Luteovariegata'. Leaves 2 cm long at most, yellow spotted. Discovered in 1864 in Russia.

'Mariesii'. Dwarf form, habit quite low and wide, branches stiff and erect, very short, dark gray; leaves circular to broadly ovate, 5–12 mm wide, somewhat concave, entire. NH 36: 16; MJ 1107 (= *I. nummularioides* Franch. & Sav.). Japan; much used there for dwarfing and topiary work; annual growth about 3 cm. Fruiting. Plate 66. ∅

'Microphylla'. Low, densely branched, branches dark brown-green; leaves densely packed, elliptic, very small, 7–13 mm long, 4–7 mm wide, often somewhat concave, dark green. NH 36: 11, 16. Korea, Japan.

var. **paludosa** (Nakai) Hara. Prostrate habit; leaves broadly elliptic, apex rounded, base obtuse (= var. *radicans* Nakai). Japan, in swampy areas. Not yet introduced into cultivation?

var. *radicans* see: var. **paludosa**

'Stokes' (Stokes 1951). Habit stiff and compact, dense; leaves small, less crenate, dull green. DB 6: 27. Introduced to the trade by W. S. Stokes, Butler, Penn., USA. US Plant Pat. 887. Very winter hardy and worthy of cultivation. Plate 61.

There exist many other cultivars of American origin which may be found listed in Wyman (Lit.).

I. dahoon see: **I. cassine**

I. decidua Walt. Deciduous shrub, 1–2 m high, also a small tree in its habitat, branches light gray, outspread; leaves partly clustered, obovate, 4–7 cm long, obtuse serrate, dark green above, glossy, with indented venation, thickish, lighter beneath, midrib pubescent; flowers solitary, May; fruits globose, 7–8 mm thick, orange-scarlet, seed pit with many ridges (important distinction from *I. verticillata* and *I. laevigata*!). BB 2360; SS 49; NH 36: 7; VT 654 (= *I. prinoides* Ait.). Southeast USA. 1760. Hardy. z5–6

I. dipyrena Wall. Large shrub, also a tree in its habitat, to 14 m, similar to *I. aquifolium,* young branches pubescent; leaves oval-elliptic to oblong, 5–10 cm long, acuminate, base cuneate to rounded, margins very scabrous dentate on young plants, only slightly dentate to entire on older plants; flowers and fruit sessile; fruits oval, 8 mm long, red, with 2(!) seed pits. DH 4: 124; NH 36: 14. Himalaya. 1840. z9 Plate 62. # ∅

ssp. *elliptica* see: **I. × beanii**

I. dubia see: **I. amelanchier** and **I. macropoda**

I. fargesii Franch. Evergreen shrub, erect, 3–5 m high, branches glabrous; leaves oblong to linear-lanceolate, 6–12 cm long, serrate on the apical half, dull green and smooth above; flowers short petioled, in clusters, May–June; fruits globose, red, 6–8 mm thick, short stalked. BM 9670; DH 130; NH 36: 17; FIO 159; ICS 3034. W. China; Hupeh and Szechwan Provinces. 1900 z7 Plate 62. # ∅

I. fortunei see: **I. crenata** f. **latifolia**

I. franchetiana Loes. Very closely related to *I. fargesii*, but leaves broader obovate-lanceolate, often nearly completely serrate, 4–10 cm long, abruptly acuminate, base obtuse cuneate; fruits globose, 8 mm thick, red, short stalked, in dense clusters, fruits abundantly. ICS 3033; FIO 160. Central China; W. Hupeh, W. Szechwan Provinces, in mountain forests, 1200–2400 m. 1907. z8? Plate 67. # ∅

I. geniculata Maxim. Deciduous shrub, 1–2 m high, thin branched, young branches glabrous, somewhat furrowed; leaves oval-elliptic, 3–5 cm long, acuminate, scabrous serrate, base round to cuneate, midrib pubescent beneath, golden-yellow fall color; flower stalk 8–15 mm long; fruits solitary, globose, 5 mm thick, vermillion-red, very attractive, on a 1.5–3.5 cm long, brown stalk, August. MJ 1104; NH 36: 17. Central Japan, in the mountains. 1926. z6 ⚥

I. georgei Comber. Compact evergreen shrub, becoming tree-like in its habitat; leaves lanceolate to ovate, 2–4.5 cm long, to 1.5 cm wide, acuminate and with a thorny tip, coarse and sparsely arranged, with up to 7 teeth on a side, base round to cordate; fruits to 5 mm thick, bright red, in clusters of 5–8, each with 1–2 nutlets. China, Yunnan Province; Upper Burma. (Caerhays; Hillier.) z10

I. glabra (L.) Gray. Inkberry. Often only semi-evergreen, occasionally to 2.5 m high, dense, thin branched shrub, usually only 1–1.5 m high, young branches pubescent; leaves oblanceolate, 2 to 5 cm long, entire or with few obtuse teeth on the apex, glossy dark green above, glabrous; flowers 5–8 parted, solitary or grouped, June; fruits globose, 6 mm thick, black, short stalked. BB 2359; HHo 13, 22; GSP 274 (= *Prinus glabra* L.). Eastern USA, on sandy to peaty acid soil. Very hardy. z5 Fig. 119. #

I. hookeri King. Evergreen, glabrous tree, to 18 m high in its habitat, branches ash-gray (!), thick, but brown in the first year; leaves elliptic to more obovate, 5–10 cm long, 2–4.5 cm wide, leathery tough, dull on both sides, quite acuminate, margins finely serrate, base obtuse to round; flowers 4 parted, glabrous, calyx not ciliate (!); fruits globose, 6 mm thick. China; Yunnan. (Hillier.) z9 # ∅

I. insignis see: I. kingiana

I. integra Thunb. Evergreen shrub, also a small tree in its habitat, 10–12 m high, conical, young branches angular, glabrous; leaves elliptic-oblong, 5–10 cm long, 2–3 cm wide, entire or (rarely) with few teeth, short and obtuse acuminate at the apex, gradually tapering to the base, dark green above, glossy, petiole 8–15 mm long; flowers like I. aquifolium, in clusters, March–April; fruits globose, 1 cm thick, dark red, solitary or several on 8–12 mm long stalks. DH 118; MJ 1094; KIF 1: 35; KO 293. Japan, Korea. 1864. z9 Plate 62. # ∅

'Accent' (I. integra × I. pernyi, 1960). Habit conical, about 2 m high and 90 cm wide in 10 years; leaves very similar to those of 'Elegance', margins with 3–4 soft thorns on either side. Introduced by the US National Arboretum as a pollinator for 'Elegance'. z8–9 # ∅

'Elegance' (I. integra × I. pernyi, 1966). Habit narrow conical, 2 m high and 60 cm wide as 10-year plants; leaves elliptic, 2–4 cm long, 1–2 cm wide, evergreen; fruits red, nearly 1 cm thick. Developed by the US National Arboretum. z8–9 # ∅

I. intricata Hook. f. Evergreen, glabrous shrub, low creeping habit; leaves obovate to elliptic, 2 cm long, 1 cm wide, leathery, rounded apex, dull green, densely arranged, very short petioles; flowers inconspicuous and in small clusters; fruits globose, 5 mm thick, bright red, solitary or grouped 2–3 together. Himalaya; Sikkim, SE. Tibet, NW. Upper Burma. 1931. z8?? # ∅

I. kingiana Cockerell. Evergreen, quite glabrous, small tree, branches thick, gray, glossy; leaves elliptic-lanceolate to ovate, 15–20 cm long, 5–6 cm wide, acuminate, generally tapering to the base, margins with small, thorn-tipped teeth, occasionally nearly entire, thick and tough leathery, deep green above, midrib raised, petiole 2.5 cm long, purple; flowers in nearly sessile clusters; fruits 5 mm long, oval, bright red (= I. insignis Hook. f.; I. nobilis Gumbleton). E. Himalaya, 1800–2400 m. Young seedlings are very different having distinctly wavy leaves, 6–8 mm long, marginal thorns pointing in every direction. (Hillier; Kew.) z9 # ∅

I. × koehneana Loes. (= I. aquifolium × I. latifolia). Occurring among the parents, resembles I. latifolia in foliage character; leaves elliptic, 11–16 cm long, 6–8 cm wide, leathery, quite glabrous, with 7–10 veins on either side, margins very thick and short thorny toothed, slightly undulate. Before 1919. z6 # ∅

I. laevigata (Dum.-Cours.) Gray. Deciduous shrub, 2–3 m high, branches erect, always quite glabrous; leaves oval-lanceolate, 3–6 cm long, appressed serrate, bright green above and glossy, yellow in fall; flowers May–June, male flowers 8–16 mm long stalked, female flowers short stalked; fruits orange, solitary, globose, 8 mm thick.

BB 2363; GSP 271; NH 36: 2. N. America, Canada to Carolina, in moist valleys. 1812. Quite hardy. z5 Plate 68.

'Hervey Robinson'. Fruits yellow. 1908 (as yet in cultivation?)

I. latifolia Thunb. Tall, evergreen shrub, 12–15 m tall tree in its habitat, young branches very thick and stiff, nearly 1 cm thick, glabrous, somewhat angular; leaves oblong, very thick, 10–20 cm long, glossy green, equally tapering to both ends, marginal teeth shallow, not thorny, yellow-green beneath, petiole 1–2.5 cm long; fruits red, globose, 8 mm thick, several in short, axillary racemes. HHo 99, 220; MJ 1105; BM 5597; DH 118; KO 291; KIF 2: 38. Japan. Not to be confused with the 'Latifolia'-forms of many species! z7 Plate 66. # ∅

I. longipes Chapm. ex Trel. Deciduous large shrub or small tree, to 5 m high, branches divaricate; leaves elliptic to obovate, 4 to 7.5 cm long, short acuminate, base cuneate, sparsely dentate, paper thin, glabrous above, midrib finely pubescent beneath; flowers solitary, in the leaf axils; fruits red, globose, about 8 mm thick, stalk 1–2 cm long. GF 3: 46; NH 1970: 181; VT 653; NBB 2: 500. Southeast USA, on gravelly river banks. 1906. Hardy. z6 Plate 62.

'Vantrompii'. Fruits yellow.

I. lawsoniana see: I. × altaclarensis 'Lawsoniana'

I. macrocarpa Oliv. Small to medium-sized deciduous tree, to about 15 m high in its habitat, young branches glabrous, green; leaves thin, elliptic, long acuminate, 7–10 cm long, 3 to 5 cm wide, flat toothed, base round to broadly cuneate, glabrous; male flowers in clusters on short side branches, female flowers solitary in the leaf axils on 15 mm long stalks; fruits like small black cherries, 15 mm wide, flat rounded, with 7–9 distinctly ribbed seed pits. BMns 72; FIO 171; ICS 3014. S. and SW. China. (Kew; Hillier and others) z8? ⚭

I. macropoda Miq. Small deciduous tree or large shrub, branches gray, glabrous; leaves thin, ovate to broadly ovate, 4 to 7 cm long, 2.5–4 cm wide, short acuminate, serrate, scattered pubescent above to glabrous, underside pubescent only on the venation, otherwise light green and glossy, petiole 1–2 cm; flowers greenish-white, like the leaves often clustered on short shoots, calyx 4–5 lobed, ciliate; fruits ellipsoid, 7 mm long, red, seed pit with a shallow furrow on the dorsal side. KIF 1: 81; ICS 3016 (= I. dubia var. macropoda [Miq.] Loes.). Japan, China, Korea. 1894. (Hillier.) z8? ⚭

I. maderensis see: I. perado

I. melanotricha Merr. Evergreen, densely foliate, tall shrub, or also a small tree, 3–11 m high, young branches stiff, glabrous (plants in the wild often with erect, 1 mm long black "hairs", actually the fruiting bodies of a fungus!), later dark red; leaves oblong-elliptic, acuminate at both ends, 7.5–10 cm long, sparse and shallowly crenate, dark green above, lighter beneath; flowers green, many in small, axillary panicles; fruits globose, 9 mm thick, scarlet-red, fruits very abundantly. BMns 84. China; Yunnan Province. Cultivated in England. z9? # ⚭ ∅

I. montana Torr. & Gray. Deciduous shrub, also a small tree in its habitat, young branches glabrous; leaves ovate to elliptic, very long acuminate, 6–16 cm long, cuneate tapered at the base, scabrous serrate, glabrous beneath, somewhat pubescent on the venation, petiole 1–1.5 cm long; flowers white, the male clustered at the ends of the previous year's short shoots, the female usually solitary, short stalked; fruits globose, light orange-red, 1 cm thick, on a 3–8 mm long stalk, September, persisting throughout the winter, seed pit much ribbed. BB 2361; NBB 2: 500; VT 656; SS 50; NH 36; 2 (= *I. monticola* Gray). Eastern USA, from New York south. Hardy. z5 Plate 68.

var. **media** (Gray) Britt. Leaves only half as large. W. Virginia. Plate 68.

I. monticola see: **I. montana**

I. mundyi see: **I. × altaclarensis 'Mundyi'**

I. myrtifolia Walt. Small evergreen tree or shrub, compact habit, branches sparse and curved; leaves linear to more elliptic-oblong or oblanceolate, 12–25 mm long, 5 mm wide, occasionally with a few tiny thorned teeth toward the apex, otherwise entire, obtuse or rounded, with a small bristle on the apex, base cuneate, margins somewhat involuted; fruits usually flat globose, 6 mm wide, red, 4 tipped on the base, with persisting calyx lobes, arranged in a square. KTF 136; HHo 99, 135. Florida, generally with *Nyssa biflora, Magnolia virginiana, Taxodium ascendens*, etc. z8–9 # ⌀

'Oriole' (*I. myrtifolia* × *I. opaca*, 1956). Compact habit, about as high as wide, 1.5 × 1.5 m; leaves 5–7 cm long, 1–1.5 cm wide, with 3–4 small teeth on either side on the apical blade half, thick, leathery tough; fruits red to orange-red, about 1 cm thick. Developed by the US National Arboretum. z8–9 # ⌀

'Tanager' (*I. myrtifolia* × *I. opaca*, 1956). Globose habit, about 2 m high and becoming somewhat wider; leaves similar to those of 'Oriole', but fruits more similar to those of *I. myrtifolia*, bright red, 7 mm thick. Developed in the US National Arboretum. z8–9 # ⌀

I. nobilis see: **I. kingiana**

I. nummularoides see: **I. crenata 'Mariesii'**

I. oldhamii see: **I. chinensis**

I. opaca Ait. American Holly. Tall, evergreen shrub, also a tree to 10 m tall in its habitat (also that tall in cultivation), crown narrow conical, branches outspread, finely pubescent at first; leaves elliptic-lanceolate, 5–10 cm long, sparsely coarsely thorny dentate, occasionally nearly entire, always dull green above (!!), yellow-green beneath; flowers white, June; fruits globose, 6 mm thick, usually solitary, September, but persisting to April. HHo 13, 24; BB 2356; GTP 269. Eastern USA, in deep fertile lowlands, also on thin soil. 1744. Hardy. Widely cultivated in the USA, berried branches used for Christmas decorations. More than 117 cultivars are described in NH 36: 21–31, which see. z6 Plate 65; Fig. 119. # ⌀

f. **xanthocarpa** Rehd. Fruits yellow. USA; Massachusetts. 1901.

I. paraguayensis St. Hil. Evergreen shrub, to 6 m high and tree-like in its habitat; leaves obovate to more elliptic, 4–12 cm long, 3–6 cm wide, leathery, flat, dark green above, lighter beneath, coarsely crenate on the apical half; flowers on the young shoots, white; fruits globose, brown-red, 6–7 mm thick. HHo 139. Paraguay and adjoining regions in Brazil and Argentina. Very important economic plant for its leaves as the source of "Yerba Mate" tea. z9 #

I. pedunculosa Mill. Evergreen shrub, also a tree in its habitat, to 10 m high, young branches glabrous; leaves ovate, 3–7 cm long, always entire (!), somewhat undulate, long acuminate, base rounded, glossy above, petiole 0.5 to 1.5 cm long; flowers in June, solitary or several together, on 1–2 cm long stalks; fruits red, globose 8 mm thick, stalk 2–4 cm long. HHo 13, 137; MJ 110; KO 290; NH 36: 13; KIF 3: 36. Japan. 1893. Hardy. z6 # ⌀

var. **continentalis** (Loes.) Bean. Differing primarily in the tougher, 8–12 cm long leaves, the apical blade half on young plants sparsely and fine appressed serrate; sepals finely ciliate. Central China; Szechwan, Hupeh Provinces. 1901. Hardy. Plate 64. ⌀

I. perado Ait. Tall, evergreen shrub or small tree, young branches glabrous or finely pubescent; leaves ovate to oblong or lanceolate or elliptic, 6–10 cm long, 3–6 cm wide, thin leathery, glossy, more or less thorny toothed or dentate to occasionally entire, petiole 5–15 mm long, appearing winged with the decurrent leaf blade; flowers in axillary clusters; fruits deep red, globose, 8 mm thick, petiole 8 mm long. DH 4 (= *I. maderensis* Lam.; *I. perado* var. *maderensis* Lam. Loes.). Canaries, Azores. 1760. Not generally cultivated; plants by this name are *I. altaclarensis*! z9 # ⌀

var. **azorica** Loes. Leaves smaller, 2.5–6 cm long, elliptic to rounded or ovate, entire or with a few forward directed teeth. Azores. z10 #

var. **platyphylla** (Webb. & Berth.) Loes. Leaves large, especially on female plants, 10–15(20) cm long, ovate, seldom oblong, entire or thorny toothed; flowers milk-white; fruits globose, red. DH 134; BM 4079; NH 1970: 164 (= *I. platyphylla* Webb.). Madeira, Teneriffe. 1842. z9

I. pernyi Franch. Upright, evergreen shrub, often tree-like in its habitat, young branches green, finely pubescent; leaves crowded, rhombic to nearly rectangular, 2–3 cm long, with 2–3 thorns on either side, dark green and glossy above, light green beneath, glabrous, blade tip long, bowed; flowers yellowish, in dense clusters on the previous year's wood; fruits ellipsoid, red, 6–8 mm thick, quite short stalked, August. HHo 133; DH 130; NH 36: 14; ICS 3030. Central and W. China. 1900. z7 Plate 62; Fig. 119. # ⌀

var. **manipurensis** Loes. Like the species, but leaves more ovate instead of rhombic, somewhat larger, the apical thorns smaller, about ⅓ the length of the blade. Yunnan, Manipur, in the mountains from 2000 to 3000 m. 1901.

var. *veitchii* see: **I. bioritsensis**

I. platyphylla see: **I. perado** var. **platyphylla**

I. prinoides see: **I. decidua**

I. purpurea see: **I. chinensis**

I. rotunda Thunb. Evergreen shrub, to 20 m high in its habitat, glabrous, closely related to *I. pedunculosa;* leaves oblong-elliptic or ovate to obovate, 4–11 cm long, entire, acuminate, base cuneate, petiole 1.2–2.8 cm long; male flowers 4–15 in umbels at the base of the young shoots, female flowers grouped 1–3, petals lilac; fruits ellipsoid, red, with 4–6 seed pits, these furrowed. KIF 2: 40; ICS 3021; HHo 13, 136; Bai 1: 35; MJ 1095; NH 36: 11 (= *I. siroki* Sieb.). Korea, Japan to SE. China. 1849. Very attractive species. z6 Plate 62 # ⊘

I. rugosa F. Schmidt. Low, sometimes procumbent, evergreen shrub, young branches pubescent, furrowed; leaves ovate to oblong, tapering to both ends, 2 to 5.5 cm long, 0.4–1.8 cm wide, apex obtuse, sparse, and rounded crenate, dark green and rugose above, lighter and distinctly veined beneath, glabrous; fruits usually ellipsoid, solitary, 6 mm thick, red, September. MJ 1108; KO ill.292; TAP 177. Japan, Sachalin. 1895. Hardy. z3 Plate 62, 67. # ⊘

I. serrata Thunb. Deciduous shrub, to 5 m high, usually only half that height in cultivation, branches pubescent, outspread, somewhat wavy; leaves elliptic, 2–5 cm long, finely serrate, dull green above, pubescent beneath; flowers 4–5 parted (!!), June; fruits globose, red, 4 to 5 mm thick, very persistent and holding their color well, fruits abundantly. HHo 138; DH 118; MJ 1102; NH 36: 18 (= *I. sieboldii* Miq.). Japan. 1893. Quite hardy. z6 Plate 68. ⊗

'Leucocarpa'. Fruits white.

'Xanthocarpa'. Fruits yellow.

I. sieboldii see: **I. serrata**

I. siroka see: **I. rotunda**

I. sugeroki Maxim. Evergreen shrub, to 2 m high, very similar to *I. yunnanensis,* but sparser and shorter pubescent; leaves larger, 2 to 6 cm long, serrate or crenate on the apical half, petioles 3–10 mm long, finely pubescent; male flowers usually in 3's on 8–10 mm long stalks; fruits solitary, red, globose, 7 mm thick, stalks 1–2 cm long. MJ 1101. Japan, Sachalin. 1914. z6 #

I. urqhuartii see: **I. × altaclarensis 'Nobilis'**

I. verticillata (L.) Gray. Deciduous shrub, to 3 m high, branches sparsely arranged, usually totally glabrous or sparsely and short pubescent; leaves broadly ovate to lanceolate, tapering to both ends, 4–7 cm long, singly or doubly serrate, pubescent beneath, especially on the venation; flowers 5–8 parted (!!), all short stalked, grouped 1–2; fruits bright red, 6 to 8 mm thick, globose,

fully ripe before leaf drop. BB 2362; HHo 13, 138; GSP 274; HN 36: 6; VT 655. Eastern N. America. 1793. Valued for the character of its fruit. (Easily confused with *I. laevigata,* but the latter with upper leaf surface always glossy, underside glabrous, long stalked male flowers and solitary fruits.) Quite variable. z3–9 Plate 68. # ⊗

'Aurantiaca'. Fruits orange-red.

'Chrysocarpa'. Fruits yellow.

'Cyclophylla'. Leaves smaller, broadly obovate to nearly orbiculate, 2–2.5 cm long, rugose above, usually clustered at the branch tips. Discovered in Illinois, USA in 1934.

var. **padifolia** (Willd.) Wats. Leaves oblong to obovate, rather thick and leathery tough, pubescent beneath, 5–12 cm long. 1920.

var. **tenuifolia** (Torr.) Wats. Leaves to 12 cm long, thin, lighter green, sparsely pubescent, fine translucent punctate (visible under a hand lens).

I. vomitoria Ait. Evergreen shrub, also a small tree in its habitat, under 10 m high, branches very sparsely arranged, finely pubescent; leaves ovate, 1–3.5 cm long, 6–15 mm wide, obtuse, tapering to the base, sparsely and shallowly dentate, bright glossy, glabrous, petiole 3–4 mm long; fruits scarlet-red, globose, grouped 1–2 together. BB 2358; HHo 13, 139; NH 36: 5 (= *I. cassine* Walt. non L.). Southeast USA. 1700. z6–7 Plate 62. #

'Pendula' (Foret & Solymosy). Tree, to 10 m high, branches very pendulous, young branches reddish; leaves 2.5–3 cm long, crenate with hornlike, brown teeth; fruits 4–6 mm thick, more or less clustered, persisting through the winter. Discovered in 1960 by S. L. Solymosy, Southwestern Louisiana Institute, Lafayette, Louisiana, USA. Greatly valued and easily propagated.

'Tricolor'. Leaves dark and gray-green, with irregularly wide, cream-white margins, often half the leaf cream-white. NH 1970: 182.

I. yunnanensis Franch. Hardy, evergreen shrub, to 4 m high, branches short and densely pubescent; leaves small, ovate, 1–3 cm long, finely crenate, leathery, glossy above, finely pubescent beneath, eventually only on the midrib; flowers June; fruits globose, 6 mm thick, red, on a 4–8 mm long stalks, solitary. NH 36: 11 (and 3 var.); FIO 158; ICS 3028. 1901. W. China. z6–7 Plate 62; Fig. 119. # ⊘

var. **gentilis** (Loes.) Bean. Branches and leaves nearly or totally glabrous.

All *Ilex* species like fertile, humusy soils; many species will thrive best in shady areas. The evergreen species will often defoliate when transplanted, but soon grow out again. The USA has several large holly plantations providing holly branches for Christmas decoration.

Each new *Ilex* cultivar, before its introduction, should be registered with the Holly Society of America at the US National Arboretum, Washington, DC. 20002.

Lit. Dallimore, W.: Holly, Box and Yew; 3–149, London 1908 ● Dengler, H. W.: Handbook of Hollies; in Nat. Hort. Mag. **36**, 1–139, 1957 (with 61 pp., ills. and index) ● Hume, H. H.: Hollies; 242 pp., New York 1953 ● Loesener, Th.: Monographia Aquifoliacearum; in Abhandl. Kais. Leop. Karol. Ak. Naturforscher, Part. 1 in Vol. **78**, 1–589, 1901; Part 2 in Vol. **89**, 1–313, 1908 ● Loesener, Th.: Über die Aquifoliaceen, besonders über *Ilex;* in Mitt. DDG **28**, 1–66, 1919 ● Thomson, B. F.: Bibliography on Holly; Bull. Holly Society of America, No. **6**, 1–26, 1955 ● Wister, J. C.: Preliminary Holly Check List; Bull. Holly Society of America, No. **8**, 1–56, 1953 ● Wyman, D.: *Ilex crenata* and its varieties; in Arnoldia **20**, 41–46, 1960 (12 forms described, 17 others listed) ● Wyman, D.: Interest in Hollies adds to list of varieties in nurseries; in Americ. Nurseryman **112**, 9, 12–13, 116–124, 1960 ● Eisenbeiss, G. K., & T. R. Dudley: International Checklist of Cultivated *Ilex;* Part I: *Ilex opaca;* US Nat. Arboretum Contr. **3**, 1–85, 1973 ● Eisenbeiss, G. K., & T. R. Dudley: Handbook of Hollies; special edition of Americ. Hort. Mag. 1970, 149–334 ● Hu, S.-Y.: The genus *Ilex* in China; in Jour. Arnold Arb. **30**, 283–344, 348–387, 1949; **31**, 39–80, 214–240, 242–263, 1950 (this is the best work on the Chinese species) ● Hu, S.-Y.: Notes on the genus *Ilex* L.; in Arnoldia **30**, 67–71, 1970 (with desc. and ills. of *I. nobilis, I.* × *meserveae* and *I. rugosa*).

ILLICIUM L. — Star Anise — ILLICIACEAE

Aromatic, evergreen shrubs or trees; leaves alternate to nearly whorled clusters, transparent punctate; flowers usually solitary, axillary, perianth with up to 30. petals; outer petals more calyx-like; stamens numerous, ovaries 3–20, whorled, unilocular, fruit a leathery follicle, grouped in a starlike ring. — About 40 species in E. and S. Asia, S. USA.

Illicium anisatum L. Shrub or small tree in its habitat, branches gray-brown, reddish when young, foliate buds thick ovate; leaves clustered at the branch tips, oblong-elliptic, acuminate at both ends, 4 to 8 cm long, glabrous, bright and glossy green above, light green beneath; flowers numerous, axillary, about 2.5 cm wide, pale greenish-yellow, not fragrant, May. BM 3965 (= *I. religiosum* S. & Z.). Japan, Korea. 1790. Hardy as *Camellia*. z7–8 Plate 37, 70. # ⟳

I. floridanum J. Ellis. Shrub, 1.5–2.5 m high; leaves oblong-lanceolate, 5–10 cm long, tapering to both ends, glabrous; flowers about 5 cm wide, solitary or several in the leaf axils, nodding on a 2.5–5 cm long stalk, petals 20–30, dark carmine to purple, May; fruit to 2.5 cm wide. BM 439; KTF 79; BS 2: 455; JRHS 97: 221. Southeast USA; often in swamps and other moist areas. 1771. z7–8 # ⟳

I. henryi Diels. A small tree in its habitat; leaves oblanceolate, long acuminate, tapering to the base, 10–15 cm long, 2.5–5 cm wide, leathery, glossy above; flowers solitary in the leaf axils on 2.5–3 cm long stalks, petals ovate to more oblong, grouped about 20 together, pink (but also carmine on plants in the wild!), carpels 8–13. FIO 5; JRHS 97: 263. W. China. (Hillier.) z8 # ⟳

I. religiosum see: **I. anisatum**

Illustration references for further species:
> *I. arborescens* LWT 57
> *I. griffithii* HA1 40
> *I. szechuanense* FIO 6

These plants prefer a moist, humus, peaty soil; winter protection is advisable. Flower in March if overwintered in a cool greenhouse.

Lit. Smith, A. C.: The families Illiciaceae and Schisandraceae; in Sargentia VII, 1–79, 1947 ● Hopkins, H.: *Illicium;* an old plant with new promise; in Jour. RHS **97**, 525–530, 1972.

INDIGOFERA L. — Indigo — LEGUMINOSAE

Deciduous shrubs, subshrubs or herbs, usually more or less densely covered with branched or simple hairs; leaves alternate, odd pinnate, sometimes trifoliate or simple; leaflets usually small; stipules awl-like, adnate to the petiole base; flowers in axillary racemes, pink, red to purple; calyx short cupulate, with 5 irregularly arranged, long teeth; standard very short clawed; keel with a claw-like appendage on either side; styles glabrous; pods quite variable, 2 to many seeded, usually cylindrical. — About 300 species in the tropics and subtropics, of these only a few winter hardy and in cultivation.

Key to the most important species

● Leaflets 13–27
  + Leaflets 13–21; flower racemes erect:
        I. heterantha
  ++ Leaflets 21–27; flower racemes pendulous:
        I. pendula

● ● Leaflets 5–13
  + Leaflets glabrous above;
    ★ Standard petal white, wings pink:
        I. decora, fortunei (white)
  ★★ Standard carmine, wings pink:
        I. hebepetala
  ++ Leaflets pubescent on both sides;
    × Flowers 1.5–2 cm long:
        I. kirilowii
    ×× Flowers 6–8 mm long; keel and wings abscising;
      > Flower stalk as long as the leaf petiole:
        I. potaninii
        >> Flower stalk much shorter than the leaf petiole:
        I. amblyantha

Indigofera amblyantha Craib. Shrub, to about 1.5 m high, branches angular, appressed white pubescent; leaves 10–15 cm long, with 7–11 leaflets, these elliptic, 1–3.5 cm long, bright green above, gray and pubescent beneath; flowers pale lilac to dark red, in erect, 6–10 cm long racemes, appearing from July–October; pods about 2.5 cm long, pubescent. ICS 2508. China. 1907. Hardy. z6 ⊕

I. decora Lindl. Shrub, only about 30(50) cm high, branches brown-red, cylindrical, glabrous; leaves 10–15 cm long, leaflets 7–13, elliptic, 3 to 7 cm long, deep green above, pale green beneath, with small bristly tips, quite lightly pubescent beneath; flowers, to about 20–40 in erect 10–20 cm long racemes, standard white with pink at the base, wings pink, July–August. BM 5063; BS 2: 144; ICS 2503 (= *I. incarnata* Willd. Nakai). Japan to China. 1846. Very attractive. z8 ⊕

f. **alba** Rehd. Like the species, but flowers totally white. Gfl 375; BC 1956. z8

I. dielsiana Craib. Open growing shrub, about 1–1.5 m high, young branches angular, appressed pubescent at first, later completely glabrous or nearly so; leaves 7–12 cm long, with 7–11 pinna, these elliptic-oblong to obovate, both ends usually rounded, 1–2 cm long, 6 to 10 mm wide, appressed pubescent on both sides, lighter beneath; flowers in 12–15 cm long, rather erect, slender, many flowered racemes, the individual flowers light pink, 12 mm long, calyx silky pubescent, lobes awl-shaped, petals soft pubescent, June to September. China; Yunnan Province, high in the mountains. z6 ⊕

I. dosua Ham. (non Hort.!). India. Only a greenhouse plant in the temperate zone. Those cultivated in the landscape are actually *I. heterantha* Wall.

I. fortunei Craib. Closely related to *I. decora;* leaflets usually 7, elliptic to ovate or broad oblong, 2–4 cm long, with reticulate venation beneath and fine appressed pubescence, eventually glabrous; flowers white, 12 mm long (= *I. reticulata* Koehne non Franch.). China. 1890. z6 Plate 69.

I. gerardiana see: **I. heterantha**

I. hebepetala Benth. Scarcely over 1 m high in cultivation, taller in its habitat; leaves 15–20 cm long, leaflets 5–11, broad elliptic, 3–5 cm long, rounded at the apex, pubescent beneath; flowers in 7–20 cm long racemes, corolla 12–15 mm long, standard dark carmine, exterior pubescent, wings and keel pink, August–September; pods 3–5 cm long, with 8–10 seeds. BM 8208. Himalayas. 1881. z8

I. heterantha Wall. ex Brandis. Only about 1 m high in the cooler regions, in milder areas to 2 m high, stoutly branched, branches lightly striped, appressed pubescent; leaves 5–10 cm long, leaflets 13–21, oval-oblong, appressed pubescent on both sides, about 12 mm long; flowers purple-pink, 1 cm long, in 7–15 cm long, dense, erect racemes, June–September; pods 3 to 5 cm long, bowed when ripe. HM 1442; SH 2: 40e–l; m–p 1(= *I. dosua* Lindl. non D. Don; *I. quadrangularis* Graham; *I. gerardiana* Wall. ex Baker). Himalayas. 1840. z7–8 Plate 67. ⊕

I. incarnata see: **I. decora**

I. kirilowii Maxim. Shrub, 0.6–0.8 m high, pubescent at first, soon glabrous; leaflets 7–11, oval-rounded, 1–3 cm long, lightly appressed pubescent on both sides; flowers in rather dense, 12 cm long racemes, corolla to 2 cm long, June; pods linear, 3–5 cm long. BM 8580; BC 1957; ICS 2500. Korea, Manchuria to N. China. 1899. z5 Plate 69. ⊕

I. pendula Franch. Shrub, broad growing, in mild areas grows to 2 m high every year (freezes back to the ground each winter); leaflets 19–27, oblong to elliptic, 2–3 cm long, appressed pubescent beneath; flowers appear continuously in the leaf axils from August to September, in 25–30 cm long, pendulous racemes, purple-pink; pods 5 cm long. BM 8745. China; Yunnan Province. 1914. z9 ✧

I. potaninii Craib. Upright, 1–1.5 m tall shrub, branches appressed pubescent at first, soon glabrous; leaves 7–15 cm long, leaflets 5–9, elliptic-oblong, 1–3 cm long, pubescent on both sides, denser and gray-green beneath; flowers in 5–12 cm long, erect racemes, corolla lilac-pink, July to September. JRHS 61: 20. China; Yunnan Province. z6 ✧

I. pseudotinctoria Matsum. Subshrub, 30–50 cm high, the many thin branches appressed pubescent; leaves 4–8 cm long, usually with 7–9 leaflets, these oblong, 8–25 mm long, 5–12 mm wide, soft pubescent on both sides, bluish beneath; flowers in short stalked, 4–10 cm long, racemes, very densely arranged, the individual flowers light pink to nearly white, only 4 mm long, flower stalk very short, July–September. ICS 2506; JRHS 90: 180. Japan, Taiwan, Central China. 1897. z6–7 ✧

I. quadrangularis see: **I. heterantha**

I. reticulata see: **I. fortunei**

Attractive, ornamental shrubs, especially valuable as summer bloomers. Need a well drained, clay soil in a sunny location. Usually freeze to the ground in winter, but resprout well in spring and flower on young shoots.

Lit. Ali, S. I.: Revision of the genus *Indigofera* L. from W. Pakistan and N. W. Himalaya; in Bot. Notiser **111** (3), 543–577, 1958 ● Craib: in Notes Roy. Bot. Gard. Edinburgh, 8 (no. 36), 1913.

IOCHROMA Benth. — SOLANACEAE

Trees or shrubs; leaves alternate, usually tomentose; flowers in clusters; corolla tubular to narrow funnelform, limb with 5 small lobes; calyx tubular to campanulate, truncate at the apex, much enlarged at fruiting time and then surrounding the round, many seeded berry with a pasty pulp. — 25 species in tropical S. America.

Iochroma coccineum Scheidw. Shrub, young branches soft pubescent; leaves oblong to ovate, 7–12 cm long, often much longer on young plants, long acuminate, margins undulate; flowers pendulous in terminal clusters of 8 or more together, 4–5 cm long, scarlet-red, 2 cm wide (measured across the narrow, slightly lobed limb), July–August. FS 1261. Central America. z10 Plate 63. ✧

Very attractive, easily grown shrubs for the landscape; only for frost free climates.

IPOMOEA L. — Morning Glory — CONVOLVULACEAE

Evergreen or deciduous perennials, very rarely shrubs or trees; leaves alternate, either simple or lobed or compound; flowers plate-like, campanulate or funnelform, often very attractive, calyx with obtuse lobes; styles with 1 capitate or 2–3 globose, but not with 2 linear or ovate stigmas; ovaries 2–4 chambered. — About 500 species in the tropics or the warm temperate zones, usually a vine or shrub.

Ipomoea arborescens Don. Shrub or tree-like, 4–6 m high, branches and foliage velvety pubescent; leaves cordate-ovate; flowers white, 2 cm long, sepals ovate, 8 mm long, velvety on both sides; seeds black, with a tuft of hair on the dorsal side. GF 7: 364. Mexico. 1880. Likes cool, dry areas; occasionally cultivated on the Canary Islands. z10 ✧

ISOPLEXIS (Lindl.) Dougl. — SCROPHULARIACEAE

Evergreen subshrubs, very closely related to *Digitalis*; leaves alternate, long narrowly toothed; flowers in terminally arranged, erect, dense racemes; corolla tubular at the base, limb 2 lipped, 5 lobed, with 4 stamens. — 3 species on the Canary and Madeira Islands.

Isoplexis canariensis (L.) Loud. Evergreen shrub, to 1.5 m high, branches thick, erect, pubescent; leaves oval-lanceolate, 12 cm long, 2 to 5 cm wide, glossy green above, sparsely pubescent beneath, finely dentate; flowers very numerous in terminal, 30 cm long racemes, very dense, corolla orange-yellow, about 3 cm long, May to June; fruit capsule longer than the calyx. BMns 559; BFCa 260, 261. Canary Islands. 1698. z10 ✛

I. sceptrum, from Madeira, is illustrated in JRHS 90: 177.

ITEA L. — ITEACEAE

Deciduous or evergreen shrubs or trees, branches with chambered pith; leaves alternate, simple, dentate; flowers small, in racemes or spikes; calyx campanulate, 5 toothed; petals 5, situated on the calyx; stamens 5; ovary oblong, 2 chambered; fruit a many seeded capsule. — Around 15 species from Himalayas to Japan and W. Malaysia; 1 species in N. America.

Itea ilicifolia Oliv. Evergreen shrub, to 3 m high, branches smooth; leaves elliptic-rounded, 6–11 cm long, thorny toothed, glossy dark green above, lighter beneath with pubescent tufts in the vein axils; flowers greenish-white, in numerous, pendulous, narrow, 15–30 cm long racemes, August. GC 50: 96; BS 2: 147. Central China. 1895. z9 Plate 68. # ✛ ⌀

I. virginica L. Evergreen shrub, narrow upright habit, bushy, 1(2) m high, branches rodlike, red, pubescent when young; leaves elliptic-oblong, 4–10 cm long, finely serrate, glabrous above, lightly pubescent beneath, fall color bright red; flowers white, fragrant, in 5–15 cm long, erect, densely pubescent racemes, May–July. BB 1864; GSP 164. Eastern N. America. 1744. Hardy. z6 Plate 70. ✪

I. yunnanensis Franch. Evergreen shrub, similar to *I. ilicifolia,* but leaves narrower, 5 to 10 cm long, margins less thorny toothed; flowers in slender, nodding, cylindrical racemes to about 15 cm long, petals dull white, July. China; Yunnan Province. 1918. z9 Plate 68.

Further illustrations of Itea species:
    I. oldhamii LWT 91
    I. omeiensis FIO 9
    I. orientalis FIO 76

Although thriving in any good garden soil, *I. virginica* is found in wet areas in its habitat. All species do equally well in a humus soil, especially the evergreen types.

IVA L. — COMPOSITAE

Deciduous shrubs or herbs; leaves opposite at the base of the branches, alternate toward the apex, entire or dentate; flowers solitary or few in racemes or panicles in the apical leaf axils; bisexual flowers sterile, female flowers fertile; involucre usually composed of only 3–5 bracts; styles widening brush-like at the apex; achene without a pappus. — About 15 species in N. and Central America and the W. Indies.

Iva frutescens L. Shrub, to 3 m high in its habitat, branches strigose pubescent; leaves lanceolate, 10–15 cm long, serrate, 3 veined, short petioled, acuminate, somewhat rough haired; flower heads greenish, 4 mm wide, usually with 5 bracts, July–September. Southeast USA. 1711. z8

var. **oraria** (Bartlett) Fern. & Griscom. Subshrub, only 1 m high; leaves elliptic, nearly linear on flowering shoots; flower heads 5–6 mm wide, with 5, occasionally 6 bracts. KD 564. USA, Massachusetts to Maryland. 1880. z7

Only for collectors. Found in salt marshes and muddy areas along the seacoast in its habitat.

JACARANDA Juss. — BIGNONIACEAE

Evergreen or semi-evergreen trees or shrubs; leaves opposite, bipinnate, occasionally only simple pinnate, leaflets small and numerous; flowers in terminal or axillary panicles, usually blue or violet; calyx small, 5 toothed; corolla tube straight or curved, somewhat 2 lipped, limb outspread and 5 parted; fruit an oblong, ovate to circular, dehiscent capsule with many winged seeds. — About 50 species in tropical America.

Jacaranda mimosifolia D. Don. Tree. To 50 m or more, leaf drop in spring; leaves about 45 cm long, fern-like finely pinnate, with about 16 paired pinnae, each pair with about 14–24 paired leaflets, these elliptic-oblong, pubescent; flowers blue, more or less erect, in loose, conical, 20 cm long panicles in spring, flowering again (less abundantly) in fall, corolla tube 6 cm long; fruit capsule circular, 5 cm wide. BM 2327; BR 631; FS 185 (= *J. ovalifolia* R. Br.). NW. Argentina. 1818. Often confused with the Brazilian *J. acutifolia* Humb. & Bonpl. z10 Plate 61, 70. ✛ ⌀

J. ovalifolia see: **J. mimosifolia**

A very popular, stately and gorgeous tree, very fast growing; flower display beyond compare. Only to be considered, however, for totally frost free climates. Sometimes used as a greenhouse plant for its fine foliage (not to be confused with *Dalbergia nigra,* also known as "Jacaranda").

JAMESIA Torr. & Gray. — PHILADELPHACEAE

Deciduous shrubs with cylindrical branches and exfoliating bark; leaves opposite, simple, gray tomentose beneath, serrate, petiolate, stipules inconspicuous;

flowers cupulate, petals 5, margin involuted or folded inward in the bud stage, sepals 5, stamens 10, filaments linear tapered, ovaries conical, unilocular, nearly totally superior, styles 3(5); fruit a 4 valved, many seeded capsule. — 1 species in N. America.

Jamesia americana Torr. & Gray. Shrub, upright, about 1 m high, occasionally taller, bark brown, exfoliating (like *Philadelphus*); leaves elliptic to rounded, 2–6 cm long, rough and pubescent above, gray-white tomentose beneath, coarsely serrate, petiolate; flowers 1.5 cm wide, white, somewhat fragrant, in 4 cm wide cymes, May–June; fruit a small capsule. BM 6142; Gfl 53: 231; SL 234. Western N. America, in the mountains. 1865. Fall color scarlet-red in sunny locations. z6 Plate 61, 68. Ø ⊕

'**Rosea**'. Flowers pink. Originated in cultivation, 1905.

Quite hardy, but only for collectors with a large rock garden.

JASMINUM L. — Jasmine — OLEACEAE

Deciduous or evergreen upright, climbing or twining shrubs, branches angular or cylindrical, occasionally with green bark; leaves opposite or alternate, odd pinnate or reduced to only 1 leaflet; flowers in terminal cymes or axillary at the branch tips, plate-like with a long, thin tube, 4–9 lobes, rolled up when in bud, calyx campanulate, with 4–9 very small or awl-shaped teeth, stamens 2, ovaries 2 chambered; fruit a 2 valved, usually black berry with 1 or 2 seeds. — About 300 species, most in the tropics and subtropics, 1 in America.

Key to the most important species

● Leaves opposite
  + Flowers axillary, solitary, yellow:
        J. nudiflorum (simple)
        J. mesnyi (double)

  ++ Flowers terminal, usually grouped;
   > Leaflets 1–3:
        J. beesianum (leaves simple, flowers dark pink)
        J. stephanense (leaflets 1–3, flowers light pink)
   >> Leaflets 3–7, flowers white
        J. officinale (flowers 10–20)
        J. polyanthum (flowers 30–40)
        (*J. azoricum* and *J. angulare* only in z9)

●● Leaves alternate, pinnate or pinnatisect
  + Leaflets and young branches glabrous;
  ★ Calyx teeth awl-shaped, as long or longer than the calyx tube:
        J. floridum (many flowered)
        J. fruticans (flowers 2–5)
  ★★ Calyx teeth shorter than the calyx tube:
        J. parkeri, humile, f. *farreri*

Jasminum affine see: **J. officinale 'Affinis'**

J. angulare Vahl. Evergreen climber, long branched, branches 4 sided; leaves opposite, pubescent, leaflets 3, circular to lanceolate, with a small tip; flowers white, about 3 cm wide, limb with 5–7 oval-lanceolate lobes, not fragrant, usually in 3's axillary. BM 6865; GC 139: 18. S. Africa. The best white flowered species, easily cultivated in the greenhouse. z9 # ⊕

J. azoricum L. Evergreen climber, with twining shoots, cylindrical, nearly smooth; leaves opposite, leaflets 3, ovate, margins undulate, base nearly cordate, middle leaflet larger; flowers white, fragrant, petals 5, as long as the tube, July–September. BM 1889. Azores. Does not flower as abundantly as *J. angulare*, but still very attractive. z9 # ⊕

J. beesianum Forrest & Diels. Deciduous shrub, to 1.5 m high, slightly twining, branches very thin, furrowed, pubescent at the nodes; leaves opposite, simple, ovate-lanceolate, 2.5–5 cm long, dull green and sparsely pubescent on both sides; flowers light to dark pink, grouped 1–3, 1.5 cm wide, fragrant, May. BM 9097; ICS 4692. China; Szechwan, Yunnan Provinces, in the mountains. 1910. z8–9.

J. blinii see: **J. polyanthum**

J. floridum Bge. Semi-evergreen, erect shrub, slightly twining, branches angular, green, glabrous, drooping; leaves alternate, usually 3 parted, occasionally 5 parted, leaflets oval-elliptic to obovate, acute, 1–3.5 cm long, glossy green above, lighter beneath, glabrous; flowers yellow, many in terminal cymes, July–August; fruits black, pea-sized. BM 6719; ICS 4684. China. 1850. z9 Plate 68; Fig. 120, 121. See also **J. humile** f. **farreri.** ⊕

J. fruticans L. Evergreen to semi-evergreen shrub, upright, branches angular, rodlike, striped, green; leaves alternate, usually 3 parted, some with 1–2 leaflets, these oblong-spathulate, somewhat leathery, 1–2 cm long, finely ciliate; flowers 2–5 on the ends of short side branches, yellow, not fragrant, calyx teeth awl-shaped, nearly as long as the corolla tube, July–September. BM 461. S. Europe, N. Africa, W. Asia. 1570. z8 Plate 68; Fig. 121. #

J. giraldii see: **J. humile** f. **farreri**

J. grandiflorum see: **J. officinale** f. **grandiflorum**

J. humile L. A quite variable species, composed of the following forms:

f. **farreri** (Gilmour) P. S. Green. Evergreen shrub, broad growing, about 1.5 m high, young branches angular, pubescent at first, reddish, later green; leaves alternate, 5–12 cm long, leaflets 3, oval-lanceolate, long acuminate, obliquely tapering to the base, middle leaflet to 10 cm long, others much smaller, dull green and rugose above, pubescent beneath; flowers yellow, not fragrant, about 7–12 in terminal cymes, June. BM 9351. Upper Burma. 1919. This plant was earlier erroneously known as *J. giraldii*, which is actually a synonym for a pubescent type of *J. floridum*. z9 # ⊕

Fig. 120. **Jasminum.** Left, *J. floridum;* middle, *J. humile* 'Revolutum'; right, *J. humile* f. *humile*
(from Bot. Reg. and Bot. Mag.)

var. *glabrum* see: f. **wallichianum**

f. **humile.** The typical form of the species. Nearly evergreen
shrub, 1(6) m high, branches angular, glabrous, green; leaves
alternate, with 3–7(9) leaflets, these oval to elliptic or oblong,
obtuse to acuminate, the terminal leaflet 2–5 cm long, the
others 1.2–3 cm long, deep green above, lighter beneath,
glabrous; flowers yellow, grouped 5–10, about 1 cm wide, often
unscented, calyx teeth quite small, rectangular, corolla lobes
reflexed, June–July. BR 350. Afghanistan and W. Pakistan to
Burma and China; Yunnan and Szechwan Provinces; SE. Tibet.
z9 Plate 68; Fig. 120. # ⊕

'Revolutum'. Nearly evergreen, glabrous shrubs; leaves with
3–5–7 leaflets, terminal leaflet 4–7 cm long, others 2.5–5 cm
long; flowers 6–12 or more together, yellow, fragrant (!),
corolla limb 2–2.5 cm wide, styles short and visible in the throat
with the tips of the 2 anthers. BM 1731; BR 178 (= *J. revolutum*
Sims; *J. humile* var. *revolutum* [Sims] Stokes; *J. reevesii* Hort. ex
Schneid.; *J. triumphans* Hort. ex Dippel). z9 Plate 68; Fig. 120,
121. # ⊕

f. **wallichianum** (Lindl.) P. S. Green. Branches more angular;
leaves with 7–13 leaflets, these ovate to lanceolate, terminal
leaflet very long acuminate, others 2–5 cm long; flowers
grouped 1–3, more or less pendulous, corolla 1.5 cm wide at the
limb. BR 1409 (= *J. wallichianum* Lindl.; *J. pubigerum* var. *glabrum*
DC.; *J. humile* var. *glabrum* DC. Kobuski). Nepal. 1812. (Also
included here is a pubescent form, *J. punigerum* D. Don not
found in cultivation.) z9 Fig. 121. # ⊕

J. mesnyi Hance. Primrose Jasmine. Evergreen shrub, to
2 m high, broad, densely foliate, branches 4 sided,
smooth; leaves opposite, leaflets 3, oblong-lanceolate,
nearly sessile, 2.5–7 cm long; flowers solitary, axillary,
3.5–5 cm wide, with 6–10 corolla lobes, bright yellow,
center somewhat darker, March–April. BM 7981; FS
168; BS 2: Pl. 14 (= *J. primuminum* Hemsl.). W. China in
cultivation but not in the wild. 1900. Best yellow species;
somewhat resembling *J. nudiflorum,* but much more
attractive. z8–9 Plate 68; Fig. 121. # ⊕

J. nudiflorum Lindl. Winter Jasmine. Deciduous shrub,
espaliered to 3 m high and wide, branches slender,
rodlike, angular, pendulous, green; leaves opposite,
trifoliate, leaflets oval-oblong, 1–3 cm long, ciliate, deep
green, otherwise glabrous; flowers solitary, axillary,
along the previous year's branches, yellow, often with
somewhat reddish exterior, corolla usually with 6 lobes,
about 2.5 cm wide, often somewhat wavy, December–
April. BM 4649; Gs 2: 248; FS 762. N. China. 1844. One
of the most beautiful winter bloomers and usually winter
hardy. z6 Fig. 121. ⊕

'Aureum'. Like the species, leaves yellow. Often reverting back
to the species.

J. officinale L. True Jasmine. Deciduous shrub, to 10 m
high in mild areas (when espaliered), branches 4 sided,
thin, green, annual growth to 1.5 m long; leaves
opposite, leaflets 5–7(9), elliptic to oblong-ovate,
acuminate, 1–6 cm long, terminal leaflet long stalked,
side leaflets sessile; flowers white, very fragrant, 2–10 in
terminal cymes, corolla nearly 2.5 cm wide, limb 4–5
lobed, calyx with 5 linear awl-shaped, 6–18 mm long
teeth, half to nearly as long as the tube, June–September.
BM 31; BC 2008; HV 14, ICS 4688. Iran to China, but
often naturalized in S. Europe. 1548. z7–8 Plate 68; Fig.
121.

'Affine'. Flowers larger, pink exterior, corolla lobes wider. BR
31: 26 (= *J. affine* Carr.). Before 1878. More prolific bloomer
than the species. ⊕

'Aureovariegatum'. Leaves yellow speckled.

f. **grandiflorum** (L.) Kobuski. More vigorous habit, coarser,
branches more nodding; leaf rachis flat to winged, leaflets 5–7,
elliptic to rounded or ovate, drawn out to a small tip; flowers to
4 cm wide, stellate, white, exterior pink, calyx teeth half as long
as the tube, June–September. MJ 3215 (= *J. grandiflorum* L.).
Subtropical Himalayas. 1629. More sensitive than the species.
z9 ⊕

Fig. 121. **Jasminum.** a *J. mesnyi*, immediately right, calyx, enlarged; b. *J. humile* f. *wallichianum;* c. *J. nudiflorum;* d. *J. floridum;* e. *J. humile* 'Revolutum'; f. *J. fruticans;* g. *J. officinale* (from Bot. Mag., Baillon, Oliver, Knoblauch)

J. parkeri Dunn. Evergreen shrub, only about 30 cm high, bushy, branches furrowed, finely pubescent at first, becoming glabrous; leaves alternate, 2–2.5 cm long, 3–5 parted, leaflets oval, obtuse, 3–6 mm long; flowers solitary, yellow, 1.5 cm wide, June; fruits 2 lobed, greenish-white, translucent. NW. India; Chambra. 1919. z8? # ✧

J. polyanthum Franch. Evergreen, occasionally only semi-evergreen climbing shrub, young branches somewhat warty, new growth brownish; leaves opposite, 7–12 cm long, leaflets 5–7, these lanceolate, middle leaflet 3–7 cm long, axillary smaller; flowers with white interior, reddish exterior, very fragrant, 30–40 in numerous axillary, panicles, summer. BM 9545; GC 139: 647; JRHS 91: 208 (= *J. blinii* Lév.). China; Yunnan Province. 1891. One of the best species of the entire genus, but only hardy in the mildest regions. z9 Plate 68, 71. # ✧

J. primulinum see: **J. mesnyi**

J. pubigerum see: **J. humile** f. **wallichianum**

J. reevesii see: **J. humile** 'Revolutum'

J. revolutum see: **J. humile** 'Revolutum'

J. sambac (L.) Ait. Arabian Jasmine. Evergreen climber with angular branches; leaves opposite or in whorls of 3, tough, glossy, glabrous, elliptic-ovate to broad ovate, round or acute at the apex, distinctly veined, entire; flowers in clusters of 3 to 12, white, turning pink, very fragrant, flowers continuously, corolla tube 12 mm long, lobes oblong to round, calyx lobes linear. BR 1; ICS 4693. India. 1665. Only for completely frost free areas. z10 # ✧

Also includes a double flowering form **'Grand Duke'** or **'Grand Duke of Tuscany'.** BM 1785.

J. × stephanense Lemoine (= *J. beesianum* × *J. officinale*). Deciduous, vigorous twining shrub, 3–5 m high, branches glabrous, somewhat angular, thin; leaves partly simple, some 3–5 parted, pinnate, leaflets dull green, somewhat pubescent beneath; flowers light pink, few in cymes, rather small and not very attractive, June–July. RH 1927: 644. Hybridized in Nancy, France by Lemoine in 1918, but later allegedly found in the wild in W. China. z7–8

J. triumphans see: **J. humile** 'Revolutum'

J. wallichianum see: **J. humile** f. **wallichianum**

The Jasmine species are not particular as to soil type when provided with the proper climate. Many of the listed species are only hardy in the mildest climates or suitable as greenhouse plants.

Lit. Kobuski, C. A.: Synopsis of the Chinese species of *Jasminum;* in Jour. Arnold Arb. **13,** 145–179, 1932 ● Kobuski, C. A.: A revised key to the Chinese species of *Jasminum;* in Jour. Arnold Arb. 1959, 385–390 ● Green, P. S.: Studies in the genus *Jasminum;* (I): Notes Roy. Bot. Gard. Edinb. **23,** 355 to 384, 1961; — (II): Jour. Arnold Arb. **43,** 109 to 131, 1962; — (III): Baileya **13,** 137–171, 1975

JOVELLANA Ruiz & Pavon — SCROPHULARIACEAE

Herbs and subshrubs, erect or creeping, leaves opposite, simple, margins more or less dentate; flowers very similar to *Calceolaria*, but with both lips nearly alike, outspread, never shoe- or pouch form like the true calceolarias.—7 species in New Zealand and Chile.

Jovellana sinclairii (Hook.) Kraenzl. Erect subshrub; leaves oval-oblong, obtuse, coarse doubly serrate, 7 cm long, 5 cm wide, somewhat pubescent above, lighter green and glabrous beneath, peticle 3–5 cm long; flowers in loose, terminal panicles, corolla white to pale lilac, purple punctate, lower lip much larger than the upper, June. DRHS 1090; BM 6597 (as *Calceolaria sinclairii*). New Zealand. 1881. z9 ✧

J. violacea (Cav.) G. Don. Semi-evergreen shrub, about 1 m high, very densely branched, branches thin, finely pubescent when young; leaves ovate, coarse and irregularly incised to lobed, dull dark green, 2–3.5 cm long, flowers in erect clusters, corolla "helmet-form", light violet, reddish speckled inside, with a yellowish spot on the throat, finely pubescent exterior, June–July. BM 4929 (= *Calceolaria violacea* Cav.). Chile. 1853. z9 Fig. 122. **H** ✧ ○

Fig. 122. *Calceolaria violaceae* and flower parts (from Bot. Mag.)

JUBAEA H. B. K. — PALMACEAE

Monotypic genus belonging to the Cocoideae group; stem short and thick, slow growing; leaves pinnate, similar to *Phoenix*, flowers in simple branched inflorescences (spadix), with fertile male and female flowers, stamens 12 to many, not connate.—Chile.

Jubaea chilensis Baill. Coquito. 10–15 m high in its habitat, stem straight, thick, smooth; leaves terminal, pinnate, 1.5–2.5 m long, pinnae numerous, linear-

lanceolate, long acuminate, stiff, reflexed, 25–30 cm long, nearly paired opposite, also pointing in different directions, petiole flat, with raised margins, densely covered with brown filaments at the base, thornless; flowers dark yellow; fruit a small coconut. SDB 2: 7; PFC 232 to 233. Chile. 1843. In its habitat, the sugary sap is boiled down to produce a "palm honey". z8 Plate 71. ✗ ∅

JUGLANS L. — Walnut — JUGLANDACEAE

Deciduous trees, very occasionally shrubs, with furrowed bark; young shoots with chambered pith, aromatic; buds with few scales, nearly naked; leaves alternate, odd pinnate, large, aromatic, leaflets serrate to entire; flowers monoecious, female in terminal, short, usually few flowered (often single flowered) spikes or clusters, male in axillary, long, pendulous, many flowered catkins; fruits (nuts) with usually thick, indehiscent outer shell, and a woody, hard inner shell, more or less furrowed; seeds edible, very high in oil content. — About 15 species from S. Europe to E. Asia, as well as N. and S. America. (Dode described 44 species, but many are excluded here.)

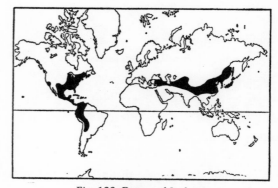

Fig. 123. Range of *Juglans*

Key to the more prominent species
(see also the ill. of the fruits)

★ Fruits totally glabrous or finely pubescent, 1–3; nut with 4 dividing walls at the base;
  + Leaflets usually 7–9, entire:
    J. regia
  ++ Leaflets 9–25;
    > Leaflets narrower than 2.5 cm;
      ○ Nuts deeply furrowed;
        × Nut to 3.5 cm thick; leaflets 9–13:
          J. major
        ×× Nut to 2 cm thick; leaflets 11–23:
          J. microcarpa
      ○○ Nuts not furrowed or indistinctly so:
          J. hindsii

    >> Leaflets 2.5 cm wide or more; nuts distinctly furrowed:
          J. nigra

★★ Fruits glutinous pubescent, in racemes; nut with 2 dividing walls at the base; leaflets stellate pubescent and glandular beneath, serrate;
  + Nuts distinctly 6–8 ribbed:
        J. cathayensis, cinerea, mandshurica
  ++ Nuts rough to nearly smooth:
        J. ailantifolia, intermedia, quadrangulata

Juglans ailantifolia Carr. Japanese Walnut. Tree, to 15 m high, with a broad crown, young branches glandular pubescent; leaves 40–50 cm long, occasionally also longer, leaflets 11–17, oblong to more elliptic, short acuminate, 7–15 cm long, 4–5 cm wide, densely serrate, base oblique rounded to somewhat cordate, both sides densely gray stellate tomentose when young, later nearly glabrous above, deep green, underside glandular pubescent especially on the venation; male catkins 10–25 cm long, very ornamental; fruits in long racemes, occasionally to 20 together, globose to ovate, shell glutinous pubescent, nut globose to more ovate, acuminate, with 2 thick, ridges drawn out at the apex, otherwise rather smooth, shallow pitted, 3 cm long. BC 2014 to 2016 (= *J. sieboldiana* Maxim. non Goepp.; *J. cordiformis* var. *ailantifolia* [Carr.] Rehd.). Japan; Sachalin. 1860. z5 Plate 72; Fig. 126. ⌀

var. **cordiformis** (Maxim.) Rehd. Habit and foliage like the species; leaflets usually somewhat more narrow; nuts however quite different, 3 cm long, compressed, distinctly 2 sided, apex 2 edged acuminate, otherwise nearly smooth, thin shelled. BC 2017; BD 1909: 34 (= *J. cordiformis* Maxim.; *J. sieboldiana* var. *cordiformis* [Maxim.] Mak.). Japan; but not known in the wild. Fig. 126 ⌀

J. alata see: **J. × quadrangulata**

J. allardiana Dode was included in *J. ailantifolia* var. *cordiformis* by Rehder. Fig. 126.

J. arizonica see: **J. major**

J. × bixbyi Rehd. (*J. cinerea × J. ailantifolia*). Intermediate between the parents, but grows more like *J. cinerea*, new growth appears early; fruit shell slightly pubescent, nut with 8 ridges, rough, but only shallowly furrowed,

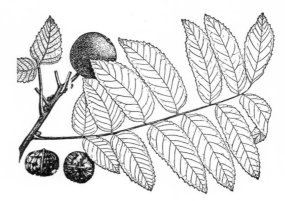

Fig. 124. *Juglans californica* (from Sudworth)

somewhat larger than *J. cinerea*. MD 42: Pl. 5 (= *J. sargentii* Sudw.). First observed in 1903.

J. californica Wats. Shrub, to 5 m high, or a small tree, round crowned, bark of the older branches ash-gray, later becoming black-brown and furrowed; leaflets 9–17, light yellow-green, eventually totally glabrous, oblong-lanceolate, acute, 2.5–6.5 cm long; fruits globose, 1–2 cm thick, hull very thin, nut indistinctly and irregularly furrowed. SPa 82; MS 69; SM 174. S. California. 1889. z9 Fig. 124.

var. *hindsii* see: **J. hindsii**

J. cathayensis Dode. Chinese Walnut. A tree in its habitat, to 23 m high, young branches (and leaf petiole, midrib and fruits) glutinous glandular pubescent; leaves to 80 cm long, leaflets 11–17, these oblong-ovate, 8–15 cm long, 4–8 cm wide, finely dentate, base oblique rounded to somewhat cordate, deep green and pubescent above, lighter and stellate pubescent beneath; female flowers in racemes, male flowers in 20–30 cm long, pendulous catkins; fruits in clusters of 6–10 on about 15 cm long stalks, ovate, 4–5 cm long, nuts with 6–8 ridges, oval, sharp acuminate, shell 3–4 mm thick. PIO 144; LWT (= *J. draconis* Dode). Central and W. China. Resembles *J. mandshurica*, but differs in the fruits. Recently (Bean, 8th ed. II–473) there has been some doubt as to whether this species can be separated from *J. mandshurica*. ℗ USSR. z5 Fig. 126. ⌀

Fig. 125. *Juglans cinerea* (from Sargent)

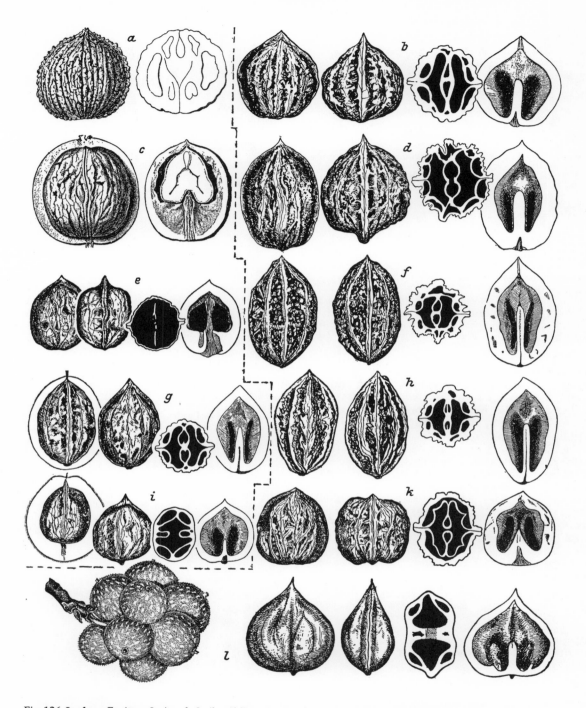

Fig. 126. **Juglans.** Fruits. a. *J. nigra;* b. *J. ailantifolia;* c. *J. regia;* d. *J. cathayensis;* e. *J. regia* ssp. *fallax;* f. *J. mandshurica;* g. *J. allardiana;* h. *J. stenocarpa;* i. *J. lavallei;* k. *J. sinensis;* l. *J. ailantifolia* var. *cordiformis,* inflorescence at left (from Dode, Sargent, Berg & Schmidt, ⅔ natural size)

J. cinerea L. Butternut. Tree, to 25 m high, bark gray, deeply furrowed, previous year's branches reddish-brown (!), branches pubescent, glandular-glutinous when young; leaves 25–50 cm long, leaflets 11–19, these oblong-lanceolate, appressed serrate, 6–12 cm long, pubescent on both sides, glandular beneath, as are the leaf petiole and rachis; fruits 2–5 together, 7–10 cm long, exterior glutinous-glandular pubescent, nut oval-oblong, black-brown, with 8 rather distinct, rough, sharp angled ridges. BB 1150; HH 50–51; GTP 117. Eastern N. America. 1633. Likes moist, fertile, lowland soil; more winter hardy than *J. nigra*. z3 Fig. 125. ∅

J. cordiformis see: **J. ailantifolia**

J. draconis see: **J. cathayensis**

J. duclouxiana see: **J. regia** ssp. **fallax**

J. fallax see: **J. regia** ssp. **fallax**

J. hindsii (Jepson) R. E. Smith. Round crowned tree, 10–15 m high, young branches densely pubescent; leaves 25–30 cm long, leaflets 15 to 19, oval-lanceolate to lanceolate, 5–10 cm long, 2–2.5 cm wide, coarsely serrate, venation pubescent beneath; fruits globose, pubescent, nuts 2.5–3 cm wide. SM 175 (= *J. californica* var. *hindsii* Jepson). N. California. 1976. Often planted as a street tree, also used as an understock for *J. regia*. z9?

J. intermedia quadrangulata see: **J. × quadrangulata**

J. × intermedia Carr. (*J. nigra × J. regia*). Tree, similar to *J. regia*, but normally with 11 leaflets, these ovate to elliptic-ovate, sparsely dentate, dark green and glabrous above, axillary pubescence beneath; fruits smooth, but deeply furrowed. Includes 2 forms:

var. **pyriformis** Carr. The type. Leaflets 9–13, finely dentate; fruits obovoid, more like *J. regia*. RH 1863: 30.

var. **vilmoreana** (Carr.) Schneid. Fruits globose, more like *J. nigra*. MD 1911: 197; BC 2018 (= *J. vilmoriniana* Meunissier). Parent tree in the Vilmorin Arboretum, Verrieres, France.

J. lavallei Dode included within *J. ailantifolia* by Rehder. Fig. 126.

J. major (Torr.) Heller. Tree, to 15 m high, narrow crowned, young branches pubescent; leaflets 9–13 (19), oblong-lanceolate to ovate, long acuminate, base cuneate or rounded, coarsely serrate, soon glabrous or with pubescent midrib beneath, 7–10 cm long, the basal pair only 3–6 cm long; stamens 30–40 (only 20 on the very similar *J. microcarpa*!); fruits globose to ovoid, 2.5–3 cm wide, thick rust-brown pubescent, nuts dark brown to black, somewhat flattened, deep longitudinally furrowed, thick shelled, nut sweet. SW 172 (= *J. rupestris* var. *major* Torr.; *J. arizonica* Dode; *J. torreyi* Dode). SW. USA. z9

J. mandshurica Maxim. A tree in its habitat, to 20 m high, corolla broad and round, branches glandular pubescent, green-yellow in the second year; leaves 75–90 cm long, to 1.2 m long on suckering shoots, with 11–19, nearly sessile leaflets, these oval-oblong, irregular and finely serrate, 7–18 cm long, 3–6 cm wide, eventually glabrous

Fig. 127. *Juglans microcarpa* (from Sargent)

above, densely glandular pubescent beneath including the midrib and rachis; fruits 6–12 on a stalk, globose-ovate, to 5 cm long, exterior glutinous pubescent, nut ovate, 3–4 cm long, with (4) 8 distinct, sharp and knotty wings. NK 20: 17; BC 2013. Manchuria, Amur River region. (The leaves, otherwise similar to those of *J. ailantifolia*, are abruptly tapered and quite short acuminate.) Ⓕ USSR. z5 Fig. 126. ∅

J. microcarpa Berl. Texas Walnut. Shrub or small tree, scarcely over 7 m high, young shoots brownish or gray, tomentose at first, later yellow-gray and pubescent; leaves 15–25 cm long, leaflets 15–23, lanceolate, finely serrate to nearly entire, 4–8 cm long, long acuminate, base oblique or round, glabrous above, somewhat pubescent beneath or eventually glabrous; fruits solitary, globose, 1.5–2 cm thick, hull thin, smooth, nut black-brown, thick shelled, with deep, irregular furrows, fruits abundantly. SS 335 (= *J. rupestris* Engelm.; *J. nana* Engelm.). USA; Texas, New Mexico. Hardy. z6–7 Fig. 127.

J. nana see: **J. microcarpa**

J. nigra L. Black Walnut. To 50 m tall, round crowned tree, bark deeply furrowed, young branches pubescent; leaves 30–60 cm long, leaflets 15 to 23, ovate to lanceolate, irregularly serrate, 6–12 cm long, glabrous above, somewhat glossy, pubescent and glandular beneath, terminal leaflet often very small or absent (an excellent distinguishing characteristic!); fruit globose, 4–5 cm thick, hull rough, very thick, nut coarse and irregularly, longitudinally furrowed, with rough ridges, thick shelled. BB 1179; SS 333–334; GTP 119. Eastern N. America. 1656. Very valuable lumber tree. Ⓕ Germany, Austria, Hungary, Romania, Yugoslavia, France, USA. z5 Plate 72, 77; Fig. 126 ∅

'Laciniata' (Hershey). Leaflets deeply incised, brought into the trade in 1937 by J. Hershey, Milton, USA. ∅

J. × notha Rehd. (*J. ailantifolia × J. regia*). Leaflets 7–9, elliptic to elliptic-oblong, glabrous above, later becoming glabrous beneath, margins quite finely and sparsely dentate; fruits very similar to those of *J. ailantifolia*, as is the nut. Developed in 1878.

J. × quadrangulata (Carr.) Rehd. (*J. cinerea × J. regia*). Tall tree, very similar to *J. regia*, but leaflets usually 9, elliptic to oblong, indistinctly and sparsely serrate, somewhat pubescent beneath; fruits only occasionally, nearly globose, about 5 cm long, nut ovoid-oblong, 4.5 cm long, acute, deeply furrowed, more or less like those of *J. cinerea*. BC 2019 (= *J. alata* Schelle; *J. intermedia quadrangulata* Carr.). First recognized in France before 1870, later more prevalent in the USA. z5

J. regia L. English Walnut, Persian Walnut. Tree, to 30 m high, broad crowned, bark silvery-gray, becoming furrowed on older trees, branches glabrous; leaflets 5 to 9, elliptic to oval-oblong, nearly entire, 6–12 cm long, glabrous; fruits globose, smooth, green, 4–5 cm long, nut ellipsoid, pointed, rather thin shelled, with 2 bulging ridges. MD 1911: 197. E. Europe, N. Asia. Cultivated for centuries for lumber and as a nut tree. ⓕ Yugoslavia, Romania, Germany, Himalaya, USSR, China, USA. z6–7 Fig. 126. ∅ ✗

The many fruit varieties will not be considered here (please see Lit.). Of the garden varieties, the following can be recommended:

'Adspersa'. Leaves whitish speckled and striped, the young branches also occasionally white striped.

f. *asplenifolia* see: **'Laciniata'**

'Bartheriana'. Nut almond form. Gfl 1935: 143. Discovered by Barthere in France. Before 1860.

'Corcyrensis'. Large leaved form; leaves 50–60 cm long, the 3 apical leaflets about 20 cm long and 12 cm wide, short stalked, the other leaflets much smaller, light green and glossy above, dull green beneath. Discovered by Sprenger on Corfu. Before 1908. ∅

ssp. **fallax** (Dode) Popov. Differing in the elliptic-ovate, short acuminate leaves and thin shelled nuts. BD 1906: 81, 83, 96 (= *J. fallax* Dode; *J. duclouxiana* Dode). S. China and Himalayas. 1917. Fig. 126.

f. *filicifolia* see: **'Laciniata'**

'Heterophylla'. Leaflets long and narrow, irregularly lobed, not as attractive as 'Laciniata'. Discovered in the vicinity of Poitiers, France in 1827. ∅

'Laciniata'. Tree, often only a tall shrub; leaves deeply incised (= f. *asplenifolia*; f. *filicifolia*; f. *salicifolia*). Attractive, better than 'Heterophylla'. ∅

'Monophylla'. Leaves usually reduced to a large terminal leaflet, occasionally with a pair of small side leaflets.

'Pendula'. Branches and twigs pendulous on a short arch. Discovered in Waterloo, Belgium around 1850 and distributed by Armand Gothier, of Fontenay-aux-Roses, France. Very rare in cultivation.

'Praepaturiens'. Shrub Walnut. Dwarf habit, often fruiting on 5-year-old plants, fruits in racemes (Lit. FS 1848: 367).

Discovered about 1830 in the Chatenay Nursery, in Doué, Maine-et-Loire, France, in a seedbed where it was fruiting as a 3-year plant. Must be vegetatively propagated as seedlings have not proven reliable in the trade. ✗

'Purpurea' (E. Schneiders). Slow growing; leaves dull red. Discovered in 1938 in the Geisenheim Research Center, W. Germany by E. Schneiders. Still present in 1976. ✗

'Racemosa'. Fruits grouped 10–15 in racemes. ✗

'Rubra'. Nut meat red, epidermis blood-red or only red speckled. Discovered in the wild before 1897 in Steiermark, Germany and distributed by the Trunner Nursery, in Ybbs, Austria, as 'Trunners Rote Donau-Walnuss' and 'Rote Mosel-Nuss'. ✗

f. *salicifolia* see: **'Laciniata'**

J. rupestris see: **J. major**

J. sargentii see **J. × bixbyi**

J. sieboldiana see: **J. ailantifolia**

J. × sinensis (DC.) Dode (*J. mandshurica* × *J. regia*). Small tree; leaves to 50 cm long, leaflets ovate to obovate, to 15 cm long and 8 cm wide, tough, acute, base round, entire or sinuate, dull green above, brownish-green beneath, only the center leaflet with long petiolule, venation brownish pubescent beneath, otherwise glabrous; nut deep pitted. BD 1906: 83. N. and E. China. z6 Fig. 126. ✗

J. stenocarpa Maxim. Tree, very similar to *J. mandshurica*, differing in the terminal leaflet, which is up to 25 cm long and obovate; and the side leaflets, which are much smaller and oblong, with coarser serration; pubescent bulge above the leaf scar absent (present on *J. mandshurica*); nut cylindrical oblong, acuminate, obtuse ridged. Manchuria. 1903. Hardy. z6 Fig. 126. ∅

J. torreyi see: **J. major**

J. vilmoriniana see: **J. × intermedia** var. **vilmoreana**

All *Juglans* species prefer a deep, fertile, but not too wet, alkaline soil.

Lit. Dode, L. A.: Contribution à l'étude du genre *Juglans;* in Bull. Soc. Dendr. France 1906, 67–98; 1909, 21–51, 165–215 (richly illustrated) ● Trelease, W.: Juglandaceae of the United States; in Ann. Missouri Bot. Gard. 1896, 25–46 (with 26 plates, most of *Carya*) ● Nitzelius, T.: *Juglans* i svenska parker; in Lustgarden 1946, 101–128 (with ills.) ● Zur Kultur der *Juglans regia*: Cronbach, W.: Die Walnuss und ihre Sorten im Schrifttum; Frankfurt (Oder) 1938 (406 listings) ● Cronbach, W.: Zierformen und Liebhabersorten der Walnuss *Juglans regia;* in Gartenflora 1935, 141–144 ● Schneiders, E.: Der neuzeitliche Walnussbau; 2nd. ed., 127 pp., Stuttgart 1947.

KADSURA Kaempf. ex Juss. — SCHISANDRACEAE

Evergreen, twining climbers; leaves alternate, simple, entire or dentate; flowers monoecious, solitary or in 2's, occasionally 2–4, usually axillary, with 7–24 perianth segments, sepals and petals generally fused; stamens numerous, short stalked, connective very wide; carpels usually with 2 seed chambers; fruit berrylike, on a short stalk. — 22 species in SE. Asia; Japan, S. Korea to Java.

Kadsura japonica (L). Dunal. Evergreen, twining shrub, to about 2.5 m high; leaves ovate to lanceolate, acuminate, 6 to 10 cm long, cuneate at the base, sparsely serrate, glabrous; flowers solitary, axillary, sulfur-yellow, about 2 cm wide, pendulous, stalks 3–4 cm long, June–September; berry scarlet-red, grouped in globose, nearly 3 cm wide heads. MJ 1613; KD 28; LWT 58 (= *Uvaria japonica* Thunb.). Japan, Korea. 1846. z9 Fig. 128. # ∅ ꧁

Illustrations of other species:
K. chinensis FIO 73
K. peltigera FIO 74

Prefers a humus soil; must be protected from frost in winter.

Lit. Smith, A. C.: The Families Illiciaceae and Schisandraceae; in Sargentia VII, 156–211, 1947.

Fig. 128. *Kadsura japonica* (from S. & Z., Baillon and others)

KALMIA L. — ERICACEAE

Evergreen, rarely deciduous shrubs; leaves alternate, opposite or whorled, entire, petiolate to sessile; flowers in terminal or axillary umbels or corymbs; corolla broad campanulate to bowl-shaped, with 10 pouch-like protuberances, in which the anthers lie until opening; anthers opening with an oblique slit on the apex; ovaries 5 chambered; fruit a 5 valved, many seeded capsule. — 8 species in N. America and Cuba.

Kalmia angustifolia L. Evergreen shrub, to 1 m high; growth tightly upright; leaves usually opposite, short petioled, oblong lanceolate, 3–6 cm long, bright green above, lighter beneath; flowers several in axillary clusters at the shoot tips, corolla about 1 cm wide, purple-red, June–July. BB 2756; GSP 391; BM 331. Eastern N. America. 1736. Grows primarily on sterile, acid soil, moist or dry; one of the more predominant plants of the swamps and peat bogs. The leaves are particularly poisonous to calves and lambs. z2 Fig. 129. # ꧁

The more commendable cultivars are:

'Candida'. White flowers.

'Nana'. Dwarf form, only 30–40 cm high. ꧁

var. **ovata** (Lodd.) Zab. Leaves ovate to elliptic or obovate. Introduced to the trade in 1953 by C. E. English of Seattle, Washington, USA.

Characteristics of the more prominent species

	K. angustifolia	K. latifolia	K. microphylla	K. polifolia
Leaves	usually opposite, 3–6 cm long	alternate, 5–10 cm long	usually 2–3 together, 0.5–2 cm long	usually 2–3 together, 2–3 cm long
Flowers	corolla 1 cm wide, purple-red, June–July	corolla 2–2.5 cm wide, pink to white, May–June	corolla 12 mm wide, pink-lilac, May–June	corolla 1–1.5 cm wide, purple-pink, May–June
Habit	1 m high	1–2 (3) m high; also tree-like in its habitat	15–20 cm high	to 1 m high

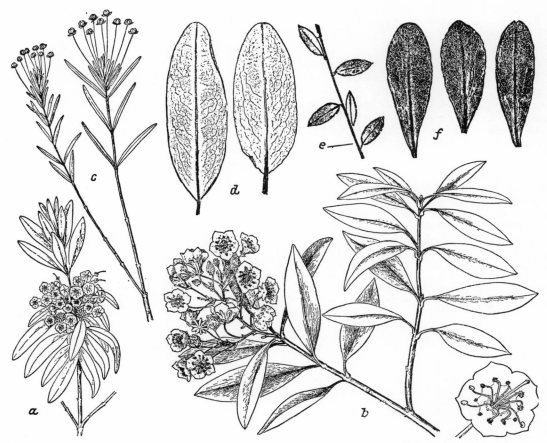

Fig. 129. **Kalmia.** a. *K. angustifolia;* b. *K. latifolia;* c. *K. polifolia;* d. *K. carolina;* e. *K. hirsuta;* f. *K. cuneata* (a–c from Cornell Ext. Bull. 538; d–f Original)

'Pumila'. Possibly var. *ovata* (?). Royal Botanic Garden, Edinburgh, Scotland.

'Rosea'. Flowers pink-red.

'Rubra'. Flowers dark purple; leaves wider. One of the more popular cultivars. ✛

K. carolina Small. Very similar to *K. angustifolia,* but the young shoots softly pubescent; leaves in whorls of 3, finely tomentose beneath; flowers purple-pink, June–July. BB 2: 684. S. Carolina to Virginia, USA in forests. 1906. z6 Fig. 129. #

K. ciliata see: **K. hirsuta**

K. cuneata Michx. Deciduous or slightly semi-evergreen shrub, about 50 cm high, young shoots glandular pubescent; leaves alternate, oblong to narrowly obovate, 2–5 cm long; flowers in axillary clusters on the apical end of the previous year's shoots, corolla shell-like, 1.5 cm wide, white with a red basal ring, occasionally pink, June–July, sepals with pink tips. BM 8319; BS 2: 176; BC 2032. Carolinas. 1820. z7 Fig. 129.

K. glauca var. *microphylla* see: **K. microphylla**

K. hirsuta Walt. Evergreen shrub, 0.25 to 0.5 m high, twiggy, young twigs pubescent; leaves alternate, lanceolate, acute, 12–15 mm long, densely pubescent at first; flowers solitary (or grouped 2–3), axillary on the previous year's shoots, crown pink, plate-like, to 2 cm wide, calyx abscises, June–July. BB 2759; BM 138 (= *K. ciliata* Bartr.; *Kalmiella hirsuta* [Walt.] Small). Southeastern USA. 1790. z9 Fig. 129. #

K. latifolia L. Mountain Laurel. Evergreen shrub, to 10 m high in its habitat; leaves alternate, elliptic-lanceolate, 5–10 cm long, dark green above, yellowish-green beneath; flowers many, in large, terminal, glandular pubescent corymbs, corolla bowl-shaped, 2 to 2.5 cm wide, pink to white, May–June. BB 2757; GSP 388. Eastern N. America. 1734. To 12 m high in the southern Appalachian Mts. Usually found as an understory plant in the oak and pine forests, in dry, rocky areas. Leaves and flowers are poisonous to livestock, but not to wildlife. The wood was once used in the manufacture of spoons (hence, "spoonwood"). Thrives only on acid soils!! z5 Plate 77; Fig. 129. # ⊘ ✛

'Alba'. Flowers white. Cultivated since 1840.

'Bettina'. Flowers campanulate-urceolate (not shell-form), purple-pink, up to 100 together in dense inflorescences. Discovered in the U.S. National Arboretum about 1950. ⊕

'Clementine Churchill' (Setford). Flowers over 2 cm wide, an indigo-purple exterior, dark pink interior. Selected by J. H. Setford in Sheffield Park, Uckfield, Sussex, England. Introduced in England in 1952. ⊕

'Fuscata'. Flowers white, with a dark purple-brown band on the inner corolla limb. Horticulture 1960: 208. Cultivated in the New York Botanic Garden. ⊕

var. **laevipes** Fern. Inflorescences and stalks glabrous. Coastal plains of SE. USA.

f. *monstruosa* see: **'Polypetala'**

f. *minor* see: **'Myrtifolia'**

'Myrtifolia'. Dense, low shrub; leaves dark green, 2–5 cm long. BS 2: 178; BC 2030 (= f. *nana*; f. *minor*). Cultivated since 1840.

f. *nana* see: **'Myrtifolia'**

var. *pavartii* see: **'Rubra'**

'Polypetala'. Corolla deeply 5 parted. BC 2031 (= f. *monstruosa* Mouillef.). 1885.

'Rosea'. Flowers monochromatic pink. Common in England.

'Rubra'. Flowers deep carmine-pink. RH 1888. Pl. 450 (= var. *pavartii* André). Not uncommon in European parks. ⊕

'Sheffield Park' (Setford). Flowers very large, deep pink, the darkest form to date. Origin the same as that for 'Clementine Churchill' (Sheffield Park, England). Introduced in England in 1952. ⊕

K. microphylla (Hook.) Heller. Evergreen shrublet, only 15–20 cm high; very similar to *K. polifolia*, but with ovate leaves, 9–20 mm long; flowers pink-lilac, 12 mm wide. RFW 265; NF 8: 169 (= *K. glauca* var. *microphylla* [Hook.]; *K. polifolia* var. *microphylla* [Hook.] Rehd.). Western N. America, Alaska. Hardy. z4 #

K. polifolia Wangh. Shrub, to 50 cm high, branches 2 sided; leaves evergreen, grouped 2–3 together, blue-green beneath, margins involuted, nearly or completely sessile; flowers, several in terminal umbels, corolla 10–15 mm wide, purple-pink, campanulate-funnelform, May–June. BM 177; BB 1957: 73; SL 239; NBB 3: 15. Northern N. America, in cold peat bogs, growing together with *Ledum*. 1767. z2 Fig. 129. # ⊕

var. *microphylla* see: **K. microphylla**

var. **rosmarinifolia** (Pursh) Rehd. Leaves narrower, more linear, more involuted, green beneath; flowers also purple-pink. # ⊕

Best used in a wooded site; *K. latifolia* is the easiest to cultivate (for cultural requirements, refer to the species description). Won't tolerate alkalinity! All species will tolerate full sun if given sufficient moisture.

Lit. Holmes, M. L.: The Genus *Kalmia*; in Baileya **4**, 89–94, 1956.

KALMIA × RHODODENDRON

As yet only one hybrid between these two genera is known and that is:

Kalmia latifolia × Rhododendron williamsianum, developed by H. Lem in Seattle, Washington, USA.

The plant which was once considered a hybrid of *K. latifolia* × *Rhododendron maximum* is at the least very doubtful. Today it is generally considered a mutation of *Rhododendron maximum*. Outside of the USA, Hillier arboretum has one plant at about 1 m high. (1976).

Lit. Pierce, L. J.: in Amer. Rhod. Soc. Quart. Bull. **28**, 45, 1974 ● Jaines, R. A.: The Laurel Book; New York 1975 (pp. 146–147, with ills.).

KALMIELLA

Kalmiella hirsuta see: **Kalmia hirsuta**

KALMIOPSIS Rehd. — ERICACEAE

Monotypic genus, classified between *Kalmia* and *Loiseleuria*; leaves evergreen, alternate; flowers in terminal, foliate racemes; corolla short, tube form with on outspread limb, lobes incised nearly to the base, calyx persistent, somewhat glandular; fruit a dry, flat-rounded, dehiscent capsule. Distinguished from *Loiseleuria* by the free stamens, from *Kalmia* in the scaleless flower buds. — 1 species in N. America.

Kalmiopsis leachiana (Henderson) Rehd. Evergreen shrublet, about 30 cm high, erect habit, well branched; leaves elliptic, acute at both ends, 3 cm long, to 1.5 cm wide, dark green above, lighter beneath; flowers in terminal racemes, March–May, grouped 6–9 together, corolla shell form, 2 cm wide, purple-pink. NF 9: 249; BS 2: 503. Oregon, USA. A chance discovery by Mrs. Lilla Leach in 1930. z7 Fig. 130 # ⊕

Some clones have been named; i.e. 'M. Le Piniec', which grows and blooms better than the species, discovered in S. Oregon, USA.

Prefers a sandy soil, rich in humus, lime free, semishade. A collector's plant without particular garden merit.

Lit. Rehder, A.: *Kalmiopsis*, a new genus of Ericaceae from Northwest America. Jour. Arnold Arb. **13**, 30–34, 1932.

Fig. 130. *Kalmiopsis leachiana* (left, according to Parry, right from Abrams)

KALOPANAX Miq. — ARALIACEAE

Deciduous tree or shrub, thick branches, shoots with short, broad prickles; leaves palmately lobed, finely serrate; flowers hermaphroditic, in umbels, the umbels grouped into large panicles composed of racemose branchlets; flowers otherwise similar to those of *Acanthopanax*, styles forming a column; ovaries 2-locular; fruits globose, with 2 flat seeds. — 1 species in E. Asia.

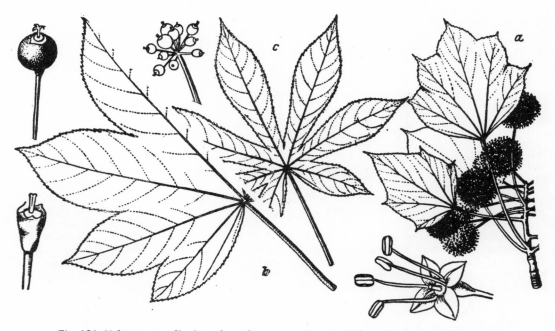

Fig. 131. **Kalopanax.** a. *K. pictus*; b. and c. var. *maximowiczii* (from Harms and Shirasawa)

Kalopanax pictus (Thunb.) Nakai. A 30 m high, wide tree in its habitat, occasionally only a shrub, sparsely branched, branches thick, with thick prickles; leaves nearly circular, 10–25 cm wide, 5–7 lobed, the lobes triangular, serrate, dark green above, lighter beneath, slightly pubescent when young; flowers white, in 20–30 cm long panicles, May; fruits globose, blue-black, 4 mm thick, with persistent style. MD 1913; 147; BC 192 (= *Acanthopanax ricinifolius* Miq.). China, E. Siberia, Korea, Japan. 1865. Hardy. Leaves quite variable. z5 Plate 37, 77; Fig. 131. ⌀

var. **magnificus** (Zab.) Nakai. Shoots without or with only a few prickles; leaves only shallowly lobed, densely pubescent beneath, lobes ovate. MD 1918: Pl. 5 and 8. China, Japan. ⌀

var. **maximowiczii** (van Houtte) Hara. Leaves deeply 5–7 lobed, incised to beneath the blade middle, lobes oblong-lanceolate, densely pubescent beneath. FS 2067; MD 1918: Pl. 5–7. China, Japan. (Hartwig suggested that var. *maximowiczii* is possibly only a juvenile form and cited an analogy to the leaf dimorphism of *Hedera;* in Mitt. DDG 1954: 170–171.) Plate 77; Fig. 131. ⌀

K. sciadophylloides see: **Acanthopanex sciadophylloides**

Cultivated like *Acanthopanax.* Very impressive as an older plant. Quite hardy.

Lit. Li, H.–L.: The Araliaceae of China, 90–93; Jamaica Plain 1942 ● Koehne, E.: *Acanthopanax ricinifolius;* in Mitt. DDG 1913, 145–150.

KENNEDIA Vent. — LEGUMINOSAE

A woody climber or herbaceous plant, partly creeping on the ground; leaves nearly always with 3 leaflets, occasionally with only 1 or 5, alternate; stipules persistent, often attractive; papilionaceous flowers paired or in clusters or racemes, axillary; with 5 stamens, of these, 4 connate; standard usually circular, wings sickle form, keel incurved; fruit a linear pod. — 15 species in Australia, most with black flowers.

Kennedia nigricans Lindl. Subshrub, strong growing, twining, more or less pubescent; leaflets in 3's, occasionally only 1, broadly ovate to rhombic, entire, obtuse or emarginate, 5–7 cm long, in one sided, axillary racemes, slender, 3 cm long, dark purple-violet to nearly black, the obovate, reflexed standard has a green spot in the center, April. BM 3652. W. Australia. z10 ⌀ ✿

KERRIA DC — ROSACEAE

Deciduous shrub with green, rodlike, glabrous twigs, buds very scaly; leaves alternate, oval, long acuminate, coarsely serrate, with pinnate venation; flowers dioecious, rarely hermaphroditic; calyx 5 lobed, petals 5 (numerous on double flowers); carpels 5–8, distinct, enclosed in the calyx base and usually noticeable only on the double flowers; styles filamentous; fruit a single seeded, eventually dry, drupe. — 1 species in China and Japan.

Kerria japonica (L.) DC. Shrub, 1.5–2 m high, broader growing when young, without a central leader, twigs green and glossy, thin, buds brownish-green; flowers solitary, golden-yellow, single, 3 cm wide, April–May and fall. BS 2: 181; DB 1959: 12. Central and W. China; in Japan only in cultivation. 1834. z5 Plate 78. ✿

f. *argenteo-marginata* see: **'Picta'**

f. *argenteo-variegata* see: **'Picta'**

'Aureovariegata'. Leaves all yellow margined; flowers yellow, single. Before 1914. ⌀

'Aureovittata'. Broad and low growing, twigs green with yellow stripes. Before 1875.

'Picta'. Slow growing; leaves more gray-green, margin white; sparsely flowering (= f. *variegata* Moore; f. *argenteo-variegata;* f. *argenteo-marginata*). Before 1844. Plate 78. ⌀

'Pleniflora'. Strictly upright habit, tight, much bushier than the wild type; flowers densely double. BM 1296; DB 1959: 12. 1804. Very attractive. ✿

f. *variegata* see: **'Picta'**

Thrives in any garden soil in a sunny location; shoots live 3 years but new shoots resprout from the rootstock. Some dieback in a cold winter.

KOELREUTERIA Laxm. — SAPINDACEAE

Deciduous, small tree or shrub; leaves alternate, simple or bipinnate; flowers yellow, in large terminal panicles; calyx 5 parted, regular, petals 4, lanceolate; stamens 5–8, with distinct, pubescent filaments; ovaries oblong, triangular, pubescent; fruit a paper-like, inflated capsule with 3 globuse, black, pea-sized seeds. — 7 species in E. Asia.

Koelreuteria bipinnata Franch. Tree, to 5 m high; leaves bipinnate, 50 cm long or longer and just as wide, pinnae

8–11, alternately arranged on the rachis, secondary leaflets oval-oblong, leathery, 5–7 cm long, finely serrate; flowers yellow, with a reddish spot at the base; fruits eventually reddish, 5 cm long. DL 2: 180; ICS 3176. Yunnan Province, China. 1888. z8 ⌀ ✿ ⚭

K. integrifolia Franch. A tree to 18 m high in its habitat, shoots light brown, glabrous, densely covered with lenticels; leaves simple to bipinnate, alternate, leaflets oval-oblong, 5–8 cm long, 3–5 cm wide, acuminate, base round, entire, glossy dark green above, silvery pubescent beneath, somewhat leathery, short stalked; flowers in terminal panicles, yellow; fruit capsule 6.5 cm long, 4 cm wide. ICS 3177. China; Chekiang. z6 Fig. 133. ✛

K. minor Hemsl. ICS 3178. Also from China. Presumably not yet cultivated.

K. paniculata Laxm. Tree, 8–15 m high, very broad crown; leaves pinnate, to 35 cm long, leaflets 7–15, ovate-oblong, 3–8 cm long, coarsely crenate, often also somewhat deeply lobed; flowers yellow, 1 cm wide, in loose, many flowered, erect panicles, July–August. BR 330; DB 1959: 5. N. China, Korea. 1763. Attractive in fall with yellow foliage and fruits. z6 Plate 79; Fig. 133. ∅ ✛ ⚭

var. **apiculata** (Rehd. & Wils.) Rehd. A low tree with a short, thick stem; leaves bipinnate especially in the central portion, often only incised on the ends, to 40 cm long, pubescent beneath; flowers light yellow, panicles 15–30 cm long, erect, August. GC 78: 307 (= *K. bipinnata* Rehd. & Wils.). China, Szechwan Province. 1904. Hardy. Not a forest tree in its habitat, rather found in hot, dry river valleys; often planted at the gravesites of Chinese aristocracy.

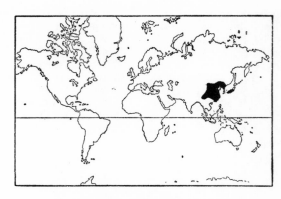

Fig. 132. Range of the genus *Koelreuteria*

'Fastigiata'. Narrow columnar form, otherwise like the species.

'September'. Always flowers in late August to early September, otherwise like the species. Discovered in 1960 in Bloomington, Indiana, USA at the Indiana University. ✛

Attractive specimen for parks; thrives in any good soil, but flowers better if given a somewhat protected location. Full sun. Good flowers, fruit, foliage and fall color.

Fig. 133. **Koelreuteria.** a. *K. paniculata* (from China); b. *K. paniculata* var. *apiculata* (from Dortmund, W. Germany); c. *K. integrifolia* (from W. Hupeh Province, China; E. H. Wilson Nr. 1609)

KOLKWITZIA Graebn — CAPRIFOLIACEAE

Monotypic genus; deciduous shrub; leaves opposite, short petioled, without stipules; flowers paired, but grouped in large terminal corymbs on short lateral shoots; sepals 5, narrow, pubescent, spreading outward; corolla campanulate, 5 lobed; stamens 4; carpels superimposed one upon the other, partially connate, 3 chambered, but with only 1 fertile locule; fruit a dry, bristly, single seeded capsule. — 1 species in China.

Kolkwitzia amabilis Graebn. Deciduous shrub, 1.5–2.5 m high, dense and upright, young shoots densely pubescent, older shoots with brown, exfoliating bark; leaves opposite, broadly ovate, 3–7 cm long, short stalked, sparsely serrate, ciliate, dull green above,
pubescent, venation pubescent beneath; flowers campanulate, 5 lobed, pink-white with a yellow throat, 1.5 cm long, May–June, the pubescent calyx persists on the fruit. BM 8563. W. China, Hupeh Province. 1901. z5 Plate 37, 79. ⊕

'Pink Cloud'. Flowers a stronger pink. Developed before 1963 in RHS Gardens, Wisley, England. ⊕

'Rosea' (Ruys). Flowers more reddish. Introduced to the trade in 1960 by Ruys of Dedemsvaart, Holland. ⊕

A splendid flowering shrub, similar to *Weigela,* but with smaller flowers and a more delicate texture. Thrives in any good garden soil.

+ LABURNOCYTISUS (LABURNUM + CYTISUS) Schneid. — LEGUMINOSAE

+ Laburnocytisus adami (Poit.) Schneid. (*Laburnum anagyroides + Cytisus purpureus*). Deciduous, 3–5 m high shrub, similar to a *Laburnum* in appearance; leaflets smaller, nearly glabrous; flowers a dull light purple, in nodding racemes, but also with pure yellow *Laburnum* flowers and pink *Cytisus* blooms on the same plant, May. BR 1965 (= *Laburnum adami* Kirchn.; *Cytisus adami* Poit.). Developed in 1825 as a periclinal Chimera on a graft of
Cytisus purpureus on *Laburnum anagyroides;* in the J. L. Adam Nursery in Vitry near Paris, France. Plate 38. ⊕

Culture and use like *Laburnum,* but less valuable.

Lit. Buder, J.: Studien an *Laburnum adami;* in Z. indukt. Abst. u. Vererb. **5,** 209–284, 1911 ● Bergann, F.: Über das Auftreten einer bisher unbekannten *Laburnum*-form on der Pfropfchimäre *Laburnum adami;* in Flora **139,** 295–299, 1952.

LABURNUM Med. — Goldenchain Tree — LEGUMINOSAE

Deciduous shrubs or small trees with green bark; leaves alternate, 3 parted, leaflets nearly sessile; flowers yellow, in simple, terminal racemes; calyx campanulate, with 5 short teeth; corolla with a rounded to broad obovate standard, obovate wings and a convex keel petal; ovaries stalked or sessile, style curved inward; fruit an oblong, compressed, several seeded pod. All parts poisonous! — 3 species in S. Europe to Asia Minor.

Laburnum adami see: **+ Laburnocytisus**

L. alpinum (Mill.) Bercht. & Presl. Alpine Goldenchain. Shrub or tree, to 5 m, twigs quite glabrous, yellow-green or gray-green; leaflets elliptic-oblong, 4–7 cm long, light green, ciliate when young, otherwise usually quite glabrous, except the midrib beneath; flowers in dense (!), pendulous, 20–30 cm long racemes, glabrous, fragrant (!), corolla light yellow, about 2 cm long, stalk 8–15 mm long, June; pods oblong, to 6 cm long, glabrous, hardly dehiscent, many seeded, distinctly stalked, seam wing-like keeled (!). HM 1315; HF 2306 (= *Cytisus alpinus* Mill). S. Europe, Alps; gravelly slopes in warm, moist areas. Ⓕ Italy (green screens). z5 Fig. 134. ⊕

'Aureum'. Leaves golden-yellow. ⊘

'Autumnale'. Flowers in fall for a second time, but less abundantly than in June (= var. *biferum* Kew Gardens). Before 1914. ⊕

'Lucidum'. Leaves conspicuously glossy. ⊘
f. **macrostachys** (Endl.) Koehne. Leaves broader; flower racemes to 40 cm long. Particularly well established in S. Tyrol; in more shaded areas. ⊕

'Pendulum'. Twigs very pendulous. Discovered in England. Before 1838.

var. *biferum* see: **'Autumnale'**

L. anagyroides Med. 5 to 6 (9) m high, twigs more rodlike, gray-green and appressed pubescent; leaflets elliptic-oblong, 3–8 cm long, dark green above, gray-green beneath and appressed pubescent; flowers in loose (!), 10(–20) cm long, pubescent racemes, not fragrant, corolla 2 cm long, light or dark yellow, standard broadly emarginate, base brown, May–June; pod to 8 cm long, silky pubescent, upper seam sharply ridged but not winged. HM 1314; HF 2306 (= *Cytisus laburnum* L.; *Laburnum vulgare* Bercht. & Presl). S. France to Romania, north to S. Germany. On limestone; in S. Europe found growing with *Quercus pubescens, Ostrya, Cotinus, Sorbus aria, Colutea, Prunus mahaleb,* etc. Cultivated for centuries. All parts poisonous, including the seeds. Ⓕ Italy (in green screens); Kashmere (reforestation). z6 Fig. 134. ⊕

var. **alschingeri** (Vis.) Schneid. Only 2–4 m high; leaflets densely white pubescent beneath, elliptic, later more leathery; flower racemes short, often erect, claw of the standard petal distinctly exserted past the calyx tube, calyx distinctly bilabiate, lower lips longer. Visiani, Fl. Dalmat. Pl. 54; Ost. Bot. Z. 1891: Pl. 4 (= *Cytisus alschingeri* (Vis.). S. Tyrol, N. Italy, Dalmatia and

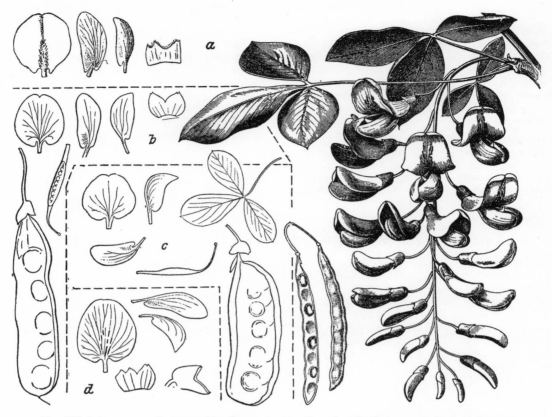

Fig. 134. **Laburnum**. a. *L. anagyroides*, flower raceme, 2 seed pods, flower parts; b. *L. alpinum*;
c. *L. caramanicum*; d. *L. anagyroides* var. *alschingeri* (from Kerner, Hempel & Wilhelm, Schneider)

Croatia; common at Lake Garda. Usually incorrectly labeled in
cultivation! Plate 80; Fig. 134.

'Aureum' (van Houtte). Leaves greenish-yellow, the new
growth more golden-yellow. FS 2247. Introduced into the
trade by van Houtte in 1875. ⌀

f. *autumnale* see: **'Serotinum'**

'Bullatum'. Leaflets "spoon-like" concave, margins curved
upward (= f. *involutum* K. Koch; f. *crispum* Ktze.; f. *monstrosum*
Nichols.). Before 1840.

'Carlieri'. Leaflets much smaller, narrower, obovate to
narrowly lanceolate on the same twig, 1–2.5 cm long, 4–5 mm
wide; flower racemes short, semi-erect, often grouped 2–3
together, light yellow; fruit with only a few seeds. Known since
1864; originated in cultivation.

'Chrysophyllum' (Späth). Leaves always golden-yellow.
Introduced to the trade by L. Späth in 1888 and touted as an
improved 'Aureum'. ⌀

f. *crispum* see: **'Bullatum'**

f. *involutum* see: **'Bullatum'**

'Incisum'. Similar to the much more common 'Quercifolium',
but with leaflets much more deeply incised.

f. *monstrosum* see: **'Bullatum'**

'Pendulum'. Branches and twigs very pendulous.

'Quercifolium'. Leaflets with 2–3 sinuate lobes on either side.
Dt. Mag. Gart. Blumenk. 1853, Pl. 45.

f. *semperflorens* see: **'Serotinum'**

'Serotinum'. Flowers again in the fall, but less abundantly in
May (= f. *autumnale* K. Koch; f. *semperflorens* Bean). Before
1840.

'Sessilifolium'. Leaves short stalked, therefore clustered and
appearing nearly sessile. Before 1847.

'Variegatum'. Leaves yellowish variegated.

L. caramanicum (Boiss.) Benth. & Hook. Erect, about 1 m
high shrub, shoots thin, long, cylindrical, glabrous;
leaves trifoliate, short stalked, leaflets gray-green,
obovate, 8–15 mm long, sessile, lateral leaflets somewhat
smaller; flowers on short side shoots on the new growth,
in erect, terminal, 10–20 cm long racemes, corolla
yellow, 2 cm long, stalk 8 mm long, with some small
prophylls, August–September. BS 2: 189; BM 7898.
Greece, Asia Minor. 1861. z9 Fig. 134.

L. vulgare see: **L. anagyroides** and **L. × watererii**

L. vossii see: **L. × waterei 'Vossii'**

L. × waterei (Wettst.) Dipp. (*L. alpinum* × *L.*
anagyroides). Hybrids between these two species have

occurred in cultivation as well as in the wild. Observed in 1856 in the wild in Tyrol and S. Switzerland among the parents. The group name is based on a plant found in the Knap Hill Nursery and introduced by this firm before 1864 as *L. vulgare watereri* (= *Cytisus* X *watereri* Wettst.). z6 ✣

'Parkesii'. Habit and particulars similar to those of *L. alpinum*, shoots always quite glabrous; leaflets oblong, often somewhat obovate, 3–6 cm long, to 3 cm wide, ciliate, dark green above and somewhat glossy, lighter beneath with thin appressed pubescence; flowers in about 30 cm long, dense, somewhat pubescent racemes, bright yellow, fragrant, appearing mid-May, standard oval, emarginate on the apex, base with some small brownish lines, somewhat longer than the calyx, calyx margin slightly pubescent, stalk nearly 2 cm long, rather glabrous; pods over 6 cm long, abscising quickly. HL 220; Gs 2: 109 (= *L. vulgare parkesii* Nichols.). Discovered in 1842 in the

Nursery of J. D. Parkes of Dartford, England; still cultivated but far surpassed by 'Vossii'.

'Vossii' (De Vos). Strong growing, slender, only pubescent on the shoot tips, more like *L. anagyroides*; leaves often with 10 cm long petioles, leaflets elliptic, 2.5–7 cm long, somewhat glossy dark green above, only the venation pubescent beneath; calyx glabrous or the limb somewhat pubescent, flower racemes to 50 cm long, fragrant; pods usually only with 1(2) seeds (!!) (= *L. vossii* Hort. ex Hendr.). Discovered about 1875 by C. De Vos of Hazerswoude, Holland. The foremost *Laburnum* in cultivation today. Plate 38. ✣

Very popular, large park trees for any good soil, but particularly fond of alkaline soil. Dislikes pruning. Trees or shrubs in full bloom are a sight to behold.

Lit. Bartrum, D.: Lilac and Laburnum, 122–150; London 1959 (Gilford) ● v. Wettstein, Untersuchungen über die Sect. *Laburnum* der Gattung *Cytisus*; in Osterr. Bot. Z. 1890, 395–399, 435–439; 1891, 17–130, 169–173, 261–265.

LAGERSTROEMIA L. — LYTHRACEAE

Deciduous or evergreen shrubs or trees; leaves opposite, the apical ones however, often alternate, usually ovate, entire; flowers in axillary or terminal panicles; petals 6, attractive, with a crispate margin and distinct, long claw; stamens numerous; ovaries sessile, with 3–6 locules, also with a long style; fruit a woody, dehiscent capsule with 3–6 valves; seeds winged. — About 30 species in S. and E. Asia and the islands to Australia.

Lagerstroemia indica L. Deciduous tree or shrub, 3–7 m high, twigs nearly quadrangular-winged, older shoots and stem with very smooth, pink-brown bark; leaves elliptic-oblong, 3–5 cm long, acute, nearly sessile; flowers pink, white or purple, 3–4 cm wide, in terminal, 15–20 cm long panicles, July to September. BC 2060; DRHS 1122; HM 2161; BS 2: Pl. 17. China, but very often planted in the tropics and subtropics (Indies, S. America,

N. Australia, S. Europe, etc.). 1759. Highly recommended for the conservatory for its ease of cultivation and abundant floral display. z10 Plate 80; Fig. 135. ✣

The U.S. National Arboretum in Washington D.C. introduced 4 cultivars to the trade in 1967 ('Catawba', 'Powhatan', 'Conestoga' and 'Potomac') and in 1969, 2 more were introduced ('Cherokee' and 'Seminole'). They are described by D. R. Egolf in Baileya 1967, 7–13, and 1969, 1–5.

L. speciosa Pers. Tall tree, leaves elliptic to oblong-lanceolate, 10–20 cm long; flowers 5–7 cm wide, with 100–200 stamens, opening to a soft pink, then turning darker purple by evening. GC 15: 77. S. China, E. India. Very attractive, but also very tender. z9 ✣

A gorgeous summer flowering shrub, widely planted in the landscape where it is hardy.

LAGUNARIA G. Don — MALVACEAE

Evergreen tree with alternate, simple leaves; flowers axillary, with a 5 toothed calyx and *Hibiscus*-like large flowers.

Lagunaria patersonii Don. Evergreen tree, to 15 m high in its habitat, a tall, narrowly conical shrub, 6 m or more in cultivation where it is hardy, young shoots and inflorescences lepidote; leaves oblong to broadly

lanceolate or oval-oblong, 7–10 cm long, half as wide, entire, underside white when young, obtuse, base cuneate; flowers solitary, axillary, mallow-like, 6 cm wide, pink, petals 5, about 3 cm long, reflexed, summer flowering, calyx 5 toothed, ovaries 5 chambered. BM 769. Australia and Norfolk Island. 1792. z10 Plate 79. # ∅ ✣

LANTANA L. — VERBENACEAE

Herbaceous or evergreen shrubs; leaves opposite or in groups of 3 together, usually rough, toothed, unpleasant smelling; flowers in axillary, hemispherical heads; corolla tube with an irregular, 4–6 or 5 incised limb, not bilabiate, tube thin, calyx very small; stamens 4, attached

in the middle of the corolla tube; fruit a berrylike drupe. — About 150 species in tropical America, the West Indies, tropical and S. Africa.

Lantana camara L. Shrub, to 3 m high in its habitat, strong growing, divaricate with prickly shoots (the species is seldom found in cultivation, the cultivated plant is a thornless selection that is only 30–100 cm high); leaves oval-oblong, acute, margin crenate, 4–6 cm long, more or less cordate at the base, rugose and rough above, softly pubescent beneath; flowers in flat heads, but later elongating spike-like, usually opening yellow or pink, later orange or scarlet, lilac to violet, depending on the cultivar (some don't change color), flowers the entire summer; fruits black, globose, pea-sized. BM 96; DRHS 1126. Tropical America, north to Texas and S. Carolina. 1692. Ⓕ Tropics; as a ground cover. z10 ✧

The following, so-called "Camara hybrids" have been named:

'Arlequin'. Dark pink and yellow.

'Fabiola'. Salmon-pink and yellow.

'Goldsonne'. Monochrome lemon-yellow.

'Naide'. White with a yellow "eye".

'Professor Raoux'. Scarlet-red and orange.

'Schloss Ortenburg'. Brick-red and salmon-yellow.

L. montevidensis (Spreng.) Briq. Procumbent and spreading, shoots softly pubescent; leaves acutely ovate, smaller than the previous species, tapering to the base; flowers in flat heads, lilac-pink, exterior somewhat lighter, flowers abundantly throughout the entire year. BM 2981; RH 1852: 461 (= *L. sellowiana* Link & Otto). S. America; Montevideo. z10 ✧

L. sellowiana see: **L. montevidensis**

LAPAGERIA Ruiz & Pav. — PHILESIACEAE

Monotypic genus; evergreen, twining shrub; leaves alternate; flowers several in the leaf axils, large, pendulous, campanulate, with 6 fleshy corolla lobes, the 3 inner ones much wider than the outer ones; fruit a many seeded, ovate berry. Chile.

Lapageria rosea Ruiz & Pav. Evergreen twining shrub, to 3 m high, shoots thin, stiff, hard, green; leaves tough, leathery, acutely cordate, 3–10 cm long, the larger ones with 5 veins, the smaller with 3 veins, glossy dark green; flowers 1–2 in the leaf axils or on the shoot tips, carmine-pink, 7 cm long, campanulate, stamen filaments 6, white, anthers yellow, July–September; fruit triangular-ovate, 5 cm long, fleshy, with numerous seeds. BM 4447; PFC 235; FS 491. Chile. 1847. z9 Plate 38; Fig. 135. # ✧

'Albiflora'. Flowers pure white, otherwise like the species. FS 2050; BM 4892. Introduced from Chile by R. Pearce. ✧

'Ilsemannii'. Flowers larger, darker carmine-pink, flowers more abundantly. Gfl 1445. Introduced in 1897. ✧

'Nash Court'. Flowers carmine-pink, somewhat darker marbled. Very attractive. Developed in England. ✧

'Superba'. Flowers larger, carmine. ✧

Hardy only in warmer climates; prefers a sandy-humus soil with sufficient moisture. The plant is inconspicuous without flowers, but gorgeous when in bloom.

Fig. 135. *Lapageria rosea* (left); *Lagerstroemia indica* (center); *Lardizabala biternata* (right) (from Baillon, Koehne, Original at right)

Plate 65

Ilex opaca in Longwood Gardens, Kennett Square, Pennsylvania, USA

Ilex cornuta in Royal Botanic Gardens, Kew, London, England

Plate 66

Ilex altaclarensis 'Atkinsonii'
in Royal Botanic Gardens, Kew, England

Ilex attenuata in the Hillier Arboretum, England

Ilex latifolia
in the Boskoop Nursery Experiment Station, Holland

Ilex crenata 'Mariesii'
in the Royal Botanic Garden, Edinburgh, Scotland

Plate 67

Ilex franchetiana
in the Hillier Arboretum, England

Ilex rugosa
in Göteborg Botanic Garden, Sweden

Indigofera heterantha
in the Dublin Botanic Garden, Ireland

Ilex altaclarensis 'Camelliifolia'
in Royal Botanic Gardens, Kew, England

Plate 68

Ilex. a. *I. montana*; b. *I. montana* var. *media*; c. *I. serrata*; d. *I. laevigata*; e. *I. verticillata.* — **Itea.** f. *I. ilicifolia*; g. *I. yunnanensis.* — h. *Jamesia americana.* — **Jasminum.** i. *J. fruticans*; k. *J. officinale*; l. (1–2) 2 different types of *J. humile* f. *humile*; m. *J. mesnyi*; n. *J. humile* 'Revolutum'; o. *J. floridum*; p. *J. polyanthum* (collected from wild plants)

Plate 69

Indigofera kirilowii
Photo: A. Purpus

Indigofera fortunei
Photo: A. Purpus

Plate 70

Illicium anisatum in its native habitat
Photo: Dr. Watari, Tokyo

Itea virginica
in the Berlin Botanic Garden, W. Germany

Jacaranda mimosifolia in the garden,
Companha Agricola-Horticola, Porto, Portugal

Plate 71

Jamesia americana in the Dortmund Botanic Garden, W. Germany

Jasminum polyanthum
in the Royal Botanic Garden, Edinburgh, Scotland

Jubaea chilensis,
on Madre Island, Lake Maggiore, Italy

Plate 72

Juglans nigra, fruits
Photo: Archiv D. B.

Juglans ailantifolia, fruits
Photo: Dr. Pilat, Prague

Juglans ailantifolia in its habitat in Japan
Photo: Dr. Watari, Tokyo

Plate 73

Magnolia soulángiana 'Norbertii'   *Magnolia soulangiana* 'Alexandrina'
both in the Hillier Aboretum, England

Magnolia virginiana   *Magnolia soulangiana* 'Picture'
both in the Hillier Aboretum

Plate 74

Magnolia stellata
in Westfalenpark, Dortmund, W. Germany

Magnolia loebneri 'Merrill'
in the Proeftuin Boskoop, Holland

Mahonia aquifolium 'Apollo'

Mahonia wagneri 'Moseri'

both in the Proeftuin Boskoop, Holland

Plate 75

Malus 'Henrietta Crosby'   *Malus* 'Oporto'
both in the Belmonte Arboretum, Wageningen, Holland

Malus floribunda   *Malus ioensis* 'Nova'
both in the Dortmund Botanic Garden, W. Germany

Plate 76

Malus 'Cowichan'   *Malus* 'Professor Sprenger'
both in the Dortmund Botanic Garden, W. Germany

Malus sieboldii 'Wintergold'   *Malus* 'Gorgeous'
both in the Dortmund Botanic Garden, W. Germany

Plate 77

Juglans nigra
in the Tervuren Garden, Brussels, Belgium

Kalmia latifolia

Kalopanax pictus
in the Dortmund Botanic Garden, W. Germany

Kalopanax pictus var. *maximowiczii* in the
Thiensen Arboretum near Elmshorn, W. Germany

Plate 78

Kerria japonica, photo taken in Japan
Photo: Dr. Watari, Tokyo

Kerria japonica 'Picta'

Ligustrum lucidum 'Tricolor'
in a park in Herceg-Novi, Yugoslavia

Plate 79

Lagunaria patersonii in the March Garden, S'Avall, Mallorca, Spain

Kolkwitzia amabilis, flowering twig in the Dortmund Botanic Garden, W. Germany

Koelreuteria paniculata as a street tree in Tashkent, USSR
Photo: K. Browicz

Plate 80

Laburnum anagyroides 'Alschingeri' in Royal Botanic Gardens, Kew, England

Lagerstroemia indica
in the Sochi Dendrarium, USSR

Laurelia serrata
in Tregothnan, SW. England

LARDIZABALA Ruiz & Pav. — LARDIZABALACEAE

Evergreen climbing shrub; leaves alternate, simple to 3 times trifoliate, entire or sinuate; flowers dioecious, purple-brown, with 6 sepals and 6 petaloid nectaries (often considered petals); male flowers in pendulous racemes, with 6 connate stamens; female flowers solitary, with 6 distinct, sterile stamens and 3 ovaries; fruit a many seeded berry. — 2 Species in Chile.

Lardizabala biternata Ruiz & Pav. Leaves bipinnate or 3 times trifoliate, usually simple trifoliate on the flowering shoots, leaflets leathery, entire or with 1–2 thorny teeth, glossy green above, lighter with reticulate venation beneath; stamens about 2.5 cm wide, purple-brown, about 15 in pendulous racemes, the white nectaries lanceolate, December; fruits sausage-shaped, sweet and fleshy, about 6 cm long. BM 4501; DRHS 1128; PFC 116; JRHS 69: 28. Chile. 1844. z9 Fig. 135. # ⌀ ✧ ⚭

LAURELIA Juss. — ATHEROSPERMATACEAE

Evergreen, aromatic trees; leaves opposite, leathery, usually serrate; flowers in axillary cymes or spikes, dioecious or polygamous, without petals; corolla or calyx of the male flowers with a short tube and 5 to 12 points in 2 or 3 rings, with 4–12 stamens; corolla of the other flowers with a narrower tube and persisting longer, 3–5 incised, stamens reduced to scales or the outermost complete; carpels numerous and distinct, each later developing into a plumose achene. — 2 species in S. America, 1 in New Zealand.

Laurelia aromatica see: **L. sempervirens** and **L. serrata**

L. novae-zealandiae A. Cunn. A tall forest tree in its habitat with a thick, white barked trunk and radially sprouting root initials, young shoots very aromatic when crushed; leaves oblong to obovate, coarsely and obtusely serrate, finely silky pubescent at first or also glabrous, 4–7.5 cm long; flowers in 2.5 cm long, axillary racemes, silky pubescent. New Zealand. z10 # ⌀

L. philippiana see: **L. serrata**

L. sempervirens (Ruiz & Pavon) Tulasne. Tall tree, very similar to the following species, but differing in: the aromatic bark (wood not fragrant); leaves elliptic to broadly lanceolate, obtuse or rounded, 6–9 cm long, 3–4.5 cm wide, base broadly cuneate to round, margins serrate nearly to the base, with appressed teeth, midrib glabrous; filaments of the stamens pubescent, as long as the anthers. PFC 137 (= *L. aromatica* Poir. non Mast.). Chile. 1926. z9 # ⌀

L. serrata Bertero. Tall, evergreen forest tree, young shoots 4 sided, not aromatic (!), wood smells unpleasantly; leaves leathery, narrowly elliptic, 6–12 cm long, 2.5–4 cm wide, tapering to both ends, coarsely serrate, dark green and glabrous above, midrib beneath with erect yellow pubescence, petiole 5 mm long,

crushed leaves scented like "Bay" (*Laurus* spp.); flowers inconspicuous; fruits brown pubescent. BM 8279; PFC 136 (= *L. philippiana* Looser; *L. aromatica* Mast. non Poir.). Chile and neighboring Argentina. z10 Plate 80. # ⌀

LAUROCERASUS

Laurocerasus officinalis see: **Prunus laurocerasus**

LAURUS L. — Laurel — LAURACEAE

Aromatic, evergreen shrubs or trees; leaves alternate, stiff, simple; flowers dioecious or hermaphroditic, small, greenish, not attractive, usually in small, axillary clusters; sepals 4, corolla 4 parted, stamens usually 12; fruit a black berry. — 2 species in S. Europe and the Canary Islands.

Laurus azorica (Seub.) J. Franco. Canary Laurel. An evergreen tree in its habitat, to 15 m high, shoots softly pubescent, reddish-brown, aromatic when crushed; leaves elliptic or ovate, 5–15 cm long, acute, very glossy above, lighter beneath with a pubescent midrib; flowers greenish-yellow, April; fruits ovate, 12 mm long, stalk 6 mm long. BFCa 15 (= *L. canariensis* Webb & Berth. non Willd.). Canaries, Azores. z10 Plate 81. # ⌀

L. canariensis see: **L. azorica**

L. nobilis L. Laurel. Evergreen, aromatic shrub or tree, 7–15 m high, conical, densely foliate, shoots black-red, glabrous; leaves narrowly elliptic, acuminate at both ends, 5–10 cm long, margins undulate, glossy dark green above; flowers greenish-yellow, in axillary clusters, March; fruits glossy black. HM Pl. 122. Original habitat is Asia Minor, but naturalized throughout the Mediterranean region and the Balkan Peninsula, north to the Ticino River and S. Tyrol. In cultivation for centuries. z9 # ⌀

'Aurea'. Leaves yellowish.

'Angustifolia'. Leaves narrowly lanceolate, about 3 to 7 cm long, but only 6–20 mm wide (= f. *salicifolia* Hort.).

'Crispa'. Leaf margin distinctly undulate (= 'Undulata').

f. *salicifolia* see: **'Angustifolia'**

'Undulata' see: **'Crispa'**

11 forms in the Nikita Botanic Garden, Jalta, Crimea, are mentioned by Th. Kagaida in Voloshin, M. P.: Review of Laurel forms in Crimea; in Trudi Gosud. Nikitsk. Bot. Sad. **29**, 85–94, 1959 (in Russian).

"Laurus tinus" see: **Viburnum tinus**

The Laurel tree is frequently cultivated in Central Europe where it is grown in tubs and overwintered in the greenhouse. Not particular as to soil type where it is winter hardy.

Lit. Hegi: Flora von Mittel-Europa, IV, 1, 11–12 (of historical significance) ● Sprenger, C.: Neue Notizen vom Lorbeerbaum; in Mitt. DDG 1914, 214–217; 1916, 99–103.

LAVANDULA L. — Lavender — LABIATAE

Aromatic, gray pubescent perennials or shrubs, shoots 4 sided; leaves decussate, usually simple and narrow, rarely pinnatisect or pinnate; flowers blue or violet, in 10–20 cm long false spikes; calyx short cupulate, with 5 teeth; corolla bilabiate, upper lip further 2 lobed, lower lip 3 lobed, with nearly equal, outspread lobes; stamens 4, inside the corolla tube; fruits composed of 4 nutlets. — About 28 species from the Canary Islands to India.

Key to the species described
(from E. Guinea)

● Leaves dentate to bi-pinnatisect;

     1. Leaves crenate-dentate to pectinate-pinnatisect; flowers 6–10 in each cyme:
                 L. dentata
     2. Leaves usually bi-pinnatisect; 2 flowers in each cyme:
                 L. multifida

●● Leaves entire;

   + Upper bracts oblong-obovate, much longer than the flowers and the other bracts and without flowers in the axil;

     1. Upper bracts white or purple, leaves tomentose:
                 L. stoechas
     2. Upper bracts green, leaves short pubescent:
                 L. viridis

   ++ All bracts similar, not longer than the flowers and all with flowers in the axil;

     1. Bracts oval-rhombic, cuspidate or long acuminate; bracteoles tiny or absent:
                 L. angustifolia
     2. Bracts linear or lanceolate; bracteoles 2–5 mm, linear or bristle-like;
       ✕ Leaves quite short and densely white tomentose when young, gray-green and less tomentose when mature; calyx with 13 veins:
                 L. latifolia
       ✕✕ Leaves densely and persistent white woolly-tomentose; calyx with 8 veins:
                 L. lantana

Lavandula angustifolia Mill. True Lavender. Evergreen subshrub, 20–60 cm high, shoots numerous, erect, 4 sided, gray; leaves densely crowded, linear to narrowly lanceolate, obtuse, sessile, entire, 2–6 cm long, margins somewhat involuted, stellate pubescent on both sides when young; inflorescences 15–20 cm long, flowers light blue to violet, white tomentose exterior, calyx 5 mm long, gray-violet, July–September. HM 3186; HF 1772 (= *L. officinalis* Chaix ex Vill.; *L. spica* L.; *L. vera* DC.). Mediterranean region, SW. Europe. Of economic use for essential oil. Includes a large number of cultivars (listed at the end of the genus). z6 #

ssp. **pyrenaica** (DC.) Guinea. Bracts exserted far past the calyx (those of the species shorter), calyx 6–7 mm long, pubescent only on the venation (= *L. pyrenaica* DC.). E. Pyrenees and NE. Spain. z9 #

L. dentata L. Shrub, 0.3–1 m high, young shoots 4 sided, leaves and shoots dark gray-green tomentose; leaves linear-oblong to lanceolate, crenate to pectinate-

pinnatisect, 1.5–3.5 cm long, gray-green above, gray tomentose beneath; inflorescence 2.5–5 cm long, upper bracts 8–15 mm long, flower stalk 7–25 cm long, July–August. BM 400. S. and E. Spain, Balearic Islands. 1597. z10 #

L. lantana Boiss. Evergreen shrub, 30–50 cm high, shoots and leaves always white woolly-tomentose; leaves 3.5–5 cm long, oblong-lanceolate to linear spathulate; inflorescence 4 to 10 cm long, flower stalk 25–40 cm long, 4 sided, flowers 8–10 mm long, lilac, exterior pubescent, July–August. S. Spain, on dry calcareous slopes. More aromatic than the other species. z9 #

L. latifolia (L. f.) Med. About 40 to 90 cm high; lower leaves on the branch base more or less clustered in a rosette, leaves oblong to lanceolate or spathulate, 3–4 cm long, to 1 cm wide, often with the margins not involuted, both sides gray; inflorescence long stalked, often also branched, with awl-shaped, greenish prophylls (!), July to September, blue-violet. HM 3186; Bai 3: 18. SW. Europe, western Mediterranean region. z9 #

L. multifida L. Shrub, 0.5–1 m high, gray-tomentose, frequently also with long, straight hairs; leaves usually bi-pinnatisect, green, sparsely pubescent; false spikes 2 to 7 cm long, bracts 2–7 mm long, cordate-ovate, long acuminate, cymes only with 2 flowers, corolla 12 mm long, blue-violet. W. Mediterranean region and S. Portugal. z10 #

L. officinalis see: **L. angustifolia**

L. pedunculata see: **L. stoechas** ssp. **pedunculata**

L. pyrenaica see: **L. angustifolia** ssp. **pyrenaica**

L. spica see: **L. angustifolia**

L. stoechas L. Shrub, 0.5–1 m high in its habitat; leaves linear to oblong-lanceolate, 1–4 cm long, entire, usually gray tomentose; inflorescence usually 2–3 cm long, fertile bracts rhombic-cordate, 4 to 8 mm long, tomentose, the uppermost bracts oblong-obovate, 1–5 cm long, usually purple with flowers in the axil, summer flowering, cymes with 6–10 flowers, these dark purple, corolla 6–8 mm long. BS 2: 537. Mediterranean region, Portugal. z9 Plate 81. # ☉

var. *albiflora* see: var. **leucantha**

var. **leucantha** Gingins de Lassaraz. Bracts and flowers white (= var. *albiflora* Bean). Discovered in 1925 in the East Pyrenees near Villefranche. #

ssp. **pedunculata** (Mill.) Samp. ex Rozeira. Quite variable, differing primarily from the type in the shorter, wider false spikes and much longer stalk, upper bracts longer than the calyx, June–August (= *L. pedunculata* Cav.). Atlantic Islands, Spain, Portugal, N. Africa, S. Balkan Peninsula and Asia Minor. z10 # ☉

L. vera see: **L. angustifolia**

L. viridis L'Hér. Like *L. stoechas,* but the shoots and leaves short pubescent; stalk of the inflorescence as long or longer than the false spike, upper bracts light green, 0.8–2 cm long, flowers white. SE. Spain, S. Portugal. z9 #

Cultivars

In 1937, Miss D. A. Chaytor established that plants of *Lavandula angustifolia* and *L. latifolia* in cultivation were part hybrids between both; therefore the name × *intermedia* Lois. An investigation by K. J. W. Hensen determined that this was the parent of the cultivars 'Dutch', Hidcote Giant' and 'Old English', while all the others probably belong to *L. angustifolia*. The × *intermedia* cultivars can be distinguished from the *L. angustifolia* cultivars by a few characteristics resembling *L. latifolia*: the foliate leaves are sometimes more than 1 cm wide, the false spikes are sometimes branched, and the leaves subtending the flowers are narrower (2–3 mm) than those of *L. angustifolia* (4 to 5 mm). In addition, the first flowers of the × *intermedia* types appear about 2 weeks later than those of the *angustifolia* types.

'Alba'. Flowering plants to 50 cm high; false spikes 5–7 cm, with 1–2 widely spaced cymes, the secondary spikes well developed, calyx green, rather woolly, corolla reddish-white, late July. Presumably also from *L.* × *intermedia*. #

'Bowles Variety' (Bowles, before 1923). Flowering plants 50–60 cm high, foliage to 40 cm high; spikes 4–5 cm long, with 6–10 flowers in a whorl, calyx and corolla light violet-gray. Not an attractive color. Fig. 136. #

'Compacta' (USA, before 1901). Flowers reaching to 50 cm high, foliage 25–30 cm high; the largest leaves linear, gray-green, to 5 cm long, 3 mm wide; spikes 5–7 cm long, with 6–10 flowers in a whorl, calyx violet, corolla lilac. #

'Dutch'. Flowering plants 75–90 cm high, foliage 35–40 cm high; the larger leaves linear-lanceolate, gray-green, to 7 cm long, 9 mm wide; spikes 7–9 cm long, dense, with 6 flowers in a whorl, calyx lilac, corolla light violet (= *L. spica* of Dutch gardens; *L. vera* Hort.; 'Dutch Lavender' in England. Fig. 136. #

'Dwarf Blue' (before 1911). Flowering plants about 50 cm high, foliage to 30 cm; larger leaves lanceolate, 6 mm wide; spikes 4–5 cm long, frequently somewhat loose, 10 flowers in a whorl, calyx violet, corolla lilac. Commendable cultivar. Fig. 136. # ⊕

'Folgate' (England, before 1933). Flowering plants to 60 cm high, foliage to 35–40 cm; the largest leaves linear, to 4 cm long, 3 mm wide; spikes 7–8 cm long, basal portion loose, 10–14 flowers in a whorl, calyx gray-lilac, corolla light violet. Good bloomer. # ⊕

'Gigantea' see: **'Grappenhall'**

'Grappenhall' (England). Flowering plants to 75 cm high, foliage to 40 cm; spikes branched, rather loose, the lowest whorl at a wider distance from the others, calyx light bluish-purple, thin woolly, corolla lavender-purple, late July (= 'Gigantea'). # ⊕

'Hidcote' (Major Lawrence Johnston, Hidcote, before 1950). Flowering plants 45–50 cm high, foliage 25–30 cm high, the larger leaves lanceolate, to 4 cm wide, gray; spikes 6–7 cm long, rather dense, with 10–18 flowers in a whorl, calyx deep lilac, corolla somewhat lighter. A first class cultivar! Fig. 136. # ⊕

'Hidcote Giant' (L. Johnston, Hidcote, before 1963). Flowering plants 70–80 cm high, loosely foliate to 35 cm; larger leaves

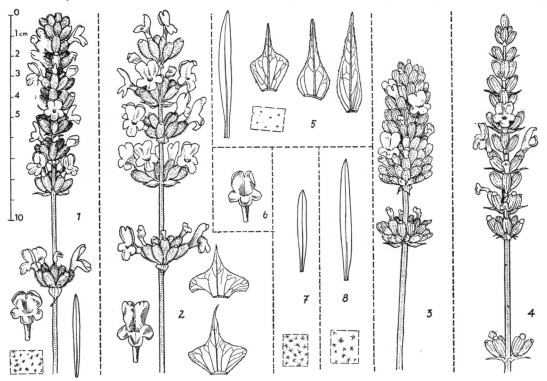

Fig. 136. **Lavandula.** *L. angustifolia* forms: 1. 'Hidcote', 2. 'Munstead'; *L. intermedia* forms: 3. 'Hidcote Giant', 4. 'Dutch', 5. 'Old English', 6. 'Bowles', 7. 'Nana Atropurpurea', 8. 'Dwarf Blue' (from K. J. W. Hensen, altered)

lanceolate to spathulate, to 6 cm long, 1 cm wide, green; spikes about 6 cm long, very dense, with 18–30 flowers in a whorl, calyx blue-purple, woolly, corolla lavender-purple, flowers very abundantly, July. A good cultivar, very aromatic, with a branched inflorescence, this often not erect. Fig. 136. # ✧.

'Hidcote Pink' (L. Johnston, Hidcote, before 1962). Flowering plants to 55 cm high, foliage to 35 cm; larger leaves linear, about 4 cm long, 3 mm wide, gray-green; spikes 5–7 cm long, loose on the basal portion, 12–18 flowers in a whorl, calyx light gray-green, corolla pale lilac-pink. #

'Loddon Blue' (Carlile, before 1963). Flowering plants about 50 cm high, foliage only 25–35 cm; larger leaves linear to lanceolate, gray, to 4.5 cm long, 4 mm wide; spikes 6–7 cm long, dense, with 10–18 flowers in a whorl, calyx and corolla purple-lilac. Somewhat lighter than 'Hidcote'. #

'Loddon Pink' (Carlile 1950). Flowering plants about 60 cm high, foliage 25–30 cm high; larger leaves linear, to 5 cm long, 5 mm wide, green; spikes 5–8 cm long, loose, 10–14 flowers in a whorl, calyx pale green, corolla pale lilac-pink. Differs only slightly from 'Hidcote Pink'. #

'Middachten' (Moerheim, before 1923). Flowering plants to 45 cm high; larger leaves linear-lanceolate, to 5 cm long, 5 mm wide, green; spikes 7–8 (10) cm long, flowers the earliest, calyx and corolla dark lilac. Rarely found in the trade because it is difficult to propagate and not winter hardy. z9 #

'Munstead' (G. Jekyll, Munstead; introduced by Barr, 1916). Flowering plants 40–45 cm high, foliage 25–30 cm high; larger leaves to 4 cm long, 3 mm wide, greenish-gray; spikes 4–6 cm long, loose, 6–10 flowers in a whorl, calyx purple, base lighter, corolla blue-lilac, late June to early July (= 'Munstead Blue'; 'Munstead Dwarf'; 'Munstead Variety'; 'Nana Compacta'). A first class cultivar; has the largest flowers of all. Fig. 136. # ✧

'Nana Alba'. Flowering plants only 10–15 cm high, very compact; larger leaves linear, 3 mm wide; spikes 3 cm long, calyx pale green, corolla white.

'Nana Atropurpurea' (before 1923). Flowering plants to 45 cm high, foliage to about 30 cm high; larger leaves lanceolate, to 4 cm long, 6 mm wide, greenish-gray; spikes 5–6 cm long, looser at the base, with 6 to 10 (14) flowers in a whorl, calyx dark lilac, lighter beneath, fewer flowers per spike and with wider leaves, greener. Fig. 136. #

'Nana Compacta' see: 'Munstead'

'Old English'. Flowering plants 100–115 cm high, foliage to 50 cm; larger leaves 6–7 cm long, to 9 mm wide; flowering shoots often covered with side shoots from the base up, these all terminating in a flower spike, spikes 5–7 cm long, dense, 12–14 flowers in a whorl, calyx lilac, corolla light violet (= L. spica Hort. Angl.). Fig. 136. #

'Rosea' (Holland, before 1949). Flowering plants to 50 cm high, foliage to 30–35 cm; largest leaves linear-lanceolate, to 5 cm long, 5 mm wide; spikes 5–7 cm long, calyx pale green, corolla pale lilac-pink. #

'Summerland Supreme' (N. May, Res. Stat. Summerland, B.C., Canada, before 1961). Flowering plants to 35 cm high, foliage to 20 cm; larger leaves linear, to 5 cm long, 3 mm wide, green-gray; spikes 3–4 cm long, dense, with 10 flowers in a whorl, calyx and corolla purple-lilac. #

'Twickel Purple' (before 1961). Flowering plants 35 cm high, rather wide; spikes 10 cm long, loose, calyx woolly, purple, as is the corolla, July. #

For sunny areas and fertile, calcareous soils; the plants can be sheared to form particularly attractive low border hedges.

Lit. De Wolf, G. P.: Notes on cultivated Labiates. 5. *Lavandula;* in Baileya 3, 47–57, 1955 ● Chaytor, D. A.: A taxonomic study of the genus *Lavandula;* in Jour. Linn. Soc. Lond. (Bot.) 51, 153–204, 1937 ● Heeger, E. F.: Handbuch des Arznei- und Gewürzpflanzenanbaues, 441–450; Berlin 1956 ● Faure, J., & B. Vercier: Les Lavandes; in Bull. Techn. d'Inform. 88, 161–174, Paris 1954 ● Barber, E.: Les lavandes et l'apiculture dans le Sud-est de la France; in Ann. Abeille 1963, 85–159 ● Hensen, K. J. W.: Het *Lavandula*-sortiment; in Groen 1974, 184–190.

LAVATERA L. — Tree Mallow — MALVACEAE

Herbaceous or soft wooded shrubs, many only annual or biennial; shoots pubescent or tomentose, leaves alternate, lobed or angular, occasionally maple-like; flowers axillary, solitary or paired or in clusters, with an involucre, 2 or 3 involucral parts connate, at least in the bud stage (all distinct on *Malva*), petals 5, colorful, but rarely yellow, stamen filament column split into many distinct filaments at the apex, stigmas 5; fruit dehiscing into numerous segments, each with 1 seed. — 25 species, most in the Mediterranean region, from the Canary Islands to Asia Minor, NW. Himalayas, N. Asia, E. Siberia, Australia and USA (California).

Lavatera arborea L. Small deciduous shrub of tree-like appearance, 1.5–3 m high, erect, with a 3–5 cm thick stem, branches outspread; leaves 5–7 lobed, 7–20 cm long and wide, the lobes unevenly crenate on the margins, base cordate, both sides densely soft pubescent, petiole long; flowers in axillary clusters, very numerous, pale purple-red, base darker veined, 5 cm wide, July–August. S. Europe, along the Atlantic coast. z9 ✧

'Variegata'. Leaves white variegated. More frequently encountered in cultivation than the species. z9

L. maritima Gouan. Evergreen shrub, 0.5–1.5 m high, shoots densely white woolly at first, but later becoming glabrous; leaves rather circular, 6–10 cm wide, but shallowly 5 lobed, both sides white stellate pubescent, but less dense above, petiole 3 cm long; flowers 1–2 in the leaf axils, petals pink with a carmine-red basal spot, obcordate, 1.5–2.5 cm long, calyx with triangular-ovate sepals, May–September; seeds black. BM 8997. Western Mediterranean region and NW. Africa. z9 ✧

× LEDODENDRON F. de Vos — ERICACEAE*

Hybrids between *Ledum* and *Rhododendron*. As far as is known this plant has only been developed in the USA and Holland. Only the following hybrid is being propagated:

'Brilliant' (*Ledum glandulosum* × *Rhododendron* 'Elizabeth'). Low, evergreen shrub, broader than high (the oldest plant in California reached 55 cm high and 90 cm wide in 5 years); leaves aromatic when crushed, thin, elliptic, acute on both ends, with a small mucro, 2–5 cm long, 1.5–2 cm wide, margins tiny glandular ciliate, bright green above, dull, much lighter and more or less scattered glandular pubescent beneath, particularly on the venation, midrib more or less brown, petiole short; flowers usually 4 in terminal umbels, corolla tubular-campanulate, 2–3 cm long, 2.5 cm wide, bright red (similar to 'Elizabeth'), 5 sepals, ovate, exterior glandular (like the stalk and the ovary), style 4–5 cm long. Developed before 1962 in the Leonard L. Brooks & Son Nursery of Modesto, California, but introduced by the Harold Greer Nursery of Eugene, Oregon USA. # ⊕

The same cross was successfully achieved at the Proefstation voor de Boomkwekerij in Boskoop, Holland; a few seedlings were produced under glass, grew to 15 cm high, flowered orange-pink, but then all died.

Donald W. Paden of Urbana, Illinois, USA crossed *Ledum glandulosum* with *Rhododendron yakusimanum* 'Mist Maiden'.

Lit. De Vos, F.: in Americ. Rhod. Soc. Quart. Bull. **16**, 272, 1962 ● Paden, D. W.: A promising innovation; in Americ. Rhod. Soc. Quart. Bull. 28, 222, 1974.

LEDUM L. — ERICACEAE

Low, densely branched evergreen shrubs; leaves very aromatic, alternate, short petioled, entire, margins often involuted, tomentose or glandular beneath; flowers small, white with a 5 part calyx, 5 part corolla, with 5 or 10 stamens; anthers opening by 2 holes at the apex; ovaries 5 locular, styles filamentous, with a 5 lobed stigma; fruit a 5 chambered, oblong, many seeded capsule. — About 10 species in the cooler regions of the Northern Hemisphere.

Ledum columbianum Piper. Erect shrub, scarcely 1 m high, shoots more or less densely pubescent and occasionally also glandular; leaves elliptic-oblong, 3–6 cm long, margins distinctly involuted, green and rough above, underside more or less whitish and finely pubescent between the glandular spots; petals white, 5–6 mm long, stamen filaments usually 5–7; fruit capsule oblong, 5–6 mm long, 3 mm wide. West coast of Canada and USA, Oregon to California, in swamps and bogs. z7 Fig. 137.

L. californicum see: **L. glandulosum**

Fig. 137. **Ledum.** From left: *L. groenlandicum; L. glandulosum; L. columbianum* (from Abrams)

L. canadense see: **L. groenlandicum**

L. glandulosum Nutt. Shrub, 0.5–1 (2) m high in its habitat, shoots finely pubescent; leaves oval, 2–5 cm long, dark green above, blue-green and very glandular-scaly beneath (not tomentose!), margins not involuted; flowers white, stamens 10, June–August. BM 7610; MS 405 (= *L. californicum* Kellogg). W. Canada to California. 1894. z7 Fig. 137. # ☼

L. groenlandicum Gunn. Shrub, to 1 m high, erect, young shoots brown tomentose; leaves elliptic-oblong, 3–5 cm long, dense red-brown tomentose beneath (!), dark green beneath the tomentum, midrib not distinguishable (!); flowers in about 5 cm wide inflorescences, white, with 5–8 (!) stamen filaments, May–June. BB 2742; GSP 395; BS 2: 12 (= *L. latifolium* Jacq.; *L. canadense* Lodd.). Northern N. America, Greenland. 1763. Its habitat is the cold peat bogs and swamps in the northern latitudes; the leaves were once used as a tea substitute (Labrador Tea) in the region, but only during hard times. Young leaves aromatic, not the mature ones. z2 Fig. 137. # ☼

'Compactum'. Lower growing; leaves somewhat shorter and wider. Plate 81.

L. hypoleucum see: **L. palustre** var. **dilatatum**

L. latifolium see: **L. groenlandicum**

L. nipponicum see: **L. palustra** var. **dilatatum**

L. palustre L. Wild Rosemary. Shrub, erect or with ascending branches, to 1 m high; leaves linear-lanceolate, 2–4 cm long, midrib normally distinctly visible (!), rusty woolly beneath; flowers in terminal corymbs, corolla 1.5 cm wide, stamens 7–10 (!), May–June. HF 2033; BB 2741. N. Europe, N. Asia; in the high moors and Pine moors. 1762. Formerly used in Norway in the brewing of beer to make the drink more intoxicating. z2 Plate 86. # ☼

var. **decumbens** Ait. Procumbent, densely branched; leaves linear, only 1–1.5 cm long, Arctic N. America and NE. Asia. 1762.

var. **dilatatum** Wahlb. Leaves wider, 2–7 cm long, not as involuted, often only the midrib brownish tomentose, the rest of the underside white tomentose. MJ 774; GC 140: JRHS 71: 107 (= *L. hypoleucum* Kom.; *L. nipponicum* Hort.). N. Europe to Japan, Sitka. 1902. ☼

'Roseum'. Flowers soft pink. Rare. ☼

For moist, sandy-bog soil, likes semishade; also likes peat; won't tolerate alkalinity. Easily forced into bloom in March.

LEIOPHYLLUM Hedw. f. — Sand Myrtle — ERICACEAE

Monotypic genus; evergreen, densely branched, usually a procumbent shrub; leaves opposite or alternate, densely crowded, small, oblong, entire; flowers in terminal, umbellate clusters; calyx and corolla 5 parted; stamens 10, anthers without an appendage, opening from a split; style filamentous; fruit a 2 to 5 valved capsule with numerous seeds. — N. America.

Leiophyllum buxifolium (Bergius) Ell. Shrub, 5–30 (50) cm high, but usually low or procumbent; leaves oblong, 3–8 mm long, dark green above, glossy, lighter beneath; flowers in terminal corymbs, white to light pink, May–June. NBB 3: 12; BB 2754. New Jersey to Florida, USA. 1736. z6 Plate 81; Fig. 141. # ☼

var. **hugeri** (Small) Schneid. More cushion-form habit, but growing to 20 cm high; leaves usually alternate, longer than those of the type; flowers pink. BM 6752 (as *L. buxifolium*). New Jersey to the Carolinas. 1884. More commonly cultivated than the species. # ☼

var. **procumbens** (Loud.). Gray. Shoots usually appressed to the ground, procumbent-ascending; leaves usually opposite, 4–8 mm long (= *L. lyonii* Sweet). Mountains of N. Carolina and Tennessee. 1912. # ☼

L. lyonii see: **L. buxifolium** var. **procumbens**

Winter hardy dwarf shrub for sandy-clay, acid soil in sunny to somewhat shady locations. Very good rock garden plant.

LEITNERIA Chapm. — Corkwood — LEITNERIACEAE

Monotypic genus; deciduous, dioecious shrub or small tree; leaves alternate, petiolate, entire, with pinnate venation, lacking stipules; flowers dioecious, not very conspicuous, catkin-like; male flowers without a perianth, with 3–12 stamens, female with a scale-form perianth, ovaries unilocular with an ovule; fruit an oblong, compressed drupe, with a thin outer shell and a hard inner shell. — N. America.

Leitneria floridana Chapm. Shrub, stoloniferous, or a small tree, scarcely over 5 m high, bark thick, gray, wood very light, broad crowned, young shoots soft pubescent;

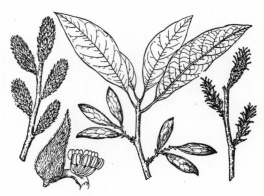

Fig. 138. *Leitneria floridana* (from Sargent)

leaves narrowly elliptic, 7–15 cm long, tapering to both ends, pubescent above at first, gray tomentose beneath; flowers in axillary, erect, about 3 cm long catkins, appearing before the leaves, March; fruit chestnut-brown, about 1.5 cm long. SS 330. Eastern USA. Winter

hardy, but lacks any particular ornamental value. z5 Fig. 138.

Occurring in swamps in its habitat; best cultivated in a cool, moist soil; easily propagated from stolons.

LEONOTIS (Pers.) R. Br. — Lion's Ear — LABIATAE

Perennials, subshrubs or shrubs; leaves opposite, petiolate, ovate, flowers in very dense, axillary whorls, orange-red to yellow, occasionally white, calyx with 8–10 veins, corolla tubular and bowed with a bilabiate limb, upper lip long and concave, pubescent exterior, lower lip bowed sharply downward, with 3 nearly equal tips, stamens 4, style 2 lobed. — About 40 species in tropical and S. Africa, 1 generally found throughout the tropics.

Leonotis leonurus (L.) R. Br. Lion's Ear. Shrub, 1–2 m

high (depending on the climate), shoots 4 sided, pubescent, erect; leaves oblong to lanceolate, 5–10 cm long, obtuse, coarsely serrate, tapering to the base, soft pubescent; flowers in many superimposed whorls in the apical leaf axils, corolla more than 3 times as long as the calyx, to 6 cm long, red-yellow or orange-red, upper lip large, lower lip small, October to December. BM 478. S. Africa. 1712. Very meritomous ornamental. z10 Plate 39. ⊕

Easily cultivated in milder regions.

LEPTODERMIS Wall. — RUBIACEAE

Deciduous shrubs; leaves opposite, entire; flowers often sessile, on the current year's shoots; corolla tubular, limb 5 lobed, stamens 5; pistil 5 rayed at the apex. — About 16 species in Himalayas, Japan and China.

Leptodermis kumaonensis Parker. Shrub, to 1 m high, usually smaller, shoots thin, reddish, glabrous, older shoots with exfoliating bark; leaves oval-lanceolate to broadly elliptic, 3.5–6 cm long, 1 cm wide, with 6–7 vein pairs, soft pubescent on both sides, dark green above; flowers 3–5 in sessile, axillary clusters, corolla tubular, 12 mm long, white to light pink, becoming darker as spent flowers, July. Central Himalayas. 1923. (Hillier.) z9 ⊕

L. lanceolata Wall. Shrub, to 1.5 m, twigs steeply erect; leaves lanceolate, quite variable in size, 2–10 cm long, 0.6–3 cm wide, long acuminate, distinctly veined, more or less pubescent beneath; flowers in terminal and axillary, stalked clusters, June–October, corolla white, funnelform, 1–1.5 cm long. N. India. 1842. z9 Fig. 139.

L. oblonga Bge. Deciduous, well branched shrub, to about 1 m high, young shoots reddish, pubescent; leaves opposite, oval-oblong, 1.5 to 2.5 cm long, tapering to the base, rough above, soft pubescent beneath; flowers few in axillary clusters, July–September, corolla tubular, violet-red, 1.5–2 cm long, pubescent inside and out (= *Hamiltonia oblonga* Franch.). N. China. 1905. Hardy. z6 Fig. 139. ⊕

L. pilosa Diels. Shrub, 1.5–2 m high, young shoots soft pubescent; leaves gray-green, ovate, 1.5–3 cm long, entire, acute, cuneate at the base, pubescent on both sides; flowers in axillary clusters at the ends of the current year's shoots, grouped into small panicles,

corolla lilac, pubescent outside, 12 mm long, funnelform, limb 5 lobed, July to September. China, Yunnan Province. 1905. z9 ⊕

L. purdomii Hutchins. Shrub, 1–2 m high, shoots very slender, pubescent at first; leaves clustered at the nodes, linear, 6–12 mm long, glabrous, obtuse, limb involuted; flowers clustered at the ends of short shoots, corolla slender tubular, 12 mm long, pink, limb 5 lobed, August to September. N. China. 1914. Hardy z6 ⊕

In its habitat grows in dry, gravelly sites in full sun which it also requires in cultivation. Flowers resemble those of many *Daphne* spp. or *Syringa persica*. *L. oblonga* and *L. purdomii* are quite hardy.

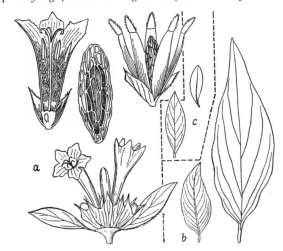

Fig. 139. **Leptodermis.**
a. and c. *L. lanceolata*; b. *L. oblonga*

LEPTOSPERMUM Forst. — MYRTACEAE

Small evergreen trees or shrubs; leaves small, alternate, entire; flowers solitary or in pairs or 3's, often on small short shoots, sessile, 1–2.5 cm wide; petals 5, outspread; stamens numerous; calyx 5 lobed; fruit a hard, persistent woody capsule. — About 50 species, mostly in Australia, Tasmania, and one in New Zealand.

Leptospermum cunninghamii see: **L. lanigerum**

L. humifusum Schauer. Evergreen, carpet-like shrub, 15–20 cm high, shoots reddish, partly upright, partly outspread; leaves elliptic to obovate, rounded, 6–9 mm long, leathery tough, dark green; flowers solitary, in the leaf axils, 12 mm wide, white, June. BS 2: Pl. 74 (= *L. rupestre* Hook. f.; *L. scoparium* f. *prostratum* Hort. non Hook. f.). Tasmania. 1930. Not to be confused with *L. scoparium* var. *eximium*, which has mucronulate leaves. z9 # ✧

L. lanigerum (Ait.) Sm. Evergreen, small tree, twigs with long, soft hairs; leaves alternate, very densely arranged, oblong to elliptic, 8–12 mm long, silvery pubescent beneath, abruptly acuminate, occasionally glossy green and glabrous above, bronze brown in fall; flowers solitary on short shoots along the branch, white, 12 mm wide, with 20–30 stamens, calyx and sepals densely white woolly, June–July (= *L. pubescens* Lam.; *L. cunninghamii* Schauer). Tasmania. 1774. (Malahide; Hillier.) z9 Plate 83. # ✧

L. pubescens see: **L. lanigerum**

L. rupestre see: **L. humifusum**

L. scoparium J. R. & G. Forst. Shrub, also a small tree in its habitat to 5 m high, densely branched and foliate, young shoots sparsely pubescent; leaves alternate, linear-oblong, 8 to 12 mm long, sharply acuminate, transparent glandular punctate, aromatic when crushed; flowers white, 12 mm long, solitary in the leaf axils, petals nearly circular, not touching, with the triangular calyx lobes between, May–June; fruit a many seeded, pea-sized, woody capsule. KF 117; BS 2: 552. New Zealand. Forms extensive stands in its habitat. 1772. z9 #

'Album Plenum'. Slender upright habit; leaves bright green; flowers white, double, about 1.5 cm wide, appearing from February to July. ✧

'Boscawenii'. Compact habit; flowers to 2.5 cm wide, buds dark pink, opening white with a pink center, flowers abundantly, May–June. Developed by A. J. Boscawen. 1912. ✧

'Chapmanii'. Upright habit, dense; leaves ovate to oval-lanceolate, 6 mm long, reddish above, green beneath; flowers pink, dark red in the center, flowers abundantly. Discovered in 1890 by J. Chapman in Dunedin, Scotland. ✧

var. **eximium** Burtt. Bushy habit, low; flowers pure white, flowers very abundantly. BM 9582. Tasmania. ✧

'Keatleyi'. Very compact habit; leaves sharply acuminate, olive-green; flowers very large, nearly 2.5 cm wide, soft pink, February–April. Quite meritorious. ✧

'Nichollsii'. Leaves bronze-brown; flowers carmine. BM 8419. 1908. ✧

var. **prostratum** Hook. f. Procumbent habit, shoots ascending; leaves ovate to rounded; flowers white. GC 132: 73: An alpine form from New Zealand, much hardier than the other forms, but less attractive.

'Red Damask' (W. E. Lammerts 1944). Narrow upright habit; leaves bright green with a reddish dorsal side and margins; flowers slightly double, an attractive carmine, 12 mm wide, flowers abundantly. GC 138: 55. Introduced into the trade in 1955 by Slieve Donard of Newcastle, Ireland. ✧

L. stellatum Cav. Shrub, 1.5–2 m high, twigs silky pubescent; leaves elliptic-oblong to more linear-lanceolate, 6–12 mm long, bright green and nearly totally glabrous, punctate with oil glands; flowers partly solitary and axillary, sessile or terminal on short shoots, white, calyx silky pubescent, May. Australia. One of the hardiest species. z9 #

All species prefer a sunny area and a light, fertile, humus soil; should only be considered for open sites in the mildest climates, otherwise for the conservatory.

Every new cultivar of *Leptospermum* should be registered with the Royal New Zealand Institute of Horticulture, P.P. Box 1368, Wellington, N.Z. before its introduction.

LESPEDEZA Mich. — Bush Clover — LEGUMINOSAE

Deciduous shrubs, subshrubs or perennials; leaves alternate, mostly trifoliate, occasionally simple, entire; stipules awl-shaped, abscising quickly; flowers in axillary racemes or heads, sometimes in terminal panicles; calyx teeth of nearly equal size or the uppermost two connate; standard petal obovate, wing claw-like, keel curved upward; pod short, ovate to elliptic, flat, single seeded, not dehiscent. — About 100 species from the Himalayas to China and Japan, in N. America and Australia; only a few of which are cultivated.

Lespedeza argyracea see: **L. sericea**

L. bicolor Turcz. Shrub, about 1.5 m high (to 2.5 m in its habitat), erect, twigs angular, somewhat pubescent at first; leaves 3 parted, leaflets broadly ovate, 2.5 cm long, somewhat emarginate at the apex, dark green above, gray-green beneath and lightly appressed pubescent; flowers violet-red or purple-pink, in 4–8 cm long racemes, these grouped into large, terminal foliate panicles, calyx pubescent, August–September. HM 1502; BC 2134. N. China to Manchuria, Japan. 1856. Hardy. Ⓕ Japan; used as a ground stabilizer in reforestation. z5 Plate 84; Fig. 140. ✧

var. *alba* see: **L. japonica**

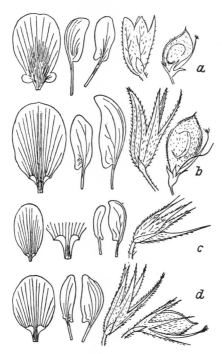

Fig. 140. **Lespedeza.** a. *L. bicolor;* b. *L. maximowiczii;* c.
L. thunbergii; d. *L. sericea* (from Schneider)

L. buergeri Miq. Similar to *L. thunbergii,* but the leaflets elliptic and more pubescent; flowers purple to white, in short racemes, July to September. SH 2: 70k; MJ 1249. Japan. z6

var. *praecox* see: **L. maximowiczii**

L. cuneata see: **L. sericea**

L. cyrtobotrya Miq. Shrub, about 1.5 m high, glabrous or occasionally somewhat pubescent, somewhat like *L. bicolor* in appearance but less attractive; leaves of the long shoots long petioled and elliptic-oblong, 2.5–3 cm long, apex round to somewhat emarginate, middle leaflet long stalked, leaves of the side shoots much smaller, densely crowded, elliptic, emarginate, 8–18 mm long, all short petioled; flowers purple, 8 mm long, in dense, axillary, only 1.5 cm long clusters or racemes, August; pods elliptic, 6 mm long. MJ 1247; KO 253 (= *L. bicolor* sensu Harms non Turcz.). Japan, Korea. 1899. z6 ⊕

L. davidii Franch. Shrub, to 3 m high in its habitat, shoots thick, erect, shaggy silvery pubescent; leaves elliptic, leaflets to 3 cm long, elliptic, obtuse, shaggy pubescent beneath; flowers purple-red, in axillary racemes, calyx and pod shaggy pubescent, September. China. About 1900. z8 Plate 84. ∅ ⊕

L. dubia see: **Campylotropis falconeri**

L. formosa see: **L. thunbergii**

L. japonica Bailey. Very similar to *L. thunbergii* (perhaps

only a white flowering form), growth upright; leaves conspicuously light green, broadly elliptic, obtuse to emarginate, rather glabrous; flowers white to reddish-white, in 3–12 cm long racemes, standard obtuse; ovaries pubescent (= *L. sieboldii* var. *albiflora* Schneid.; *L. bicolor* var. *alba* Bean). Japan. 1900. z6 ⊕

L. macrocarpa see: **Campylotropis macrocarpa**

L. maximowiczii Schneid. To 4 m high in its habitat, shoots cylindrical, finely pubescent at first; leaflets oval-elliptic to ovate, 2.5–5 cm long, acute, base round to broadly cuneate, silky pubescent beneath; flowers purple, in 3 to 8 cm long racemes, calyx teeth awn-like; July–August, pod 1–1.5 cm long. SH 2: 70f, 71h–i; Nakai l.c. 34 (= *L. buergeri* var. *praecox* Nakai). Korea. 1907. Hardy. z5 Fig. 140.

L. sericea (Thunb.) Miq. non Benth. Subshrub, to 70 cm high; leaves crowded, rather short petioled, leaflets linear-oblong to linear, 8–18 mm long, short stalked, strigose pubescent beneath, also above, but shorter pubescent; flowers white, nearly sessile, 6 mm long, in dense, axillary clusters or short, about 2.5 cm long racemes, calyx with 5 awl-shaped teeth, August–September. MJ 1251 (= *L. cuneata* [Dum.-Cours.] G. Don; *L. argyracea* S. & Z.). China, Japan. Ⓕ Japan and S. USA, as a pioneer species in reforestation. z6 Fig. 140.

L. sieboldii see: **L. japonica** and **L. thunbergii**

L. thunbergii (DC.) Nakai. Shrub, to 2 m high and wide, often only a subshrub in cultivation which freezes back to the ground every year (but then resprouts, reaching 1.5 m each summer), twigs long nodding, channeled, finely pubescent at first; flowers purple-pink, in small racemes, but these grouped into 60–80 cm long, terminal panicles, September to October. BC 2135; FS 1888; BM 6602 (= *L. sieboldii* Miq.; *L. formosa* Koehne; *Desmodium penduliforum* Oudemans). China, Japan. z6 Fig. 140. ⊕

Most very attractive fall bloomers; require full sun in light, dry soils (in heavy or moist soil the plant continues growing too long and blooms too late).

Lit. Nakai, T.: *Lespedeza* of Japan and Korea; in Bull. Forest Exp. Sta. Chosen **6**, 1–101, 1927.

LEUCADENDRON Berg. — Silver Tree — PROTACEAE

Evergreen trees and shrubs; leaves alternate, entire, glabrous or silvery silky pubescent; flowers dioecious, in cone-like heads; male flowers usually many, each solitary with a bract enlarged after flowering, the inflorescence often surrounded by a wide, hull-like, colorful bract, anthers oblong to linear, sessile; female flowers with a deeply incised perianth, lobes reflexed. — 70 species in S. Africa.

Leucadendron argenteum R. Br. Silver Tree. Evergreen, densely foliate tree, 5–8 cm high; leaves sessile, oblong-lanceolate, 5–15 cm long, 1.2–3 cm wide, long acuminate, entire, base rounded, both sides covered with tightly appressed, long silky hairs; flowers in terminal, 2.5–3 cm wide heads, yellowish, surrounded by the silvery involucral leaves, these much longer than the flower heads, July to October; fruit a woody cone, like *Cedrus* in form and size, with a silvery silky exterior. RPr 92–93. S. Africa; found wild only on Table Mountain near Cape Town. 1693. z10 Plate 38, 82. # ∅

Cultivated in many botanic gardens but usually short lived; easily propagated from seed, but the young plants are very sensitive, easily diseased and difficult to transplant.

Lit. under *Protea*.

LEUCOPOGON R. Br. — EPACRIDACEAE

Evergreen shrubs, very closely related to (and often included in) *Cyathodes*, erect or prostrate habit; leaves alternate; flowers small, partly solitary, but usually in slender, occasionally dense, axillary or terminal spikes, each flower with a subtending leaf and 2 small bracts near the calyx; calyx 5 lobed, corolla tubular at the base, the 5 erect or outspread lobes with a distinct white pubescent tuft; fruit a berrylike drupe. — About 150 species, most in Australia, otherwise in Malaysia and New Caledonia.

Leucopogon colensoi see: **Cyathodes colensoi**

L. forsteri see: **Cyathodes juniperina**

L. fraseri see: **Cyathodes fraseri**

LEUCOSPERMUM R. Br. — PROTEACEAE

Evergreen shrubs; leaves alternate, very densely crowded, very often dentate at the apex (the similar genera *Protea* and *Leucadendron* entire), tough-leathery,

usually pubescent; flowers many in small terminal heads, these solitary and surrounded by a ring of bracts; perianth (of the individual flowers) tubular, with a limb of 2 or 4 segments, brightly colored, pistil markedly exserted, often 5–7 cm long, colorful and attractive; fruit a whitish, hard shelled nut. — About 40 species, all in S. Africa, rare in cultivation. Flowering shoots are occasionally seen in floral bouquets.

Leucospermum cordifolium (Salisb. ex Knight) Fourcade. Shrub, about 1 m high, shoots finely pubescent; leaves ovate to oblong or elliptic, 2.5–7 cm long, rounded at the apex and entire or with 2–4 teeth, base cordate, rather blue-green; flower heads solitary, 5–7 (10) cm wide, carmine-pink, perianth 6–8 mm long, glabrous, pistil 5 cm long, pink with a yellow stigma. RPr 51 (= *L. nutans* R. Br.). S. Africa; Cape Province. z10 Plate 82. # ✧

L. lineare R. Br. Low shrub, growth erect or with procumbent twigs, shoots glabrous; leaves loosely arranged, linear, tapering to the base, sessile, 7–10 cm long, 3–6 mm wide; flowers in conical, 8 cm high and wide heads, mostly solitary, terminal on a 3 cm long stalk, perianth normally greenish-yellow, occasionally dark red, August–September. RPr 53. S. Africa; Paarl and French Hoek. 1774. z10 Plate 82. # ✧

L. nutans see: **L. cordifolium**

L. reflexum Buek. ex Meisn. Tall shrub, to 3 m, twigs usually erect, bluish-gray woolly; leaves directed upward, oblong to oblanceolate, 2–3 cm long, 6–10 mm wide, becoming smaller toward the shoot tip, leaf apex either entire or with 3 teeth; flowers in 10 cm long, 5 cm wide heads, solitary or in terminal pairs, carmine, style carmine with a yellow stigma, eventually curved vertically downward. RPr 53. S. Africa. z10 # ✧

Lit. Rourke, J. P.: Taxonomic studies on *Leucospermum*; Suppl. Vol. 8 from Jour. South Afric. Bot. 1954. — See also *Protea*.

LEUCOTHOE D. Don — ERICACEAE

Evergreen or deciduous shrubs; leaves alternate, short petioled, normally serrate; flowers in axillary or terminal racemes or panicles; calyx 5 parted, imbricate; corolla ovate to tubular, white, stamens 10; anthers with 2 or 4 short or long awn (bristle) tips; fruit a 5 loculed capsule; locules with numerous, small seeds. — About 40 species in N. and S. America, 4 in E. Asia.

Leucothoe axillaris (Lam.) Don. Evergreen shrub, about 1.5 m high, shoots long, arching outward, finely pubescent when young; leaves elliptic-lanceolate, 5–10 cm long, abruptly (not gradually!) short acuminate, sparsely serrate, very glossy above, lighter beneath; flowers in 2–7 cm long, axillary racemes, white, corolla narrowly ovate, 8 mm long, sepals narrowly ovate, May–June. NBB 3: 17; BB 2763 (= *Andromeda axillaris* Michx.). Southeast USA. 1765. z6 Plate 85; Fig. 141. # ∅ ✧

L. catesbaei see: **L. fontanesiana**

L. davisiae Torr. Narrowly upright, evergreen shrub, 0.3–1 m high; leaves ovate-oblong, 2–7 cm long, finely serrate; flowers in erect (!) racemes at the shoot tips, corolla urceolate, white, 8 mm long, sepals 5 to 6, somewhat glandular ciliate, June. BS 2: 558; BM 6247. Southeast USA. 1853. z9 Plate 85; Fig. 141. # ✧

L. fontanesiana (Steud.) Sleumer. Evergreen shrub, to 2 m high (often only 1 m in cultivation), fast growing and luxuriant, always branched to the ground, shoots bowed downward, reddish and pubescent when young; leaves ovate-lanceolate (!), 6–15 cm long, long acuminate, appressed ciliate and serrate, glossy green above, lighter and finer brownish punctate beneath, glabrous; flowers in 4–6 cm long, axillary racemes, clustered at the branch tips, corolla nearly cylindrical, 8 mm long, white, anthers without an appendage, sepals triangular-ovate, April–

May. BB 2764; NF 9: 72 (= *L. catesbaei* Gray). Southeast USA. Foliage often a good red in late fall and during the entire winter. z5 Plate 83; Fig. 141 # ⌀ ✧

'Multicolor' see: **'Rainbow'**

'Rainbow'. Leaves pink marbled and speckled, with white spots, coppery and yellow new growth. Horticulture 1960: 349 (= 'Multicolor'). Developed by Girard Bros., Geneva, Ohio, USA in 1949. ⌀

'Rolissonii' (Bean). Leaves narrower, 5–10 cm long, only 1–1.5 cm wide. Apparently selected in Kew Gardens. 1914. ⌀ ✧

'Trivar'. Stronger growing; leaves red with cream-yellow and green, somewhat larger than 'Rainbow', but paler. Developed in 1947 in De Wilde's Rhodo-Lake Nursery, Bridgeton, New Jersey, USA. ⌀

L. grayana Maxim. Usually deciduous, erect shrub, about 0.5(0.7) m high, slow growing, stiff, shoots reddish; leaves reddish on the new growth, broadly elliptic to oval-oblong, 5–8 cm long, short acuminate, base round, entire and ciliate, often rough above, fall foliage scarlet-red; flowers urceolate-campanulate, 4 to 5 mm long, ivory-white, spent flowers reddish, in terminal, 7–10 cm long, nodding racemes, May–June. MJ 740. N. and Central Japan. 1890. Hardy. z6 Plate 88. ⌀ ✧

L. griffithiana Clarke. Evergreen shrub, about 1 m high, twigs widely nodding; leaves lanceolate, somewhat caudate tipped, base tapered to nearly round; margins finely serrate to nearly entire, 7–12 cm long; inflorescences in elongated, axillary racemes, anthers drawn out to an awn tipped apex, June (= *Pieris cavaleriei* Lévl. & Vaniot). Bhutan, China, Yunnan and Kweichow Provinces. 1921. z9? Plate 88. # ⌀

L. keiskei Miq. Evergreen, scarcely over 30 cm high (!), twigs usually procumbent, red when young; leaves acutely ovate, 4–8 cm long, leathery, somewhat bristly pubescent beneath; flowers in pendulous racemes, axillary at the branch tips, corolla cylindrical, about 1.5 cm long (the largest flowers of the winter hardy species), limb tips recurved, white, July. MJ 741; KO ill. 389. Japan. 1915. Very attractive. z7 # ⌀ ✧

L. populifolia (Lam.) Dipp. Evergreen shrub, 1 m high, twigs widely arching; leaves ovate-lanceolate, entire (!) or indistinctly serrate, 5–10 cm long; flowers in laterally arranged corymbs, corolla tubular, white, about 15 mm long, June. DL 1: 234 (= *Andromeda populifolia* Lam.; *A. lucida* Jacq. non Lam.; *A. acuminata* Ait.). Southeast USA. 1765. z9 Plate 85; Fig. 141. # ⌀ ✧

L. racemosa (L.) Gray. Deciduous shrub, 1–2.5 m high, twigs erect or outspread, one-sided racemes on the apical portion of the previous year's shoots, corolla tubular, 9 mm long, white, anthers with 4 appendages, May–June; capsule nearly globose, seeds not winged. BS 2: 216; Hy WF 133; GSP 402 (= *Lyonia racemosa* D. Don; *Andromeda racemosa* L.). Eastern N. America. 1736. z6 Plate 85; Fig. 141. ✧

L. recurva (Buckl.) Gray. Deciduous shrub, 1(2.5) m high, shoots bowed outward, somewhat pubescent when young; leaves elliptic to lanceolate, serrate, 4–10 cm long, venation pubescent beneath, fall foliage a fiery red; flowers in 2–10 cm long, downward curving racemes, corolla like that of *L. racemosa*, anthers with only 2 appendages, April–June. BC 2141. Southeast N. America; in dry areas. 1880. Hardy. z6 Plate 83, 85. ⌀ ✧

	Height	Leaves	Flowers	Corolla
● **Evergreen Species**				
axillaris	to 1.5 m	5–12 cm long, abruptly acuminate	2.5–4 cm long racemes with 8–30 flowers	narrowly ovate
fontanesia	to 1.5 m	6–15 cm long, long acuminate	4–6 cm long racemes with 20–60 flowers	nearly cylindrical
'Rollisonii'	to 1 m	5–10 cm long, narrow-lanceolate	4–6 cm long racemes	nearly cylindrical
davisiae	0.3–1 m	2–7 cm long, oval-oblong	in erect racemes	urceolate
griffithiana	to 1 m	8–12 cm long, lanceolate	in axillary racemes, 5–12 cm long	
keiskei	0.2–0.3 m	4–8 cm long, acutely ovate	in pendulous racemes, July	tubular
populifolia	1–2 m	5–10 cm long, oval-lanceolate	in lateral corymbs	tubular
●● **Deciduous Species**				
racemosa	1–3 m erect	2–7 cm long, acutely elliptic	racemes straight, 3–8 cm long	tubular
recurva	1–3 m wide	4–10 cm long, acutely oval-elliptic	racemes curved 2–10 cm long	tubular

Nearly all the above species are winter hardy, but relatively little known; the evergreen species prefer a moist, humus soil and do best in a wooded setting; the deciduous species are best in dry, sunny areas.

Lit. Sleumer, H.: Studien über die Gattung *Leucothoe* D. Don; in Bot. Jb. **78**, 435–480, 1959 ● Green, P. S.: *Leucothoe fontanesiana*; in Arnoldia **23**, 93 to 99, 1963 (contrasted with *L. axillaris* by illustration) ● Ingram, J.: Studies in the cultivated Ericaceae; 1. *Leucothoe*. Baileya **9**, 57–66, 1961 (with ills.).

Fig. 141. **Leucothoe.** a. *L. davisiae;* b. *L. axillaris;* c. *L. populifolia;* d. *L. racemosa;* e. *L. fontanesiana.* **Leiophyllum.** f. *L. buxifolium* (from Drude, Guimpel, Jacquin, Dippel and B. M.)

LEYCESTERIA Wall. — CAPRIFOLIACEAE

Deciduous shrubs with hollow branches; leaves opposite, simple, petiolate; flowers in terminal and axillary, nodding spikes surrounded by large, colorful bracts; calyx lobes 5, small, 1–2 often more prominent, persistent; corolla tube funnelform, base bulging on one side, corolla limb usually 5 lobed; stamens 5, ovaries 5 locular; fruit a many seeded berry.—6 species in the Himalayan Mts. to W. China.

Leycesteria crocothyrsos Airy-Shaw. A shrub in its habitat, 1–2 m high, young twigs hollow, thin, sparsely glandular-pubescent; leaves ovate, long acuminate, 5–15 cm long, somewhat serrate, base rounded, dull blue-green above, finely pubescent beneath with distinctly reticulate venation, stipules kidney-shaped, 1–2 cm wide; flowers in nodding, 12–18 cm long racemes, grouped in whorls of 6, corolla yellow, pubescent exterior, calyx green, April; fruit a small (gooseberry-like) berry with a persistent calyx. BM 9422; BS 2: 563. Assam, in the mountains at 2000 m. 1928. z9 ✤

L. formosa Wall. Shrub, to 2 m high, narrowly upright, multi-branched from the base, young shoots bluish pruinose; leaves broadly cordate-ovate, often caudate tipped, 5–17 cm long, entire or dentate, finely pubescent at first; flowers in 3–10 cm long, pendulous spikes, corolla reddish-white to violet, narrowly campanulate, to 2 cm long, surrounded by purple-violet bracts, August–September. BM 3699; DL 1: 183. Himalayas. 1928. z7 Plate 39, 84, 88. ✤

'Rosea'. Flowers pink. ✤

Only the latter species somewhat hardy and ornamental; often freeze back to the base, but resprout well in spring. Not particular but prefer a moist, humus soil.

Lit. Airy-Shaw, H. K.: A revision of the genus *Leycesteria*; Kew Bull. 1932, 241–245.

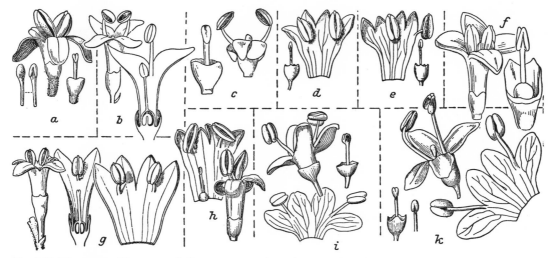

Fig. 142. **Ligustrum.** Flowers and flower parts, enlarged. a. *L. vulgare;* b. *L. japonicum;* c. *L. lucidum;* d. *L. delavayanum;* e. *L. henryi;* f. *L. strongylophyllum;* g. *L. massalongianum;* h. *L. ovalifolium;* i. *L. sinense;* k. *L. sinense* var. *stauntonii* (from Koehne, Shirasawa, Sargent, Hort. Then. and B. M.)

LIGUSTRUM L. — Privet — OLEACEAE

Evergreen or deciduous shrubs or small trees; leaves opposite, entire, short petioled, simple; flowers hermaphroditic, white, usually small, but often in large, terminal panicles, similar to *Syringa;* corolla funnelform, with a short or long tube, corolla lobes 4, stamens 2; fruit a 1–4 seeded, normally black berry. — About 50 species, most in E. Asia, from China and Japan to the Malaysian Archipelago, only 1 species in N. Europe to N. Africa.

Division of the Genus

Sect. I. **Euligustrum** Rehd.
　　L. vulgare

Sect. II. **Ibota** Koehne
　Ser. 1. **Robusta** Mansfeld
　　　L. chenaultii, compactum, delavayanum, henryi, japonicum, lucidum, massalongianum, strongylophyllum

　Ser. 2. **Sinensia** Mansfeld
　　　L. acutissimum, amurense, ibolium, ibota, indicum, obtusifolium, ovalifolium, purpusii, quihoui, sinense, tschonoskii, vicaryi

Ligustrum acuminatum see: **L. tschonoskii**

L. acutissimum Koehne. Deciduous shrub, resembling *L. obtusifolium,* broad growing, to 3 m high, twigs pubescent; leaves elliptic to lanceolate, 1–7 cm long, acuminate, base cuneate, dark green above, light green with a pubescent midrib beneath, often also sparsely pubescent on the blade; flowers in 2–5 cm long, narrow panicles, June; fruits oval, 8–9 mm long, blue-black. ICS 4681. Central China. 1900. z6 Plate 85; Fig. 143.

L. amurense Carr. Deciduous, erect shrub, to 3 m high, twigs pubescent; leaves elliptic-oblong, 3–6 cm long, dull green, finely ciliate, only the midrib finely pubescent beneath; flowers in loose, floccose pubescent, 4–5 cm long panicles, corolla about 7–9 mm long, June–July; fruit rounded, 6–8 mm thick, black, slightly pruinose. MD 1904: 72; RH 1861: 352. N. China. 1860. Extraordinarily frost hardy. z3 Plate 85; Fig. 143, 147.

L. angustifolium see: **L. massalongianum**

L. brachystachyum see: **L. quihoui**

Fig. 143. **Ligustrum.** Flowers and flower parts, enlarged. a. *L. tschonoskii;* b. *L. ibota;* c. *L. acutissimum;* d. *L. amurense;* e. *L. obtusifolium;* f. *L. obtusifolium* var. *regelianum* (from Koehne)

Fig. 144. **Ligustrum.** a. *L. compactum;* b. *L. indicum;* c. *L. quihoui* (from B. M. and Schneider)

L. chenaultii Hickel. Semi-evergreen in its habitat, a shrub to 6 m tall, upright, but with divaricate side branches, young shoots glabrous, glossy brown-red, with numerous white lenticels; leaves lanceolate, 15–20 cm long on long shoots, somewhat boat-shaped (navicular), light green above, dull, both sides reddish on the new growth, leaf drop often early without a noticeable fall color; flowers in 15–20 cm long and equally wide panicles, June; fruit violet-blue. BD 1925: 52. SW. China. 1908. z8 Plate 85. ∅ ✤

L. cilatum see: **L. ibota** and **L. tschonoskii**

L. compactum Brandis. Shrub, semi-evergreen, very closely related to *L. chenaultii,* also a tree in its habitat, to 7(10) m high, young shoots pubescent, likewise the leaf petiole; leaves ovate-lanceolate, long acuminate, 7–17 cm long, glabrous; flowers yellowish-white, in 15–20 cm long and equally wide panicles, anthers pink, June–July; fruits black, pruinose. MD 1915: Pl. 3, 1 (as *L. yunnanense*); DL 1: 77; ICS 4676 (= *L. yunnanense* L. Henry; *L. simonii* Carr.; *L. longifolium* Carr.). SW. China, Himalayas. 1877. z9 Fig. 144. ∅ ✤

L. confusum Dcne. Deciduous shrub, but semi-evergreen or evergreen and tree-like in very mild regions, young shoots soft pubescent; leaves lanceolate, 4–9 cm long, 1.2–2.5 cm wide, light glossy green, totally glabrous, petiole channeled above; flowers in pubescent panicles on the previous year's shoots, white, the individual flowers 4 mm wide, nearly sessile, calyx cupulate, glabrous, lobes flat-triangular, filaments white, anthers pink, June–July; fruits black, to 12 mm long, 8 mm wide, bluish pruinose. Himalaya; E. Nepal to Bhutan and in the Khasi Hills of India. 1919. Very attractive, but also very sensitive to frost. z9 ✤ ⚬

L. coriaceum see: **L. japonicum 'Rotundifolium'**

L. delavayanum Hariot. Evergreen shrub, to 2 m high, widely branched, young shoots finely pubescent; leaves ovate-elliptic, occasionally obovate, 1–3 cm long, acute, glossy green above, lighter beneath, midrib pubescent; flowers in 3 to 5 cm long, cylindrical panicles, pubescent and foliate at the base, corolla tube 5 mm long, anthers violet, June. BMns 60; ICS 4679 (= *L. prattii* Koehne; *L. ionandrum* Diels). W. China, Yunnan. 1890. z8 Fig. 142. # ∅

L. excelsum aureum see: **L. lucidum 'Aureovariegatum'**

L. henryi Hemsl. Evergreen shrub, very graceful when young, to 3 m high in its habitat, young shoots very pubescent; leaves oval to oval-lanceolate, acuminate, 2–3 cm long, glossy black-green above; flowers white, fragrant, in terminal, 5–15 cm long panicles, corolla 6 mm long, calyx and stalk glabrous; fruit black, 8 mm long. BS 2: Pl. 79. Central China. 1901. Very attractive. z7 Plate 85; Fig. 142. # ∅

L. × ibolium Coe (= *L. obtusifolium* × *L. ovalifolium*). Semi-evergreen shrub, similar to *L. ovalifolium,* but with pubescent twigs, also the inflorescence and leaf underside, anthers about as long as the corolla lobes. Developed in the USA about 1910 at the Elm City Nursery, Connecticut. Frequently used as a hedging plant. z5 ∅

L. ibota S. & Z. Deciduous shrub, to 2 m high, usually lower, divaricate, twigs somewhat pubescent, long shoots usually glabrous; leaves rhombic-ovate to elliptic-oblong, 1.5 to 5 cm long, finely ciliate, dull green above, lighter beneath with a pubescent midrib; flowers 4–8 in 1–1.5 cm long, nearly capitate inflorescences, corolla whitish, to 8 mm long, June. MD 1904: 73; DL 1: 82 (= *L. ciliatum* Sieb. ex Bl.). Japan. 1870. Hardy, but of no particular ornamental value. z6 Fig. 143.

Fig. 145. **Ligustrum.** a. *L. lucidum;* b. *L. massalongianum;* c. *L. obtusifolium* (from B. M., Dippel)

L. indicum (Lour.) Merrill. Evergreen shrub, a tree in its habitat, young shoots gray-yellow, rough, tomentose; leaves ovate-oblong, 4–8 cm long, acuminate, glossy green above, yellowish-green beneath; flowers in terminal and axillary, 10–18 cm long and equally wide panicles, small, fragrant, May–June; fruits blue-black. DL 1: 72 (= *L. nepalense* Wall.; *L. spicatum* Hamilt.). Himalaya, Indochina. Should be overwintered in a cool greenhouse! z9 Fig. 144. # ∅ ✛

L. insulare see: **L. vulgare 'Insulense'**

L. insulense see: **L. vulgare 'Insulense'**

L. ionandrum see: **L. delavayana**

L. italicum see: **L. vulgare var. italicum**

L. japonicum Thunb. Evergreen shrub, 1 to 2 m high, in its habitat to 3(6) m high, twigs finely pubescent (at least when young), later glabrous, with lenticels; leaves broadly ovate to oval-oblong, 4–10 cm long, short acuminate or acute to obtuse, margins and midrib often reddish, with 4–5 distinct vein pairs; flowers in 6–15 cm long, pyramidal panicles, corolla lobes somewhat shorter than the tube, July to September. BS 2: 571; MJ 660; NT 1: 386; BM 7519. Japan, Korea. 1845. z8 Fig. 142, 146. # ✛

'Revolutum'. Growth narrowly upright, like 'Rotundifolium', scarcely over 1 m high, shoots very densely foliate; leaves only half as large as those of 'Rotundifolium', narrower, hemispherically recurved. (Minier et al, Nurseries in Angers, France). #

'Rotundifolium'. Low shrub, to 2 m high only in exceptional cases, twigs stiff, erect, internodes very short; leaves broadly ovate to nearly rounded, 3–6 cm long, obtuse to emarginate at the apex, leathery tough; flowers in crowded, 5–10 cm long clusters. BM 7519; DL 1: 79 (= *L. coriaceum* Nois.). From horticulture, Japan. 1860. z7 Plate 86. # ∅

'Variegatum'. Leaves white variegated on the margin and speckled. In English gardens.

L. longifolium see: **L. compactum** and **L. massalongianum**

L. lucidum Ait. f. Evergreen shrub, a tree in its habitat and other mild climates, to 10 m, young shoots glabrous (!), outspread, with lenticels; leaves ovate, long acuminate, 8–12 cm long, glossy dark green above, lighter beneath, with 6–8 (!) distinct veins on both sides; flowers in 10 to 20 cm long and equally wide panicles, corolla white, lobes as long as the tube, August–September; fruit blue-black, oblong, 1 cm long. DL 1: 78; MD 1915: Pl. 1, 4; BM 2566; ICS 4675 (= *L. japonicum macrophyllum* Hort.; *L. magnoliifolium* Hort.). China, Korea, Japan. 1794. z8–9 Fig. 142, 145. # ∅ ✛

'Alivonii'. Young shoots totally finely pubescent; leaves ovate-lanceolate, 7–17 cm long, long acuminate, often yellow variegated when young, not so tough and less glossy than the species; fruits black. Kew Gardens. 1886. (Possibly belongs elsewhere?) ∅

'Aureovariegatum'. Leaves yellow variegated, but not very attractive (= *L. excelsum aureum* Hort.). 1900.

'Excelsum Superbum'. Strong growing; leaves dark yellow and cream-white speckled and margined. Very attractive. In English gardens (Hillier). ∅

'Tricolor'. Also strong growing; leaves smaller than those of the species, yellow and white speckled, also pink when young. Around 1900. z9 Plate 78. ∅

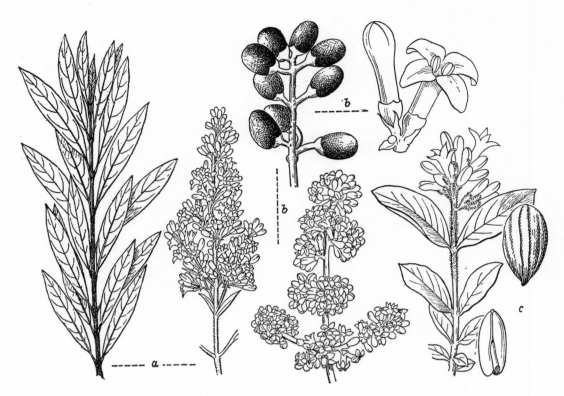

Fig. 146. **Ligustrum**. a. *L. massalongianum;* b. *L. japonicum;* c. *L. obtusifolium*

L. macrocarpum see: **L. tschonoskii** var. **macrocarpum**

L. magnoliifolium see: **L. lucidum**

L. massalongianum Vis. Evergreen, scarcely 1 m high, upright shrub, shoots slender, papillate and very pubescent; leaves linear-lanceolate, 3–8 cm long, tapered to both ends, 1 cm wide, glabrous (abscising in late fall in cooler climates); flowers stalked, in 6 to 8 cm long, many flowered, branched panicles, abundantly flowering, anthers half as long as the corolla lobes, June–July; fruit ovoid, blue (= *L. angustifolium; L. rosmarinifolium; L. longifolium* Hort.). Himalayas. 1877. z9 Fig. 142, 145, 146. # ⊘

L. medium see: **L. ovalifolium**

L. nepalense see: **L. indicum**

L. obtusifolium S. & Z. Deciduous, broad growing shrub, 2–3 m high, twigs short shaggy pubescent, bowed outward; leaves elliptic to oblong or oblong ovate, some only 2 cm, some to 9 cm long, acute or obtuse, deep green, glabrous above, totally pubescent or only on the midrib beneath; flowers in 5 cm long, cylindrical, nodding panicles, corolla 8 to 10 mm long, anthers about as long as the corolla lobes, June; fruits 6 mm thick, lead-gray to black. DL 1: 80; ICS 4682; BC 1861 (= *L. ibota* S. non S. & Z.!). Japan. 1860. z3 Plate 85; Fig. 143, 145, 146.

var. **reglianum** (Koehne) Rehd. Growth broader than the species, to about 2 m high, twigs more divaricate and horizontally spreading, short, rough pubescent; leaves 2 ranked, oblong to obovate, 5–7 cm long, bright green, pubescent beneath; flowers more numerous, in short, dense panicles along the twig, June–July; fruits 4–5 mm thick, globose, black. DL 1: 83; MD 1904: 70; NT 1: 365 (= *L. regelianum* Koehne). Japan. 1885. Very popular for dense, wide hedges. Plate 87; Fig. 143. ⊘ ⚭

L. ovalifolium Hassk. Deciduous to semi-evergreen shrub, to 5 m high under favorable conditions, tightly upright, young shoots glabrous; leaves elliptic-oblong, 3–7 cm long, glossy dark green above, yellowish-green beneath; flowers in crowded, 5–10 cm long panicles, corolla yellowish-white, 8 mm long, anthers as long as the corolla lobes, July. DL 1: 84 (= *L. medium* Franch. & Sav.). Japan. 1847. One of the most commonly used hedging plants, but hardy only to z6. Fig. 142, 147. ⊘

'Argenteum'. Leaves white margined, occasionally somewhat pale on the new growth. ⊘

'Aureum'. Leaves broadly golden-yellow margined or also totally yellow (= *L. ovalifolium aureum elegans; L. ovalifolium elegantissimum; L. ovalifolium aureomarginatum* Hort.). Before 1862. Plate 86. ⊘

'Multiflorum'. Especially floriferous. GC 50: 237.

'Tricolor'. New growth pink, developed leaves yellow to white variegated.

Fig. 147. **Ligustrum**. a. *L. sinense* var. *stauntonii*; b–c. *L. ovalifolium*; d. *L. amurense*;
e. *L. vulgare* 'Insulense'; f. *L. strongyphyllum* (from B. M., Sargent, Dippel)

"Walkeri" of some nurseries is nothing more than *L. ovalifolium*, distributed from France around 1910 and still occasionally found by this name. It has nothing to do with the tropical species *L. walkeri* Dcne. (WCT 326).

L. prattii see: **L. delavayanum**

L. purpusii Hoefk. Semi-evergreen, broad growing shrub, similar to *L. quihoui*, but the twigs lighter, less reddish, finely pubescent; leaves usually elliptic, 1–3 cm long, thin leathery, acute to obtusish; flowers in dense pyramidal panicles about 15 cm long and 10 cm wide, corolla lobes less outspread, corolla yellowish-white, August. MD 1915: Pl. 5. China. 1916. z7 ∅

L. patulum see: **Syringa patula**

L. quihoui Carr. Deciduous shrub, broad and divaricately spreading, to 2 m high, twigs often long, occasionally with thorn-like short shoots, brown-red, softly pubescent; leaves elliptic to obovate, 2–5 cm long, tough, abscising very late, dark green and somewhat glossy above, glabrous, petiole finely pubescent; flowers in up to 20 cm long, loose, narrow panicles, fragrant, lateral peduncles somewhat pendulous or spreading horizontally, calyx light green, corolla pure white, anthers crosswise in arrangement, September; fruits ovate, black-purple, numerous. DL 1: 73; BM 9209; ICS 4678 (= *L. brachystachyum* Dcne.). China. 1862. Much valued for its late and abundant flowers. z7 Plate 88; Fig. 144. ∅ ✧

L. regelianum see: **L. obtusifolium** var. **regelianum**

L. rosmarinifolium see: **L. massalongianum**

L. sempervirens see: **Parasyringa sempervirens**

L. simonii see: **L. compactum**

L. sinense Lour. Deciduous shrub, bushy and upright, to 4 m tall, twigs densely gray-yellow pubescent; leaves elliptic to elliptic-oblong, 3–7 cm long, dull green above, lighter beneath, midrib pubescent; flowers in 6–10 cm long panicles, whitish, fragrant, peduncles pubescent, July; fruit globose, 4 mm thick, reddish-black, persistent. BS 2: Pl. 19; DL 1: 75; MD 1915: Pl. 1, 2; ICS 4677 (= *L. villosum* May). China 1852. Very hardy. Valued for its abundant flowers. z7 Plate 85; Fig. 142. ✧

'Multiflorum' (Bowles). Abundantly flowering selection with red-brown anthers (yellowish on the species). Before 1911.

var. **stauntonii** (DC.) Rehd. Lower growing, wider, twigs violet, soft pubescent; leaves elliptic to ovate, usually obtuse, 3–4 cm long, dark green, lighter beneath and scattered pubescent; flowers in broad, loose panicles, flowers abundantly in July–August; fruits like the species. DL 1: 76; GF 3: 213; SL 212 (= *L. stauntonii* DC.). Central China. Around 1863. Plate 88; Fig. 142, 147. ✧

L. spicatum see: **L. indicum**

L. stauntonii see: **L. sinense** var. **stauntonii**

L. strongylophyllum Hemsl. Evergreen shrub, tree-like in its habitat, twigs thin, divaricately spreading, finely gray-yellow pubescent; leaves rounded to broadly ovate, 1–2.5 cm long, thick, leathery, glossy dark green above, light green beneath; flowers in pyramidal, loose, 5–10 cm long panicles, corolla tube white, 7 mm long, July; fruits oblong. MD 1915: Pl. 6; BM 8096. Central China. 1879. Easily recognized by the very small leaves. z9 Plate 85; Fig. 142, 147. # ∅

L. tschonoskii Dcne. Deciduous shrub, upright, to 2 m high, twigs divaricate to broadly arching; leaves rhombic-ovate to more lanceolate, acuminate, 3–8 cm long, finely ciliate and pubescent above near the margins; flowers in 3–6 cm long, pubescent, panicles, short stalked, corolla tube 1 cm long, stamens somewhat exserted, anthers crosswise, June; fruit ovate, 8 mm long, black, glossy. MD 1904: 74; NT 1: 383; NK 10: Pl. 16; MJ 659 (= *L. acuminatum* Koehne; *L. ciliatum* Rehd. non Bl.; *L. medium* Hort.). Japan. 1888. z6 Plate 85; Fig. 143.

var. **glabrescens** Koidz. Leaves smaller, quickly becoming totally glabrous. Japan. Plate 85.

var. **macrocarpum** (Koehne) Rehd. More upright growing; leaves bright green; panicles like those of the species, corolla tube somewhat smaller; fruits thicker, to 1 cm long. MD 1904: 75; ST 1: 281 (= *L. macrocarpum* Koehne). Japan. Before 1900.

L. × vicaryi Rehd. (= *L. ovalifolium* 'Aureum' × *L. vulgare*). Deciduous shrub, broad growing, compact, branches divaricate; leaves golden-yellow, especially in a sunny area, later more green, somewhat wider than *L. ovalifolium*, more oval; inflorescence short pubescent (= *L. ibota aureum vicaryi* Beckett). Developed about 1920 by Beckett in Aldenham House, England. ⌀

L. villosum see: **L. sinense**

L. vulgare L. Common Privet. Deciduous shrub, dense, to 5 m high, ascending, young twigs finely pubescent; leaves obovate-oblong to lanceolate, 3–6 cm long, glabrous; flowers in 4–6 cm long, pubescent panicles, fragrant, corolla tube white, June to July; fruits pea-sized, glossy black. HM Pl. 213; HW 3: 122. Europe, N. Africa, Asia Minor. Cultivated for centuries. z5 Fig. 142. ⌀

Includes numerous cultivars:

Differing in habit
  a) dense and low:
      'Densiflorum', 'Lodense', 'Rupicolum'
  b) weeping:
      'Pendulum'

Different leaves
  a) colored:
      'Argenteovariegatum' (white variegated)
      'Aureovariegatum' (yellow variegated)
      'Aureum' (yellow)
      'Glaucum' (gray-green, white margined)
  b) other forms or lengths of persistence:
      'Buxifolium' (1–3 cm long)
      'Microphyllum' (8–15 mm long)
      'Laurifolium' (laurel-like)
      'Insulense' (narrow lanceolate)
      'Atrovirens' (dark green, persisting late)
      var. *italicum* (nearly evergreen, light)
      'Triphyllum' (n whorls of 3)

Differing in flower color:
      'Auriflorum' (light yellow)

Differing in fruit color:
      'Chlorocarpum' (green fruited)
      'Densiflorum' (greenish-yellow, leaves nearly evergreen)
      'Leucocarpum' (whitish)
      'Xanthocarpum' (yellow)

'Argenteovariegatum'. Leaves normal, white speckled, England. Around 1770.

'Atrovirens' (Späth). Growth narrowly upright, lateral shoots short and outspread; leaves broadly elliptic to ovate, 4–6 cm long, deep green, somewhat metallic, deep brown in winter and persistent. Introduced by Späth in 1880. Not identical to 'Italicum', as often suggested by Rehder. ⌀

'Aureovariegatum'. Leaves yellow speckled. England. 1770.

'Aureum'. Entire leaf blade awash with yellow. Introduced in Germany, 1884.

'Auriflorum'. Leaves often more rounded; flowers dirty yellow (= f. *lutescens*; fl. *luteo* Hort.). Cultivated in England.

'Buxifolium'. Semi-evergreen; leaves ovate, 1 to 3 cm long. From England.

'Chlorocarpum'. Fruits greenish-yellow. Discovered in England in 1838.

'Densiflorum'. Compact habit, erect; flowers very densely crowded; fruits greenish-yellow.

'Glaucum'. Leaves 5–6 cm long on long shoots, otherwise much smaller, effectively gray-green above from the very thick cuticle, very narrowly white limbed. Developed by Späth in Berlin, 1883. ⌀

'Insulense'. Young shoots soft pubescent at first; leaves narrowly lanceolate, 5–10 cm long, acuminate, pendulous, yellowish-green; flowers and fruits larger than those of the species (= *L. insulare* Dcne.; *L. insulense* Dcne.). Fig. 147.

var. **italicum** (Mill.) Vahl. Leaves lanceolate, light green, very persistent to nearly evergreen; flowers greenish-white; fruits greenish-yellow (= f. *sempervirens* Ait.; *L. italicum* Mill.). Presumably from Italy. # ⚤

'Laurifolium'. Strong upright grower, densely branched; leaves broadly oval, laurel-like, dark violet colored in winter. ⌀

'Leucocarpum'. Fruits whitish. 1838.

'Lodense' (Jackson & Perkins). Very low and dense growing, scarcely higher than 50 cm; leaves narrowly elliptic, deep green, 3–5 cm long, bronze-brown in winter, persistent (= f. *nanum* Rehd.). Discovered about 1924 by Kohankie & Sons, Painesville, Ohio, USA; but introduced by Jackson & Perkins. ⌀

f. *lutescens* see: **'Auriflorum'**

'Microphyllum'. Leaves only 8–15 mm long, 6 to 8 mm wide. Discovered on a dry, sunny site in Vorarlberg, Austria.

f. *nanum* see: **'Lodense'**

'Pendulum'. Twigs outspread and pendulous (still cultivated?).

'Pyramidale' (Späth). Compact habit, many branched, narrowly upright, shoots stiff, lateral shoots nearly whorled. Introduced into the trade in 1893 by Späth. ⌀

'Rupicolum'. Dwarf habit; leaves narrower and tougher than those of the type. Found in dry areas of the species habitat, in Switzerland. (cultivated?)

f. *sempervirens* see: var. **italicum**

'Triphyllum'. Leaves always in whorls of 3.

'Xanthocarpum'. Fruits pure yellow. Discovered in France about 1811.

L. yunnanense see: **L. compactum**

The Asiatic, evergreen species (at least those listed here) should be somewhat protected or planted in milder regions. The deciduous species are, however, quite winter hardy. All species will tolerate regular shearing, therefore quite suitable as hedging material. As a flowering shrub only *L. sinense* should be considered.

Lit. Hoefker, H.: *Ligustrum vulgare* und seine Varietäten; in Mitt. DDG 1911, 219–226 ● Hoefker, H.: Übersicht über die Gattung *Ligustrum;* in Mitt. DDG 1915, 51–66; addendum in the report for 1930, 31–35 ● Mansfield, R.: Vorarbeiten Monogr. *Ligustrum;* in Bot. Jahrb. **59**, Supplement 132, 19–75, 1924.

LIMONIA

Limonia aurantifolia see: **Citrus aurantifolia**

LIMONIASTRUM Moench — PLUMBAGINACEAE

A shrubby genus, very closely related to *Limonium;* leaves alternate, with leaf sheaths and salt glands; corolla tube about as long as the lobes, styles connate on the basal half. — 10 species in the Mediterranean region, but only the following in culture.

Limoniastrum monopetalum (L.) Boiss. Densely branched and foliate shrub, 0.5–1.2 m high; leaves oblanceolate to linear-spathulate, 2–3(8) cm long, 0.5(1.5) cm wide, drawn out to a broad stem-clasping sheath at the base, blue-green, fleshy; flowers pink-lilac, in 5–10 cm long spikes, corolla 1 to 2 cm wide, drying to violet. Mediterranean coast of Portugal, Spain, France, and Italy in salt marshes and on dunes. z9 Plate 89 # ✿

LINDERA Thunb. — Spicebush — LAURACEAE

Deciduous or evergreen, aromatic trees or shrubs; leaves alternate, simple to three lobed, venation pinnate or three veined at the base; flowers dioecious, 4–6 in sessile or stalked umbels; perianth short tubular, deeply 6 parted, male flowers with 9 stamens in 3 rings; female flowers with 6–9 staminodes and globose or ovate ovaries with short or filamentous styles; fruit a globose or ovate, berrylike drupe. — About 100 species, most in E. Asia, only 2 in America.

Lindera benzoin (L.) Bl. Deciduous shrub, broad growing, 3–6 m high, bark gray-brown; leaves broadly elliptic, thin, acute, 7–12 cm long, glabrous when young, later pubescent, bright green above, lighter beneath, golden-yellow in fall, very aromatic; flowers greenish-yellow, sessile, 2–5 on the previous year's wood, 5 mm wide, March to April; fruits ellipsoid, 1 cm long, scarlet. BB 1656; GSP 144 (= *Benzoin aestivale* Nees; *B. odoriferum* Nees). Southeastern USA. 1683. Hardy! z5 Plate 88; Fig. 148. ✿

'Xanthocarpa'. Fruits yellow.

L. cercidifolia Hemsl. Deciduous tree, bark gray, young shoots gray, glabrous; leaves rounded, with an abrupt, short, obtuse apex, 4–10 cm long, occasionally 3–5 lobed, base cordate, 3 veined, glossy green above, bluish beneath, leathery tough, petiole 2–4 cm long; flowers yellow, in dense, terminal and axillary clusters; fruits red, about 8 mm long. ICS 1725; BS 2: Pl. 80; BMns 492. China; Hupeh, Yunnan, Szechwan Provinces; E. Tibet; at altitudes from 1500–3000 m. 1907. z8?

L. glauca Bl. Large, deciduous shrub; leaves oblong to elliptic, often also ovate or obovate, acute at both ends, green and glabrous above, whitish and quickly becoming glabrous beneath, petiole 4 mm long, fall color an attractive purple, orange and red; flower stalk 13–15 mm long, thickened at the apex; fruits globose, black, 6–7 mm thick. Japan, Korea, China, Taiwan. z9 ⊘

L. membranacea see: **L. umbellata**

L. megaphylla Hemsl. Evergreen shrub, a tall tree in its

Fig. 148. *Lindera benzoin*

habitat, young shoots glabrous, reddish; leaves oblong-lanceolate, 10–20 cm long, dark green and glossy above, blue-green and dense red-brown pubescent beneath, leathery tough, with pinnate venation, petiole 2–3 cm long; flowers greenish, in dense, axillary umbels; fruits fleshy, black, 2 cm long. LF 149; ICS 1714; LWT 78 (= *Benzoin touyunense* [Lévl.] Rehd.). China; Yunnan, Szechwan, Hupeh Provinces. 1900. z9 # Ø ⚭

L. obtusiloba Bl. Deciduous shrub or tree, to 10 m, twigs glabrous, gray-yellow, occasionally reddish or yellow, with scattered lenticels; leaves broadly ovate, 6–12 cm long, obtuse, usually 3 lobed, 3 veined at the base, gray-green beneath, silky pubescent on the venation; flowers yellow, in nearly sessile, silky pubescent clusters, April; fruits globose, pea-sized, black. KO 182; MJ 597; ICS 1726; KIF 3: 14 (= *Benzoin obtusilobum* Ktze.). Japan, Korea, China. 1880. z6 Plate 88. Ø

L. praecox (S. & Z.) Bl. Deciduous shrub or also a small tree, shoots brown, papillate, glabrous, lenticels white, buds with several glossy scales; leaves ovate-elliptic, 4–9 cm long, acuminate, with pinnate venation, bluish beneath, fall color golden-yellow, petiole 1.5–2.5 cm long; flowers greenish-yellow, in small umbels, April; fruit 1.5–2 cm long, rather globose, yellow to brown. MJ 1595 (= *Benzoin praecox* S. & Z.; *Parabezoin praecox* [S. & Z.] Nakai). Japan; China; Anhui Province. 1891. z9 ⚭

L. sericea (S. & Z.) Bl. Shrub, tree-like in its habitat, young shoots red-brown, more or less glabrous; leaves on the new growth dense silky pubescent, but later totally glabrous, elliptic-oblong, 3–6 cm long, 2–4 cm wide, acuminate at both ends, gray-green beneath and white pubescent; flowers yellow, abundant, March–

April. KO 75; KIF 3: 15. Japan, Korea. Hardy. z6

L. triloba Bl. Deciduous shrub, to 6 m high; leaves very similar to *L. obtusiloba*, but the lobes deeper and acute, nearly of equal size; fruits globose, yellow-green, about 1 cm thick. MJ 1594 (= *Parabenzoin trilobum* [Bl.] Nakai; *Benzoin trilobum* S. & Z.). Japan. 1915 z7 Plate 88. Ø

L. umbellata Thunb. Deciduous shrub, to 3 m high, shoots dark red, without lenticels, smooth; leaves obovate-elliptic to oblong, 5–10 cm long, short acuminate, base cuneate, somewhat bluish beneath, midrib pubescent; flowers yellow, appearing with the leaves (!), in short stalked, pubescent, about 2.5 cm wide umbels, April–May; fruits globose, black, 8 mm thick, on 1.5–2 cm long stalks. MJ 1596; ICS 1707; KIF 3: 16 (= *Benzoin umbellatum* Ktze.; *L. membranacea* Maxim.). Japan, Central and W. China. 1892. z8 Plate 88.

Illustrations of further *Lindera* species in ICS:

L. glauca 1705	*L. angustifolia* 1711
L. communis 1706	*L. chienii* 1712
L. reflexa 1708	*L. kwangtungensis* 1713
L. erythrocarpa 1709	*L. rubronervia* 1715
L. latifolia 1710	*L. fruticosa* 1716
L. chunii 1717	*L. thomsonii* 1722
L. caudata 1719	*L. strychnifolia* 1723
L. fragrans 1720	*L. hemsleyana* 1724
L. tonkinensis 1721	

Attractive, aromatic plants which flower early, but generally of only slight garden merit; *L. obtusiloba* should be considered for its splendid fall color. Prefer a moist, acid, humus soil in a semishaded, protected area (woodland).

Lit. Lee, Sh.-Ch.: Forest Botany of China, 552 to 564 (23 species covered).

LINDLEYA Kunth—ROSACEAE

Small evergreen trees or shrubs, leaves alternate, with stipules, simple, serrate; flowers, few in clusters or solitary, with 5 petals; carpels connate forming a 5 chambered capsule; seeds winged (= *Lindleyella* Rydb.).—2 species in Mexico.

Lindleya mespiloides Kunth. Evergreen shrub or also a small tree in its habitat, densely branched, shoots gray, glabrous; leaves oblong-lanceolate to more oblanceolate, 1–2 cm long, acute to obtuse, margin tiny and obtuse glandular dentate, base cuneate, glossy green and glabrous on both sides, stipules tiny, abscising quickly; flowers solitary, terminal, 2 cm wide, petals 5, obovate-round, white, stamens about 20, July; fruit an oval, 5 sided, 8 mm long, eventually woody capsule. DRHS 1185; RH 1854; 81 (= *Lindleyella mespiloides* [Kunth] Rydb.). Mexico; in the mountains. z10 Fig. 149. #

Fig. 149. *Lindleya mespiloides*
(from Schneider, altered)

LINNAEA Gronov. — Twinflower — CAPRIFOLIACEAE

Procumbent, evergreen, dwarf shrublet, shoots filamentous; leaves opposite, rounded; flowering shoots upright; flowers paired, terminal; corolla funnelform-campanulate, rather regularly 5 lobed, stamens 4; ovaries 3 chambered; fruit single seeded, leathery, indehiscent, dry. — Monotypic, quite variable species in the cooler regions of the Northern Hemisphere.

Linnaea borealis L. Evergreen shrublet, procumbent, shoots 30–120 cm long, pubescent; leaves opposite, oval-rounded, 6–25 mm long, with some crenate teeth, ciliate; flowers paired, terminal on a long stalk, light pink, with darker markings, campanulate, 6–9 mm long, fragrant, June–August; fruits 3 mm thick, ocher-yellow, indehiscent. HF 2921; Hm Pl. 250. N. Europe, Siberia.

1762. Named for Linnaeus by Gronovius. z2 Plate 87. #

var. **americana** (Forbes) Rehd. Leaves glabrous, only ciliate at the base; corolla tube longer, 8–15 long, longer than the calyx, darker pink-red. BB 3450 (= *L. americana* Forbes). N. America.

var. **longiflora** Torr. Leaves somewhat larger than those of the type; corolla more funnelform, to 16 mm long. British Columbia to California.

L. spaethiana see: **Abelia × grandiflora**

Pretty plants, but not easy to maintain; they perform best on a moist, peat or humus area, or under conifers and birches. Winter protection by conifer branches advisable.

LINUM L. — Flax — LINACEAE

Annuals, perennials or subshrubs, occasionally shrubs; leaves sessile, blue-green, oblong-obovate, single veined or parallel veined; flowers 5 parted, sepals entire, petals clawed, longer than the sepals, stamens 5, intermittently arranged with the tooth-like staminodes, filaments connate at the base; capsule dehiscing with 10 valves, often short beaked, seeds flat. — 230 species in the temperate and subtropical zones, particularly in the Mediterranean region. Only 1 species will be mentioned here:

Linum arboreum L. Evergreen shrub, glabrous, to 1 m high; leaves spathulate, about 3–5 cm long, 3–10 mm

wide, thickish, single veined, margins cartilaginous-like, often clustered in dense rosettes; inflorescence 7–15 cm long (on plants in cultivation), in erect, few flowered panicles, sepals lanceolate, acute, 5–8 mm, petals golden-yellow, 12–18 mm long, flowers about 3 cm wide, appearing continuously from May to the end of July/August. Crete (Greece), SW. Anatolia (Turkey), Rhodes (Greece). 1788. Usually only short-lived in cultivation, but easily propagated from cuttings. z10 # ✧

Prefers a calcareous soil in warm, dry, very sunny sites.

LIPPIA see: **ALOYSIA**

LIQUIDAMBAR L. — Sweetgum — HAMAMELIDACEAE

Tall or medium-sized deciduous trees with fragrant sap, often with corky bark on the younger shoots; leaves alternate, long stalked, palmately lobed, with 3–7 lobes, serrate, stipules small; flowers unisexual, capitate, yellow, capitulae surrounded by 4 bracts; male flowers lacking a calyx and corolla, densely crowded, individual flowers not distinguishable; female flowers with a fused calyx, also lacking a corolla; ovaries of the individual flowers connate; fruit heads globose, hardened, composed of many capsules, stiff beaked by the long and persistent styles. — 6 species, one in N. America, one in SW. Asia Minor, 2 in China.

Liquidamber acerifolia see: **L. formosana**

L. formosana Hance. 20–40 m high in its habitat, straight trunked, strong branched crown, young shoots often corky winged; leaves 3 lobed (occasionally 5 lobed), 7–15 cm wide, underside often pubescent, fall foliage a gorgeous wine-red. LF 152–153; FIO 104; LWT 98 (= *L. acerifolia* Maxim.; *L. maximowiczii* Miq.). Central China, Formosa. 1884. ⓕ China. z7 Plate 92. ∅

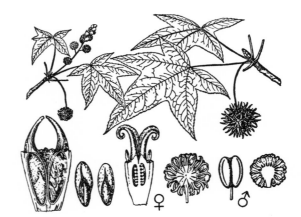

Fig. 150. *Liquidambar styraciflua*. 2 twigs above (left male flower, fruit at right); below (from left to right) fruit capsule, 2 seeds, male and female flowers and their respective inflorescences (from Sargent)

var. **monticola** Rehd. & Wils. Leaf underside always totally glabrous, leaves always 3 lobed, base often cordate, petiole longer, to 10 cm long; fruit heads smaller, to 2.5 cm wide, thorny. China; W. Hupeh, E. Szechwan Provinces. Definitely hardier than the species. z7? Plate 92. ∅

L. imberbe see: **L. orientalis**

L. maximowiczii see: **L. formosana**

L. orientalis Mill. Small tree, to 7 m high, or only a shrub, more delicately textured than *L. styraciflua*, young shoots glabrous; leaves usually 5 lobed, 5–7 cm wide, the sinuses reaching past the blade middle, lobes coarsely dentate, margins finely glandular, otherwise usually quite glabrous; flowers greenish, in globose heads, appearing with the new growth. NF 1: 173 (= *L. imberbe* Ait.). Asia Minor. 1750. z8 Plate 92. ∅

L. styraciflua L. Tall tree, to 45 m high, twigs red-brown, often with corky ridges, habit rather conical; leaves palmately lobed, with 5–7 lobes, 12–15 cm wide, the lobes triangular, finely serrate, glossy green and glabrous above, with axillary pubescent tufts on the venation beneath, fall color a beautiful carmine with yellow, green and often somewhat violet (a pure carmine-red fall color is occasionally seen; such as a specimen in the RHS Gardens, Wisley, generally considered the most beautiful in all of England); flowers greenish-yellow, in globose, 2 cm wide heads, March–May. BB 1880; GTP 217. Southeast USA; in moist regions, valley bottoms and along the seacoast. 1681. The wood contains the pleasant smelling and tasty cinnamic acid; produces a resin (*Storax liquidus*), which is considered a medicine for the common cold, also used as an additive to chewing gum. ⓕ USA. z6 Plate 89; Fig. 150. ∅

'Burgundy' (Saratoga). Early leafing; fall color appears 1 month later than the other forms and persists longer, deep red. ∅

'Festival' (Saratoga). Especially upright habit; fall color particularly light, mostly yellow or also with pink and peach tones. ∅

'Palo Alto' (Saratoga). Selection; leaves all coloring orange-red at the same time in the fall. NH 38: 234. Introduced to the trade by the Saratoga Horticultural Foundation, Saratoga, California, USA. ∅

'Pendula'. Upright growing, rather columnar, shoots pendulous.

'Rotundiloba'. Lobes not acute, rather rounded. JA 1931: 70.

'Variegata'. Leaves bright yellow marbled. Occasionally found in cultivation but not particularly ornamental.

Very attractive trees for specimen use in lawn areas, most effective in fall. Prefer a moist soil, tolerate standing water, otherwise not particular as to soil.

Lit. Samorodova-Bianki, G.: De genere *Liquidambar* L. notulae systematicae; in Botanitscheski Materialny **18**, 77–89, 1957 (in Russian and Latin) ● Harms, H.: in Mitt. DDG **44**, 21–24, 1932 ● Thomas, J. L.: *Liquidambar*; Arnoldia **21**, 59–66, 1961 (6 cultivars described).

LIRIODENDRON L. — MAGNOLIACEAE

Deciduous trees; buds covered by 2 pruinose stipules fused at the margin (also during elongation of the new growth); leaves alternate, usually 4 to 6 lobed, truncate at the apex; flowers terminal, attractive, solitary, campanulate, with 3 outspread sepals and 6 upright petals; stamens numerous, with long filaments, directed outward, linear anthers; ovaries numerous, on a spindle-form, elongated column; fruit cone-like, brown, composed of winged, single seeded achenes. — 2 species one in N. America and China respectively.

Liriodendron chinense (Hemsl.) Sarg. Tall tree, about 15 m high, bark gray; leaves usually larger than those of the North American species, much more deeply lobed (to the blade middle or deeper), normally with only one pair of basal lobes, bluish beneath with papillae; flowers smaller, petals 3 to 4 cm long, green exterior, interior yellow, May to June, filaments 5 mm long (!); fruit cones 7 to 9 cm long, fruit segments obtuse at the wing apexes. ST 52; GC 44: 429. Central China; W. Hupeh, Kiangsi Provinces, in the mountains at 1000–1500 m. 1901. z8 Fig. 151. ∅

L. tulipifera L. Tulip Tree. Large tree, fast growing, 40(60) m high, crown spreading or conical; leaves nearly rectangular in outline, 4–10 cm long and wide, base usually rounded, with 1–2 large, short acuminate lobes on either side, apexes usually truncate, bright green above, lighter beneath or bluish, petiole 5–10 cm long, coloring a good golden-yellow in fall; flowers tulip-form, 4–5 cm long, sepals oval-lanceolate, outspread, greenish-white, petals obovate-oblong, greenish-yellow outside, with a broad orange band inside near the base, erect, filaments 1 cm long, May–June; fruit cones 6–8 cm long, fruit segments with an acute wing apex. SS 13; BS 2: 35; EH 24 to 27. N. America, Massachusetts to Florida and Mississippi. 1663. ⓕ USA, W. Germany, Hungary, USSR. z5 Plate 87; Fig. 151. ∅ ✧

Includes the following cultivars:

'Aureomarginatum'. Slow growing; leaves yellow variegated on the margin. FS 2025: 2181. Other variegated forms are known. ∅

'Compactum'. Compact habit, nearly globose, densely branched.

f. *contortum* see: **'Crispum'**

'Crispum'. Leaves broader than long, occasionally to 16 cm wide and 9 cm long, deeply crenate on the apex, lobes undulate, base cuneate (= f. *contortum* Hort.). Plate 92.

'Fastigiatum'. Strictly conical upright (= f. *pyramidale* Hort.). ∅

'Integrifolium'. Juvenile form with nearly rectangular, unlobed leaves. Not unusual. ∅

'Medio-pictum'. Center yellow and cream colored, margin broad and normal green. Plate 93.

'Obtusilobum'. Leaves with one round lobe on either side of the base.

f. *pyramidale* see: **'Fastigiatum'**

For deep based sandy-clay, moist soil; a good park specimen. An important forest tree in the USA. Best transplanted in the spring because of the easily damaged, fleshy roots.

Lit. Schwerin, F.: Angeblicher Atavismus bei *Liriodendron*; in Mitt. DDG 1919, 135–142.

Fig. 151. **Liriodendron.** Left, 2 leaves of *L. tulipifera*, the large leaf is from a young plant (1 m high, collected from the wild in Pennsylvania); right, 2 leaves of *L. chinense*, the large leaf is from a young plant (Original)

LITHOCARPUS Bl. — FAGACEAE

Oak-like evergreen trees of the tropics and subtropics; buds with a few, leaflike scales; leaves leathery, entire or dentate; male flowers in erect (!), simple or branched spikes, ovaries atrophied; female flowers at the base of male spikes or in special catkins; styles 3, cylindrical, stigma only on the tip of the style; fruit a very hard shelled, acorn-like nut in a calyx cup; these with either imbricate, distinct or concentric rings on connate scales. — Around 100 species, one in western N. America, the others in S. and E. Asia and Indonesia. The species covered here are quite hardy for milder temperate climates. (i.e. London).

Lithocarpus cleistocarpus Rehd. & Wils. About a 12 m high, upright, but spreading tree in its habitat, glabrous; leaves oblong to narrowly elliptic, long acuminate, 10–20 cm long (also to 25 cm on young plants), 3–6 cm wide, entire, gray-green, quite glabrous, with 9–12 vein pairs; acorns to 2.5 cm thick, in dense clusters on stiff, 5 to 7 cm long spikes. CQ 379; FIO 117 (= *Quercus cleistocarpa* Seemen; *Quercus wilsonii* Seemen). China; W. Hupeh, Szechwan Provinces. 1901. Slow growing, but one of the most beautiful evergreen oaks. z9 # ∅

L. densiflorus (Hook. & Arn.) Rehd. Tree, to 20 m high or more in its habitat, young shoots densely white woolly; leaves elliptic to oblong, acuminate, 5–10 cm long, base round to broadly cuneate, stiff and leathery, with 12–14 vein pairs, these terminating in as many sharp teeth, with loose stellate pubescence above at first, later glabrous and glossy, densely white tomentose beneath at first, later brownish, eventually nearly totally glabrous and gray-green; male flowers in erect, slender, 5–10 cm long spikes; acorns grouped 1–2, about 2–2.5 cm long. SPa 148 to 150; BM 8695; CQ 444; HI 380 (= *Pasania densiflora* Oerst.; *Quercus densiflora* Hook. & Arn.). N. America; California, Oregon. 1865. Threatened with extinction because of commercial exploitation for its high tannin content. z9 # ∅

var. *echioides* see: var. **montanus**

var. **montanus** (Mayr) Rehd. Shrub form, leaves only 3–5 cm long, obtuse. SS 488; MS 56 (= var. *echioides* [R. Br.] Abrams; *Quercus echioides* R. Br.). Western USA; Mt. Shasta and the Siskiyou region of California and Oregon. z9 # ∅

L. edulis (Mak.) Nakai. Small tree, to 7 m high, or only a shrub, young shoots glabrous (!); leaves narrowly

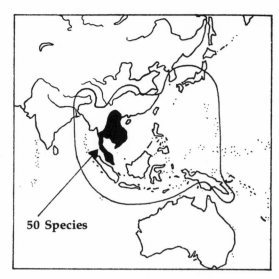

Fig. 152. Range of the genus *Lithocarpus*
(the center in black)

elliptic to oblanceolate, entire, leathery, 7–15 cm long, tapered toward the apex, obtuse, glossy yellowish-green above, with 9–11 vein pairs, dull gray-green beneath; acorns about 2.5 cm long, 8 mm thick, usually grouped 2–3 together. CQ 473 to 474; MJ 1961 (= *Quercus edulis* Mak.). Japan. 1842. z9 Plate 90. # ∅

L. glaber (Thunb.) Nakai. A tree in its habitat, to 7 m high, or only a shrub, young shoots always tomentose (!); leaves elliptic-oblong to lanceolate, widest in the middle, 7–12 cm long, 2.5–4 cm wide, leathery tough, gradually acuminate, occasionally with a few marginal teeth at the apex, glossy green and glabrous above,

whitish tomentose beneath (!) when young, later becoming glabrous, with about 8 very distinct vein pairs only on the underside; acorns sessile and pubescent, 5–12 cm long spikes, 1.5–2 cm long, ovoid, cup flat. CQ 474; MJ 1962; FIO 120 (= *Quercus glaber* Thunb.; *Quercus sieboldiana* Bl.). Japan, E. China. Very often confused with *L. edulis* and *Quercus acuta*. z6 Plate 89. # ∅

L. henryi (Seemen) Rehd. & Wils. A 12 m high tree in its habitat, rounded crown, young shoots gray-brown, with white lenticels, soft pubescent only when young; leaves elliptic-oblong, acute, 8–12 cm long, 2 to 4.5 cm wide, entire, base cuneate, glossy green above, lighter green beneath, glabrous, leathery thick; acorns usually well developed, 2 cm long, many on stiff, 15 cm long, erect spikes. CQ 488; FIO 119 (= *Quercus henryi* Seemen). China; Hupeh, Yunnan, Szechwan Provinces, in the mountains at 1200–1400 m. 1901. z9 # ∅

L. pachyphyllus (Kurz) Rehd. Small evergreen tree, with outward spreading branches (a tall tree in its habitat), young shoots soft pubescent; leaves elliptic to more lanceolate, 10–20 cm long, drawn out to a caudate apex, base cuneate, entire, leathery tough, glossy dark green above, silvery-green stellate pubescent beneath, petiole 6–12 mm; fruit spikes thick, to 15 cm long, with many lenticels, about one third of the fruits sessile with the acorn cups fused into about a 3.5 cm thick mass, acorns wider than high, nearly totally enclosed within the cup. E. Himalayas. (A 12 m tall tree grows in Cornwall at Caerhays Castle garden.) z10 # ∅

Soil requirements somewhat like those for *Quercus*; a deep, fertile soil; but this genus likes the heat, and therefore, may only be considered for the warmest climates.

Lit. Camus, A.: Les Chênes; vol. 3; Monograph of the genus *Lithocarpus*, 511–1188; Atlas vol. 3, Pl. 351 to 522 + LXXIV–XCVII, 1948–1954.

LITHODORA Griseb. — BORAGINACEAE

Evergreen or semi-evergreen shrubs or subshrubs; leaves alternate, entire, usually rough on both sides; flowers 1–10 in short, loose corymbs, never in a cincinnus (!); corolla blue or purple, funnelform; nutlets constricted just above the base, later breaking at that point leaving a persistent, cupulate appendage on the nutlet (!). — 7 species in W. and S. Europe, N. Africa, Asia Minor.

Lithodora diffusa (Lag.) Johnst. Evergreen, procumbent subshrub, 10–25 cm high, dense mat-form habit, shoots bristly pubescent; leaves linear-oblong to lanceolate, 1.5 to 2 cm long, obtuse, somewhat involuted, sessile, bristly pubescent on both sides; flowers in terminal, foliate spikes, corolla tube exterior pubescent, deep blue, somewhat reddish-violet striped, 12 mm wide, limb lobes rounded, anthers inserted, of equal length (!), May–June; nutlets striped. Bai 6: 17A (= *L. prostrata* [Loisel.] Griseb.; *Lithospermum diffusum* Lag.). S. Europe; Pyrenees, W. France. z6 # ✛

'Alba' (Ingerwersen). Leaves lighter green; flowers white. 1918. Plate 91. ✛

'Grace Ward' (H. Ward). More compact; leaves somewhat larger than those of 'Heavenly Blue'; flowers darker blue, to 2 cm wide. Developed in England about 1931. ✛

'Heavenly Blue' (D. H. Lowe). Larger than the wild type in all respects; flowers dark blue. Developed in England about 1907. Most widely distributed, but surpassed by 'Grace Ward'. ✛

L. fruticosa (L.) Griseb. Prostrate to upright subshrub, to 25 cm high; leaves elliptic to linear, very involuted, 1 to 2 cm long, rough haired on both sides; flowers in foliate, few flowered racemes, purple-blue, corolla tube with a glabrous exterior (!), anthers all inserted and of equal length, the tube limb exserted, May–June. Bai 6: 17D (= *Lithospermum fruticosum* L.). S. Europe, in the mountains. z6 1683. ✛

L. oleifolia (Lapeyr.) Griseb. Procumbent, evergreen shrub, shoots ascending, scarcely higher than 15 cm;

leaves elliptic-oblong, margins not involuted (!), clustered at the branch tips, 1–1.5 cm long, rough haired, silvery beneath, silky pubescent, green above; flowers in terminal clusters, violet (or pink to blue), corolla tube not pubescent, anthers of the same length, surpassed slightly by the corolla tube (!), May–June. BM 8994, 9559; Bai 6: 17C (= *Lithospermum oleifolium* Lapeyr.). Pyrenees Mts. (Europe) 1900. Difficult to cultivate. The flowers of this species are sometimes heterostylous. z6 # ⊕

L. prostrata see: **L. diffusa**

L. rosmarinifolia (Ten.) Johnst. Upright, evergreen shrub, 25–50 cm high; leaves lanceolate, 2.5–4 cm long, margins involuted; flowers light blue with white lines, in terminal clusters, corolla tube pubescent on the exterior, anthers all of equal length, exserted past the tube limb, December. DRHS 1192; Bai 6: 17B. Central Italy. z9 # ⊕

L. zahnii (Heldr. ex Halácsy) J. M. Johnston. Densely branched shrublet, 40 cm high, but to 90 cm wide, shoots erect, the older ones black and leafless, the younger ones densely foliate and silky pubescent; leaves linear-oblong, 2–4 cm long, 2–4 mm wide, leathery, obtuse, green above or more gray with appressed bristles, dense gray bristly beneath, margins conspicuously involuted; flowers grouped 1–3, blue or white, salver-shaped, 13 mm wide, March–April, very attractive. BMns 530 (= *Lithospermum zahnii* Heldr. ex Halácsy). S. Greece; Peloponnesus. z10 ⊕

Best cultivated in a sunny site in the rock garden, the more sensitive species are best handled in an alpine conservatory; all tolerate limestone except *L. diffusa* and particularly its cultivars.

Lit. Ingram, J.: Studies in the Cultivated Boraginaceae; 1. *Lithospermum* and related genera; in Baileya 6, 91–100, 1958.

LITHOSPERMUM L. — Gromwell — BORAGINACEAE
(including BUGLOSSOIDES Moench)

Annual plants, perennials or subshrubs; leaves alternate, sessile, entire, rough bristly pubescent or silky; flowers blue or violet (on *Buglossis*) or white or yellow, in curved spikes or racemes, occasionally heterostylous (not *Buglossis*); calyx 5 parted, corolla funnel- or salverform, tube cylindrical, straight, 5 lobed, stamens 5, inserted in the tube; nutlets not constricted at the base. — About 50 species in the Northern Hemisphere (of these about 7 belong to *Buglossoides*).

Lithospermum diffusum see: **Lithodora diffusum**

L. froebelii see: **Moltkia × intermedia 'Froebelii'**

L. fruticosa see: **Lithodora fruticosa**

L. gastonii Benth. A subshrub, upright, 20–30 cm high; leaves to 7 cm long, oval-lanceolate, with rough appressed pubescence; flowers deep blue with a white "eye", in terminal clusters, June–August; nutlets yellow. BM 5926; NF 4: 129; Bai 6: 19C (= *Buglossoides gastonii* [Benth.] Johnst.). Pyrenees Mts. Likes a calcareous soil. z6 ⊕

L. oleifolium see: **Lithodora oleifolia**

L. purpureo-coeruleum L. Subshrub with long shoots, rooting easily at the tips, flowering shoots erect, to 30 cm high; leaves lanceolate, acute, 3–4 cm long, limb involuted; flowers red at first, then blue, in a terminal pair of corymbs, June. Bai 6: 19B (= *Buglossoides purpureo-coeruleum* [L.] Johnst.). Europe. z6 ⊕

L. zahnii see: **Lithodora zahnii**

L. zollingeri A. DC. Like the previous species, but the flowering shoots only 10–20 cm high, arising from the previous year's wood (those of the above species arise from the rhizomes!); leaves ovate to oblanceolate; flowers light blue with a white "eye", in unbranched, terminal corymbs, June. Bai 6: 19A (= *Buglossoides zollingeri* [A. DC.] Johnst.). The Orient. z6 ⊕

No special cultural requirements; for sunny areas in the rock garden.

Lit. Johnston, J. M.: A survey of the genus *Lithospermum*; in Jour. Arnold Arb. 33, 299–366, 1952 ● Johnston, J. M.: Suppl. notes; in Jour. Arnold Arb. 34, 1–16, 1953.

LITSEA Lam. — LAURACEAE

Evergreen, dioecious trees; leaves alternate, occasionally nearly opposite, usually with pinnate venation; flowers in umbels, sessile or stalked, surrounded by 4–6 involucral leaves; perianth lobes 6, tube short or ovate; stamens 9–12, the first two rings of stamens not glandular; fruit a berry, often surrounded by the swollen perianth base. — About 400 species in the warmer parts of Asia (north to Korea and Japan), Australia and America. See also *Neolitsea*.

Litsea aciculata see: **Neolitsea aciculata**

L. glauca see: **Neolitsea sericea**

L. japonica (Thunb.) Juss. Evergreen tree, young shoots yellow limbed; leaves leathery tough, oblong or narrowly oblong, 7–15 cm long, 2–5 cm wide, obtuse, margins somewhat recurved, glabrous above, yellowish beneath, with 8–12 vein pairs, petiole 1.5–4 cm long, densely woolly; inflorescence short stalked, stamens 9; fruit ellipsoid, to 18 mm long, 12 mm thick, purple-blue, October–November. KO 193; MJ 1605; KIF 2: 25. Japan, S. Korea. z9 Plate 92. # ∅

LOISELEURIA Desv. — ERICACEAE

Monotypic genus of dwarf evergreen shrubs; leaves usually opposite, leathery, small, entire; flowers very small, broadly campanulate, calyx and corolla 5 parted; stamens 5, not exserted; fruit a 2–3 chambered, many seeded capsule. — The higher mountains of Central and Northern Europe, N. Asia and N. America as well as the Arctic tundra.

Loiseleuria procumbens (L.) Desv. Habit grasslike or mat-form, twigs very thin; leaves densely crowded, oval-oblong, 4–8 mm long, involuted, tough, bluish-white beneath; flowers few in clusters at the tips of the previous year's growth, in the leaf axils, corolla about 5 mm long and equally wide, pink to white, flowers abundantly, calyx deeply 5 lobed, lilac-red, April–May. HF 2028; HM Pl. 206; NBB 3: 12 (= *Azalea procumbens* L.). Habitat as mentioned above. Plate 90. # ✧

Very difficult to cultivate; does best in an acidic, humus soil, full sun, with a well drained, gravelly base and preferably covered with snow in winter.

LOMATIA R. Br. — PROTACEAE

Evergreen trees or shrubs; leaves opposite or alternate, simple or dentate or pinnate, occasionally with many forms on the same plant; flowers hermaphroditic, usually yellowish-white, in pairs, but grouped in terminal or axillary panicles or racemes; corolla tube oblique, split into 4 linear, twisted lobes at the limb. — 12 species in Australia, Tasmania, Chile.

Lomatia dentata R. Br. Evergreen shrub of medium height in cultivation; leaves nearly *Ilex*-like, elliptic to obovate, coarsely toothed nearly to the base, deep glossy green above, lighter or bluish beneath; flowers greenish-white. PFC 170. Chile. (Hillier; Wakehurst 1963.). z9

L. ferruginea R. Br. Shrub (a tree to 7 m high in its habitat), often multistemmed, narrowly upright, shoots rust-brown silky pubescent; leaves usually pinnate, the individual pinnae deeply pinnatisect, about 20 cm long, 10 cm wide (see Plate 92), dull dark green above, white tomentose beneath when young, later brown; flowers in axillary, 4–5 cm long racemes of about 12 individual flowers, these brownish-yellow with pink-red, but inconspicuous, July. PFC 169; DRHS 1201; BM 8112. Chile. 1851. z9 Plate 91, 92. # ∅

L. hirsuta (Lam.) Diels. Evergreen shrub or a small tree, 6(18) m high, young shoots slightly pubescent; leaves alternate, ovate, leathery tough, 3.5–10 cm long, 2 to 6 cm wide, obtuse, margins coarsely crenate, base cuneate or rounded, new growth brownish pubescent, later totally glabrous and glossy deep green, petiole brownish, 1–2.5 cm long; flowers light greenish-yellow, in axillary, 5–7 cm long panicles, May. Without particular ornamental value. PFC 168; BMns 335 (= *L. obliqua* [Ruiz & Pavon] R. Br.). Chile, Peru, Argentina, Ecuador. 1902. z9 Plate 93. # ∅ ✧

L. longifolia see: **L. myricoides**

L. myricoides (Gaertn.) Dorrien. Evergreen shrub, broad, 1.8–2.5 m high, young shoots angular, somewhat brown pubescent; leaves narrowly linear to oblong-lanceolate, 7–15 cm long, 6–12 mm wide, coarsely and sparsely dentate on the apical half or also nearly entire, acute or obtuse, base gradually tapering to a short petiole or sessile, glabrous on both sides; flowers cream-white or yellowish, very fragrant, in 7–15 cm long racemes, terminal or in the apical leaf axils, June–July. Very decorative. BM 7698 (= *L. longifolia* R. Br.). SE. Australia. z9 # ∅ ✧

L. obliqua see: **L. hirsuta**

L. silaifolia (Sm.) R. Br. Dwarf shrub, only 0.5–1 m high, resembles *L. tinctoria*, but with more coarsely parted leaves, young shoots finely pubescent or glabrous; leaves twice or 3 pinnate, 10–20 cm long and equally wide, the outermost segments linear–lanceolate, acute, becoming broader at the base; flowers in large panicles, cream-white, July. BM 1272. SE. Australia. z9 # ∅ ✧

L. tinctoria (Labill.) R. Br. Low shrub, only 0.5–0.7 m high, often with stolons, then developing a dense thicket, usually quite glabrous; leaves usually pinnate or somewhat bipinnate, rarely simple, 5–8 cm long, the segments linear and parallel, dark green; flowers in long, spreading racemes on the shoot tips, 10–20 cm long, light yellow, very fragrant, July–August. BM 4110. Tasmania. 1822. z9 # ∅ ✧

Can be used in the landscape only in the mildest regions; otherwise overwinter in a cool greenhouse.

LONICERA L. — Honeysuckle — CAPRIFOLIACEAE

Deciduous or evergreen, upright or climbing shrubs, shoots hollow or with a solid pith; leaves opposite, short petioled, occasionally nearly sessile, normally simple, occasionally lobed; flowers in axillary pairs on the upright growing species, stalked, with 2 large bracts and 4 smaller prophylls, flowers of the climbing species usually in 6 flowered whorls; corolla usually long or short tubular, with 5 lobes or distinctly bilabiate; stamens 5; fruit a black, red, yellow or white many-seeded berry. — About 200 species over the entire Northern Hemisphere.

Outline of the subgenera and sections (from Rehder, expanded)

● Subgenus I. **Chamaecerasus** L.
   Upright or climbing shrubs; leaves always distinctly separate; flowers in axillary pairs, occasionally in racemose clusters at the branch tips or solitary

   + Section 1. **Isoxylosteum** Rehd.
      Upright, rarely procumbent shrubs; twigs with a solid, white pith; secondary buds absent; leaves usually small, simple or folded in the bud stage; corolla tubular or campanulate, regularly 5 lobed; bracts usually leaflike; prophylls always present, usually all 4 fused into a involucre; style glabrous;

      Subsection 1. **Microstylae** Rehd.
      Stamens inserted to the middle of the corolla tube; style as long or longer than the corolla tube; ovaries 2–3 locular; fruits red or blue-black:
         L. angustifolia, myrtilloides, myrtillus, rupicola, syringantha, thibetica, tomentella

      Subsection 2. **Spinosae** Rehd.
      Stamens exserted at the mouth of the corolla tube, the pistle likewise exserted; ovaries 2–3 locular; fruit light violet to whitish:
         L. alberti, spinosa

   ++ Section 2. **Isaka** Rehd.
      Upright, rarely prostrate shrubs, twigs with a solid pith, later hollow, secondary buds often present; leaves rolled or tightly clustered in the bud stage; corolla more or less widening sack-like or bulging at the base;

      Subsection 3. **Purpurascentes** Rehd.
      Ovaries 2 locular, rarely 3 locular; corolla tube with a nearly regular limb or indistinctly bilabiate; prophylls absent or present, but not tightly enclosing the ovaries; fruits orange-yellow to scarlet-red, seldom dark blue; corolla tube pink or purple to yellowish; seeds small, 2–4 mm long:
         L. canadensis, gracilipes, microphylla, obovata, purpurascens, szechuanica, tangutica, tenuipes, utahensis

      Subsection 4. **Coeruleae** Rehd.
      Ovaries 2 locular; prophylls hull-like adnate to the ovaries forming a blue, pruinose mock fruit; corolla with a nearly regular limb; winter buds with 2 pairs of outer scales:
         L. coerulea, villosa

      Subsection 5. **Cerasina** Rehd.
      Ovaries 2 locular; corolla distinctly bilabiate, upper lip erect, with short, broad lobes; prophylls fused into a 4 lobed involucre, half as long as the ovary; fruit red, with 2 smooth, about 6 mm long seeds:
         L. cerasina

      Subsection 6. **Pileatae** Rehd.
      Calyx with a collar-like continuation at the base, this covering the prophyll limbs; bracts usually subulate; stipules absent; corolla small, nearly regular or indistinctly bilabiate; leaves evergreen to semi-evergreen (except on *L. gynochlamydea*):
         L. gynochlamydea, nitida, pileata

      Subsection 7. **Vesicariae** Komar.
      Calyx without a collar-like continuation at the base; corolla distinctly bilabiate, with reflexed, bristly hairs; involucre adnate to the calyx base and also adnate to the ovary forming a rather dry fruit:
         L. ferdinandii, vesicaria

      Subsection 8. **Chlamydocarpi** Jaub. & Spach.
      Calyx without the collar-like continuation at the base; corolla soft pubescent; involucre not adnate to the calyx; ovaries growing out of the involucre, eventually red:
         L. iberica

      Subsection 9. **Fragrantissimae** Rehd.
      Prophylls only partly connate; the ovaries about half connate; bracts narrow; corolla bilabiate; fruits red; twigs without terminal buds:
         L. fragrantissima, purpusii, standishii

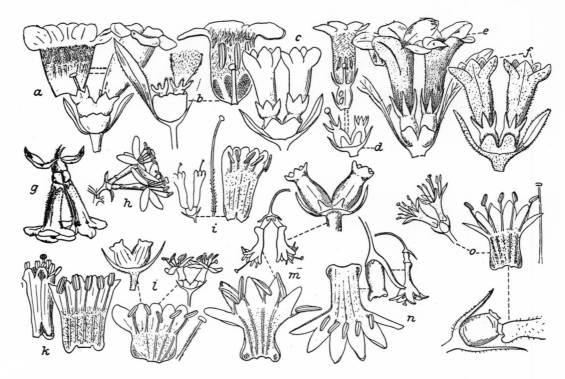

Fig. 153. **Lonicera.** Flowers and floral parts, most enlarged. a. *L myrtillus*; b. *L. myrtilloides*; c. *L. syringantha*; d. *L. syringantha* var. *wolfii*; e. *L. thibetica*; f. *L. rupicola*; g. *L. tomentella*; h. *L. alberti*; i. *L. tangutica*; k. *L. obovata*; l. *L. microphylla*; m. *L. canadensis*; n. *L. gracilipes*; o. *L. coerulea* (from Koehne, Rehder, Schneider, B. M.)

Subsection 10. **Bracteatae** Hook. f. & Thoms.
  Like Subsection 9, but with the ovaries completely separate, bracts usually very large and wide; corolla with a nearly regular limb, seldom bilabiate:
    L. altmannii, bracteolaris, chaetocarpa, hispida, praeflorens, setifera, strophiophora

Subsection 11. **Pyrenaicae** Rehd.
  Twigs with a fully developed terminal bud; prophylls all distinct, half as long as the ovary, ovate, acute; ovaries separate, 3 locular; corolla campanulate, with an outspread limb, white:
    L. pyrenaica

Subsection 12. **Distegiae** Rehd.
  Twigs with a terminal bud; corolla nearly regular; ovaries distinct; prophylls and bracts large, surrounding the ovaries and the corolla base, glandular; corolla yellow to scarlet-orange, base bulging or sack-like; fruits black, surrounded by the large red prophylls:
    L. involucrata, ledebourii, propinqua

Subsection 13. **Oblongifoliae** Rehd.
  Corolla bilabiate; bracts and prophylls small, inconspicuous; leaves oblong, obtuse, finely pubescent on both sides; corolla yellowish, exterior pubescent; fruits red; no part of plant glandular:
    L. oblongifolia

Subsection 14. **Alpigenae** Rehd.
  Corolla bilabiate; flowers appearing after the leaves; ovaries usually 3 locular; fruits red, with 3–6 cm long, yellowish, smooth seeds; winter buds ovate; flower stalk long; leaves often glandular:
    L. alpigena, glehnii, heteroloba, heterophylla, tatsienensis, webbiana

Subsection 15. **Rhodanthae** Maxim.
  Like Subsection 14, but with the winter buds oblong-acute, sharply quadrangular; flower stalk long or short; calyx teeth lanceolate, persistent; fruits red or black, with 2–4 mm long, brownish, granular seeds; leaves not glandular:
    L. caucasica, chamissoi, discolor, kesselringii, maximowiczii, nervosa, nigra, orientalis

+++ Section 3. **Coeloxylosteum** Rehd.
  Twigs very quickly becoming hollow; flower corolla always bilabiate; ovaries always distinct; prophylls paired (connate) above the bracts or distinct; fruits red, yellow and white;

  Subsection 16. **Tataricae** Rehd.
    Flowers pink to white, never yellowing; flower stalks always longer than the leaf petiole; calyx short 5 toothed; berries red or yellow:
      L. amoena, floribunda, korolkowii, tatarica, xylosteoides

  Subsection 17. **Ochranthae** Zab.
    Flowers white or yellowish, occasionally with a trace of red, changing to yellow; flower stalk longer or shorter than the leaf petiole; calyx often cupulate or campanulate; berries red, yellow or white:
      L. arborea, bella, chrysantha, deflexicalyx, demissa, maackii, minutiflora, morrowii, muendeniensis, muscaviensis, notha, nummulariifolia, pseudochrysantha, prostrata, quinquelocularis, ruprechtiana, trichosantha, vilmorinii, xylosteum

++++ Section 4. **Nintooa** (Sweet) Maxim.
  Climbing shrubs; twigs hollow; leaves usually semi-evergreen; flowers in axillary pairs, often grouped into terminal spikes or panicles; fruits black, seldom white or yellow; corolla always bilabiate;

  Subsection 18. **Calcaratae** Rehd.
    Corolla with a long spur; ovaries 5 locular, connate; prophylls connate; fruits yellow:
      L. calcarata (not described in this work)

  Subsection 19. **Breviflorae** Rehd.
    Corolla withour a spur, ovaries and prophylls distinct; ovaries 3 locular; corolla scarcely longer than 3 cm, red, orange or yellowish-white; tube about as long as the limb, sometimes bulging beneath the middle; styles with erect pubescence; fruits black;
      L. alseuosmoides, giraldii, henryi

  Subsection 20. **Longiflorae** Rehd.
    Corolla without a spur, ovaries and prophylls distinct; ovaries 3 locular; corolla 4–16 cm long, white or yellow, turning yellow later; tube slender, longer than the limb; style glabrous or short soft pubescent; fruits black, seldom white:
      L. affinis, biflora, confusa, hildebrandiana, japonica, similis

● ● Subgenus II. **Caprifolium** (Adans.) Dipp.
  Twining shrubs; twigs hollow; leaf pairs under the inflorescences usually disk-like connate (except *L. periclymenum*); flowers usually in whorls of 6 at the branch tips, sessile; fruits red;

  Subsection 21. **Phenianthi** Rehd.
    Corolla with a short, or nearly regular bilabiate limb; tube more or less bulging beneath the middle or slightly expanded; stamens inserted beneath the throat:
      L. arizonica, brownii, ciliosa, sempervirens, subaequalis

  Subsection 22. **Cypheolae** Raf.
    Corolla deeply bilabiate; stamens inserted in the throat; corolla 1.5–2.5, rarely to 3.5 cm long, yellow or yellow-white, often reddish; tube usually bulging or expanded below the middle, pubescent interior; style pubescent or glabrous:
      L. dioica, flava, glaucescens, hirsuta, hispidula, prolifera, yunnanensis

  Subsection 23. **Eucaprifolia** Spach.
    Corolla deeply bilabiate, stamens inserted in the throat; corolla 4–8 cm long, occasionally shorter, usually white or yellow-white; tube slender, gradually becoming narrower toward the base, longer than the limb, glabrous interior (except *L. implexa* and *L. tragophylla*); styles glabrous (except *L. implexa*):
      L. americana, caprifolium, etrusca, heckrottii, implexa, periclymenum, splendida, tellmanniana, tragophylla

  Subsection 24. **Thoracianthae** Rehd.
    Prophylls at each flower pair cup-form connate, as large as the ovaries; leaf pairs all distinct:
      L. griffithii

Parentage of the more prominent cultivars

(names such as 'Rosea', 'Alba', 'Nana' etc. are frequently encountered but more collective in nature and not included here)

'Arnold Red'	→ *L. tatarica*	'Dropmore'	→ *L. bella*
'Arnoldiana'	→ *L. amoena*	'Dropmore Scarlet'	→ *L. brownii*
'Aureo-reticulata'	→ *L. japonica*	'Elegant'	→ *L. nitida*
'Aurora'	→ *L. korolkowii*	'Ernest Wilson'	→ *L. nitida*
'Baggesen's Gold'	→ *L. nitida*	'Fuchsioides'	→ *L. brownii*
'Belgica'	→ *L. periclymenum*	'Goldflame'	→ *L. heckrottii*
'Clavey's Dwarf'	→ *L. xylosteum*	'Graziosa'	→ *L. nitida*

'Hack's Red'	→ *L. tatarica*
'Halliana'	→ *L. japonica*
'Louis Leroy'	→ *L. tatarica*
'Morden Orange'	→ *L. tatarica*
'Pauciflora'	→ *L. caprifolium*
'Plantierensis'	→ *L. brownii*
'Praecox'	→ *L. caprifolium*
'Punicea'	→ *L. brownii*
'Redgold'	→ *L. tellmanniana*
'Serotina'	→ *L. periclymenum*
'Superba'	→ *L. sempervirens*
'Yunnan'	→ *L. nitida*
"Zabelii"	→ *L. korolkowii*

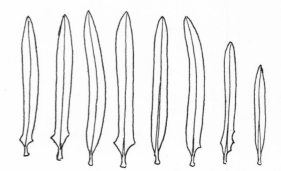

Fig. 154. *Lonicera spinosa* var. *alberti*
(actual size; Original)

Lonicera acuminata see: **L. henryi**

L. affinis Hook. & Arn. Twining shrub, closely related to *L. similis*, but with normally glabrous shoots; leaves ovate to oval-oblong, acuminate, glabrous; the lower flower stalks to 1.5 cm long; flowers white, corolla 3.5–4.5 cm long. MJ 3303. Liukiu-Archipelago; Japan; Cochin China; Yunnan China. Quite variable in leaf form and pubescence. z6 Fig. 156, 162.

L. alberti Reg. Deciduous shrub, low, scarcely 0.5 m high, twigs prostrate or nodding; leaves linear, sessile, 2–3 cm long, entire or with 1–2 teeth at the base on either side, blue-green above, whitish beneath; flowers lilac-pink, fragrant, in short stalked axillary pairs, May; fruits purple, to nearly white, pruinose. BM 7396; Gfl 1065; DL 1: 168 (= *L. spinosa* var. *alberti* [Regel] Rehd.). Turkestan, Tibet. 1880. Differing from the similar *L. spinosa* in the thornless twigs, longer corolla, oblong limb lobes, filaments twice as long as the anthers. z6 Plate 95, 102; Fig. 154, 153. ⊕

L. alpigena L. Deciduous shrub, narrowly upright, about 1 m high, shoots green, slightly glandular and pubescent; leaves elliptic to oblong, acuminate, 5–10 cm long, dark green and somewhat glossy above, lighter beneath and pubescent at first, ciliate; flowers yellow or greenish, with a trace of brown, bilabiate, 1 to 1.5 cm long, on erect, 2–5 cm long stalks, May; fruits cherry-like, ovate-globose, glossy dark red, to 1.3 cm long, hanging on thin stalks. HF 2920; HM Pl. 252; POe 511 to 512. Central and Southern Europe; in the mountains of Bosnia (Yugoslavia) to the Pyrenees (Spain). z6 Fig. 155, 169. ⊗

var. *glehnii* see: **L. glehnii**

'Macrophylla'. Leaves larger than those of the species, totally glabrous, prophylls small; flowers with a glabrous exterior.

f. **nana** (Carr.) Nichols. Dwarf form, only half as high; pubescent on the flower stalks and leaf undersides; flowers dark red, but small. AB 22: 65. Before 1889.

L. alseuosmoides Graebn. Evergreen, twining shrub, very closely related to *L. henryi*, but differing in the 3–6 cm long leaves, tapering or rounded base, margins ciliate, otherwise with the leaves glabrous beneath; flowers at the shoot tips in the apical leaf axils, developing short, broad panicles, corolla funnelform, tube longer than the limb, about 1.2 cm long, with a

purple and pubescent interior, a yellow, glabrous exterior, July–October; fruits black, globose, purple pruinose. BS 2: 596. W. China. 1904. z6 Plate 98; Fig. 162. #

L. altaica see: **L. coerulea** var. **altaica**

L. altmannii Regel & Schmalh. Deciduous shrub, upright, twigs reddish, stiff haired; leaves broadly ovate to oval-elliptic, acute, rarely obtuse, 1.5–5 cm long, base round to slightly cordate, bluish-green above and more or less rough haired, ciliate, lighter and soft pubescent beneath, petiole 3–5 mm long; flowers paired, bracts lanceolate, 6 mm long, corolla yellowish-white, 12 mm long, exterior loosely pubescent and glandular, tube interior pubescent, distinctly expanded at the base,

Fig. 155. *Lonicera alpigena*, twig with fruits
(Original)

Fig. 156. **Lonicera.** Left *L. affinis* var. *pubescens;* center *L. chrysantha;* right *L. fragrantissima*
(from ICS)

stamens and styles glabrous, April–May; fruits oval-
rounded, orange-red, 8 mm thick (= *L. tenuiflora* Regel &
Winkl.). Turkestan. 1899. Hardy! z5 Plate 96; Fig. 168.

var. **hirtipes** Rehd. Shoots and leaf petioles dense bristly
pubescent; leaves rough haired on both sides, broadly ovate, 3–
5 cm long; ovaries often pubescent. Turkestan, Alatau Mts.
1910. Hardy.

var. **pilosiuscula** Rehd. Twigs and petioles finely pubescent and
sparsely bristled, often glandular; leaves ovate to elliptic, 2–4
cm long, sparsely pubescent above, glabrous beneath or with
pubescent venation; corolla usually totally glabrous, as are the
ovaries. Alatau Mts. 1900.

var. **saravshanica** Rehd. Shoots finely pubescent to glabrous,
bristles scarce or absent; leaves ovate to elliptic, obtusish, 1.5–
2.5 cm long, densely gray tomentose beneath. Saravshan, W.
Buchara.

L. × americana (Mill.) K. Koch (= *L. caprifolium* × *L.
etrusca*). Deciduous shrub, medium high climber, closely
related to *L. caprifolium*, the apical leaf pairs similarly
cup-form connate, the basal ones more acuminate,
shoots reddish, glabrous; leaves broadly ovate to elliptic,
to 8 cm long, bluish beneath, glabrous; flowers in
densely packed whorls, often in 25 cm long and 20 cm
wide panicles, corolla tube slender, yellow, but usually
more or less reddish on the exterior and glandular
pubescent, fragrant, June–August; fruits red. FS 1120;
DL 1: 167 (= *L. caprifolium major* Carr.; *L. italica* Schmidt;
L. grata Ait.). Presumably a natural hybrid; found in the
wild in S. France and in Istria (NW. Yugoslavia); not
from the USA as might be inferred from the name. 1730.
z6 Fig. 158. ✣

'Atrosanguinea'. Flowers with a dark red exterior. Before 1870.
✣

'Quercifolia'. Leaves open emarginate, occasionally yellow
margined or reddish striped (= *L. caprifolium variegata; L.
caprifolium erosa* Hort.). Before 1870.

'Rubella'. Flowers pale purple outside, darker in bud. Before
1830.

L. amherstii see: **L. strophiophora**

L. × amoena Zab. (= *L. korolkowii* × *L. tatarica*). Tall,
strong growing, deciduous shrub, about 2–3 m high;
leaves ovate, 3–4 cm long, 1.5–2.5 cm wide, base
somewhat cordate, gray-green, to 15 mm wide, short
stalked; flowers very numerous at the shoot tips, lighter
pink, only 18 mm wide, prophylls very small, distinct or
connate in pairs, flowering time between that of *L.
korolkowii* and *L. tatarica*. Originated as a chance seedling
before 1895 in the garden of H. Zabel of Gotha, W.
Germany. z5

'Alba' (Zabel). Shrub, rounded, shoots more pubescent;
flowers white, turning yellowish, very fragrant, flowers
abundantly. ✣

'Arnoldiana' (Arnold Arboretum). Graceful, open habit,
shoots drooping; leaves oblong-lanceolate, 2–3 cm long, gray-
green; flowers white with a trace of pink, later totally white,
nearly 3 cm wide (!), flowers abundantly. Quite valuable. 1899.
Plate 96. ✣

'Rosea' (Zabel). Flowers fleshy pink to a bright, light pink, pale
pink inside, very fragrant, quickly becoming yellow.

L. angustifolia Wall. ex DC. Deciduous shrub, 2–2.5 m
high, rounded, twigs gracefully nodding; leaves oval-
lanceolate, 2–5 cm long, acuminate, base round or
tapered, bright green and glabrous above, lighter
beneath and somewhat pubescent, at least on the midrib,
petiole 2 mm long, pubescent; flowers in axillary pairs
on the young shoots, pendulous on 1–2 cm long stalks,
fragrant, corolla tube form, whitish-pink, tube 8 mm
long, lobes of equal size, 3 mm long, style short, totally
enclosed, May to June; fruits red, in connate pairs,
edible. DL 1: 167; FS 408b. Himalayas; Kashmir to
Sikkim. Hardy. z5 Fig. 157.

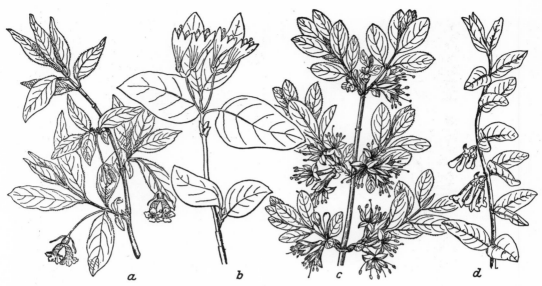

Fig. 157. **Lonicera**. Flowering shoots. a. *L. angustifolia*; b. *L. canadensis*; c. *L. coerulea*; d. *L. tomentella* (from Guimpel, B. M., F. S., Torrey)

L. arborea Boiss. Upright, deciduous shrub, to 2 m, also occasionally a small tree in its habitat, branches gray-white, shoots 4 sided, gray-white pubescent; leaves oval to elliptic, 2.5–3.5 cm long, apex rounded to somewhat emarginate, base round to somewhat cordate, both sides gray pubescent at first, eventually more glabrous, gray above, more white-gray beneath; flowers very short stalked, corolla 1.5 cm long, reddish, softly pubescent exterior, interior pubescent or glabrous, tube only slightly expanded at the base; berries distinct, orange-yellow, often only one fruit develops. S. Spain, N. Africa. 1900. Rare in cultivation outside of a few botanic gardens. z7–8 Plate 95; Fig. 166.

var. *persica* see: **L. nummulariifolia**

L. arizonica Rehd. Medium high, climbing, deciduous shrub, similar to *L. ciliosa* (tube more slender, style glabrous, leaves smaller) and *L. pilosa* (leaves thin, obtuse, glabrous, but ciliate, tube more slender, prophylls shorter), new growth brownish; leaves oval-rounded, to 5 cm long, ciliate, bright green, whitish beneath; corolla 3–4.5 cm long, exterior turning brick-red, limb lobes regular. ST 23. Arizona, New Mexico USA. 1900. Very attractive shrub. z6 Plate 100; Fig. 162. ⊕

L. × bella Zab. (= *L. morrowii* × *L. tatarica*). Deciduous shrub, upright, intermediate between the parents; leaves more acute than those of the parents, soon becoming glabrous; flowers more reddish, but later becoming yellow (= *L. "morrotarica"* Lemoine). Originated from seed obtained from the Leningrad Botanic Garden. USSR. z5 Plate 96.

'Albida' see: **'Candida'**

'Atrorosea' (Zabel). Flowers dark pink, May; fruits dark red, pea-sized. ⊕ ⊗

'Candida' (Zabel). Flowers white, greenish in bud (= 'Albida').

'Dropmore' (F. L. Skinner). Shrub, to about 2 m high, with an attractively weeping habit; flowers very abundantly, white, turning yellowish, mid-May; good fruiter, red. Developed in Dropmore, Manitoba, Canada and very popular in the USA for its winter hardiness. z2 Plate 93. ⊕

'Polyantha' (Zabel). Flowers light carmine-pink, flowers especially abundant. The most attractive cultivar. ⊕

'Rosea' (Zabel). Flowers light pink, soon becoming totally white, eventually yellow, small flowered, only about 12 m wide. AB 20: 60. Often confused with 'Atrorosea'.

L. biflora Desf. Semi-evergreen, twining shrub, similar to *L. confusa*, but with glabrous ovaries; leaves elliptic to nearly oblong, often acute, 3–5 cm long, ciliate, pubescent above at first, later gray-green, glabrous, margins ciliate, soft gray-white pubescent beneath; flowers in terminal, raceme-like inflorescences, fragrant, corolla 3–4 cm long, white at first, a turbid yellow later, exterior glutinous pubescent, tube slender, widening clavate-like at the apex, limb half as long as the tube, calyx short 5 toothed, pubescent; fruits globose, blue-black. DL 1: 139 (= *L. canescens* Schousboe). Spain, Sicily, N. Africa. 1889. z9 Fig. 160. #

L. bracteata see: **L. hispida** var. **bracteata**

L. bracteolaris Boiss. & Buhse. Upright, deciduous shrub, glabrous, closely related to *L. altmannii*, but the leaves ovate to oval-oblong, 2.5–5 cm long, obtuse or somewhat acute, glabrous or somewhat ciliate; flowers white, corolla glabrous, tube about as long as the limb,

Plate 81

Laurus canariensis
in Caerhays Garden, Cornwall, England

Lavandula stoechas
in the Lausanne Botanic Garden, Switzerland

Ledum groenlandicum 'Compactum'
in the Arnold Arboretum, USA

Leiophyllum buxifolium
in the Royal Botanic Garden, Edinburgh, Scotland

Plate 82

Leucadendron argenteum in its native habitat at the Barlow Reservation, Stellenbosch, South Africa

Leucospermum lineare in its native habitat at the Barlow Reservation, Stellenbosch, South Africa

Leucospermum cordifolium in the Caledon Gardens, Cape Province, South Africa

Plate 85

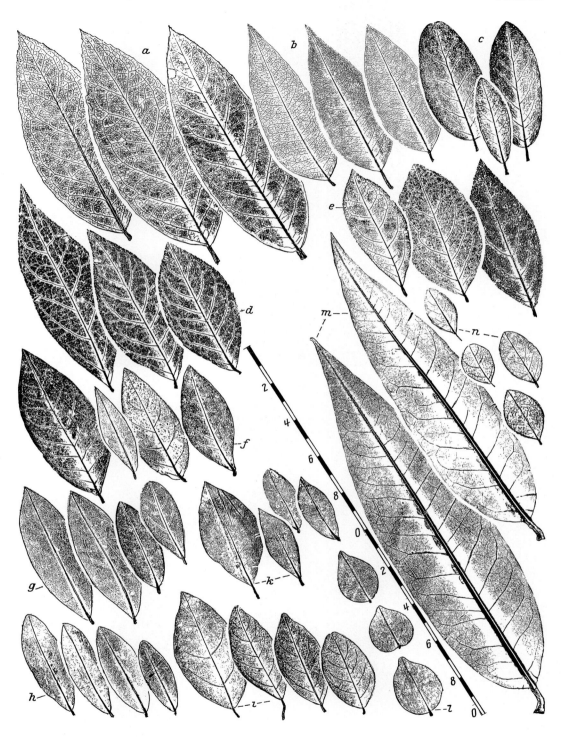

Leucothoe. a. *L. axillaris;* b. *L. populifolia;* c. *L. davisiae;* d. *L. racemosa;* e. *L. recurva.* — **Ligustrum.** f. *L. acutissimum;* g. *L. obtusifolium;* h. *L. amurense;* i. *L. sinense;* k. *L. tschonoskii* (large leaf), var. *glabrescens* (3 small leaves); l. *L. henryi;* m. *L. chenaultii;* n. *L. strongylophyllum* (collected from plants in the wild)

Plate 86

Ledum palustre
Photo: C. R. Jelitto, Berlin

Ligustrum japonicum 'Rotundifolium'
in the Dortmund Botanic Garden, W. Germany

Ligustrum ovalifolium 'Aureum'
in the Dortmund Botanic Garden, W. Germany

Plate 87

Linnaea borealis
Photo: C. R. Jelitto, Berlin

Liriodendron tulipifera, flowers
Photo: C. R. Jelitto, Berlin

Ligustrum obtusifolium var. *regelianum* in the Dortmund Botanic Garden, W. Germany

Plate 88

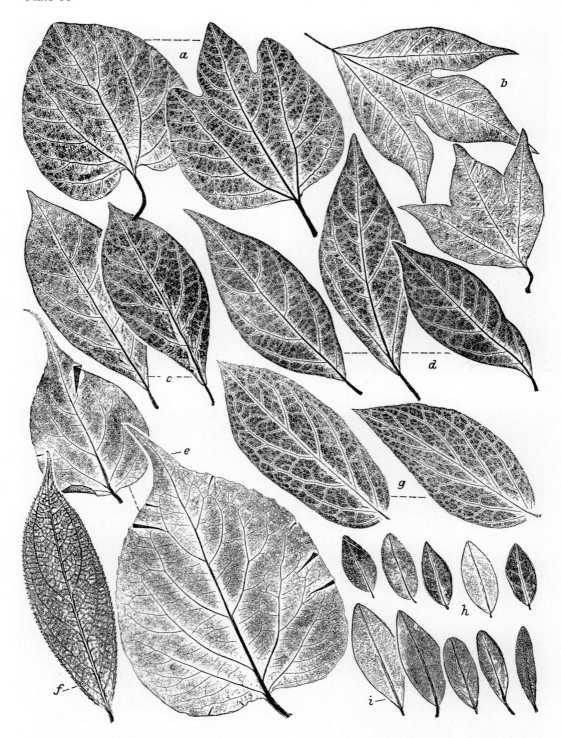

Lindera. a. *L. obtusiloba;* b. *L. triloba;* c. *L. benzoin;* d. *L. umbellata.* e. *Leycesteria formosa.* — **Leucothoe.** f. *L. griffithiana;* g. *L. grayana.* — **Ligustrum.** h. *L. sinense* var. *stauntonii;* i. *L. quihoui* (from material collected in the wild)

Plate 89

Liquidambar styraciflua,
young tree with the typical corky bark

Lithocarpus glaber
in the Fota Island Arboretum near Cork, Ireland

Lupinus arboreus
in the Hillier Arboretum, England

Limonastrum monopetalum
in a garden on Iviza, Balaeric Islands, Spain

Plate 90

Lithocarpus edulis
Photo: Dr. Watari, Tokyo

Loiseleuria procumbens in its native habitat in the Alps
Photo: C. R. Jelitto, Berlin

Plate 91

Lithodora diffusa 'Alba' in the Royal Botanic Garden, Edinburgh, Scotland

Lomatia ferruginea in Ashbourne, Ireland

Plate 92

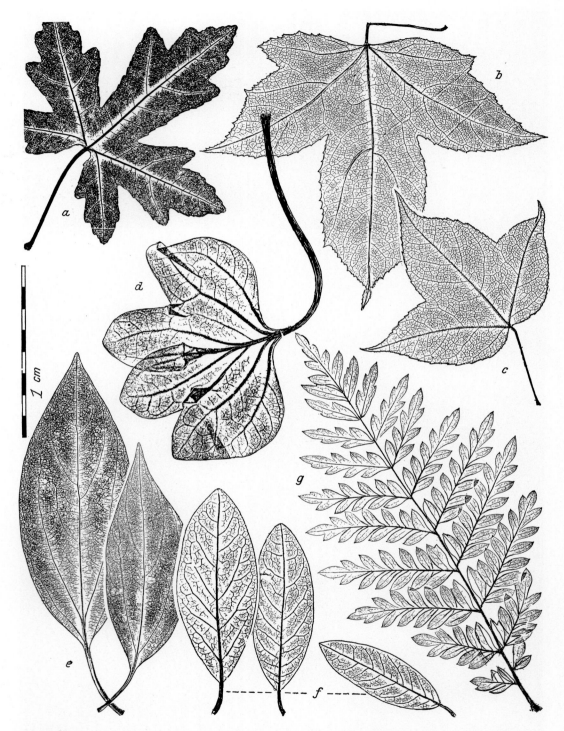

Liquidambar. a. *L. orientalis;* b. *L. formosana;* c. *L. formosana* var. *monticola.* — d. *Liriodendron tulipifera* 'Crispum'. — e. *Neolitsea glauca.* — f. *Litsea japonica.* — g. *Lomatia ferruginea* (only c. is from a wild plant)

Plate 93

Luzuriaga radicans
in Malahide, Ireland

Liriodendron tulipifera 'Mediopictum'
in the Sochi Dendrarium, USSR

Lomatia hirsuta
in Crarae, Scotland

Lonicera bella 'Dropmore'
in the Wageningen Arboretum, Netherlands

Plate 94

Lonicera pileata   Lonicera nitida

Lonicera nitida 'Elegant'
Photos: Dortmund Botanic Garden

Plate 95

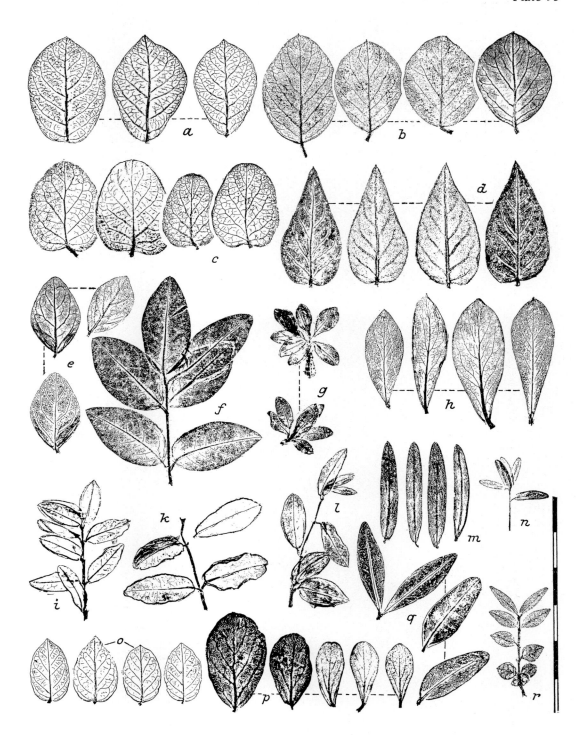

Lonicera. a. *L. chamissoi;* b. *L. arborea;* c. *L. iberica;* d. *L. ferdinandi;* e. *L. prostrata;* f. *L. myrtilloides;* g. *L. myrtillus;* h. *L. pyrenaica;* i. *L. syringantha;* k. *L. syringantha* var. *wolfii;* l. *L. thibetica;* m. *L. alberti;* n. *L. spinosa;* o. *L. tomentella;* p. *L. microphylla;* q. *L. pileata;* r. *L. nitida* (from material collected in the wild; actual size)

Plate 96

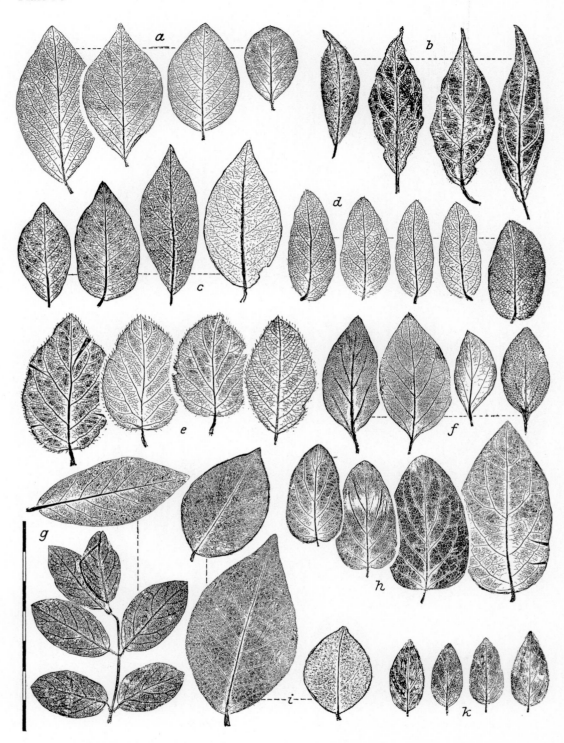

Lonicera. a. *L. quinquelocularis;* b. *L. heteroloba;* c. *L. nervosa;* d. *L. hispida;* d. *L. altmannii;* f. *L. korolkowii;* g. *L. morrowii;* h. *L. bella;* i. *L. tenuipes;* k. *L. amoena* 'Arnoldiana' (from wild plants)

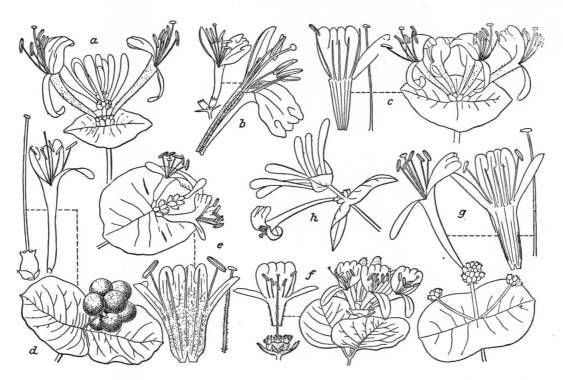

Fig. 158. **Lonicera.** Flowers and flower parts, enlarged. a. *L. americana*; b. *L.* × *brownii* 'Fuchsioides'; c. *L. caprifolium*; d. *L. tragophylla*; e. *L. prolifera*; f. *L. griffithii*; g. *L. etrusca*; h. *L. implexa* (from Rehder, Koehne, Dippel)

stamens somewhat shorter, styles somewhat longer than the tube. RLo 12. Transcaucasus. 1910. Hardy. z5 Fig. 168.

L. brachypoda see: **L. japonica** var. **repens**

L. brachypoda aureo-reticulata see: **L. japonica 'Aureo-reticulata'**

L. × **brownii** (Reg.) Carr. (= *L. hirsuta* × *L. sempervirens*). Deciduous, climbing shrub, to about 3 m high; leaves elliptic, bluish beneath and somewhat pubescent, petiole glandular, the apical leaf pairs connate; flowers in stalked heads, not fragrant, corolla usually distinctly bilabiate, orange-scarlet, somewhat tubercled at the base, longer than the limb, glandular pubescent outside, May–August. FS 1133; Gfl 38 (*L. sempervirens brownii* Lav.; *L. etrusca brownii* Regel). Developed before 1850. Plate 100.

'Dropmore Scarlet' (F. L. Skinner). Long lasting flowers, strong growing; flowers bright red, June–October. Supposedly very winter hardy. Developed by F. L. Skinner, Dropmore, Manitoba, Canada. Before 1950. z2 Plate 39. ⊕

'Fuchsioides'. Very much resembling *L. sempervirens*, but the flowers distinctly bilabiate, tube exterior scarlet-red. GF 9: 496 (= *L. fuchsioides* Hort. ex Koch). Difficult to cultivate, poor grower. Fig. 158. ⊕

'Plantierensis' (Simon Louis). Flowers larger, corolla with a brownish-orange exterior, less deeply bilabiate. IH 18: 86.

Developed before 1871. The most attractive form. ⊕

'Punicea'. Slower growing than the species; flowers less abundantly, flower exterior orange-red (= *L. punicea* Hort. ex Koch). Before 1964. Somewhat sensitive in cultivation. ⊕

'Youngii'. Strong grower; flowers deeper orange-red. Before 1872. ⊕

L. bungeana see: **L. microphylla**

L. canadensis Bartr. Deciduous shrub, upright but broad growing, to 1.5 m high, twigs glabrous, red-brown; leaves ovate to oval-oblong, 4–8 cm long, light green, ciliate; flowers yellowish-white, turning reddish, to 2 cm long, corolla tubular to funnelform, exterior glabrous, always in slender stalked pairs, April to May; fruits light red, connate at the base. BB 3465; GSP 454; DL 1: 164; NBB 3: 299 (= *L. ciliata* Muehl.). N. America; Canada to Michigan. z2 Plate 98; Fig. 153, 157. ⊕

L. caprifolium L. Medium high, twining, deciduous shrub, glescent only when young, the apical pair fused into an acute-elliptical disk; flowers in 6's in the axil of the apical leaf pair, other 6 parted flower whorls situated in the axils of the next lower 1–2 leaf pairs, corolla tube yellowish-white, often reddish outside, 4–5 cm long, interior pubescent or glabrous, fragrant, upper lip erect, later reflexed, with 4 sections, lower lip ligulate, May–June; fruits coral-red. HF 2913; NBB 3: 300; BB 3455; POe 491–494. Central and W. Europe, from France to the

Caucasus and Asia Minor. See also: **L. × americana.** z6 Fig. 158. ⊕

'Pauciflora'. Corolla with a purple exterior, interior yellowish-white, tube 2–3 cm long (= *L. caprifolium rubra* Tausch). ⊕

'Praecox'. Leaves lighter, more gray-green; flowering a few weeks earlier, corolla pale red or only cream-white, later becoming a turbid yellow. ⊕

L. caucasica Pall. 2 m high, bushy, deciduous shrub, winter buds like those of *L. nigra*, shoots totally glabrous; leaves elliptic to ovate, 3–10 cm long, 6–30 mm wide, acute, base broadly cuneate or round, green above, more gray-green beneath, glabrous; flowers pink, 12 mm long, paired in the leaf axils of the current year's wood, somewhat fragrant, corolla bilabiate, with a very short tube, inflated on one side, pubescent interior, May–June; fruits black, paired and connate along the inner margin (= *L. orientalis* var. *caucasica* [Pall.] Rehd.). Caucasus. 1825. Without particular garden merit. z6 Fig. 169.

var. *longifolia* see: **L. kesselringii** and **L. orientalis** var. *caucasica*

L. chaetocarpa Rehd. Upright, deciduous shrub, 1.5–2 m high, twigs stiffly glandular pubescent, winter buds 1 cm long; leaves oblong-ovate to oblong, 3–8 cm long, often convex, rough haired, especially dense on the gray-green underside; flowers yellowish-white, 3–4 cm long, tubular, nodding, June, bracts 2.5 cm long, nearly circular, pubescent; fruits light orange-red, surrounded by the eventually nearly white, persistent bracts. BM 8804; BS 2: 600 (= *L. hispida* var. *chaetocarpa* Batal.). China, Kansu, Szechwan Provinces; E. Tibet. 1904. Gorgeous in fruit. Quite hardy. z5 Plate 99; Fig. 168. ⊕ ⚭

L. chamissoi Bge. Deciduous shrub, about 1 m high, erect, glabrous; leaves nearly sessile, oval-elliptic, 2.5–5 cm long, usually obtuse, base rounded, distinctly veined beneath; flower stalks 6–14 mm long, bracts and prophylls very small, glabrous, calyx teeth triangular, corolla dark violet outside, 12 mm long, tube somewhat tubercled, exterior glabrous, stamens and styles shorter than the limb, glabrous, May–June; fruits connate, red. MJ 3298. E. Asia; Sachalin, Manchuria, Kamchatka, Kuril, Japan. 1909. Hardy. z5 Plate 95; Fig. 169.

L. chrysantha Turcz. Erect, deciduous shrub, 2–4 m high, twigs stiff haired; leaves short petioled, ovate-rhombic to oval-lanceolate, 6–12 cm long, acuminate, bright green and glabrous above, lighter beneath and soft pubescent; flowers yellowish-white, eventually yellow, about 2 cm long, somewhat pubescent on the exterior, May–June, corolla upper lip incised to the midpoint, prophylls oblong to rounded, glandular and ciliate, half as long as the ovaries; fruits coral-red. DL 1: 141; Gfl 12: 404; NT 1: 642 (= *L. gibbiflora* Maxim. non Dipp.). NE. Asia to Central Japan. 1880. z6 Fig. 156, 165.

L. ciliata see: **L. canadensis** and **L. ciliosa**

L. ciliosa (Pursh.) Poir. Leaves short stalked, ovate to oblong-elliptic, 5–10 cm long, ciliate, tapered to both ends, blue-green beneath, pubescent when young, the

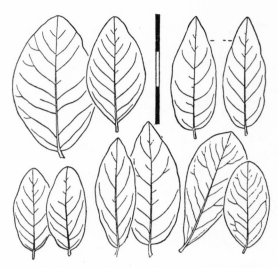

Fig. 159. Forms of *Lonicera coerulea*, 2 leaves of each. Upper row: left var. *altaica*, right var. *angustifolia*; Lower row: left var. *glabrescens*, middle var. *dependens*; right *L. villosa* (from Rehder)

uppermost pair fused into an acute elliptic disk; flowers in short stalked heads, corolla tube yellow, turning red, 3–4 cm long, June. FS 1133; DL 1: 136 (= *L. ciliata* Dietr.). West. N. America, British Columbia to California, Montana, Utah. 1824. z5 Plate 101; Fig. 162. ⊕

var. *occidentalis* (Hook.) Nichols. Flowers somewhat larger, corolla tube glabrous and dark orange on the exterior. BR 1457 (= *L. occidentalis* Hook.).

L. coerulea L. Deciduous shrub, upright, often also divaricate, usually very densely branched, 1–1.5(2) m high, bark eventually red-brown and exfoliating, winter buds spreading horizontally, often with 3–4 superimposed at each node; leaves ovate rounded to oblong, 2–8 cm long, bright green, eventually totally glabrous; flowers yellowish-white, tubular-funnelform, 15 mm long, on short, nodding stalks, April–May; fruit globose, 1 cm thick, black-blue, light blue pruinose. HF 2919; HM Pl. 252; DL 176; GPN 815; RLo 1. Northern Hemisphere, on a humus soil in forests, thickets and among low brush. Quite variable and widely distributed species with numerous named varieties occasionally occurring in cultivation. z2 Fig. 153, 157. ⊕

Outline of the forms

● Winter buds spreading more or less horizontally, twigs usually spreading at an angle greater than 45°; corolla tube funnelform, usually longer than the limb, usually pubescent on the exterior;

★ Leaves glabrous or pubescent when young; shoots glabrous or nearly so, rarely finely tomentose;

> Leaves usually somewhat pubescent, rather membranous:

var. *glabrescens* (= the type), 'Praecox', 'Sphaerocarpa', 'Globosa', var. *dependens*, 'Viridifolia', 'Angustifolia'

Fig. 160. **Lonicera.** Flowers and flower parts, enlarged. a. *L. confusa;* b. *L. biflora;* c. *L. japonica;* d. *L. hildebrandiana* (from Rehder, Andrew, Watson, Desveaux)

>> Leaves glabrous, not ciliate, occasionally somewhat pubescent while developing, elliptic, rather tough, with distinctly reticulate venation beneath:
var. *venulosa*
★★ Leaves and shoots more or less pubescent;
+ Leaves mostly elliptic; stamens not exserted past the corolla; fruit nearly globose:
var. *altaica,* f. *emphyllocalyx*
++ Leaves mostly oblong to lanceolate; stamens exserted past the corolla; fruit ovoid:
var. *edulis,* var. *tangutica*
● ● Winter buds erect; usually spreading at less than a 45° angle; corolla usually campanulate, glabrous exterior; fruit nearly globose:
L. villosa! which see (= *L. coerulea* var. *villosa* Torr. & Gray)

var. **altaica** (Pall.) Sweet. Upright, shoots bristly; leaves oblong-elliptic, obtusish to acute, 4–7 cm long, with persistent pubescence on both sides; corolla pubescent outside, stamens not exserted; fruit globose (= *L. altaica* Pall.). N. Europe, N. Asia to Japan. Fig. 159.

var. **angustifolia** Regel. Shoots finely soft pubescent; leaves oblong to more lanceolate, 2–4 cm long, finely pubescent at first; corolla small, slightly pubescent. SH 2: 242c. Turkestan. Fig. 159.

var. **dependens** (Dipp.) Rehd. Broad growing, twigs red-brown, thin; leaves elliptic, 1.5–3 cm long, bristly ciliate, somewhat pubescent at first; flowers rather small, slender, pubescent (= f. *graciliflora* Dipp.). Turkestan. Fig. 159.

var. **edulis** Regel. Twigs pubescent; leaves oblong-lanceolate, pubescent; stamens exserted; fruits oblong, supposedly edible. E. Siberia, Tibet.

f. **emphyllocalyx** Rehd. Differing from var. *altaica* in the ovate leaves, truncate to round at the base, somewhat exserted stamens and dense bristly-shaggy pubescent shoots. Japan, Hondo Island.

var. **glabrescens** Rupr. Twigs pubescent to glabrous, yellow-brown; leaves oblong to more elliptic, base rounded, 4–6 cm long, pubescent at first, or at least ciliate, later becoming glabrous above, only the venation pubescent beneath; corolla with a glabrous exterior, tube short, rather thick, glabrous on the exterior (= var. *praecox* Dipp.). Europe to Northeast Asia. The type of the species. Fig. 159.

'**Globosa**'. Fruit nearly globose (= f. *sphaerocarpa* Dipp. non Regel).

f. *graciliflora* see: var. **dependens**

var. *praecox* see: var. **glabrescens**

'**Sphaerocarpa**'. Fruit oval-globose (= f. *sphaerocarpa* Regel non Dipp.).

f. *sphaerocarpa* see: '**Globosa**'

var. **tangutica** Maxim. Differing from var. *edulis* in the more lanceolate leaves, smaller, about 8 mm long, greenish, glutinous, pubescent flowers and only slightly exserted anthers. China; Kansu.

var. **venulosa** Rehd. Leaves glabrous, tough, eventually with distinctly reticulate venation beneath; corolla glabrous. Japan; S. Europe; Croatia.

var. *villosa* see: **L. villosa**

'Viridifolia.' Shrub, stiffly upright, shoots red; leaves elliptic to obovate, rounded at both ends, 1.5–3 cm long, bright green above, lighter beneath and short, rough haired; flowers with a thicker, about 1 cm longer tube. ⊕

L. canescens see: **L. biflora**

L. confusa (Sweet) DC. Twining shrub, twigs brown, short soft pubescent; leaves deciduous to semi-evergreen, elliptic to oblong, thickish, 4–7 cm long, both sides pubescent and ciliate when young, later glabrous above, deep green, gray-green beneath; flowers in short, dense panicles at the shoot tips, axillary and terminal, corolla about 4 cm long, white at first, later yellow, very fragrant, with a glandular pubescent exterior, slender, June–September; fruit black. DL 1: 138. E. China. Often confused with *L. japonica*, but distinguished by the awl-shaped bracts. z8 Fig. 160.

L. coerulescens see: **L. × xylosteoides**

L. deflexicalyx Batal. Bushy, deciduous shrub, to 1.5 m high, twigs spreading, somewhat nodding, brown-red, finely pubescent; leaves oblong-lanceolate, 4–8 cm long, 1.5 to 2.3 cm wide, acuminate, dark green above and somewhat rugose, light gray-green beneath, both sides somewhat pubescent, especially on the venation beneath, bracts small, awl-shaped; flowers very numerous, on the upper side of the branch, yellow, later dark yellow, 15 mm long, appressed pubescent exterior, corolla tube short, bulging, June; fruits brick-red, globose. BM 8536. W. China, Tibet. Hardy and flowers very abundantly. z6 Plate 99; Fig. 166. ⊕

var. **xerocalyx** (Diels) Rehd. Leaves very narrow-lanceolate, 6–10 cm long, base round to truncate; prophylls fused into a bilabiate, calyx cup reaching past the ovaries, calyx about 4 mm long (= *L. xerocalyx* Diels). SE. China. 1915. ⊕

L. delavayi see: **L. similis** var. **delavayi**

L. demissa Rehd. Deciduous shrub, to 4 m high, twigs short pubescent, widely spreading, winter buds small; leaves obovate, elliptic on long shoots, 1.5–3 cm long, obtuse, dull green above and appressed pubescent, gray-green beneath and densely pubescent; flowers whitish, later yellow, tube to 1 cm long, upper lip with short, ovate lobes, May–June; ovaries somewhat glandular pubescent, fruits scarlet red, globose, pea-sized. IM 3297. Japan. 1914. Hardy. Attractive in fruit. z6 ⊕ ⚬

L. depressa see: **L. myrtillus** var. **depressa**

L. dioica L. Deciduous shrub, slightly climbing or only bushy, about 1.5 m high, glabrous; leaves elliptic to oblong, 5–9 cm long, bright blue-green beneath, the apical pair disk-like connate, margins often somewhat undulate; flowers in terminal, sessile or short stalked whorls, corolla tube yellowish or greenish white, often reddish outside, about 1.5 cm long, rather cylindrical, bilabiate, May to June; fruits red. BB 3458; HyWF 191; DL 1: 133; Gs 13: 193 (= *L. glauca* Hill; *L. parviflora* Lam.). Northeastern N. America, in moist thickets and forests, occasionally in swamps or on dunes. 1776. z5 Plate 100; Fig. 163. ⊕

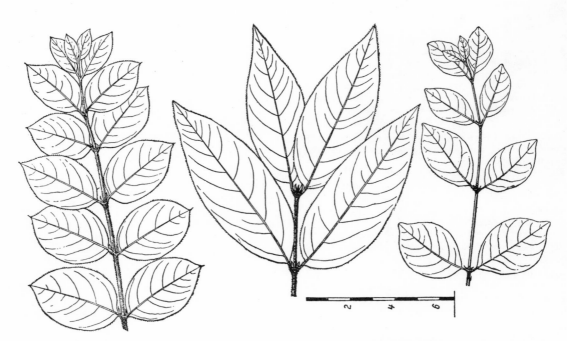

Fig. 161. **Lonicera.** Branches. Left, *L. fragrantissima;* middle, *L. standishii;* right, *L. gracilipes*
(from Dippel, ½ actual size)

L. discolor Lindl. Deciduous shrub, upright, closely related to *L. orientalis*, to 2 m high, glabrous, twigs thin, young shoots red-brown; leaves oblong to elliptic, 5–8 cm long, base round, deep green above, light blue-green beneath, glabrous on both sides; flowers yellowish-white, often somewhat reddish outside, 2–3 cm long, brown stalks, bracts glandular pubescent, tube short, widened pouch-like at the base, interior pubescent, exterior glabrous, May to June; fruit globose, black, 8–10 mm thick. DL 1: 156 (= *L. orientalis* var. *discolor* Clarke). Kashmir to Afghanistan. 1847. z6 Fig. 165.

L. diversifolia see: **L. quinquelocularis**

L. douglasii see: **L. glaucescens**

L. etrusca Santi. Semi-evergreen, high climbing shrub, young shoots glabrous, reddish; leaves short stalked, obovate to broadly elliptic, 3–8 cm long, obtuse or acute, usually glabrous above, blue-green and pubescent beneath, the uppermost pair connate; flowers in dense, many whorled, terminal spikes on a stalk up to 4 cm long, often 3 spikes together, corolla 4–5 cm long, yellowish-white, exterior often reddish, tube very slender, exterior glabrous to glandular pubescent, corolla limb to 2.5 cm wide, stamens widely exserted, June–July; fruits red. HF 2915; JRHS 84: 46; HM 6/1: 142; POe 499–503. Mediterranean region. 1750. z8–9 Plate 100; Fig. 158. See also: **L. × brownii**. ⊕

'Superba'. More vigorous than the species, shoots reddish; flowers in large, terminal panicle-like inflorescences, corolla cream-yellow at first, eventually orange. BM 7977 (= *L. gigantea* Carr.). More attractive than the species, but more sensitive in cultivation! z9 ⊕

L. ferdinandi Franch. Deciduous shrub, upright, to 3 m high, strongly branched, young shoots usually bristly pubescent; leaves brownish on the new growth, ovate to lanceolate, 3–5 cm long, acute, base round to somewhat cordate, dark green above, bristly ciliate, lighter and rough haired beneath, with large, stem-clasping stipules on long shoots, bracts to 1 cm long; flowers in axillary pairs, crown yellowish, bilabiate, 1.5–2 cm long, tube widened pouch-like, densely glandular and short haired outside, May–June; fruits red, separate, surrounded for some time by the opened hull. Mongolia, N. China. 1910. z6 Plate 95. ᗡ

var. **leycesterioides** (Graebn.) Zab. Twigs often nearly glabrous; leaves larger, 4–7 cm long, oval-oblong to lanceolate; corolla not or only slightly bristly pubescent, but not glandular. SH 2: 443c–e (= *L. leycesterioides* Graebn.). Mongolia. ᗡ

L. flava Sims. Deciduous shrub, slightly climbing, 2–2.5 m high, often also bushy, glabrous, twigs gray-green pruinose; leaves short stalked, broadly elliptic to elliptic, 4–8 cm long, bright green above, blue-green beneath and densely pruinose, the uppermost leaf pair fused into a long-rounded disk, apices acute; flowers usually in 2–3 superimposed whorls at the branch tips, yellow at first, later orange, very fragrant, corolla about 3 cm long, tubular, not bulging, May–June. BB 3460; DL 1: 131; BM 1813. Southeast USA. Very hardy, rarely true in cultivation; small flowered and of only slight garden merit. z6

Plate 101; Fig. 170. ⊕

L. flavescens see: **L. involucrata** var. **flavescens**

L. flexuosa see: **L. japonica** var. **repens**

L. floribunda Boiss. & Buhse, non Zab., non Rehd. Deciduous, low growing shrub, closely related to *L. korolkowii*, densely branched, shoots more or less finely tomentose; leaves broad ovate to elliptic, 1.5–3 cm long, obtuse to acute, base truncate to somewhat cordate, bright green and slightly pubescent above, gray-green beneath, densely and soft pubescent; flowers reddish, corolla widened pouch-like at the base, bracts awl-shaped, abundantly flowering; fruits yellow-red. RLo Pl. 3, 18, 19; DL 1: 146. Iran. See also: **L. korolkowii** 'Floribunda' z8 ⊕

L. fragrantissima Lindl. & Paxt. Semi-evergreen shrub, very open habit, to 2 m high, similar to *L. standishii*, but the shoots longer, more bowed, never bristly pubescent (!), pruinose when young; leaves elliptic to broadly ovate or obovate, 3–7 cm long, acute (but not long and drawn out!), tough, dark green and glabrous above, underside more blue-green, midrib bristly, new growth appears very early; flowers cream-white, very fragrant, in axillary pairs, tube short, glabrous outside, December–March; fruits dull-red, long-rounded. BM 8585; DL 1: 144. China, but only known in cultivation. 1845. z5 Plate 99; Fig. 156, 161. # ᗡ

L. fuchsioides see: **L. × brownii** 'Fuchsioides'

L. gibbiflora see: **L. chrysantha**

L. gigantea see: **L. etrusca** 'Superba'

L. giraldii Rehd. Evergreen, twining shrub, usually not over 2 m high, shoots stout, covered with soft yellowish hairs, hollow; leaves oblong-lanceolate, 6 to 7 cm long, acute, base cordate, dull green, often somewhat brownish, both sides densely pubescent; flowers in capitate clusters at the branch tips, corolla purple-red outside, yellow pubescent, 2 cm long, tube slender, slightly bulging, June to July; fruits violet-black, pruinose. RLo Pl. 3; BM 8236. NW. China. 1909. Much more pubescent than *L. henryi*, with which it is often confused. z6 Plate 100; Fig. 162. #

L. glauca see: **L. dioica**

L. glaucescens Rydb. Deciduous, twining shrub, shoots glabrous, very closely related to *L. dioica*, but the leaves pubescent beneath and with only the uppermost pair connate, 4–8 cm long, elliptic to oblong, bluish beneath; flowers light yellow somewhat reddish inside, not fragrant, about 2 cm long, tube slightly widened, interior somewhat short haired, style pubescent, May–June; fruits light red. BB 3457; RWF 358–359 (= *L. douglasii* Koehne non DC.). Northeast N. America, Canada to Nebraska. z3 Fig. 162.

L. glehnii F. Schmidt. Deciduous shrub, very closely related to *L. alpigena*, shoots glandular pubescent; leaves, however, pubescent beneath, ovate to obovate or oblong-oval-cordate; corolla greenish-yellow (not red!),

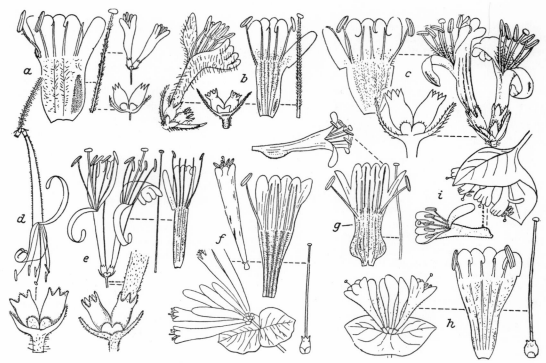

Fig. 162. **Lonicera**. Flowers and flower parts, enlarged. a. *L. alseuosmoides*; b. *L. giraldii*; c. *L. henryi*; d. *L similis*; e. *L. affinis*; f. *L. arizonica*; g. *L. ciliosa*; h. *L. subaequalis*; i. *L. glaucescens* (from Rehder)

anthers yellow, filaments glabrous. MJ 3301 (= *L. alpigena* var. *glehnii* [F. Schmidt] Nakai). Sachalin, Japan, Hondo, Hokkaido. z6

L. gracilipes Miq. Deciduous shrub, upright to 2 m high, twigs spreading, glabrous to finely pubescent; leaves short stalked, broadly ovate to elliptic, acute, base rounded, upper side bright green with a reddish margin, slightly ciliate at first, bluish beneath, somewhat pubescent only at first; flowers usually solitary, on 2–3 cm long stalks, pendulous, funnelform, about 1.5 cm long, carmine-pink, glabrous exterior, style somewhat shorter than the limb, April–May; fruits ellipsoid, scarlet-red, attractive. DL 1: 170; GF 10: 34; MJ 299 to 300; YWP 1: 8 (= *L. uniflora* Bl.). Japan. 1870. z6 Fig. 153, 161. ⊕ ⦵

'Alba'. Flowers white.

var. *glandulosa* see: **L. tenuipes**

L. grata see: **L. × americana**

L. griffithii Hook. f. & Thoms. Deciduous, twining shrub, to 5 m high, shoots glabrous; leaves broadly oval-oblong or rounded, often oak-like, deeply lobed, 3–5 cm long, glabrous, the pair beneath the inflorescence circular, petiolate, bluish; flowers in terminal, stalked clusters of 2–3 whorls, tube about 2.5 cm long, pink to whitish, bilabiate, finely glandular pubescent exterior, calyx with 5 outspread, lanceolate, persistent lobes, May. BM 8965; ST 24; JRHS 66; 121–122. Afghanistan. 1900. Fig. 158. z9

L. gynochlamydea Hemsl. Deciduous shrub, upright, related to *L. pileata* (but with totally different leaves!), shoots reddish, glabrous; leaves oblong-lanceolate to narrowly lanceolate, 5–10 cm long, acuminate, base rounded, midrib pubescent on both sides; flowers erect, short stalked, in pairs, corolla bilabiate, 8 to 12 mm long, white, reddish, tube short, base much widened, pubescent exterior, calyx like that of *L. pileata*, May; fruits reddish to white, separate. Seeds nearly black. SH 2: 441p–r, 442g. W. China. Hardy. z6

L. × heckrottii Rehd. (= ? *L. americana* × *L. sempervirens*). Deciduous shrub, climbing only slightly, often more bushy, shoots glabrous; leaves nearly sessile, elliptic to more oblong, 3–6 cm long, acute, bluish beneath; flowers in elongated, stalked spikes, composed of a few distinctly separate whorls, corolla 3.5–5 cm long, exterior an intensive pink-red, not pubescent, somewhat glandular, interior yellow and slightly pubescent, fragrant, flowers abundantly, buds carmine, prophylls half as long as the ovaries, July–August. Developed in the USA before 1895, but the exact origin is unknown. Sometimes confused with *L. periclymenum* 'Serotina' whose flowers are dark carmine outside, interior light yellow. But the latter is a similarly excellent, long flowering form. z6 Plate 39, 101. ⊕

'**Goldflame**' (Willis Nurseries in Ottawa, Kansas, USA) is a selection of the above with darker flowers; also the leaves are somewhat deeper green. Considered a first rate improvement. ⊕

Fig. 163. **Lonicera.** Flowering shoots and flower parts, partly enlarged.
a. *L. hirsuta;* b. *L. periclymenum;* c. *L. dioica;* d. *L. splendida* (from Boissier, B. M., Nose)

L. henryi Hemsl. Strong growing, evergreen, twining shrub, but also often procumbent, shoots very pubescent and hollow; leaves oblong-lanceolate to lanceolate, 4–9 cm long, acuminate, base often cordate, dark green and ciliate above, underside only pubescent on the midrib, otherwise glabrous, light green and glossy; flowers 2 cm long, usually paired axillary and terminal, reddish or yellow-red, June–July; fruits black. BM 8375; LWT 359 (as *L. acuminata*) (= *L. acuminata* Wall.). China; Hupeh Province. Hardy. z5 Plate 98; Fig. 162. # ❂

L. heteroloba Batal. Deciduous shrub, upright, closely related to *L. tatsiensis,* but with smaller leaves, more densely pubescent, especially on the midrib; stamens shorter than the anthers. RLo Pl. 15. NW. China. 1911. z6 Plate 96.

L. heterophylla Decne. Deciduous shrub, upright, closely related to *L. webbiana,* but the twigs glabrous; leaves glabrous or nearly so, elliptic to oblong-lanceolate, 5–9 cm long, occasionally also lobed (on long shoots); corolla yellowish, turning reddish, exterior glandular and pubescent. Himalayas. Only its variety found in cultivation.

var. **karelinii** (Bge). Rehd. Leaves not lobed, 4 to 8 cm long, elliptic to oblong-lanceolate, somewhat thickish, usually with both sides glandular pubescent on the venation, dark green above, lighter beneath; flowers yellowish-white, turning reddish, 1.5 cm long, corolla usually glandular on the exterior; fruit pea-sized, red (= *L. karelinii* Bge.). Central Asia. 1906. z5 Fig. 169.

L. hildebrandiana Collet & Hemsl. Strong growing, evergreen, twining shrub, twigs glabrous, cylindrical; leaves ovate to elliptic, 7–12 cm long, short acuminate, bright green above, lighter beneath with a few scattered, large, brown glands; flowers in pairs in the leaf axils, but also grouped in larger inflorescences, corolla tube 10–16 cm long, narrow, limb bilabiate, 7–10 cm wide, cream-white at first, later becoming orange, very fragrant, June to August; fruits to 2.5 cm long. BM 7677. Burma, Thailand, China. 1888. The largest and most attractive of all the climbing species, but only for warmer regions or a greenhouse. z9 Fig. 160. # ❂

L. hirsuta Eat. Strong growing, high climbing, deciduous species, twigs glandular and pubescent; leaves short petioled, elliptic, 5–10 cm long, dark green above, gray-green beneath, pubescent on both sides, both apical leaves fused into an acute-elliptic disk; flowers in several dense whorls grouped into a terminal spike, corolla tube orange-yellow, 2–3 cm long, bilabiate, glandular pubescent on the exterior, tube bulging beneath the middle, June–July; fruits yellow-red. BM 3103; DL 1: 132; GF 9: 344; NBB 3: 300 (= *L. pubescens* Sweet). N. America, in moist forests from Canada to Michigan. z3 Plate 101; Fig. 163, 170. ❂ &

L. hispanica see: **L. periclymenum** var. **glauco-hirta**

L. hispida Roem. & Schult. Upright, deciduous shrub, to 1.5 m high, twigs bristly pubescent, winter buds surrounded by a long, conical hull; leaves elliptic to oval-oblong, 3–8 cm long, bristly ciliate, dark green and glabrous above, bluish and bristly pubescent on the venation beneath; flowers in axillary pairs, pendulous on 1 cm long stalks, yellowish-white, corolla funnelform, 2–3 cm long, surrounded by 2 large, about 2–2.5 cm long, ovate, whitish bracts, April–May; fruits oblong, 1.5 cm

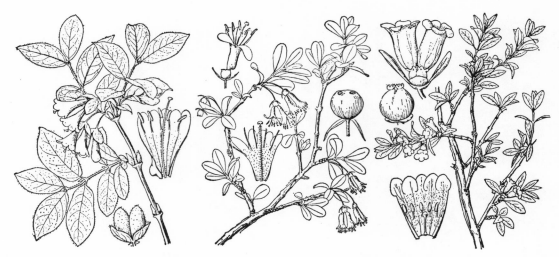

Fig. 164. **Lonicera.** Left *L. hispida;* middle *L. microphylla;* right *L. myrtillus* (from ICS)

long, bright red, framed by the white bracts. DL 1: 174. Turkestan to W. China. 1878. z6 Plate 96; Fig. 164, 168. ⊕ ⚬

var. **bracteata** (Royle) Rehd. ex Airy-Shaw. Leaves oblong-lanceolate to oblong, apex long and drawn out, soft pubescent. BM 9360 (= *L. bracteata* Royle). SE. Tibet. 1924.

var. **hirsutior** Regel. Leaves bristly pubescent on both sides, base normally rounded. Turkestan, Tibet.

var. **setosa** Hook. f. & Thoms. Leaves bristly pubescent on both sides, base slightly cordate. Himalaya; Sikkim.

L. hispidula Dougl. Deciduous shrub, usually procumbent, occasionally also climbing, twigs usually with glandular bristles; leaves oval-oblong, 3–6 cm long, acute or abruptly acuminate, base rounded to nearly cordate, ciliate, slightly pubescent to glabrous above, soft pubescent beneath, the uppermost leaf pair fused into an acute disk; flowers usually in several closely spaced whorls forming a long stalked spike, corolla bilabiate, tube about 1.5 cm long, exterior whitish, turning red, glabrous or somewhat pubescent, stamens and styles as long as the limb, June–July. DL 1: 134; BR 1761. Western N. America. 1830. z6 Fig. 170.

var. *chaetocarpa* see: **L. chaetocarpa**

var. **vacillans** Gray. Stronger growing; leaves larger, more oval-oblong, 3–8 cm long, acute, glabrous above, pubescent beneath; inflorescence longer, often panicle-like, corolla tube 1.8 cm long. MS 647–648. British Columbia to S. California. 1880.

L. iberica Bieb. Upright, deciduous shrub, dense, 1–2 m high, twigs brown-yellow, pubescent; leaves rounded ovate, 2–4 cm long, apex rounded, base cordate, pubescent on both sides, ciliate, gray-green; flowers yellowish-white, bilabiate, 1.5 cm long, pubescent, in axillary pairs, bracts oval-oblong, bracteole hull urceolate, densely pubescent and glandular, June; fruits bright red. 6–8 mm thick. BS 2: 608. Transcaucasia, N. Iran. 1824 z6 Plate 95; Fig. 168. ⊕ ⚬

'Erecta'. Growth narrowly upright, to 3 m high; leaves 1.5–2 cm long; otherwise like the species.

'Microphylla'. Prostrate habit, twigs nodding; leaves 1.5–2 cm long; more pubescent, more blue-green.

L. implexa Sol. Evergreen shrub, climbing to 2 m high (or more), shoots reddish, thin, soft bristly pubescent; leaves oblong to oval-oblong, sessile, 2.5–7 cm long, dark green above and eventually totally glabrous, blue-white pulverulent beneath, leaf pairs on flowering shoots nearly all connate forming rhombic disks; flowers in the axils of the 3 uppermost leaf pairs, corolla 3–4.5 cm long, yellowish-white, often turning red, flowers very abundantly, fragrant, tube pubescent inside, style pubescent on the upper half, June–August. HF 2914; DL 1: 130. S. Europe, N. Africa. 1772. z9 Fig. 158. # ⊕

f. **balearica** DC. Leaves elliptic to obovate or oblong, the lower leaves truncate to cordate on the base; flowers in few flowered whorls. Found throughout the range of the species and more common. #

L. involucrata (Richards.) Banks. Upright, deciduous shrub, usually not over 1 m high, twigs glabrous, somewhat angular; leaves thin, elliptic-ovate to oblong-lanceolate, 5 to 10 cm long, bright green, base tapered, glabrous or somewhat pubescent when young; flowers in axillary pairs, corolla yellow, tubular, 1–1.5 cm long, limb erect, stamens about as long as the limb, May–June; fruits glossy black-red, surrounded by the outspread, eventually reflexed bracts. DL 1: 172; NBB 3: 299; MS 641. Western N. America, Alaska to Mexico. 1824. z6 Plate 99; Fig. 169. ⊕

var. **flavescens** Rehd. Leaves oblong-lanceolate, 5 to 12 cm long, light green, glabrous; corolla bulging, but not widened sack-like. DL 1: 173 (= *L. flavescens* Dipp.). Northwestern USA. ⊕

f. **humilis** Koehne. Shrub, to about 60 m high, young shoots finely pubescent; leaves oval-oblong, acuminate, 4–6 cm long, ciliate; corolla 12–15 mm long, yellow or somewhat reddish, exterior pubescent, style widely exserted. Colorado. Mountain form. ⊕

Fig. 165. **Lonicera.** Flowers and flower parts, enlarged. a. *L. discolor;* b. *L. nervosa;* c. *L. tatarica;* d. *L. ruprechtiana;* e. *L. xylosteum;* f. *L. chrysantha;* g. *L. morrowii;* h. *L. korolkowii* (from Rehder, Schneider, Koehne)

f. serotina Koehne. Shrub, about 1 m high or more; leaves rather large, elliptic-oblong, glabrous; flowers 15–20 mm long, orange-yellow, turning to scarlet-red, July–August (!), bracts outspread when in fruit, not reflexed (!). Colorado. 1903. Most attractive form. ⊕

L. italica see: **L. × americana**

L. japonica Thunb. Semi-evergreen shrub, strong growing, climbing to 6 m high, shoots pubescent; leaves ovate to oblong, 3 to 8 cm long, both sides pubescent at first, eventually glabrous above; flowers white, turning pink, spent flowers yellow, very fragrant, particular at night, tube pubescent and glandular on the exterior, 3–4 cm long, prophylls a third as long as the ovaries, June–July; fruit black. Hy WF 192; GSP 450; MJ 298. Japan, Korea, China; naturalized in the southeastern USA. 1806. z5 Fig. 160. #

'Aureo-reticulata'. Like var. *repens,* but the leaves have golden-yellow reticulate venation (= *L. brachypoda aureo-reticulata* Jakob-Makoy). Around 1862. Colors best in full sun. Often used as a potted plant. z9 #

var. **chinensis** (Wats.) Bak. Leaves nearly glabrous, but usually ciliate, often with the venation pubescent beneath, young leaves frequently reddish; corolla with a carmine exterior, upper lip incised to the midpoint, prophylls narrower than the ovaries. SL 1: 137 (as *L. japonica*), # ⊕

'Halliana'. Leaves pubescent on both sides when young; corolla totally white, occasionally somewhat reddish, spent flowers yellowish, upper lip barely split to the middle,

prophylls as wide as the ovaries; fruits black. MG 16: 609. Around 1862. Easily naturalized in the garden. # ⊕

var. **repens** (Sieb.) Rehd. Young shoots red; leaves ovate-oblong, often lobed oak-like, 6–12 cm long, bright green above, gray-green beneath, venation often reddish; flowers milk-white, spent flowers light yellow, upper lip divided into short, ovate sections. DL 1: 137 (as *L. japonica*); BM 3316 (= *L. brachypoda* DC.; *L. flexuosa* Thunb.; *L. japonica* var. *chinensis* [Wats.] Bak.; *L. japonica* var. *flexuosa* [Thunb.] Nichols.). China. Not as strong growing as 'Halliana'. #

L. karelinii see: **L. heterophylla** var. **karelinii**

L. kesselringii Reg. Shrub, very similar to *L. caucasica,* but with oblong or elliptic-lanceolate leaves, 3–6 cm long, 1–2 cm wide; flowers smaller, pink, the corolla tube only slightly inflated, paired on 8 mm long stalks. Gfl 40: 44 (= *L. caucasica* var. *longifolia* Dipp.). Kamchatka. Introduced around 1888. z6

L. korolkowii Stapf. Deciduous shrub, upright, to 3 m high, broad growing, finely branched, shoots finely pubescent; leaves broadly ovate to elliptic, acute to obtuse, 1–2.5 cm long, gray-green on both sides, lightly pubescent above, more dense beneath; flowers whitish-pink, 1.5 cm long, in axillary pairs, corolla tube narrow, as long as the narrow sections of the upper lip, base hardly bulging, June; fruits bright red. RLo Pl. 3; GF 7: 4; MD 1910: 115. Turkestan, Buchara (USSR). 1880. z6 Plate 96; Fig. 165. ∅ ⊕

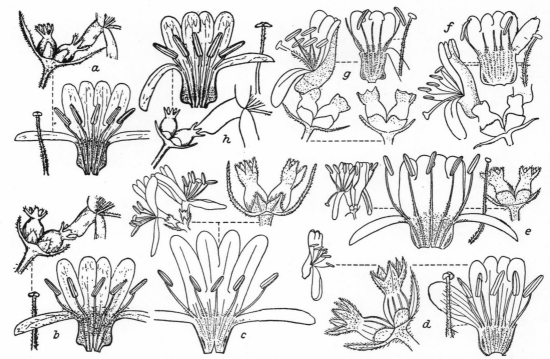

Fig. 166. **Lonicera.** Flowers and flower parts, enlarged. a. *L. × muscaviensis;* b. *L. × muendeniensis;* c. *L. maackii;* d. *L. arborea;* e. *L. quinquelocularis;* f. *L. trichosantha;* g. *L. deflexicalyx;* h. *L. korolkowii* var. *zabelii* (from Rehder, Schneider, Zabel, Koehne)

'Aurora'. An elegant shrub, to 2.5 m high, twigs gracefully nodding; leaves narrowly ovate, 1.5 to 2.5 cm long, to 4 cm long on long shoots, acute, velvety short haired beneath, gray-green on both sides; flowers 18 mm wide, bright pink, tube 7–8 mm long, glabrous; fruits small, orange-red. 1893. Flowers abundantly. Ø ✣

'Floribunda'. Leaves more broadly ovate, more obtuse, only broadly cordate-rounded on long shoots, velvety short haired; flowers white, corolla 15 mm long, tube distinctly bulging. MD 1910: 115 (= *L. floribunda* Zab. non Boiss. & Buhse). 1901. Ø ✣

var. **zabelii** (Rehd.) Rehd. Shrub, to 2 m high; leaves ovate, 3–3.5 cm long, 2–2.5 cm wide, acute to obtuse, totally glabrous; corolla about 1.2 cm long, white with a trace of pink or totally pink, distinctly broad, pouch-like at the base. DM 1910: 10 (= *L. zabelii* Rehd.). Buchara (USSR). Obtained by the Späth Nursery of Berlin in 1893 from Kesselring of Petersburg (now Leningrad). Fig. 166. ✣

L. ledebourii Esch. Deciduous shrub, upright, to 2 m high, twigs long, rodlike, glabrous or somewhat pubescent; leaves thickish, oblong to ovate-lanceolate, 6–12 cm long, very dark green and somewhat glossy above, lighter and soft pubescent beneath, ciliate, base round to tapered; flowers deep yellow, turning orange or · scarlet, interior yellow, 1.5–2 cm long, tube funnelform, limb somewhat outspread, exterior glutinous pubescent, on erect, 2–4 cm long stalks, June–July; fruit pea-sized, purple to black, always surrounded by the persistent, red bracts. BM 8555; DL 1: 171; GF 2:

289. California. 1838. z6 Fig. 169. ✣ ⚇

L. leycesterioides see: **L. ferdinandi** var. **leycesterioides**

L. ligustrina see: **L. nitida**

L. maackii (Rupr.) Maxim. Deciduous shrub, upright, wide, to 5 m high, young shoots pubescent; leaves ovate-elliptic, to ovate-lanceolate, 5–8 cm long, dark green above, lighter beneath, usually slightly pubescent on both sides; flowers in axillary pairs, white, old flowers yellow, fragrant, 2 cm long, exterior usually glabrous, tube very short, slightly bulging, upper lip ⅔ incised, stamens and styles usually much shorter than the corolla limb, June; fruits dark red, distinct. DL 1: 153; Gfl 1162; NT 1: 641. Manchuria, China, Japan. 1880. z6 Plate 98; Fig. 166. ✣ ⚇

'Erubescens'. Like var. *podocarpa* but with a pink blush on the flowers. ✣

var. **podocarpa** Franch. Growth broader; leaves wider, ovate to elliptic, abruptly acuminate, darker green and more pubescent; flowers somewhat smaller, often in 2 rows on the horizontal twigs, flowering somewhat later, ovaries with short but distinctly stalked bracteoles (!), stamens and pistils as long as the corolla limb; fruit red, more attractive than that of the species. Gs 1921: 36; AB 22: 61. China. 1900. Plate 98. ⚇ ✣

L. maximowiczii (Rupr.) Maxim. Deciduous shrub, upright, to 3 m high, twigs reddish, glabrous; leaves elliptic or ovate to oval-oblong, acute, 3–7 cm long, dark green and glabrous above, light green and pubescent beneath; flowers violet-red, 1 cm long, exterior glabrous, on 2 cm long, glabrous stalks, tube very expanded at the base, May–June; fruits pea-sized, red. DL 1: 160; Gfl 17: 597. Manchuria, China. 1878. z5 Plate 99; Fig. 169.

var. **sachalinensis** F. Schmidt. Leaves reddish on new growth, broad and less acute, deep green and glabrous above, distinctly bluish beneath; flowers larger, 13–15(18) mm long, dark purple, calyx teeth acuminate; fruits totally connate, dark red (= *L. sachalinensis* E. Wolf). Sachalin, Japan; Hondo. 1917. More attractive than the species. ⊕ ⚭

L. microphylla Roem. & Schult. Deciduous shrub, upright, to 1 m high, abundantly branched, shoots glabrous to finely pubescent; leaves short petioled, obovate to oblong, 1–2.5 cm long, obtuse, tapered at the base, dull green above, finely pubescent on both sides; flowers yellowish-white, about 1 cm long, exterior glabrous to finely pubescent, calyx indistinctly 5 toothed, anthers slightly exserted past the limb, May; fruits usually connate to the apex, orange-red. DL 1: 161 (= *L. bungeana* Ledeb.). Afghanistan, Turkestan, Himalaya, Tibet, Mongolia. 1818. Hardy. Often incorrectly labeled in cultivation. z5 Plate 95; Fig. 153, 164.

L. × minutiflora Zab. (*L. morrowii* × *L. xylosteoides*). Deciduous shrub, upright, twigs spreading outward, finely pubescent; leaves oval-oblong to oblong, obtuse, 2–3 cm long, finely pubescent beneath; prophylls oval-oblong, about as long as the ovaries, corolla whitish, 12 mm long, limb outspread, upper lip incised past the middle, stamens half as long as the limb, May–June; fruits red. Before 1878. z6

L. "morrotarica" see: **L. × bella**

L. morrowii Gray. Deciduous shrub, to 2 m high, broad arching habit, young twigs soft pubescent; leaves elliptic to obovate-oblong, 3–5 cm long, slightly soft pubescent above when young, more pubescent beneath; flowers white at first, spent flowers yellow, 1.5 cm long, pubescent exterior, tube slender, expanded, upper lip incised to the base, forming 12 mm long, narrow sections, stamens shorter than the limb, glabrous, May–June; fruits dark red. DL 1: 142; MJ 303; Gs 10: 215; YWP 1: 9. Japan. z6 Plate 96, 97; Fig. 165. ⊕ ⚭

'Xanthocarpa'. Fruits yellow. Ad 17: Pl. 153. ⚭

L. × muendeniensis Rehd. (*L. bella* × *L. ruprechtiana*). Deciduous shrub, upright, young shoots pubescent; leaves ovate to oval-lanceolate, acuminate, 3–7 cm long, base round or tapered, dark green above, pubescent beneath; prophylls broadly ovate, ⅓ as long as the ovaries, ciliate, corolla white to yellowish-white, occasionally somewhat reddish, upper lip erect, incised nearly to the base, filaments pubescent on the basal half, May; fruits red. Gfl 42: 101. Developed in the Munich Botanic Garden, W. Germany. Before 1883. z5 Fig. 166. ⚭

L. muscaviensis Rehd. (*L. morrowii* × *L. ruprechtiana*). Deciduous shrub, upright, new growth pubescent;

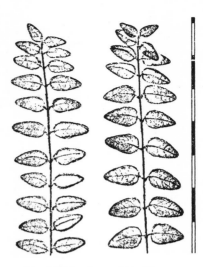

Fig. 167. *Lonicera nitida* 'Elegant' (= *L. pileata* "yunnanensis")

leaves ovate to oval-oblong, acuminate, dark green and scattered pubescent above, densely pubescent beneath, 3–5 cm long; prophylls ovate, ciliate, half as long as the ovaries, corolla white, upper lip erect, incised to somewhat under the middle, filaments glabrous, May–June; fruits bright red. Gfl 42: 101. Developed before 1893 in the Muskau Nursery, Silesia (now N. Czechoslovakia and SW. Poland) from seed obtained from St. Petersburg (Leningrad) USSR. z5 Fig. 166. ⚭

L. myrtilloides Purpus (perhaps *L. angustifolia* × *L. myrtillus*). Deciduous shrub, finely branched, about 1.5 m high, young shoots glandular pubescent; leaves elliptic to narrowly oblong, 1–3 cm long, somewhat pubescent, at least on the midrib; flower stalk 1 cm long, flowers nodding, corolla white, reddish on the base, exterior pubescent, fragrant, May–June; fruits connate, red. MD 1907: 255; SH 2: 436a–c, 437b to c. Himalaya. 1907. z6 Plate 95; Fig. 153.

L. myrtillus Hook. f. & Thoms. Deciduous shrub, rounded, dense, finely branched, 0.5–1 m high, twigs glabrous or also pubescent when young; leaves very short stalked, ovate to oblong, 6–25 mm long, dark green above, gray-green beneath, glabrous; flowers tubular-campanulate, 6 to 8 mm long, yellowish-white, glabrous, fragrant, May to June; ovaries connate, 2 locular, fruit orange-red, 6 mm thick. DL 1: 166; ST 1: Pl. 44. Afghanistan to Himalayas, Sikkim. 1879. z6 Plate 95; Fig. 153, 164.

var. **depressa** (Royle) Rehd. Bracts larger, wider, usually elliptic (narrowly oblong on the species); flower stalk as long as the leaves (= *L. depressa* Royle). Himalayas; Nepal, Sikkim. 1910.

L. nepalensis see: **L. × xylosteoides**

L. nervosa Maxim. Deciduous shrub, upright, glabrous, to 3 m high, twigs erect and spreading, young shoots usually red-brown; leaves short stalked, elliptic to oval-oblong or rhombic, 3–6 cm long, acute at both ends,

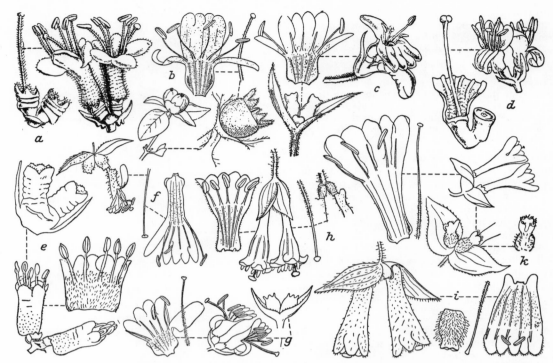

Fig. 168. **Lonicera.** Flowers and flower parts, most enlarged. a. *L. pileata*; b. *L. iberica*; c. *L. standishii*; d. *L. fragrantissima*; e. *L. nitida*; f. *L. altmannii*; g. *L. bracteolaris*; h. *L. hispida*; i. *L. chaetocarpa*; k. *L. strophiophora* (from Rehder, Koehne, B. M.)

bright green above with reddish venation, bluish beneath, glabrous; prophylls fused into a usually deeply lobed cupule, ovaries distinct, corolla 1 cm long, light pink, lip somewhat longer than the tubercled tube, style pubescent nearly to the apex, May–June; fruits distinct, black. RLo 16. China; Kansu Province. 1872. z6 Plate 96; Fig. 165.

L. nigra L. Deciduous shrub, upright, to 1.5 m high, twigs glabrous; leaves elliptic to oval-oblong, 4–6 cm long, bright green and glabrous above, bluish beneath and pubescent along the midrib, also often totally glabrous; flowers dull pink, bract half as long as the ovaries, prophylls equally long, ciliate, May–June; fruits blue-black, only connate at the base. HF 2918; DL 1: 145; HM Pl. 251. Central Europe to Serbia (Yugoslavia). Ⓕ Czechoslovakia, in windbreaks. z6 Plate 99.

L. nitida Wils. Evergreen shrub, upright, to 2 m high in its habitat, abundantly branched, young twigs usually regularly decussate, purple at first, densely pubescent; leaves ovate-oblong, 6–12 mm long, glossy dark green above, lighter beneath; flowers cream-white, 8–10 mm long, in short stalked pairs on small side shoots, May; fruits connate, 5 to 6 mm thick, purple, glossy, usually hidden beneath the leaves, hardy visible. BM 9352 (= *L. pileata* f. *yunnanensis* [Franch.] Rehd.; *L. ligustrina yunnanensis* Franch.). W. China. 1908. z7–8 Plate 94, 95; Fig. 168. #

Several clones have been selected and are being propagated today:

'Aurea'. Narrowly upright; leaves golden-yellow. Origin unknown; offered in 1957 in the catalogue of Duncan & Davies Ltd. Nurseries of New Plymouth, New Zealand. Prefers full sun. # ⌀

'Baggesen's Gold'. Low, broad growing; leaves yellow, more yellow-green in summer. Developed by J. H. Baggensen, Pembury, Kent, England. 1967. # ⌀

'Elegant'. Shrub, to 1 m high, twigs loose, often spreading nearly horizontally or gracefully nodding, fast growing; leaves mostly distichous, ovate to oval-rounded, 15 mm long, obtuse, dull green above; flowers and fruit like those of the species. Origin unknown. Cultivated since 1935. Plate 94, 95; Fig. 167. # ⌀

'Ernest Wilson'. Lateral shoots directed downward; leaves usually oval-lanceolate to triangular-ovate, less than 12 mm long, glossy green. DB 1964: 26. Probably developed from Wilson's Nr. 833. #

'Fertilis' (Hillier). Stiffer, more upright habit, growing to over 2 m high; flowers and fruits abundantly. DB 1964: 31. Selected by Hillier, Winchester, England. #

'Graziosa' (Juergl). Growth broader, denser; leaves somewhat smaller; flowers and fruits sparsely. DB 1957: 59. Originated from seed/of "*L. pileata yunnanensis*" in 1952 by Jürgl in Sürth, near Cologne, W. Germany. #

'Hohenheimer Findling'. Growing to 1.25 m wide with nodding twigs; leaves bright green, 10–18 mm long, narrowly oval. Discovered by Dietrich in Stuttgart-Hohenheim, W. Germany in 1940 but first recognized many years later. Very similar to *L. nitida* 'Elegant' but more frost tolerant and remaining green later in the fall.

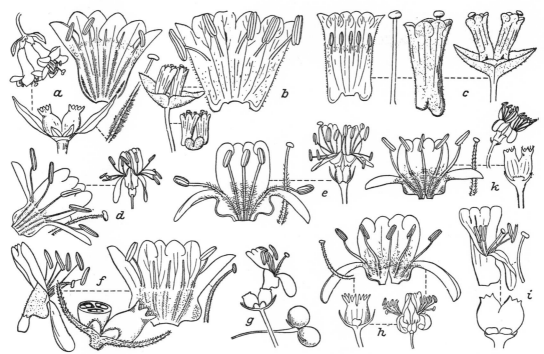

Fig. 169. **Lonicera.** Flowers and flower parts, most enlarged. a. *L. pyrenaica;* b. *L. involucrata;* c. *ledebourii;* d. *L. oblongifolia;* e. *L. alpigena;* f. *L. webbiana;* g. *L. heterophylla* var. *karelinii;* h. *L. caucasica;* i. *L. chamissoi;* k. *L. maximowiczii* (from Rehder, Schneider, Koehne)

'Yunnan'. Similar to 'Ernest Wilson', but the lateral shoots are shorter and more erect; leaves somewhat larger and usually not distichous; a good bloomer. DB 1964: 28 (= *L. pileata yunnanensis* in British nurseries). #

L. × notha Zab. (*L. ruprechtiana* × *L. tatarica*). Strong growing, upright, deciduous shrub, young shoots glabrous or nearly so; leaves ovate to oval-lanceolate, seldom elliptic, 3–6 cm long, acuminate, base round to truncate, slightly pubescent beneath to nearly glabrous, prophylls ovate, ⅓ as long as the ovaries, glabrous to glandular ciliate; corolla 18 mm long, white, yellowish or pink, upper lip erect, divided to the middle or nearly to the base, tube expanded, May–June; fruit red. Developed in 1878. z4

'Alba'. Flowers white.

'Carneorosea'. Flowers bright red.

'Gilva'. Flowers yellowish-white, pink margined.

'Grandiflora'. Flowers large, yellowish, turning pale pink, May; fruit dark red. ⊕

'Ochroleuca'. Flowers yellowish; fruit orange.

L. nummulariifolia Jaub. & Spach. Very similar to *L. arborea,* but the leaves smaller, ovate to circular, often somewhat pubescent above; prophylls shorter than the ovaries, corolla tube more slender, not expanded. SL 1: 147 (= *L. arborea* var. *persica* [Jaub. & Spach] Rehd.; *L. persica* Jaub. & Spach; *L. turcomanica* Fisch. & Mey.). Turkestan; as well as Southern Greece and Crete. 1880.

Hardier than *L. arborea.* z6

L. oblongifolia (Goldie) Hook. Deciduous shrub, upright, to 1.5 m high, twigs finely soft pubescent; leaves nearly sessile, oblong to oblanceolate, 3–8 cm long, blue-green above, gray-green beneath, short pubescent on both sides, bracts small, prophylls indistinct; flowers yellowish, deeply bilabiate, 1–1.5 cm long, tubular, on 2.5 cm long, erect stalks, May; fruits red. DL 1: 162; NBB 3: 298; GSP 457. Northern N. America, in swamps and bogs. 1823. z3 Plate 99; Fig. 169.

L. obovata Royle. Deciduous shrub, upright, bushy, to 2 m high; leaves obovate, 5–12 mm long, base cuneate, whitish beneath, bracts scarcely reaching past, or as long as the totally connate ovaries; flowers yellowish-white, 1 cm long, glabrous outside, stamens as long as the limb, style longer, May; fruits blue-black, short ellipsoid. SH 2: 439a–d, 440e (= *L. parvifolia* Edgew. non Hayne). Himalayas to Afghanistan. 1894. Hardy. z5 Fig. 153.

L. occidentalis see: **L. ciliosa** var. **occidentalis**

L. orientalis Lam. Upright, deciduous shrub, 2–3 m high, twigs glabrous; leaves ovate to oval-lanceolate, 4–10 cm long, acute, base usually round, dark green above, gray-green beneath and glabrous except for the venation; flowers tubular, 1–1.2 cm long, pink to violet, stalk 1 cm long, bracts subulate, ¼ as long as the ovaries, May–June; fruits connate, black. DL 1: 157. Asia Minor. According to Rehder, the species is not yet in cultivation, only the two varieties.

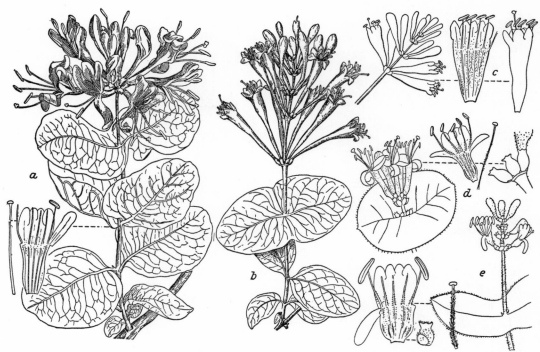

Fig. 170. **Lonicera.** Flowering shoots and flower parts, partly enlarged. a. *L. flava;* b. *L. sempervirens;* c. *L. sempervirens* 'Minor'; d. *L. hirsuta;* e. *L. hispidula* (from Schmidt, B. M., Rehder)

var. *caucasica* see: **L. caucasica**

var. *discolor* see: **L. discolor**

L. parviflora see: **L. dioica** and **L. obovata**

L. periclymenum L. Deciduous climbing shrub, 3–4 m high, young twigs pubescent or glabrous; all leaves always distinct, the uppermost pair never connate (!), ovate to elliptic or oval-oblong, 4–6 cm long, dark green above, blue-green beneath, usually totally glabrous, somewhat pubescent only when young; flowers in stalked, 3–5 whorled spikes, yellowish-white, exterior turning red, very fragrant, 4–5 cm long, exterior glandular glutinous, May–June; fruits round, bright red, with persistent calyx lobes. POe 504–507; HF 2916. Central and W. Europe, N. Africa; at the forest's edge, brushy slopes, hedge and fence rows. z6 Fig. 163. ⊕

Includes the following cultivars:

'**Aurea**'. Leaves yellow variegated, never "oak-like" sinuate. IH 59. Before 1871.

'**Belgica**'. Habit more stocky; leaves glabrous, tough, elliptic to oblong, longer petioled; flowers pale purple on the exterior, usually turning yellowish. BS 2: 253. Before 1789. ⊕

var. **glauco-hirta** Kunze. Leaves pubescent on both sides, bluish beneath, ovate to elliptic, acute (= *L. hispanica* Boiss. & Reuter). Spain, Morocco. z9

'**Quercina**'. Leaves "oak-like" sinuate, occasionally also somewhat white variegated. Lu 1935: 249. Before 1770.

'**Serotina**'. Leaves narrower; flowers deep purple outside, turning paler in time, interior yellow, flowers abundantly, June–September. Gn 45: 307. ⊕

L. persica see: **L. nummulariifolia**

L. pileata Oliv. Evergreen shrub, broad growing, procumbent, scarcely higher than 30 cm, twigs thin, pubescent; leaves distichous in arrangement, oblong-lanceolate, obtuse, glabrous, 12–25 cm long, deep green and glossy above, lighter beneath, short petioled; flowers in sessile pairs, bracts lanceolate-subulate, usually as long as the ovaries, calyx with a collar-like appendage covering the cupule margin, corolla 6–8 mm long, funnelform, light yellow, pubescent exterior, fragrant, base very tubercled, May; fruits globose, purple-violet, 5 mm wide. BM 8060. China; W. Hupeh, W. Szechwan Provinces. 1900. z6 Plate 95; Fig. 168. # ∅

"f. *yunnanensis*" is *L. nitida* 'Elegant'. See also: *L. nitida* 'Yunnan'

L. pilosa see: **L. strophiophora**

L. praeflorens Batal. Deciduous shrub, to 2 m high, closely related to *L. altmannii;* leaves short petioled, broadly ovate to oval-elliptic, 4 to 6.5 cm long, acute, pubescent on both sides; flowers appear before the leaves, very short stalked, corolla about 15 mm long, yellowish-white, interior and exterior glabrous, limb regular, scarcely bilabiate, tube short, hardly expanded, anthers purple, March–April. RLo 13–14. Manchuria, Korea. 1917. Hardy. z5 ⊕

Fig. 171. **Lonicera.** Left, *L. tatsienensis;* middle, *L. thibetica;* right, *L. tragophylla* (from ICS)

L. prolifera (Kirchn.) Rehd. Deciduous shrub, almost never climbing, usually bushy, about 1–1.5 m high, very blue pruinose, twigs glabrous; leaves sessile to short petioled, elliptic to oblong-obovate, 5–9 cm long, bright green above, often pruinose, blue-green beneath and often finely pubescent, the uppermost 2–4 pairs fused into rounded, flat, tough disks, very pruinose and emarginate on the apex; flowers usually in 4 whorls, superimposed on each other, corolla tube yellow, exterior glabrous, interior pubescent, 3 cm long, tube somewhat expanded, June–July; fruits scarlet-red, fruits abundantly. GF 3: 34; BC 2207; NBB 3: 301 (= *L. sullivantii* Gray). N. America, in moist forests and thickets. 1840. z5 Plate 101; Fig. 158. ✣ ⚭

L. × propinqua Zab. (*L. alpigena* × *L. ledebourii*). Deciduous shrub, developed in 2 forms, one more like the former parent, the other resembling the latter; corolla yellowish-brown, bilabiate, very expanded, prophylls about as long as the partly connate ovaries, densely glandular like the almost twice as large bracts. Developed in 1844 in the H. Münden Botanic Garden. z6 ✣

L. prostrata Rehd. Related to *L. trichosantha,* deciduous shrub, mat-like procumbent, twigs often creeping; leaves ovate to elliptic, 2–3 cm long, acute, soft pubescent and ciliate; flowers pale yellow, in axillary pairs, corolla 15 mm long, pubescent, bilabiate, June; fruits ovate, red, 6–8 mm long. China. 1904. z6 Plate 95.

L. × pseudochrysantha Barun. (= *L. chrysantha* × *L. xylosteum*). Similar to *L. chrysantha,* but with prophylls broad and about half as long as the ovaries, ciliate. DL 1: 140 (as *L. regeliana* Dipp.). Origin unknown. Before 1889.

L. pubescens see: **L. hirsuta**

L. punicea see: **L. × brownii 'Punicea'**

L. purpurascens Walp. Deciduous shrub, related to *L. tangutica,* but to 3 m high, shoots pubescent; leaves ovate to oblong or obovate-oblong, 2–4 cm long, obtuse to acute, densely pubescent beneath, less so above; flowers nodding on small stalks, corolla tubular-funnelform, 1.5 cm long, turbid red, exterior pubescent, base expanded, prophylls glandular ciliate, ⅓ as long as the ovaries, anthers as long as the limb, style somewhat longer, May, (= *L. sericea* Royale). Himalayas; Kashmir; Afghanistan. 1864. z6

L. × purpusii Rehd. (= *L. fragrantissima* × *L. standishii*). Semi-evergreen shrub, upright, 2–3 m high, densely branched, twigs broadly arching, glabrous, light bristly pubescent only on the long shoots; leaves ovate-elliptic, 5–10 cm long, glabrous on both sides except for the pubescent venation beneath and the bristly margin; flowers 2–4 in axillary clusters, cream-white, very fragrant, exterior totally glabrous, filaments glossy white, anthers golden-yellow; December–April; fruits red. Gfl 75: 165; BMns 323; Gs 1941: 93. Developed before 1920 in the Darmstadt Botanic Garden. Hardy. z6 Plate 97. # ✣

L. pyrenaica L. Small, deciduous shrub, upright, to 1 m high, densely branched, shoots glabrous, somewhat angular; leaves oblong-ovate to oblong-oblanceolate, 2–4 cm long, bright green above, blue-green beneath, glabrous; flowers funnelform-campanulate, 2 cm long, nodding on 1–2 cm long stalks, corolla yellowish-white with a reddish trace, base tubercled, prophylla oval-lanceolate, bracts leaf-like, May; fruits globose, red, 6 mm thick. BM 7774; DL 1: 163; BS 2: 619; NF 4: 129. Pyrenees Mts. and the Balearic Islands (Spain). 1739. z7 Plate 95; Fig. 169 ⊘ ✣

L. quinquelocularis Hardw. Upright, deciduous shrub, 3–4 m high, branches spreading broadly, young shoots reddish, pubescent; leaves broad ovate to elliptic or oblong-ovate, 3–5 cm long, glabrous above, gray-green and pubescent beneath; flowers short stalked in axillary pairs, cream-white, later turning yellow, corolla bilabiate, 1.5–2 cm long, exterior with densely appressed pubescence, tube slender, only slightly expanded, somewhat shorter than the limb, June; fruits translucent, whitish, with violet seeds, fruits abundantly and attractively. BR 30: 33; Fgl 1551 (= *L. diversifolia* Wall.; *L. royleana* Wall.). Himalayas to Afghanistan and Beluchistan. 1840. z6 Plate 96; Fig. 166. ⚬

f. **translucens** (Carr.) Zab. Shrub, usually not over 2.5 m high; leaves ovate to oval-oblong, longer acuminate, base round to somewhat cordate; flowers somewhat smaller, tube distinctly expanded on one side; fruits translucent white, seeds violet. Gs 1928: 376; GC 130: 67 (= *L. translucens* Carr.). Around 1870. ⚬

L. 'Redgold' see: *L.* × **tellmanniana**

L. regeliana see: *L.* × **pseudochrysantha**

L. royleana see: *L.* **quinquelocularis**

L. rupicola Hook. f. & Thoms. Dense, bushy, deciduous shrub, to 1 m high, twigs arching, glabrous; leaves ovate-oblong, 1–2.5 cm long, obtusish, bluish above, gray-green beneath, pubescent to glabrous, base round to somewhat cordate; flowers pale lilac, in axillary pairs, fragrant, corolla limb with 5 ovate lobes, tube pubescent outside, June; fruit red. SH 2: 436k–l. Himalayas, Tibet. 1888. Quite variable in habit and degree of pubescence; usually short branched in its mountain habitat, long branched in cultivation. z8 Fig. 153.

L. ruprechtiana Regel. Upright, deciduous shrub, 2–3 m high, twigs somewhat nodding, somewhat pubescent when young; leaves oblong-obovate to lanceolate, 6–10 cm long, dark green and glabrous above, lighter and pubescent beneath; flowers in axillary pairs, each pair on a 1.5–2 cm long stalk, white, turning yellow, corolla with a glabrous exterior, about 1.5–2 cm long, upper lip half or ²⁄₃ incised, prophylls small, glabrous or with marginate glands, like the calyx, May; fruits orange to dark red. DL 1: 150; Gfl 645. NE. Asia; Manchuria, China. 1880. z6 Plate 98; Fig. 165. ⊕ ⚬

'Xanthocarpa'. Leaves densely pubescent; flowers smaller, fruit yellow. Developed in the Botanic Garden of Hanover-Münden, W. Germany. ⚬

L. sachalinensis see: *L.* **maximowiczii** var. **sachalinensis**

L. sempervirens L. Semi-evergreen, high climbing shrub, totally glabrous, shoots not pruinose; leaves elliptic or ovate to oblong, 3 to 8 cm long, nearly sessile, both uppermost pairs fused into round disks, dark green and glabrous above, blue-green, often pubescent beneath; flowers in 3–4 superimposed whorls, corolla 4–5 cm long, orange-yellow and scarlet-red, interior yellowish, without fragrance, tube with 5 nearly equal sized lobes, May–fall; fruits bright red. BB 3461; JRHS 89: 194; DL 1: 135; MJ 306. N. America. 1656. z8 Plate 100; Fig. 170. See also *L.* × **brownii**. # ⊕

coccinea superba see: '**Superba**'

'Dreer's Everblooming' see: '**Superba**'

'Magnifica' see: '**Superba**'

var. **minor**. Ait. Leaves more conspicuously semi-evergreen, elliptic to oblong-lanceolate; flowers smaller, more slender, orange-red to scarlet, flowers very abundantly. BM 1753; BR 556. Before 1789. More tender than the species since it is found in the southern part of the species range. z8–9 Fig. 170. # ⊕

'Red Coral' see: '**Superba**'

'Red Trumpet' see: '**Superba**'

'Rubra' see: '**Superba**'

var. *speciosa* see: '**Superba**'

'**Sulphurea**'. Flowers golden-yellow. Gfl 2: 38. # ⊕

'**Superba**'. Leaves quite glabrous above and on the margins, ovate-elliptic, usually only deciduous; flowers more scarlet-red. FS 1128 (= *L. sempervirens coccinea superba* Dipp.; var. *speciosa* Carr.). Before 1856. Found under many names in American nurseries, such as 'Magnifica', 'Dreer's Everblooming', 'Red Coral', 'Red Trumpet' and 'Rubra'. ⊕

L. sericea see: *L.* **purpurascens**

L. setifera Franch. An attractive, medium high shrub, with very bristly pubescent shoots; leaves oblong-lanceolate, 4.5–7.5 cm long, usually coarsely dentate, soft pubescent on both sides; flowers tubular, straw-yellow to pink, with a sweet fragrance, appearing sparsely in clusters before the leaves, interior and exterior bristly, ovaries glandular and bristly, February–April; berries red, bristly. Himalayas; Assam; China. 1924. z8? Fig. 172. ⊕

L. similis Hemsl. Semi-evergreen, climbing shrub. SH 2: 457, 458. Central and W. China. Not in cultivation. Fig. 162. #

var. **delavayi** (Franch.) Rehd. Glabrous variety of the species. Evergreen, climbing shrub, shoots glabrous; leaves broadly lanceolate, acute, 4–6 cm long, base round to somewhat cordate, bright green and glabrous above, short white tomentose beneath; flowers in axillary pairs, grouped into large racemes at the shoot tips, corolla 6–8 cm long, white, spent flowers turning yellow, tube slightly expanded, slender, limb bilabiate, stamens and pistils widely exserted, August; fruit black, ovate. BM 8800 (= *L. delavayi* Franch.). SW. China. 1901. z9 Plate 100; Fig. 172. # ⊕

L. spinosa (Decne.) Walp. Deciduous shrub, thorny branched, often nearly leafless, low, differing from *L. alberti* (which is often considered a variety of this species) in the much smaller leaves, only 2–3 cm long, narrowly oblong, glabrous, occasionally with 2 teeth; corolla tubular-funnelform, lilac-pink, with a slender, thin tube and an outspread limb, exterior glabrous, interior pubescent, limb tips oval (!), anthers not exserted past the limb tips (!). NW. Himalayas, Tibet, E. Turkestan, in dry regions. Not in cultivation! z6 Plate 95.

var. *alberti* see: *L.* **alberti**

L. splendida Boiss. Evergreen, climbing shrub, strong growing, annual growth occasionally to 1.5 m; leaves sessile, ovate to oblong, 2.5–5 cm long, very blue-green, the apical pair connate at the base; flowers in terminal,

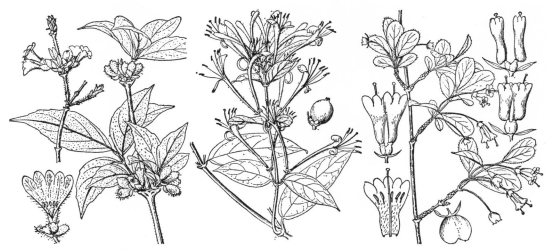

Fig. 172. **Lonicera.** Left, *L. setifera;* middle, *L. similis* var. *delavayi;* right, *L. tangutica* (from ICS)

sessile clusters of densely compound whorls, corolla 3 to 5 cm long, bilabiate, whitish-yellow, turning red, exterior finely pubescent, limb 2.5 cm wide, June to August. FS 1130; DL 1: 129; BM 9517. Spain. z9 Fig. 163. # ⟐

L. standishii Jacq. Semi-evergreen, upright shrub, to 2 m high, young shoots bristly pubescent (!); leaves ovate-oblong to lanceolate, 5–10 cm long, bristly ciliate, both sides rough haired, especially dense beneath, slender acuminate (!); flowers in axillary pairs, cream-white with a trace of pink, exterior usually pubescent (!), very fragrant, bracts pubescent and ciliate, linear-lanceolate, prophylls barely developed, March–April, often flowering as early as December. BM 5709; DL 1: 143. China: Szechwan, Hupeh Provinces. 1845. z6 Fig. 161, 168. # ⟐

f. **lanceifolia** Rehd. Like the species, but with much narrower leaves, only 1–2.5 cm wide; otherwise hardly differing. GC 67: 101; RH 1916: 24. Szechwan Province. 1908. Equally meritorious. # ⟐

L. strophiophora Franch. Deciduous shrub, upright, closely related to *L. hispida,* winter buds scaly, twigs long pubescent at first, later glabrous; leaves oval-elliptic, acuminate, 4–8 cm long, short petioled, bright green on both sides, loosely pubescent, often only on the underside; flowers on nodding, somewhat glandular, pubescent stalks, bracts broadly ovate, to 15 mm long, corolla tubular-funnelform, 2 cm long, whitish, expanded at the base, limb lobes oblong, glabrous on both sides, March to April; fruits nearly globose, red, pubescent. DL 1: 175; MJ 304; SH 2: 447 (= *L. pilosa* Maxim. non Willd.; *L. amherstii* Dipp.). Japan. 1915. z6 Fig. 160.

L. subaequalis Rehd. Deciduous, climbing shrub, entire plant glabrous, except for the inflorescence; leaves elliptic to oblong-obovate, 7–10 cm long, obtuse, base tapered to a short petiole, the uppermost pair fused into an elliptic disk; flowers in sessile whorls, corolla funnelform, 2.5 to 3 cm long, somewhat curved, exterior glandular, interior pubescent, stamens inserted just beneath the tube opening (!). RLo 4. China, Szechwan Province. z6 Fig. 162.

L. sullivantii see: **L. prolifera**

L. syringantha Maxim. Deciduous shrub, upright and finely branched, 2–3 m high, very densely branched, glabrous; leaves short stalked, oblong, 1–2.5 cm long, obtuse, bluish-green, glabrous; flowers whitish pink to lilac, tube campanulate, to 8 mm long, interior pubescent, fragrant, calyx teeth lanceolate, May–June; fruits red. BM 7978; RH 1907: 281; Gfl 41: 115–116. China, Kansu Province; Tibet. 1890. z5 Plate 95; Fig. 153. ⟐

var. **minor** Maxim. Prostrate habit; leaves smaller; flowers shorter stalked, bracts and prophylls ciliate. China, Kansu Province; Tibet.

var. **wolfii** Rehd. Habit like that of the species, but the twigs mostly procumbent; leaves elliptic to narrowly oblong 1.5–3.5 cm long, dull green above, blue-green beneath; calyx teeth ciliate and connate at the base, corolla to 14 mm long, carmine-pink, glabrous, very fragrant. SH 2: 436n–o; 437h (= *L. wolfii* Hao). Central China. 1900. Plate 95, 102; Fig. 153. ⟐

L. szechuanica Batal. Differs only slightly from *L. tangutica;* leaves obovate, 8–25 mm long, obtuse and quite glabrous, bluish beneath; corolla slender, 12 mm long, bracts shorter than the ovaries, style glabrous. RLo 8; SH 2: 439h–i, 440g. W. China: Szechwan, Kansu, Yunnan Provinces. z6

L. tangutica Maxim. Deciduous shrub, low, broad growing, young shoots glabrous; leaves obovate to elliptic or obovate-oblong, 1.5 to 3 cm long, acute to obtuse, base cuneate, ciliate, usually somewhat pubescent above, glabrous beneath and whitish; flowers yellowish-white, turning pink, on thin, 1.5–3 cm long pendulous stalks, corolla tubular-funnelform, 1–1.4 cm long, somewhat bulging at the base, stamens shorter than the limb, style longer and glabrous, May–June; fruits usually half connate, red, pendulous, attractive. SH 2: 439m to o, 440i; Gfl 40: 580. China: Kansu, Szechwan, Hupeh, Yunnan Provinces. 1890. z6 Fig. 153, 172. ⊕ ⊛

L. tatarica L. Deciduous shrub, upright, 3–4 m high, twigs gray-brown, glabrous; leaves ovate to lanceolate, 4–6 cm long, dark green above, light bluish-green beneath; flowers in axillary pairs, each pair on a 1–2 cm long stalk, corolla tube to 2.5 cm long, limb to 2 cm wide, white to pink, June; fruits globose, light red. BM 8688; NBB 3: 298; BR 31. S. Russia to Altai and Turkestan. 1752. Ⓕ USA, USSR, in windbreaks and screens. z4 Fig. 165.

Includes many cultivars:

Outline

Dwarf habit:
    'Louis Leroy', 'Nana'

Very strong growing:
    'Grandiflora'

  Leaves yellow variegated:
    'Fenzlii'

  Leaves smaller, narrower:
    'Angustifolia', 'Sibirica'

    Flowers light pink:
      'Alborosea', 'Discolor', 'Latifolia', 'Punicea', 'Rosea', 'Splendens'

    Flowers dark red:
      'Arnold Red' (the darkest), 'Hack's Red'

    Flowers white:
      'Alba', 'Grandiflora', 'Virginalis' ('Gracilis' with a trace of pink)

      Fruits yellow or orange»
      'Lutea', 'Morden Orange'

'Alba'. Flowers medium sized, pure white. Gfl 627. Before 1801. First observed in Austria. Little to no garden merit.

'Alborosea' (Späth). Flowers light pink, large. Späth, 1880.

f. *angustata* see: **'Angustifolia'**

'Angustifolia'. Leaves oval-oblong to oval-lanceolate, 4 to 6 cm long, 1–2 cm wide; flowers light red, 12 mm wide, corolla lobes with a wide border, small flowers (= f. *angustata* Hort.).

'Arnold Red' (Arnold Arboretum). Flowers 2.5 cm wide, dark red, the darkest form of the group; fruits deep red, 9 mm thick. AB 20: 8; 22: 6. Originated as a chance seedling in the Arnold Arboretum in 1947, introduced in 1954. ⊕

f. *bicolor* see: **'Discolor'**

'Discolor'. Corolla with a light pink interior, margin and the dorsal side carmine; fruits orange (= f. *bicolor* Hort.).

f. *elegans* see: **'Latifolia'**

'Fenzlii'. Leaves yellow variegated, speckled and striped.

'Gracilis' (Späth). Flowers large, white with a trace of soft pink. ⊕

'Grandiflora'. Strongest growing of all the cultivars, always twice as high as the other cultivars in the nursery, new growth light green; flowers pure white, very large (= f. *virginalis grandiflora* Dauvesse). Developed in France in 1847. ⊕

f. *grandiflora rubra* see: **'Latifolia'**

'Hack's Red' (Hack). Flowers deep purple-pink, but not as dark as 'Arnold Red'. Developed by Hack's Nursery, near Winnipeg, Manitoba, Canada. ⊕

'Latifolia'. Shoots thick, always somewhat curved; leaves to 10 cm long and 5 cm wide; flowers light pink, darker striped (= f. *elegans*; f. *grandiflora rubra*; f. *pulcherrima*; f. *rubrissima*; f. *speciosa* Billard). ⊕

f. *leroyana* see: **'Louis Leroy'**

'Louis Leroy'. Shrub, slow growing, very low, rounded, new growth appears very early; leaves bluish-green; flowers nearly 3 cm wide, corolla lobes purple-pink, white bordered; fruits orange (= f. *leroyana* Hort.).

'Lutea'. Fruits yellow (= f. *xanthocarpa* Hort.). ⊛

'Morden Orange' (Morden). Flowers pale pink; fruits orange. Developed in the Dominion Experiment Station, Morden, Manitoba, Canada. ⊛

'Nana' (Billard). Dwarf form; flowers pink, small flowers. Developed in France about 1825.

f. *pulcherrima* see: **'Latifolia'**

'Punicea'. Similar to 'Rosea', but the flowers are a pink monochrome. ⊕

f. *purpurea* see: **'Sibirica'**

'Rosea'. Flowers large, light pink, about 2 cm wide; fruits dark scarlet-red.

f. *rubra* see: **'Sibirica'**

f. *rubriflora* see: **'Sibirica'**

f. *rubrissima* see: **'Latifolia'**

'Sibirica'. Small leaves; flowers 2 cm wide, corolla lobes dark red, broad white bordered. BM 2469 (= f. *purpurea*; f. *rubra*; f. *rubriflora*). Developed in England. Before 1822.

f. *speciosa* see: **'Latifolia'**

'Splendens' (Späth 1883). Buds dark red, old flowers pink-red, limb somewhat lighter. ⊕

'Virginalis'. New growth green; flowers large, pure white; berries scarlet-red.

f. *virginalis* see: **'Grandiflora'**

f. *xanthocarpa* see: **'Lutea'**

L. tatsienensis Franch. Deciduous, glabrous, upright shrub, to 2 m high; leaves ovate, obovate or oblong-lanceolate, 2.5–5 cm long, occasionally deeply lobed on long shoots, occasionally pubescent on both sides; flowers in axillary, slender stalked pairs, peduncle nearly 3 cm long, corolla 12 mm long, dark purple, tube short, May; fruits connate, nearly cherry sized, very conspicuous. Tibet; K'wang-ting. 1910. Hardy. z6 Plate 99; Fig. 171. ⊕ ⊛

L. × tellmanniana Magyar (*L. sempervirens* × *L. tragophylla*). Deciduous, strong growing, high climbing shrub, twigs glabrous, new growth olive-brown; leaves elliptic to ovate, later deep green above, whitish pruinose beneath, 5 to 10 cm long, the uppermost pair fused into an elliptic disk; flowers in terminal whorls, deep orange-yellow, tube 4 to 4.5 cm long, buds reddish, June–July. BS 2: 20; GC 90: Pl. 421 (= *Lonicera* 'Redgold' in the USA). Developed by J. Magyar in Budapest, Hungary around 1920, described in 1926. Introduced into the trade by L. Späth in 1927. One of the more popular climbing loniceras today. z8 ⊕

L. tenuiflora see: **L. altmanii**

L. tenuipes Nakai. Upright deciduous shrub, to 2 m high, twigs yellow to red-brown, pubescent at first to nearly glabrous; leaves elliptic to obovate-oblong, 3–6 cm long, pubescent above, dense brown pubescent beneath; flowers usually solitary, red, exterior pubescent, base expanded, to 18 mm long, on 1–2 cm long stalks, April to May; fruits ellipsoid, red. MJ 301; KO ill. 450 (= *L. gracilipes* var. *glandulosa* Maxim.) Japan. z6 Plate 96. ⊕ ⊗

L. thibetica Bur. & Franch. Deciduous shrub, to about 1.5 m high, twigs thin, partly procumbent, loosely tomentose; leaves often in groups of 3, oblong-lanceolate, 1–3 cm long, acute, deep green, and glossy above, white tomentose beneath; flowers light purple, tubular-campanulate, 15 mm long, exterior pubescent, bracts linear-lanceolate, involucral leaves as long as the ovaries, glandular ciliate, occasionally incised, June–July; fruits light red, ellipsoid. BS 2: 625; ST 1: Pl. 45; RH 1902: 449. W. China; Tibet. 1897. z6 Plate 95; Fig. 153, 171. ⊕

L. tomentella Hook. f. & Thoms. Deciduous shrub, divaricate, 1.5(3) m high, stiff branched, twigs thin, usually densely tomentose; leaves nearly distichous, short petioled, oval-oblong to elliptic, 1.5–3 cm long, acute, base round, dull green and finely pubescent to glabrous above, lighter and more or less tomentose beneath; flowers short stalked in axillary pairs, corolla 16 mm long, narrowly funnelform and finely pubescent outside, base not expanded, white, the horizontally spreading limb tips reddish, bracts linear-oblong, June; fruits blue-black, pea-sized, globose. DL 1: 169; BM 6486. Himalayas; Sikkim, in the mountains. 1849. z5 Plate 95; Fig. 153, 157.

L. tragophylla Hemsl. Deciduous, strong growing, twining, glabrous shrub, flower shoots 15–20 cm long; leaves short petioled to sessile, thin, oblong, 5–10 cm long, obtuse to rounded, blue-green beneath with a pubescent midrib, the apical 1–3 leaf pairs fused into elliptic to rhombic disks; flowers 10–20 in short stalked, terminal heads, corolla orange to yellow, upper lip often reddish on the dorsal side, tube somewhat bowed, not fragrant, June. BM 8064; DRHS 1206; DL 1: 129; BS 2: Pl. 85. China; Hupeh Province, in thickets. One of the most attractive climbing species. z6 Fig. 158, 171. ⊕

L. translucens see: **L. quinquelocularis f. translucens**

L. trichosantha Bur. & Franch. Deciduous shrub, divaricate, upright, to 1.5 m high, twigs slender, glabrous; leaves ovate, obtuse, with a protruding, fine tip, 3–5 cm long, pubescent beneath; flowers in axillary pairs, pale yellow, spent flowers becoming darker, tube

1.5 cm long, expanded at the base, exterior appressed pubescent, June; fruits bright red. RLo Pl. 20. W. China, Tibet. 1908. Resembling *L. deflexicalyx*, but the leaves not so acute and less pubescent. z6 Fig. 166. ⊕

L. uniflora see: **L. gracilipes**

L. utahensis S. Wats. Deciduous shrub, upright similar to *L. canadensis*, but the leaves are broadly ovate to oblong, rounded at both ends, 3–6 cm long, not ciliate or only at the base; corolla shorter and somewhat expanded. SH 2: 440n. Western N. America. z5

L. vesicaria Komar. Deciduous shrub, closely related to *L. ferdinandi*, differing in the ovate to oval-oblong, 5–10 cm long leaves, these ciliate with pubescent venation beneath, more leather-like, petiole 3–10 mm long, bracts ovate-oblong, acuminate, 1.5 cm long; corolla with recurved bristles, involucre eventually broken apart by the emerging red fruit. RLo PL. 1 and 10. Korea. 1924. z5

L. villosa (Michx.) Roem. & Schult. Deciduous shrub, closely related to *L. coerulea*, but the shoots and winter buds are ascending at acute angles (less than 45°), secondary buds and stipule cushions absent, young shoots densely short tomentose; leaves densely pubescent on both sides; corolla tubular-campanulate, exterior densely pubescent, tube as long or shorter than the limb; fruits blue, edible. GSP 457 (= *L. coerulea* var. *villosa* Torr. & Gray). N. America; Newfoundland to Alaska. z2 Fig. 159.

var. **solonis** (Eat.) Fern. Shrub, to 75 cm high, bark exfoliating in narrow strips, brown, young shoots soft, pubescent, not tomentose; leaves 2–4 cm long, strigose pubescent to glabrous above, shaggy pubescent beneath; flowers paired in the apical leaf axils, corolla pale yellow, narrowly campanulate, exterior glabrous or slightly pubescent, interior shaggy pubescent, April–June; fruits ovate, blue-black, edible. RMi 387. N. America; Newfoundland to Minnesota; in coniferous forests, swamps and in rocky meadows.

L. × vilmorinii Rehd. (*L. deflexicalyx* × *L. quinquelocularis*). Deciduous shrub, usually similar to *L. quinquelocularis*, but with smaller leaves, these wider and less acuminate; calyx distinctly 5 toothed, inflorescence shorter; fruits yellowish-pink, with tiny red dots. Developed about 1899 by Vilmorin of Les Barres, France. z6

L. webbiana Wall. Deciduous shrub, upright, to 3 m high, young shoots scattered glandular pubescent to nearly glabrous; leaves elliptic to oval-oblong or oblong-lanceolate, 5–12 cm long, acuminate, base cuneate, both sides glandular and pubescent, often rather glabrous above, bracts and prophylls glandular ciliate; flowers bilabiate, tube short, expanded at the base, exterior yellowish-green with a reddish trace, interior whitish to yellow-green, pubescent on both sides, stamens as long as the limb, style shorter and pubescent, April–May; fruits short oval, red, distinct. ST 69. SE. Europe, Afghanistan, Himalayas. 1885. z6 Plate 98; Fig. 169. ⊕

L. wolfii see: **L. syringantha var. wolfii**

L. xerocalyx see: **L. deflexicalyx var. xerocalyx**

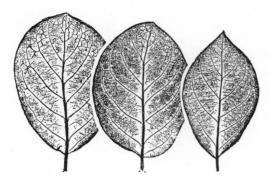

Fig. 173. *Lonicera xylosteum* 'Clavey's Dwarf'

L. × xylosteoides Tausch. (= *L. tatarica* × *L. xylosteum*). Deciduous shrub, narrowly upright and strongly branched, 1–2 m high, branches gray-yellow, reddish when young, stiffly pubescent; leaves ovate to broadly oval-elliptic or obovate, 3–6 cm long, roughly ciliate, short pubescent on both sides, both sides bluish-green, somewhat lighter beneath; flowers light red, pubescent, small, tube short, expanded at the base, flower stalks thin, 1–1.5 cm long, glabrous, stamens shorter than the limb, densely shaggy at the base, style somewhat shorter, shaggy pubescent, May; fruits connate only at the base, yellow-red. DL 1: 149 (= *L. nepalensis* Kirchn.; *L. coerulescens* Dipp.). Origin unknown, but planted in Prague, Czechoslovakia before 1838. z6

L. xylosteum L. Deciduous shrub, broadly upright, to 3 m high, twigs glabrous to pubescent; leaves broadly ovate, 3–6 cm long, pubescent on both sides, dark gray-green above, lighter beneath; flowers yellowish-white, old flowers becoming yellow, pubescent on both sides or glabrous on the interior, 1 cm long, stamens pubescent only at the base, style pubescent to the apex, May–June; fruits flat globose, dark red. BB 3466; GNP 814; HF 2917. Found nearly throughout Europe, the Caucasus, Siberia, the Amur River region of China; in forests, thickets and hedges. Likes a dry soil. Ⓕ Czechoslovakia, in screen plantings. z5 Plate 98; Fig. 165.

Including 16 forms, none of which are particularly meritorious for the garden (see Rehder, synopsis).

'Clavey's Dwarf' (Clavey). Slow growing, about 1 to 1.5 m high, nearly globose and densely branched, branched to the ground; foliage bluish-green; flowers white, inconspicuous. AB 22: 61. Developed in Clavey's Ravinia Nurseries, Deerfield, Ohio, USA. Introduced into the trade about 1950. Has recently gained much popularity in the USA as a hedging plant. Fig. 173.

'Compacta'. To 1.5 m high, hemispherical, denser and more compact than the species. Selected and named in 1931 by A. Wroblewski, Kornik Arboretum, Poland. Probably identical or similar to the more recently introduced 'Nana' from the USA (= 'Emerald Mound'). Cultivated in the Morton Arboretum of Lisle, Illinois for 30 years.

'Emerald Mound' see: **'Compacta'**

'Nana' see: **'Compacta'**

L. yunnanensis Franch. Evergreen shrub, to 4 m high, climbing, shoots glabrous; leaves oblong or narrowly ovate to obovate, 3 to 12 cm long, glabrous above, bluish-white beneath and somewhat pubescent to glabrous, the uppermost pair connate; flowers in short stalked heads of few to many whorls, corolla 2–2.5 cm long, yellow, glabrous, tube somewhat bulging, interior pubescent. SW. China. 1900. z9 #

var. **tenuis** Rehd. Smaller, more bushy, shoots thinner; leaves only 3 cm long, somewhat pubescent beneath; flowers in single whorls, corolla 2 cm long, white, turning yellow. SW China. 1900. z9 #

L. zabelii see: **L. korolkowii** var. **zabelii**

Despite the diversity in the species of this genus, most will thrive in any garden soil, preferably in light shade but also doing well in full sun. The evergreen species, however, prefer an open forest location; *L. pileata* and *L. nitida* in particular will rapidly develop as understory plants. Many of the climbing species are tender.

Lit. Rehder, A.: Synopsis of the genus *Lonicera*; in Report of the Missouri Botanical Garden 1903, 27–232 (with 20 plates) ● Schneider: Handbuch 2 ● Wyman, D.; Shrub Honeysuckles with pink to red flowers; in Arnoldia **20**, 29–32, 1960 ● Yeo, P. F.: *Lonicera pileata* and *L. nitida* in cultivation; Baileya **12**, 56–66, 1964.

LORANTHUS Jacq. — Mistletoe — LORANTHACEAE

Deciduous shrub, parasitic on trees, with cylindrical, brown twigs; leaves opposite, occasionally alternate; flowers hermaphroditic, rarely dioecious, perianth 4–6 parted; stamens 4–6, present as staminodes on female flowers; fruits berrylike, pear-shaped, small. — About 600 species, primarily in the tropics; only one species occurs in Europe.

Loranthus europaeus Jacq. Mistletoe. Deciduous, shrubby plant, parasitic on trees, about 20–40 cm high, well branched, branches brittle, brown; leaves obovate-

oblong, thick, stalked, 4–6 cm long; flowers dioecious, in small, terminal racemes, light yellow, May–June; fruits about 1 cm long, berrylike, pear-shaped, with a persistent style. HM 513. SE. Europe, west to Italy, north to Saxony. Grows on *Quercus, Casanea* and *Olea*.

Cultivated only occasionally in botanic gardens; otherwise not cultivated, rather it is considered a pest!

Lit. Baillon, H.: Deuxième mémoire sur les Loranthacées; in Adansonia **3**, 50–128, 1862 (with 3 plates).

LOROPETALUM R. Br. — HAMAMELIDACEAE

Evergreen shrub, similar to *Hamamelis*, but differing in the stamens with 4 pollen sacs, dehiscing by means of 2 winged valves, with horned, inferior ovaries; leaves evergreen; flowers whitish, 6–8 in terminal heads; fruit a woody, 2 pointed, broadly ovate capsule, dehiscing with 2 valves. — 3 species from E. Himalaya to S. China.

Loropetalum chinense (R. Br.). Oliv. Small evergreen shrub, seldom over 1 m high in cultivation, but to 3 m in its habitat, dense, stellate pubescent; leaves thin, ovate, somewhat asymmetrical, 2–4 cm long, acute, entire, bristly ciliate, stellate pubescent on both sides, denser beneath; flowers whitish to greenish, the 4 petals strap-like, about 18 mm long, February–April. BM 7979; MD 1932: 2. SE. China. Plants occurring in NE. India (Assam and Khasia) are considered an independent species. *L. indicum* Tong. 1880. z9 Plate 102; Fig. 174. # ⊕

Cultivated in the open landscape only in the mildest regions; soil preference about like that of *Hamamelis*. An attractive shrub for the cool greenhouse. Not advisable for cooler regions since the plant must be covered in the winter, when it normally flowers.

Fig. 174. *Loropetalum chinense* (actual size)

LUCULIA Sweet — RUBIACEAE

Deciduous shrub to a small tree; leaves opposite, obovate to lanceolate, acuminate; flowers in terminal corymbs; calyx with 5 awl-shaped lobes; corolla plate-like with a long, very narrow corolla tube, limb 5 lobed, stamens 5, inserted within the tube, hardly protruding. — 5 species from Himalaya to SW. China.

Luculia grandifolia Chose. Shrub, to 2 m high; large leaved, fall color intense; flowers in very large corymbs, white, fragrant, June. Discovered by F. Kingdon-Ward in Bhutan in the mountains at 2500 m. Probably much more winter hardy than the other species. z9 ⊕

L. gratissima Sweet. A small tree in its habitat, or a shrub, 3–7 m high, shoots reddish, soft haired; leaves oval-oblong, acuminate, 10–20 cm long, 5–10 cm wide, smooth above with 9–15 vein pairs, soft pubescent beneath; flowers soft pink in large corymbs (resembling hydrangeas), globose, 10–15 cm wide, very fragrant, corolla tube very slender, corolla 2.5–3 cm wide across the 5 lobed limb, flowering September–October (!). BM 3946. Himalayas. 1816. z10 Fig. 175. ⊕

L. pinceana Hook. Shrub, to 2 m high, semi-evergreen; leaves elliptic-lanceolate, 10–15 cm long, 5–7 cm wide, long acuminate; flowers cream-white, turning pink, 3.5–5 cm, exterior reddish, with 2 small, warty tubercles at the base of each incision between the limb lobes (!), very fragrant, in 10–20 cm wide cymes, loose, May–September. BM 4132; DRHS 1210. Assam; Khasia Mts. 1843. z9 ⊕

Fig. 175. *Luculia gratissima* (from Jour. RHS)

LUETKEA Bong. — ROSACEAE

Subshrub of grasslike habit, twigs procumbent, with stolons; leaves 2–3 times trilobed; flowers hermaphroditic, in racemes, 5 parted, sepals valvate, petals twisted in the bud stage, stamens about 20, filaments subulate, connate at the base; stigmas 4 to 6(usually 5), distinctly separate; fruit a leathery follicle, dehiscing on both seams; seeds lanceolate. — One species in N. America.

Luetkea pectinata (Pursh) Ktze. Procumbent, evergreen dwarf shrublet, flowering shoots 5–15 cm high, glabrous, foliate; leaves divided into linear lobes, 1–1.5 cm long; flowers in 1–1.5 cm long racemes, petals white, rounded, 3–3.5 mm long, sepals 2 mm long, July–September; fruits 4 mm long. MS 203 (= *Spiraea pectinata* Torr. & Gray). Northwest N. America, Alaska to Oregon, in the mountains. z6 Fig. 176. #

These *Saxifraga*-like plants require an acid soil, sufficiently moist; suitable only for collectors.

Fig. 176.
Luetkea pectinata
(from Abrams)

LUPINUS L. — Lupine — LEGUMINOSAE

Perennials or shrubs; leaves alternate, palmately divided; flowers hermaphroditic, irregular, in racemes or also often whorled, calyx composed of 5 sepals, but bilabiate; corolla with 5 petals (standard, 2 wings and 2 keel petals); stamens 10; fruit a pod dehiscing along both seams, with 2 to 12 seeds. — About 200 species, most in the western USA and in the Mediterranean region, but also found on every continent except Australia.

Lupinus arboreus Sims. Shrub, to 3 m high, woody at the base, bushy, twigs with appressed silky pubescence; leaves 7–11 parted, leaflets oblanceolate, abruptly short acuminate, about 3 cm long, glabrous above, pubescent beneath; flowers in terminal, 25 cm long racemes of widely spaced whorls, corolla normally sulfur-yellow, occasionally blue to violet, May–September. BM 682; NF 3: 275. California. 1793. z8 Plate 89. ⊕

Includes the following cultivars:

'Golden Spire', gold-yellow; **'Mauve Queen'**, lilac; **'Snow Queen'**, white; **'Yellow Boy'**, yellow. Brought into cultivation by Esveld, Boskoop, Holland, 1974.

LUZURIAGA Ruiz & Pav. — PHILESIACEAE

Evergreen, ground covering subshrub or somewhat climbing; shoots glabrous, green; leaves alternate, sessile, oblong-elliptic, distinctly 3 veined or parallel veined; flowers solitary in the leaf axils, white, broadly campanulate to stellate, with 6 petals; fruit a pea-sized, many seeded berry. — 3 species from Peru to Tierra del Fuego, as well as in New Zealand. — Only the following species is cultivated.

Luzuriaga radicans Ruiz & Pav. Shrublet, evergreen, the thin wiry shoots lying on the ground or climbing to about 1 m high by means of tendrils, green, totally

L. chamissonis Esch. Upright shrub, 0.3 to 0.7 m high, twigs and leaves pubescent to tomentose; leaves 6–9 parted, leaflets 12–25 mm long, silky pubescent on both sides; flowers blue to lilac, standard with a yellow center, keel usually not ciliate, May–July. MS 267. California. z9

L. excubitus Jones. Upright shrub, 0.5–1 m high, but often 1.5 m wide, total plant densely silky pubescent; leaves 5–8 parted, leaflets 2–3 cm long, petiolules 2.5–10 cm long; flowers blue, lilac or whitish, standard with a yellow center, but later becoming reddish; May to June; pods densely pubescent. California; in dry, gravelly river beds. z9

Easily propagated from seed; *L. arboreus* will flower in the first year. Requires a very sandy, gravelly, dry soil in full sun.

Fig. 177. *Luzuriaga radicans* (from Hooker)

glabrous; leaves distichous, alternate, oval-oblong to narrowly linear, 1.5–5 cm long, 0.5–1.5 cm wide, acute at both ends, entire, with 6–12 longitudinally parallel veins, bright green above, blue-green beneath; flowers white, stellate in form, stamens yellow, conically grouped in the center of the flower, summer flowering (but only a few flowers); fruit pea-sized, orange or scarlet. Chile, Argentina. Before 1850. (Malahide) z10 Plate 93; Fig. 177. # ✧

A very interesting ground cover for milder regions in a moist soil.

LYCIUM L. — Matrimony-vine, Box-thorn — SOLANACEAE

Deciduous shrubs; branches usually long, rodlike, nodding, thin, thorny or thornless; leaves alternate or clustered, entire; flowers 1–4 in axillary clusters; calyx campanulate, most 5 toothed; corolla funnelform, with a normally 5 lobed limb, reddish, violet, whitish or greenish, limb lobes outspread; stamens 5, rarely 4, enclosed; ovaries 2 locular; fruit usually a red berry with many or few seeds. — About 80–90 species in the temperate and subtropical zones of the world, often in dry areas; the tropical species are found mostly in S. America.

Key to the more prominent species
(from W. T. Stearn, in Flora Europaea)

- Leaves usually widest beneath the middle, the largest at least 10 mm wide; corolla lobes about as long or longer than the corolla tube;
  - \+ Corolla tube with a 2–3 mm long tubular base; leaves usually widest in the middle:
    - *L. barbarum*
  - \++ Corolla tube with a 1.5 mm long tubular base; leaves widest beneath the middle:
    - *L. chinense*
- ● Leaves 0.5–10 mm wide, most widest above the middle; corolla lobes no more than ⅓ as long as the corolla tube;
  - \+ Calyx 5–7 mm; corolla 20–22 mm:
    - *L. afrum*
  - \++ Calyx 1.5–4 mm; corolla 8–18 mm;
    1. Stamens enclosed;
       a) Leaves 20–50 mm; calyx 2 to 3 mm; corolla 11–13 mm:
          - *L. europaeum*
       b) Leaves 3–15 mm; calyx 1.5 to 2 mm; corolla 13–18 mm;
          - *L. intricatum*
    2. Stamens exserted;
       a) Leaves 0.5–1.5 mm wide; filaments pubescent at the base; fruit black:
          - *L. ruthenicum*
       b) Leaves 3–10 mm wide; filaments glabrous; fruit reddish:
          - *L. europaeum*

Lycium afrum L. Shrub, 1–2 m high, twigs stiff, very thorny, thorns thick; leaves 10 to 23 × 1–2 mm, quite narrowly oblanceolate; calyx 5–7 mm, deeply 5 toothed, corolla nearly tubular, purple-brown, 20–22 mm long, lobes 2 mm, stamens enclosed, filaments with dense hair fascicles at the base, flowers abundantly, May to June; fruit red at first, later purple-black, ovate 8 mm, with a persistent calyx at the base. N. Africa; cultivated and naturalized in Spain, Portugal and Italy. 1712. z9 Fig. 179.

L. barbarum L. Chinese Box-thorn. Shrub, to 2.5 m high, twigs and branches bowed, thorns thin; leaves 2–10 cm × 6–30 mm, very narrowly elliptic to narrow lanceolate, usually widest in the middle; calyx 4 mm, bilabiate, corolla 9 mm, funnelform, the tube 2.5–3 mm long at the base, narrowly cylindrical, purple at first, later brownish, lobes 4 mm, axillary, usually in groups of 2–3, stamens widely exserted, filaments with pubescent tufts at the base, May–July; fruit ellipsoid to ovate, scarlet to orange, 2–2.5 cm long. HLy 15; SL 1: 9; HM Pl. 231; NBB 3: 202 (= *L. halimifolium* Mill.; *L. vulgare* Dun.; *L. europaeum* Hort. non L.). China. 1772. This species was cultivated for many years under the erroneous name of *L. europaeum*. z6

L. chilense Bert. Small shrub, more or less pubescent, shoots pendulous or procumbent; leaves cuneate, 1.5–5 cm long, gray-green, obtuse, glandular pubescent and ciliate, in clusters; flowers solitary, pubescent, exterior yellowish, interior reddish, June–September; fruit orange-red. HLy 19 (= *L. grevilleanum* Gillies ex Miers). Mountain valleys of Chile. 1890. Quite variable. z9 Fig. 178.

L. chinense Mill. Very similar to *L. barbarum*, but the leaves are 10–14 cm × 5–60 mm, the basal leaves much larger than the apical ones, lanceolate to ovate, usually widest beneath the middle; calyx 3 mm, corolla 10–15 mm, broadly funnelform, the tube narrowly cylindrical for 1.5 mm from the base, lobes 5–8 mm. DL 1: 10; NBB 3: 202; LWT 335 (= *L. chinense* var. *ovatum* [Veill.] Schneid.; *L. rhomifolium* Dipp.). China, but cultivated in Europe before 1709 and naturalized in many parts. z6 Fig. 178.

L. depressum Stocks. Upright shrub, twigs thin, thorny; leaves lanceolate, 2–7 cm long, thick; flowers similar to those of *L. europaeum*, but grouped 2–4 together, nodding, corolla tube pink, 8–10 mm long, twice as long as the limb tips, filaments pubescent, June–August; fruits small, globose, red (= *L. turcomanicum* Turcz.). W. and Central Asia. z5

L. europaeum L. Shrub, 1–4 m high, twigs stiff, with very stout thorns, thorns thick; leaves 2–5 cm × 3–10 mm, usually oblanceolate; flowers solitary or grouped 2–3 together, calyx 2–3 mm, 5 toothed or bilabiate, corolla 11 to 13 mm, narrowly funnelform, pink or white, limb lobes 3–4 mm, stamens usually exserted, filaments not pubescent, somewhat uneven; fruits reddish. DL 1: 14

Fig. 178. **Lycium**. a. *L. europaeum;* b. *L. ruthenicum;* c. and d. *L. chilense;* e. *L. pallidum;* f. *L. barbarum;*
g. *L. chinense* (from G. F., Miers, Pallas, Poiret, Sibthorp, Thomé, Wettstein)

(= *L. mediterraneum* Dun.). Mediterranean region and Portugal. z9 Fig. 178.

L glaucum see: **L. ruthenicum**

L. grevilleanum see: **L. chilense**

L. halimifolium see: **L. barbarum**

L. horridum H. B. K. Small, strongly branched, thorny shrub, branches thick, leafless, lateral shoots foliate at the base, drawn out to a thorny tip; leaves clustered, obovate, 2.5–4 cm long (on long shoots), glossy dark green above, lighter beneath; flowers solitary, calyx hemispherical, campanulate, 5 toothed, glabrous, corolla violet, cylindrical, tube somewhat curved, limb outspread, stamen filaments glabrous; fruits red, enclosed within the calyx (= *Lycoplesium horridum* Miers). Peru, Andes Mts. z9 Plate 103.

L. intricatum Boiss. Shrub, 0.3–2 m, strongly branched, very thorny, thorns stiff and very thick; leaves oblanceolate, 3–15 mm × 1 to 6 mm, thickish; flowers solitary or in clusters of 2–3, calyx 1–5–2 mm, shallowly 5 toothed, corolla narrowly funnelform, 13–18 mm long, blue-violet, purple, pink, lilac or white, limb tips 2–3 mm, stamens inserted, filaments glabrous. S. Europe,

from Crete to Portugal, N. Africa, SW. Asia. z6 Plate 103.

L. mediterraneum see: **L. europaeum**

L. pallidum Miers. Upright, loosely branched shrub, shoots spreading, often twisted, thorny, glabrous; leaves usually in clusters, lanceolate to oblanceolate, 2–5 cm long, gray-green, thickish; flowers grouped 1–2, nodding, greenish-yellow with a reddish trace and darker veins, funnelform, 2 cm long, May–August; fruits globose, red, about 1 cm thick. BM 8440; HLy 18; BS 2: 634; GF 1: 341. Southwestern N. America. 1886. z6 Fig. 178. ✧ ⚭

L. rhombifolium see: **L. chinense**

L. ruthenicum Murray. Shrub, 0.5–2 m high, twigs with thin thorns; leaves very narrowly lanceolate and somewhat fleshy, 7 to 30 mm × 0.5–1.5 mm, calyx 3–4 mm, bilabiate, corolla 8–10 mm, narrowly funnelform, purple above, whitish beneath, lobes 2–3 mm, stamens exserted, filaments pubescent at the base; fruits black. DL 1: 12 (= *L. glaucum* Miers). Central and SE. Asia, W. Kasachstan. 1804. z6 Plate 103; Fig. 178.

L. turcomanicum see: **L. depressum**

L. vulgare see: **L. barbarum**

All species, where winter hardy, suitable for planting on gravelly slopes. Prefer a sunny location and calcareous soil. Easily cultivated.

Lit. Hitchcock, C. L.: A monographic study of the genus *Lycium* of the Western Hemisphere; in Ann. Mo. Bot. Gard. **19**; 179–364, 1932 ● Feinbrunn, N., & W. T. Stearn: Typification of *Lycium barbarum* L., *L. afrum* L. and *L. europaeum* L.; Israel J. Bot. **12**, 114–123, 1964 ● Stearn, W.T.: *Lycium*; in Flora Europaea 3, 193–194, 1972.

LYCOPLESIUM

Lycoplesium horridum see: **Lycium horridum**

LYGOS

Lygos sphaerocarpa see: **Genista sphaerocarpa**

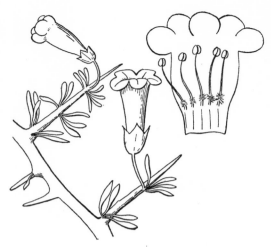

Fig. 179. *Lycium afrum* (from Miers)

LYONIA Nutt. — ERICACEAE

Deciduous or evergreen shrubs, twigs cylindrical or angular; leaves alternate, short petioled, entire or slightly serrate or dentate; flowers urceolate or tubular-campanulate, with short corolla lobes, in axillary clusters or racemes or terminal panicles; fruit a capsule with thickened seams. — 30 species in E. Asia, Himalaya, N. America and in the Antilles.

The genera *Lyonia* and *Pieris* are so closely related that a careful examination of all characteristics is necessary for positive identification.

● Leaves evergreen
  Undersides rust-brown lepidote:
        L. ferruginea

  .Finely black punctate beneath, glossy above:
        L. lucida

●● Leaves deciduous
  Flowers in terminal panicles:
        L. ligustrina

  Flowers in axillary racemes;
  + Leaves glabrous, base cuneate:
        L. mariana

  ++ Leaves finely pubescent above, base cordate or round:
        L. ovalifolia

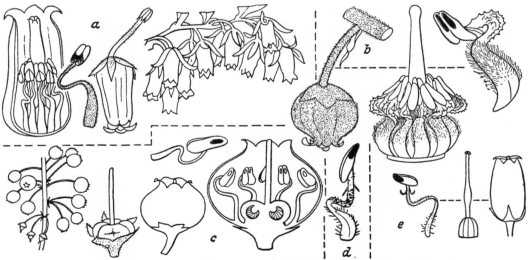

Fig. 180. **Lyonia.** Flowers and flower parts. a. *L. lucida*; b. *L. ligustrina*; c. *L. ferruginea*; d. *L. mariana*; e. *L. ovalifolia* (from Drude, Jacquin, Sargent, Schneider)

Fig. 181. **Lyonia.** From left: *L. mariana; L. ferruginea; L. lucida; L. ligustrina* (from J. Ingram)

Lyonia calyculata see: **Chamaedaphne calyculata**

L. ferruginea (Walt.) Nutt. Evergreen shrub, also a small tree in its habitat, to 5 m high, entire plant with rusty scales, especially the leaf undersides; leaves ovate, obovate or lanceolate, 3–6 cm long, entire, limb slightly involuted; flowers in axillary, nodding clusters, corolla white, globose-urceolate, 3 mm long, limb tips outspread, short, acute, sepals acutely ovate, rusty, February–March. Bai 11: 10; KTF 162 (= *Andromeda ferruginea* Walt.). Southeastern N. America. 1784. z9 Fig. 180, 181. #

L. ligustrina (L.) DC. Deciduous shrub, to 4 m high in its habitat, abundantly branched, twigs glabrous or somewhat pubescent; leaves elliptic-oblong to oblong-lanceolate, 3–7 cm long, entire to finely serrate, tapering to both ends; flowers in dense, pubescent, 8–15 cm long, terminal panicles, corolla globose to oval-urceolate, whitish, pubescent, 4 mm long, corolla tips wide, small, reflexed, sepals ovate triangular, acute, pubescent, May to June. BB 2771; GSP 400; BC 1110; Bai 11: 11 (= *Andromeda ligustrina* Muehlb.). Eastern N. America. 1748. Hardy. Poisonous to horses. z6 Plate 101; Fig. 180, 181. ⊕

L. lucida (Lam.) K. Koch. Evergreen shrub, 0.5–1.5 m high, twigs sharp angular, glabrous; leaves elliptic-oblong to obovate, acuminate, base abruptly tapering, entire, glabrous and glossy above, limb somewhat involuted, finely black punctate beneath, leathery; flowers in axillary clusters at the branch tips, grouped into a terminal, foliate raceme, corolla oval-urceolate, white, often reddish, 3–5 mm long, limb tips small and usually erect, calyx tips small, lanceolate, reddish, June–July. NBB 3: 19; HyWF 134; BM 1095; Bai 11: 11 (= *Andromeda lucida* Lam.; *Pieris lucida* Rehd.). SE. USA.

1765. z9 Fig. 180, 181. # ⊕

'Rubra' Flowers dark pink. Bot. Cab Pl. 672. 1822. z9 ⊕

L. macrocalyx (Anthony) Airy-Shaw. Evergreen or semi-evergreen shrub, young shoots gray-brown, glabrous; leaves ovate to more oblong or oval-lanceolate, 5–10 cm long, 3–5 cm wide, slender acuminate, entire, base round, glabrous and bright green above, blue-green beneath, with fine, appressed, brown bristly pubescence; flowers in axillary, pendulous, 7–10 cm long racemes at the branch tips, corolla yellowish-white, urceolate, 1 cm long, with 5 tiny, triangular teeth at the apex, exterior lightly reddish pubescent, July, calyx with erect, 6 mm long sepals. BM 9490 (= *Pieris macrocalyx* Anthony). China; NW. Yunnan, SE. Tibet. 1925. z9 ⊕

L. mariana (L.) D. Don. Deciduous shrub, 0.5–1.5 m high, loose, twigs cylindrical, glabrous; leaves elliptic-oblong, 3–6 cm long, leathery, entire, fall color red, glabrous above, brown glandular punctate beneath; flowers in axillary, nodding clusters, grouped into leafless, terminal racemes, corolla ovate-cylindrical, white to light pink, 7 to 9 mm long, calyx green, red speckled, May–June. HyWF 134; BB 2770; GSP 402; NBB 3: 19; Bai 11: 9 (= *Pieris mariana* Benth. & Hook.). Eastern N. America, on moist, peaty forest soil. 1736. z6 Fig. 180, 181. ⊕

L. ovalifolia (Wall.) Drude. Deciduous to semi-evergreen shrub, also a tree in its habitat, to 10 m high; leaves ovate, oval-oblong to elliptic, 6–12 cm long, acute, base round to cordate, finely pubescent on both sides, more bristly beneath, especially on the venation; flowers in terminal and axillary, pubescent racemes, corolla urceolate, whitish, pubescent, 9 mm long, limb tips small, erect, calyx lobes lanceolate, May–June. Bai 11: 12; SWT 286; FIO 83 (= *Andromeda ovalifolia* Wall.; *Pieris ovalifolia* D. Don). Himalayas, W. China. 1825. z6 Plate 101; Fig. 180. ⊕

var. **elliptica** (S. & Z.) Hand.-Mazz. Leaves longer and thinner; flowers in short racemes.

var. **lanceolata** (S. & Z.) Hand.-Mazz. Leaves elliptic-oblong to lanceolate, base usually cuneate (seldom rounded). Himalayas, W. China.

L. racemosa see: **Leucothoe racemosa**

Lit. Ingram, J.: Studies in the cultivated Ericaceae. 2. *Lyonia; Baileya* **11**, 28–35, 1963.

LYONOTHAMNUS A. Gray — ROSACEAE

Monotypic genus; characteristics described below.

Lyonothamnus floribundus A. Gray. Evergreen tree, 5–15 m high in its habitat, bark reddish-brown, peeling; leaves opposite, simple, oblong-lanceolate, 10–15 cm long, entire or crenate or somewhat lobed beneath, short petioled, dark green and glossy above, tomentose or glabrous beneath; flowers white, about 1 cm wide, in 10–20 cm wide, many flowered cymes, the individual flowers with 5 persistent sepals, 5 petals, without a claw, stamens 15, ovaries 2, style thick, stigma small, May–June; fruit composed of a pair of follicles. SM 336; SPa 154 to 155. S. California; Catalina Island. z9

var **aspleniifolius** (Greene) Brandegee. The only plant in cultivation. Leaves pinnately partite, with 3–7 leaflets, these oblong-lanceolate, 8–10 cm long, pinnatisect with broad, oblique lobes, narrow sinuses between; otherwise like the species. 1900. z9 Fig. 182 # ∅

Fig. 182. *Lyonothamnus floribundus* var. *aspleniifolius* (Original)

LYTHRUM

Lythrum verticillatum see: **Decodon verticillatus**

MAACKIA Rupr. & Maxim. — LEGUMINOSAE

Deciduous shrubs or small trees; leaves odd pinnate, pinnae nearly opposite; flowers in erect racemes, these usually grouped into panicles; calyx campanulate, 5 toothed. — 10 species in E. Asia.

Very closely related to *Cladrastis*, but easily distinguished by the distinct winter buds (those of *Cladrastis* are hidden beneath the petiole base in the vegetative period), the erect floral racemes and the opposite leaflets.

Maackia amurensis (Rupr. & Maxim). K. Koch. Tree, to 15 m high in its habitat, twigs glabrous, but finely pubescent when young; leaves 20–30 cm long, leaflets 7–11, ovate, rather obtuse, 3–7 cm long, glabrous; flowers small, greenish-white, in 10–15 cm long, stiff, upright racemes on the shoot tips, July–August; pods linear, flat, 3–5 cm long. BM 6561; BS 2: 64; LF 183.

Manchuria, China. 1864. Ⓕ Ukraine, for reforestation. z4 Plate 104. ⊕

var. **buergeri** (Maxim.) Schneid. Shorter, shoots and leaves more pubescent; flower smaller and denser. MJ 1287 (= *M. buergeri* Tatewaki). Japan. 1892

M. buergeri see: **M. amurensis** var. **buergeri**

M. chinensis Takeda. Tree, 5–12 m high in its habitat, shoots dark gray, rough, young shoots dense, silky brown pubescent; leaves with 7 to 17 leaflets, rachis cylindrical, pubescent, leaflets oblong, about 5 cm long, 2 cm wide, acute, base round, dark green above, densely silky pubescent beneath; flowers in dense, pubescent, 15–20 cm long panicles, July–August (= *M. hupehensis* Takeda). China, Hupeh Province. 1908. z5 ⊕

M. fauriei (Levl.) Takeda. Small shrub, about 1.5 m high, young shoots glabrous; leaves 20 cm long, with 9–17 leaflets, these 3 to 5 cm long, 15–20 mm wide, only pubescent when young; flowers in dense, slender, 7–10 cm long panicles on the shoot tips, August. Korea. 1917. z5

M. honanensis Bailey. Tall shrub, young shoots pubescent; leaves 10–14 cm long, with 5 leaflets, terminal leaflet elliptic, short acuminate, 6–7 cm long, petiole 1.5–2 cm long, the lateral leaflets sessile, oval-oblong, narrower and smaller, the lowest pair much smaller, venation pubescent beneath; flowers few in short racemes. GH 1: Pl. 9. China; Honan Province. z6

M. hupehensis see: **M. chinensis**

M. tashiroi (Yatabe) Mak. Shrub, leaflets 11–15, elliptic-ovate, 2–4 cm long, slightly pubescent beneath; flowers 7 mm long; pods 2 to 3.5 cm long. Japan; Liukiu Island. LT 467. 1919. Hardy. z6

M. tenuifolia Hand.-Mazz. Shrub, to 1.5 m high, shoots thin, glabrous, brownish; leaflets 3 to 7, oblong-ovate, about 8 cm long and 5 cm wide, acute or obtuse, base broadly cuneate, glossy green above, likewise beneath but lighter, glabrous on both sides, rachis green, white silky pubescent, flowers white, in axillary racemes. China; Kiangsi Province. z6

Not particular as to soils or location; thrives in any garden soil.

Lit. Takeda: *Cladrastis* and *Maackia*; in Not. Bot. Gard. Edinburgh 8, 95–104, 1963 (Pl. 26–27, Fig. 16 to 27).

MACHILUS Nees — LAURACEAE

Evergreen trees with an aromatic scent, leaves alternate, pinnately veined; flowers hermaphroditic or polygamous, in large terminal panicles; perianth 6 parted, the outer parts equal to or somewhat smaller than the inner petals and usually all persisting until fruit set; stamens 9, 6 of these with a short stalked gland on either side of the filament base; fruit a globose, fleshy berry. — About 35 species in tropical and subtropical Asia, 27 of these in China. Quite rare in cultivation.

Machilus ichangensis Rehd. & Wils. Evergreen tree, 6–12 m high, stem 10–20 cm thick, bark gray-brown, smooth, but occasionally very rough, young shoots red-brown, glabrous, with rounded leaf scars; leaves oblong-lanceolate, 10–15 cm long, 2–5 cm wide, short acuminate, base narrowly cuneate, deep glossy green above, glabrous, bluish beneath, lightly silky pubescent, thick and leathery, petiole 1–2 cm; flowers in short,

axillary panicles, cupulate, white, 5 mm wide, sepals lightly silky pubescent, oblong-lanceolate, May–June; fruit a pea-sized, black berry with a persistent, 6 lobed, brownish perianth tube. ICS 1651 (= *M. thunbergii* S. & Z.). China; W. Hupeh Province. Around 1901. (Wakehurst Gardens, England). Not suited for alkaline soil. Ⓟ India. z9 #

Descriptions of the following species found in ICS:
* *M. microcarpa* 1645
* *M. oreophila* 1646
* *M. bournei* 1647
* *M. chinensis* 1648
* *M. leptophylla* 1649
* *M. velutina* 1650
* *M. pauhoi* 1652
* *M. thunbergii* 1653
* *M. yunnanensis* var. *duclouxii* 1654

× MACLUDRANIA (CUDRANIA × MACLURA) André — MORACEAE

Bigeneric hybrid; known only in the following combination:

× **Macludrania hybrida** André (*Cudrania tricuspidata* × *Maclura pomifera* 'Inermis'). Small, deciduous tree, bark yellowish, furrowed, twigs dark brown, with short, straight, woody thorns; leaves alternate, ovate, long acuminate, glabrous, violet beneath to 15 cm long, not lobed. RH 1905: 138. z6

MACLURA Nutt. — Osage Orange — MORACEAE

Deciduous trees or shrubs with axillary thorns; leaves alternate, simple, entire; flowers dioecious, axillary, female flowers in dense heads, male in short spikes or racemes; male flowers with a 4 part perianth and 4

Fig. 183. Fruit of *Maclura pomifera*
(⅔ actual size; from Illick)

stamens; female flowers with long, filamentous styles; fruits united into a large, globose, rugose multiple fruit. — 12 species, mostly in the warmer regions of America, Asia and Africa.

Maclura pomifera (Raf.) Schneid. Small tree, to 20 m high in its habitat, fast growing, open crowned, twigs thorny, olive-green; leaves ovate, acuminate, 5–12 cm long, glossy above, with a milky latex; flowers inconspicuous, May–June; fruit an 8–14 cm thick, globose, orange-yellow multiple fruit of nearly orange-like appearance, not edible. BB 1259; GTP 201; SM 302 (= *M. aurantiaca* Nutt.). N. America; Arkansas to Texas. 1818. Ⓕ USSR, Yugoslavia, India. z5 Fig. 183. ⚭

'Inermis'. Twigs totally thornless. Only the female known to be in cultivation.

'Pulverulenta' (Hesse). Leaves white pulverulent. Introduced into the trade by H. A. Hesse of Weener, W. Germany.

An attractive, fast growing tree or shrub, creating an impenetrable hedge. Fruits can be a nuisance. No soil preference.

MADDENIA Hook. f. & Thoms. — ROSACEAE

Deciduous trees or shrubs, very similar to *Prunus*, but with 10 (instead of 5) sepals; winter buds many scaled; leaves alternate, serrate, with stipules; flowers dioecious, appearing with the leaves, short stalked, in racemes, petals absent; sepals 10, very small; stamens 25–40; ovaries with a long style and 2 ovules; the staminate flowers occasionally have 2 atrophic styles with sessile stigmas; fruit a single seeded drupe. — 4 species in Himalaya and China.

Maddenia hypoleuca Koehne. Often a shrub in cultivation, a small tree in its habitat, to 6 m high, young twigs dark brown, glabrous; leaves ovate-oblong, 4–7(12) cm long, long acuminate, base round, double-serrate, glabrous and dark green above, blue-white beneath, with 14–18 vein pairs; flowers in 3–5 cm long, dense racemes, red-brown at first, later green, February–March; fruits black, ellipsoid, 8 mm long. DB 11: 71. Central and W. China. 1907. Quite hardy. z5 Fig. 184. ∅ ◑

Not difficult to cultivate; thrives in any garden soil.

Fig. 184. *Maddenia hypoleuca* (Leaf and inflorescence at ½ actual size, individual flower enlarged) (Original)

MAGNOLIA L. — MAGNOLIACEAE

Deciduous or evergreen trees or shrubs; shoots with a white pith; leaves alternate, simple, not lobed, entire, enclosed within a sheath in bud; flowers hermaphroditic, solitary, terminal, often enclosed by 2 large, often very pubescent scales when in bud, sepals 3 (usually petaloid), petals 6–15, free; stamens many; ovaries arranged around an ovate or oblong axis, elongated in fruit; fruit usually a woody follicle, dehiscing along the dorsal side; seeds later hanging from long filaments. — About 80 species in E. Asia, N. and Central America and the Himalayan Mts.

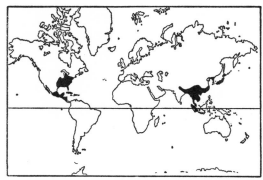

Fig. 185. Range map of the genus *Magnolia*

Outline of the Genus (without hybrids) (from Dandy, altered)

● Subgenus **Magnolia**

Anthers introrse (opening from the floral axis forward); flowering either before the leaves or flowers with a much reduced (calyx-like) outer whorl of petals; leaves evergreen or deciduous; fruits variable;

+ Stipules adnate to the petiole, leaving leaf scars on the petiole surface;

   ✕ Leaves evergreen; flower buds surrounded at first by one to several sheath-like involucral leaves, these leaving scars on the flower stalk;

    Section **Gwillima** DC.
        Fruit carpels short beaked, the beak flattened on its dorsal side; including:
           M. championii, coco, delavayi

Section **Lirianthe** (Spach) Reichenb.
    Fruit carpels long beaked:
        —not described in this work

   ✕✕ Leaves deciduous (occasionally evergreen on the American species in the Section *Magnoliastrum*); flower buds at first enclosed in a sheath-like involucre, the solitary stigma persists on the flower stalk;

Section **Rhytidospermum** Spach
    Leaves in mock whorls clustered at the ends of the annual growth, usually large or very large; including:
        M. ashei, fraseri, hypoleuca, macrophylla, officinalis, pyramidata, rostrata, tripetala
    Leaves not in mock whorls, clustered at the shoot tips;

Section **Magnoliastrum** DC.
 § Anthers with the connective drawn out into a small, acute appendage; leaves evergreen or deciduous:
       M. virginiana

Section **Oyama** Nakai
 §§ Anthers with the connective obtuse or truncated, normally not drawn out to an appendage; including:
       M. sieboldii, sinensis, wilsonii

  ++ Stipules not adnate to the petiole, petiole without stipular scars; leaves evergreen;
    x Petals nearly alike in texture; fruits ellipsoid to oblong, occasionally twisted; leaves with a more or less elongated petiole;

Section **Theorhodon** Spach
    Gynoecium sessile:
        M. grandiflora

Section **Gynopodium** Dandy
    Gynoecium on a short stalk; plants totally glabrous:
        M. nitida

    xx Petals of the outer whorl much finer in texture than those of the inner whorl; fruit more or less cylindrical, usually twisted; leaves relatively short petioled;

Section **Maingola** Dandy
        —not described in this work

●● Subgenus **Pleurochasma** Dandy

Anthers laterally or nearly laterally opening; flowers appear before the leaves and/or with a much reduced (calyx-like) outer whorl of petals; leaves deciduous; fruits cylindrical to oblong, usually more or less twisted;

Section **Yulania** (Spach) Reichenb.
    Petals all nearly alike; flowers appear before the leaves, white to pink or purple-pink; including:
        M. campbellii, cylindrica, dawsoniana, denudata, sargentiana, sprengeri

    Petals very unequal, the outer whorl much shorter and appearing as a calyx:

Section **Buergeria** (S. & Z.) Baill.
    + Flowers appear before the leaves; the inner (larger) petals white, often also pink and purple toned:
        M. biondii, kobus, loebneri, salicifolia, stellata

    ++ Flowers appear with or before the leaves; the inner (larger) petals purple or green to yellow:
        M. acuminata, cordata, liliiflora

Outline of the specific origins of the cultivars

(Only fancy names are listed, not those of Latin origin such as 'Alba', 'Rosea', 'Grandiflora', etc.)

'Alba Superba'	— *soulangiana*
'Alexandrina'	— *soulangiana*
'Amabilis'	— *soulangiana*
'André Leroy'	— *soulangiana*
'Ann'	— *liliiflora* hybrids
'Betty'	— *liliiflora* hybrids
'Brozzonii'	— *soulangiana*
'Burgundy'	— *soulangiana*
'Charles Coates'	— *sieboldii* hybrids
'Charles Raffill'	— *campbellii* × *mollicomata*
'Darjeeling'	— *campbellii*
'Dark Raiment'	— *liliiflora* hybrids
'Else Frye'	— *salicifolia*
'Ethel Hillier'	— *campbellii*
'Evamaria'	— *brooklynensis*
'Exmouth'	— *grandiflora*
exoniensis	— *grandiflora*
'Ferruginea'	— *grandiflora*
'Freeman'	— *grandiflora* × *virginiana*
"Galissoniensis"	— *grandiflora*
'Galissonière'	— *grandiflora*
'Georg Henry Kern'	— *stellata*
'Goliath'	— *grandiflora*
'Grace McDade'	— *soulangiana*
'Halleana'	— *stellata*
'Highland Park'	— *soulangiana*
'Isca'	— *veitchii*
'Jane'	— *liliiflora* hybrids
'Jermyns'	— *salicifolia*
'Judy'	— *liliiflora* hybrids
'Kewensis'	— *salicifolia* × *kobus*
'Kew's Surprise'	— *campbellii* × *mollicomata*
'Lanarth'	— *campbellii* × *mollicomata*
'Lennei'	— *soulangiana*
'Lennei Alba'	— *soulangiana*
'Leonard Messel'	— *loebneri*
'Liliputian'	— *soulangiana*
'Lombardy Rose'	— *soulangiana*
'Maryland'	— *grandiflora* × *virginiana*
'Melanie'	— *soulangiana*
'Merrill'	— *loebneri*
'Michael Rosse'	— *sargentiana*
'Nannetensis'	— *grandiflora*
'Neil McEacharn'	— *loebneri*
'Norbertii'	— *soulangiana*
'Norman Gould'	— *stellata*
'Peppermint Stick'	— *liliiflora* hybrids
'Picture'	— *soulangiana*
'Pinkie'	— *liliiflora* hybrids
'Princess Margaret'	— *campbellii*
'Randy'	— *liliiflora* hybrids
'Raspberry Ice'	— *liliiflora* hybrids
'Rouged Alabaster'	— *soulangiana*
'Royal Crown'	— *liliiflora* hybrids
'Rustica Rubra'	— *soulangiana*
'Saint Mary'	— *grandiflora*
'Samuel Sommer'	— *grandiflora*
'San José'	— *soulangiana*
'Sayonara'	— *veitchii*
'Sidbury'	— *campbellii* × *mollicomata*
'Slavin's Snowy'	— 'Kewensis'
'Snowdrift'	— *loebneri*
'Speciosa'	— *soulangiana*
'Spring Beauty'	— *soulangiana*

'Susan'	— *liliiflora* hybrids
'Trewithen'	— *liliiflora*
'Trewithen Dark Form'	— *campbellii*
'Trewithen Light Form'	— *campbellii*
'Triumphans'	— *soulangiana*
'Undulata'	— *grandiflora*
'Vanhouttei'	— *soulangiana*
'Verbanica'	— *soulangiana*
'Vin Rouge'	— *liliiflora* hybrids
'Wada's Memory'	— 'Kewensis'
'Wakehurst'	— *sprengeri*
'Waterlily'	— *stellata*
'Woodsman'	— *brooklynensis*

Magnolia acuminata (L.) L. Cucumber Tree. Deciduous tree, growth vigorous, to 25 m high in its habitat, young trees conical, branches regularly arranged around the stem, bark gray, smooth, fissured only on older trees; leaves elliptic, 10–20 cm long, dark green above, lighter and soft pubescent beneath, golden-yellow in fall, stipules adnate to the petiole; flowers not showy, bluish-green, about 5 cm long, sepals 3, 2.5–3 cm long, reflexed, petals 6, oblong, June to July; fruit cones cylindrical, 5–8 cm long, eventually red. GTP 204; RWF 135; BB 1541. E. USA. An attractive, quite hardy park tree. 1736. z4 Fig. 186. ∅

var. *subcordata* see: **M. cordata**

'Variegata'. Leaves very white variegated, speckled and striped. A large tree stands in Kew Gardens, London.

M. ashei Weatherby. Deciduous shrub (or tree, to 8 m); leaves obovate, 25 to 60 cm long, thin, base deeply cordate, light green and glossy above, silvery-white beneath, margins sinuate; flowers appear with the

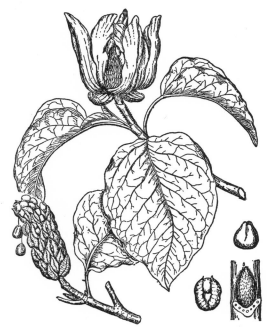

Fig. 186. *Magnolia acuminata* (from Illick)

leaves, erect, cupulate, sepals gray-white, petals 6, white, 15 cm long, oblong, eventually outspread (flowers then 30 cm wide!), very fragrant. WT 70; JRHS 72: 146; 78: 90; KTF 74. NW. Florida. 1933. Closely related to *M. macrophylla*, but flowering as a young plant. z7 ✥

M. aulacosperma see: **M. biondii**

M. auriculata see: **M. fraseri**

M. biondii Pamp. Deciduous tree in its habitat, to 12 m, twigs ascending, ash gray, bark somewhat aromatic, winter buds large, silvery pubescent; leaves oblong-elliptic, short acuminate, to 14 cm long and 7 cm wide, base cuneate, leathery tough, dark green and glossy above, lighter beneath, petiole 1–1.5 cm long, glabrous; fruits to 13 cm long, 3 cm thick, red-brown (= *M. aulacosperma* Rehd. & Wils.). China; W. Hupeh, Shensi, Honan Provinces. Closely related to *M. salicifolia*. z8

M. × brooklynensis G. Kalmbacher (*M. acuminata × M. liliiflora*). This new hybrid was developed because the Asiatic magnolias are susceptible to frost damage in many parts of the USA as a result of their early leafing and flowering. Both parents were natural tetraploids crossed in the Brooklyn Botanic Garden and the University of Illinois in Urbana. A few clones have been named and propagated in the USA.

'Evamaria'. Flowers purple, overlain by light green and ocher-yellow. Developed in the Brooklyn Botanic Garden in 1970. Protected by US plant patent. z5 ✥

'Woodsman'. Flower color a mixture between the parents, the 3 outer petals green, the 3 middle petals dark purple on both sides, exterior with a trace of green, the 3 inner petals light pink, later white, 10 cm long when opening, but soon extended to 13 cm, 5 cm wide, not fragrant. Developed by J. C. McDaniel of Urbana, Illinois, USA. 1974. z5 ✥

M. campbellii Hook. f. & Thoms. Deciduous tree, to 10(30) m high in its habitat; leaves elliptic, 20–30 cm long; flowers very large, appearing before the leaves, pink, carmine or white, 20–25 cm wide; fruits conical, about 8 cm long, 2 cm thick. Himalayas, Tibet.

This species has recently been subdivided as follows:

ssp. **campbellii.** Leaves elliptic, 15–25 cm long, glabrous above, not rugose, appressed pubescent beneath; flowers cupulate, later outspread, then 15 to 25 cm wide, petals 10–12 cm long, darker or lighter carmine-pink exterior, occasionally also white, interior somewhat lighter than the exterior, fleshy, buds ovate, February–March. BM 6793; JRHS 91: 142, 143. E. Nepal, Assam. 1868. Although a wonderful tree in bloom, the plant must be 20–25 years old before it flowers! The large flowers are sensitive to frost and wind. Ⓕ India. z9 Plate 109, 111. ✥

f. **alba** Hort. Flowers white, exterior with a soft pink middle line. JMa 4 (as 'White Form'; JRHS 91: 145). Before 1945. Occurs within the range of the species. z9 ✥

'Darjeeling' (Hillier). Flowers especially attractive, dark pink. Original tree stands in Lloyd Botanical Gardens in Darjeeeling, India. z9 ✥

'Ethel Hillier' (Hillier). Very strong grower; flowers white, exterior turning soft pink at the base, very large flowered. Selected as a seedling by Hillier. z9 ✥

'Princess Margaret' (Descendant of *M. campbellii* f. *alba*). Flowers about 25 cm wide, petals 9–12 cm long, 5–7 cm wide, exterior purple-red, interior cream-white with a trace of purple (darker than 'Charles Raffill' and petals less rounded), bud scales long silky pubescent. JRHS 99: 125. Developed in Windsor Great Park, England in 1957. ✥

'Trewithen Dark Form' (Johnstone). Flowers carmine on the exterior, interior dark pink, an attractive cupulate form. JMa 2. ✥

'Trewithen Light Form' (Johnstone). Flowers with a dark pink exterior, interior whitish-pink. JMa 3. Both selected by G. H. Johnstone, Trewithen, Cornwall, England. Before 1950. ✥

ssp. **mollicomata** (W. W. Sm.) Johnstone. Young shoots pubescent at first; leaves broadly elliptic to obovate, 15–25 cm long, base rounded to cordate, glabrous and dull green above, soft pubescent beneath, with 10–14 vein pairs, petiole 2–5 cm long, pubescent; flowers cupulate, 9 cm long, later outspread, then 18–20 cm wide, exterior lilac-pink, darker at the base, interior nearly white, flowers 10 days earlier than ssp. *campbellii*. JMa Pl. 4 (= *M. mollicomata* W. W. Sm.). Upper Burma; SE. Tibet; China, W. Yunnan Province. 1920. Flowers at 7 to 8 years of age! z9 Plate 107. ✥

'Lanarth' (M. P. Williams). Leaves more tough and leathery, distinctly rugose above; flowers monochromatic inside and out, deep purple, opening about 20 cm wide, the last internode before the flower long pubescent. JMa Pl. 5; JRHS 97: 161 (= *M. campbellii* ssp. *mollicomata* convar. *williamsiana* Johnstone). Developed by M. P. Williams in Lanarth, Cornwall, England from seed collected by G. Forrest (Nr. 25 655; collected in NW. Yunnan Province, China). Flowered first in 1943. z9 ✥

M. campbellii ssp. **campbellii** × ssp. **mollicomata.** This cross was artificially developed, but there probably exist natural hybrids in cultivation in England. Two cultivars have been named to date:

'Charles Raffill'. Very strong growing; more similar to ssp. *mollicomata* in the early flowering character, flower buds dark pink-red at first, exterior purple-pink when fully open, interior white with a purple-pink blush on the margin. JRHS 1963; 173; 91: 123. Cross developed by Charles Raffill at Kew; cultivated and first flowered in Windsor Great Park, England in 1959. ✥

'Kew's Surprise'. Flowers larger than those of 'Charles Raffill', exterior also somewhat darker purple-pink. JRHS 93: 45. Also developed by C. Raffill of Royal Botanic Gardens, Kew, England; flowered for the first time at Caerhays, Cornwall in 1967. ✥

'Sidbury'. Flowers similar to those of *M. campbellii*, but appearing earlier. Developed in 1946 by Sir Charles Cave of Sidbury Manor, Devon, England. Propagated by Hillier. ✥

M. coco (Lour.) DC. Evergreen, low growing shrub, 0.5–1.5 m high; leaves elliptic, tapering to both ends, glabrous, venation reticulate, to 10 cm long; flowers nodding, cream-white, very fragrant at night. NH 39; 122; JRHS 89: 296 (= *M. pumila* Andr.). Java. 1786. z9 #

M. conspicua see: **M. denudata**

M. cordata Michx. Small deciduous tree or a shrub, scarcely over 7 m high; leaves obovate-oblong to elliptic, 8–15 cm long, acuminate or obtuse, base broadly ovate to round, only rarely cordate (!), somewhat pubescent

Plate 97

Lonicera purpusii in the Späth Arboretum, Berlin
Photo: L. Späth, Archive

Lonicera morrowii in flower

Lonicera morrowii with fruit in its native habitat
Photo: Dr. Watari, Tokyo

Plate 98

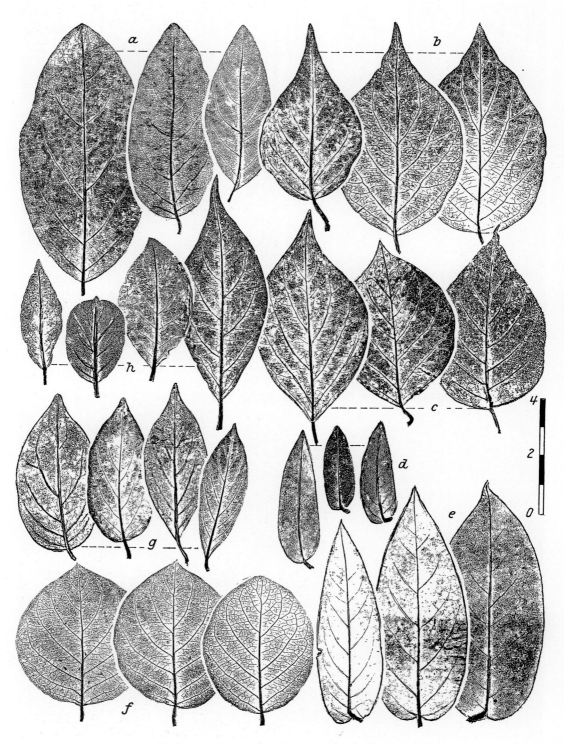

Lonicera. a. *L. canadensis*; b. *L. maackii*; c. *L. maackii* var. *podocarpa*; d. *L. alseuosmoides*; e. *L. henryi*; f. *L. xylosteum*; g. *L. ruprechtiana*; h. *L. webbiana* (from wild plants)

Plate 99

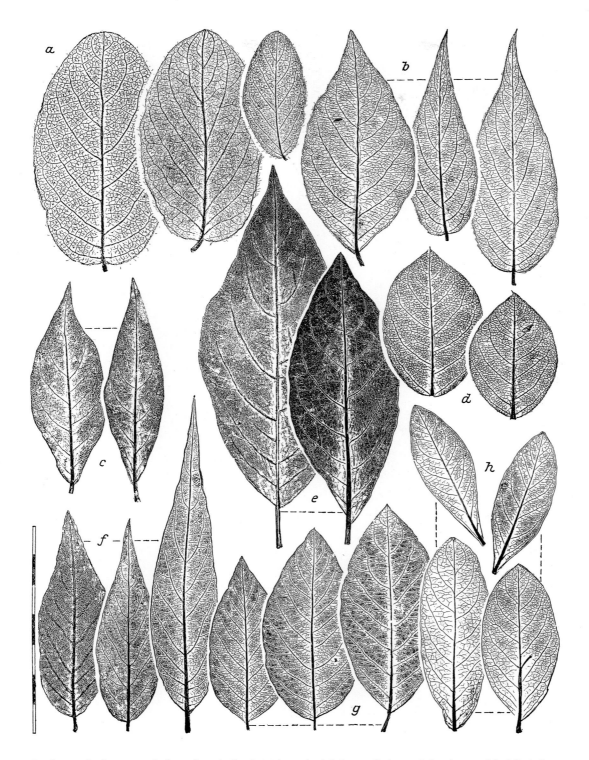

Lonicera a. *L. chaetocarpa*; b. *L. maximowiczii*; c. *L. tatsienensis*; d. *L. fragrantissima*; e. *L. involucrata*; f. *L. deflexicalyx*; g. *L. nigra*; h. *L. oblongifolia* (from wild plants)

Plate 100

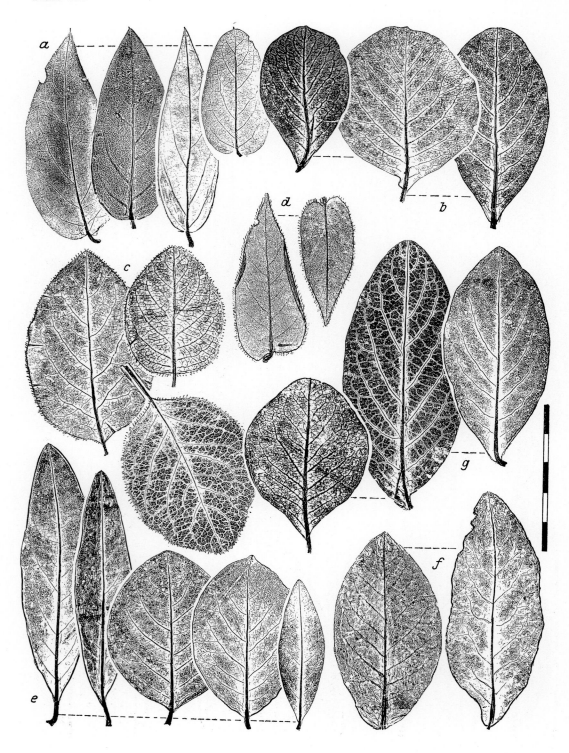

Lonicera. a. *L. similis* var. *delavayi;* b. *L. etrusca;* c. *L. arizonica;* d. *L. giraldii;* e. *L. sempervirens;* f. *L.* × *brownii;* g. *L. dioica* (from wild plants)

Plate 101

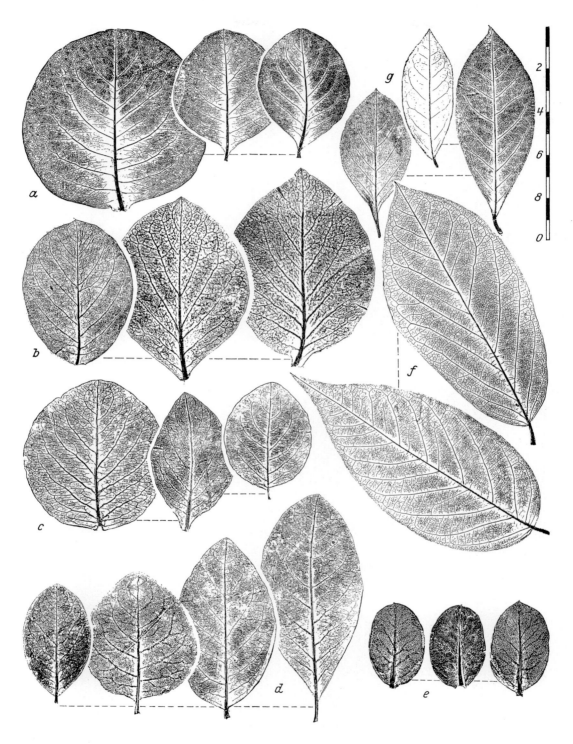

Lonicera. a. *L. prolifera;* b. *L. hirsuta;* c. *L. flava;* d. *L. ciliosa;* e. *L.* × *heckrottii.*
Lyonia. f. *L. ovalifolia;* g. *L. ligustrina* (from wild plants)

Plate 102

Lonicera syringantha var. *wolfii*
in the Pruhonice Arboretum

Lonicera alberti
in a park in Prague, Czechoslovakia

Loropetalum chinense in a garden in Minusio, Tessin, Switzerland
Photo: Dr. Anliker, Wädenswil, Switzerland

Plate 103

Lycium ruthenicum
in the Copenhagen Botanic Garden, Denmark

Lycium intricatum
in its native habitat on Mallorca Island, Spain

Lycium horridum in the Lisbon Botanic Garden, Portugal

Plate 104

Maackia amurensis, fruits, in the Banska Stavnice Arboretum, Czechoslovakia
Photo: Dr. Pilat, Prague

Magnolia grandiflora
in the U.S. National Arboretum, Washington, D.C.

Magnolia macrophylla
in Longwood Gardens, USA

Plate 105

Magnolia delavayi in Birr Castle Park, England

Magnolia cylindrica
in the Hillier Arboretum, England

Magnolia soulangiana 'Brozzonii' in the Palmer Garden,
Rosemoor, Torrington, Devon, England

Plate 106

Magnolia × *soulangiana* 'Alexandrina'
Photo: Gartenbau-Blatt, Switzerland

Magnolia × *soulangiana* 'Lennei'
Photo: Gartenbau-Blatt

Magnolia × *soulangiana* 'Speciosa'
Photo: J. Jenni, Bern (Gartenbau-Blatt Switzerland)

Magnolia × *soulangiana* 'Norbertii'
Photo: Gartenbau-Blatt

Plate 107

Magnolia. a. *M. denudata;* b. *M. globosa;* c. *M. sprengeri* var. *elongata;* d. *M. sinensis;* e. *M. sargentiana;*
f. *M. campbellii* var. *mollicomata* (most material from wild plants)

Plate 108

Magnolia cordata in Royal Botanic Gardens, Kew, England

Magnolia grandiflora × *virginiana* in the U.S. National Arboretum in Washington, D.C.

Plate 109

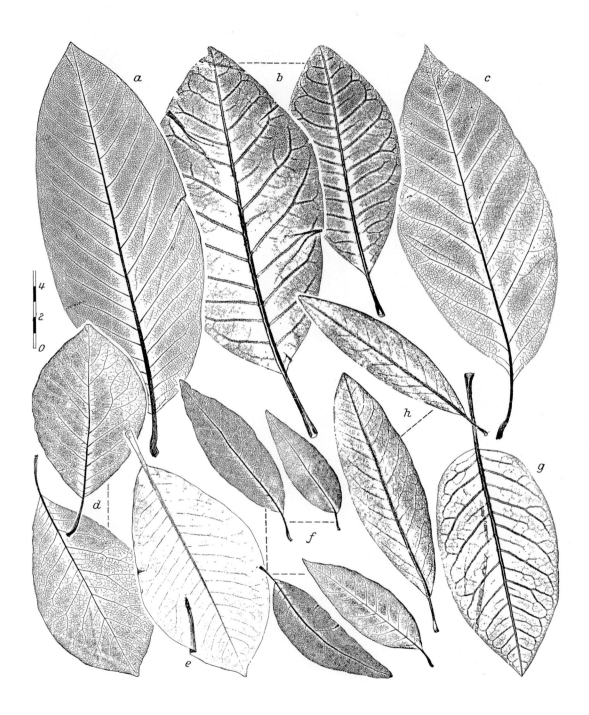

Magnolia. a. *M. campbellii*; b. *M.* × *watsonii*; c. *M. tripetala*; d. *M. kobus*; e. *M. grandiflora*;
f. *M. salicifolia*; g. *M. wilsonii*; h. *M. virginiana* (most from wild plants)

Plate 110

Magnolia stellata
Photo: Dortmund Botanic Garden

Magnolia kobus
Photo: C. R. Jelitta, Berlin

Magnolia loebneri
Photo: Dortmund Botanic Garden

Plate 111

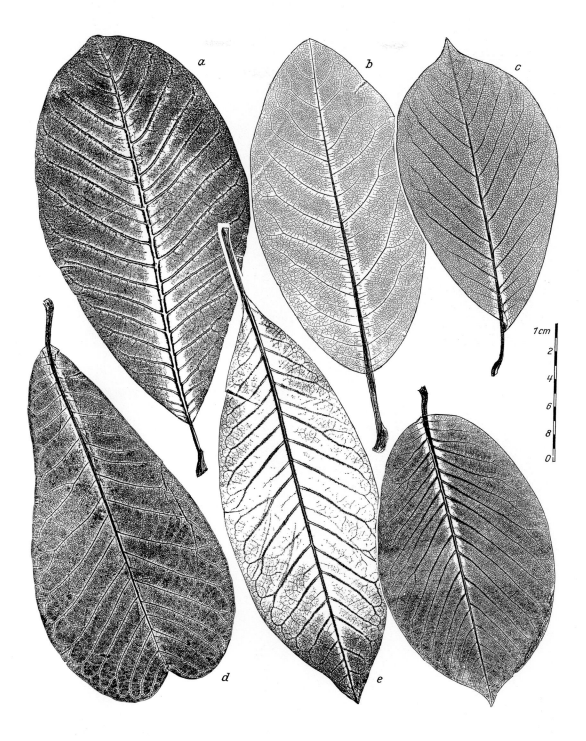

Magnolia. a. *M. hypoleuca;* b. *M. delavayi;* c. *M. veitchii;* d. *M. officinalis;* e. *M.* × *thompsoniana;* f. *M. campbellii* (material from wild plants)

Plate 112

Magnolia tripetala
in the Dortmund Botanic Garden, W. Germany

Magnolia hypoleuca
in the Berlin Botanic Garden, W. Germany
Photo: C. R. Jelitto, Berlin

Magnolia sinensis
in the Birr Castle Park, Ireland

Magnolia liliiflora 'Nigra'
Photo: Gartenbau-Blatt, Switzerland

Fig. 187. *Magnolia cordata* (from Sargent)

beneath; flowers on a thick stalk, cupulate, light yellow (sometimes only greenish-yellow), 4–7 cm long, sepals ovate, soon becoming reflexed, slightly fragrant, somewhat pleasant scented, May–June; fruits red, 2.5–3.5 cm long. SM 311; RWF 136 (= *M. acuminata* var. *subcordata* [Spach] Dandy). Eastern North America. 1801 z6 Plate 40, 108; Fig. 187.

M. cylindrica Wils. Deciduous shrub or small tree, to 9 m high in its habitat, young shoots reddish-brown, silky pubescent; leaves obovate to elliptic-obovate, 10–16 cm long, 4.5 to 9.5 cm wide, obtuse or with a short, protruding, obtuse tip, base cuneate, dark green above, glabrous, with distinctly reticulate venation, pale gray-green beneath, primary veins with somewhat appressed silky pubescence; flowers appear before the leaves, similar to those of *M. denudata*, with 5 petals, white, oblong-spathulate, exterior with a soft pink middle band, 10 cm long, 4 cm wide, stamens with pink filaments, April; fruits cylindrical, 5–7 cm long, 2–3 cm wide, China; Anhui Province. 1936. Apparently good winter hardiness. z6 Plate 105. ✧

M. dawsoniana Rehd. & Wils. Deciduous tree, 6–10 m high, but more commonly a broad, open shrub, young shoots glabrous; leaves obovate or more elliptic, 12–17 cm long, obtuse to short acuminate, glabrous and dark green above, glossy, gray-green and finely pubescent beneath, with 8–10 vein pairs, both sides with distinctly reticulate venation; flowers spreading horizontally or nodding, with 9–12 petals in 3 rings, narrowly oblong, about 10 cm long, exterior light pink, darker and striped at the base, interior white with a trace of pink, about 25 cm wide when fully open, petals eventually pendulous, March to April. JMa 6; NF 12: 13; BM 9678. China; E. Sikang Province. 1910. Similar to *M. sargentiana*, but bushier, flowers more abundantly, leaves with reticulate venation. z9 ✧

M. delavayi Franch. Evergreen shrub, to 5 m high, occasionally a tree in its habitat, to 10 m, bark ash-gray, eventually nearly black, rough; leaves ovate to oblong, 15–30 cm long, half as wide, leathery, dull green, base round, pubescent above only when young, softly pubescent beneath; flowers with 3 sepals, green at first, later white, then reflexed, petals 6, in 2 rings, fleshy, ivory-white, about 10 cm long, to 7 cm wide, July–September; fruit cones erect, 10–20 cm long, ovoid-cylindrical. BM 8282; MMa 104. China; Yunnan Province. 1899. z9 Plate 105, 111. # ✧ ⚭

M. denudata Desr. Deciduous tree, 2 to 4 m high (to 10 m in its habitat), abundantly branched and wide arching, shoots bent, young shoots pubescent at first, later quite glabrous; leaves obovate-elliptic to obovate-oblong, with a protruding tip, 8–15 cm long, base round to cuneate, eventually quite glabrous above, with scattered soft pubescence beneath; flowers numerous, erect (!), large, campanulate, sepals and petals of equal form (tepals), 9 in number, white, oblong-obovate, tapering to the apex, the outer 3 somewhat reddish at the base, pleasantly scented, April–May; fruit cones spindle-form, to 15 cm long, erect. JMa 1 to 2 ; LF 132; MMa 16, 48, 58, 112 (= *M. precia* Corr. ex Vent.; *M. conspicua* Salisb.; *M. yulan* Desf.). E. and S. China, widely distributed. 1789. Good winter hardiness. z6 Plate 107; Fig. 188. ✧

M. diva see: **M. sprengeri** var. **diva**

M. foetida see: **M. grandiflora**

M. fraseri Walt. Deciduous, broad crowned tree, to 10 m high, shoots glabrous; leaves obovate-spathulate, 15–30 cm long, obtuse, base distinctly cordate and usually with conspicuous auricles, thin, glabrous on both sides, abscising green in the fall; flowers appear after the leaves, campanulate, 20 to 25 cm wide, light yellow at first, then milky-white, pleasantly scented, with 6–9 petals, the 3 sepals more greenish, soon becoming reflexed and then abscisisng, the others obovate, compressed beneath the middle, May–June; fruits 10–12 cm long, cone-like, pink. BB 1537; SS 11–12; NF 6: 235 (= *M. auriculata* Bart.). SE. USA. 1786. Hardy. z6 Plate 40; Fig. 189. ∅ ✧ ⚭

M. glauca see: **M. virginiana**

M. glauca var. *major* see: **M.** × **thompsoniana**

M. globosa Hook. f. & Thoms. Deciduous shrub or a small tree in its habitat, 2.5–5 m high, young shoots velvety brown pubescent; leaves ovate to elliptic, with a short apex, 10–20 cm long, glossy dark green above, somewhat bluish pruinose while developing, later brown pubescent on the midrib, gray-green with silvery-brown pubescence beneath, especially the 12 vein pairs; flowers cream-white, nodding, about 6–7 cm

Fig. 188. *Magnolia denudata* (from Baillon)

Fig. 189. *Magnolia fraseri* (from Sargent)

wide, more or less globose, petals 9–10, in 3 rings, obovate, anthers dark pink, opening in June after the leaves; fruits pendulous, carmine, 5 cm long. BM 9467; JMa 17; NF 10: 272 (= *M. tsarongensis* W. W. Sm. & Forrest). Burma, Himalyas. 1920. z9 Plate 107. ⌀ ⊕ ⦻

var. *sinensis* see: **M. sinensis**

M. gracilis see: **M. liliiflora 'Gracilis'**

M. grandiflora L. Evergreen tree, to 25 m high, often an attractive conical form from the ground up, young shoots and buds rust-brown tomentose; leaves obovate-elliptic, 12–20 cm long, obtuse, base usually cuneate, glossy dark green above, rust-brown beneath, short tomentose, leathery tough, abscising in the 2nd year, stipules not adnate to the petiole; flowers very large, cream-white, 20–30 cm wide, fragrant; petals usually 6(9–12), obovate, fleshy, sepals 3, petaloid, May–August; ovaries woolly, fruit ovoid, 7–10 cm long, brown tomentose. SM 312; SS 1–2; RWF 133–134 (= *M. foetida* Sarg.). Southeast N. America. For mild climates. z6–7 Plate 104, 109. # ⌀ ⊕

This species has no less than 165 cultivars (including some synonyms) listed in the Magnolia Register. Most cultivated only in the USA.

'Angustifolia'. Leaves narrow, about 15–20 cm long, only 3–5 cm wide, underside only slightly or totally lacking rusty brown coloration. # ⌀ ⊕

'Exmouth'. Habit narrowly conical; leaves narrower than those of the species, brown tomentose beneath; flowers appearing on young plants, 20–25 cm wide. BM 45 (= var. *exoniensis* Lodd.; var. *lanceolata* Ait.). Developed by Sir John Colliton in Exmouth, England before 1737.

var. *exoniensis* see: **'Exmouth'**

'Ferruginea'. Compact habit, erect; leaves elliptic, obtuse, densely red-brown tomentose beneath, dark green and glossy above. A clone selected by Hillier in 1973. #

'Galissonière'. Especially regular habit, conical; leaves red-brown beneath (= 'Galissoniensis'; 'Galissonieri'). Brought into France from the USA between 1741 and 1749 by Roland Michel Baron de la Galissonière and disseminated by A. Leroy after 1856 as the "hardiest form". Still propagated in France (Minier). # ⊕

'Goliath'. Stiff branched; leaves oval-oblong, 12–25 cm long, margins sinuate, apex distinctly rounded, not brown tomentose beneath; flowers with 8–10 petals, nearly 30 cm wide when in full bloom, flowering continuously June to

September. GC 136: 193. Developed by C. Smith, Caledonia Nursery, Guernsey Island, England. Before 1910. One of the most popular forms in England today. # ⊕

var. *lanceolata* see: **'Exmouth'**

'Nannetensis'. Flowers double, flowers very abundantly. Introduced around 1865 by the Delaunay Nursery in Angers, France. (The name should be corrected to read 'Namnetensis', derived from Namnetes, the old Roman name for Nantes, France.) # ⊕

'Saint Mary'. Selection with especially dark brown leaf undersides; flower on young plants. Introduced to the trade by the Glen St. Mary Nursery, Glen Saint Mary, Florida, USA. Before 1930. # ⊕

'Samuel Sommer'. Narrow upright habit; leaves very large, very glossy above with raised venation, rust-brown pubescent beneath; flowers to 35 cm wide, with 12 petals in 3 layers. Introduced in 1961 by the Saratoga Horticultural Foundation in Saratoga, California and widely used in the USA today. Patented. # ⊕

'Undulata'. Foliage differing from the norm, leaves elliptic-oblong to oblong-obovate, margins very undulate, very glossy above and distinctly veined, green beneath, also on young plants. Probably = 'Longifolia Undulata' from Leroy, Angers, France. 1850. #

M. grandiflora × M. virginiana Freeman. Closely resembles *M. grandiflora* in habit and foliage; leaves evergreen, 12–20 cm long, only 3–7 cm wide, obtuse, green beneath; flowers small, like those of *M. virginiana*. NH 16: 161. Developed in the U. S. National Arboretum, Washington D.C. 1930. Plate 108. # ⊕

2 of these so-called 'Freeman Hybrids' have been selected to date:

'Freeman'. Habit a particularly attractive conical form; flowers significantly larger than those of *M. virginiana*, very fragrant. 1967. # ⊕

'Maryland'. Habit less regular than the above selection; foliage lighter green. Very easily propagated from cuttings. # ⊕

M. halleana see: **M. stellata**

M. × highdownensis Dandy (*M. sinensis × M. wilsonii*). Tall, deciduous shrub, young shoots pubescent; leaves elliptic-oblong, 10–19 cm long, 5–10 cm wide, acute (!), dark green above, gray pubescent or finely tomentose beneath; flowers with 9 petals, white, 7 cm long, nodding, fragrant, anthers red, June. JRHS 1950: 82; NF 10: 33. Developed by J. C. Williams of Caerhays Castle, Cornwall, England. Before 1938. Possibly, however, only an acute leaved form of *M. sinensis*. ⊕

M. hypoleuca S. & Z. Deciduous tree, to 30 m high, habit broadly pyramidal, often only a tall shrub in cultivation, twigs glabrous, reddish, thick, buds blue-black; leaves obovate, 20–40 cm long, abruptly narrowed at the base, apex obtuse, bluish and pubescent beneath at first, abscising in September, all leaves spreading umbrella-like at the branch tips; flowers shell-like, 10–15 cm wide, white, fragrant, petals 6–9, leathery, filaments scarlet-red, anthers yellow, June; fruit "sausage-like", 15–20 cm long, red. BM 8077; MD 1904: 1; MMa 148, 150 (= *M. obovata* Thunb.). Japan. 1865. z6 Plate 111, 112; Fig. 192. ⊕ ⌀ ⦻

M. × **'Kewensis'** Hort. (*M. salicifolia* × *M. kobus*). Small tree, narrowly conical habit, habit intermediate between the parents, bark of the young shoots with a strong lemon scent (like *M. salicifolia*), buds with fine silky pubescence; leaves narrowly obovate to elliptic, tapering to a narrow apex, base cuneate, 8–12 cm long, 3 to 5 cm wide, smooth above and somewhat glossy, bluish-green beneath; flowers appear before the leaves, white, petals 6–7 cm long, very fragrant. Originated as a chance seedling in Kew Gardens, 1952.

'Kew Clone'. The name of the original plant.

'Slavin's Snowy'. Deciduous, dense, conical tree, very strong growing; the 3 outer petals lanceolate, 3 cm long, green at first, later white, the 6–9 inner petals white with red basal spots, 8–9 cm long, flowers abundantly, before the leaves. NH 1954: 11 (= *M.* × *slavinii* Harkness). Developed by Harkness in Highland Park, Rochester, New York, USA in 1917 from *M. salicifolia* × *M. soulangiana*. The correct classification of this hybrid is still unclear; it is found in the "check list" as a *M.* × *kewensis*. ✧

'Wada's Memory' is probably best included here. Upright habit, dense, compact; young leaves somewhat reddish; flowers nodding, white, to 16 cm wide, flowering on young plants, fragrant. Developed in the Arnold Arboretum from *M. kobus* seed obtained from K. Wada of Yokohama, Japan. These flowers are, however, much larger than those of *M. kobus* and more abundant. Before 1959. ✧

M. **kobus** DC. Deciduous tree or shrub, habit broad or conical or rounded; leaves elliptic-obovate, 6–17 cm long, 3–11 cm wide, most abruptly short acuminate, cuneate tapered to the base; flowers appear before the leaves, erect, white, exterior occasionally light reddish, about 10 cm wide when open, sepals 3, petaloid, petals 6, spathulate, in 2 rings, much larger than the early abscising, much smaller sepals, flower buds thick and pubescent; ovaries green. Japan. z5

var. **borealis** Sarg. Strong growing variety, to 25 m high, habit more conical, lateral twigs and long shoots stiff and thick, over 5 mm thick at the base; leaves on the outward directed shoots normally longer than 10 cm; flowers like the species. Bai 5: 3; GF 6: 11. Japan; N. Honshu, Hokkaido. Flowering on young plants. z4 ✧

var. **kobus**. The type of the species. Tree, to 10 m high, often lower, long shoots and lateral shoots slender, less than 4 mm thick at the internodes; leaves elliptic-oblong, 10 cm long; flowers with 6–9 petals, these white, occasionally with a reddish middle stripe, the 3 sepals 10–15 mm long, quickly abscising. NK 20: 24; BM 8428. Japan; Central Honshu and southward. The type flowers only on older plants. Plate 109, 110. ✧

var. *stellata* see: **M. stellata**

M. **liliiflora** Desr. Deciduous shrub, to 3 m high, branches limp; leaves green on both sides, obovate, occasionally nearly elliptic, 10–15 cm long, with short tips, scattered pubescent above, pubescent venation beneath; flowers large, campanulate, exterior purple, interior usually white, petiole short and thick, petals 6, obovate-oblong, obtuse, 6–7 cm long, sepals greenish, only ⅓ as long, soon abscising, appearing together with the leaves, May–June; fruits oblong. JMa Pl. 11; LF 133; BM 390 (= *M. purpurea* Curt.; *M. denudata* Schneid. non

Fig. 190. *Magnolia liliiflora*
(from Nouveau Dehamel)

Desr.). China; much planted in Japan. 1790. Hardy. z6 Fig. 190. ✧

'Gracilis'. Smaller in all respects, lower; leaves smaller; flowers dark purple (= *M. gracilis* Salisb.). Introduced from Japan in 1804. ✧

'Nigra'. Flowers much larger, to 12 cm long, exterior dark purple, interior light purple, appearing with the leaves, late May–June. Gs 1921: 79 (= *M. soulangiana nigra* Nichols.). Introduced from Japan in 1861. Plate 112. ✧

'Trewithen'. Flowers very large, to 12 cm long, exterior dark carmine, interior soft pink with darker veins and a darker throat. JMa Pl. 11. ✧

Included here are a group of triploid hybrids of *M. liliiflora* with *M. stellata*, developed in the U.S. National Arboretum in Washington, D.C. in 1955–1956. They can generally be described as multistemmed shrubs, 2 to 3 m high with a globose or conical habit, flowers superior to those of the parents in size, color, fragrance and number. The following clones have been developed to date by Dudley and Kosar:

'Ann' (*M. liliiflora* 'Nigra' × *M. stellata*). Early flowering, flowers small, 5–10 cm wide, buds red-purple with 6–8 petals, these erect when in full bloom, narrowly obovate, 2.5–6.5 cm long, 12–20 mm wide, interior and exterior purple, stamens 50–60. ✧

'Betty' (*M. liliiflora* 'Nigra' × *M. stellata* 'Rosea'). Middle-early, flowers large, to 20 cm wide, buds purple-red, with 12–19 petals, these spathulate or ligulate to awl-shaped, obtuse to acuminate, 5–7.5 cm long, 2–3 cm wide, limp when in full bloom, narrowed to a claw-form base, exterior red-purple, interior white, stamens 70–90, gray-purple. ✧

'Jane' (*M. liliiflora* 'Reflorescens' × *M. stellata* 'Waterlily'). Late flowering, flowers small, 7–10 cm wide, buds erect, slender, red-purple, 8–10 petals, these obovate-spathulate, 5–6 cm long, exterior red-purple to purple, interior white, stamens 90 to 110, red-purple. ✧

'Judy' (*M. liliiflora* 'Nigra' × *M. stellata*). Middle-early, flowers small, 5–7.5 cm wide, buds erect, red-purple, petals cuneate and somewhat outspread, red-purple, interior cream-white, stamens 40–50, red-purple. ✧

'Pinkie' (*M. liliiflora* 'Reflorescens' × *M. stellata* 'Rosea'). Late flowering, flowers large, 12–18 cm wide, buds thick, red-purple, petals 9–12, obovate-spathulate, red-purple, interior white, stamens 50–60, reddish purple. ✧

'Randy' (*M. liliiflora* 'Nigra' × *M. stellata*). Flowering middle-early, upright to columnar habit; flowers cupulate, 7–12 cm wide, buds erect, red-purple, petals 9–11, awl-shaped to spathulate, 5–7 cm long, 1.5–2 cm wide, red-purple, stamens 50–70, red-purple. ⊕

'Ricki' (*M. liliiflora* 'Nigra' × *M. stellata*). Middle-late, flowers 10–15 cm wide, buds erect, slender, red-purple, petals 10–15, twisted, obovate-spathulate, 5–8 cm long, 1.5–2.5 cm wide, exterior red-purple, interior white to red-purple, stamens 80–100, red-purple. ⊕

'Susan' (*M. liliiflora* 'Nigra' × *M.stellata* 'Rosea'). Middle-early, flowers 10–15 cm wide, buds erect, red-purple, petals 6, slightly clawed, red-purple, stamens 60–70, gray-purple. ⊕

Also included here are the 'Gresham Hybrids', developed in 1955 by Dr. Todd Gresham in Santa Cruz, California, USA.

'Dark Raiment' (*M. liliiflora* × *M. veitchii*). Leaves dark green, tough, attractive fall color; flower buds red-violet, 12 cm long, petals 12, red-violet, the outer 8 cupulately arranged and later reflexed, the inner 4 remaining erect and surrounding the ovaries. ⊕

'Peppermint Stick' (*M. liliiflora* × *M. veitchii*). Flowers white, violet on the base, petals with a purple midstripe, inner petals erect, outer ones eventually reflexed. JRHS 98: 131. ⊕

'Raspberry Ice' (*M. liliiflora* × *M. veitchii*). Flowers campanulate, petals 12, red-violet on the base, gradually becoming whiter toward the apex. JRHS 89: 130. ⊕

'Royal Crown' (*M. liliiflora* × *M. veitchii*). Flowers dark red-violet, buds 13 cm long, 5 cm wide, petals 12, the outer ones reflexed, appearing crown-like. JRHS 89: 128. ⊕

'Vin Rouge' (*M. liliiflora* × *M. veitchii*). Young leaves bronze-red, venation and stipules red; flowers dark wine-red, very fleshy. ⊕

M. × loebneri Kache (*M. kobus* × *M. stellata*). Tall shrub, 6–8 m high and wide with age, but usually lower; leaves similar to those of *M. stellata*, but more obovate and larger; flowers also larger, normally with 12 petals, white, very abundant. Developed before 1920 by Max Loebner, Dresden, E. Germany. The opinion that this plant is simply a variety of *M. kobus* has not been generally accepted. z5 Plate 110.

Includes the following cultivars:

'Leonard Messel' (Messel). Strong growing; flowers with about 12 petals, these oblanceolate, about 7 cm long, exterior pink (HCC 30/2), interior white, grouped calyx-like at first, eventually spreading quite flat. JRHS 80: 104; 96: 218. Developed before 1950 by L. Messel, Nymans, Handcross, Sussex, England. Received an Award of Merit in 1955. ⊕

'Merrill' (Arnold Arboretum). Strong growing; flowers with 15 petals, somewhat larger than those of *M. stellata*, petals also twice as wide. AB 20: 25. Bred in the Arnold Arboretum in 1939. Flowers as a 5-year-old plant. Hardy. Plate 74. ⊕

'Neil McEacharn' (*M. kobus* × *M. stellata* 'Rosea'). Grows tree-like; flowers white with a trace of soft lilac, flowers very abundantly. Developed from seed in Windsor, England by N. McEacharn, then proprietor of the Taranto Gardens in Pallanza, Italy on Lake Maggiore. ⊕

'Snowdrift' (Hillier). Clone, developed from one of the original seedlings of *M. loebneri*; leaves somewhat larger; flowers larger with about 12 petals. Named by Hillier. ⊕

M. macrophylla Michx. Small, deciduous tree, to about 10 m high, crown broad and round, branches stiff, thick,

Fig. 191. *Magnolia macrophylla* (from Sargent)

young shoots brown tomentose at first; leaves obovate, 30–80 cm long, obtuse, base cordate to auricled, dark green above, bluish and finely pubescent beneath; flowers shell-form, 20–30 cm wide, petals 6, white, with a half-moon shaped, pink basal spot, June; fruit oval-oblong, 6–8 cm long, woolly, pink. BB 1538; BM 2981; SS 7–8. USA, Kentucky to Arkansas. Hardy, but only for very rich, fertile soil in park-like settings, out of the wind. z6 Plate 191. ∅ ⊕ ⊛

M. mollicomata see: **M. campbellii** ssp. **mollicomata**

M. nicholsoniana see: **M. sinensis** and **M. wilsonii** var. **taliensis**

M. nitida W. W. Sm. Evergreen tree or shrub, 5–7 m high, young shoots glabrous; leaves elliptic to ovate, with short tips, 6–11 cm long, 2–5 cm wide, leathery, glossy green above, lighter and glabrous beneath; flowers pale yellow, exterior reddish on the base of the sepals, 5–7 cm wide, petals 9, narrowly obovate, 5 cm long; fruits 5–7 cm long. JMa Pl. 14; BMns 16. Tibet; China, Yunnan Province. z9 # ⊕

M. officinalis Rehd. & Wils. Deciduous tree, to 22 m high in its habitat, bark ash-gray, young shoots yellow-green and finely pubescent at first, later glabrous; leaves clustered at the branch tips, predominantly elliptic-obovate, 30–38 cm long, 15–17 cm wide, obtuse or occasionally emarginate at the apex (!), base cuneate, with 27–30 prominent veins, conspicuously apple-green above, glabrous, more gray-green and loosely pubescent beneath, petiole light green; flowers cupulate, 15–20 cm wide, sepals 3, pale green, with a reddish trace, quickly abscising, petals 6–9, white, fleshy, stamens numerous, red, June; fruits cone-like, oblong, red, 10–12 cm long. W. China. 1900. z8 Plate 111. ∅ ⊕ ⊛

var. **biloba** Rehd. & Wils. Leaves always deeply incised at the apex, otherwise hardly differing from the type. W. China. The bark of both plants has medicinal uses in China. ∅

M. precia see: **M. denudata**

M. × proctoriana Rehd. (*M. kobus* × *M. salicifolia*). Deciduous shrub, very similar to *M. salicifolia*, but differing in the short pubescent foliar buds, leaves widest above the middle, green beneath; flowers with 6–12 petals. Developed in the USA in 1928. Slight garden merit.

M. parviflora see: **M. sieboldii**

M. pumila see: **M. coco**

M. purpurea see: **M. liliiflora**

M. pyramidata Bartr. Deciduous tree, to 10 m high, twigs ascending, glabrous; leaves obovate, 14–22 cm long, short acuminate (!), base auricled, underside blue-gray, petiole 2–4 cm long; flowers 8–12 cm wide (!), cream-white, petals oblanceolate, gradually tapering to the base, sepals much shorter, June; fruits oblong, 5 to 7 cm long, tips of the ripe carpels curved (!). SM 317; ST 51; BR 407. SE. USA. 1825. Occasionally confused with *M. fraseri*, but flowers only half as large. z8 ∅ ⊛

M. rostrata W. W. Sm. Deciduous tree, 10–20 m high in its habitat, young shoots reddish, glabrous; leaves obovate, 40–50 cm long, to 30 cm wide, apex rounded, tapered to slightly cordate at the base, young leaves rust-brown tomentose and reddish, later glabrous above, bluish beneath, venation reddish pubescent, about 30 vein pairs, petioles 3–7 cm long; flowers solitary, terminal, sepals 3, greenish and pink, abscising quickly, petals usually 8, chalk-white, fleshy, to 14 cm long, appearing after the leaves develop, June; fruit cones erect 12 cm long, red, later brown, carpel apices beaked. JMa Pl. 13. China, Yunnan Province; SE. Tibet; Upper Burma. 1917. Attractive for the very large leaves, flowers insignificant. z9 ∅ ⊛

M. salicifolia (S. & Z.) Maxim. Large, deciduous, conical shrub or tree, to 7 m, shoots somewhat flexuose, young shoots glabrous; leaves elliptic-ovate to oblong-lanceolate, 7–12 cm long, long acuminate, base round to cuneate, with 10–12 vein pairs, dull green above, bluish beneath, petiole 1–2 cm long; flowers white, 7–10 cm wide, fragrant, sepals 3, much shorter than the petals, occasionally reddish at the base, petals 6, narrowly obovate, 5–6 cm long, appearing before the leaves, April–May; fruit cones cylindrical, 4–7 cm long, erect, dark pink. BM 8483; GF 6: 67. Japan. 1892. Hardy. z6 Plate 109. ⊕

var. **concolor** Miq. Broader growing, shoots thicker, bark distinctly aromatic; leaves oval-oblong, larger, to 12 cm long and 6 cm wide, with 12 or more vein pairs; flowers 7–10 cm wide, pure white, flowering 2 weeks later than the species. Japan. Introduced by Veitch in 1892.

'Else Frye'. Upright habit, shoots slender, glabrous, yellow-green to brown; leaves elliptic-lanceolate, 7 to 15 cm long, 3–6 cm wide, green above, becoming bluish beneath, sparsely pubescent; flowers large, 8–9 cm long, with 9 petals, white, turning purple-pink at the base, stamen filaments pink. Developed by W. B. Clarke, San Jose, California. Before 1962. ⊕

'Fasciata'. Narrow upright habit, with many dark shoots (like a bundle of branches). MMa 212. Frequently found in England, but unattractive. A typical tree grows in Tilgate, Sussex, England (1927). ⊕

'Jermyns' (Hillier). Less vigorous, shrubby; leaves wider, underside more bluish; flowers larger, flowering later. Described by Hillier as one of the best clones. ⊕

M. sargentiana Rehd. & Wils. Slender, deciduous tree, narrowly upright, to 15 m high, never a shrub, shoots slender, yellowish, later gray; leaves obovate, 10–15 cm long, apex rounded, occasionally with a protruding apex or emarginate, 6–12 cm wide, light green above, reticulate venation beneath, gray pubescent, leathery; flowers spreading horizontally or nodding, to 20 cm wide, petals 10–14, normally 12, obovate-spathulate, interior white, exterior purple-pink, particularly at the base, stalk thick, yellow, pubescent, before the leaves, April–May. JMa Pl. 7–8; NF 5: 45. China; Szechwan Province. Flowers usually only at the top of the plant. z9 Plate 40, 107. ⊕

'Michael Rosse'. Tall tree; flowers soft purple. The original tree is in Nymans, S. England. Presumably a seedling of the var. *robusta*. ⊕

var. **robusta** Rehd. & Wils. Large shrub to (rarely) a broad crowned tree (to 10 m), twigs outspread from the base up; leaves oblong-obovate, 12–20 cm long, emarginate at the apex on older plants, 5–8 cm wide, cuneately tapered at the base; flowers nodding, 20–30 cm wide, with 12–16 petals, these 10–15 cm long, elliptic, overlapping, exterior lilac-pink, interior usually pure white, evenly distributed over the entire plant, March–April. JMa Pl. 1 (and 2 forms at Pl. 8); JRHS 91; 144; 97: 189. China; W. Szechwan Province, in forests. Seedling plants flower in about 10 years. ⊕

M. sieboldii K. Koch. Deciduous shrub hardly over 4 m high in cultivation (a small tree to 7 m in its habitat), young shoots thin, pubescent; leaves broadly elliptic, 10–15 cm long, abruptly acuminate, base rounded, dark green above, glabrous, blue-green beneath and finely pubescent at first, later only on the midrib, with 6–10 vein pairs; flowers cupulate, 7–10 cm wide, appearing after the leaves, fragrant, on 3–6 cm long, pubescent stalks, nodding, petals usually 6, obovate, concave, 5 cm long, sepals 3, somewhat smaller, exterior reddish, abscising quickly, filaments an attractive carmine-red, June–July; fruit ovoid, 3–4 cm long. JMa 17; NK 20: 23; BM 7411 (= *M. parviflora* S. & Z. non Bl.). Japan, Korea. 1865. z7–8 ⊕ ⊛

'Charles Coates' (*M. sieboldii* × *M. tripetala*). Small tree or large shrub; leaves similar in form to those of *M. tripetala*, but smaller, with long hairs on the midrib beneath; flowers appear with the leaves, similar to those of *M. sieboldii*, but erect, cream-white, very fragrant, stamens intensely pink, May–June. Chance seedling, originated and named in Royal Botanic Gardens, Kew, England, 1958. ⊕

'Semiplena'. Flowers partly semidouble, but also partly single on the same plant. Origin unknown; cultivated in Dutch gardens before 1949. ⊕

M. sinensis (Rehd. & Wils.) Stapf. Deciduous shrub or small tree, to 5 m high in mild areas, closely related to *M. sieboldii*, young shoots silky pubescent at first; leaves obovate to elliptic, 7–15 cm long, light green above, blue-green and silky pubescent beneath; flowers shell-form, 12–15 cm wide, pendulous when in full bloom (!), sepals and petals total 9, oblong-obovate, 2.5–5 cm wide, fragrant, appearing after the leaves, June; fruits cylindrical, 7 cm long, carmine-pink, pendulous. BM 9004; JMa 18; BS 2: Pl. 21; NF 5: 13 (= *M. globosa sinensis*

Rehd.; *M. nicholsoniana* Hort. non Rehd. & Wils.). China; W. Szechwan Province. 1920. z8–9 Plate 107, 112. ⟡ ⚭

"M. sinensis × wilsonii" see: **M. × highdownensis**

M. × slavinii see: **M. × 'Kewensis' 'Slavin's Snowy'**

M. × soulangiana Soul.- Bod. (*M. denudata × M. liliiflora*). Broad, deciduous shrub, seldom over 3–4 m high and wide; leaves obovate to more elliptic, 10–15 cm long, tapering to the apex, underside more or less pubescent; flowers erect, campanulate, interior usually white, exterior more or less reddish, appearing before the leaves, flowers very abundantly, flowers sometimes reappearing sporadically, form of the petals varies with the cultivar, March–June. BC 2300; BR 1164. Developed in 1820 by Soulange-Bodin at Fromont, near Paris, France; the first plant flowered in 1826. z5

Many cultivars are known today:

Outline of the more prominent cultivars

1. Flowers nearly totally white on the exterior
   + Early cultivars:
     'Alba Superba', 'Amabilis'
   ++ Late cultivars:
     'Lennei Alba'

2. Flowers with a predominantly light pink exterior
   + Early cultivars:
     'Alexandrina'
   ++ Late cultivars:
     'Brozzonii', 'Speciosa', 'Verbanica'

3. Flowers with a predominantly deep pink-red exterior
   + Early cultivars:
     'Burgundy', 'San Jose'
   ++ Late cultivars:
     'Lennei', 'Lombardy Rose', 'Norbertii', 'Rustica Rubra', 'Triumphans', 'Vanhouttei'

4. Leaves variegated:
     'Variegata'

'Alba' see: **'Amabilis'**

'Alba Superba' (Cels). Shrub, to 5 m high, rather erect, not so broad growing; flowers white, exterior with a faintly pink line on the mid-line, 9 petals, early April. MMa 22 (= 'Superba'). Developed by J. Cels, Montrouge, near Paris, France around 1825. ⟡

'Alexandrina' (Cels). Shrub, strong growing, upright, to 5 m high; flowers large, to 10 cm long, petals 9, nearly 5 cm wide, exterior soft pink, becoming darker toward the base, the basal ¾ streaked with dark purple lines, interior pure white, early April, appearing with the leaves, flowering time lasts nearly 3 weeks. MMa 80: 22; RH 1912: 370. Developed about 1825 by Cels of Montrouge, France. Plate 73, 106. ⟡

'Amabilis'. Slow growing; flowers white, medium size, exterior slightly white striped, filaments dark brown-red (= 'Alba'). Before 1900.

'André Leroy' (Leroy). Flowers distinctly cupulate, exterior dark pink to purple (similar to 'Verbanica'), interior white. Introduced before 1892 by A. Leroy of Angers, France.

'Brozzonii' (Leroy). Very fast growing and vigorous (10 m high

in Les Barres Arboretum, France), broad; flowers very large, to 25 cm(!) wide when fully open, petals soft reddish outside, inside pure white, flowering as quite young plants, April–late May. MMa 86; NF 4: 19. Introduced to the trade in 1873 by A. Leroy, France. ⟡

'Burgundy' (W. B. Clarke). Flowers dark purple-red, very early. Introduced in 1930 by W. B. Clarke & Co., San Jose, California, USA. ⟡

'Grace McDade' (McDade). Flowers white, exterior reddish at the base of the petals. Introduced in 1945 by C. McDade of Semmes, Alabama, USA.

'Highland Park'. Small tree, flowers abundantly every year, flowers cupulate, petals about 6 × 6 cm large, brownish-purple, early, fragrant. Developed in Highland Park, Rochester, N.Y., USA, before 1961. ⟡

'Lennei' (A. Topf). Leaves broadly ovate; flowers reverse campanulate, petals obovate, very fleshy, exterior quite dark purple-red to magenta, 10–15 cm long, to 10 cm wide, interior pure white, flowers continuously from April through May, often a few flowers appear in the fall; fruits in long green cones. Developed by Conte Guiseppe Salvi, in Florence, Italy, but introduced to the trade in 1854 by Alfred Topf, of Erfurt, E. Germany. Plate 106. ⟡

'Lennei Alba' (Froebel). Flower form like that of 'Lennei', but pure white, without a pink tone, flowers poorly in some years. Bk 1954: 26. Developed by Froebel of Zurich, Switzerland, 1905, but first introduced in 1930 by Keessen, Aalsmeer, Holland. ⟡

'Liliputian'. Slow growing form, small flowers; otherwise very similar to the species. Developed about 1950 in the Semmes Nurseries, Crichton, Alabama, USA.

'Lombardy Rose' (McDade). Seedling of 'Lennei', exterior dark pink, interior white, flowering period lasts for several weeks. Introduced to the trade before 1957 by C. McDade of Semmes, Alabama, USA.

'Melanie' (*M. liliiflora* 'Darkest Purple' × *M. soulangiana* 'Lennei'). Interior and exterior equally dark purple, 9 petals, about 7 cm long. Developed by F. B. Galyon, Knoxville, Tenn., USA. Before 1970. ⟡

"Nigra" see: **M. liliiflora 'Nigra'**

'Norbertii' (Cels). Slow growing; flowers purple-pink outside, becoming lighter toward the apex, somewhat darker than 'Alexandrina', petals narrow. Developed in 1800 by Cels, Montrouge, France. One of the latest flowering cultivars; of little merit. Plate 73, 106.

'Picture' (Wada, Japan, around 1925). Strong growing, long branches, upright, large leaves; flowers to 25 cm wide, erect, exterior deep wine-red, interior white, very fleshy, flowers well on young plants. Preeminent cultivar. Plate 73. ⟡

'Rouged Alabaster'. Bud scales roughly pubescent, dark brown-black, flowers to over 25 cm wide, stalks shaggy gray, petals to 15 cm long and 10 cm wide, pink-red. JRHS 89: 132. Developed from *M. soulangiana* 'Lennei Alba' × *M. veitchii* by Gresham. ⟡

'Rustica Rubra'. Presumably a seedling of 'Lennei', but the flowers more pink-red, smaller, more open cupulate, interior white when fully developed, 9 petals, these to 10 cm long, 7 cm wide, early (= 'Rustica'). Developed about 1892 in Boskoop, Holland and distributed by Wezelenburg, Hazerswoude, Holland. ⟡

'San José' (W. B. Clarke). Very large flowers, exterior dark purple, very early, fragrant. Developed about 1938 by W. B. Clarke of San Jose, California, USA. Somewhat lighter than 'Lennei', but grows more vigorously.

'Speciosa' (Cels). Flowers with a white exterior, somewhat like 'Brozzonii', but the petals erect at first, then with the apical half spreading more outward, exterior red striped, interior white, fragrant, large flowered. 1825, developed by Cels. One of the latest blooming cultivars. Plate 106. ⊕

'Spring Beauty' (Cross of 'Lennei' with 'Lennei Alba'). Leaves light green, glossy; flowers very large, exterior lilac-pink, interior white. Introduced in 1971 by H. den Ouden, Boskoop. ⊕

'Superba' see: **'Alba Superba'**

'Triumphans' (Rinz). Flowers white inside, exterior red on the base, gradually becoming paler toward the apex. Little different and not as good as 'Rustica'.

'Vanhouttei' (Van Houtte). Probably a 'Lennei' seedling, but the flowers much smaller, exterior black-purple, early May. Developed by L. Van Houtte, Ghent, Belgium. Before 1900.

'Variegata' (Walraad). Flowers yellowish variegated. Before 1900.

'Verbanica'. The inner and middle 3 sepals narrowly kidney-shaped and usually evenly pink on the exterior, the 3 outer petals nearly as long as the 6 inner ones, also pink outside, but becoming paler toward the tips. Flowers very late, with 'Brozzonii'. ⊕

M. sprengeri Pamp. Deciduous tree, 5 to 10 m high, young shoots glabrous, with short internodes, thick, flower buds long gray pubescent; leaves quite variable in size and form, oblong-spathulate to oval-oblong, 15 to 20 cm long, 6–15 cm wide, apex round or with a protruding tip, base broad or narrowly cuneate, dark green and glossy above, dark green with reticulate venation beneath, leathery tough; flowers large, erect, on short stalks, 3 sepals, 9 petals, interior white, pink striped, exterior pink, filaments short, pink-red, appearing before the leaves, April; fruit cones 7 cm long, dark red, erect at first, later nodding. China, Hupeh, N. Honan Provinces. 1901. Hardy in England. z9 ⊕ ⊛

var. **diva** Stapf. Leaves 10–17 cm long, obovate, widest above the middle; flowers erect, very large, to 20 cm wide, petals 12 or more, exterior carmine-pink, interior soft pink to whitish, with thin pink lines, late March, early April. JMa Pl. 9; BM 9116; JRHS 91: 139 (= *M. diva* Stapf). China, W. Honan, W. Hupeh, W. Szechwan Provinces. 1901. ⊕

var. **elongata** (Rehd. & Wils.) Stapf. Twigs olive-green; leaves narrowly obovate; flowers erect, 7 to 10 cm wide, petals 12, narrow, much shorter, white, exterior somewhat reddish at the base, March–early April. JMa Pl. 10 (2 forms). China, W. Hupeh, Changyang-hsien Provinces, in forests and on open land. Plate 107. ⊕

M. stellata (S. & Z.) Maxim. Shrub, hardly over 3 m high in cultivation, slow growing, as wide as high, shoots densely silky pubescent when young; leaves obovate, 4–10 cm long, apex obtuse or rounded, base cuneate, margins usually undulate; flowers appear before the leaves, about 7–8 cm wide, white, fragrant, with 12–15 narrow petals, March–April. BM 6370; RH 1878: 270 (=

M. kobus var. *stellata* [S. & Z.] Blackburn; *M. halleana* Parsons). Central Japan; Owari and Mikawa Region, East of Nagoya. 1862. Only found in cultivation elsewhere in Japan. z4 Plate 74, 110. ⊕

'George Henry Kern'. Rather small flowered, but an attractive form, 8–10 petals, exterior lilac-red, interior white. Developed from a cross of *M. stellata* with an unknown *M. soulangiana* form by Carl E. Kern, Wyoming Nurseries, Cincinnati, Ohio, USA. Patented (1949). ⊕

'Halleana'. The type of the plant brought from Japan in 1862 by George R. Hall. ⊕

'Norman Gould'. A triploid form developed by colchicine treatments from *M. stellata* with which it is very similar in leaf and habit, as are the flowers. Developed in the RHS Gardens, Wisley by E. K. Janaki Ammal and given an Award of Merit in 1967. ⊕

'Rosea' (Veitch). Buds pink-red, petals usually white inside, the entire flower soon becoming white. Introduced to the trade in 1902 by Veitch, England.

'Royal Star' (Vermeulen). Fast growing; flowers pure white, with 25–30 petals, flowers 10 days later than the species. Introduced to the trade by J. Vermeulen & Son, Neshanic Station, N.J., USA. ⊕

'Rubra' (Kluis). Improvement on 'Rosea'. Abundantly branched, plants about 1.5–2 m high; leaves 10 cm long; petals 15, purple-pink, about 6 cm long, 1.5 cm wide, March–April. Developed by Kluis Brothers, Boskoop, Holland before 1947. A pink flowering type is grown in Japan. ⊕

'Waterlily' (Greenbrier). Habit more upright, more stiffly branched; buds pink, opening white, petals somewhat longer, but narrower than the species. Developed by Greenbrier Farms, Inc., of Norfolk, Virginia, USA. Allegedly as var. *stellata* × *M. soulangiana*, which is unlikely. ⊕

M. × thompsoniana (Loud.) Vos. (*M. tripetala* × *M. virginiana*). Deciduous, tree-like shrub, similar to *M. virginiana*, grows irregularly, but the leaves much larger; leaves elliptic-oblong, 12–20 cm long, acute, dark green above, bluish and somewhat pubescent beneath, in an umbrella-form arrangement at the shoot tips; flowers cream-white, 12–15 cm wide, fragrant, not so globose as *M. virginiana*, June–July. BM 2164 (= *M. glauca* var. *major* Smith). Developed in the Thompson Nursery at Mile End, near London, England, 1808. Plate 111; Fig. 192. ∅ ⊕

M. tripetala (L.)L. Umbrella-tree Magnolia. Deciduous tree, to 12 m high with a broad crown, twigs glabrous; leaves spreading umbrella-like from the branch tips, obovate, 30–60 cm long, acuminate at both ends, underside pubescent when young, petiole 2–5 cm long; flowers appear after the leaves, cream-white 20–25 cm long; unpleasant smelling, petals 6–9, the inner ones narrower; fruit cones pink, 10 cm long. BB 1539; GTP 208; SS 9–10 (= *M. umbrella* Lam.). W. and SW. USA. 1752. Hardy. z5 Plate 109. ∅ ⊕ ⊛

M. tsarongensis see: **M. globosa**

M. umbrella see: **M. tripetala**

M. × veitchii Bean (*M. campbellii* × *M. denudata*). Small, deciduous tree, young shoots reddish, appressed

Fig. 192. Left, *Magnolia* × *thompsoniana*; right, *M. hypoleuca* (from G. F.)

pubescent at first, later brown; leaves obovate to oblong, 15–30 cm long, half as wide, eventually dark green above (reddish when young, particularly beneath), abruptly acuminate, venation gray-pubescent beneath, petiole 1–2.5 cm long; flowers soft pink, 15 cm long, sepals 3, petals 6, widest near the apex, tapered toward the base, appearing before the leaves, April. GC 107: 104 to 105; GC 135: 223. Developed by Peter C. M. Veitch in Exeter, England. 1907. z7 Plate 111. ⊕

'Isca'. Flowers about the same size as those of the species, but nearly white, turning soft pink, appearing 1 week earlier than the species. Developed before 1950 by Robert Veitch & Son, Exminster Nurseries, Exeter, England. ⊕

'Peter Veitch'. The original type of the cross, 1907; flowered for the first time in 1917. ⊕

'Sayonara' (*M.* × *veitchii* × *M. liliiflora*). Branches bowed outward; flowers very large, white, rather globose, petals fleshy, white, pink-red at the base, not tapered claw-like, late flowering. JRHS 89: 129. Developed by D. Todd Gresham, Santa Cruz, California, USA. Before 1964. ⊕

M. virginiana L. Deciduous shrub, scarcely over 5 m high (tree-like and to 15 m tall only in Florida), twigs glabrous, thin; leaves elliptic to oblong-lanceolate, 7–12 cm long, bright green above, bluish-green beneath, silky pubescent at first, occasionally semi-evergreen; flowers solitary, nearly globose, white, fragrant, 5–7 cm wide, petals and sepals alike, 9–12, obovate, June–July, often a few flowers appearing to September; fruit ellipsoid, 4–5 cm long, dark red, glabrous. BB 1540; BMns 457; SM 313; NF 6: 234 (= *M. glauca* L.). E. to SE. USA. 1688. z5 Plate 73, 109; Fig. 193. ⊕

M. × **watsonii** Hook. f. (*M. obovata* × *M. sieboldii*). Deciduous shrub, somewhat stiffly branched, upright, to 5 m high, twigs glabrous; leaves obovate to more oblong, 10–18 cm long, rather leathery tough, dark green above, blue-green and finely pubescent beneath, with 10–15 vein pairs; flowers shell-form, petals 6, obovate, white, sepals 3, reflexed, exterior pink, filaments carmine, stalk thick, about 2.5 cm long, appearing after the leaves, June–July. BM 7157; BB 2: 288. Presumably imported from Japan. z6 Plate 109. ⊕

M. wilsonii (Finet & Gagnep.) Rehd. Deciduous shrub or tree, to 8 m, shoots and buds pubescent; leaves oval-oblong to oblong-lanceolate, 6–12 cm long, acute to acuminate, base usually round, glabrous above, silky pubescent beneath, petiole 1.5–4 cm long; flowers shell-form, 10–12 cm wide, fragrant, on 1.5–2 cm long stalks, sepals and petals alike, pure white, totalling 9, obovate, filaments carmine, appearing after the leaves, June; fruits cylindrical, 6 cm long. JMa Pl. 12; NF 12: 124; BM 9004. W. China. 1904. z7 Plate 109. ⊕

f. **taliensis** (W. W. Sm.) Rehd. Like the species, but the leaves are eventually nearly glabrous beneath and become bluish, except for the brown pubescent midrib (= *M. nicholsoniana* Rehd. & Wils. non Hort.). China; Yunnan, Tali Mt. Range. 1910.

M. yulan see: **M. denudata**

All the magnolias need a fertile, humus soil with sufficient moisture; many species prefer to be planted in a wooded setting, particularly the large leaved and early flowering species. Magnolias benefit greatly from a spring fertilizer application. Never dig around magnolia plants as damage to the shallow root system is likely.

Fig. 193. *Magnolia virginiana* (from Sargent)

Lit. Blackburn, B. C.: Early flowering Magnolias of Japan; in Baileya **5**, 3–13, 1957 ● Dandy, J. E.: A survey of the genus *Magnolia* together with *Michelia* and *Manglietia;* in RHS Camellia and Magnolia Conference Rep., 64–81, London 1950 ● Dandy, J. E.: *Magnolia hypoleuca;* in Baileya **19**, 44, 1973 ● Fogg, J. M.: The temperate American magnolias; in Morris. Arb. Bull. **12**, 51–58, 1961 ● Fogg, J. M., & J. C. McDaniel: Check List of the cultivated Magnolias; Americ. Hort. Soc. 1975, 74 pp. ● Gresham, D. T.: Deciduous magnolias of Californian origin; Morris Arb. Bull. **13**, 47–50, 1962 ● Graebener: Die in Deutschland winterharten Magnolien; in Mitt. DDG 1905, 366–382, and 1920, 73–74 ● Johnstone, G. H.: Asiatic Magnolias in Cultivation; London 1955 ● McDaniel, J. C.: Modern Magnolias; in Amer. Nurseryman from 11/15/1969, 56–58 ● McDaniel, J. C.: Breeding the 'Woodsman' *Magnolia;* in Amer. Nurseryman from 8/1/1974, 88–90 ● Millais J. G.: Magnolias; London 1927 ● Murray, E.: *Magnolia* species descriptions; in Kalmia **5**, 1–17, 1973 ● Spongberg, S. A.: A tentative key to the cultivated Magnolias; in Arnoldia **34**, 1–11, 1974 ● Wyman, D.: Magnolias hardy in the Arnold Arboretum; in Arnoldia **20**, 17–28, 1960.

× MAHOBERBERIS Schneid. (BERBERIS × MAHONIA) — BERBERIDACEAE

Semi-evergreen or evergreen shrubs; distinguished from *Berberis* by the thornless branches, from *Mahonia* by the simple, occasionally 3 toothed leaves; flowers like *Mahonia.*

× **Mahoberberis aquicandidula** Krüssm. (*B. candidula* × *M. aquifolium*). Small, evergreen shrub, habit somewhat divaricate; leaves densely arranged, ovate to oval-oblong, 2.5–4 cm long, with 2–4 large, thorned, acuminate teeth on either side, normally on an even plane, glossy green above, bluish-white pruinose beneath, quite short petioled, flat toothed at the shoot base; flowers yellow. DB 1950: 301. Developed about 1943 by Holger Jensen, Ramlösa, Sweden. z6 Plate 113. # ∅

× **M. aquisargentii** Krüssm. (*B. sargentiana* × *M. aquifolium*). Strong growing, evergreen shrub, dense, presumably reaching 2 m high, shoots erect; leaves quite variable: like those of *Ilex aquifolium* on young shoots, with very stout thorns, ovate to elliptic on older wood, about 5–7 cm long, occasionally trifoliate, then the side leaflets are much smaller than the terminal leaflet, often becoming brown in winter; flowers sparsely arranged in terminal racemes; fruits black. DB 1950: 301. Developed at the same time as the previous hybrid by H. Jensen. z6 Plate 113. # ∅

× **M. miethkeana** Melander & Eade (*B. julianae?* × *M. aquifolium*). Strong upright grower, much more vigorous and stout than the very similar × *M. aquisargentii;* leaves quite variable, stout with coarse thorns on the young shoots, somewhat like *Ilex aquifolium,* elliptic on older shoots, simple or trifoliate, the lateral leaflets much smaller than the terminal leaflet, slightly dentate, glossy green above, lighter beneath, bronze-brown toned in winter, the younger leaves also bronze; flowers light yellow, in terminal racemes; fruits blackish. NH 33: 257. Discovered in a seedbed of *M. aquifolium* at H. O. Miethke Nursery, near Tacoma, Washington, USA. 1940. Usually considered identical to × *M. aquisargentii,* but different and presumably (according to v.d Laar, in Dendroflora 1975: 33) hybrids with *B. julianae,* not *B. sargentiana.* z6 Plate 114. # ∅

× **M. neubertii** (Lem.) Schneid. (*B. vulgaris* × *M. aquifolium*). Evergreen or semi-evergreen, open growing shrub, to 1 m high, shoots long, stiff, erect; leaves on long shoots oval-oblong, 3–7 cm long, tough, sinuately toothed, base cuneate, glossy green, leaves on short shoots and at the base of the long shoots often 3 (5) parted; flowers unknown (= *Berberis neubertii* Baumann; *B. neubertii* var. *ilicifolia* Schneid.). Developed about 1850 by Baumann in Bollweiler, Alsace, France; named for Prof. Neubert, University of Tübingen, W. Germany. z6 Plate 113. # ∅

'Latifolia'. Wider growing, more open, to 2 m high; leaves wider on the long shoots, obtuse, gray-green (!), base broad cuneate (= *Berberis latifolia* Hort.). Origin the same as above.

Uses and cultural requirements like those of *Mahonia.*

Lit. Jensen & Krüssmann: Zwei neue *Mahoberberis*-Hybriden; in Deutsche Baumschule 1950, 300, ills. ● Wyman, D.: Two new *Mahoberberis* Hybrids in Arnoldia 1958, 9–12.

MAHONIA Nutt. — BERBERIDACEAE

Low or tall evergreen shrubs, some also becoming tree-like; twigs thornless; leaves alternate, odd pinnate, rarely trifoliate; leaflets usually thorny toothed; flowers in clustered racemes or panicles at the shoot tips, many flowered, yellow; fruits usually blue and pruinose, occasionally red or whitish. — About 70 (to over 100) species, 58 in Asia, from Japan to Sumatra, the others in N. and Central America.

Outline of the species described in this work

★ Inflorescences in clusters of spike-form racemes;

- \+ Bracts persisting at the base of the inflorescences, 1.5–4 cm long; leaves thick, stiff, usually faintly veined:
  - Includes all the asiatic species: *M. acanthifolia, bealii, fortunei, japonica, lomariifolia, magnifica, napaulensis* and *siamensis,* also the American species *M. nervosa*
- ++ Bracts at the base of the inflorescence abscising, 3–10 mm long; leaves much thinner, with reticulate venation, veins indented above:
  - Here are the Aquifoliatae: *M. aquifolium, dictyota, eutriphylla, gracilis, piperiana, pinnata, repens, sonnei,* and × *wagneri*

★★ Flowers in loose panicles, simple racemes, in umbels or fascicles;

- > Inflorescence many flowered, in a loose panicle or simple raceme:
  - Included here are only the species from Mexico and Central America; only *M. arguta* described in this work
- >> Inflorescence few (3–7) flowered, umbellate or fascicled; leaves thick, stiff:
  - Including the Horridae from Mexico and SW. USA: *M. fremontii, haematocarpa, nevinii, swaseyi,* and *trifoliata.*

Mahonia acanthifolia Wall. ex. G. Don. Tall shrub, 3–5 m high in its habitat, but lower in cultivation; leaves to 50 cm long, with about 27 leaflets, these ovate, to 10 cm long, with 2–5 (!) teeth on either side, sinuately toothed, glossy and distinctly veined above; flowers light yellow (HCC 602) in dense, terminal, to 30 cm long, 2 cm thick, outspread (!), fascicled racemes. JRHS 84: 60; 94: 193. Nepal z9 # ∅ ⊕

M. aquifolium (Pursh) Nutt. Erect shrub, seldom over 1 m high, without stolons (not typical of plants in cultivation!); leaves 5 to 11 parted, leaflets ovate to elliptic, 4–8 cm long, base rounded, with 5–12 teeth on either side, glossy green above, greenish beneath, young leaves usually reddish, the older leaves also often reddish in winter, lowest leaflet pair 2.5–5 cm from the petiole base; flowers in panicled, erect racemes, yellow, occasionally somewhat reddish toned, April–May; fruits black purple, blue pruinose, with a reddish juice, stalk absent. HM 718; MS 113; BR 1425 (= *Berberis aquifolium* Pursh). Western N. America, British Columbia to N. California. 1823. z5 Fig. 194. ∅ ⊕ &

Includes a few cultivars:

'Apollo' (Brouwers 1973). Growth broad and low, to 60 cm high, many branched; leaves rather large, with 5–7 leaflets, these rather flat, dull glazed, brownish in fall and winter, as is the new growth; flowers in very large, somewhat loose, paniculate racemes, very numerous, deep yellow, April–May. Selected by Brouwers, Groenekan, Holland, 1973 (patented in Holland). Plate 74. # ⊕

'Atropurpurea'. Growth broad and low, to about 60 cm high, many branched; leaves medium size, with 5–7 leaflets, these deep green, dull glazed, dark bronze brown in winter; flowers in numerous small racemes, April–May. Selected in Holland before 1915 and still a very common form. # ∅

'Donewell' (Donewell Nurseries, USA). Very broad, arching habit, to 1 m high; leaves 10 to 20 cm long, usually with 7 leaflets, these narrowly oval-elliptic, somewhat undulate, long thorned, glossy dark green above, blue-gray beneath; rachis conspicuously red, flowers yellow, April–May. Attractive foliage. # ∅

'Forescate'. Growth to about 1 m high, many branched; leaves rather large, usually with 7 leaflets, these coarse and sharply thorned, dark green and glossy above, blue-green beneath, bronze-red-brown in winter and spring, as is the new growth; flowers from March–April. Selected in 1975 by van Alphen, Voorschoten, Holland. # ∅

'Moseri' see: **M. wagneri 'Moseri'**

'Orangee Flame'. Growth broad and low, to 60 cm high, many branched; leaves rather large, with 5–7 leaflets, dark green, young leaves yellow-green at first, later becoming luminescent bronze-orange, then becoming duller, purple-brown in fall, wine-red in winter; flowers golden-yellow, March–April. Selected in 1965 by Charles Szerszen, Fairview Floral Nursery, Westlake, Ohio, USA. Unfortunately quite susceptible to rust and powdery mildew! # ⊕ ∅

'Undulata' see: **M. wagneri 'Undulata'**

M. arguta Hutch. Shrub, 1–1.5 m high, quite glabrous, young shoots reddish; leaves pinnate, to 25 cm long, 9–13 pinnae, these lanceolate, thorny, acuminate, 2.5–9 cm long, 6–20 mm wide, base cuneate, entire or with a few sharp teeth, stiff, distinctly reticulate veined, sessile; flowers in 30–40 cm long, bowed panicles, with 3–4 flowers on a branch of the panicle, light yellow, May; fruits globose, dark blue, 6–8 mm. BM 8266. Mexico (?). (Glasnevin, Ireland) z9 #

M. bealei (Fort.) Carr. Shrub, to 4 m high in mild regions, shoots erect, thick; leaves 30–40 cm long, leaflets 9–15(17), the lowest pair are very small, and close to the petiole base, terminal leaflet much larger than the laterals, oval-rounded to more oblong, 5–12 cm long, very oblique at the base, with 2–5 large marginal thorns on either side, the terminal leaflet truncate at the base to slightly cordate, tough and stiff-leathery, dull blue-green above, yellowish on the thorns, bluish to yellowish green beneath; flowers light yellow, in 8–15 erect racemes, 7–15 cm long and grouped into a cluster, scented like lily-of-the-valley, (February) May–June; fruits blue-black, bluish pruinose, with a distinct stalk. BM 4846; FS 6: 79 (= *M. japonica* var. *bealei* Fedde).

Fig. 194. **Mahonia.** Left row (from top to bottom) *M. nervosa, M. piperiana, M. fremontii;* middle row *M. pumila, M. repens, M. dictyota;* right row *M. aquifolium, M. sonnei, M. pinnata, M. nevinii* (½ actual size; from Abrams)

China. 1845. A very attractive species, but often confused with *M. japonica* in cultivation. z7–8 Plate 117. # ⌀ ⊕

M. dictyota (Jeps). Fedde. Erect shrub, 0.5–2 m high, rather sparsely foliate; leaflets 5–7, broad-oblong to nearly circular, margins very undulate with 3–5 thorny teeth on either side, leathery tough, with raised venation on both sides, pale green above, bluish beneath; flowers in fascicles; fruits very blue pruinose. California, USA, in chaparral. z9 Fig. 194.

M. eutriphylla Fedde. Low, slow growing shrub; leaves small, 6–11 cm long, 4 to 5 cm wide, with only 1 pair of leaflets, these 3–7 cm above the petiole base, leaflets oval-oblong, about 3 × 1.5 to 4 × 2 cm, acute, margins thorny serrate, with 5–6 thorns on either side, terminal leaflet with an 8–12 mm long petiolule, glossy above, finely reticulate veined; flower racemes only 1–2 cm long, April–May; fruits black. Mexico. z9 ⌀

M. fascicularis see: **M. pinnata** and **M. wagneri 'Pinnata'**

M. fortunei (Lindl.) Fedde. Shrub, erect, about 1 m high, with fewer branches; leaves 15 to 25 cm long, leaflets 7–13, narrowly oblong, 6 to 12 cm long, sessile, margins finely undulate and with fine marginal thorns, usually dull dark green above, somewhat lighter beneath; flowers in 6–8 cm long racemes, yellow, October–November; fruits blue-black, pruinose, globose. MJ 1638 (= *M. trifurca* Fedde). W. China. 1846. z8 Plate 117. # ⌀

M. fremontii (Torr.) Fedde. Erect shrub, 0.7(2.5) m high, multistemmed, occasionally a small tree in its habitat; leaves 3–7 cm long, with 3–5 leaflets, these ovate to angular in outline, 12–25 mm long, leathery tough, light gray-green on both sides, either edge with 1 to 4 stiff, thorny teeth, the terminal leaflet only slightly larger than the others, petiole very short, whitish; flowers in 2.5–3 cm long, erect racemes with 3–9 flowers, light yellow, fragrant, June; fruits globose, dark blue, on erect stalks. MS 121; BC 1917. SW. USA, in the desert regions. 1895. z9 Fig. 194 # ⚭

M. glumacea see: **M. nervosa**

M. gracilis (Benth.) Fedde. Closely related to *M. aquifolium*, to 1.5 m high; leaflets 5 to 11, ovate to oval-lanceolate, 3.5–6 cm long, entire or indistinctly crenate, glossy above; fruits with a distinct stalk. Mexico. 1900. z9 #

M. haematocarpa (Woot.) Fedde. Low shrub (1.5–3 m high in its habitat), young shoots glabrous, often gray; leaves about 7 cm long, with 5–7(9) leaflets, these elliptic, lanceolate or ovate, 1.2–4 cm long, with 3–5 thorns on either edge, apex thorny, margins undulate, distinctly blue-green; flowers about 6 in slender, stalked racemes, light yellow; fruits globose, red, 8 mm thick, without a stalk. New Mexico to California, USA. z9 # ⚭

M. herveyi see: **M. repens 'Rotundifolia'**

M. heterophylla see: **M. × 'Heterophylla'**

M. × 'Heterophylla'. Shrub, about 1 m high, young shoots reddish; leaves with 5–7 widely spaced, lanceolate, 3–7 cm long, partly sessile, partly stalked leaflets, base cuneate, margins undulate, with 6–12 coarse teeth, very glossy on both sides, bright green above, lighter beneath; flowers in 3–4 fascicled racemes, of little merit; fruits blue, pruinose, but rarely set (= *M. heterophylla* [Hort.] Schneid.; *M. toluacensis* Hort. non J. J.). Origin unknown. z9 Plate 115. #

M. japonica (Thunb.) DC. To 3 m high (taller only in exceptional cases), shoots stiffly erect, with only a few leaves at the apex; leaves 30–40 cm long with 7–13 leaflets, these obliquely ovate, very hardy and stiff, 5–12 cm long, dull green above (not blue-green!), with 4–6 large thorny teeth on either edge, the lowest leaflet pair attached nearly at the petiole base; flowers sulfur-yellow, fragrant, in 10–20 cm long, very loose, nodding to outspread racemes, winter to May; fruits dark purple, pruinose, ellipsoid. MJ 1637; DB 1951: 154. Japan, only known in cultivation. z7 Plate 114, 117. # ⌀ ⊕ ⚭

var. *bealei* see: **M. bealei**

'Hiemalis'. Vigorous erect growth, to 1.5 m high, wide; leaves 35–50 cm long, with 13–15(17) broad elliptic leathery leaflets, these with more or less white-yellow markings at the base, terminal leaflet often somewhat larger, dull gray-green, light yellow-green beneath, slightly pruinose; flowers in up to 35 cm long, light yellow racemes, obliquely ascending, later more outspread, flowers very abundant, December–February. Dfl 1975: 27. Developed in England, exported to Holland and further disseminated from Boskoop. Considered one of the best of the winter hardy, large leaved forms. z8 # ⊕ ⌀

M. lomariifolia Tak. Multistemmed shrub, shoots thick, steeply erect, 7–10 m high in its habitat, much lower in cultivation; leaves only at the branch tips, 30–70 cm long, leaflets in 10–20 pairs, lowest pair nearly on the petiole base, linear-lanceolate, 3 to 10 cm long, base oblique, with 2–6 thorny teeth on either edge, stiff, leathery, glabrous, new growth an attractive bronze-red; flowers in erect, candle-like, 10–20 cm long spikes, 18–20 in spikes at the shoot tips, light yellow, fragrant, November–March; fruits ovate, blue-black, 1 cm long. BM 9634; JRHS 94: 194; MCL 75. China, NW. Yunnan Province; Burma. 1934. A very attractive *Mahonia*. z9 Plate 115. # ⌀ ⊕ ⚭

M. magnifica Ahrendt. Very similar to the previous species but more attractive; leaflets narrower. CG 136: 117. Manipur (India). Not yet cultivated. # ⌀ ⊕

M. × media Brickell. (*M. japonica* × *M. lomariifolia*). Medium tall to tall (in mild areas) shrub; leaves with numerous pinnae; flowers in clusters of loose racemes, terminal, in late fall and winter.

Includes the following cultivars:

'Buckland'. A particularly attractive clone, conspicuous for its 60 cm wide inflorescences, composed of 13–14 branched racemes, light yellow, fragrant. Developed by Lionel Fortescue, Buckland, Monachorum, Devon, England. 1971. z9 # ⊕

'Charity'. Strong growing, 2–4 m high, erect shrub, intermediate between the parents; leaves to 50 cm long, leaflets to 21, narrowly ovate, to 10 cm long, deep green above, lighter beneath; flowers in narrow, to 30 cm long, partly erect, partly obliquely spreading racemes, these grouped into clusters, light yellow (HCC 603), January–February in mild regions. JRHS 94: 192. Discovered as a chance seedling in Windsor Great Park and named around 1955. Considered quite meritorious. z9 # ⊘ ⊕

Another clone of the same cross as 'Buckland' is **'Lionel Fortescue'** (1971) with thick, erect, to 40 cm long racemes, fragrant. **'Faith'** (1972) and **'Hope'** (1966) have been developed from seed of 'Charity' in Windsor Great Park, both hard to find in the trade. Another cultivar **'Winter Sun'** was developed in 1970 at the Slieve Donard Nursery, Newcastle, N. Ireland.

M. napaulensis DC. A shrub scarcely over 2 m high in cultivation, to 5 m high in its habitat; similar to *M. japonica*, but the leaflets in 5–11 pairs, lanceolate, 3–10 cm long, more acuminate, consistent, margins with 4–10 teeth on either edge, tough, very glossy above, the lowest pair much smaller, rounded, near the petiole base; flowers in narrow, 15–30 cm long racemes, these 5–7 at the branch tips on rather thick axes, yellow, March–April; fruits ovate, 6 mm thick, blue-white pruinose. SNp 17 (= *M. nepalensis* Spreng.). Nepal. Before 1850. z9 #

M. nepalensis see: **M. napaulensis**

M. nervosa (Pursh) Nutt. Low, stoloniferous shrub, 25–40 cm high (!); leaves about 40 cm long, with 11–23 leaflets, these obliquely ovate, 4–7 cm long, with 8–12 teeth on either edge, base rounded; flowers in erect, to 20 cm long racemes, April–June; fruits ellipsoid, blue, gray pruinose, 1 cm long. BM 3949; MS 114 (= *M. glumacea* DC.). Western N. America, British Columbia to California. 1826. z6 Fig. 194. # ◐ ⊕

M. nevinii Fedde. Shrub, very similar to *M. fremontii*, but with red fruit, 0.7–1.5 m high, erect; leaves 3–7 cm long, leaflets 3–5(7), the lateral leaflets oblong to ovate, about 2.5 cm long, terminal leaflets broad lanceolate, to 4 cm long, gray-blue on both sides (!), margins undulate, with 5–16 fine thorned teeth; flowers in loose racemes of 5–7 flowers, March–May; fruits globose, 6 mm thick, dark red, juicy, pruinose. MS 121; GF 9: 415; MCL 174. California, USA. z9 Plate 113; Fig. 194. # ⚭

M. pinnata (Lag.) Fedde. Erect shrub, about 1 m high (to 4 m in very mild regions) leaves 5–12 cm long, leaflets 5–9, very closely spaced and normally overlapping, oval-elliptic to oblong, 3–5 cm long, margins very undulate and with 10–20 finely thorned teeth on either edge, both sides very glossy, the lowest leaflet pair close to the petiole base; flowers in 6–8 cm long racemes, along the twig (!), May; fruits oval, bluish pruinose, 6 mm long. MS 115; BM 2396; JRHS 94: 195 (= *M. fascicularis* DC.). North America; California to Mexico. 1819. z9 Fig. 194. #

var. *wagneri* see: **M. wagneri**

M. piperiana Abrams. Erect shrub, 20 to 50 cm high (to 2.5 m in cultivation!); leaves 10–20 cm long, leaflets 5–9, ovate, 2.5 to 6 cm long, with 7–10(20) marginal thorns on either side, glossy green above and finely reticulate veined, gray-green and papillose beneath; flowers in 4–10(20) cm long, erect racemes, at the shoot tips and along the branch, March–April; fruit blue-black, ellipsoid, 6 mm long. MS 118. California. A similar, but somewhat superior plant to *M. aquifolium*. z7 Fig. 194. # ⊘ ⊕

M. pumila (Greene) Fedde. Stem upright, 20–40 cm high, simple or branched; leaflets 5–9, broadly oval-oblong, obtuse at the apex, dull green above with very reticulate venation, blue-green beneath, margins undulate and with 5 to 9 thorny teeth on either edge; flowers many in clusters; fruits blue-black, pruinose, 6 mm long. Western N. America, from Oregon to California. z9 Fig. 194.

M. repens (Lindl.) G. Don. Shrub, 30–60 cm high, stoloniferous (!) and creeping, often only 10–20 cm high in its habitat; leaves 10–20 cm long, leaflets nearly always 5, broadly ovate or oval-oblong, base oblique to slightly cordate, 2.5–6 cm long, nearly totally flat, dull green (!), lighter and distinctly papillose beneath, margins with 8–20 bristly teeth on either edge; flowers in 3–7 cm long racemose clusters at the branch tips, golden-yellow, May; fruits blue-black, pruinose, ovate. MS 117; NF 10: 130; RWF 139. Western N. America. 1822. This plant is often confused in cultivation with *M. aquifolium* when the plant in question is stoloniferous; differing however in the dull green leaves and lack of a red fall color. z6 Plate 115; Fig. 194. # ⊕

'Rotundifolia' (Hervé) Shrub, over 1 m high, stoloniferous, densely foliate; leaflets usually 5, oval, base cordate, entire to very slightly dentate, dull blue-green above, lighter beneath, green or light bronze in winter, new growth yellow-green, then pink; similar to *M. wagneri* 'Moseri', flowers more abundantly (= *M. herveyi* Hort.; *M. rotundifolia herveyi* Hort.). Developed in Versailles, France around 1875 by Hervé. #

M. rotundifolia see: **M. repens 'Rotundifolia'**

M. schiedeana see: **M. trifoliolata**

M. siamensis Takeda. Shrub, to 4 m high, old stems with a deeply furrowed, corky bark, young shoots green or reddish; leaves to 70 cm long, with 11–17 widely spaced leaflets, these asymmetrically lanceolate to ovate, very tough, leathery, dull green above, much lighter beneath, 14–17 cm long, 5 to 8 cm wide, margins flat and thorny toothed, often *Ilex*-like undulate; inflorescence a bundle of 6–10 more or less erect, 12–25 cm long racemes, flowers pure yellow, very fragrant, petals about 1 cm long, the outermost with the apex split for up to ⅓ of their length, January–February; fruits blue-black, 5 mm, pruinose. BMns 605. Siam; E. Burma; China, Yunnan Province. 1931. Most attractive species of the entire genus, but only rarely found in cultivation. z10 # ⊘ ⊕

M. sonnei Abrams. Shrub, 20–50 cm high; leaflets 5, oval-lanceolate, 4–8 cm long, glossy green above, lighter beneath, but not gray, with loosely scattered papillae; flowers in dense racemes, 4–7 cm long, March–May; fruits blue-black, 6 mm long. North America, California; in the mountains. z9 Fig. 194.

Fig. 195. *Mahonia trifoliolata* (from B. R.)

M. swaseyi (Buckl.) Fedde. Closely related to *M. haematocarpa*. Shrub, to 1.5 m high; leaflets 5–11, elliptic to oblong-lanceolate, very thin, with reticulate venation, blue-green, rachis and venation red; inflorescences with 2 broadly ovate, 5 mm long bracts at the base; fruits whitish-yellow at first, later red, 1 cm long. Texas. 1907. z9 # ⚭

M. toluacensis see: **M. × 'Heterophylla'**

M. trifoliolata (Moric.) Fedde. Shrub, erect, divaricate, 1(2) m high, very easily recognized by the trifoliate leaves; leaflets acutely ovate, 2.5–5 cm long, 8–15 mm wide, long acuminate, with 1–2 large marginal thorns on either edge, distinctly blue-white beneath; flowers in short corymbs; fruits red. FS 56 (= *M. schiedeana* Wats.). New Mexico, USA; in the hills. 1839. z9 Fig. 195. # ∅

M. trifurca see: **M. fortunei**

M. wagneri (Jouin) Rehd. (*M. aquifolium* × *M. pinnata*). Shrub, erect, to 2.5 m high; similar to *M. aquifolium*, but the leaves are nearly sessile, reddish when young, leaflets 7–11, usually oval-oblong, deep green above and somewhat glossy, lighter beneath, margins with 4–7 teeth on either edge, petiole 5–30 mm long; flowers and fruits like those of *M. pinnata* (= *M. pinnata* var. *wagneri* Jouin). Known before 1863, from Simon Louis Frères in Metz. z8 # ∅

Includes the following cultivars:

'Aldenhamensis'. Strong, erect growing, to 1.5 m high; leaves rather large, leaflets usually 9, slightly dentate, dull gray-green above, light blue-gray beneath, dull green in winter, young leaves bronze; flowers yellow, April–May (= *M. × aldenhamensis* Hort.; *M. aquifolium* 'Aldenhamensis'). Developed by Vicary Gibbs, Aldenham, England, before 1931. Garden merit only slight. #

'Fireflame'. Growth broadly erect, to 1.25 m high, very widespreading; leaves medium size, most with 7 leaflets, these thin, flat to dull glazed, blue-green, light gray-green beneath, margins finely thorny dentate, rachis conspicuously red, winter color bronze, new growth bronze-brown-red; flowers deep yellow, numerous, April–May. Discovered about 1955 by Boot & Co., Boskoop, Holland. Introduced in 1965. Somewhat susceptible to powdery mildew. # ∅

'King's Ransom'. Growth narrowly upright, branches rather stiff, to 1.6 m high; leaves medium size, dark green above, lighter beneath, margins undulate and dentate (soft to the touch), new growth bronze, young leaves dark blue-green and pruinose, green to blue-green beneath and somewhat pruinose; flowers light yellow, April–May; fruits blue-black, pruinose. Developed in the USA. # ∅

'Moseri'. Erect habit, to about 80 cm high; leaves medium size, most with 7–9 leaflets, dull yellow-green to orange-yellow, later an attractive pink-red and finally pale green; flowers numerous, yellow, April–May; fruit blue-black, pruinose (= *M. aquifolium* 'Moseri'; *M.* × 'Moseri'). Developed by Moser, of Versailles, France in 1895. Collector's plant. Plate 74. # ∅

'Pinnata'. Erect habit, growth strong, to 1.5 m, many branched; leaves medium size, with 9–11 leaflets, dull bluish gray-green above, lighter beneath, sharp thorny dentate, slightly bronze colored in winter, young leaves an attractive coppery-brown, dull; flowers numerous, pure yellow, April–May; fruits not produced, since the inflorescences abscise after flowering (= *M. pinnata* Hort. non Fedde; *M. fascicularis* Hort. non DC.). 1930. #

'Undulata'. Strong, upright grower, to 1.5 m high; leaves large, usually with 7, occasionally 5 or 9 leaflets, dark green, very glossy, margins conspicuously undulate and sharply dentate, blue-green beneath, light bronze in winter, young leaves an attractive red-brown and very glossy; flowers dark yellow, in dense racemes, very numerous, April–May (= *M. aquifolium* 'Undulata'; *M.* 'Undulata'). Collected in a park in N. Ireland around 1930 and later disseminated by Notcutt. A very attractive form. # ∅

'Vicaryi'. Broad, erect habit, many branched, to 1 m; leaves rather small, most with 9 leaflets, these dull bluish gray-green, light gray-green beneath, sharp thorny dentate, some of the leaves coloring red in fall, more bronze in winter, young leaves dull light green; flowers not numerous, in small racemes, dense, yellow, March–April; fruits blue-black, pruinose. Introduced into the trade by Vicary Gibbs of Aldenham, England in 1931. # ∅

Nearly all mahonias prefer a semishaded area; the species with small, blue-green leaves come from desert regions and thrive in a light, sandy soil in full sun. The others need a moist, humus soil. A number of the Asiatic species will flower in mild areas from October to March and are very well suited to conservatories in cooler climates.

Lit. Fedde, F.: Versuch einer Monographie der Gattung *Mahonia*; in Bot. Jahrb. **31**, 30–133, 1901 ● Jouin, E.: Die in Lothringen winterharten Mahonien; in Mitt. DDG 1910, 86–91 ● Piper, C. V.: The identification of *Berberis aquifolium* and *B. repens*; in Contr. U.S. Nat. Herb. **2**, 437–451, 1922 (3 Pls.) ● Ahrendt, L. W. A.: *Berberis* and *Mahonia* (see under *Berberis*) ● Hensen, K. J. W.: Het *Mahonia*-sortiment; in Mededel. Bot. Tuin. Belmonte Arb. Wageningen Vol. 7 (1), 1963 ● Li, Hui-Lin: The cultivated Mahonias; in Morris Arb. Bull. **14** (3), 1963 ● van der Laar, H.: *Mahonia* en *Mahoberberis*; in Dendroflora, 11–12, 19–33, 1975.

MALLOTUS Lour. — EUPHORBIACEAE

Deciduous, mostly tropical trees or shrubs; leaves alternate, opposite on some species, wide, simple, palmately veined, petiolate, with stipules; flowers dioecious, small, in spikes or panicles; calyx valvate or imbricate, petals absent, stamens numerous, styles 3, fruit a capsule. — About 140 species in the tropics of the Old World, 1 in Japan.

Mallotus japonicus (Thunb.) Muell.-Arg. Deciduous tree or shrub, 3–4 m high, shoots rather thick and with a white pith, dense reddish stellate pubescent when young, later gray; leaves rounded-obovate to broadly ovate, 10–20 cm long, 6–15 cm wide, long acuminate, entire, 3 veined, yellowish-green beneath with yellow,

sessile, translucent glands, 2 glands on the base, petiole very long; flowers inconspicuous, in 8–15 cm long panicles, white pubescent, fall. ICS 2929; KIF 2: 35 (= *Croton japonicum* Thunb.; *Rottlera japonica* [Thunb.] Spreng.). Japan, China, Korea. z9 Plate 114. ⌀

Illustrations of other species in ICS:

M. apelta 2927	*M. barbatus* 1931
M. paniculatus 2928	*M. philippinensis* 1932
M. tenuifolius 1930	*M. repandus* 2933

Only rarely found in cultivation. For full sun, well drained, calcareous soil; only suited for very mild climates. Very attractive for its large leaves.

× MALOSORBUS Browicz See: **MALUS florentina**

MALUS Mill. — Apple — ROSACEAE

Deciduous trees or shrubs, the lateral twigs thorned; leaves alternate, serrate or (rarely) lobed, rolled up or folded in the bud stage; flowers hermaphroditic, white, pink to carmine, in corymbs; petals 5, usually nearly circular to obovate, stamens usually 20, very rarely 15; carpels 5, occasionally 3–4, always connate at the base, basal portion partly shaggy; fruit an "apple" (pome), usually without hard stone or grit cells like those found in *Pyrus*; endocarp usually tough and parchment-like; each locule with 2 ovules, later each with 1–2 seeds; sepals persistent or abscising. — About 35 species in Europe and Asia, some in N. America. For practical purposes the cultivars of *Malus* are arranged alphabetically following species' descriptions.

Outline of the genus
(from Rehder, simplified)

★ Leaves rolled up in the bud stage, always unlobed;
Section I. **Eumalus** Zabel
Characteristics as above;

Series 1. **Pumilae** Rehd.
Calyx persistent:
M. astracanica, magdeburgensis, micromalus, prunifolia, pumila, purpurea, silvestris, spectabilis

Series 2. **Baccatae** Rehd.
Calyx abscising (also persistent on some hybrids):
M. adstringens, atrosanguinea, baccata, halliana, hartwigii, hupehensis, robusta, rockii, sikkimensis

★★ Leaves folded in the bud stage, sharply serrate and more or less lobed, at least on the long shoots;
Section II. **Sorbomalus** Zabel
Calyx abscising, except series 6;

Series 3. **Sieboldianae** Rehd.
Style base shaggy pubescent; calyx and flower stalk glabrous or only slightly pubescent:
M. arnoldiana, brevipes, floribunda, gloriosa, sargentii, scheideckeri, sieboldii, sublobata, zumi

Series 4. **Florentinae** Rehd.
Style base shaggy pubescent; calyx and flower stalk always tomentose; leaves always lobed:
M. florentina

Series 5. **Kansuensis** Rehd.
Style glabrous:
M. dawsoniana, fusca, honanensis, kansuensis, toringoides, transitoria

Series 6. **Yunnanensis** Rehd.
Calyx persistent, styles 5, fruit with a cupulate depression at the apex; with hard stone cells; leaves not or only shallowly lobed:
M. prattii, yunnanensis

Fruit with the core apex open; style shaggy pubescent at the base;

Section III. **Chloromeles** (Dcne.) Rehd.
Fruits without stone cells, usually compressed with an indented calyx; flowers white to pink, stalk thin:
M. angustifolia, bracteata, coronaria, glabrata, glaucescens, heterophylla, ionensis, lancifolia, platycarpa, soulardii

Fruits with stone cells; calyx not indented; flowers pure white, calyx tomentose;

Section IV. **Eriolobus** (DC.) Schneid.
Leaves deeply lobed, flowers white, grouped 6–8, thin stalked; fruits with stone cells, 1.5 cm thick:
M trilobata

Section V. **Docyniopsis** Schneid.
Leaves not lobed or only slightly so; flowers white, grouped 2–5; style base shaggy pubescent; fruits with stone cells, to 3 cm thick:
M. tschonoskii

Malus acerba see: **M. silvestris**

M. × adstringens Zabel (*M. baccata* × *M. pumila*). Group of hybrids; leaves soft pubescent beneath; flowers usually pink, calyx and flower stalk short shaggy pubescent; fruits usually nearly globose, 4–5 cm thick, rather short stalked, red, yellow or only green, calyx not always abscising from the fruit.

The following hybrids have originated from this cross, including the red leaved cultivars from *M. baccata* × *M. pumila* var. *niedzwetzkyana*;

'Almey', 'Crimson Brilliant', 'Helen', 'Hopa', 'Irene', 'Nipissing', 'Osman', 'Red Silver', 'Robin', 'Simcoe', 'Timiskaming', 'Transcendent', 'Wabiskaw' and others.

M. angustifolia (Ait.) Mich. Shrub or tree, 5–7(10) m high, twigs thin, somewhat pubescent when young or glabrous; leaves lanceolate-oblong to oval-oblong, obtuse to acute, 3–7 cm long, base usually broadly cuneate, coarsely serrate to nearly entire, more coarsely serrate and somewhat lobed on the long shoots, light green above, glabrous to tomentose beneath, leathery and nearly semi-evergreen in mild areas in the fall; flowers few in corymbs, corolla 2.5 cm wide, calyx lobes narrowly acuminate, petals narrowly obovate, whitish-pink, fragrant, late flowering, style shaggy pubescent only at the base; fruits slightly pear-shaped, 1.5–2.5 cm wide, indented at both ends, yellow-green, fragrant. BB 1978; SS 169; BFC 163 (= *M. sempervirens* Desf.). S. USA. 1750. z6 Plate 121, Fig. 197. (#)

M. apetala see: **M. pumila 'Apetala'**

M. × **arnoldiana** (Rehd.) Sarg. (*M. baccata* × *M. floribunda*). Shrub, about 2 m high, similar to *M. floribunda*, twigs widely arching, somewhat pubescent at first to glabrous; leaves elliptic to oval-oblong, acuminate, base tapered to round, 5–8 cm long, uneven and nearly double serrate, lightly pubescent at first, later only the leaves on the long shoots with persistent pubescence on the venation beneath; flowers 4–6 in cymes, stalks slender, reddish, calyx and ovaries red, glabrous, calyx lobes 5 mm long, lanceolate, buds carmine-red, petals narrowly elliptic, pink, later nearly white, distinctly clawed, to 3 cm long, styles usually 3(4), connate and pubescent on the basal ⅓, May; fruits globose, yellowish, to 15 mm thick, calyx abscising, September–October. MG 1909: 27; BFC 72 (= *M. floribunda* var. *arnoldiana* Rehd.). Developed in the Arnold Arboretum in 1883. Attractive but a poor grower. z5 ✧ ⚭

M. × **astracanica** Dum.-Cours. (*M. prunifolia* × *M. pumila*). Differing from *M. pumila* in the coarser and more sharply serrate leaves, longer stalked, bright red flowers and pruinose fruits. Known in European cultivation since 1700 but presumably originating in Asia. Not significant as an ornamental but includes some of the fruit cultivars.

M. asiatica see: **M. prunifolia** var. **rinki**

M. atropurea see: cv. **'Jay Darling'**

M. × **atrosanguinea** (Späth) Schneid. (*M. halliana* × *M. sieboldii*). Growth somewhat like *M. floribunda*, divaricately spreading, twigs nodding; leaves ovate, serrate, dark green, and glossy, each side with a basal lobe only on the long shoots, usually however without these basal lobes on the normal shoots, flowers simple, buds deeply carmine, not fading; fruits globose, 1 cm thick, red or yellow with a red cheek, consistently produces fruit but of little ornamental merit. BFC 78. Introduced into the trade by Späth, Berlin in 1898, first as "*M. floribunda atrosanguinea*". z5 ∅ ✧

M. baccata (L.) Borkh. Tree or shrub, to 5 m high, young shoots glabrous (!), thin; leaves ovate, acuminate, 3–8 cm long, fine and scabrous serrate, never lobed, glossy above, light green, normally with both sides quite glabrous, petioles 3–5 cm long; flowers white, 3–3.5 cm wide, cream-white, in small corymbs, April to May, calyx glabrous, calyx teeth long acuminate; fruits more or less globose, about 1 cm thick, yellow with a red cheek, calyx abscising (!). BS 2: 278; ICS 2198 (= *M. sibirica* Borkh.). NE. Asia to N. China. The plant often found by this name in cultivation is often incorrectly labeled and belongs to *M. robusta*. z2 Plate 121; Fig. 196. ∅ ✧ ⚭

'Columnaris'. Columnar growing, no other *Malus* cultivar grows so narrowly upright; flowers pure white; fruits yellow with a large, red cheek, bears fruit every year. BFC 88. Developed before 1927.

'Gracilis'. Shrubby habit, very slow growing, thin branched, dense; leaves rather small, narrow, long acuminate; flowers small, white, stellate, buds pink; fruits pea-sized, red, slightly pear-shaped. BFC 98. Developed in the Arnold Arboretum in 1910 from Chinese *M. baccata* seed.

var. **himalaica** (Maxim.). Schneid. Leaves broad elliptic, coarsely serrate, venation pubescent beneath; flowers 3 cm wide, buds pink; fruits 1–1.5 cm thick, yellow with a red cheek. W. Himalayas, SW. China. Around 1919.

var. **jackii** Rehd. Growth wider than that of the species; leaves broad elliptic; flowers 3.5 cm wide, pure white; fruits 1 cm thick, bright red, glossy. BFC 119. Korea. 1915. ⚭

'Macrocarpa' (Wroblewski). Fruits to 3 cm wide, glossy, yellowish with a trace of red, the dry part of the calyx often persisting on the fruits.

var. **mandshurica** (Maxim.) Schneid. Leaves broad elliptic, sparsely fine serrate, pubescent beneath when young, petiole pubescent; flowers fragrant, 4 cm wide, pure white, calyx and flower stalks pubescent, styles hardly as long as the stamens; very early, flowers before other *Malus*, April; fruits to 12 mm thick, broadly ellipsoid, bright red, coloring early. BFC 133; BM 6112; JRHS 94: 184; ICS 2199 (= *M. cerasifera* Spach). Central Japan to the Amur region and central China. Before 1825. ✧

M. bracteata Rehd. Small tree, to 7 m high, corolla broad, twigs later becoming glabrous; leaves oval-elliptic to long ovate, serrate or incised-serrate, but less deep and less lobed than the very similar *M. ioensis*, more lobed on the long shoots, lobes somewhat recurved, young leaves pubescent, soon glabrous; flowers 3–5 in corymbs, pink, stalk with awl-shaped prophylls, these abscising right after the flowers; fruits 3 cm thick, yellow. SM 344. S. USA. 1812. Hardy. z6

M. brevipes (Rehd.) Rehd. Very compact habit, stiffly branched, twigs stiff, usually only a shrub; leaves like those of *M. floribunda*, 5–7 cm long, finely and densely serrate, not lobed; flowers very numerous, nearly pure white, 3 cm wide, flower stalk glabrous; fruits 1.5 cm thick, nearly globose, bright red, somewhat ribbed, on a stiff, erect, reddish stalk, fruit calyx usually abscising. BFC 139 (= *M. floribunda* var. *brevipes* Rehd.). Habitat unknown. Before 1883. z6 ♑

M. cerasifera see: **M. baccata** var. **mandshurica**

M. communis see: **M. pumila**

M. coronaria (L.) Mill. Small tree, scarcely over 7 m high, broad growing, twigs outspread, stiff, white woolly at first, later glabrous, with many, somewhat thorny short twigs; leaves ovate to more oblong, 5–10 cm long, acute, irregularly serrate and shallowly lobed (more coarsely lobed on the long shoots), floccose-tomentose at first, later glabrous, scarlet-red and orange; flowers 4–6 in corymbs, to 4 cm wide, soft pink, petals nearly circular, very late, often appearing as late as early June, fruits flat globose, to 4 cm thick, greenish, somewhat ribbed at the calyx end, fragrant. BM 2009; BR 651; BFC 177; BB 1979; GTP 222. Eastern N. America. 1724. z5 Plate 119; Fig. 196. ∅

'Charlottae' (de Wolf). Fall foliage red and orange; flowers double, with about 18 petals, to 5 cm wide, soft pink, scented like violets; fruits like those of the type. BFC 80. Discovered in 1902 by De Wolf in Illinois, USA. Fig. 198. ∅ ✥

var. **dasycalyx** Rehd. Leaves lighter on the underside, venation pubescent beneath on long shoots; flowers only 3.5 cm wide, pink, very fragrant, calyx pubescent (!); fruits 4 cm thick, yellow-green. North America, Ontario to Indiana. 1920. ✥

var. **elongata** (Rehd.) Rehd. Leaves longer, oval-oblong to narrowly triangular, occasionally nearly oval-lanceolate, only slightly lobed; flowers 3.5 cm wide, pink; fruits 3 cm thick, green. New York to Alabama, USA. 1912.

'Nieuwlandiana' (Slavin). A shrub, about 3 m high and wide; flowers double, with 13–27 petals, 3–4 cm wide, somewhat more intensely pink than 'Charlottae', limb often somewhat fringed and dentate, very fragrant, in large, eventually pendulous clusters; fruits to 4.5 cm wide, yellow-green, bluish pruinose. BFC 137. Developed by B. H. Slavin of Rochester, N.Y., USA in 1931. ✥

M. crataegifolia see: **M. florentina**

M. dasyphylla see: **M. pumila**

M. × dawsoniana Rehd. (*M. fusca* × *M. pumila*). Habit similar to that of *M. fusca*; leaves usually wider, more elliptic, 6–8 cm long, acute, base round, more scabrous serrate, occasionally lobed; flowers simple, 2.5–3.5 cm wide, pale pink at first, later white to whitish, very late; fruits to 4 cm long and 2.5 cm thick, ellipsoid, yellow, calyx small, persistent. ST 2: 3; BFC 92. Developed in the Arnold Arboretum in 1881. Easily distinguished from all other species by the very long fruits. z5 ∅ ♑

M. × denboerii Kruessm. (*M. ioensis* [female] × ?*M. purpurea*). Differing from *M. ioensis* in the stronger, more erect growth habit; leaves on the long shoots lobed like those of *M. ioensis*, purple to bronze, elliptic on short shoots and normally unlobed, fall color a bright yellow, light brown and deep purple; flowers pink-red, about 2.5 cm wide; fruits red, globose, 2.5 cm wide, pulp reddish.

BFC 104. Developed about 1939 by Arie Den Boer, but first named in 1953. z4

'Evelyn' (Den Boer). The type of the cross. Also included here is 'Lisa'. For description see list of cultivars.

M. dioeca see: **M. pumila 'Apetala'**

M. diversifolia see: **M. fusca**

M. domestica see: **M. pumila** and **M. silvestris** var. **domestica**

M. florentina (Zuccagni) Schneid. Small, round crowned tree, 8 m × 6 m in size, a conical shrub when young, twigs shaggy pubescent; leaves broadly ovate, 5–7 cm long, lobes always incised (*Crataegus*-like), serrate, dull green above, gray-yellow beneath, tomentose, fall color orange-scarlet; flowers white, 2 cm wide, 2–6 in shaggy pubescent corymbs, late, flowers abundantly, May–June; fruits red, 1 cm thick, broadly ellipsoid. BM 7423; BFC 118; DRHS 1239 (= *M. crataegifolia* Koehne). Italy. Considered by K. Browicz to be a bigeneric hybrid of *Malus silvestris* × *Sorbus torminalis* and named × *Malosorbus florentina* (Zuccagni) Browicz. z6 Plate 121. ∅

M. floribunda Van Houtte. Shrub or tree, 4(10) m high, crown dense, broadly vaulted, twigs usually nodding to widespreading, young shoots slender, reddish on the sunny side, pubescent; leaves folded in the bud stage, ovate, 4–8 cm long, acuminate, sharply serrate, more coarsely serrate on long shoots and somewhat lobed; flowers very numerous, deep carmine in bud, opening pink, but soon fading, interior white, 2.5–3 cm wide, May, along the entire branch; fruits pea-sized, yellow. FS 1585; BFC 120. Japan? Habitat not exactly known. Introduced from Japan in 1862. One of the most popular species of all time. z5 Plate 75, 117. ✥ ♑

var. *arnoldiana* see: **M. × arnoldiana**

var. *atrosanguinea* see: **M. × atrosanguinea**

var. *brevipes* see: **M. brevipes**

M. formosana (Kawakami & Koidz.) Kawakami & Koidz. Deciduous tree or shrub, to 12 m, young shoots often thorny, gray pubescent; leaves ovate to oblong, acute, 8 to 15 cm long, 2.5–5.5 cm wide, irregularly serrate, base round or cuneate, with 9–11 vein pairs, pubescent when young, soon becoming glabrous; flowers 2.5 cm wide, few together, white pubescent, calyx lobes lanceolate, white pubescent, petals 5, white, obovate, 12 mm long, styles with 4–5 incisions; fruit broadly oval, 4 to 5 cm thick, with a persistent calyx. LWT 102. S. China, Hainan Province; Taiwan, distributed throughout the mountain forests. z8–9

M. fusca (Raf.) Schneid. Shrub or a small tree 5–7(10) m high, young shoots more or less pubescent; leaves oval-lanceolate to oblong elliptic, acute to acuminate, 3–10 cm long, often 3 lobed on long shoots, sharply serrate, both sides pubescent at first, later glabrous above, light green; flowers grouped 6–12 together, flower stalk thin and pubescent, petals whitish-pink, later white, May; fruits ellipsoid, 1 cm thick, yellow or red, calyx abscising. SPa

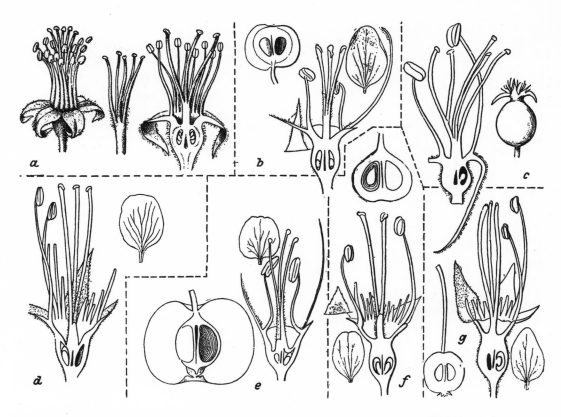

Fig. 196. **Malus.** Flowers, flower parts and fruits. a. *M. pumila;* b. *M. baccata;* c. *M. prunifolia;* d. *M. ioensis;* e. *M. coronaria;* f. *M. halliana;* g. *M. spectabilis* (from Koehne, Schneider, Berg & Schmidt)

160; SS 170; BM 8798 (= *M. rivularis* M. J. Roem.; *M. diversifolia* M. J. Roem.). Western N. America. 1836. z6 Plate 117.

M. glabrata Rehd. Very closely related to *M. glaucescens,* but differing in the green underside of the leaves, not bluish, these otherwise thinner, light green, glabrous, deeply lobed, base distinctly cordate, the basal vein pair arising directly from the base; flowers 3 cm wide, pink, calyx glabrous, reddish, petals rounded to broadly ovate, abruptly short-clawed, styles 5; fruits flat globose, yellow-green, 3 cm wide, distinctly ribbed on the indented calyx end. ST 188; SM 380. N. Carolina to Alabama, USA. 1912. z6 Plate 121.

M. glaucescens Rehd. Small tree or shrub, broad growing, branches occasionally thorned, young twigs glabrous; leaves oval to triangular, acute, 5–8 cm long, with short, triangular lobes, deeper lobed on long shoots, thinly tomentose at first, later totally glabrous, dark green above, blue-green beneath, yellow and dark purple in fall; flowers white to pink, 5–7 together, 3.5 cm wide, flower stalk thin, calyx with a slightly shaggy pubescent exterior, lobes lanceolate, interior tomentose, petals ovate, gradually tapering to a claw; fruits flat globose, indented at both ends, yellow, waxy, fragrant,

3–4 cm thick. ST 157; SM 381; BC 3297–99. E. USA. 1902. z5

M. × gloriosa Lemoine (*M. pumila* 'Niedzwetzkyana' × *M. scheideckeri*). More similar to *M. scheideckeri,* but the young leaves are bronze-red; flowers light purple-red, 4 cm wide, few flowers; fruits about 3 cm wide, bright red (= *M. hybrida gloriosa* Lemoine). Developed by Lemoine of Nancy, France before 1931. ∅

M. halliana Koehne. Shrub, 2–4 m high, occasionally also a small tree, open crowned; branching erratically, young shoots quickly becoming glabrous; leaves reddish on the new growth, oval-oblong, glabrous, leathery, 4 to 8 cm long, acute, crenate, glossy dark green above, lighter beneath, totally glabrous, petiole and venation often reddish; buds deep red, flowers dark pink, 4–7 together, on thin, reddish stalks, 3–4 cm wide, single to semidouble, often pendulous, styles usually 4, calyx lobes often obtuse, May; fruits obovoid, 6 to 8 mm thick, red-brown, ripening very late in fall, of little ornamental value, seeds large. BC 3289; MJ 1408. Japan, China. Not well known in cultivation. 1863. z6 Fig. 196.

'Parkmanii'. Shrub; leaves oval-lanceolate, margins finely crenate, 4–8 cm long; flowers double, with 15 petals, about 3 cm

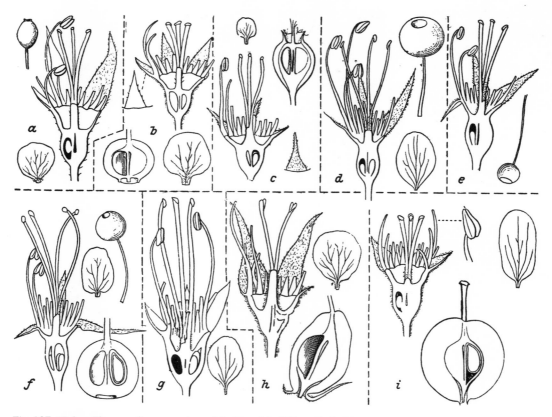

Fig. 197. **Malus.** Flowers, flower parts and fruits. a. *M. sikkimensis;* b. *M. yunnanensis;* c. *M. prattii;* d. *M. zumi;* e. *M. sieboldii;* f. *M. sargentii;* g. *M. angustifolia;* h. *M. trilobata;* i. *M. tschonoski* (most from Schneider)

wide, soft pink, in pendulous clusters; fruits globose, 1.5 cm thick, red. Add 134; BFC 142. Japan, in cultivation. Imported to the USA in 1861 by Hall & Parkman and distributed from there. An excellent cultivar. Plate 121; Fig. 199. ∅ ⊕

var. **spontanea** (Mak.) Koidz. Low, densely branched shrub, nearly vase-like habit, broader than high; leaves smaller, elliptic to obovate, 3 to 4 cm long, with a small apex; flowers smaller, nearly white, 3 cm wide, flowers heavily only every other year, styles usually 4; fruits 1 cm thick, yellow-green. MJ 3558. Japan. 1910. ∅

M. × hartwigii Koehne (*M. baccata × M. halliana*). Growth broad columnar, upright, twigs dark brown; leaves ovate, acute, unlobed, 6–8 cm long; flowers slightly double, 4 cm wide, dark pink at first, later nearly totally white, calyx red; fruits yellow-green, globose to pear-shaped, 1 cm thick, fruit calyx abscising leaving a light brown ring-like scar. BFC 108. Introduced into the trade in 1906 by K. G. Hartwig of Lübeck, W. Germany, but the origin is unknown. z5 Plate 120; Fig. 198. z5

M. × heterophylla Spach. (*M. coronaria × M. pumila*). Similar to *M. soulardii* in appearance, but the leaves are wider, less pubescent, petioles pubescent; flower buds pink, opening white, 4 cm wide; fruits green, to 6 cm wide. 1832.

M. 'Hillieri' see: List of Cultivars

M. honanensis Rehd. Shrub, closely related to *M. kansuensis*, but grows more slowly, twigs thinner; leaves broadly ovate, seldom oval-oblong, 6–8 cm long, with 2–5 pairs of broadly ovate, serrate lobes, pubescent beneath, scarlet-red in fall; flowers white, 2 cm wide, in clusters of up to 10, inflorescences glabrous, styles 3–4; fruits globose, 1 cm thick, yellow-green, punctate. BFC 111; ICS 2209. NE. China. 1921. z6 ∅

M. hupehensis (Pam.) Rehd. Shrub or tree, 5–7 m high, twigs rather stiff, outspread, pubescent at first, soon glabrous; leaves ovate to more oblong, 5–10 cm long, acuminate, sharply serrate, base round or slightly cordate, somewhat pubescent beneath, new growth reddish; flower exterior pink at first, later pure white (as are the buds), 4 cm wide, fragrant, in clusters of 3–7, calyx and calyx lobes reddish, pubescent; fruits globose, about 1 cm thick, green-yellow with reddish cheeks. BM 9667; ICS 2200 (= *M. theifera* Rehd.). China; Assam. Plants occasionally found in cultivation with thorny shoots and small leaves are not this species. z5 Plate 121. ⊕

'Rosea'. Flowers an attractive pink, abundant, resembling a flowering cherry from a distance; prolific fruit, these yellow, reddish on the sunny side. JRHS 86: 45. ⊕ ⊗

M. hybrida gloriosa see: **M. × gloriosa**

M. ioensis (Wood) Britt. Small tree with an open crown, shoots densely tomentose at first, later red-brown, glabrous; leaves oblong-ovate, 5–10 cm long, acute, coarse to incised-serrate, dark green above, yellow-green beneath, tomentose, dark red and yellow in fall; flowers white with a trace of soft pink, 4 cm wide, scented like violets, late May–early June; fruits globose to broadly ellipsoid, occasionally somewhat angular, green, waxy, 3 cm thick, with a persistent calyx. BB 1980; SM 278; BM 8488; BFC 114. Central USA. 1885. z2 Plate 121; Fig. 196.

var. **bushii** Rehd. Differing from var. *palmeri* in the more oblong-lanceolate, acute leaves, these more deeply lobed than the species, later becoming glabrous. Missouri, USA.

var. **creniserrata** Rehd. Slender tree, twigs not thorny, shaggy pubescent when young; leaves elliptic-ovate to oval-oblong, crenate to entire, somewhat double serrate only on the long shoots; calyx tomentose.

'Fimbriata' (Slavin). Leaves narrower than those of the type; flowers pink, densely double, with about 34 petals, limb fringed (fimbriate). Introduced into the trade in 1931 by Slavin of Rochester, N.Y., USA, where it was developed. ⊕

'Nova' (Augustine). Presumably a mutation of 'Plena' with darker pink flowers, these 5 cm wide, with 18–35 petals, sterile. Introduced around 1950 by Augustine of Normal, Illinois, USA. More attractive than 'Plena'. Plate 75. ⊕

var. **palmeri** Rehd. Small tree; otherwise differing in the smaller, more oblong, thinner pubescent leaves with rounded apexes, margins crenate on leaves of the flowering shoots. Missouri, USA. 1910.

'Plena' (Bechtel). Leaves like those of the species; flowers very large, densely double, with about 33 petals, soft pink, flowers abundantly every year; fruits green, 3 cm thick. BFC 74. Found in the wild in Illinois, USA before 1850 and introduced into the trade by A. Bechtel in 1888. Gorgeous cultivar. Plate 119, 120. ⊘ ⊕

var. **spinosa** Rehd. Small shrub, 1.5–2.5 m high, twigs thin, thorny; leaves smaller than those of the species, serrate to finely serrate, somewhat lobed only on the long shoots; flowers pink, 3.5 cm wide; fruits green, 3 cm thick. Missouri, USA.

var. **texana** Rehd. Shrub, densely branched, to 4 m high, twigs densely tomentose, usually glabrous in the second year; leaves smaller, but much wider than those of the species, not or only slightly lobed, with a persistent, dense tomentum. Texas, USA.

M. kaido see: **M. × magdeburgensis** and **M. × micromalus**

M. kansuensis (Batal.) Schneid. Shrub or a small tree, to 5 m high, young twigs finely pubescent, soon glabrous and red-brown; leaves broadly ovate, 3(5) lobed, the lobes triangular, densely and tightly serrate, 5–8 cm long, deep green above, lighter and pubescent beneath, at least on the venation, petiole 1.5–4 cm long; flowers white, 1.5 cm wide, grouped 4–10, calyx shaggy pubescent, May; fruits ellipsoid, 1 cm long, yellow to purple, lighter punctate. BFC 123; ICS 2206. NW. China. 1904. z5 Plate 121.

f. **calva** Rehd. Leaves, calyx and flower stalks totally glabrous, as is the new growth; fruits more yellowish. BMns 251. NW. China. 1911.

M. lancifolia Rehd. Shrub to a small tree, to 6 m high, twigs spreading outward, thorny, young shoots somewhat pubescent, later becoming glabrous; leaves oval-lanceolate to oval-oblong, 3–7 cm long, acute, finely to coarsely double serrate, occasionally somewhat lobed; flowers 3–6 in corymbs, 3 cm wide, soft pink, calyx and flower stalk glabrous, styles 5, late May–June; fruits globose, 3 cm thick, green, waxy, pendulous on thin stalks. SM 342; ST 158; BC 3296; BFC 69. E. USA. z5. 1912.

M. × magdeburgensis Hartwig (*M. pumila* × *M. spectabilis*). A small, round crowned tree or shrub, similar to *M. spectabilis*, but with wider leaves, more elliptic, tapering to both ends, 6–8 cm long, pubescent beneath; flowers bright pink, 4.5 cm wide, with 5–12 petals, buds bright red, calyx and flower stalks pubescent; fruits yellow-green with a red cheek, globose, 3 cm thick. MG 20: 254; BFC 131; JRHS 86: 41 (= *M. kaido* Dipp.). Discovered before 1898 by Schoch in Magdeburg, E. Germany, possibly originated as early as 1850. z5 Plate 118; Fig. 199. ⊕

M. × micromalus Makino (*M. baccata* × *M. spectabilis*). Upright, small tree or shrub, to 4 m high and nearly 3 m wide; twigs long, dark brown, pubescent at first, soon becoming glabrous; leaves elliptic-oblong, acuminate, base cuneate, 5–10 cm long, finely serrate, tough, glossy above, eventually glabrous beneath; flowers grouped 3–5 together, pink, not fading (!), 4.5 cm wide, simple, very early, May; fruits globose, 1.5 cm thick, yellow (very similar to those of *M. scheideckeri*), somewhat angular, calyx abscising, calyx end indented. MJ 1409; NK 6: 13; BFC 136; ICS 2204 (= *M. kaido* Pardé; *M. spectablilis kaido* Sieb.). Origin unknown. Introduced from Japan around 1856, but not occurring there in the wild. z4 ⊕

M. × moerlandsii Doorenbos (= *M. purpurea* 'Lemoinei' × *M. sieboldii*). Tall growing shrub; leaves brown-green, glossy, partly lobed; flowers dark red to pink, flowers very abundantly; fruits globose, 1–1.5 cm wide, purple. A group of hybrids developed before 1938 by S. G. A. Doorenbos of Den Haag, Holland. z5

Including the cultivars **'Liset'** and **'Profusion'**. See List of Cultivars.

M. niedzwetzkyana see: **M. pumila 'Niedzwetzkyana'**

M. × platycarpa Rehd. (? *M. coronaria* × *M. pumila*). Tree, to 6 m high, twigs outspread, thornless, young shoots tomentose at first, soon glabrous; leaves ovate to elliptic, rounded at both ends, base occasionally slightly cordate, 5–8 cm long, otherwise scabrous and usually biserrate, often with a few short triangular lobes on either side of the long shoots, pubescent on the venation beneath; flowers white, 3.5 cm wide, buds pink, calyx and flower stalks glabrous; fruits yellowish-green, 4–5 cm wide, flat globose, indented at both ends, waxy, fragrant, occasionally somewhat reddish. SM 384; ST 189; BFC 107. N. Carolina, USA. 1912. z6 Plate 121.

'Hoopesii'. Leaves only slightly lobed or unlobed, usually with only a few large teeth on the leaves of the long shoots; flowers pink-white, 4 cm wide; fruits to 5 cm wide, green, otherwise like the species. BFC 112; Add 523. Cultivated since 1867. Not known in the wild.

M. prattii (Hemsl.) Schneid. Upright shrub or tree, to 7 m high, young shoots pubescent at first, soon becoming glabrous; leaves ovate or elliptic to oblong, long acuminate, base usually round, 6–15 cm long, finely and doubly serrate, with 8–10 vein pairs, loosely pubescent beneath; flowers white, 2 cm wide, 7–10 in cymes, petals rounded, styles 5, glabrous; fruits ovate to globose, 1–1.5 cm thick, red or yellow, punctate, with a persistent calyx, on strong stalks. LF 170; ICS 2208. China, Szechwan, Hupeh Provinces. 1904. z6 Plate 119; Fig. 197. ∅

M. prunifolia (Willd.) Borkh. Small tree, 5(10) m high, twigs soft pubescent when young; leaves elliptic or ovate, 5 to 10 cm long, crenate, with a scattered pubescence beneath; flowers 3 cm wide, petals always pure white, buds pink, grouped in clusters of 6–10, calyx white tomentose, calyx lobes lanceolate, April; fruits ovate, 2 cm thick, yellow-green to red, always with a persistent calyx (!), fruits abundantly. BM 6158; BFC 145; ICS 2203. Introduced from NE. Asia around 1750 but not yet found in the wild. z5 Fig. 196.

See also List of Cultivars.

'Fastigiata'. Tightly upright when young, but the branches later bowed under the weight of the fruits and then remain bowed; flowers white, 4.5 cm wide; fruits ellipsoid, 2 cm thick, yellow and red (= *M. prunifolia fastigiata bifera* Dieck). Distributed by Dieck of Zöschen, E. Germany before 1906.

var. **rinki** (Koidz.) Rehd. Leaves pubescent beneath; flowers pink, to 5 cm wide, calyx shaggy pubescent; fruits 1.5–3 cm thick, yellow-green, edible, but bittersweet. BM 8265; BC 3290–91 (= *M. ringo* Carr.; *M. asiatica* Nakai; *M. yezoensis* Koidz.). China; in cultivation. Introduced around 1850.

M. pumila Mill. Short trunked, round crowned tree, 5–7(15) m high, not as high as *M. silvestris*, more open, twigs usually thornless, young shoots and short shoots somewhat tomentose, buds pubescent; leaves more elliptic to ovate, acute to obtuse, base cuneate, 4–10 cm long, crenate, both sides pubescent at first, later glabrous above, petiole to 3 cm long; flowers white, turning pink, to 5 cm wide, calyx lobes acuminate, longer than the calyx cup, calyx and flower stalks pubescent, styles pubescent on the basal half, April; fruits globose, 2–6 cm thick, green, indented on both halves. HW 2: 76; SH 1: 397c–h (= *M. communis* Poir.; *M. domestica* Poir.; *M. dasyphylla* Borkh.). Europe, Asia Minor, particularly in the Caucasus and Turkestan. This species, through hybridization with other species, is a parent of many modern day crabapple cultivars. z5 Fig. 196.

'Apetala'. Flowers small, greenish, the petals stunted, stamens absent, with 10–15 styles; fruits to 5 cm thick, green, open at the calyx end. BC 328, 4–85 (= *M. dioeca* Loisl.; *M. apetala* Poit. & Turpin). 1770; has since occurred repeatedly in cultivation.

'Niedzwetzkyana'. Shrub, to 4 m high, upright twigs outspread, new growth an attractive red; leaves later more bronze-brown, elliptic to ovate, eventually only with only the midrib and venation reddish; flowers 4–7 in clusters, 4.5 cm wide, dark red, petals white clawed (!), flowers sparsely, May; fruits somewhat conical, 5–6 cm thick, with a dark red exterior, interior also more or less reddish, seeds pink, sweetish. BM 7975; RH 1906: 232 (= *M. niedzwetzkyana* Dieck). SW. Siberia, Turkestan. 1891. Discovered by the Russian botanist Niedzwetzky in Kashgan, Turkestan; Dr. Dieck of Zöschen, E. Germany obtained seed & named the resulting plants from which all cultivated plants today are descended. This cultivar has often been used in hybridization, resulting in today's dark red flowering, red leaved forms.

var. **paradisiaca** Schneid. Paradise Apple. This is the name widely assigned in the nursery trade to *Malus* forms used as understocks in grafting. See especially the Lit. (Maurer).

'Translucens'. Flowers double, to 3.5 cm wide, with about 15 petals, buds pink, opening white; fruits only 2 cm thick, yellow-red (= *M. pumila plena*).

M. × purpurea (Barbier) Rehd. (*M. atrosanguinea* × *M. pumila* 'Niedzwetzkyana'). A large shrub to small tree, strong growing, twigs very long, bark black-red; leaves ovate, acute, 8–9 cm long, crenate, 1 or 2 leaves on the long shoots with a few small lobes, brown-red at first, later dark green, glossy; flowers 3–4 cm wide, purple-red at first, but fading after a few days, very early; fruits globose (!), purple-red, 1.5–2.5 cm thick, with a persistent fruit calyx (!), long stalked. BFC 150. Developed by Barbier of Orléans, France before 1900. z5 ∅

Includes the following cultivars (for descriptions see List of Cultivars)

'Aldenhamensis', 'Eleyi', 'Eleyi Compacta', 'Lemoinei', 'Amisk', 'Hoser', 'Jadwiga', 'Jay Darling', 'Kobendza', 'Szafer', 'Sophia' and 'Wierdak'.

M. ringo see: **M. prunifolia** var. **rinki**

M. rivularis see: **M. fusca**

M. × robusta (Carr.) Rehd. (*M. baccata* × *M. prunifolia*). Strong grower, upright, broadly conical shrub to a small tree, branches outspread-nodding; leaves elliptic, acute, 8–11 cm long, somewhat narrower than those of *M. baccata*, bright green, crenate, base rounded; flowers 3–8 in corymbs, white, occasionally somewhat pink, 3–4 cm wide, April–May; fruits globose to ellipsoid, 1–3 cm thick, yellow or red, occasionally also bluish pruinose, calyx abscising or presistent, stalk long, thin. BFC 83. Around 1815. z3

Includes numerous cultivars (see List of Cultivars)

'Erecta'. Strictly upright only when young, later becoming more broad and rounded, bushy; leaves broadly ovate, acuminate, 6–10 cm long; flowers white, with a light pink exterior, usually semidouble (!), 4 cm wide; fruits 2 cm thick, yellow with a red cheek or otherwise red and bluish pruinose. BFC 87. Cultivar from Japan. 1905.

var. **persicifolia** Rehd. A large shrub, branches long, thin;

leaves oval-lanceolate, 5–10 cm long, finely crenate; flowers pure white, soft pink in bud, 4 cm wide, flowers abundantly; fruits 2 cm thick, globose or more oblong, bright red, prolific fruiter. BFC 144. N. China. 1910 ⊕ ⚭

M. rockii Rehd. Closely related to *M. baccata*, but with the young shoots shaggy pubescent; leaves elliptic to more ovate, to 12 cm long, finely and densely serrate, base round, soft pubescent beneath with reticulate venation; flower stalk and calyx shaggy pubescent, flowers 2.5 cm wide, white, buds reddish, fruits ovoid to nearly globose, often acuminate at the apex, yellow, reddish on the sunny side, calyx abscising very late. W. China. z5 1922. Plate 119.

M. sargentii Rehd. Shrub, to 2 m high, often wider, twigs spreading horizontally, very dense, often thorny; leaves ovate, 5–8 cm long, sharply serrate, usually 3 lobed, bright green, fall color orange-yellow; flowers along the entire twig, buds white or a very soft pink, opening white, 2.5 cm wide, usually flowering and fruiting only every other year, May; fruits globose, scarcely 1 cm in diameter, dark red, long stalked, calyx abscising. BM 8757; BFC 157; JRHS 86: 46. Fruits often persist until spring and are quite decorative. One of the best *Malus* species. z5 Plate 117; Fig. 197. ∅ ⊕ ⚭

'Rosea'. Somewhat taller than the type; flowers to 3.5 cm wide, white, but with distinctly dark pink buds; fruits like those of the type. Discovered before 1921 in Rochester, N.Y., USA, Park System.

M. × scheideckeri Späth ex Zabel (*M. floribunda* × *M. prunifolia*). Shrub, scarcely taller than 3 m, occasionally a small tree, growth narrowly upright, young twigs pubescent; leaves ovate, sharply serrate, occasionally with a large incision, bright green, lighter and pubescent beneath; flowers light pink, 4–5 cm wide, semidouble, with 10 petals, buds darker pink, mid-May; fruits cherry-sized, yellow to orange, long stalked, calyx usually persistent. BFC 158. Developed by Scheidecker, Munich, W. Germany, but introduced into the trade by Späth, Berlin in 1888. An excellent hybrid. z5 Plate 115, 117; Fig. 199. ⊕

M. sempervirens see: **M. angustifolia**

M. sibirica see: **M. baccata**

M. sieboldii (Regl.) Rehd. Shrub, to 4 m high, branches widely spaced and nodding, black-brown; leaves of cultivated plants larger than those of plants in the wild, ovate-elliptic, acuminate, 3–6 cm long, coarsely dentate, particularly on the long shoots 3(5) lobed, pubescent on both sides, deep green, lighter beneath, fall color red and yellow; flowers 2 cm wide, light pink, eventually nearly white, darker in bud, petals obovate, May; fruits globose, pea-sized, red to yellow-brown, persisting to December. BC 2: 3295; MJ 1407; KIF 4: 14; ICS 2205 (= *M. toringo* Sieb.). Japan. 1856. According to Rehder this is a mountain form taken into cultivation. z6 Plate 121; Fig. 197. ⊕ ⚭

var. **arborescens** Rehd. Tree, to 10 m high; leaves larger, not so deeply lobed, less pubescent; flowers nearly white, to 3 cm

wide; fruits 1 cm thick, yellow-red. Japan, Korea. 1892. The wild form. ⚭

"*calocarpa*" see: **M. zumi** var. **calocarpa**

'Fuji'. Broad, low habit (original tree is 8 m high and 16 m wide after 40 years); foliage green; flowers greenish-white, occasionally with a trace of purple, the outer 5 petals flat-outspread, the inner 8 to 10 petals directed upward in a loose cluster, *Anemone*-like; fruits orange, 12 mm. US Plant Introduction Station, Glenn Dale, Md. 1968. ⊕

M. sikkimensis (Wenzig) Koehne. Closely related to *M. baccata*; tree, 5–7 m high, shoots woolly pubescent; leaves ovate to oval-oblong, acuminate, 5–7 cm long, sharply serrate, underside more or less woolly; flowers 4–9 in clusters, 2.5 cm wide, white, exterior somewhat pink, calyx woolly, May; fruits pear-shaped, yellow-red, to 1.5 cm wide, punctate. BC 7430; BS 2: 294. N. India. 1849. z6 Plate 119. Fig. 197.

M. silvestris (L.) Mill. Tree or shrub, to 7 m high, short shoots more or less thorny, leaf buds woolly; leaves more oval-rounded, crenate or irregularly serrate, 4–8 cm long, nearly glabrous or only slightly pubescent, petiole half as long as the blade; flowers pink-white on the exterior, 4 cm wide, open flowers with a pink interior, anthers yellow, flower stalks glabrous, styles not or slightly connate, glabrous, April–May; fruits 2–4 cm wide, globose, yellow-green with a reddish cheek, sour, stalk shorter than the fruit. HM 4(2): 744 (= *Pirus malus* L. p. p.; *M. acerba* Mérat). Europe. Rarely cultivated. Ⓕ W. Germany, USSR; in windbreaks and screens. z3

var. **domestica** (Borkh.) Mansf. (= *M. domestica* Borkh.). Most of the cultivated fruit apples are included here and are, therefore, not covered in this text. Please refer to the fruit literature.

'Plena'. Medium size tree, often only a tall, spreading shrub; leaves ovate to oval-elliptic, acute, 4–7 cm long, finely serrate, rather glabrous; flowers double, soft pink in bud, opening pure white, to 4 cm wide, with 13–15 petals; fruit 4 cm in diameter, yellow with red to totally red, sweet, stalk short, not thickened. BFC 96, 194. Almost always found listed as *M. spectabilis alba plena*; but, according to Von Den Boer, this is not an asiatic species but definitely a double form of the native European wild apple! Fig. 199 ⊕

M. × soulardii (Bailey) Britt. (*M. ioensis* × *M. pumila*). Tall shrub, usually more closely resembling *M. ioensis*; leaves broadly elliptic, 5–8 cm long, irregularly crenate, somewhat lobed, tough, rugose above, densely pubescent beneath; flowers pink, fading lighter, 3.5 cm wide, on short stiff stalks, very attractive; fruits flat globose, yellow-green, to 5 cm wide, with a reddish cheek. BB 1981; BC 3286. Occasionally occurs in the wild in the USA. z4

'Soulard'. The type of this cross.

M. spectabilis (Ait.) Borkh. Tall shrub, or a small tree, to 8 m high, habit usually conical when young, branches later more outspread, forming a broader crown, twigs sparsely pubescent at first, later red-brown; leaves elliptic to oblong, 5–8 cm long, short acuminate, serrate, teeth appressed, dark green above, glossy, lighter beneath, pubescent; flowers light pink, 4–5 cm wide, single to semidouble, buds dark pink, stalk 2–3 cm long, glabrous (as is the calyx) to somewhat pubescent, calyx lobes triangular-ovate; fruits globose, 2–3 cm thick, yellow, sour, not indented at the stalk end, stalk thickened on the tip. BC 3292; JRHS 86: 42; GF 1: 272; BM 267. China. Not observed in the wild! Cultivated in Europe since 1750. z5 Fig. 196.

"Alba Plena" see: **M. silvestris 'Plena'**

'Plena'. Flowers double, with about 15 petals, buds a strong pink, soft pink when fully open, 3–5 cm wide, flowers abundantly; fruits globose, 2 cm in diameter, yellow, calyx end often with a second, deformed seed core, stalk thickened. BFC 86, 200, 210.

'Riversii'. Leaves broadly ovate (BFC 155); flowers distinctly double, 4.5–6 cm wide, with 9–20 petals, pink; fruits yellow, 2.5–3.5 cm thick. Developed by Thomas Rivers of Sawbridgeworth, England, around 1872. Considered the hybrid with the largest flowers, but frequently not true to name in cultivation. ✥

M. spectabilis kaido see: **M. micromalus**

M. × sublobata (Dipp.) Rehd. (*M. prunifolia* × *M. sieboldii*). Conical tree, young twigs pubescent; leaves narrowly elliptic, 4 to 8 cm long, wider on long shoots with 1 to 2 lobes, tomentose at first, eventually pubescent only beneath; flowers pale pink, to 4 cm wide, stalk and calyx shaggy, styles 4–5, rarely 3, May; fruits nearly globose, 1.5 cm thick, yellow, calyx persistent or abscising. Gs 4: 181 (= *M. ringo sublobata* Dipp.). Japan around 1892. z6

M. theifera see: **M. hupehensis**

M. toringo see: **M. sieboldii**

M. toringoides (Rehd.) Hughes. Shrub or tree, to 8 m, twigs somewhat hairy only at first, later glabrous; leaves ovate, 3–8 cm long, usually with 2 crenate lobes on either side, some scattered leaves simple and lanceolate, eventually only the venation pubescent beneath; flowers 3–6 in nearly sessile umbels, white 2 cm wide, calyx tomentose, May; fruits globose to slightly pear-shaped, yellow, reddish on the sunny side, 1.5 cm long, very abundant and persistent. BM 8948; BFC 91 (= *M. transitoria* var. *toringoides* Rehd.). W. China. 1904. One of the most decorative fruiting species! z6 Plate 121. ⚭

M. transitoria (Batal.) Schneid. Very closely related to *M. toringoides*, but smaller, habit more slender, young twigs tomentose; leaves broad ovate, with deeper, narrower lobes, 2–3 cm long (usually somewhat larger on plants in cultivation), more pubescent; flowers to 2 cm wide, white, petals broad oblong; fruits 1.5 cm wide, light red. ICS 2207. NW. China. 1911. Less valuable than the previous species. z5 Plate 121.

M. trilobata (Labill.) Schneid. Upright growing shrub or tree; leaves deeply 3 lobed, 5–8 cm long, serrate, glossy green above, pubescent only at first beneath, soon becoming glabrous, light green, a gorgeous red in fall; flowers white, 3.5 cm wide, grouped 6–8 together, May; fruits ellipsoid, red, 2 cm thick. BM 9305 (= *Eriolobus trilobata* Roem.). W. Asia. 1877. Hardy. z6 Fig. 197. ∅

M. tschonoskii (Maxim.) Schneid. Tree, to 12 m, conical at least when young, twigs white tomentose, winter buds red; leaves ovate-elliptic to more oblong, 7–12 cm long, irregularly serrate, often shallowly lobed, white tomentose at first, later glabrous and dark green above, thin tomentose beneath, an attractive orange and red in fall; flowers white, 3 cm wide, grouped 2 to 5, May; fruits globose, 2–3 cm thick, yellow-green with a reddish cheek. BC 3294; MJ 1411; BM 8179; KIF 2: 31 (= *Eriolobus tschonoskii* Rehd.). Japan. 1892. z6 Plate 119; Fig. 197. ∅

M. yezoensis see: **M. prunifolia** var. **rinki**

M. yunnanensis (Franch.) Schneid. Tree, to 10 m high, narrow and narrowly upright when young, young shoots tomentose; leaves broadly ovate, 6–12 cm long, base round to somewhat cordate, scabrous and double serrate, partly with 3–5 pairs of short, broad lobes (but mostly unlobed), tomentose beneath, red and orange in fall; flowers white, 1.5 cm wide, many in 4–5 cm wide, dense corymbs, stalk and calyx shaggy tomentose, styles 5, May; fruits nearly globose, 1–1.5 cm thick, red punctate, calyx reflexed. ICS 2210. W. China. 1908. z6 Plate 119; Fig. 197. ∅

var. **veitchii** Rehd. Leaves ovate, base always distinctly cordate, all leaves with small distinct lobes, the lobes short acuminate, underside eventually glabrous; flowers only 1.2 cm wide; fruits to 1.3 cm thick, red, white punctate. BM 8629 (= *Pirus yunnanensis* Bean p. p.) Central China. 1901. Plate 119. ∅

M. × zumi (Matsum.) Rehd. (*M. baccata* var. *mandshurica* × *M. sieboldii*). Small tree, conical habit, young twigs somewhat pubescent; leaves ovate, acuminate, 5–9 cm long, crenate, slightly lobed on the long shoots, 4–9 cm long; pubescent beneath when young, later glabrous; flowers pink in bud, opening white, 3 cm wide, petals elliptic, styles 4–5, calyx with a somewhat shaggy exterior (villous), May; fruits 1 cm thick, globose, red. ST 91. Japan. 1892. z6 Plate 116; Fig. 197. ∅

var. **calocarpa** Rehd. Spreading habit; leaves smaller on the flowering shoots, entire, more deeply lobed on the long shoots; flowers somewhat smaller, white, styles 3–4; fruits red to orange-red, 1–1.3 cm thick, in dense clusters, fruits abundantly, persisting to December (= *M. sieboldii* var. *calocarpa* Rehd.). Japan. 1905. ∅ ✥ ⚭

List of Cultivars

Flowering time is based on plants observed in the Dortmund Botanic Garden, W. Germany (at about 51° N latitude but in zone 6); grown in other zones, flowering will occur earlier or later depending upon zone. The earliest to flower is always *Malus baccata* var. *mandshurica*, together with *Magnolia kobus* var. *stellata*. Following quickly thereafter, are *M. baccata*, its varieties and hybrids, *M. purpurea*, *micromalus*, *prunifolia*, *magdeburgensis*, etc. The middle flowering time coincides with the Japanese flowering cherries (*Prunus serrulata*), and is represented by *M. arnoldiana*, *floribunda*, *halliana*, *hupehensis*, *purpurea*-types, *pumila*-types, *robusta*, *scheideckeri*, *spectabilis*, *zumi* and its descendents. The late species and cultivars flower with *Caragana arborescens*; they are *M. bracteata*, *ioensis*, *sargentii*, *sieboldii*, 'Dorothea', 'Elise Rathke', 'Marshall Oyama', 'Profusion', 'Wynema', etc. The latest group to flower comprises the American and Asiatic species: *M. coronaria* and varieties,

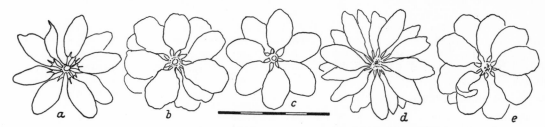

Fig. 198. **Malus.** Types with double flowers. a. 'Aldenhamensis'; b. *M. coronaria* 'Charlottae'; c. *M. hartwigii*; d. 'Katherine'; e. 'Dorothea' (from Den Boer)

florentina, ioensis-types, *kansuensis, platycarpa, sikkimensis, toringoides, transitoria, tschonoskii* and *yunnanensis.*

The average flowering period from bud opening to petal fall is about 10 days, but can last longer. Collectively, the flowering period of the genus covers 4–5 weeks. It should be noted that many species and cultivars flower and fruit abundantly only every other year. This characteristic is noted in the cultivar descriptions.

Of the many breeders who have developed a large number of beautiful crab apples since the end of the 19th century, only the more notable will be mentioned here; for a complete list, see Wyman.

Aldenham House (Vicary Gibbs), Elstree, Hertfords., England
Arnold Arboretum, Jamaica Plain, Mass., USA
E. A. Bechtel, Staunton, Ill., USA
Central Experimental Farm, Ottawa, Canada
J. Cheal & Sons, Crawley, Sussex, England
A. F. Den Boer, Des Moines, Iowa, USA
Dr. Georg Dieck, Zöschen, near Merseburg, W. Germany
S. G. A. Doorenbos, The Hague, Holland
Charles Eley, East Bergholt, Suffolk, England
Niels E. Hansen, Agr. Exp. Station, Brookings, S. Dakota, USA
Heinrich Henkel, Darmstadt, W. Germany
Humm, J., Christchurch, New Zealand
Kornik Arboretum, Kornik, near Poznan, Poland
V. Lemoine & Fils, Nancy, France
W. R. Leslie, Dominion Exp. Station, Morden, Manitoba, Canada
William Oakes, Miami, Manitoba, Canada
Miss Isabella Preston, Central Exp. Farm, Ottawa, Canada
Dr. Karl Sax, Arnold Arboretum, USA
J. P. Scheidecker, Munich, W. Germany
B. H. Slavin, Park System, Rochester, N.Y., USA
James G. Soulard, Galena, Ill., USA
L. Späth, Berlin, W. Germany
J. A. Tanner, Palo, Iowa, USA
G. P. Van Eseltine, N.Y. State Exp. Sta., Geneva, N.Y., USA
Robert Veitch & Son, Exeter, England
Wayside Gardens, Mentor, Ohio, USA
Hayward Wright, Auckland, New Zealand (since 1910)
Dr. D. Wyman, Arnold Arboretum, USA

'Aldenhamensis' (Gibbs) (*M. purpurea*). Shrub to a small tree, 2.5–3.5 m high, usually lower than the other *purpurea* types, twigs thinner and shorter; leaves like those of *M. purpurea* in form and color, reddish to bronze, never lobed; flowers purple-red, to 2.5 cm wide, with 6–10 petals, these narrow in spring, wider on premature flowers in fall, flowers abundantly every year, middle-early (about 10 days after 'Eleyi'), buds deep red; fruits flat-globose, brownish-red, to 2.5 cm thick, with a persistent calyx, pulp reddish. BFC 68. Chance seedling, found in 1922 by Hon. V. Gibbs. Plate 116; Fig. 198. ⊕ ⚭

'Aldenham Purple' (Gibbs). (Originated from *M. pumila* 'Niedzwetzkyana' and not to be confused with 'Aldenhamensis'!). Flowers single, lilac-red; fruits like a normal apple, purple-red, to 6.5 cm in diameter. Before 1925. Very susceptible to canker. ⚭ ✖

'Alexis' (*M.* × *robusta*). Flowers white, single, 4 cm wide, buds opening pink; fruits bright red to dark carmine, bluish pruinose, 3 cm wide. Developed in 1897 by N. E. Hansen from Russian seed. Similar to 'Dolgo'.

'Almey' (W. R. Leslie) (*M. adstringens*). Erect growing, rounded, medium size, scarcely over 3–4 m high; leaves purple on the new growth, later more bronze-green; buds chestnut-brown, opening deep purple, to 5 cm wide, base of the petals lighter (forming a whitish "star", like 'Hopa'), petals characteristically folded backward, flowers early and abundantly; fruits globose, distinctly angular, 2 cm thick, orange with a large red cheek to totally red, September–October. BFC 71; JRHS 86: 50 (= 'Sunglory'). Developed around 1945. ⊕ ⚭

'Amisk' (Preston) (*M. purpurea*). Flowers simple, buds carmine, fading to a dull pink, to 5 cm wide, middle-early, flowering fully every other year; fruits red and yellow, 3 cm thick, conical. Introduced in 1930.

'Barbara Ann' (Seedling of 'Dorothea', presumably crossed with a *purpurea* clone; first flowered in 1960). Leaves reddish during the entire growth period; flowers semidouble, with 12–15 petals, purple-pink, then somewhat fading, 4–5 cm wide, fruits purple, 12 mm thick. Developed in the Arnold Arboretum by Dr. Karl Sax, named by Dr. Donald Wyman in 1964. ⊕

'Beauty' (Hansen) (*M. robusta*). Growth narrow, nearly tightly conical-upright; flowers white, single, 4–5 cm wide; fruits 3 cm thick, bright red, of economic use. Developed in 1919 from Russian seed. ⊕ ⚭

'Blanche Ames' (Sax) (*M. spectabilis*). Shrub, 3–4 m high, twigs delicate; buds carmine, opening to a white interior, exterior carmine-pink, semidouble, 3.5 cm wide; fruits 8 mm thick, globose, yellow, soon abscising. 1947. Quite meritorious. ⊕

'Bob White' (*M. zumi*). Habit like that of the species; flowers white, buds light pink, middle-early, 2.5–3 cm wide, single, flowers fully only every other year; fruits globose, 1–1.5 cm thick, yellow to dull brown, persisting on the plant the entire winter. Discovered in Massachusetts in 1876. ⚭

'Cheal's Crimson' (Cheal) (*M. prunifolia*). Habit upright-rounded, 3–5 m high; leaves broad oval-elliptic, short acuminate; buds pink, opening white, single, 3 cm wide, petals usually circular; fruits orange-yellow with a bright red cheek, 2.5 cm wide, fruits abundantly, calyx persistent or abscising. BFC 82 (= *M. prunifolia* 'Fructu coccineo'). Introduced in 1919. ⚭

Fig. 199. **Malus.** Types with double flowers. a. *M. scheideckeri;* b. *M. halliana* 'Parkmanii';
c. *M. magdeburgensis;* d. *M. silvestris* 'Plena; e. 'Crimson Brilliant' (from Den Boer)

'Cheal's Golden Gem' (Cheal) (*M. prunifolia*). Flowers white, single; fruits oval-rounded, yellow, 1.5 cm in diameter, fruits abundantly, but abscising in September. Introduced in 1929. ⊛

'Chilko'. (Preston) (*M. purpurea*). Flowers single, purple-pink, 5 cm wide, opening dark purple-red; fruits bright red to carmine, 5 cm wide, fruits well every other year. Developed in 1920, named in 1930. ✧

'Cowichan' (Preston) (Descendant of *M. pumila* 'Niedzwetz-kyana'). Foliage with an effective fall color; flowers soft lilac to nearly white, opening pale pink, single, to 4.5 cm wide, middle-early, flowers every year; fruits purple-red and light yellow to ivory-yellow, 4 cm thick, attractive. Has the lightest flower color of the "Rosyblooms", but very attractive. Plate 76. ✧ ⊛ ✂

'Crimson Brilliant' (Den Boer) (*M. adstringens*). Shrub, rather slow growing, medium high, twigs outspread; leaves purple or bronze; flowers single to semidouble, with 5–10 petals, carmine with a white "star", 4 cm wide; fruits about 2 cm wide, dark purple, somewhat "rusty". BFC 89. Developed in 1939; introduced into the trade in 1952 by Wayside Gardens (patented). Fig. 199. ✧

'Dartmouth' (*M. pumila*). Strong growing, broad upright; flowers white, single, 3.5 cm wide, attractive; fruits medium size, somewhat angular, otherwise rounded, dark red and yellow, to 4 cm thick, very bluish pruinose, edible, mildly sour. Discovered about 1883 in New Hampshire, USA. ✧ ⊛ ✂

'Dolgo' (Hansen) (*M. baccata*). Shrub, medium size, upright; flowers white, to 4.5 cm wide, single, early, flowers fully every other year; fruits oval-globose, 3 cm thick, bright red, pulp yellow-white. BFC 93. Developed from Russian seed in 1897. ✧ ⊛ ✂

'Dorothea' (Wyman) (*M. scheideckeri*). Shrub, low to medium high, slow growing; flowers abundantly as a young plant, opening carmine, then pink, does not fade, flowers semi-double to double, to 5 cm wide, with about 16 petals, late, flowers every year; fruits globose, about 1.3 cm thick, yellow. BFC 95. Discovered as a chance seedling in the Arnold Arboretum in 1943. Fig. 198. ✧ ⊛

'Dorothy Rowe' (Seedling of *M. spectabilis* 'Riversii', first flowered in 1962). Flowers white to cream-white, single to semidouble, with many yellow anthers; fruits glossy red, 2.5 cm thick. Named by Arie Den Boer, Des Moines, Iowa, USA. 1964. ⊛

"Echtermeyer" see: **'Oekonomierat Echtermeyer'**

'Eleyi' (Eley) (*M. purpurea*). Large shrub, upright, twigs outspread; leaves darker red than the species; flowers dark purple, somewhat darker than the species, about 3.5 cm wide, single; fruits normally ovate, about 2.5 cm long, 1.5 cm thick, on long, thin stalks, purple-red, persisting to November. BS 1: Pl. 25; GC 104: 243 (= *M. eleyi* Hesse). Around 1920. In the USA this cultivar is variable, often producing flat-globose fruits on short stalks; then named 'Jay Darling'. See Den Boer, l. c.: 100–101. ✧ ⊛

'Eleyi Compacta' (Doorenbos) (Seedling of 'Eleyi'). Shrub, very slow growing, short branched; leaves purple-red; flowers appearing on young plants, purple-red, single, 3.5 cm wide; fruits 1 cm in diameter, purple-red. Around 1950.

'Elise Rathke' (Döring) (*M. pumila*). Actually nothing more than a common apple with umbrella-form, pendulous branches; flowers pink-white, late every other year; apples 5–7 cm long, green with a red cheek. Developed by L. A. Doering in Elbing; introduced into the trade around 1885 by F. Rathke, Praust near Gdańsk, Poland. ✂

'Evelyn' (Den Boer) (*M. denboerii*). Upright, strong growing shrub; leaves nearly like those of *M. ionensis* in form and size, but red or bronze, more deeply lobed on long shoots, fall color of the foliage red; flowers single, pink-red (darker than on *M. ionensis*), 3–3.5 cm wide, middle-early, fragrant, flowers abundantly; fruits greenish-yellow and red, 3 cm thick, pulp reddish. BFC 104. Discovered as a natural hybrid by Den Boer in 1939. ✧ ∅

'Exzellenz Thiel' (Späth) (*M. scheideckeri*). Known only as a grafted plant on a high standard, the branches weeping nearly vertically, somewhat stiff; leaves oval-elliptic, acuminate, scabrous serrate; buds pink, opening white, 4.5 cm wide, middle-early, blooms fully every other year; fruits 2 cm thick, angular, yellow and red, without a calyx. BFC 166. 1909. ✧

'Fairy' (*M. robusta*). Strong upright grower; flowers single, white, 3.5 cm wide; fruits red, 4 cm thick. Developed about 1870 in England. Economically valuable. ✂

'Flame' (parentage unknown). Strong growing, upright; flowers white, buds pink, 4 cm wide, early, flowers every other year, attractive; fruits bright red, 2 cm in diameter. Developed at the University of Minnesota State Fruit Farm in Excelsior, Minn., USA around 1920. ✧ ⊛

'Florence' (Gideon). Large shrub; buds pink, flowers white, 4 cm wide, single; fruits yellow with red, 4–5 cm thick. Same origin as 'Flame', around 1886. Of economical value. ✧ ⊛ ✂

"Frau Luise Dittmann". This cultivar, developed before 1909 by H. Henkel of Darmstadt, W. Germany, is actually (according to Den Boer) *M. spectabilis* 'Plena'.

'Fryderyk Chopin'. Shoots glabrous, particularly in the second year, purple; leaves nearly glabrous, mostly green, only the youngest red to red-brown; flowers large, to 6.5 cm wide, pink inside and out, petals broadly elliptic, 32 × 23 mm, overlapping, stalk 3–4 cm long, glabrous, purple, like the calyx;

fruits to 27 mm wide, 24 mm long, dark purple, pulp more or less reddish, calyx usually abscising. From Kornik, Poland.

'Fuji' see: **M. sieboldii**

'Garry' (*M. pumila* 'Niedzwetzkyana' × *M. baccata* ?). Upright habit, shoots thin; flowers chestnut-purple at first, then dark pink-red; fruits carmine, with a thick, waxy pruinose, 2 cm thick, persisting the entire winter. Developed in Morden, Manitoba, Canada; introduced in 1962. ⚮

'Geneva' see: '**Van Eseltine**'.

'Gibbs Golden Gage' (Gibbs). Flowers white, buds pink; fruits yellow, 2.5 cm thick, persisting throughout the entire winter. Discovered as a chance seedling before 1923. ⚮

'Gold'. Flowers white, 4 cm wide, single; fruits yellow to red, 2.5 cm in diameter. Discovered in Illinois, USA, around 1910. ⚮

'Golden Hornet' (Waterer) (Seedling of *M. zumi* 'Calocarpa'). Flowers white; fruits usually in groups of 4, globose to ovate, yellow, 3 cm, persisting until the New Year. JRHS 82: 8. ⚮

'Gorgeous' (*M. sieboldii* × *M. halliana*). Medium high shrub or small tree; flowers white, 3 cm wide, opening soft pink; fruits carmine to orange-red, oval, 2.5 cm thick, very abundant and present every year. Hybridized by Hayward R. Wright of Avondale, Auckland, New Zealand. Before 1925. (Often found under the erroneous name "Striped Beauty"). Plate 76. ⚮

'Guiding Star'. Tree, quite narrow, conical habit; with very attractively formed double flowers, in clusters of 5 or more, buds pink, opening pure white, very fragrant; fruits small, yellow. Developed by Arie Den Boer, Des Moines, USA; introduced by Wayside Gardens of Mentor, Ohio in 1963.

'Helen' (Den Boer) (*M. adstringens*). Shrub, medium high, twigs spread horizontally, leaves narrowly oval-elliptic, acuminate, light red (!), color quite different from all the other red foliage forms; flowers purple, 4–5 cm wide, single, attractive; fruits purple-red, 2 cm thick. BFC 109. Developed by Den Boer in 1939. Seedling of 'Jay Darling'. ⚘

'Henrietta Crosby' (Sax) (*M. arnoldiana* × *M. pumila* 'Niedzwetzkyana'). Medium strong grower, graceful habit; flowers single, pink, 4.5 cm wide; fruits bright red, 2.5 cm thick, fruiting every year, but abscising early. 1939. Plate 75. ⚘ ⚮

'Henry F. Dupont' (Sax) (*M. arnoldiana* × *purpurea* 'Eleyi'). Spreading habit; flowers single to semidouble, pink, 4 cm wide, buds dark purple-red; fruits 1.2 cm thick, brownish-red, persistent. 1946. ⚘

'Henry Kohankie' (selected from seedlings of *M. sieboldii* from Japanese germ plasm, 1938). Fruits elliptic-oblong, 2.5–3.5 cm long, 2–3 cm wide, bright red, the pulp a light orange, in clusters of 2–4, pendulous on the twigs and persisting throughout the winter. Developed in the Kohankie Nursery, Painesville, Ohio, USA. 1946.

'Hillieri' (Hillier) (*M. scheideckeri*). Shrub, strong growing; leaves broadly oval-elliptic, coarsely serrate, occasionally slightly lobed; flowers semidouble, light pink, 2.5 cm wide, 5–9 petals; fruits 1.5–2 cm thick, globose, yellow and orange. JRHS 86: 40. Selected out of a shipment of plants from Holland by Hillier, Winchester, England; 1928. Plate 117. ⚘

'Hopa' (Hansen) (*M. adstringens*). Small, broad and loosely upright growing tree, to 4 m high; leaves oval-elliptic, with a short apex, scabrous serrate, bronze-red; flowers lilac-red, middle-early, very abundant, 4–5 cm wide; fruits globose, 2 cm thick, bright red, without a purple tone, occasionally somewhat

translucent orange, edible, calyx persistent or abscising, pulp reddish. BFC 113 (= 'Pink Sunburst'). 1920. ⚘ ⚮ ✖

'Hoser' (Wroblewski) (*M. purpurea*). Buds dark purple, flowers pink, 4–4.5 cm wide, petals elliptic, 2 cm long; fruits globose, purple, bluish pruinose. 1938 in the Kornik Arboretum, Poland. ⚘

'Hyslop' (Downing) (*M. prunifolia* ?). Small tree, strong growing, broad crowned, twigs somewhat nodding; fruits globose, about 4 cm thick, yellow, largely carmine-red striped and mottled, bluish pruinose, pulp dense, yellowish, juicy, harshly sour. SpB 121; RH 1892: Pl. 420. Before 1869. ⚮ ✖

'Irene' (Den Boer) (*M. adstringens*). Medium size, slow growing shrub to a small tree; leaves oval-elliptic, obtuse to acute, finely serrate, reddish; flowers purple, 4.5 cm wide, single, flowers very abundantly; fruits dark purple, about 2 cm thick, pulp red. BFC 116. 1939. ⚘

'Jadwiga' (Wroblewski) (*M. purpurea*). Broad crowned, branches pendulous; buds large, purple-brown, flowers to 7 cm wide, petals broadly elliptic, to 3 cm long, rich pink, outspread; fruits 4–5 cm long, conical, raspberry-red, juicy, calyx persistent, stalk 1.5 cm long. Developed in 1938 in the Kornik Arboretum, Poland. ⚘ ⚮ ✖

'Jay Darling'. Practically identical to 'Eleyi' (which see), except for the fruits which are globose to flat-globose, 2.5 cm thick, purple-red, stalk short. BFC 122 (= *M. atropurpurea* Croux & Fils, Chatenay). 1904. ⚘ ⚮

'Joan' (Dunbar) (*M. robusta*). Flowers white, single, early, to 5 cm wide; fruits red, 3.5 cm wide, present every 2 years. Discovered in 1918 by J. Dunbar in New York State. ⚮ ✖

'John Downie'. Strong grower, tall shrub; flowers white, 5 cm wide, buds pink, flowers fully every other year; fruits 3 cm wide, orange with a red cheek, abundant. Hybridizer, Edward Holmes, Whittingdon Nursery, Lichfield, England. 1875. Economically valuable. ✖

'Justyna'. Except for the flower form, generally resembling the previous cultivar, flowers to 6 cm wide, with narrower petals (to 28 × 15 mm); fruits nearly as wide as long (to 23 mm), with somewhat shorter stalks. Developed in Warsaw, Poland, 1963. Named in honor of Chopin's mother, Justyna Krzyzanowska.

'Katherine' (Slavin) (*M. hartwigii*). Small to medium size, erect shrub, twigs soon spreading outward; leaves narrowly oval-elliptic, acuminate, finely serrate; flowers densely double, with 15–24 petals, opening pink, quickly becoming pure white, middle-early, flowers fully only every other year (but then prolific!), 5.5 cm wide; fruits yellow, occasionally somewhat reddish, ripening late, 1 cm thick, not conspicuous. BFC 125. Developed around 1928. Very valuable. Plate 118; Fig. 198. ⚘

'Kelsey' ("Rosybloom" type). Flowers semidouble, purple-pink, very persistent. Introduced in 1970 from Morden, Manitoba, Canada; notable as the first winter hardy crabapple with double flowers. ⚘

'Kingsmere' (Preston) *M. pumila* 'Neidzwetzkyana' × *M. sieboldii*). Small tree, twigs outspread, habit like a common apple tree; leaves oval-elliptic, acute, finely serrate, purple to bronze, but young leaves distinctly bluish-lilac toned (!); flowers purple, early, single, 5.5 cm wide, flowers abundantly every year; fruits 3 cm thick, carmine and brownish, greenish on the shaded side, edible. BFC 126. 1920. ⚘ ⚮ ✖

'Kobendza' (Wroblewski) (*M. purpurea*). Buds small, purple-brown, flowers 4 cm wide, purple-brown exterior, interior pink, petals elliptic, 1.5 cm long, stalk thin, 4–5 cm long; fruits globose, 1.5 cm thick, purple-red. Developed in the Kornik Arboretum, Poland, 1938. Best of the Polish hybrids. ⊕

'Kola' (Hansen) (*M. coronaria* 'Elk River' × 'Oldenburg'). Growth conical when young, later broader, round crowned, branched to the ground; leaves broadly ovate, base cordate, large lobed, the lobes finely serrate, with a blue-gray shimmer (!); flowers pink, fragrant, 4–5 cm wide, late; fruits flat-globose, 5 cm wide, green-yellow, fragrant, waxy. BFC 127, 198. 1922. Tetraploid. ⊕ ⊛ ✕

'Lady Northcliffe' (Aldenham) (*M. baccata*). Buds soft pink, opening white, flowers very abundantly; fruits orange-yellow, 1.5 cm thick. Before 1929.

'Lemoinei' (Lemoine) (*M. purpurea*). Shrub, medium high, strong, erect, twigs not outspread; leaves elliptic to ovate, often with 1–2 lobes, dark purple, later bronze or more dark green; flowers single to slightly double, purple, 4 cm wide, appearing every year, but less abundant on young plants and then only on the periphery of the crown; fruits dark purple, 1.5 cm thick. BFC 129; JRHS 68: 10. 1928. ⊕ ⊛

'Linda' (*M. arnoldiana* descendent). Flowers carmine at first, pale pink when fully open, 4.5 cm wide, single; fruits carmine, about 3 cm wide. Developed and named in 1958 by A. Den Boer. ⊕

'Lisa' (Den Boer) (*M. denboerii*). Very similar to 'Evelyn', but much less vigorous; flowers pink-red to light carmine, fragrant; fruits orange and carmine or totally carmine, 2.5 cm thick.

'Liset' (Doorenbos) (*M. moerlandsii*). Habit very similar to that of 'Lemoinei', as are the leaves, leaf color, flowers and fruit, but growing taller; leaves later in the year becoming predominantly dark green and glossy; flowers single, 4 cm wide, pure purple-red, middle-early, flowers abundantly even on young plants; fruits dark brown, 1.5 cm thick. Developed before 1938, but first recognized in 1949. Very meritorious plant! ⊕

'Makamik' (Preston) ('Niedzwetzkyana' descendent). Small tree, crown tall, rounded, twigs erect, somewhat outspread; leaves elliptic, not lobed, reddish on the new growth, later bronze-green; flowers dark red at first, then purple-pink, later fading, with a white star in the middle, 5 cm wide, flowers fully every year; fruits nearly globose, 2 cm thick, light red, often more orange on the shaded side. BFC 132. 1921. Considered by Miss Preston, her best "Rosybloom" type. ⊕ ⊛

'Marshall Oyama'. Strong growing, tightly upright, broadly conical, very long branched, nodding with age from the weight of the fruits; leaves oval-elliptic, acuminate, sharply serrate; flowers light pink, buds darker, 3.5 cm wide, late and flowering every year; fruits often to 3.5 cm wide, carmine, often yellow on the shaded side. BFC 134. 1930. ⊕

'Mary Potter' (Sax) (*M. atrosanguinea* × *M. sargentii* 'Rosea'). Low, broad growing shrub, similar to *M. sargentii*, but grows more vigorously; flowers white, buds pink, 2.7 cm wide; fruits red, 1 cm thick, abscising early. 1939. Considered a very promising hybrid. ⊕

'Montreal Beauty' (Cleghorn) (*M. pumila*). Large shrub or small tree; flowers white, single, 5 cm wide, buds pink; fruits green with a red cheek, 4 cm thick. Developed before 1833 in Quebec, Canada; introduced into the trade by Robert Cleghorn, near Montreal, Canada. ⊛ ✕

'Neville Copeman' (Seedling of 'Eleyi'). Resembles 'Eleyi' in habit and flower, but more vigorous; fruits more ornamental, orange-red to glossy carmine, flat-globose to ellipsoid. Breeder, T. N. S. Copeman of Norfolk, England; introduced in 1954 by Notcutt, Woodbridge, England. ⊕ ⊛

'Nicoline' (*M. moerlandsii*). Tall shrub; foliage purple-green during the entire summer; flowers in clusters, rather small, purple-red. Hybridized by S. G. A. Doorenbos. Susceptible to apple scab; surpassed by 'Liset'.

'Nipissing' (Preston) (*M. adstringens*). Large shrub; leaves light purple; flowers carmine-pink at first, then dark pink, quickly fading, about 4 cm wide, flowers every other year; fruits dark red with orange, bronze-green on the shaded side, 3.5 cm thick, sour. 1920; first named in 1930. ⊕

'Nova' see: **M. ionensis**

'Oekonomierat Echtermeyer' (Späth) (*M. gloriosa*). Weeping form, twigs nodding in a wide arch, graceful; leaves brown-red at first, quickly becoming an attractive bronze-green, irregular and coarsely serrate, also lobed on long shoots; flowers carmine, single, 4 cm wide, flowering every year; fruits red-brown, about 2.5 cm thick, pulp reddish. BFC 100. 1914. ⊕ ⊛

'Oporto' ('Niedzwetzkyana' seedling). Medium tall shrub, open growing, twigs slightly nodding; leaves purple-red at first, later dark reddish-green; flowers deep purple, a duller color, not contrasting well with the foliage; fruits 1.5–2 cm thick, purple-red. England. Plate 75. ⊕

'Orange' (*M. baccata*). Large shrub; buds pink, opening white, flowers 3.5 cm wide; fruits 3 to 4 cm thick, orange-yellow. Known in the USA since before 1869. ⊛ ✕

'Ormiston Roy' (Den Boer). Similar to *M. floribunda* in habit and foliage, but flowering later; flowers soft pink, 4 cm wide, single, appearing every year; fruits orange-yellow and reddish, 1 cm thick, in large numbers, persisting throughout the winter. Origin unknown; named by Den Boer in 1954. ⊕ ⊛

'Osman' (Saunders) (*M. adstringens*). A large shrub; flowers pink-white, single, 5 cm wide; fruits orange to red, 4 cm thick. Developed by W. Saunders before 1904, but first named in 1911. ⊕

'Patricia' (Den Boer) (*M. adstringens*; seedling of 'Hopa'). Round crowned, upright tree, to 5 m high; leaves larger than those of 'Hopa', darker green, often somewhat bronze; flowers purple-red, single, 4–5 cm wide, early; fruits dark red, 2.5 (5) cm in diameter, pulp reddish. 1953. ⊕ ⊛ ✕

'Peachblow' (*M. floribunda*). Habit more upright than that of the type; flowers white, 3 cm wide, buds dark pink; fruits 1 cm thick, red, coloring much earlier than the species. In USA, before 1930. ⊕

'Pink Beauty' (*M. adstringens*). Flowers single, pink; fruits red, 2.3 cm wide. Developed in Morden, Manitoba, Canada; but named before 1958 by Simpson Orchard Co., Vincennes, Indiana, USA.

'Pink Giant' (C. Hansen) (*M. adstringens*). Flowers pink at first, light pink to pale lilac when fully open, to 5 cm wide, simple; fruits orange to red, 2 cm thick. 1939. ⊕

'Pink Sunburst' see: **'Hopa'**

'Prince Georges' (Glenn Dale Station) (*M. angustifolia* × *M. ioensis* 'Plena'). Small, erect shrub, about 2.5 m high, conical, twigs somewhat outspread, very densely foliate; leaves narrowly oval-elliptic, slightly lobed, coarsely lobed on long shoots, last of the crabapples to drop its leaves in the fall, base cuneate, very much resembles *M. ioensis* 'Plena'; flowers pink, 5 cm wide, with 50–60 petals, flowers well every year, very late, fragrant; fruitless. BFC 147. Developed in the Glenn Dale Station from seed collected in the Arnold Arboretum, USA. Named at the Arnold Arboretum in 1930. ✿

'Professor Sprenger' (Doorenbos) (see *M. zumi*). Leaves and flowers scarcely differing from the species; fruits orange, 1 cm thick, persisting until late winter. Before 1950. Plate 76. ✿ ⚘

'Profusion' (Doorenbos) (*M. moerlandsii*). Shrub to a small tree, about 3–4 m high, upright, twigs nodding; leaves oval-elliptic, lobed on long shoots, young red or reddish, later more bronze-green, very similar to 'Lemoinei'; flowers carmine-red, but quickly becoming lighter, single, 4 cm wide, late; fruits red-brown ("oxblood red"), flat-globose, deep red, often somewhat pruinose, 1.5 cm thick, somewhat angular, long stalked. BFC 148; JRHS 86: 51. Developed before 1938, but first named in 1946. ✿ ⚘

'Purple Wave' (Den Boer) (*M. adstringens* ?) Shrub to a small tree, 3–4 m high, twigs outspread; leaves broad ovate to obovate, often about 8 cm long and 6 cm wide, new growth reddish, later becoming much darker (then somewhat like *Fagus silvatica* 'Atropurpurea'!) and remaining so until leaf drop; flowers purple, single, 4 cm wide; fruits 2 cm thick, purple. BFC 151. 1939, named in 1951. ✿

'Radiant' (L. E. Longley) (seedling of 'Hopa'). Compact habit, erect; young leaves an attractive reddish and contrasting well with the older green foliage; flower buds deep red, opening dark pink, single, medium size, flowers abundantly every year, effective for 10 days; fruits cherry-red, 13 mm thick, persisting into the winter. Selected by Dr. L. E. Longley at the University of Minnesota, USA, around 1940. ✿ ⚘

'Redfield' (*M. pumila*; 'Niedzwetzkyana' descendent). Flowers purple, single, 4 cm wide; fruits red, 3.5 cm thick, pulp red. Developed in the New York Experiment Station, USA, around 1924. Flowers and fruits attractive. ✿ ⚘

'Redflesh' (Hansen) (*M. soulardii*). Leaves bronze-red; flowers single, pale purple, buds carmine, flowers every 2 years; fruits red, 4 cm thick. Developed by N. E. Hansen, 1928. Flowers and fruits attractive. ✿ ⚘ ✗

'Redford' (*M. pumila*; 'Niedzwetzkyana' descendent). Leaves red-brown; flowers single, 5 cm wide, very similar to those of 'Redfield', but lighter, flowers fully every other year; fruits yellow with a red cheek, 5 cm in diameter. Developed in 1921 at the New York Exp. Station, USA; but not named until 1938. ✿

'Red Jade' (Reed) (allegedly a seedling of 'Excellenze Theil'). Twigs long, arching, thin, often touching the ground; buds pink, flowers pure white, 4 cm wide, single; fruits ovate, 1.5 cm long, bright red, in large numbers and persisting long into the winter, produced abundantly every year. Developed by Dr. G. M. Reed of the Brooklyn Botanic Garden, USA, in 1935. (Patented.) ✿

'Red River' (Yeager) ('Dolgo' × 'Delicious'). Tree; flowers pink, 5 cm wide, single; fruits bright red, 5 cm thick, edible. Developed in the USA before 1938. Very attractive in flower and fruit. ✿ ⚘ ✗

'Red Sentinel' (*M. robusta*). Flowers white, early May; fruits rather flat-globose and somewhat ribbed, dark red and glossy, persisting long into winter, calyx abscising or persisting. Introduced into the trade in 1959 by Notcutt. ✿ ⚘

'Red Silver' (C. Nansen) (*M. adstringens*; 'Niedzwetzkyana' descendent). Medium size tree or a tall shrub; leaves bronzered to chestnut-red, dense silvery pubescent on the new

growth, ovate, irregularly serrate, more or less deeply incised and lobed on the long shoots; flowers dark pink at first, soon fading, 4 cm wide, single; fruits purple, 2 cm thick. BFC 152. Developed in 1928. ✿ ⚘ ✗

'Red Splendor' (Descended from 'Red Silver'). Flowers pink-red at first, lighter pink when fully open, 4.5 cm wide; fruits red, 1.5 cm wide. Selected in 1948 by M. Bergeson of Fertile, Minnesota, USA. Very resistant to disease.

'Red Tip' (N. E. Hansen) (*M. heterophylla* ?). Strong growing; young leaves with red apexes (hence the name!), fall foliage orange; flowers solitary, a dark pink monochrome, 4 cm wide; fruits 4 cm thick, yellow-green. 1919. ✿ ⚘ ⊘

'Robin' (Saunders) (*M. adstringens*). Flowers dark pink, quickly fading, but attrative, 4 cm wide; fruits 4 cm thick, orange-yellow and reddish, evenly ribbed, attractive, edible. 1911. ✿ ⚘ ✗

'Rosilda' (Saunders) (*M. pumila*). "Apple tree" with pink-white flowers and 5 cm thick, greenish-yellow and red apples, edible. Developed from 'Prince' × 'McIntosh'. 1920. ✿ ⚘ ✗

'Rosseau' (Preston) (seedling of 'Neidzwetzkyana'). Shrub, similar to 'Makamik'; leaves bronze-red in spring, red in fall; flowers purple, 4.5 cm wide, middle-early, flowers abundantly every year; fruits 2 cm thick, bright red. Named in 1930. One of the best "Rosybloom" cultivars. ✿ ⊘

"Rosybloom". A collective name for the Canadian hybrids of Miss Isabella Preston; 'Niedzwetzkyana' descendents ('Amisk', 'Makamik', 'Scugog', 'Wabiscaw', 'Simcoe'); this name was also used later for the Leslie hybrids ('Oakes', 'Almey', 'Sundog').

'Royalty' ("Rosybloom" form). Leaves red-brown; flowers carmine to nearly purple, single; fruits dark red, about 1.5 cm thick. Selected in 1958 by W. L. Kerr, Sutherland, Saskatchewan, Canada; introduced in 1962. ✿ ⚘

'Rudolph' (presumably a *M. baccata* hybrid) Buds carmine, pink when fully open, 5 cm wide; otherwise similar to 'Almey', but more winter hardy. Introduced in 1954 by F. L. Skinner of Dropmore, Manitoba, Canada.

'Scugog' (Preston) ('Niedzwetzkyana' descendent). Large shrub; flowers dark pink, soon becoming lighter, single, 5 cm wide, middle early, flowers abundantly every 2 years; fruits dark carmine to brown-red, 4 cm in diameter, edible. 1930. Attractive flowers and fruits. ✿ ✗

'Seafoam' (*M. sieboldii*). Nodding habit; leaves deeply incised; flowers solitary, pink-red to carmine at first, white with a trace of pink when in full bloom, 3.5 cm wide; fruits yellow, 1.5 cm. Selected in 1940 by A. Den Boer; introduced in 1952. ⚘ ✿

'Selkirk' (*M. baccata* × *M. pumila* 'Niedzwetzkyana'; another "Rosybloom" form). Strong grower, round crowned; flowers in clusters along the shoot, appearing in rings, flat, pure pink, with scarlet-red fruits in early August. Introduced into the trade in Morden, Manitoba, Canada, 1963. First flowered in 1939. ✿ ⚘

'Simcoe' (Preston) (*M. adstringens*'; 'Niedzwetzkyana' descendent). Flowers 5 cm wide, dark pink, but soon fading, single, flowers only moderately every other year; fruits orange-red, 2.5 cm thick. 1930. ✿

'Sissipuk' (Preston) (*M. adstringens*; 'Niedzwetzkyana' descendent). Flowers only 3 cm wide, very dark red and attractive, eventually light pink, single, late; fruits purple, 2.5 cm thick. 1930. ✿

'Snowdrift' (Cole). Leaves tough, strong, green; flowers in masses, pure white, over a very long period, buds soft pink;

fruits about 10 mm thick, glossy orange-red. Origin unknown; first appeared in the Cole Nursery, Painesville, Ohio in 1965. ⊕ ⊕

'Sophia' (Wroblewski) (*M. purpurea*). Buds somewhat oblong, dark purple, flowers gradually turning pink, 4 cm wide, on long, thin stalks; fruits pea-sized, globose, purple, bluish pruinose. Developed in 1918 in Kornik, Poland. ⊕

'Spring Snow'. Seedling of and similar to 'Dolgo', but flowers better with larger flowers, white; fruitless! Discovered in 1963 as a 25-year-old tree in Parkside, Saskatchewan, Canada and introduced in 1965. ⊕ ⊕

'Striped Beauty'. Plants cultivated by this name in Europe are nearly always 'Gorgeous'! Both come from New Zealand but the former has red and yellow striped fruit.

'Sundog'. Tree, narrowly upright when young, opening more vase-like with age; flowers solitary, buds pink-red, opening pink, gradually fading with a trace of lilac; fruits dark scarlet-red, 2.5 cm wide. Developed in 1947 as *M. pumila* 'Niedzwetzkyana' × *M. baccata* in Morden, Manitoba, Canada. ⊕ ⊕

'Sunglory' see: **'Almey'**

'Szafer' (Wroblewski) (*M. purpurea*). Buds somewhat oblong, violet-pink, flowers pale lilac, to 5.5 cm wide, with a trace of pink, stalk 2.5 cm long; fruits pea-sized, globose, purple, glossy, 12 mm long. Developed in the Kornik Arboretum, Poland in 1938. ⊕

'Tanner's Variety' (Tanner). Flowers pure white, 4 cm wide; fruits red, 1.5 cm thick, persisting on the plant nearly the entire winter. Before 1931. The parentage attributed to *M. baccata* is uncertain since the ripe fruits become tough and the seeds differ in appearance. ⊕ ⊕

'Thunderchild' (a "Rosybloom" type). Foliage more green during the flowering period, dark coppery bronze after mid June; flowers carmine-pink; fruits purple-red, 2 cm. Percy Wright, Saskatchewan, Canada. Supposedly quite winter hardy. ⊕ ⊕

'Timiskaming' (Preston) (*M. adstringens*; 'Niedzwetzkyana' descendent). Foliage red in fall; flowers 5 cm wide, dark purple, single, flowers fully every other year, middle early; fruits 2 cm thick, purple. 1930. ⊕ ⊘

'Trail' (Saunders) (*M. pumila*). Flowers white, buds pink, 4 cm wide, attractive; fruits orange-red, 4 cm thick, edible. 1913. Very valuable. ⊕ ⊕ ✕

'Transcendent' (Downing) (*M. adstringens*). A large shrub or small tree, broad growing, twigs somewhat nodding; flowers white, 4 cm wide, buds pink; fruits obtusely conical, 3–4 cm high, somewhat angular, striped on a dark yellow background to nearly totally reddish, pulp dark yellow, firm, hard, sour, edible. Before 1844. ✕

'Van Eseltine' (G. P. van Eseltine). Very similar to *M. spectabilis*, but quite narrow growing, nearly columnar; flowers a strong pink, darker than those of *M. spectabilis*, with about 15 petals, about 4.5 cm wide, middle early; fruits flat-rounded, 2 cm thick, yellow with a large red cheek to totally yellow, with a larger calyx scar, without a thickened stalk. Developed in 1930; first introduced erroneously under the name 'Geneva' by the New York Agr. Exp. Station, Geneva, N.Y. in 1941. Hybrid between *M. arnoldiana* × *M. spectabilis*. Quite meritorious. ⊕

'Vanguard' (selected from seedlings of 'Hopa'). Erect habit, narrow, but often opening vase-form from the weight of the fruits on older trees; young leaves reddish, but soon green; flowers abundantly as a young plant, buds large, pink, opening light pink, single; fruits bright red, 2 cm, persisting into winter. Selected by Dr. E. Longley at the University of Minnesota, USA. ⊕ ⊕

'Veitch's Scarlet' (Veitch) (*M. pumila*). A large shrub, twigs very thorny; flowers white, 3.5 cm wide, buds pink; fruits ovate, 4.5 cm thick, bright red. Before 1905. ⊕

'Wabiskaw' (Preston) (*M. adstringens*; 'Niedzwetzkyana' descendent). Growth narrowly upright, strong growing, nearly columnar when young; leaves oval-elliptic, narrow, acute, reddish when young, later bronze or more green; flowers single to semidouble, dark reddish-purple, early, anthers golden-yellow; fruits slightly angular, carmine-red, somewhat yellow or brownish on the shady side, flat-rounded, 3 cm wide, bearing every year. BFC 174. 1930. ⊕ ⊕

'White Angel' (presumably a seedling of *M. sieboldii*). Twigs somewhat nodding from the weight of the fruits; flowers pure white, about 2.5 cm wide, in clusters of 5–6 together along the shoots, mid May; fruits 12 mm thick, scarlet-red, very numerous and persisting until spring. Discovered in 1955 in Beno's Nurseries, Youngstown, Ohio, USA; introduced in 1962. ⊕ ⊕

'Wierdak' (Wroblewski) (*M. purpurea*). Flower buds conspicuously oblong (!), deep purple, flowers 5–6 cm wide, lilac-pink, petals narrowly elliptic, 3 cm long, flower stalks 2–2.5 cm long; fruits globose to somewhat oblong, 1.5 cm thick, purple-red, glossy. Developed in the Kornik Arboretum in 1938. ⊕

'Wintergold' (Doorenbos) (*M. sieboldii*). Shrub to small tree, erect, twigs somewhat outspread; leaves broadly oval-elliptic, often 3–5 lobed, irregularly serrate; flowers white, buds pink, middle-early; fruits globose, 1.2 cm thick, golden-yellow, persisting until December. BFC 178. 1946. Plate 76, 117. ⊕

'Wisley Crab'. Similar to 'Niedzwetzkyana', but stronger growing, more erect; leaves large, bronze-red; flowers strong carmine-pink, 5 cm wide; fruits wine-red, similar to those of 'Aldenham Purple', large. Developed around 1924 in Wisley Gardens of the Royal Horticultural Society, England. ⊕ ⊕

'Wynema' (*M. soulardii*). Low, open shrub, 3–4 m high, twigs outspread; leaves green, large, obtusely lobed, especially on the long shoots; flowers soft pink, turning darker, late, 4 cm wide, flowers fully every other year; fruits green with a red cheek to totally red, 5 cm thick, edible. Discovered around 1920 in Iowa, USA; presumably a natural hybrid. Meritorious. ⊕ ⊕ ✕

The ornamental crab apples require the same culture as fruit apples; a good, fertile, medium heavy soil, in an open site and pruned occasionally to maintain the shape of the plant. They will benefit from regular fertilization and spraying for disease and insects as needed.

Lit. Asamai, Y.: The Crab Apples and Nectarines of Japan; 1–89, 1929 (pp. 1–50, fig. 1–50; in Japanese) ● Den Boer, A. F.: Ornamental Crab Apples; 226 pp., Chicago 1959 (ill.) ● Browicz & Bugala: The new varieties of *Malus purpurea* Rehd. bred in the Kornik's Arboretum by A. Wroblewski; in Act. Soc. Bot. Polon. **21**, 719 to 724, 1952 (in Polish) ● Seeliger, R.: Beobachtungen an *Malus*-Arten; in Mitt. DDG 1934, 1–22 ● Wyman, D.: Crab Apples for America; 1st ed., 81 pp., Jamaica Plain 1943; 2nd ed., 63pp., 1955 ● Wyman, D.: Crab Apples of merit; in Arnoldia 1959, 5–22 ● Grootendorst, H. J.: *Malus* — Sierappels; in Dendroflora 1, 3–15, 1964 ● Jefferson, R. M.:

History, Progeny, and Locations of Crab-apples of documented authentic origin; U.S. Nat. Arboretum Contr. **2**, 1–107, 1970 ● Browicz, K.: "*Malus florentina*", its history, systematic position and geographical distribution; Fragm. Floristica et Geobotan. Ann. **16** (I), 59–83, 1970 ● Huckins, C. A.: Flower and Fruit Keys to ornamental Crabapples cultivated in the United States (*Malus,* Rosaceae); Baileya **15**, 129 to 164 (with index).

MALVAVISCUS Adans. — MALVACEAE

Mostly glutinous shrubs or tall growing herbs; leaves alternate; flowers axillary, red, petals erect and inclined together or somewhat outspread on the apical half, style column very long and thin, widely exserted, with the anthers near the apex, the filaments connate and tubular around the styles; fruits berrylike at first and fleshy, later dry and disintegrating. — 3 species in Central and S. America.

Malvaviscus arboreus Cav. Shrub, to 3 m high; leaves broadly ovate, 6–11 cm long, 4 to 8 cm wide, mostly 3 lobed and coarsely dentate, base cordate, the lobes long acuminate, rough above, softly pubescent beneath; flowers usually solitary in the leaf axils, corolla bright red, 2.5 cm long and wide, petals folded in bud, opening totally later, calyx 2 cm long, flower stalk 3–7 cm long, summer and fall. BM 2305 (as *Achania malvaviscus*); BC 2312. Mexico. 1713. z10 ⊕

Good summer bloomer, but only totally winter hardy in frost free climates; needs a good soil, has no insect pests and flowers during the entire year.

MANDEVILLA Lindl. — APOCYNACEAE

Tall growing, climbing plants with a milky sap, shoots woody, leaves opposite, pinnately veined; flowers in simple, often one sided, dense or loose racemes, often reduced to only 2–3 flowers; corolla funnelform with a long, narrow tube and a broad, 5 lobed limb, yellow or white, seldom violet; calyx 5 parted, with a few glands on the inside or 5 scales; stamens 5, enclosed at the mouth of the tube; fruit usually composed of 2 follicles, these erect or outspread, 30–40 cm long, cylindrical or angular. — 114 species in Central and tropical South America, only a few of which are found in cultivation.

Mandevilla suaveolens Lindl. Deciduous climber, to 3 m high or more, young shoots very thin, glabrous, hollow; leaves cordate, 5–8 cm long, 3–5 cm wide, drawn out in a long, fine apex, entire, dark green above, glabrous, lighter beneath with tufts of pubescence in the vein axils, petiole 3–5 cm; flowers 3–6 in the leaf axils, corolla white to ivory-white, 5 cm long, 3 cm wide, the 5 ovate limb lobes overlapping, sweet smelling, calyx green, with 5 about 1 cm long awl-shaped lobes, June–September; follicles paired, 30–40 cm long, 6 mm thick. BM 3797 (= *M. tweediana* Gadeceau & Stapf). Argentina. 1837. z9 ⊕

M. laxa (Ruiz & Pavon) Woodson, from Peru, is different and preferably not included with the previous species, as recommended in Bean, II, p. 718.

MANGLIETIA Blume — MAGNOLIACEAE

Evergreen trees, closely related to *Magnolia,* but differing in the 6 or more ovules in each carpet as opposed to 1–2 in *Magnolia;* leaves often with a somewhat swollen petiole base; flower parts in 3–4 rings, sepals or petals usually 9, occasionally 10–11 or more; carpels 25–32. — About 30 species in SE. Asia to Sumatra.

Manglietia forrestii W. W. Sm. Small tree, to 6 m high, shoots short, thick, with ring-form leaf scars, buds cylindrical, with 2 hard, brown, pubescent scales; leaves oblong, ovate, quite variable in size, from 12–25 cm long and 6–10 cm wide, long acuminate, base narrowly cuneate, light green above, glossy, lighter beneath, leathery thick, glabrous on both sides, petiole base widened; flowers white, sepals and petals 4.5 cm long, 3.5 cm wide, fleshy; fruits large, globose, purple. E. Tibet; China, Yunnan Province; 1500 m. z9 # ⊕ ∅

M. hookeri Cubitt & W. W. Smith. Tree, to 15 m high, conical, young shoots thick, gray, glabrous, buds 2 cm long, thick, gray silky pubescent; leaves oblong, 15–30 cm long, 5 to 10 cm wide, acute to obtuse, entire, base narrowly cuneate, glossy dark green above, lighter beneath, leathery thick, glabrous on both sides; flowers solitary, terminal, 10 cm wide, sepals 3, obtuse, oblong, cream-white, petals 9, white, May; fruits oval, 6 cm long. Tibet; SW. China, 2100–2400 m. 1912. z9 # ⊕ ∅

M. insignis (Wall.) Bl. Tree, 12 m or taller, young shoots, buds and leaf petioles pubescent at first, later glabrous, shoots ringed at the buds; leaves oblong-elliptic, 10–20 cm long, 4–6.5 cm wide, acute to short acuminate, base narrowly cuneate, glossy green above, lighter beneath, somewhat bluish; flowers solitary, terminal, 7 cm wide, appearing after the leaves, sepals 3, petals 9, all obovate, white, fragrant, May; fruits like an elongated cone, about 9 cm long, 3 cm thick, purple. JRHS 89: 206; BMns 443. Himalaya; Upper Burma; China, Yunnan Province, 2400–2800 m. 1912. z9 # ⊕ ∅

Cultivated like *Magnolia;* prefers an acid soil.

MARGYRICARPUS Ruiz & Pavon — ROSACEAE

Low evergreen, densely branched, divaricate shrubs; leaves alternate, clustered, odd pinnate, the pinnae eventually abscising and leaving a thorned rachis; flowers small, inconspicuous, 1–3 axillary, without petals; calyx 3–5 parted; stamens 1–3, carpels 1, stigma brush-form; fruit a small berry or drupe. — 10 species in the Andes of S. America, from Ecuador southward.

Margyricarpus pinnatus (Lam.) E. Ktze. Evergreen shrub, scarcely over 30 cm high, twigs densely foliate; leaves 1–2 cm long, leaflets usually 9, linear, margins involuted, 3 to 10 cm long, leaf rachis only slightly thorny; dark red, April to summer; fruits white, 3–4 mm thick, persisting into winter (= *M. seticosus* Ruiz & Pavon). Habitat as above. 1829. z9 # ⚭ ✥ ○

Only for mild climates. Very attractive for a warm site in the rock garden. Needs a sandy-gravelly soil.

MARLEA See: ALANGIUM

MARSDENIA R. Br. — ASCLEPIADACEAE

Deciduous or evergreen, mostly tropical shrubs or subshrubs, with a milky sap, erect or twining; leaves opposite, usually broad; flowers in cymes; calyx cupulate, corolla campanulate, funnel-shaped or rotate, with 5 deep, erect sections; stamens 5, pollen in each pollen sac in a coherent mass (pollinium); stigma a follicle, seeds with hair fascicles. — About 5–10 species in the tropics and subtropics.

Marsdenia erecta (L.) R. Br. Deciduous shrub, twining to 8 m high in its habitat, scarcely over 1 m high in cultivation, shoots glabrous; leaves cordate-ovate, 4–8 cm long, entire, glabrous on both sides, blue-green above, more gray-green beneath; flowers white, at the branch tips, in corymbs, fragrant, inflorescences 3–7 cm wide, May–July; fruit capsule spindle-form, 5–7 cm long, seeds with 2.5 cm long hair fascicles (= *Cynanchum erectum* L.; *Cionura erecta* [L.] Griseb.). SE. Europe, Asia Minor. 1597. The milky sap of the plant is very poisonous. z7–8 Plate 129; Fig. 200. ✥

Rather hardy, but of little ornamental value. Needs a good garden soil and a sunny area.

Fig. 200. *Marsdenia erecta*. Twig, flower enlarged (without corolla), fruit capsule, seed (from Baillon, and Jacquin)

MAYTENUS Molina — CELASTRACEAE

Evergreen trees or shrubs; leaves alternate, often 2 ranked (distichous), petiolate, leathery, serrate; flowers axillary, small, solitary or in groups of a few, white, yellow or red; calyx 5 parted; petals 5, stamens 5, the latter inserted under the circular, undulate limbed disc; styles absent or columnar; fruit a capsule, dehiscing by 2–3 valves (like *Celastrus*). — About 225 species in S. America and the W. Indies.

Maytenus boaria Molina. Evergreen shrub, 2–3 m high (in its habitat a tree to 25 m), twigs thin, green, young shoots pendulous or nodding; leaves small, oval-lanceolate, 2.5–5 cm long, finely serrate, thin, leathery, glabrous, glossy; flowers greenish, tiny, inconspicuous, May; fruits with a red aril. BC 2340; PFC 47. Chile. 1822.

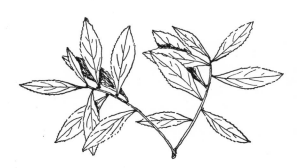

Fig. 201. *Maytenus boaria* (Original)

A valuable street tree in its habitat, with a very hard wood. z8–9 Fig. 201. # ⌀ ⚬

M. chubutensis (Speg.) Lourteiz, O'Donnell & Sleumer. Evergreen, low, dense shrub; leaves broadly ovate, 10–12 mm long, 6–8 mm wide, very thick, rough above from short, stiff hairs; flowers small, reddish, solitary or in clusters in the leaf axils. BMns 635. Argentina; Neuquen

Province. 1926. z9

M. magellanica (Lam.) Hook. f. Differing from *M. boaria* in the lower and narrowly upright habit; leaves broader, thicker, more dentate, leathery, 3–6 cm long, 18–30 mm wide; flowers mostly solitary in the leaf axils; fruits with an orange-yellow aril and a red seed coat. PFC 48; BMns 669. S. Chile. z9

MEDICAGO L. — LEGUMINOSAE

Mostly herbaceous, only occasionally shrubs; leaves small, 3 parted, petiolate, the leaflets finely dentate; flowers small, purple or yellow, in heads or short racemes or 1–2 axillary; stamens in 9's and 1's; fruits spiraling, not dehiscent, with one or several seeds.— About 50 species in Europe, Asia, and Africa. Only the following generally found in cultivation.

Medicago arborea L. Evergreen shrub, 1–2 m high,

densely foliate, twigs fine gray pubescent, wood black, hard; leaves 2.5–3 cm long, 3 parted, leaflets obovate, 6–18 mm long, silky pubescent beneath, finely toothed on the apex; flowers yellow, in short, axillary racemes, May–September; pods spiraled, with 1–2 seeds. SE. Europe. 1596. z9 Plate 125, 130. # ⚬

To be considered only for warm climates; not particular as to soil types.

MELALEUCA L. — MYRTACEAE

Evergreen trees or shrubs; leaves alternate, occasionally opposite, simple, lanceolate or linear, flat or nearly cylindrical, with 1–3 or many veins, with oil glands beneath on many species; flowers in cylindrical spikes or globose to ovate heads, very ornamental; flower tube hemispherical, sepals 5, corolla outspread and abscising, stamens numerous, in connate bundles of 5 opposite the petals (distinction from *Callistemon!*); fruit capsules densely clustered on the branches, surrounded by the woody flower tube. — About 100 species in Australia and Tasmania; only 1 species, *M. leucadendron*, distributed from Australia through the Pacific Islands to the East Indies and the Philippines.

Key to the mentioned species

● Inflorescence a cylindrical spike (seldom capitate like *M. decussata* and *M. ericifolia*);
   × Flowers red or scarlet:
         M. fulgens, hypericifolia, wilsonii

   ×× Flowers white, blue, lilac or pink;
    ○ Leaves usually opposite:
         M. decussata, linearifolia

    ○○ Leaves usually alternate or spiraled;
     + Leaves usually longer than 12 mm:
         M. leucadendron, styphelioides

     ++ Leaves usually shorter than 12 mm:
         M. ericifolia, huegelii

●● Inflorescence terminal, globose or capitate, the rachis occasionally growing past the flower:
         M. nesophila

Melaleuca decussata R. Br. Broad, open shrub, to 6 m high; leaves mostly opposite, lanceolate to oblong, 6–12 mm long, very densely arranged (up to 28 in 2.5 cm), directed upward or curved inward; flowers lilac, in cylindrical 2.5 cm long spikes or, in globose 12–15 mm wide heads, stamens purple, in fascicles of 10–15,

connate only for a short length at the base, June. BM 2268; BC 2346. Australia. z9 # ☙

M. ericifolia Sm. Shrub, 1.5–3 m, also a small tree in its habitat, glabrous or pubescent, bark very thick and quite soft; leaves alternate, narrowly linear, 6–12 mm long, rather obtuse, usually with the apical half recurved; flowers yellowish-white, in 1.3–2.5 cm long spikes, 1 cm wide, rachis soon elongating, the sterile flowers in terminal, globose heads, stamens 6–8 mm long; fruits with a wide base, partly imbedded in the bark. Australia, Tasmania. 1790. z9 # ☙

M. fulgens R. Br. Open growing, glabrous shrub, to 1.5 m high or more; leaves mostly opposite, linear to somewhat lanceolate, acute, furrowed above with dark glands, 2–3 cm long, 1.5–2 mm wide; flowers in attractive, 3.5–7 cm long spikes, bright red, stamen bundles 3 cm long, June. BR 103. W. Australia. z9 Plate 123. # ☙

M. huegelii Endl. Shrub, erect, 1.5–3(4.5) m high, stiff, nearly glabrous; leaves spirally arranged and overlapping, sessile, appressed, ovate to lanceolate, acute, 3–5 mm long, striped with 3–7 veins; flowers white, buds occasionally pink, in slender, cylindrical, 2.5–12 cm long spikes, the rachis growing on before the flowers open, stamens 7–11 in a bundle; fruits globose. W. Australia. z9 # ☙

M. hypericifolia (Salisb.) Smith. Tall, glabrous shrub, 1.5–6 m high; leaves opposite, very densely arranged, oblong to lanceolate, 2 to 4 cm long, obtuse or acute, midrib raised beneath, with glands; flowers in dense, 3.5–7 cm long and 3–5 cm wide spikes, bright red, stamens red, 2 cm long, May–October. BC 2346. New South Wales, Australia. 1792. z9 # ☙

Plate 113

Mahoberberis. *M. aquisargentii* (above) *M. aquicandidula* (middle beneath) *M. neubertii* (right)

Mahonia nevenii in the Dublin Botanic Garden, Ireland

Plate 114

Mallotus japonica
in the Suchumi Botanic Garden, USSR

Mahoberberis miethkeana
in the Hillier Arboretum, England

Mahonia japonica in Ashbourne, Ireland

Plate 115

Malus × *scheideckeri*
Photo: Archiv D.B.

Mahonia lomariifolia
in Royal Botanic Gardens, Kew, England

Mahonia. Large leaf *M.* × 'Heterophylla';
small leaf *M. repens*

Plate 116

Malus ✕ *purpurea* 'Aldenhamensis'
Photo: Archiv D.B.

Malus ✕ *zumi*
Photo: Archiv D.B.

Plate 117

Malus. Leaves. a. *M. sargentii;* b. *M. × scheideckeri;* c. *M. sieboldii* 'Wintergold'; d. *M.* 'Hilleri'; e. *M. floribunda;* f. *M. fusca.* — **Mahonia.** g. 2 leaflets of *M. bealei;* h. *M. fortunei;* i. 2 leaflets of *M. japonica* (material mostly from wild plants)

Plate 118

Malus ✕ *zumi* in RHS Gardens, Wisley, England

Malus ✕ *magdeburgensis*
in Royal Botanic Gardens, Kew, England

Malus 'Katherine'
Photo: Arnold Arboretum, USA

Plate 119

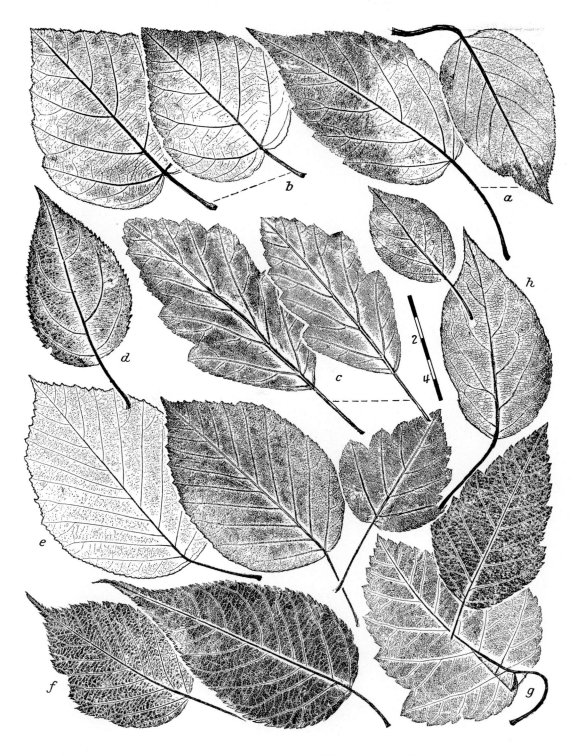

Malus. Leaves. a. *M. yunnanensis;* b. *M. yunnanensis* var. *veitchii;* c. *M. ioensis* 'Plena'; d. *M. sikkimensis;* e. *M. tschonoskii;* f. *M. prattii;* g. *M. coronaria;* h. *M. rockii* (material mostly from wild plants)

Plate 120

Malus ioensis 'Plena'
Photo: Archiv D.B.

Malus × *hartwigii*
Photo: Archiv D.B.

Plate 121

Malus. Leaves. a. *M. ioensis;* b. *M. florentina;* c. *M. baccata;* d. *M. sieboldii;* e. *M. transitoria;* f. *M. toringoides;* g. *M. glabrata;* h. *M. kansuensis;* i. *M. hupehensis;* k. *M. angustifolia;* l. *M. halliana* 'Parkmanii'; m. *M.* × *platycarpa* (material mostly from wild plants)

Plate 122

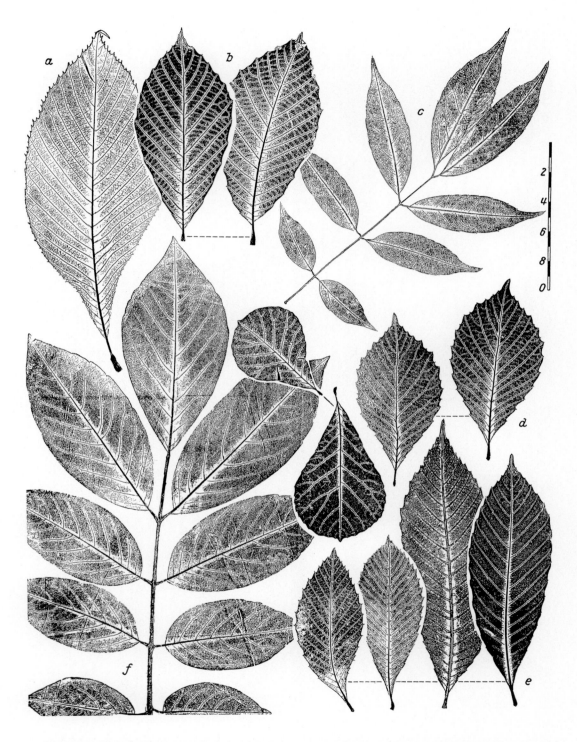

Meliosma. a. *M. dilleniifolia;* b. *M. flexuosa;* c. *M. oldhamii;* d. *M. dilleniifolia;* e. *M. cuneifolia;* f. *M. veitchiorum* (The large leaf & both small leaves in the middle are from *M. parviflora.*) (material from Nymans and Kew Gardens, England)

Plate 123

Melianthus major
in its native habitat in South Africa

Melaleuca fulgens
Photo: Australian News & Information Bureau

Myrica rubra
in the Forestry Trial Station, Kyoto, Japan

Meliosma veitchiorum
in Royal Botanic Gardens, Kew, England

Plate 124

Menispermum canadense
in the Malonya Arboretum, Czechoslovakia
Photo: Dr. Pilat, Prague

Microglossa albescens
in Royal Botanic Gardens, Kew, England
Photo: Dr. Pilat, Prague

Mespilus germanica in its native habitat in the Strandsha Mts., Bulgaria
Photo: K. Browicz

Plate 125

Medicago arborea
in a park on Mallorca, Spain

Musa basjoo
on Mainau Island, W. Germany

Myrtus luma
at Dunloe Castle, Ireland

Nicotiana glauca
in a garden in Spain

Plate 126

Nyssa sinensis
with fall foliage in the Hillier Arboretum, England

Nothofagus pumilio
in Nymas Gardens, England

Paeonia suffruticosa 'Rock's Var.' and *Paeonia suffruticosa* 'Argosy'
in an English garden

Plate 127

Pandorea pandorana
in a garden on Mallorca, Spain

Phaedranthus buccinatorius
in a Spanish garden

Penstemon davidsonii

Penstemon laetus var. *roezlii*
both in the Royal Botanic Garden, Edinburgh, Scotland

Plate 128

Pieris japonica 'Flamingo'
in a nursery in Boskoop, Holland

Photinia fraseri 'Robusta'
in the Hillier Arboretum, England

Polygala chamaebuxus var. *grandiflora*
in the Pygmy Pinetum, Devizes, England

Pieris 'Wakehurst'
in the Hillier Arboretum

M. leucadendron L. Cajuput Tree (*kaju* is Malayan for wood; *putih* is white, for the white, peeling bark). Large tree, trunk with very thick, spongy, brownish-white bark exfoliating in layers, shoots glabrous or silky pubescent; leaves elliptic to oblong, acutely tapering at both ends, 5–10 (occasionally to 15) cm long, 12–30 mm wide, with 3 to 7 parallel veins; flowers cream-white, in 3 to 10 cm long, 3 cm wide spikes, the rachis growing on after flowering, stamens 12 mm long, June–October. Australia to Malaysia. 1796. The indigenous peoples of its habitat have many uses for the bark; cajuput oil, obtained from the leaves, is medicinal. Ⓕ Indonesia. z9 # ⌀

M. linearifolia Smith. Tall shrub, a tree in its habitat, 15–18 m high, young shoots and inflorescences pubescent; leaves opposite, broadly linear, stiff, long acuminate, mostly 2.5–3 cm long, 2–3 mm wide, midrib raised on the underside; flowers paired in 2.5–3 cm long spikes, white, stamens in 12 mm long fascicles, rachis continues to grow, July. LAu 94; BM 9493; MCL 26. Australia. z9 # ⌀

M. nesophila F. v. Muell. Glabrous shrub, to a 10 m high tree in its habitat, bark thick, spongy, exfoliating in wide strips; leaves alternate, thick, obovate-oblong, obtuse or with a mucro, 12–30 mm long, 6 mm wide, indistinctly 1–3 veined; flowers in about 3 cm wide heads, pink, May–September. LAu 139. Australia. z9 # ⌀

M. styphelioides Sm. A stately tree, with a thick, papery bark, 5–10 m high, 15–25 m high in its habitat, young shoots and inflorescences silky pubescent; leaves ovate, sessile, with a broad base, stiff, acuminate, often somewhat twisted, 12–20 mm long, 6 mm wide, many veined; flowers cream-white, in dense, 3–5 cm long spikes, stamens in 6 mm long bundles, May–July; fruits globose, with persistent calyx teeth. New South Wales, Australia. 1793. z9 # ⌀

M. wilsonii F. v. Muell. Erect shrub, usually not over 1.5 m high, glabrous or slightly pubescent; leaves opposite, imbricate on the twigs, linear-subulate, 6–12 mm long, stiff, acute, mostly erect; flowers solitary or in loose, small clusters along the shoot, red, stamens in bundles of 15–20, red or pink, 12 mm long, June to July. BM 6131. Australia. z9 # ⌀

Evergreen shrubs for warm climates, full sun, acid soil; usually very well suited for coastal areas. *M. leucadendron* and *M. ericifolia* are particularly salt tolerant.

MELIA L. — MELIACEAE

Deciduous or semi-evergreen trees or shrubs; leaves alternate, bipinnate, leaflets entire or serrate, stellate tomentose when young; flowers in large, usually axillary panicles, hermaphroditic, attractive; calyx 5 to 6 parted, corolla with 5–6 distinct petals, overlapping in bud; stamens 10–12, the filaments connate forming a narrow tube; ovaries 5–8 locular, style slender, stigma knob-like; fruit a drupe, surrounded by a fleshy hull. — About 10 species in S. Asia to Australia.

Melia azadirachta L. Medium high, tropical tree, 5–12 m high, bark very bitter (with medicinal properties); leaves simple pinnate, to 40 cm long, with 9–15 leaflets, these oval-lanceolate, long acuminate, 3–8 cm long, uneven sided, 1–3 cm wide, dentate or lobed, glabrous; flowers in axillary, 15–25 cm long panicles, white, with a honey-scent, 12 mm wide, petals ciliate, May–June; fruit ellipsoid, 1.5–2 cm long, purple (= *Azadirachta indica* A. Juss.; *Antelaea azadirachta* [L.] Adelbert). E. Indies. Ⓕ Indies, Sudan, Cameroon, Ghana, Chad, used in reforestation. z10 ⌀ ⌀

M. azedarach L. Chinaberry, Bead-tree. Deciduous tree, to about 15 m high with furrowed bark, young shoots green to reddish, glabrous; leaves 25–35 cm long (to 80 cm long in mild areas!), the leaflets ovate to elliptic, 2–5 cm long, acute, sharply serrate to lobed, dark green above, lighter beneath, glabrous; flowers in 20–25 cm long, loose panicles, corolla lilac, 2 cm wide, the filament tube violet, style violet, stigma yellow, May; fruits globose, light yellow, 1.5 cm thick. BM 1066; LF 197; HM 1719. Himalayas. In cultivation for centuries. Ⓕ In China, Indies, S. and Central America, widely used in reforestation. z9 ⌀ ⌀

'Umbraculifera'. Crown flat spreading; leaves smaller and narrower. GF 7: 20; BC 2348. Allegedly found near San Jacinto, Texas, but the origin is unknown. A widely used street tree in Texas, USA. ⌀

Very fast growing trees with attractive flowers and fruits. Tolerant of low temperatures, but not below freezing! No soil preference.

Lit. Candolle, D. C.: Meliaceae; in Monogr. Phanerog. I, 399-752, 1878 ● Harms, H.: Meliaceae; in Engler & Prantl, Nat. Pfl. Fam., 2nd ed., 19b: 1–172, 1940.

MELIANTHUS L. — Honeybush — MELIANTHACEAE

Evergreen subshrubs; unpleasantly scented when crushed; leaves alternate, odd pinnate, with stipules; flowers in terminal and axillary racemes, with 5 sepals and petals, secreting a great deal of nectar, 4 stamens; calyx laterally compressed, with or without a sack-like bulge at the base, interior with nectaries; fruit a papery, 4 winged capsule, seeds black, glossy — 6 species in S. Africa.

Melianthus major L. Evergreen, large leaved subshrub, spreading habit, 1.5 m high or more, shoots hollow; leaves gray-green, 30–45 cm long, leaflets 9–11, very coarsely serrate, 7–10 cm long, 5 cm wide, rachis winged between the leaflets, base with stem-clasping, connate stipules; flowers very densely arranged, dark red-brown, 3 cm long, in terminal racemes, about 30 cm long or longer, summer flowering. BR 45. S. Africa. Found in many Botanic Gardens. Outstanding plant with beautiful foliage, but hardy only in the mildest areas. z10 Plate 123. # ⌀ ⌀

MELICYTUS J. R. & G. Forst. — VIOLACEAE

Shrubs or small trees; leaves alternate, dentate; flowers often unisexual, small, regular, in few flowered clusters at the nodes; sepals, petals and stamens each 5, anthers distinct, the subtending connective is developed as a wide membrane with a scale on the back side; fruit a berry — 6 species, all in New Zealand.

Melicytus ramiflorus J. R. & G. Forst. Deciduous, tall shrub or also a small tree, twigs glabrous; leaves oblong or more lanceolate, 5–15 cm long, 2–5 cm wide, coarsely serrate, tapering to both ends, dark green above, petiole 1.5–2 cm long; flowers 3–9 in clusters at the nodes of the previous year's growth, 5 mm wide, yellowish-green, flower stalk 5 to 10 mm long, June; fruits violet, 5 mm thick, globose. BM 8763; DRHS 1281. New Zealand. z9 ⚘

MELIOSMA Bl. — SABIACEAE

Deciduous or evergreen trees or shrubs; leaves alternate, simple, with numerous parallel straight vein pairs, or odd pinnate with opposite leaflets; flowers hermaphroditic, small, but in large, axillary or terminal panicles; sepals 5, occasionally only 4; petals 5, uneven, the 3 outer ones large, nearly circular and concave, the 2 inner ones often reduced to scales; fruit normally a single seeded drupe, black or red. — About 100 species, most in tropical E. and S. Asia and S. America; the winter hardy species are all deciduous and indigenous to China and Japan.

- Leaves Pinnate:
    M. beaniana, oldhamii, veitchiorum
- • Leaves simple:
    M. cuneifolia, dillenifolia, flexuosa, myriantha, parvifolia, tenuis

Meliosma beaniana Rehd. & Wils. Tree, 10–20 m high in its habitat, young shoots somewhat brown tomentose at first; leaves pinnate, 15–30 cm long, with (5) 9 (13) leaflets, these ovate to elliptic-lanceolate, acuminate, tapered to the base, sparsely and finely dentate to entire, the basal pair being the smallest, the more apical leaflets becoming larger, 5–12 cm long; flowers cream-white, in axillary (!!), 10–16 cm long, outspread to pendulous panicles, appearing before the leaves, May; fruit globose, black, 6 mm thick. ICS 3196. China; Hupeh, Szechwan Provinces. 1907. z8 ∅

M. cuneifolia Franch. Shrub, to about 5 m high, twigs erect, glabrous or nearly so; leaves simple, 7–17 cm long, 3–7 cm wide, obovate, acute or abruptly short acuminate, narrowly cuneate at the base, with 20–25 straight vein pairs, sinuate, rough to the touch above, light green beneath with pubescent venation; flowers yellowish-white at first, soon becoming pure white, very fragrant, in erect panicles, to 25 cm high and equally wide, July; fruits globose, black, 6 mm thick. BS 2: 315; BM 8357; ICS 3193. W. China. 1905 z9 Plate 122; Fig. 203. ⊕

M. dilatata see: **M. parviflora**

M. dilleniifolia (Wight & Arn.) Walp. Small tree, to 6 m high, shoots pubescent; leaves obovate to oblanceolate, 15–30 cm long, awn-like serrate, slightly rough above, finely pubescent beneath; flower panicles 20–30 cm long, bracts abscising. Himalayas. 1924. z9 Plate 122. ∅ ⊕

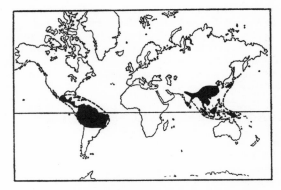

Fig. 202. Range of the genus *Meliosma*
(from Van Beusekom)

M. flexuosa Pamp. Shrub, erect, to 4 m high, twigs gracefully arching, young shoots reddish, pubescent; leaves simple, mostly obovate to elliptic, usually long acuminate, gradually tapering to the base, 5–15 cm long, with 12–20 vein pairs, sparsely pubescent above, especially on the midrib, lighter beneath and more pubescent; flowers in terminal, pendulous, 10–20 cm long, panicles, white, fragrant, July. ICS 3194 (= *M. pendens* Rehd. & Wils.). China; W. Hupeh Province. 1907. Similar to *M. cuneifolia*, but the leaves without hair fascicles in the vein axils beneath. z8 Plate 122. ∅ ⊕

M. myriantha S. & Z. Shrub to a small tree, to 5 m high, spreading habit; leaves simple, elliptic-lanceolate, 7–20 cm long, short acuminate, sharp and regularly dentate, base broadly cuneate to round, with 24–30 vein pairs, midrib and petiole red-brown pubescent, without pubescent tufts in the vein axils; flowers greenish-yellow, in terminal panicles, 15–20 cm long and equally wide, with ascending branches, the rachis brown pubescent, individual flowers very small, very numerous, fragrant, June–July; fruits globose, 5 mm thick, red. LVP 197; GC 31: 30; KIF 4: 26; ICS 3192 (= *M. stewardii* Merrill). Japan. z9 ∅ ⊕

M. oldhamii Miq. A tree in its habitat, 15(20) m high, but usually much lower, young shoots glabrous, gray; leaves pinnate, 15–30 cm long, leaflets 7–13, elliptic-oblong to oblong-lanceolate, 4–10 cm long, the basal pair about 2.5 cm long, 2 cm wide, oval, the apical pair largest at 7–12 cm long, 3–5 cm wide, acuminate, sparsely finely serrate, rather glabrous; flowers in terminal panicles, these 20–30 cm high and wide, abundantly branched, finely pubescent, pure white, July. BS 1: Pl. 28; ICS 3198 (= *M. rhoifolia* Maxim.; *M. sinensis* Nakai). Central China to Korea. z9 Plate 122. ∅ ✣

M. parviflora Lecomte. Deciduous tree, about 8 m high in its habitat, shoots covered with short, erect hairs; leaves obovate, 6–8.5 cm long, 2.5–4 cm wide, broadly truncate on the apex and with a short, protruding tip, drawn out to a very narrow, cuneate base, both sides glabrous, with 8–12 vein pairs, branching before they reach the leaf margin and terminating in the acute marginal teeth, both sides nearly glabrous, petiole 12 mm long, with erect, brown pubescence; flowers very small, white, in sparsely branched, terminal or axillary panicles, 20—30 cm long, August; fruits red, globose, 5 mm thick. ICS 3190 (= *M. dilatata* Diels; *M. parvifolia* Hort. Kew.). Western and Central China. 1936. z9 Plate 122. ∅

M. pendens see: **M. flexuosa**

M. rhoifolia see: **M. oldhamii**

M. sinensis see: **M. oldhamii**

M. stewardii see: **M. myriantha**

M. tenuis Maxim. Shrub, young shoots reddish at first, winter buds elongated; leaves simple, ovate to obovate, abruptly short acuminate, 5–10 cm long, dark green above with scattered pubescence, lighter beneath and occasionally pubescent on the 10–15 vein pairs; flowers yellowish-white, in loose, slender panicles, about 15 cm long, July–August; fruits 5 mm thick, black, reddish pruinose at first. Japan. 1915. z9

M. veitchiorum Hemsl. Tree, 7–10 m high, branches thick, stiff, upright, young shoots pubescent at first, soon becoming glabrous, punctate; leaves pinnate, 40–80 cm long, leaflets 9–11, ovate to oval-oblong, 8–15 cm long,

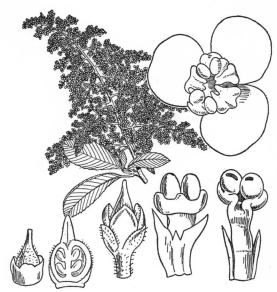

Fig. 203. *Meliosma cuneifolia*,
flower and flower parts
(from B.M.)

half as wide, obtuse to short acuminate, entire or (rarely) somewhat crenate, glabrous, only the midrib pubescent beneath; flowers yellow, 5 mm wide, in pendulous, 40 cm long, 20 cm wide, loose panicles, May, fragrant; fruits globose, 6 mm thick, black. BS 2: Pl. 29; ICS 3199. Central China. 1901. z8–9 Plate 122, 123. ∅ ✣

Only for milder climates if not noted otherwise; needs a good garden soil and a protected, but sunny area.

Lit. How, F. C.: Revision of the Chinese *Meliosma*; in Acta Phytotax. Sin. 1955, 415–452 ● Van Beusekom, C. F.: Revision of *Meliosma* (Sabiaceae), Section *Lorenzanea* excepted, living and fossil, geography and phylogeny; in Blumea **19**, 355–529, 1971.

MENISPERMUM L. — Moonseed — MENISPERMACEAE

Deciduous, twining shrubs; leaves alternate, long petioled, 3–4 lobed, peltate; flowers dioecious; male plants: sepals 6, petals 6, involuted on the sides, stamens 12–24, filaments distinct from the base up or at the apex, erect, outspread or bowed inward; female plants: sepals 4–8, petals 4–8, shorter; style attached very near the base of the fruit, stigma sessile, carpels 2–5, fruits 2–3 compressed together, kidney- to horseshoe-shaped, laterally indented. — 2 species in N. America and from Siberia to Japan.

Menispermum canadense L. Twining to 4 m high, young shoots pubescent (!); leaves oval-rounded, 10–15 cm long, pentangular, soft pointed, obtusely lobed, petiole 5–15 cm long, attached at the blade base or slightly indented, base often somewhat cordate; flowers yellow-

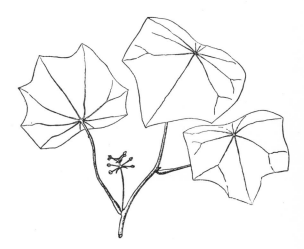

Fig. 204. *Menispermum dauricum* (Original)

green, in loose, 2–6 cm long panicles, stalk about 7 to 10 cm long, May–June; fruits 8 mm thick, blue-black, like small bunches of grapes, but poisonous! BM 1910; BC 2357; GSP 137. N. America, along river banks and forests. z5 Plate 124.

M. dauricum DC. Twining to 3 m high, with stolons, very similar to the previous species, but young shoots glabrous, oval-rounded, 6–12 cm long, totally glabrous, more or less 3 to 5 lobed, bluish beneath, petiole

attached shield-like (peltate), usually about 1 cm from the margin (!); flowers in small, umbellate, short stalked panicles, June; fruits about 1 cm thick, blackish, in small clusters. GF 5: 233; BC 2358; DL 1: 48. E. Asia. 1883. z5 Fig. 204.

Easily cultivated, but of little garden merit; well suited as a foliage screen. Prefers a moist soil. Both species quite hardy.

Lit. Diels, L.: Menispermaceae; in Engler, Pflanzenreich **46**, 1–345, 1910.

MENZIESIA J. E. Sm. — ERICACEAE

Low, deciduous shrubs, very closely related to *Rhododendron*, but with 4 parted flowers; leaves alternate, petiolate, entire; flowers 4(5) parted, long stalked, in terminal clusters, appearing with the flowers; corolla campanulate to urceolate; stamens 5–10, occasionally exserted, without an appendage, drawn out in a tube, anthers opening with a slit on the apex; ovaries 4(5) locular, many seeded; fruit a 4–5 valved, leather capsule; seeds linear, acuminate on both ends or caudate.— 7 species in E. Asia and N. America.

Menziesia ciliicalyx (Miq.) Maxim. Low shrublet, 30–60 cm high, rarely higher; leaves ovate to obovate, tapering to both ends, with a small tip, 2–5 cm long, margins and midrib bristly; flowers in pendulous, umbellate clusters

at the ends of the previous year's shoots, tubular to urceolate, pink to red, with a lilac bloom when ripe, limb often glandular, with 4–5 sections, calyx undulate, not lobed, May to June. NT 1: 46; BMns 35 (the non-glandular form); MJ 747; JRHS 92: 37. Japan. 1915. z6 Plate 129; Fig. 205. ⊕ ◑

var. **multiflora** (Maxim.) Mak. Taller, 0.7 to 1 m high, but very slow growing; calyx with distinct, but also very small, triangular lobes, flowers grouped 6–10 together, nodding to pendulous, corolla pink to nearly cream-white, flower stalks glabrous. GC 104: 457 (= *M. multiflora* Maxim.). Japan. 1915. ⊕ ◑

var. **purpurea** Mak. Leaves partly glabrous above, partly long haired; flowers purple-pink, slightly constricted at the mouth, buds blue (!), flower stalk hairs partly without glands. BS 2 Pl.

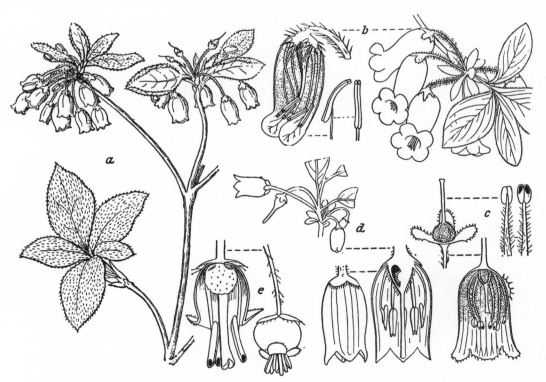

Fig. 205. **Menziesia.** a. and d. *M. pilosa*; b. *M. ciliicalyx*; c. *M. purpurea*; e. *M. pentandra* (from Guimpel, Maximowicz, Hort. Then.)

100 (= *M. lasiophylla* Nakai). Usually confused with *M. purpurea* Maxim., which see. The most attractive of the entire genus. ✛

M. ferruginea Sm. Shrub, about 1(2) m high, more or less upright; leaves elliptic to obovate, 2–5 cm long, upper surface with rust-brown appressed pubescence, glandular ciliate, less pubescent beneath; flowers less numerous, in terminal clusters of 2–5, nodding, corolla ovate-urceolate, whitish with pink, limb 4 parted, outspread, calyx with acuminate, pubescent lobes, June. DL 1: 258. Western N. America, Alaska to Oregon. 1811. z6 Plate 129. ◖

var. **glabella** (Gray) Peck. Rarely over 1 m high; leaves obtusely obovate, finely crenate and ciliate, only slightly pubescent to nearly glabrous, 2.5–6 cm long; flowers in small, pendulous clusters at the end of the previous year's shoots, corolla ovate-campanulate, 7–9 mm long, cream-white, limb 4 parted, calyx lobes acutely ovate, ciliate, May–June. BB 2753; RWF 261 (= *M. glabella* Gray). N. America, British Columbia to Idaho. 1885. z5 Plate 129. ◖

M. glabella see: **M. ferruginea** var. **glabella**

M. globularis see: **M. pilosa**

M. lasiophylla see: **M. ciliicalyx** var. **purpurea**

M. multiflora see: **M. ciliicalyx** var. **multiflora**

M. pentandra Maxim. Upright shrub, 0.7(1.5) m high; leaves elliptic to oblong, tapered to both ends, 1.5–4 cm long, acute, margins ciliate, bristly above, less bristly beneath; flowers in small umbels, pendulous, corolla globose urceolate, 6 mm long, greenish-white, limb 5 lobed, stamens 5, calyx lobes 5, ciliate, May–June. NT 1:

22; MJ 748. Sachalin, Japan. 1905. z6 Plate 129; Fig. 205. ✛ ◖

M. pilosa (Michx.) Juss. Upright shrub, 1 to 2 m high, azalea-like in habit, twigs somewhat pubescent; leaves elliptic to obovate, 2 to 5 cm long, abruptly finely acuminate, ciliate, slightly appressed pubescent above, bluish and less pubescent beneath; flowers few, pendulous, campanulate, 6–7 mm long, yellowish-white to orange-pink, on a glandular pubescent, 1–2 cm long stalk, corolla limb 4 parted, stamens 8, enclosed, style as long as the corolla, May–June; fruits capsule glandular bristly. BB 2752; NBB 3: 12; BS 2: 733 (= *M. globularis* Salib.). Eastern N. America, in the mountain forests from Pennsylvania to Alabama. 1806. Quite variable! z5 Plate 129; Fig. 205. ✛ ◖

M. purpurea Maxim. Upright, broad growing shrub, 0.7(1.5) m high; leaves oval-elliptic to oblong or obovate, obtuse and with fine tips, tapering to the base, 3–4 cm long, sparsely strigose pubescent above, as is the midrib beneath; flowers in numerous, nodding clusters, corolla tubular-campanulate, 12–15 mm long, red to purple, limb tiny, never constricted, 4 parted, calyx lobes oval-oblong, glandular ciliate, May–June. NT 1: 44. Japan. 1915. Very rare! Usually confused with *M. ciliicalyx* var. *purpurea*, which is more commonly found in cultivation. z6 Plate 129; Fig. 205. ✛ ◖

Cultural requirements exactly like that of azaleas; a moist, acid soil, fertile with much humus, preferably somewhat shaded; also for equivalent site in the rock garden. All species have good winter hardiness.

MERATIA See: **CHIMONANTHUS**

MESPILUS L. — Medlar — ROSACEAE

Monotypic genus; tree-like, thornless or rarely a thorny shrub, twigs light brown tomentose when young; leaves alternate, very short stalked, oblong; flowers large, usually solitary, terminal on short shoots, large, white, appearing after the leaves; calyx usually with bracts at the base, calyx lobes to 4 times as long as the calyx cup; petals 5; stamens numerous, usually 25–40; fruits attractive, brown when ripe, with a cupulate disk at the apex; the seed pit completely covered with fruit pulp at the apex.

Mespilus germanica L. Medlar. Shrub to small tree, 2–5 m high; leaves oblong-lanceolate, 6–12 cm long, short acuminate, dull green above, tomentose beneath, margins glandular and finely serrate, fall color yellow and red-brown; flowers 1–2 on short stalks, white, 4–5 cm wide, calyx tomentose, May–June; fruits top-shaped,

dirty brown-green at first, later brown, crowned with the 5 persistent, leaflike and inward inclined calyx lobes, edible only after frost, then sweet. HM 1963. SE. Europe: Pontus, Armenia, Transcaucasia, N. Persia. In cultivation for centuries. z6 Plate 124. ✛ ⚭ ✄

'**Apyrena**'. Seedless *Medlar*. Fruits without a seed pit. Supposedly found in the 18th century on the Balkan Peninsula. ⚭ ✄

'**Macrocarpa**'. Large fruited cultivar, fruits abundantly, fruits 3–4 cm thick, with a seed pit. DB 6, 132. ⚭ ✄

Refer to the literature for the large fruited cultivars.

Prefers a fertile, clay soil, with sufficient moisture and full sun to light shade.

Lit. Evreinoff, V. A.: Notizen über Ursprung, Biologie und Sorten der Mispel; in Deutsche Baumschule 1954, 260–265.

METAPLEXIS R. Br. — ASCLEPIADACEAE

Deciduous, twining shrubs; leaves opposite, simple; flowers in long stalked corymbs arising from between the leaf bases; calyx 5 parted; corolla nearly rotate, with 5 lobes pubescent on the interior; corona simple, with 5, short convex lobes alternating with the stamens; filaments fused into a tube; anthers each with 10 short stalked, pendulous pollen masses; fruit a follicle. — 6 species in E. Asia.

Metaplexis stauntonii Roem. & Schulte. Deciduous, twining shrub, shoots thin, pubescent when young; leaves ovate, base deeply cordate, long acuminate, 5–11 cm long, petiole 3–7 cm long; flowers a turbid pink-white, about 12 mm wide, in 3–15 cm long, often also nearly umbellate panicles, long stalked, July–September; fruit capsules to 10 cm long, seeds with hair fascicles. MJ 611 (= *M. japonica* Mak.) China, Japan. 1862. z6

METROSIDEROS Banks ex Gaertn. — Iron Tree — MYRTACEAE

Evergreen trees or shrubs, some climbing by means of aerial roots; leaves opposite, glandular punctate, leathery, flowers mostly in terminal, rarely axillary, racemes or cymes, petals 5, very small, round, outspread, stamens numerous, 2.5 cm long or longer, red, carmine or white, sepals 5, obtuse, wide; fruit a 3 locular, leathery capsule. — 60 species in S. Africa, Australia, New Zealand, Polynesia and E. Malaysia. Differing from the similar genus *Callistemon* in the opposite leaves.

Metrosideros citrinus see: **Callistemon citrinus**

M. diffusa (Forst. f.) Smith. Evergreen shrub with aerial roots, climbing into the crowns of tall trees in its habitat, in open areas a small, compact, stiff, creeping shrub, bark rough, shoots quadrangular; leaves oblong-elliptic, 1.5–3 cm long, 1–1.5 cm wide, thin, glossy, nearly sessile; flowers mostly in terminal, small racemes, about 3 cm wide, petals pink, stamens carmine, anthers yellow, April–May. BM 8628; HI 569. New Zealand. 1910. z9 # ✿

M. excelsa Sol. ex Gaertn. A broad, 9–20 m high tree in its habitat, often only shrubby in cultivation, shoots white tomentose; leaves quite variable, from lanceolate to broadly oblong, 3 to 10 cm long, dark green and glossy above, limb involuted, with white pubescence beneath; flowers in terminal, 5–10 cm wide cymes, bright carmine, buds white, woolly, petals small, stamens about 3 cm long, scarlet-red, flowering in winter (hence the New Zealand name "Christmas Tree"). BM 4488; MCL 27 (= *M. tomentosa* A. Rich.). New Zealand. 1840. z9 # ✿

M. florida see: **M. robusta**

M. linearis see: **Callistemon linearis**

M. lucida see: **M. umbellata**

M. robusta. A. Cunn. Evergreen, to 30 m high forest tree in its habitat with an irregular stem, often only a shrub to 1.5 m high in cultivation, shoots softly pubescent; leaves elliptic to more lanceolate, obtuse, 2.5–3.5 cm long, base cuneate, leathery, glabrous, with 1 pair of marginal veins; flowers in many, large, terminal cymes, the solitary flowers small, scarlet-red, as are the 2.5 cm long stamens, May. JRHS 93: 162; BM 4471 (incorrectly as *M. florida*). New Zealand. 1845. Called "Rata" in New Zealand. z9 # ✿

M. salignus see: **Callistemon salignus**

M. semperflorens see: **Callistemon citrinus**

M. speciosa see: **Callistemon speciosa**

M. tomentosa see: **M. excelsa**

M. umbellata Cav. Evergreen tree, 9–18 m high in its habitat, but often also only a shrub, 1.5–6 m high, young shoots and leaves appressed silky pubescent, later glabrous; leaves lanceolate to narrowly elliptic, 3 to 7 cm long, both ends acute, glossy; flowers in terminal, 4–5 cm wide cymes, petals and the 2.5 cm long stamen filaments bright scarlet-red, late summer. JRHS 48: 4 (= *M. lucida* [Forst. f.] A. Rich.). New Zealand. z9 # ✿

Some species may be found in most botanic gardens, usually as tub plants, in the landscape in the warmer climates.

Lit. Oliver, W. R. B.: The New Zealand species of *Metrosideros*; in Transact. and Proc. N. Z. Inst. **59**, 1928.

MICHELIA L. — MAGNOLIACEAE

Evergreen or deciduous trees or shrubs; leaves oblong, elliptic or oval-oblong, entire or auricled, leathery, acuminate, base round to cuneate, petiole without stipules; flowers hermaphroditic, axillary; petals 6–9, in 2–3 rings of 3 each; stamens numerous; carpels with 2 ovules each; gynoecium stalked; fruits in the form of a loose or dense spike with leathery carpels opening on the dorsal side, similar to *Magnolia*. — About 45 species, most in tropical Asia, about 17 in China.

Michelia compressa (Maxim.) Sarg. Evergreen shrub, a tree in its habitat, to 12 m high, very slow growing, bark dark, smooth; leaves oblong to narrowly obovate, obtuse or short acuminate, base tapered to a long petiole, leathery, entire, 7–10 cm long; flowers pale yellow, about 5 cm wide, sepals and petals narrowly obovate, about 2.5 cm long, base reddish on the exterior, very fragrant, April; fruit cones 5 cm long. GF 6: 77; MJ 1618; KIF 3: 11. Japan. z9 # ✿

M. doltsopa Buch.-Ham. Shrub, a tree in its habitat, to 20 m, conical at first, later broad growing, shoots pubescent at first; leaves elliptic to oblong, 10–18 cm long, acute, base broadly cuneate, leathery, finely pubescent at first, later rather glabrous, glossy green above, bluish beneath; flowers white to yellowish, base green toned, globose, 5 cm wide, sepals and petals about 12–16, oblong-lanceolate, very fragrant, April. BM 9645; NH 38: 117; MCL 35. Nepal; China, W. Yunnan Province; Tibet. 1920. z9 # ⌖

M. figo (Lour.) Spreng. Shrub, to 5 m high, but usually much lower in cultivation, twigs thin, ascending, densely yellow-brown pubescent when young; leaves oval-elliptic, 3–9 cm long, acute, cuneately tapered at the base, leathery thick, entire, dark green and glossy above, lighter beneath, petiole 5 mm long, densely brown pubescent; flowers yellowish-white with a trace of purple, about 3 cm wide, cupulate, very fragrant, April–June. HKS 69; BM 1008; LT 64; DRHS 1296 (= *M. fuscata* [Andr.] Bl.). China. 1789. z9 Plate 134. # ⌖

M. fuscata see: **M. figo**

17 species are described in Lee, Forest Botany of China, some of which are illustrated, *M. champaca* JRHS 89: 297; *M. velutina* JRHS 89: 297.

A conservatory plant in all but the mildest climates; delightfully fragrant in flower. Prefers a sandy-clay soil with a good covering of leaf mold.

MICROGLOSSA DC. — COMPOSITAE

Erect or climbing subshrubs or shrubs; leaves simple, oblong, pubescent, usually entire, alternate; flowers similar to those of *Aster* or *Erigeron,* in heads; ray florets 10–20, narrowly ligulate, disc florets tubular, short 5 toothed, dioecious; involucre imbricate, without an appendage; receptable flat or vaulted, usually pitted; hair crown bristly, later fox-red; achenes compressed, pubescent — 9–10 species in E. Asia and Africa.

Microglossa albescens (DC.) Clarke. Subshrub, about 0.5 m high, twigs slightly angular, reddish, pubescent; leaves lanceolate; 5–12 cm long, entire to finely dentate, dark green above and short pubescent, gray-green beneath and more densely pubescent; flowers *Aster*-like, bluish to whitish, 8 mm wide, many in branched, terminal 7–15 cm wide corymbs, flowers very abundantly, July. BM 6672; SH 2: 474; SNp 90 (= *Amphirapis albescens* DC.; *Aster cabulicus* Lindl.). Himalayas, China; in the mountains to 4000 m. 1840. z8 Plate 124; Fig. 206. ⌖

No particular soil requirements; needs only a good garden soil. Best suited for a sunny area in the rock garden. Good winter protection is advisable since it easily freezes back to the ground.

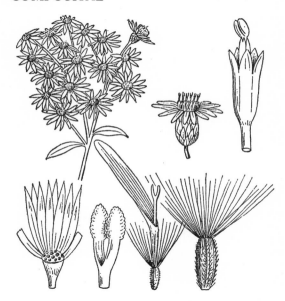

Fig. 206. *Microglossa albescens,* flowers and fruits (from B. M.)

MICROMELES

Micromeles japonica see: **Sorbus japonica** var. **calocarpa**

M. keissleri see: **Sorbus keissleri**

M. rhamnoides see: **Sorbus rhamnoides**

MIMULUS L. — Monkey Flower — SCROPHULARIACEAE

Herbaceous, occasionally subshrubs; leaves opposite, simple, entire or dentate; flowers axillary, solitary, stalked, the apical ones occasionally racemose; corolla bilabiate, the upper lip erect or reflexed, 2 lobed; lower lobe 3 lobed; stigma with 2 sensitive appendages which close when touched; fruit a 2 chambered capsule. — About 100 species in western N. America, Asia, Australia, Africa.

Mimulus aurantiacus Curt. Shrubby, 0.5(1.5) m high, rather glabrous, but the entire plant is glutinose; leaves narrowly oblong to linear, entire to somewhat toothed, 8 to 10 cm long, margins somewhat involuted; flowers solitary, axillary, to 5 cm long, tube narrowly funnel-shaped, limb indistinctly bilabiate, nearly 5 lobed, the lobes finely toothed or incised, orange to salmon-pink, summer. BM 354; FS 883 (= *M. glutinosus* Wendl.). California. 1796. Quite variable. z9 Fig. 207. ⊕ ○

var. **puniceus** (Steud.) Gray. Flowers carmine-red. RWF 329 (= *Diplacus puniceus* Don.). California; dry hillsides. ⊕ ○

Best rooted from cuttings each year or overwintered in pots. Occasionally found in botanic gardens. No particular cultural requirements, drought tolerant.

Fig. 207. *Mimulus aurantiacus* (from B. M.)

MITCHELLA L. — RUBIACEAE

Dwarf evergreen shrub of mat-like habit; leaves opposite, small, simple; flowers paired, small; calyx 4 toothed, corolla funnelform, with 4 outspread limb lobes, these with pubescent tufts; stamens 4; ovaries of both corollas connate; fruit a double berry. — One species each in N. America and Japan.

Mitchella repens L. Mat-like habit, creeping, shoots thin; leaves ovate-rounded, 0.5–1 cm long, often with white venation, petiolate, stipules tiny; flowers paired, on a small, terminal stalk, white, interior often somewhat reddish, May–June; fruits about 6 mm thick, nearly globose, red, edible, but tasteless, persisting the entire winter. BB 3404; HyWF 191; GSP 445. N. America. z5 Fig. 208. # ⊘ ⊕ ⅋ ◑

'**Leucocarpa**'. Fruits white.

M. undulata S. & Z. Like the above; flowers also white, but with a longer corolla, the limb lobes crispate, bracts larger, naviculately lobed. MJ 342 (= *M. repens* var. *undulata* [S. & Z.] Mak.). Japan. z6 # ⊕ ⅋

Hardy, attractive ground cover for semishady, moist wooded areas; easily propagated by division.

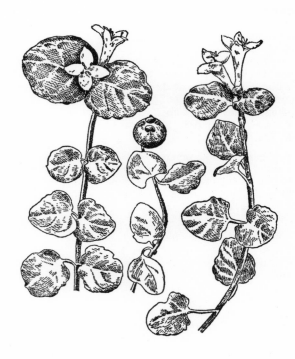

Fig. 208. *Mitchella repens*

MITRARIA Cav. — GESNERIACEAE

Monotypic genus; evergreen, procumbent or somewhat climbing, thin branched shrub; leaves opposite; flowers attractive; calyx distinct, 4 or 5 parted; corolla tube much longer than the calyx, somewhat narrowed at the mouth and with an evenly 5 lobed, outspread limb; stamens 4, exserted; fruit a berry. — 1 species in Chile.

Mitraria coccinea Vav. Evergreen, thin trunked shrub, procument or twining to 2 m high or more in thickets; leaves opposite, ovate to elliptic, acute, base round, 1.5–2.5(4) cm long, with a few teeth on either edge, leathery tough, dark green and glossy above, bluish beneath with a pubescent midrib; flowers solitary in the leaf axils, long stalked, corolla tubular-inflated, 2.5–3 cm long, scarlet-red, finely pubescent, June to fall; fruit a 1 cm long berry, with the 3 cm long style at the apex. BM 4462; DRHS 1309; FS 385; PFC 106; JRHS 100; 82. Chile; Chiloe Island. 1846. z7 Plate 134; Fig. 209. # ✧ ◗

For semishady wooded areas, will not tolerate drought or too much sun. Very attractive in flower.

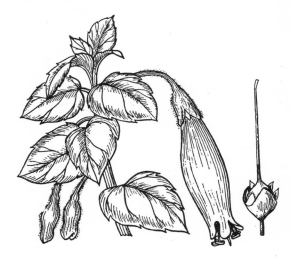

Fig. 209. *Mitraria coccinea* (from Dimitri)

MOLTKIA Lehm. — BORAGINACEAE

Perennials or subshrubs, rough haired; flowers in a many flowererd cincinnus; corolla funnelform, blue to purple, tube always glabrous on the exterior (!), anther exserted or not; nutlets either smooth and glossy or dull and rough, abscising cleanly at the base. — 3 species in S. Europe and the Near East.

Moltkia graminifolia see: **M. suffruticosa**

M. × intermedia (Froeb.) Ingram (*M. petraea × M. suffruticosa*). Growth taller and more shrubby than *M. suffruticosa*; leaves broader; flowers a strong blue, in outspread inflorescences. Developed before 1906 by Froebel in Zurich, Switzerland.

'Froebelii' (Sündermann). Differs only slightly from the species, growth upright, about 15 cm high; leaves linear-lanceolate, dark green; flowers azure-blue, in forked cincinnae (= *Lithospermum froebelii* Sündermann). Before 1906. Developed by Sündermann in Lindau, W. Germany.

M. petraea (Tratt.) Reichb. Subshrub, bushy, woody at the base, 15–30 cm high; leaves narrowly linear to oblong-lanceolate, about 3 cm long, light green, tough; flowers in dense, terminal, occasionally forked cincinnae, corolla deep violet-blue, somewhat pink-blue in bud, anthers blue, shorter than the corolla tips, June. BM 5942; Bai 6: 18B. Greece. 1845. z7 Plate 130. ✧

M. suffruticosa (L.) Brandt. Subshrub, more bushy and dense, 15–45 cm high, entire plant (except the flowers) with a gray appressed, silky pubescence; leaves linear, acute, the basal leaves to 10 cm long, only 5 cm long on the shoots, silvery beneath; flowers in branched cincinnae to 5 cm wide, blue, buds more purple, anthers yellow (!), not exserted past the corolla limb tips, June–August. Bai 6: 18A; BMns 394 (= *M. graminifolia* Nyman). N. Italy. z7 ✧

Most suitable for rock gardens or tub planting; for dry sunny sites, likes a well drained, gravelly soil.

Lit. As for *Lithodora*; which see.

MORUS L. — Mulberry — MORACEAE

Deciduous trees or shrubs producing a milky sap; leaves alternate, undivided or lobed, serrate or dentate; flowers 1 or 2 dioecious, both sexes in stalked, pendulous, cylindrical to ovate spikes; male flowers with a deeply 4 parted perianth, stamens and atrophic ovaries; female flowers with a 4 part perianth, ovaries sessile; fruits blackberry-like; the individual fruits enclosed by the fleshy perianth. — About 10 species in the northern temperate and subtropical zone.

Morus acidosa see: **M. australis**

M. alba L. White Mulberry. Tree, to 15 m high, round crowned, sparsely branched, young twigs glabrous or lightly pubescent, later gray to gray-yellow; leaves broadly ovate, acuminate, 6 to 12 cm long, often quite variably lobed, light green above and rather smooth, only the venation pubescent beneath, coarsely dentate; flowers inconspicuous, mono- or dioecious, styles absent or very short, May; fruits quite variable, usually narrow, 1.5–2.5 cm long, white, red to black-red, on a 1–2.5 cm long stalk, sweet, but bland tasting. HM 3: 87.

China. In cultivation for ages. The foliage is economically important as food for silkworms. Ⓕ USSR and India for reforestation, and Czechoslovakia in screen plantings. z5 Plate 131.

Outline of the cultivars

Differing in leaf color:
'Aurea' (leaves yellow)

Differing in growth habit:
'Nana' (dwarf)
'Pendula' (branches pendulous)
'Pyramidalis' (conical habit)

Forms with small leaves, habit normal:
M. tatarica (leaves small, lobed)
'Constantinopolitana' (leaves medium-sized, usually not lobed, deep green)

Forms with particularly large leaves:
'Microphylla' (leaves to 20 cm long, blade flat)
'Multicaulis' (leaves to 30 cm long, blade blistery)

'Aurea' (Rothe). Leaves and bark yellow.

'California Giant'. Similar to 'Multicaulis'; from the Marlborough Nurseries in Salisbury, Rhodesia.

'Constantinopolitana'. Growth tree-like, but compact and thickly branched; leaves ovate, 9–15 cm long, tough, coarse serrate, base broad cordate, dark green and glossy above, lighter beneath; fruits dark to black-red. DL 3: 2. A large leaved form cultivated in Turkey.

f. fastigiata see: 'Pyramidalis'

'Fegyvernekiana' (Rosenthal). Dwarf habit; flowers not yet observed.

f. globosa see: 'Nana'

'Laciniata'. Leaves regularly and deeply incised, the lobes very narrow and long acuminate, deeply serrate. SH 1: 151b (= f. skeletoniana Schneid). ∅

'Macrophylla'. Leaves very large, 15–20 cm long, usually totally unlobed, blade always flat (!), coarse serrate, base cordate; fruits light red. 1836. ∅

'Multicaulis'. Strong growing, mostly multistemmed shrubs, 5–6 m high eventually; leaves very large, pendulous, 8–25 cm long, with blistery new growth, usually unlobed, light green above, obtuse and somewhat rough, teeth obtuse; fruits white at first, then red, eventually nearly black (= M. cucullata Bonefous; M. tookwa Sieb.; M. chinensis Loud.). Chinese cultivar for the silk industry. Imported from Manila to Senegal in 1821 by Perrotet and later introduced into France. z8 Plate 133. ∅

'Nana'. Dwarf habit, shrubby, globose; leaves regularly lobed, medium-sized (= f. globosa Hort.). Plate 130.

var. nervosa see: 'Venosa'

'Pendula'. Branches hanging umbrella-like, rather thin; leaves usually lobed. MG 17: 27.

'Pyramidalis'. Growth tightly and narrowly conical, branches twisted, rough (= f. fastigiata Hort.).

f. skeletonia see: 'Laciniata'

var. tatarica (Pallas) Ser. Shrub or small tree with a dense crown; leaves small, 4–8 cm long, lobed or unlobed; fruits small, 1 cm long, deep red, occasionally also white. BC 2400 (= M. tatarica Pallas).

'Venosa'. Leaves rhombic, distinctly yellowish-white veined, acute to acuminate, occasionally also rounded, irregularly serrate, base cuneate. BC 2397 (= var. nervosa Loud.; M. urticifolia Hort.). Plate 131. ∅

M. australis Poir. Shrubby habit, also a small tree in its habitat, shoots glabrous; leaves ovate, lobed or unlobed, 6–15 cm long, often long acuminate, base cordate, crenate or sharply serrate, somewhat rough above, quite finely pubescent beneath or eventually nearly totally glabrous and without axillary pubescence in the vein axils; fruits very numerous, 1–1.5 cm long, dark red, edible, serrate and juicy. LWT 39; YWP 2: 76; KIF 2: 19 (= M. acidosa Griff.; M. japonica Bailey; M. bombycis Koidz.). China, Korea, Japan. 1907. z5 Plate 131. ⚤ ✂

M. bombycis see: **M. australis**

M. cathayana Hemsl. Shrub, 3–5 m high, occasionally a small tree, to 10 m high, densely branched, crown broad-rounded, young shoots glabrous, often densely white punctate; leaves quite variable, unlobed to lobed, 5–18 cm long, oval, acute, coarsely serrate, teeth broad and short, base round to cordate, glossy green and rough above, lighter and soft pubescent beneath, leathery, stalk 2–3 cm long; male flowers in up to 8 cm long spikes, female flowers 2 cm long; fruits narrowly cylindrical, 2–3 cm long, 7 mm thick, white, red or black, sweet, edible. ICS 959. Central and E. China. 1907. z6 Plate 131. ∅ ⚤ ✂

M. celtidifolia see: **M. microphylla**

M. chinensis see: **M. alba 'Multicaulis'**

M. cucullata see: **M. alba 'Multicaulis'**

M. japonica see: **M. australis**

M. kagayamae Koidz. Small tree, rather slow growing; leaves oval-oblong, usually deeply lobed, the middle lobes much larger than the side lobes, acuminate, 12–15 cm long, 6–10 cm wide, obtusely serrate, often simply crenate, the sinuses entire, somewhat rough above, glabrous beneath, a gorgeous golden-yellow in fall. GC 127: 66; MJ 3645. Japan. 1918. z6 Plate 131. ∅

M. microphylla Buckl. Very similar to M. rubra, but small leaved; small tree, only rarely over 6 m high; leaves ovate, 3–7 cm long, unlobed to 3 lobed, serrate or crenate, base round to slightly cordate, rough above, usually soft pubescent beneath; fruits short-ovate to nearly globose, 1–1.5 cm long, deep red to nearly black, sweet, edible. SM 301; SS 321 (= M. celtidifolia Sarg.) North America; Texas and Arizona to Mexico. 1926 z8–9 ⚤ ✂

M. mongolica (Bge.) Schneid. Small tree, 3 to 5 m high, developing thickets, bark gray-brown, finely channeled, young shoots gray-brown, glabrous, punctate; leaves oblong-ovate, rarely lobed, about 8–16 mm long, caudately tipped, coarse and awn-like serrate, smooth above and nearly glabrous on both sides, paper thin; fruits cylindrical, 1 cm long, 5 mm thick, red, later black. NK 19: 33; ICS 958. Manchuria, China, Korea. 1907. Hardy. z5 Plate 131. ∅

var. **diabolica** Koidz. Young shoots silky pubescent; leaves often finely and deeply lobed, glossy green above, white tomentose beneath. NK 19: 33B. China. 1923.

var. **vestita** Rehd. Young shoots glabrous, dark gray, densely punctate; leaves oval-oblong, 5 cm long, acuminate, bristly serrate, base slightly cordate, tough, light green above and somewhat pubescent, densely silky pubescent beneath; male flowers in 3–4 cm long, yellow-green, pubescent catkins. China; Yunnan Province.

M. nigra L. Black Mulberry. Tree, to 10 m, crown dense and rounded, often only a shrub or espaliered in cooler climates, twigs pubescent, eventually brown; leaves broadly ovate, 6–12(20) cm long, acute, base deeply cordate, coarsely serrate, dark green above and rough, lighter and pubescent beneath; flowers mono- or dioecious, May; fruits oval-oblong, 1–3 cm long, sweetly aromatic, deep red. BS 2: 329, Pl. 30. Orient. In cultivation for ages. z7 Plate 131, 133. ∅ ⚘ ✄

M. rubra L. Tree, to 20 m high, often lower, crown broadly rounded, bark brown, scaly, young shoots pubescent at first; leaves quite variable, usually broadly ovate or oval-oblong, usually not lobed, short acuminate, base cordate, dense and sharply serrate, to 20 cm long on long shoots and also 3–5 lobed, finely scabrous above, soft pubescent beneath, fall color golden-yellow; fruits 2–3 cm long, eventually dark purple, sweet, juicy, edible. SS 320; BB 1257; GTP 148. Eastern USA. 1629. z5 Plate 131. ⚘ ✄

M. tartarica see: **M. alba** var. **tartarica**

M. tookwa see: **M. alba 'multicaulis'**

M. urticifolia see: **M. alba 'Venosa'**

All species prefer a sunny, warm site in a good, deep, fertile, alkaline soil. The dwarf forms are attractive in small gardens. The fruits of some species are of economic use.

Lit. Hotta, T.: Taxonomic study of cultivated mulberry in Japan; Bot. Inst. Fac. of Textile Fibers, Kyoto 1954 ● Koidzumi, G.: Monograph of the genus *Morus*; Tokyo 1917 (in Japanese) ● Koidzumi, G.: Synopsis specierum generis Mori; in Bull. Sericult. Exp. Sta. Japan **2**, 1–45, 1923 (pl. 1–11) ● Tsen, M.: Révision des variétés du murier blanc observées dans le midi de la France; in Bull. Soc. Hist. Nat. Toulouse **68**, 283–334, 1935.

MUEHLENBECKIA Meissn. — POLYGONACEAE

Small or large, erect or climbing shrubs or subshrubs, often procumbent or erratically branched; leaves alternate, petiolate, small or large, occasionally also totally absent; stipules small, often sheath-like; flowers polygamous or dioecious, small, whitish or greenish, axillary or terminal and clustered; perianth deeply 5 lobed, stamens 8; styles 3, short; fruit a 3 sided nutlet surrounded by the fleshy perianth. — About 15 species in Australia and S. America.

Muehlenbeckia axillaris (Hook. f.) Walp. Deciduous, mat-like shrublet, 3–5 (in cultivation to 25) cm high, shoots wiry thin, finely pubescent; leaves oval-rounded to nearly circular, 3–8 mm long, glabrous, flat, punctate beneath; flowers small, greenish, 1–2 in a leaf axil, July; fruits black, triangular, glossy (= *M. nana* Hort.). New Zealand, Tasmania, Australia, in the mountains. z8 Fig. 210.

M. complexa (Cunn.) Meissn. Deciduous shrub, similar to *M. axillaris* in appearance, but taller, forming a broad bush or also climbing to 3 m high, young shoots warty with short, stiff hairs; leaves quite variable on the same plant, 5–20 mm long, ovate to obovate or circular, base cordate or round, acute or emarginate at the apex, often with fiddle-shaped sinuses on the sides; flowers greenish-white, in axillary and terminal, 2.5 to 3 cm long spikes, often of only 2–3 flowers, July; fruits waxy white, juicy, 5 mm thick, seeds black. BM 8449. New Zealand, in the mountains to 700 m. 1842. z9 Plate 132.

f. **trilobata** Colenso. Leaves to 2.5 cm long, broadly ovate, acuminate, base round to cordate, conspicuously sinuate on the sides, and so appears trilobed. New Zealand. Plate 132.

Fig. 210. *Muehlenbeckia axillaris*
(⅔ actual size; Original)

M. ephedroides Hook. f. Procumbent shrublet, shoots broad and chaotically outspread, nearly always totally leafless, reed-like, thin, brown, deeply channeled, glabrous; leaves linear to sagittate, 8–25 mm long, but rarely seen; flowers in axillary clusters or small spikes; fruits fleshy or dry. New Zealand. z9

M. nana see: **M. axillaris**

M. platyclada see: **Homalocladium platycladum**

The species described are all well suited to sunny areas in the rock garden; for warmer climates. More interesting than attractive.

MUSA L. — Banana — MUSACEAE

Huge, tropical, perennial plants with an occasionally bulbous base; leaves originating from the base of the rootstock, spirally arranged, rolled up when young, twisting to the right, usually very large, oblong-elliptic, entire, parallel-pinnately veined; leaf sheaths forming a false stem; inflorescence a terminal raceme, the flowers in semi-whorled clusters, each cluster with a large bract, the lower flower clusters are female, the uppermost, male; fruit juicy or leathery, not dehiscing. — 35 species in the tropics of the Old World, some species cultivated in the tropical regions of the entire world.

Although the *Musa* species have no woody stem, one is seemingly formed from the leaf sheaths. Only a few species will be mentioned here.

Musa acuminata Colla. Dwarf Banana. Stoloniferous, to 2 m high; leaves 6–7 together, spreading in a dense rosette, oblong, 70–100 cm long, 30 cm wide, bluish-green; bracts red, leathery; fruits 6 sided, yellow, 15 cm long, sweet, edible (= *M. cavendishii* Lamb. & Paxt.; *M. nana* Lour.). S. China. 1829. Cultivated throughout the tropics and subtropics. Fruits well in the conservatory. z10 ⌀ ⚭ ✗

M. basjoo S. & J. Japanese Fibre Banana. To 4 m high or more, with many root shoots, stem reddish; leaves to 3 m long, 60 cm wide, both sides glossy green; inflorescence a dense, pendulous raceme, bracts, male flowers abscising, the first ones shell-form, leathery, reddish-yellow; fruits 7 cm long, 2.5 cm thick, triangular. BM 7182; DRHS 1328 (= *M. japonica* Hort.). Japan. 1890. z10 Plate 125. ⌀

M. cavendishii see: **M. acuminata**

M. ensete Gmel. Ornamental Banana, Abyssinian Banana. Stem conical, 10–13 m high; leaves to 6 m long and 1 m wide, midrib red (!); inflorescence globose, bracts dark red; fruits leathery, dry, seeds large, black. BM 5223; FS 1418 (= *M. ventricosa*). In open mountain forests to 2500 m. 1853. Now more botanically correct as *Ensete ventricosum* (Welw.) Cheesem. z10 ⌀

M. nana see: **M. acuminata**

M. × paradisiaca L. (*M. acuminata × M. balbisiana*). Pisang; Banana. Plants with many sprouts from the rootstock, stem round, to 8 m high; leaves to 2.5 m long, 60 cm wide; inflorescences pendulous, 1–1.5 m long. z10 ⌀ ⚭ ✗

This is the more typical banana plant of which there are many varieties. These are divided into 4 groups, including:

var. **normalis** O. Ktze. Vegetable banana or plantain. Fruits 30 cm long, green, only edible when cooked.

var. **paradisiaca** (L.) Bak. Fruit banana. Fruits yellow, sweet, edible raw, seedless.

MUTISIA L. f. — COMPOSITAE

Evergreen, usually subshrubs or shrubs, climbing by means of tendrils; leaves alternate, simple or pinnate, the midrib usually drawn out to a long tendril; flower heads usually solitary and terminal, large, stalked, usually red or yellow; involucre large, cylindrical to campanulate, many scaled; ray florets in a ring, usually red or yellow; fruits with a stiff pappus. — About 60 species in S. America, mostly in Chile.

Mutisia clematis L. f. Climbing, evergreen shrub, 5–7 m high, young shoots thin, channeled, white tomentose; leaves pinnate, 8–10 cm long, leaflets 6–10, oval-oblong, entire, acute, base round, 1.5–3 cm long, nearly sessile, both sides tomentose at first, later only beneath, as well as on the rachis, the flower stalk and the outer involucral bracts, leaf rachis ending in a branched tendril; flower heads solitary, terminal, pendulous, the involucre 4–5 cm long, ray florets 9–10, rotate in arrangement, bright orange-red, total flower about 5 cm wide, May–October. BM 8391; DRHS 1336. S. America, Andes of Ecuador and Columbia. 1859. z9 or in a conservatory. # ✧

M. decurrens Cav. Shrub, 2.5–3 m high, climbing, glabrous, sparsely branched; leaves sessile, narrowly oblong, 7–12 cm long, to 2.5 cm wide, base somewhat decurrent on the stem on both sides, apex terminating in a tendril; flower heads solitary, terminal, 10–12 cm wide,

Fig. 211. *Mutisia retusa* (⅔ actual size)

orange to scarlet-red, nearly like a small, simple dahlia flower, June–August. BM 5273; FS 2408. Chile. 1859. The hardiest of all the species. z9 # ✧

M. retusa Remy. A high climbing shrub, becoming 2.5–5 m high in the wild; leaves sessile, elliptic-oblong, 3–5 cm long, base lobed, apex truncate and somewhat emarginate, entire or triangular toothed for the entire length, but often with only 1–2 teeth at the apex, deep green and glabrous above, underside glabrous or woolly; flowers solitary, terminal, stalk to 7 cm long,

midrib terminating in a long, unbranched tendril, ray florets pink, pretty, summer. Chile. 1868. z9 Fig. 211. # ✧

'Alba'. Ray flowers white. ✧

Very attractive plants, but only for warmer climates or as a cool greenhouse plant. Easily cultivated in a fertile soil.

Lit. Comber, J.: The Hardier Mutisias; in Jour. R.H.S. **74**, 241–245, 1949 ● Ingram, J.: *Mutisia* in cultivation. Baileya **18**, 33–39, 1971 (with 3 ills.).

MYOPORUM Banks & Soland. ex Forst. f. — MYOPORACEAE

Evergreen trees or shrubs, many heather-like in appearance, partly erect and tall, partly loose, glabrous or glutinous; leaves alternate, occasionally opposite, entire or dentate, translucent glandular punctate; flowers axillary, usually in clusters, small or medium size, mostly white; calyx with 5 incisions or 5 parted, corolla somewhat campanulate or funnel-shaped, the tube either long or quite short, usually with 5 lobes; stamens 4, occasionally 5–6; fruit a small, more or less fleshy drupe. — 32 species from Australia and New Zealand, the Pacific Islands, New Guinea, Mauritius and E. Asia.

Myoporum acuminatum R. Br. Upright shrub, quite variable in height, also variable in form and size of the leaves and flowers; leaves alternate, elliptic-oblong to lanceolate or linear, long acuminate, to 7 cm long, entire or with very short teeth; flowers in clusters of 2–4 or also solitary, white, corolla nearly campanulate, 8 mm long, inerior with pubescent tufts, the lobes much shorter than the corolla tube, April; fruit nearly globose, 6 mm. Australia. 1812. z10 # ⊘

M. laetum Forst. f. Shrub to a small, round crowned, 4.5–9 m high tree, branch tips glutinous; leaves lanceolate to obovate-lanceolate, acute to long acuminate, 5–10 cm long, finely serrate above the middle, glabrous, rather fleshy, dense and finely punctate from translucent oil glands; flowers in clusters of 2–6, white, purple punctate, lobes rounded, interior pubescent; fruits reddish-purple, oblong. New Zealand. z10 # ⊘

M. serratum R. Br. Upright, occasionally somewhat loose shrub or a round crowned, small tree, usually glabrous, extraordinarily variable; branch tips not glutinous (!!); leaves elliptic-oblong to lanceolate, obtuse or acute, more or less serrate or also entire, 3 to 6 cm long; flowers white, purple punctate, 2–6 in the leaf axils, interior pubescent, corolla lobes as long as the tube, obovate, May. Australia. z10 # ⊘

Frequently planted in warm climates, particularly *M. laetum*; very tolerant of sea winds.

MYRICA L. — Bayberry — MYRICACEAE

Deciduous or evergreen, aromatic trees or shrubs; leaves alternate, simple, often dentate or crenate or entire, often with resin glands; stipules absent; flowers mono- or dioecious, inconspicuous, without sepals and petals, usually with 2–8 stamens, in small, dense catkins; fruit a small, globose drupe, often with a waxy coating. — About 35 species in the temperate and subtropical zones of both hemispheres, but absent from Australia.

Myrica californica Cham. & Schlechtd. Upright, evergreen shrub, to 2.5 m, in its habitat a 10 m high tree, shoots soft pubescent; leaves lanceolate or more oblong, acute, base cuneate, 5–10 cm long, sparsely serrate, dark green and glossy above, glabrous or pubescent beneath, with tiny black spots; catkins about 2.5 cm long, May–June; fruits globose, about 4 mm thick, purple, with a white waxy coating. SPa 83–84; MS 70; SS 461. N. America, the Pacific Coast region from W. Washington to S. California. 1848. z9 Plate 134, 135. #

M. carolinensis see: **M. pensylvanica**

M. cerifera L. (non Hort.!!). Evergreen, slender, upright tree, 10–12 m high, rather glabrous; leaves oblong-oblanceolate, acute, 3–7.5 cm long, base cuneate, coarsely serrate on the apical half, deep green above, somewhat lighter beneath and glabrous or somewhat pubescent, both sides with tiny, golden-yellow resin glands; flowers dioecious, March–April; fruits gray-white, 2–3 mm thick, stiff with a white waxy coating. BB 1160; SS 459; GF 7: 476; BC 2422. N. America; Coastal region from S. New Jersey to Florida and Texas. 1669. Often confused with *M. pensylvanica* as mentioned by Rehder in BC 2: p. 2092. z9 Plate 134. # ⚯

M. curtissii see: **M. heterophylla** var. **curtissii**

M. faya Ait. Evergreen shrub or small tree, young shoots glabrous; leaves oblong-lanceolate, acute, 5–10 cm long, entire or serrate, base cuneate; male flowers with 4 stamens, female flowers and fruits in elongated spikes; fruits red, 6 mm thick, edible, exterior rough, waxy. BFCa 3. Canary Islands. z10 # ⚭ ✖

M. gale L. Bog Myrtle. Deciduous, erect, multistemmed shrub, to 1.5 m high, twigs rodlike, brown, glabrous; leaves oblanceolate, 3–5 cm long, obtuse, occasionally acute, base cuneate, serrate toward the apex, deep green and glabrous above, usually somewhat pubescent beneath; flowers dioecious, in about 1.5 cm long, abundant, brownish, cone-like catkins on the older wood, March–April; fruits ovate, thick, golden, glandular punctate. Europe, N. Asia, N. America. z1 Plate 134.

var. **subglabra** (Chevalier) Fern. Leaves totally or nearly totally glabrous. N. America.

var. **tomentosa** DC. The young shoots and leaves densely pubescent or tomentose on both sides, particularly beneath. Siberia, Japan. Plate 134.

M. hartwegii Wats. Deciduous shrub, low, 0.7–1.5 m high, shoots rather thin, pubescent, later glabrous and deep red-brown; leaves appearing after the flowers, oblanceolate, 3–7.5 cm long, 1.2–2.5 cm wide, thin, light green and slightly pubescent above, lighter with persistent pubescence beneath, serrate on the apical half, base cuneate and entire, petiole nearly 1 cm long; flowers dioecious, male flowers in 2 cm long, cylindrical catkins, female flowers in 4 mm thick, globose inflorescences, June–July; fruits compressed flat. MS 71. North America; California. z9(?)

M. heterophylla Raf. Evergreen shrub, or also a small tree, to 5 m high, young twigs black brown and soft pubescent; leaves obovate to oblong (lanceolate), 4–7 cm long, 1.5–2.5 cm wide, usually round on the apex; entire or sharply serrate, both flowers dioecious; fruits gray-white, 3 mm thick, with a heavy waxy coating. E. USA, from S. New Jersey to Florida and Louisiana. 1903. z8 # ⚭

var. **curtissii** (Chevalier) Fern. Young shoots glabrous, red-brown; leaves glabrous above, glandular beneath (= *M. curtissii* Chevalier). Maryland to Florida and Louisiana. #

M. nagi see: **M. rubra**

M. pensylvanica Loisel. Deciduous shrub, dense, about 2 m high, shoots gray pubescent and glandular; leaves obovate to oblong, 4–10 cm long, shallowly dentate or entire toward the apex, pubescent on both sides, glandular punctate beneath; young fruits pubescent, later 3.5–4.5 mm thick, with a gray-white waxy coat, persisting deep into the winter. GSP 93; BC 2423 (as *M. carolinensis*) (= *M. cerifera* sensu Bigel. and Hort., non L.). NE. coast of the USA, in dry and sandy areas. 1725. Hardy. Widely used in the USA for erosion control. z2 Plate 134. ⌀ ⚭

M. rubra S. & Z. Evergreen shrub or small tree, very densely branched, young shoots warty and slightly pubescent; leaves oblong-lanceolate, acute, entire on the apical half or serrate, 7–15 cm long, base cuneate, deep green above, lighter beneath, glabrous; fruits globose to oval, warty, 12–25 mm thick, purple, juicy, sour, edible, seed pit 8 mm long, flat. BM 5727; YWP 2: 100; KIF 1: 29 (= *M. nagi* DC. non Thunb.). S. China, Japan. 1864. Cultivated in China as "Chinese Arbutus" where the fruits are canned for export. Ⓕ Japan, for erosion control. z10 Plate 123. ⚭ ✖

Strongly recommended, where winter hardy, for the aromatic foliage and the gray-white fruits. Easily satisfied with a dry, sandy soil (for *M. pensylvanica*) or a moist, marshy soil (for *M. gale*).

Lit. Chevalier, A.: Monographie des Myricacées; in Mém. Soc. Sci. Nat. Cherbourg **32**, 85–340, 1901 to 1902 (8 plates) ● Youngken, H. W.: The comparative morphology, taxonomy and distribution of the Myricaceae in the eastern United States; in Ann. Jour. Pharmacy **87**, 391–398, 1915 ● Youngken, H.W.: (the same title); in Contrib. Bot. Lab. Univ. Penn. **4**, 339–400, 1920.

MYRICARIA Desv. — False Tamarisk — TAMARICACEAE

Deciduous shrubs or subshrubs; leaves small, scale-like, imbricately overlapping; flowers in long, terminal and axillary, narrow racemes; calyx and corolla 5 parted, stamens 10, the filaments connate at the base or to past the middle (!), ovaries with 3, nearly sessile stigmas; fruit a conical, acuminate, 3-valved, dehiscent capsule; seeds with pubescent tufts. — About 10 species in S. Europe, Central Asia, China and Siberia.

Very similar to *Tamarix*, but differing in the 10 connate stamens (*Tamarix* has 4–8 distinct stamens).

Myricaria davurica (Willd.) Ehrenb. Very similar to the following species, but better known; the flower racemes usually axillary, bracts oblong-ovate, obtuse, with an encircling, membranous margin, shorter than the flower buds, stamen filaments often only ⅓ connate. SH 2: 230h

to k. Dzungaria Transbaikal, USSR. 1816. Hardy.

M. germanica (L.) Desv. Shrub, narrowly upright, 1–2 m high, shoots gray-brown, the young shoots blue-green or gray-green, later more yellowish, very densely foliate; leaves scale-like, blue-green, linear-lanceolate, imbricate; flowers bright red, in 10–15 cm long, spike-like racemes, mostly terminal, occasionally some lateral, bracts oval-oblong, long acuminate, with a broad membranous margin, stamens connate to about the mid-point, May to August. DL 3:5; HM pl. 184 (= *Tamarix germanica* L.). Central and S. Europe, usually in gravelly river beds and on the banks of the alpine streams. z6 Plate 136; Fig. 212.

Cultivated like *Tamarix*, but less valuable in cultivation.

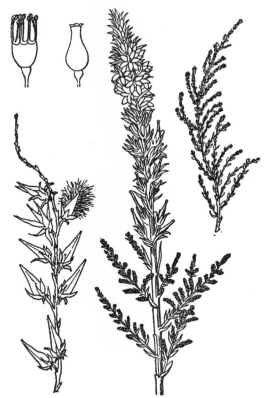

Fig. 212. *Myricaria germanica,* fruits at left (from Lauche, Dippel, Nose; all about actual size)

MYRSINE L. — MYRSINACEAE

Evergreen shrubs or trees; leaves alternate, leathery tough, usually entire; flowers small, sessile or in stalked, axillary or lateral clusters, dioecious-polygamous; flowers 4 or 5 parted, the corolla lobes imbricately arranged in bud, anthers short and usually obtuse; fruit a pea-sized drupe, dry or fleshy, with 1 pit, seeds globose. — 7 species, in the Azores and from Africa to China.

(The genus **Rapanea** Aubl., of around 200 species from the tropics and subtropics, is often included with *Myrica*. It is dealt with separately in this work.)

Myrsine africans L. Evergreen shrub, 0.5–1.5 m high, very densely foliate, young shoots angular, soft pubescent; leaves alternate, elliptic to narrowly obovate, rounded or truncate at the apex, 6–20 mm long, tapered to the base, 6–12 mm wide, somewhat dentate on the apical half, very glossy above, glabrous on both sides, petiole 2 mm long; flowers unisexual, very small, light brown, 3–6 in sessile clusters in the leaf axils, May; fruits globose, 6 mm thick, blue-lilac. BM 8712; RCA 100; DRHS 1342 (= *M. retusa* Ait.). Himalayas, China, Azores as well as in the mountains of E. and S. Africa. 1691. z9 # ⚭

M. chathamica see: **Rapanea chathamica**

M. nummularia see: **Rapanea nummularia**

M. retusa see: **M. africans**

MYRTUS L. — Myrtle — MYRTACEAE

Evergreen shrubs, occasionally trees; leaves opposite, aromatic, short stalked or sessile, entire and pinnately veined; flowers solitary and axillary or in few flowered cymes; calyx tube top-shaped, with (4) 5 sepals and petals respectively, the latter white, occasionally also pink; stamens numerous, in several rings, distinct, longer than the petals; fruit a globose to ovate, usually black berry with a persistent calyx. — About 100 species in the warmer temperate zones and the tropics.

Myrtus bullata Banks & Soland. Strong growing, upright shrub, 1–2 m high, to 9 m high in its habitat; easily distinguished from all the other species in the consistently red-brown (!), bullate, metallic-glossy leaves, these ovate to elliptic, 2 to 4 cm long, petiolate; flowers white, solitary in the leaf axils, 2 cm wide; fruits dark red, ovate, 8 mm long, edible. KF 131; BM 4809. New Zealand. 1854. z9

M. cheken Spreng. Upright shrub, tree-like in its habitat; leaves elliptic, acute, 2 to 4 cm long, aromatic, very short petioled, dull green; flowers white. BM 5644. Chile. Introduced by Veitch. z9 Plate 135; Fig. 213.

M. communis L. Common Myrtle. Evergreen shrub, 3–5 m high in its habitat; leaves opposite, occasionally in 3's, oval to lanceolate, entire, smooth, glossy, glabrous, very aromatic when crushed, leathery, translucent punctate, 2.5–5 cm long, petiole very short or absent; flowers white, 2 cm wide, fragrant, always solitary and axillary, July–August; fruit a globose, purple-black, 12 mm long berry. HM 2183. S. Europe to W. Asia; widely distributed in the Mediterranean region. 1597. z8 Fig. 213. #

var. **acutifolia** L. Upright habit, shoots reddish; leaves lanceolate, long acuminate, base cuneate, 2.5–4 cm long, 1–2 cm wide, leaf stalk reddish; sepals large, obtuse. Portugal.

var. **italica** Mill. Branches and shoots narrowly upright; leaves oval-lanceolate, rather small, about 3 cm long and 1 cm wide. Central Italy.

var. **latifolia** Willk. & Lge. Leaves oval-oblong to oblong-lanceolate, acuminate, 1.8–3 cm long, 0.8 to 1.5 cm wide. Spain, Tarragona. ⊘

'Leucocarpa'. Fruits white.

var. **romana** Mill. Leaves broadly ovate, sharply acuminate,

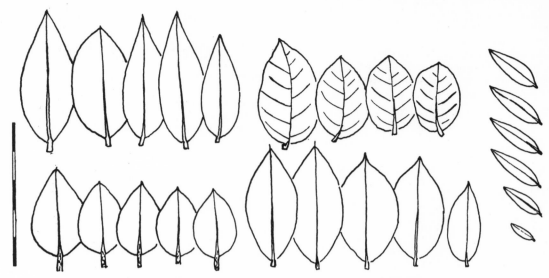

Fig. 213. **Myrtus.** Above *M. communis* at left, middle *M. cheken;*
below *M. lechleriana* left, *M. luma* middle; right *M. communis* var. *tarentina* (Original)

light green, 3–4.5 cm long, 1 to 1.5 cm wide, often in whorls of 3–4. Spain.

var. **tarentina** L. Short branched, young shoots and petioles densely pubescent; leaves closely spaced, decussate, in 4 rows, narrowly ovate, 1.2–2 cm long. Mediterranean region. Plate 135; Fig. 213 ⌀ ✛

'Variegata'. Leaves white variegated. ⌀

M. lechleriana (Miq.) Sealy. Tall shrub, also a small tree in its habitat, to 7 m, densely branched and foliate to the ground, shoots gray pubescent; leaves ovate to broadly elliptic, 1.2–3 cm long, abruptly acuminate, purple to golden-brown on the new growth, later dark green, glossy, and glabrous, petiole remaining pubescent;

flowers white, 12 mm wide, grouped 4–10 in the axils of the apical leaves, petals circular, May; fruit a 6 mm thick berry, red at first, then black. BM 9523. Chile. 1927. A gorgeous shrub in bloom. z9 Fig. 213, 214. # ⚇ ✛

M. luma Molina. Tall shrub to a small tree, 5–6 m high and equally wide, stem cinnamon-brown with an exfoliating bark, shoots with a fine reddish pubescence; leaves elliptic, 1.2–2.5 cm long, sharp and short acuminate, tapering to the base, dull, dark green above, lighter beneath; flowers solitary, white, 2 cm wide, sepals and petals 4, stamens grouped into a ring, anthers pink, July–October; fruits black, sweet, edible. BM 5040 (= *Eugenia apiculata* DC.). Chile. 1843. z9 Plate 125, 135; Fig. 213, 214. # ⌀ ✗ ✛

Fig. 214. **Myrtus.** Left, *M. lechleriana;* right *M. luma* (from Dimitri)

M. ugni Molina. Upright, abundantly branched shrub, 0.7–1.5 m high, bark dark brown, twigs nearly 4 sided, the younger shoots compressed, with a brown strigose pubescence; leaves oval-oblong, short petioled, leathery, thick, 2–3 cm long, acuminate, margins usually somewhat involuted, glossy dark green above, flowers white, pink toned, solitary and axillary, petals 5, rounded and concave, globose clustered, May; fruits dark purple at first, then blue-black, juicy, fragrant, good flavor. PBL. 2: 171; BM 4626 (= *Eugenia ugni* Hook.). Chile. 1844. z9 ✗ ∅ ⚭ ✧

Very easily cultivated in mild climates; prefers a moist, humus soil in a lightly shaded area.

Lit. Sennen, F., & Teodoro: Formes du *Myrtus communis* sur le Territoire de Tarragone; in Bull. Soc. Dendrol. France **69**, 6–19, 1929.

NANDINA Thunb. — NANDINACEAE

Monotypic genus of evergreen shrubs; upright, leaves alternate, bi- to tripinnate; flowers in terminal panicles, white; sepals in several rings, becoming increasingly large toward the center, petals 3–6; stamens 6, anther dehiscing by long slits; ovaries with a short conical stigma; fruit a globose berry with short style remnants. — 1 species in China and Japan.

Nandina domestica Thunb. Upright, evergreen shrub, usually unbranched and multistemmed, stems usually finger thick; leaves mostly bi- to tripinnate, 30–60 cm long, base often nearly globose vaulted, leaflets elliptic-lanceolate, 3–6 cm long, red-brown on new growth, later a rich green, lighter beneath, leathery tough, entire, usually purple in fall; flowers white, small, but in 20–30 cm long, erect panicles, June–July; fruits bright red, nearly pea-sized, persisting for some time. BM 1109; DRHS 1345; MJ 1632; HKS 73. Central China; cultivated in Japan. 1804. "Sacred Bamboo" of Japan whose fruiting twigs are commonly sold at the markets in winter to decorate the home and altar. z8–9 Fig. 215. # ∅ ⚭

Many cultivars:

'Alba'. Fruits white (= var. *leucocarpa* Makino). 1877. ⚭

'Flava'. Fruits light yellow. ⚭

f. *heterophylla* see: **'Purpurea'**

var. *leucocarpa* see: **'Alba'**

'Longifolia'. Leaflets oblong-lanceolate, 5–10 cm long.

'Purpurea'. Low growing, more compact; leaves always reddish, more intensely red in fall (= f. *heterophylla* Hort.).

'Variegata'. Leaflets white variegated.

Also occasionally cultivated is a form with only 1–2 cm long, more acutely ovate leaflets (this however, as yet unnamed). Fig. 215.

In Uehara (I: 1013–1036) no less than 56 cultivars are listed! Unfortunately, they are known only by Japanese names and are not generally cultivated in the West.

Only for very mild climates; prefers a moist, humus soil in a protected area; cultivated somewhat like the large leaved *Mahonia* spp.

Fig. 215. *Nandina domestica*. The normal form is at left, at right a form with small leaflets (Original)

NEILLIA D. Don. — ROSACEAE

Deciduous shrubs, similar to *Physocarpus*, but the flowers are in simple or compound racemes; fruit capsules not inflated and dehiscing only at the ventral seam; leaves alternate, usually lobed and biserrate, stipules large, quickly abscising; sepals 5, petals 5, stamens 10–30, carpels 1–2, styles terminal; fruits usually with 5 glossy seeds. — About 12–15 species in China, Himalayas, Korea, and Java.

Outline of the more common species

- Young shoots pubescent
  Shoots angular:
    N. longiracemosa

  Shoots rather cylindrical:
    N. thibetica
- ● Young shoots smooth
  Shoots angular:
    N. thyrsiflora

  Shoots rather cylindrical:
    N. sinensis

Neillia affinis Hemsl. Very closely related to *N. longiracemosa*, to 2 m high, young shoots glabrous, angular; leaves ovate to oval-oblong, long acuminate, 5–9 cm long, base cordate, small lobed, both basal lobes often long acuminate, venation somewhat pubescent beneath, petiole 1–2.5 cm long; flowers in 3–8 cm long racemes (!), pink, calyx tube campanulate, finely pubescent and with stalked glands, as long as the sepals, May–June. W. China. 1908. z6

N. longiracemosa Hemsl. Shrub, erect, 2 to 3 m high, young shoots soft pubescent, angular, thin; leaves petiolate, ovate to broadly ovate, 4–9 cm long, long acuminate, base cuneate to round, the lowest leaves often larger, 8–12 cm long and more distinctly lobed

beneath, dark green and glabrous above, lighter beneath with pubescent venation; flowers in simple, axillary, 11–18 cm long racemes, calyx cup tubular-campanulate, yellowish-pink, 5–7 mm long, petals nearly circular, 2.5 mm long, ciliate, May–June; ovaries only pubescent at the apex. BMns 3. China; Sikiang Province. 1904. z6 Fig. 216, 217. ✦

N. malvacea see: **Physocarpus malvaceus**

N. ribesioides see: **N. sinensis var ribesioides**

N. sinensis Oliv. Shrub, 1–2 m high, twigs glabrous, brown, with exfoliating bark; leaves oval-oblong, long acuminate, 5–8 cm long, incised-serrate and lobed, teeth sharp, venation pubescent beneath at first, both sides eventually totally glabrous, light green; flowers about 12–20 in nodding, 3–6 cm long, simple racemes, calyx tube-form, whitish-pink, glabrous, 1–1.2 cm long, with a few glandular bristles, sepals triangular and long acuminate, May–June. BC 2454; NH 1945: 159; BS 2: 345. Central China. 1901. z6 Fig. 216. ✦

var. ribesioides (Rehd.) Vidal. Leaves only 3–5 cm long, with small incised lobes, usually somewhat pubescent only on the venation beneath, stipules oblong, entire, shorter than the petiole; calyx tube glabrous or finely pubescent, sepals triangular; ovaries pubescent only at the apex (= *N. ribesioides* Rehd.). W. China. 1930.

N. thibetica Franch. Shrub, to 2 m high, twigs nearly cylindrical, finely pubescent; leaves ovate, long acuminate, base nearly cordate, 5–8 cm long, biserrate and small lobed, eventually glabrous above, venation beneath with a fine and dense, persistent pubescence, petiole 8–15 mm long, stipules ovate and serrate (!); flowers in short, dense, 4–8 cm long racemes, calyx tube-form, finely pubescent, June; ovaries silky pubescent. W. China. 1910. z6

Fig. 216. **Neillia**. Leaves. Left, *N. thyrsiflora*; *N. sinensis* in the center; right, *N. longiracemosa*
(Original)

N. thyrsiflora D. Don. Shrub, erect, 1(2) m high; shoots angular and glabrous, often reddish; leaves ovate or more oblong, 4 to 10 cm long, long acuminate, 3 lobed (particularly on the long shoots), double incised-serrate, glabrous beneath or with pubescent venation; flowers in 3 to 7 cm long racemes in the upper leaf axils, calyx tube campanulate, whitish, pubescent, 8 mm long, petals ovate, white, August; ovaries nearly totally glabrous. RH 1888: 416. Himalayas. Around 1850. Of only slight ornamental merit. z8 Fig. 216.

N. torreyi see: **Physocarpus malvaceus** and **P. monogynus**

All species cultivated somewhat like *Spirea,* but more tender; they will occasionally freeze back to the ground in a hard winter, but resprout well in spring. The best are *N. sinensis* and *N. longiracemosa.*

Lit. Cullen, J.: The genus *Neillia* (Rosaceae) in Mainland Asia and in Cultivation; in Jour. Arnold Arb. **52**, 137–158, 1971.

Fig. 217. *Neillia longiracemosa* (drawn by A. Olsen)

NEMOPANTHUS Raf. — Mountain Holly — AQUIFOLIACEAE

Monotypic genus; deciduous shrub; leaves alternate, thin petioled, simple to finely dentate; stipules small, quickly abscising; flowers polygamous-dioecious, small, axillary, grouped 1–4, white, without particular ornamental merit; petals 4–5, totally distinct, linear, sepals 4–5, quickly abscising; stamens 4–5, long stalked; ovaries with 4–5 stigmas; fruit a globose berry with 4–5 seed pits.

Nemopanthus canadensis see: **N. mucronatus**

N. mucronatus (L.) Treal. Deciduous shrub, to 2 m high, young twigs thin, reddish at first, later gray; leaves elliptic-oblong, 2.5–3.5 cm long, entire or slightly dentate, with small tips, dark green above, gray-green beneath, fall color yellow, petiole 1 cm long; flowers small, inconspicuous, whitish, the female solitary, males grouped 1–4 together, filamentously thin stalked; fruits pea-sized, red, with 4 furrows, stalk 3–4 cm long. BB 2364; GSP 275; DL 1: 249 (= *N. canadensis* DC.). Atlantic N. America. 1802. z4 Fig. 218.

Found in the wild in or near *Sphagnum* peat bogs or on the shores of very cold, deep lakes. Quite hardy, but without particular ornamental merit; scarcely fruiting or fruitless.

Fig. 218. *Nemopanthus mucronatus* (from Michaux)

NEOLITSEA (Benth.) Merrill — LAURACEAE

Evergreen, dioecious trees; leaves alternate, petiolate, entire, normally 3 veined from the base up, occasionally pinnately veined, often silky pubescent when young; flowers in nearly sessile umbels, surrounded by an involucre when young; flowers unisexual, the 4 part perianth abscising, tube very short; stamens in the male flowers 6(8), all fertile, in 3 rings, the outer 4 without glands, the 2 inner ones with one gland on either side; fruit a red or black berry. — About 80 species in E. and SE. Asia and Indomalaysia.

Neolitsea aciculata (Bl.) Koidz. Evergreen tree; leaves leathery tough, oblong to oval-oblong, 5–12 cm long, 2–4 cm wide, obtuse, 3 veined from the base up, white beneath and glabrous or with appressed silky pubescence, petiole 8 to 15 mm; flowers reddish, in spring; fruits ellipsoidal, somewhat longer than the stalk, with a small cupulate perianth tube at the base, black. KIF 3: 17 (= *Litsea aciculata* Bl.). Japan, Korea. z9 # ∅

N. glauca see: **N. sericea**

N. latifolia see: **N. sericea**

N. sericea (Bl.) Koidz. Evergreen tree, young shoots green; leaves leathery tough, oblong or oval-oblong, 8–18 cm long, 4–7 cm wide, obtuse, 3 veined from the base, densely yellow-brown pubescent when young, later glabrous and white beneath, petiole 2–3 cm long; flowers yellow, in fall; fruit ellipsoid, 12–15 mm long, red. KIF 2: 26 (= *Litsea glauca* Sieb.; *Neolitsea glauca* [Sieb.] Koidz.; *N. sieboldii* [E. Ktze.] Nakai; *N. latifolia* Koidz. non S. Moore). Japan, Korea, China, Taiwan. z9 Plate 92. # ∅

N. sieboldii see: **N. sericea**

Illustrations of other species in ICS:

N. aurata 1698	*N. zeylanica* 1702
N. confertifolia 1699	*N. ellipsoidalis* 1703
N. ferruginea 1700	*N. chuii* 1704
N. levinei 1701	

NEOPANAX Allan — ARALIACEAE

Monoecious or dioecious shrubs or trees with evergreen, palmately compound leaves, later reduced to simple leaves; flowers 5 parted, in compound, occasionally simple umbels; calyx very small, truncate or dentate, ovaries 2 locular, styles 2, connate at the very base; fruits more or less appressed, with 1 seed in each locule. — 6 species, all endemic to New Zealand.

Neopanax arboreum (Murr.) Allan. Evergreen, small crowned, rounded tree, 3–6 m high in its habitat, totally glabrous, young shoots stiff; leaves palmately compound, leaflets 3–7, elliptic-oblong to narrowly obovate, coarsely and obtusely dentate, tapering at both ends, 7–20 cm long, 2.5–7 cm wide, dark green and glossy above, lighter beneath, petiolules of the leaflets 1.5–2.5 cm long, petiole to 20 cm long, flared out at the base and surrounding the stem; flowers greenish-brown, in terminal, compound umbels. BM 9280 (= *Panax arboreum* L. f.; *Nothopanax arboreum* [Forst. f.] Seem.; *Pseudopanax arboreum* W. R. Philipson). New Zealand. 1820 z8 Fig. 219. # ∅

N. laetum (Kirk) Allan. Shrub or tree, to 5 m high, glabrous in all respects; leaves 5–7 parted, the leaflets obovate to cuneate-oblong, very large, the terminal leaflets 12–25 × 5–10 cm in size, thick and leathery, coarsely serrate, green above, lighter beneath, venation conspicuous on both sides, stalk to 25 cm long, purple-red (!); flowers in racemose, compound umbels, terminal (= *Nothopanax laetum* [Kirk.] Cheesem.). New Zealand. z9 Fig. 219.

See also *Nothopanax* and *Pseudopanax.*

Fig. 219. **Neopanax.** *N. arboreum* at left; right, *N. laetum* (Original)

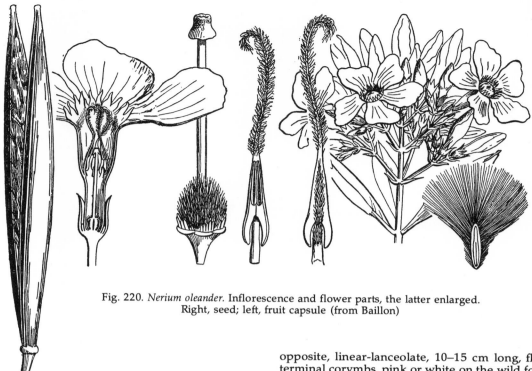

Fig. 220. *Nerium oleander*. Inflorescence and flower parts, the latter enlarged.
Right, seed; left, fruit capsule (from Baillon)

NERIUM L. — Oleander — APOCYNACEAE

Evergreen shrubs, upright, glabrous; leaves in whorls of
3, occasionally 2 or 4, narrow and leathery; flowers in
terminal cymes, short petioled; corolla funnelform, limb
deeply 5 parted, throat apex with 5 toothed or slitted
scales, the 5 limb lobes twisted in the bud stage; stamens
attached at the tube apex, but not exserted, anthers
nodding together around the stigma and attached to it,
drawn out to a filamentous, pubescent point at the base;
ovaries 2; fruit a follicle, fruits oblong, pubescent. — 3
species from the Mediterranean region to Japan.

Nerium indicum see: **N. odorum**

N. odorum Soland. Fragrant Oleander. Upright shrub,
1.5–2 m high; leaves in whorls of 3, linear-lanceolate, 12–
20 cm long, acute, thick, leathery, midrib very thick,
petiole very short; flowers pink-white on the wild type, 3
to 5 cm wide, often 50–80 flowers together in large,
terminal corymbs, fragrant, June–August. MJ 621; BM
2032; HKT 74; ICS 4836 (= *N. indicum* Mill.). Iran to
Japan and China. 1683. z9 # ∅ ✿

Includes a few known cultivars:
'Leucanthum', flowers white;
'Lutescens', flowers yellowish;
'Plenum', flowers pink, double (BM 1799).

N. oleander L. Common Oleander. Erect, bushy shrub,
2–3(5) m high and wide; leaves in whorls of 3 or
opposite, linear-lanceolate, 10–15 cm long, flowers in
terminal corymbs, pink or white on the wild form, 3 cm
wide, not scented, June–October. Mediterranean region.
1596. z9 Fig. 220. # ∅ ✿

Includes many cultivars:

Simple flowering types:
'Album Grandiflorum', white with a trace of red;
'Album Maximum, pure white;
'Coccineum Simplex', light red;
'Conte Pusteria Cortesia', apricot-yellow;
'Etna', bright purple;
'Imperio', salmon-orange;
'Solfatare', sulfur-yellow;
'Suor Lisa', dark red;
'Virginie', bright pink.

Double flowering types:
'Agnes Darac', dark red;
'Claude Blanc', bright red;
'Jean Gallen', pink;
'Louis Pouget', flesh-pink;
'Luteum Plenum', light yellow;
'Madoni Grandiflorum', white;
'Pierre Rondier', carmine-red;
'Prof. Blanchon', orange-reddish;
'Rayanat', apricot-yellow;
'Splendens', pink.

Semidouble forms:
'Album Plenum', white;
'Flavescens', light yellow;
'Pierre Gallen', pink.

Used in the open landscape only in the warmest climates. Full
sun is recommended for best flowering. Plants cultivated in
tubs must, however, be kept cool in winter, about 5–8°C (41–
46°F).

NESAEA

Nesaea salicifolia see: **Heimia salicifolia**
N. verticillata see: **Decodon verticillatus**

NEVIUSIA Gray — Snow Wreath — ROSACEAE

Monotypic genus; deciduous shrub; leaves alternate, petiolate, biserrate, with stipules; flowers hermaphroditic, solitary or in clusters; sepals 5, green, petal-like, outspread and incised-serrate; petals absent; stamens numerous, longer than the sepals, persistent; ovaries 2–4, styles slender, fruit a small achene, surrounded by the persistent calyx. — N. America.

Neviusia alabamensis Gray. Erect, broad growing, exceedingly stoloniferous shrub, 1–1.5 m high, shoots delicate, terete, finely pubescent when young, soon glabrous; leaves nearly distichous, oval or oblong, acuminate, 3–7 cm long, double serrate, finely pubescent when young, later glabrous; flowers light green, in groups of 3–8, occasionally also solitary, about 2–2.5 cm wide, filaments white and ornamental, anthers yellow, June–July. BM 6806. Found in the wild only on the sandbank of the Warrior River near Tuscaloosa, Alabama, USA. Ⓕ Lebanon, used for reforestation. z6 Fig. 221. ✣

For warm, protected sites in clay soil; quite hardy, but not particularly ornamental and more suitable for collectors.

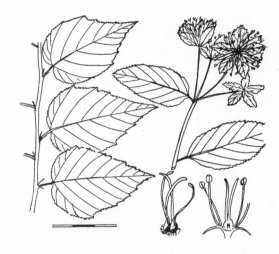

Fig. 221. *Neviusia alabamensis* (Original)

NICOTIANA L. — Tobacco — SOLANACEAE

Annual or perennial plants, very occasionally woody, strongly scented, usually glutinous-pubescent; leaves alternate, simple, usually sessile; flowers hermaphroditic, in terminal racemes or panicles, usually opening at night and then very fragrant, corolla funnelform, with a long tube, limb 5 lobed, calyx 5 parted, persistent, stamens 5, enclosed, fruit a 2 or 4 locular capsule. — 66 species, 45 in the non-tropical regions of N. and S. America, 21 species in Australia and Polynesia.

Nicotiana glauca Graham. Deciduous, upright shrub, totally glabrous, 3–6 m high, all plant parts blue-green; leaves long petioled, cordate-ovate, somewhat uneven, acute, 12–20 cm long, 7–10 cm wide, entire or somewhat sinuate; flowers in loose, terminal racemes, corolla tube opening greenish, soon turning yellow, slightly curved, slightly inflated, somewhat constricted at the throat, limb short and cupulate, August–October. BM 2837. Argentina, Paraguay, Bolivia; naturalized in the Mediterranean region and southern California. 1827. z10 Plate 125. ✣

NIEREMBERGIA Ruiz & Pavon — SOLANACEAE

Mostly herbaceous plants, only the following species is shrubby; leaves alternate, simple, flowers solitary at the tips of the young shoots; calyx tubular or campanulate, 5 parted, sepals outspread, corolla tube long, narrowed at the base, limb more or less outspread, plate-like or cupulate, with 5 lobes; fruit a bivalved capsule. — 35 species in Mexico and in subtropical S. America.

Nierembergia frutescens Dur. Small shrub, 0.3–0.8 m high, abundantly branched; leaves linear, 3–5 cm long; flowers very numerous, 2.5 cm wide, limb plate-like, light blue, gradually becoming whiter toward the margin, throat yellow, May–September (= *N. fruticosa* Hort.). Chile. 1876. z9 ✣

'Albiflora'. Flowers white.

'Atroviolaceae'. More compact habit; flowers dark violet.

'Grandiflora'. Like the species, but the flowers to 3 cm wide.

The plants are most easily grown from seed as annuals.

NITRARIA L. — ZYGOPHYLLACEAE

Divaricate branching, thorny or thornless deciduous shrubs; leaves alternate on the young twigs, clustered on the older shoots, simple, fleshy, stipules small; flowers small, yellowish or white, in small, terminal corymbs; calyx small, persistent, fleshy, 5 parted; petals 5, hollow, hooded on the apex; stamens 15 (occasionally 10–14); ovaries sessile, oblong-conical, with short styles and 2–6 nodding stigmas; fruit an ovate drupaceous berry, the exterior of the seed pit is deeply pitted; seed with a dry membranous shell. — 8 species from S. Russia to Mongolia, W. China and Iran and from Asia Minor to Saudi Arabia and N. Africa; in salt plains.

Nitraria schoberi L. Thorny, divaricate shrub, 1–1.5 m high, twigs gray-white; leaves obovate to linear spathulate, entire, fleshy, silky pubescent when young, 1–1.5 cm long, dark blue-green above, gray-green beneath; flowers in about 2 cm wide corymbs, petals spathulate, hollow, stamens as long as the petals, June; fruit red or black, 8 mm long, LAu 375. S. Russia to Mongolia, W. China and Iran. z5 Fig. 222.

Generally cultivated only in botanic gardens; somewhat difficult; needs an alkaline soil. Must be given salt in a normal garden soil.

Fig. 222. *Nitraria schoberi* (from Watson)

NOLINA Michx. — AGAVACEAE

Upright, evergreen trees, mostly short stemmed with a tuft of foliage; stem often somewhat swollen at the base; leaves in a compact cluster, long linear, very finely serrate or entire; flowers in loose, large, terminal panicles, the individual flowers rather small, perianth parts distinct, anthers cordate; ovaries 3 locular; fruit 3 sided or 3 winged, with 1–3 globose seeds. — About 30 species in North America, Texas, California and Mexico.

Nolina longifolia (Karw.) Hemsl. Stem (in cultivation), scarcely over 2 m high, flared at the base, with a thick,

corky bark, usually with several short branches on the upper part of the plant, these with tufts of about 2 m long and 2.5 cm wide, pendulous, tight, thin, long acuminate, leaves, finely scabrous on the margin; inflorescence to 2 m high, in large, branched panicles, the individual flowers small, white, appearing in summer; seeds globose, 8 mm thick. RH 1911: 206 (= *Yucca longifolia* Schult.). Mexico. 1868. z9 Plate 138. # ∅ ✧

Occasionally found in botanic garden conservatories, frequently planted in the landscape in warmer climates; needs full sun and a fertile soil.

NOLTEA Reichb. — RHAMNACEAE

Evergreen shrubs; leaves alternate, simple, serrate; flowers in axillary or terminal, few flowered corymbs; calyx with 5 incisions, the lobes erect or outspread, triangular ovate, acute; petals 5, sessile, rounded, hollow; stamens 5, as long as the petals; disc thin, lined by the calyx cup; ovaries sub-inferior, 3 lobed, with 3 single seeded locules, styles 3 sided, 3 parted on the apex; fruit an ovate, 3 lobed, drupaceous berry surrounded by the calyx tube on the basal half, with knobby, ventrally dehiscent seed pits. — 2 species in S. Africa.

Noltea africana (L.) Reichb. Evergreen shrub, erect, glabrous in all respects; leaves elliptic, obtuse, 4–6 cm long, serrate, dark green and glossy above, stipules auricle-like, margin with reddish glands; flowers white, in small, axillary- and terminal, 1–1.5 cm wide panicles, May; fruits globose, dry, 6 mm thick (= *Ceanothus africanus* L.; *Willemetia africana* Brogn.). S. Africa. 1798. z9 # ∅

Attractive, but suitable only for very mild areas.

NOTELAEA Vent. — OLEACEAE

Evergreen shrubs or trees; leaves opposite, simple, entire; flowers in short, usually axillary racemes; calyx small, 4 lobed or 4 toothed; petals 4, wide, obtuse, distinct or paired from the insertion of the stamens; fruit a globose or ovate drupe. — 7 species in Australia, 1 on the Canary Islands.

Notelaea excelsa Webb & Bert. Tall, evergreen shrub, but a 12 to 18 m high tree in its habitat, shoots smooth, gray, flattened; leaves oval-lanceolate, acute, 7.5–12.5 cm long, 2–3 cm wide, tapering to the 1.5–2 cm long petiole; flowers grouped 6–12 in axillary and terminal racemes, 3–5 cm long, white, fragrant, February–July; fruits ovate, 2 cm long, red at first, eventually dark violet-blue. Canary Islands. 1784. z9 Plate 138.

Completely hardy and tree-like in southern England; often found in the botanic gardens of milder regions.

NOTHOFAGUS Bl. — Southern Beech — FAGACEAE

Evergreen or deciduous trees or shrubs, resembling the true beech, but differing in the much smaller, densely crowded and short petioled leaves, evergreen on many species; male flowers solitary or in axillary groups of 3; fruits beechnut-like but much smaller, fruit cup (cupule) with entire or dentate scales. — 35 species in Antarctic S. America, Australia and New Zealand.

Outline of the more prominent species
(from Bean, expanded)

● Leaves evergreen

N. apiculata: Leaves entire to crenate, 18–25 mm long, with small tips.
N. betuloides: Leaves crenate, punctate beneath.
N. blairii: Leaves entire, 12–18 mm long, yellow-brown beneath.
N. cliffortioides: Leaves entire, 4–16 mm long, white beneath.
N. cunninghamii: Leaves crenate, not punctate beneath.
N. dombeyi: Leaves finely and unevenly dentate, to 3 cm long.
N. fusca: Leaves coarsely dentate, to 3 cm long.
N. menziesii: Leaves doubly crenate.
N. moorei: Leaves sharply dentate, to 7 cm long.
N. solandri: Leaves entire, obtuse, 6–18 mm long, white beneath.

●● Leaves deciduous

N. antarctica: Leaves usually less than 3 cm long, shoots densely tomentose.
N. obliqua: Leaves 5–7 cm long, shoots glabrous to lightly pubescent.
N. procera: Leaves to 10 cm long, shoots with brown pubescence.

Nothofagus alessandrii Espinosa. Deciduous, conical tree, 25–30 m high in its habitat, bark gray, young shoots pubescent; leaves ovate, obtuse, to 13 cm long and 8–9 cm wide, base broadly rounded to lightly cordate, bright green and pubescent above, blue-green and somewhat glandular beneath, margins toothed, with 11–13 raised vein pairs beneath, venation pubescent, petiole 5–8 mm long, pubescent, blade pubescent. PFC 94. Chile z7

N. alpina see: **N. procera**

N. antarctica (Forst.) Oerst. Deciduous tree, to 35 m high in its habitat, narrow upright grower, young shoots densely and finely pubescent; leaves very closely spaced, ovate, 2–3 cm long and equally wide, apices

Fig. 223. New Zealand and Australian range of the genus *Nothofagus*

Fig. 224. South American range of the genus *Nothofagus*

rounded, base cordate to truncate, margins finely undulate and irregularly crenate; flowers inconspicuous, May; fruits in 3's in a 4 lobed calyx cup. BM 8314. Tierra del Fuego to Chile. 1830. Ⓕ New Zealand, on a trial basis. z7 Plate 137; Fig. 225. ∅

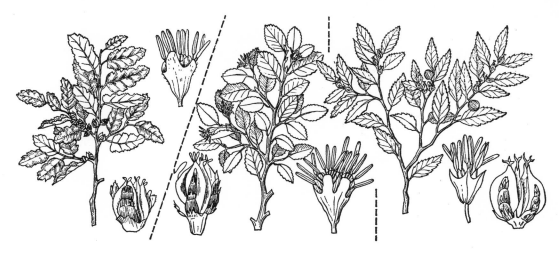

Fig. 225. **Nothofagus.** Left, *N. antarctica;* center, *N. betuloides;* right, *N. dombeyi* (from Dimitri)

'Benmore'. Dwarf form, very broad and low growing. Developed in Benmore, Scotland (the original plant is 4 m wide and 1 m high [1976]).

var. **uliginosa** (A. DC.) Distinguished by its fine, short, erect, pubescence on both sides.

N. apiculata Col. Evergreen tree, to 10 m high in its habitat, young shoots soft pubescent; leaves petiolate, 18–25 mm long, oval-oblong to elliptic-oblong, with short tips, base cuneate, entire or very finely and irregularly crenate, glabrous, venation not very conspicuous; nutlets pubescent, 2–3 winged. KF 135. New Zealand. z9 Fig. 227. # ∅

N. betuloides (Mirbel) Blume. Tall, evergreen tree, young shoots finely pubescent and glutinous (!); leaves very densely arranged, only 6 mm apart along the twig, ovate to nearly rounded, 12–25 mm long, acute or obtuse, base cuneate to round, with very small and irregular teeth, dark green above and glossy (as if

lacquered), with a fine, reticulate venation and dark punctate spots beneath, petiole 3 mm long; flowers appear in May; fruit cup 4 lobed. GC 33: 3, 11. Chile. 1830. z9 Fig. 225, 227. # ∅

N. blairii (Kirk) Krasser. Evergreen tree, 10–15 m high in its habitat, young shoots and petioles pubescent; leaves petiolate, ovate, acute or acuminate, 16–18 mm long, base rounded, entire, leathery, glabrous above, with a yellow-brown tomentum beneath; fruit cup 6–8 mm long, ovate, glabrous, 4 lobed, the lobes with 3–4 lamellae. KF 57. New Zealand, in the mountains to 600 m. Considered by Cockayne to be a "polymorphic hybrid" with *N. solandri, cliffortioides* and *fusca* as the probable parents. z9 Fig. 227. # ∅

N. cliffortioides (Hook. f.) Oerst. Evergreen tree, 5–10 m high, occasionally higher, often ony a shrub in its mountain habitat, twigs outspread, often distichous, particularly on young plants, young shoots soft

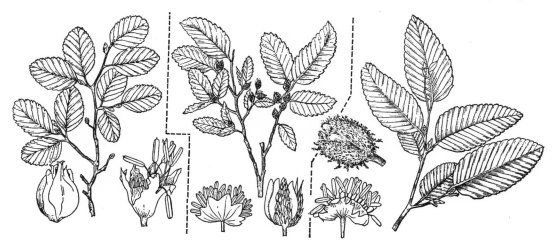

Fig. 226. **Nothofagus.** Left, *N. pumilio; center, N. obliqua;* right, *N. procera* (from Dimitri)

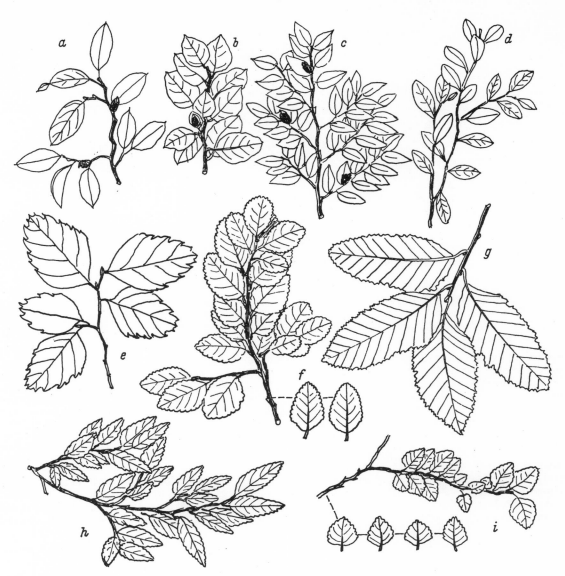

Fig. 227. **Nothofagus.** a. *N. apiculata;* b. *N. blairii;* c. *N. cliffortioides;* d. *N. solandrii;* e. *N. fusca;*
f. *N. betuloides;* g. *N. obliqua;* h. *N. dombeyi;* i. *N. cunninghamii* (from material collected in the wild)

pubescent; leaves short petioled, distichous, oval-
oblong to more rounded, 4–16 mm long, acute, rarely
obtuse, base round or cordate, entire, very leathery
tough, glabrous with reticulate venation above, more or
less densely appressed, white haired, beneath, margin
thickened, often somewhat inward curving. KF 101–
101A. New Zealand. z9 Fig. 227. # ⌀

N. cunninghamii (Hook. f.) Oerst. Evergreen, a very
large tree in its habitat, young shoots wiry-thin, dark
pubescent; leaves ovate to rhombic or nearly triangular,
6–15 mm long, half as wide, apical half acuminate, with
obtuse teeth, lower half broadly cuneate, entire, both
sides glabrous, petiole very short, pubescent; fruit cup 4

lobed, each lobe with a short scale, scales terminating in
a small, globose gland. BM 8584; DRHS 1380. Tasmania.
z9 Fig. 227. # ⌀

N. dombeyi (Mirbel) Blume. Evergreen or semi-
evergreen tree, very large with a thick trunk in its habitat,
young shoots densely tomentose; leaves regularly and
densely arrange, oval-lanceolate, 1.8–3 cm long, finely
and unevenly dentate, leathery-tough, deep green and
glossy above, lighter beneath and black punctate (occasionally punctate on both sides, then more densely so
beneath), major veins indistinct, petiole 2 mm long;
stamens bright red. NF 11:8. Chile (Chiloe Island) and
Argentina. 1916. In England only semi-evergreen in

hard winters. Ⓕ Most valuable tree in the Patagonian pampas (Southern S. America); England; experimentally in W. Germany. z9 Plate 138; Fig. 225, 227. # ∅

N. fusca (Hook. f.) Oerst. Evergreen tree, to 30 m high in its habitat, growth more columnar when young, shoots on young plants very wavy-curving, finely pubescent; leaves broadly ovate to rounded, 18–35 mm long, base cuneate to truncate, margins coarsely dentate, the teeth ciliate particularly on the incisions, thin, leathery, but tight, finely pubescent above when young, glandular beneath, later glabrous; fruit cup glutinous pubescent, 4 lobed. KF 91. New Zealand. z9 Fig. 227. #

var. **colensoi** (Hook. f.) Oerst. Leaves more leathery-tough, teeth smaller, more obtuse. KF 90, 2 (= *N. truncata* [Col.] Cock.). New Zealand. #

N. glauca (Phil) Krasser. Tree with a very rough trunk, young shoots bluish pruinose, densely rough haired; leaves ovate to more elliptic, obtuse, to 8 cm long, 5–6 cm wide, doubly incised crenate, base cordate, blue-green on both sides, particularly beneath, with 8–10 vein pairs, very short petioled. PFC 96. Chile; Andes. Easily distinguished from all the other species in the large blue-green leaves. z7 ∅

N. leonii Espinosa. Deciduous tree, young shoots pubescent; leaves ovate, obtuse, 3 to 10 cm long, 2–4 cm wide, margins double serrate, the teeth very small, base broadly cuneate to nearly truncate, with 7–13 vein pairs, these raised on the glabrous underside, petiole about 1 cm long. Chile. z7

N. menziesii (Hook. f.) Oerst. Evergreen tree, 18–20(30) m high in its habitat, bark silvery-white, particularly on young trees, young twigs yellow-brown pubescent; leaves short petioled, broadly triangular ovate or rhombic to nearly circular, obtuse, base unevenly cuneate, 8–12 mm long, thick leathery, irregularly and double crenate, margins thickened, glabrous on both sides, except for the 2 pubescent pits on the underside near the petiole, these 2 mm long and pubescent. KF 89. New Zealand. z9 # ∅

N. moorei Maiden. Evergreen tree, 30–45 m high in its habitat; leaves ovate to oval-lanceolate, 2.5–7.5 cm long and 1.8–3 cm wide, acuminate on both ends, sharply serrate, tough, with 9–15 distinct vein pairs, glabrous on both sides, but the midrib above, leaf petiole and young shoots are brownish pubescent, stipules very narrow, 8–12 mm long, Australia; New South Wales. 1892. z9 # ∅

N. nitida (Phil.) Krasser. Tall, evergreen tree, young shoots short yellow pubescent; leaves rhombic-ovate, 2.5–3.5 cm long, acute to acuminate, with 5–6 vein pairs, margins serrate, dark green and very glossy above, without resin papilla beneath (as compared to *N. dombeyi*!), leathery, glabrous except for the petiole and the yellowish pubescent midrib. PFC 92. Chile; Valdiva, Chiloe Island, on marshy soil. z7 #

N. obliqua (Mirbel) Blume. Deciduous tree, to 30 m high in its habitat, very fast growing, young shoots long, glabrous; leaves distichous, ovate to oblong, obtuse, 3–7.5 cm long, 2.4 cm wide, base round to broadly cuneate, oblique, margins irregular and finely serrate, teeth triangular, dark green above, blue-green and glabrous beneath, petiole 3 mm long; male flowers solitary in the leaf axils, composed of the calyx and 30–40 stamens. Child. 1849. Ⓕ England, W. Germany. z7 Fig. 226, 227. ∅

N. procera (Poepp. & Endl.) Bl. Deciduous tree, 20 m high and more in its habitat, young shoots with short brown pubescence, winter buds acuminate, 6 mm long; leaves oblong to narrowly ovate, obtuse at the apex or rounded, 3–10 cm long, 2–4 cm wide, base round to broadly cuneate, quite finely dentate, with 14–18 vein pairs, yellow-green above and finely pubescent, particularly on the midrib, lighter with pubescent venation beneath, petiole 2–4 cm long, pubescent. Chile. 1913. (Pizzarro has suggested that the correct name for this plant is *N. alpina* [Poepp. & Endl.] Oersted.) Ⓕ England, W. Germany. z7 Plate 137; Fig. 226. ∅

N. pumilio (Poepp. & Endl.) Krasser. Deciduous shrub or small tree, but also growing to 18 m high in protected areas, young shoots reddish, with white lenticels and dense yellow pubescence; leaves ovate, 2.5–3 cm long, with 5–7 vein pairs, margin regularly and deeply bicrenate, always with 2 teeth between each vein pair. PFC 100. Chile; in the mountains past the timber line and south to the Strait of Magellan. Often confused with *N. antarctica*, but the leaves and fruits are totally different. z7 Plate 126; Fig. 226.

N. solandri (Hook. f.) Oerst. Tall, evergreen tree, 12–24 m high in its habitat, bark black and furrowed on older trees, young twigs densely pubescent; leaves short petioled, linear-oblong to elliptic-oblong, 6 to 18 mm long, obtuse, base cuneate and oblique, entire, leathery, glabrous above with reticulate venation, gray-white appressed pubescent beneath, margins somewhat curved. KF 56. New Zealand. 1914. z9 Fig. 227. # ∅

N. truncata see: **N. fusca** var. **colensoi**

Outside of Ireland, England and the warmer zones of North America, where nearly all the species thrive, *Nothofagus* is not yet sufficiently tested for winter hardiness. It appears certain that at least the deciduous species will prove hardy in zone 7 in protected areas. Soil preference somewhat like that of beech.

Lit. Krasser, F.: Bemerkung zur Systematik der Buchen; in Ann. K. K. Naturhist. Hof-Mus. **11**, 159 to 163, 1896 ● Espinosa, M. R.: Dos especies nuevas de *Nothofagus*; in Rev. Chil. Hist. Nat. **32**, 171–197, 1928 ● Pizarro, C. Munoz: La justification del nombre *Nothofagus alpina* (Poepp. et Endl.) Oersted, para el rauli; Bol. Univ. Chile **52**, 59–61, 1964 ● Philippi, R. A.: Plantarum novarum Chilensium Centuria quarta; Linnaea **29**, 43–44, 1857–1858.

NOTHOPANAX Miq. — ARALIACEAE

Evergreen trees and shrubs; leaves alternate, simple or palmate; flowers small, unisexual or hermaphroditic, in small, 1.5–2.5 cm wide umbels, but these are grouped into large double or multiple compound umbels; the individual flowers are dull colored and not very conspicuous; fruits and seeds compressed. — About 15 species in New Zealand and China.

Nothopanax arboreum see: **Neopanax arboreum**

N. davidii (Franch.) Harms. Evergreen, glabrous, 3–12 m tall tree; leaves either simple or (occasionally) bifoliate, or (often) trifoliate, the simple leaves and the leaflets are rather equal in form and size, usually narrowly lanceolate, 7–15 cm long, tapered to both ends, long and narrowly acuminate, sparsely dentate, glossy dark green above; flowers small, greenish-yellow, in rounded or conical, 7–15 cm long, compound umbels, July–August; fruits black. JRHS 56: 54 (= *Panax davidii* Franch.). China; Hupeh, Szechwan Provinces. 1910. z9 Fig. 228. # ⌀

N. delavayi (Franch.) Harms. Shrub, 1–5 m; leaves usually 3–5 part pinnate, leaflets oblong-lanceolate, leathery tough, 6 to 12 cm long, 1–2.5 cm wide, entire to finely crenate, sessile; flowers in 15 cm long compound panicles of small, racemose umbels (= *Panax delavayi* Franch.). China: Yunnan Province. z9

N. guilfoylei Merrill. Upright shrub, to 5 m high, only sparsely branched with conspicuous lenticels; leaves

Fig. 228. *Nothopanax davidii* (from ICS)

pinnate, petioles to 40 cm long, leaflets 3–7, oblong-elliptic, 5–7 cm long, terminal leaflets 12–15 cm long, often irregularly lobed, thorny toothed, usually white margined and gray speckled. RH 1891: 225 (= *Polyscias guilfoylei* [Bull.] Bailay). Polynesia. 1873. z10

N. laetum see: **Neopanax laetum**

Cultivated only in the warmer climates.

NOTOBUXUS Oliv. — BUXACEAE

Both South African species have been considered an independent genus; but botanists now inclined to the view that they belong in the genus *Buxus* since the characteristics are so similar to the latter genus.

Notobuxus macowanii Oliv. Cape Box. A small tree in its habitat, to 9 m high, with a slender, limbless, greenish-brown trunk, shoots angular, somewhat pubescent only when young, later glabrous; leaves nearly sessile, broadly lanceolate, obtuse or rounded, to 2.5 cm long, base cuneate, leathery tough, deep green, smooth on both sides, venation hardly visible; flowers in the leaf axils, appearing in winter; fruit a small horned capsule, with black, glossy seeds. PPT 1184 (= *Buxus macowanii* [Oliv.] Phillips). S. Africa. z10 Plate 138. #

NOTOSPARTIUM Hook. f. — LEGUMINOSAE

Leafless shrubs (only the juvenile plants foliate!), similar to *Carmichaelia*, but differing in the fruits; twigs compressed, lateral twigs pendulous; flowers small, in lateral racemes; calyx campanulate, 5 toothed, teeth short, rather equal in form; standard obovate-obcordate, tapering to a short claw, wing petals oblong, shorter than the keel petal, auricled at the base; keel hatchet-shaped, obtuse; pods linear, straight or sickle-form, very knotty, not dehiscing, short beaked, 1 seed in each knot. — 3 species in New Zealand; very rare in its habitat.

Notospartium carmichaeliae Hook. f. Shrub, 1.3 m high, twigs somewhat compressed to nearly cylindrical, glabrous, pendulous, furrowed; leaves simple (only on young plants), obcordate to circular, entire to emarginate, 6 mm long; flowers purple-pink, in 3–5 cm long racemes with 8–20 flowers, calyx silky pubescent, July; pod 2–2.5 cm long. BM 6741; GC 129: 83; DRHS 1383. New Zealand, S. Island. 1883. z9 ✛

Only for very mild regions.

NUTTALLIA See: OSMARONIA

NYSSA L. — Tupelo, Black Gum — NYSSACEAE

Deciduous trees; leaves alternate, entire or dentate; flowers inconspicuous, dioecious-polygamous; in stalked, axillary clusters; male flowers many in stalked clusters, calyx disk-form or cupulate, 5 toothed; female flowers in groups of 1–2, sessile, occasionally short petioled, with a campanulate calyx, limb 5 toothed; ovaries 1–2 locular; fruit an oblong drupe; seed pit ribbed or winged. — 4 species in N. America, 2 in Asia.

- Male flowers in sessile heads:
  N. ogeche

- - Female flowers in stalked umbels or racemes;
    1. Female or hermaphroditic flowers 3–6 in stalked umbels or short racemes:
       N. sinensis

    2. Female or hermaphroditic flowers solitary, axillary or grouped in heads of 2–7;
       ★ Female flowers grouped 2–8; fruits 1–2 cm long:
       N. silvatica

       ★★ Female flowers solitary, axillary; fruits 2–3 cm long:
       N. aquatica

Nyssa aquatica L. Tree, 15 to 20(30) m high in its habitat, trunk often markedly swollen at the base, crown narrow, young shoots pubescent, soon becoming glabrous and red; leaves oval-oblong, 12–16(20) cm long, acuminate, entire to sparsely dentate, glossy dark green above, blue-green beneath, both sides densely pubescent when young, later only beneath, fall color blue-red; pistillate flowers solitary, March–April; fruits dark purple, 2–3 cm long, seed pit with 8–10 ribs. MD 1931: 10; BB 2733; SS 200 (= *N. uniflora* Wang.). SE. USA, in river basins where it is in standing water for a large part of the year. Before 1735. z7 Plate 137. ∅ ⚲

N. biflora see: **N. silvatica** var. **biflora**

N. candicans see: **N. ogeche**

N. capitata see: **N. ogeche**

N. multiflora see: **N. silvatica**

N. ogeche Bartr. ex Marsh. Multistemmed shrub or tree, 8–10 m high, narrow crowned, young shoots greenish, densely pubescent, later brown and glabrous; leaves oval-elliptic to obovate-oblong, 10–14 cm long, acute to obtuse, eventually glossy above and yet somewhat pubescent, bluish beneath, brownish pubescent on the venation or also glabrous; pistillate flowers solitary, on short, shaggy pubescent stalks; fruits about 1.5 cm long, red, sour to the taste, seed pit with 10–12 ribs. MD 1931: 11; SS 219; EP IV, 220a:1 (= *N. capitata* Walt.; *N. candicans* Michx.). USA; Florida to E. Georgia, in swamps or along rivers. 1806. z8 Plate 137.

N. silvatica Marsh. Tree, 25–30(35) m high, crown rather narrowly cylindrical, rounded, twigs often hanging to the ground, young shoots eventually glabrous and reddish; leaves obovate to elliptic, 5–12 cm long, acute, entire, occasionally with a few large teeth, glossy green

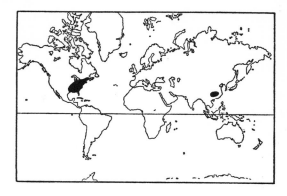

Fig. 229. Range map of the genus *Nyssa*

above, bluish beneath, fall foliage a beautiful scarlet-red to blue-red; flowers greenish, appearing soon after the leaves; fruits grouped 1–3 together, blue-black, 1–1.5 cm long, bitter tasting, seed with 10–12 ribs. BB 2721; GTP 308; MD 1923: Pl. 1 (= *N. multiflora* Wangh.). Eastern USA, in moist and dry forests and swamps, on hills and mountain slopes. Before 1750. z3 Plate 137; Fig. 230. ∅ ⚲

var. **biflora** (Walt.) Sarg. Smaller than the species, about 12(18) m high, trunk thin, but much swollen at the base, crown conical to somewhat rounded, shoots short pubescent at first, later glabrous and dark brown; leaves narrowly elliptic, obtuse, 5–8 cm long; pistillate flowers in groups of 2; fruits dark blue, seed pits ribbed. BB 2722; SM 574; MD 1931: 14 (= *N. biflora* Walt.). SE. USA. 1793. z9 Plate 137.

N. sinensis Oliv. Tree, to 15 m high, young shoots short, thick, pubescent, later glabrous; leaves oval-oblong, 8–

Fig. 230. *Nyssa sylvatica*, with fruits (from Illick)

12 cm long, acute to acuminate, base round to broadly cuneate, deep green and glossy above, lighter beneath with pubescent venation, petioles 2–4 cm long; fruits about 1.2 cm long, bluish, 1–3 on a common, to 5 cm long stalk. LF 246; CIS 39. Central China. 1902. z9 Plate 126, 137.

N. uniflora see: **N. aquatica**

With the exception of *N. silvatica,* which thrives equally well on dry soil, all species prefer a moist site. All like a fertile soil. Somewhat difficult to transplant.

Lit. Uphof, Th.: Die amerikanischen *Nyssa*-Arten; in Mitt. DDG 1931, 2–16 (with 4 plates.) ● Wangerin, W.: Nyssaceae; in Engler, Pflanzenreich **41**, 1–20, 1910.

OCHNA L. — OCHNACEAE

Evergreen and deciduous trees and shrubs; leaves alternate, leathery, margins finely serrate; flowers in racemes or solitary, mostly yellow, occasionally greenish; sepals 5, colored, overlapping, persistent; petals 5 to 10; stamens numerous; ovaries deeply 3–10 lobed, the lobes later developing into single seeded drupes, these sessile and surrounding a central disc; seeds black. — 85 species in tropical and S. Africa as well as tropical Asia, but only the following is generally cultivated.

Ochna atropurpurea DC. Low, evergreen shrub, occasionally to 6 m high in its habitat, twigs thin, bark smooth brown, but very densely covered with lenticels and therefore nearly scaly in appearance; leaves elliptic to

oblong or ovate, 3–7 cm long, 1 to 1.5 cm wide, thin, obtuse, margins finely dentate, teeth erect; flowers solitary or in 2's, at the tips of short shoots, sepals 5, golden-yellow, bright red at fruiting, petals 5, golden-yellow, abscising, May; fruits, 5–6 globose, nearly pea-sized, black drupes, pendulous on the bright red receptacle, the similarly red sepals are reflexed (being curved upward). BM 9042; DRHS 1391; MCL 80. Natal. 1860. z10 # ⊕ ⊛

The plant cultivated in Europe as *O. serrulata* (Hochst.) Walp. and originally collected in the forest near Natal Bay (S. Africa) is, according to South African botanists, a dubious species and probably identical to *O. atropurpurea* DC. A revision of the genus by D. J. B. Killick may be found in Vol. 22 of Flora of Southern Africa.

OCOTEA Aubl. — LAURACEAE

Evergreen trees or shrubs; leaves alternate, simple, with 2 large glands in the vein axils at the midrib beneath, pinnately veined; flowers small, in axillary, stalked panicles, corolla tube short or campanulate, with 6 segments, greenish; fruit an elliptic berry. — 300 to 400 species in the tropics and subtropics, 1 on the Canary Islands.

Ocotea foetens (Ait.) Benth. An evergreen tree, to 20 m

high in its habitat; leaves broad lanceolate or elliptic to ovate, to 20 cm long and 5 cm wide, acute to acuminate, dark green above, lighter beneath, with 2 large, pubescent glands, usually in the vein axils of the basal vein pair; flowers in about 10 cm long, stalked panicles, white, 1 cm wide; fruits nearly acorn-like in appearance, in a cup, fleshy, 2 cm long, black-red, glossy. KGC 13; BFCa 17. Canary Islands. z10 # ⊛

ODONTOSPERMUM

Odontospermum sericeum see: **Astericus sericeus**

OLDENBURGIA Less. — COMPOSITAE

Low shrubs with very thick, woolly shoots, leaves alternate and very densely arranged, large, resembling a large leaved *Rhododendron* (see Plate 136); flowers hermaphroditic, solitary or a few at the shoot tips, ray florets radiating, corolla bilabiate; fruits silky-shaggy. — 3 species in South Africa.

Oldenburgia arbuscula DC. Shrub, to 2 m high in its

habitat and 3 m wide, young shoots finger-thick, white woolly; leaves clustered in a rosette at the shoot tips, obovate to oblong, 20–40 cm long, 10–20 cm wide, both sides white woolly on the new growth, later dark green above, glossy, rather stiff; flower heads solitary at the shoot tips, large, reddish and white, calyx thick and woolly. BM 7942. South Africa. 1830. A beautiful shrub, but rarely found in culture. z10 Plate 136.

OEMLERIA See: **OSMARONIA**

OLEA L. — Olive Tree — OLEACEAE

Thorny or thornless, evergreen trees or shrubs; leaves opposite, entire, rarely dentate; flowers hermaphroditic or dioecious and polygamous, in axillary, rarely terminal, clusters or 3 forked branched panicles; calyx short, 4 toothed or 4 parted; corolla with a short tube and 4 lobed (or absent) limb; ovaries 2 locular, locules 2 seeded, style short; fruit an ovate, ellipsoid or globose drupe, usually single seeded. — About 20 species in the Mediterranean region, N. and S. Africa, tropical and central Asia, Australia, New Zealand and Polynesia.

Olea europeae L. Olive Tree. Small, evergreen tree, usually not over 5–6 m high, slow growing and becoming very old, trunk of cultivated plants usually short, thick, furrowed and irregular; leaves silver-gray, leathery tough; flowers yellowish-white, fragrant, June–August; fruits globose to plum-shaped, eventually blue-black. z9 #

var. **oleaster** DC. The wild olive tree. Shrub or small tree, divaricate branching habit, thorny, twigs angular; leaves elliptic to oblong; fruits small, more globose, not oily, not edible (= *O. oleaster* Hoffm. & Link). z9 # ⚭

var. **sativa** (Hoffm. & Link) Rouy. The cultivated olive tree. Taller growing, twigs more cylindrical, not thorny; leaves lanceolate, willow-like; fruits either plum-shaped or more globose according to the cultivar, dark blue or yellow, edible, very oily. HM 2921; BC 2576 (7 forms). z9 Plate 139. # ∅ ⚭ ✕

Italian nurseries grow a number of olive cultivars distinguished by their oil content or culinary use. A few of the nearly 150 cultivars are described here:

Cultivated for olive oil production

'Coratina'. Self fertile tree, upright habit; fruits large, not evenly colored.

'Corregiola' see: **'Frantoio'**

'Frantoiana' see: **'Frantoio'**

'Frantoio' ('Frantoiana', 'Corregilo', 'Razzo'). Shoots more pendulous; fruits large, ovate, fleshy. Needs a warm site.

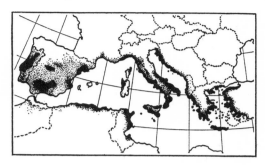

Fig. 231. Range map of *Olea europaea* (from Allbaugh & Soule)

'Leccino' ('Leccio'). Strong growing, twigs pendulous; fruits large, fleshy, violet. Suitable for less favorable climates.

'Maurino'. Fruits medium-sized, abundantly produced.

'Moraiolo' ('Morinello'). Strong growing; fruits medium-sized. Suitable for less favorable climates.

'Razzo' see: **'Frantoio'**

Cultivated for culinary uses

'Ascolana'. Yellow-green, very large, oval, fleshy.

'Cucco'. Black, oval, medium-sized, abundantly produced.

'Santa Caterina'. Fruits large, ovate.

'Uovo di Piccione'. Medium-sized, black-red.

O. oleaster see: **O. europaea** var. **oleaster**

Common as an ornamental in mild climates, and normally found in botanic gardens as one of the oldest cultivated plants. Suitable for conservatory use in cooler climates.

Lit. Sprenger, C.: Ölbaum und Oleaster; in Mitt. DDG 1916, 103–110 ● Hegi, G.: Fl. v. Mitteleuropa, V (3), 1935–1944, 1927 ● Condit, I.: Olive Culture in California; Cal. Agr. Ext. Circ. **135**, 1–36, 1947 (18 ills.).

OLEARIA Moench. — COMPOSITAE

Evergreen shrubs or small trees; leaves alternate, occasionally opposite, entire or dentate, underside usually tomentose; flower heads large or small, solitary or in corymbs or panicles; involucre ovate-campanulate, imbricate; ray florets white or reddish to blue, narrowly ligulate; disc florets tubular, short 5 toothed, hermaphroditic; ovary finely pitted, flat or somewhat convex; pappus bristly, in two rows, outer row often composed of very short hairs; achenes round or compressed, ribbed or striped. — About 100 species in New Zealand, Australia and New Guinea.

Outline of the species described

I. Flower heads large, solitary on stalks with bracts; leaves large:

O. angustifolia, chathamica, phlogo-pappa, semidentata

II. Flower heads small, 4–12 mm wide, in panicles or corymbs (solitary on *O. nummulariifolia*); with 6–24 individual flowers (per head);
> Leaves opposite:

O. buchananii, traversii,

>> Leaves alternate, 3–12 cm long, scabrous and coarsely dentate:
O. macrodonta

>>> Leaves alternate, linear, 10–20 cm long;
O. lacunosa

>>>> Leaves alternate, small, entire, 6–30 mm long:
O. haastii, nummulariifolia

III. Flower heads small, 4–5 mm long, narrow, tubular in panicles with 1–5 individual flowers, occasionally more; leaves large, 3–12 cm long, alternate:
O. albida, avicenniifolia, forsteri

IV. Flower heads small, 4–6 mm long, solitary or in clusters; leaves opposite or in opposing clusters, small, 5–30 mm long:
O. lineata, odorata, solandri, virgata

Olearia albida (Hook. f.) Hook. f. Shrub, 1.5–2.5 m high, to 5 m high in its habitat; leaves alternate, entire, oblong to oval-oblong, 5–10 cm long, leathery, somewhat mealy pulverulent above when young, later glabrous, white tomentose beneath; flowers white, in terminal and axillary clusters, 5–7 cm wide, heads with 3–6 individual flowers, ray flowers 1–3, July to August. New Zealand, North Island. z9 Plate 141. # ⌀ ✛

O. alpina see: **O. lacunosa**

O. angustifolia Hook. f. Stiffly branched shrub, about 1.5 m high (to 5 m high in its habitat), twigs, leaf undersides and all stalks soft white tomentose; leaves linear-lanceolate, 5–12 cm long, 0.8–1.8 cm wide, very stiff, leathery tough, glossy green and glabrous above, finely crenate with callous tips; flower heads 3 to 5 cm wide, white, disc dark purple, July. KF 138. New Zealand. z9 # ⌀ ✛

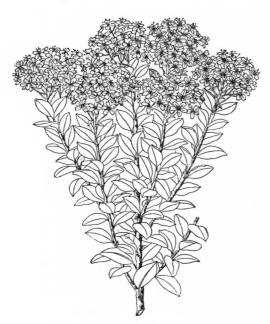

Fig. 232. *Olearia haastii* (from Gartenflora)

O. arborescens. Shrub, or also a small tree in its habitat, to 4 m, shoots angular; leaves alternate, broad ovate to more elliptic, 2–6 cm long, 2–4 cm wide, acute, leathery, eventually glabrous above, silky pubescent below, margins distinctly dentate to nearly entire, petiole 1–2 cm long; flowers white, *Aster*-like, in terminal and axillary, 10–15 cm wide, panicles, May–June (= *O. nitida* Hook. f.). New Zealand. z9 Plate 141. # ✛

O. avicenniifolia Hook. f. Shrub, 1–2 m high, twigs striped and angular, finely white tomentose; leaves oval-lanceolate, 5–12 cm long, leathery, glabrous above, whitish or brownish tomentose beneath, both sides with distinct reticulate venation; flower heads small, but many in 5–7 cm wide, dense cymes, white, axillary and terminal, August. KF 111. New Zealand. z9 # ⌀ ✛

O. buchananii Kirk. Upright shrub, also a small tree in its habitat, young shoots 5 mm thick, reddish, glabrous; leaves opposite, 5–10 cm long, elliptic-lanceolate, obtuse, entire, thin whitish pubescent beneath; flowers very small, in loose axillary cymes, about as long as the leaves. New Zealand, N. Island. z9 # ⌀

O. chathamica Kirk. Stiffly branched, upright, 0.7–1.5 m high shrub, young shoots channeled, white tomentose as are the leaf undersides and petiole; leaves alternate, quite variable in form, lanceolate, oblanceolate to oblong-obovate, 2.5–12 cm long, thick, leathery, densely serrate with short, calloused teeth, green and glabrous above; flower heads *Aster*-like, 4–5.5 cm wide, solitary on 10–15 cm long stalks, lilac, disk violet, May–June. BM 8420. Chatham Island. 1910. z9 Plate 141. # ⌀ ✛

O. erubescens Dipp. Shrub, 0.7–1.2 m high, young shoots long and thin, brownish and glossy pubescent like the leaf undersides; leaves alternate, sub-sessile, narrowly ovate to oblong, 1.2–3 cm long, stiff and leathery, acute, base round or tapered, distinctly serrate, dark green and glabrous above, inflorescences branched, with several (occasionally only 1) head, these 2.5 cm wide, white, disk yellow, May–June. Tasmania, S. Australia. 1840. z9 # ✛

O. fasciculifolia see: **O. soldandri**

O. forsteri see: **O. paniculata**

O. gunniana see: **O. phlogopappa**

O. haastii Hook. f. Bushy, rounded shrub, 1–2 m high, young shoots densely gray-white tomentose; leaves alternate, densely crowded, oblong to elliptic, tapering to both ends, obtuse, entire, 12–25 mm long, leathery thick, dark green above, glossy and glabrous, white tomentose beneath; flowers in axillary cymes in the apical leaf axils, developing a flat, terminal, 5–7 cm wide cyme, white, disk florets yellow, the individual flower heads 8 mm wide, July–August. BM 6592. New Zealand. 1858. The hardiest species, but still only to z9. Plate 141; Fig. 232. # ⌀

O. hectori Hook. f. A shrub to 5 m high in its habitat, shoots thin, furrowed, glabrous, bark dark red-brown; leaves in opposite clusters of 2–4 on short shoots, from

Plate 129

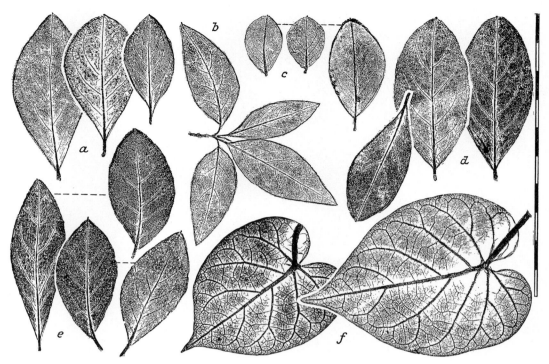

Menziesia. a. *M. ciliicalyx;* b. *M. pentandra;* c. *M. purpurea;* d. *M. pilosa;* e. *M. ferruginea* —
f. *Marsdenia erecta* (c, d, e. from wild plants)

Menziesii ciliicalyx in the Royal Botanic Garden, Edinburgh, Scotland

Plate 130

Medicago arborea
in the S'Avall Garden, Mallorca, Spain

Moltkia petraea
Photo: L. Späth, Archiv

Morus alba 'Nana' (= f. *globosa*) in the Copenhagen Botanic Garden, Denmark

Plate 131

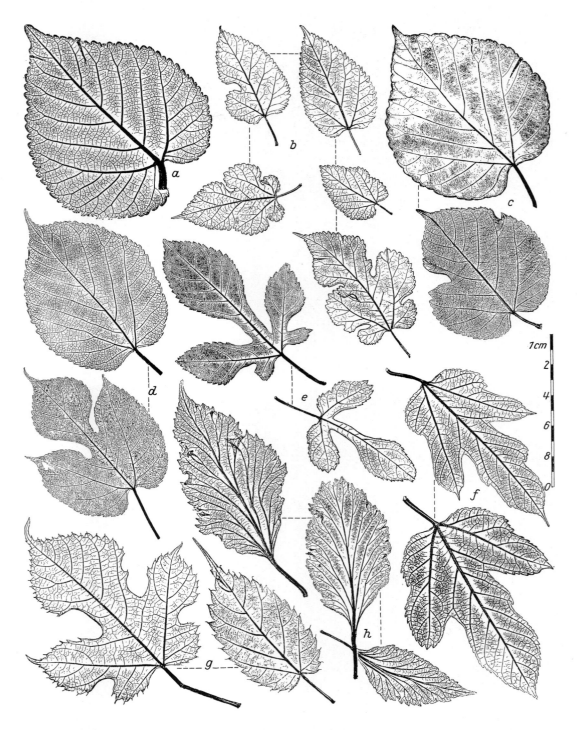

Morus. Leaves. a. *M. nigra;* b. *M. alba;* c. *M. cathayana;* d. *M. rubra;* e. *M. australis;* f. *M. kagayamae;* g. *M. mongolica;* h. *M. alba* 'Venosa' (most material from Royal Botanic Gardens, Kew, England)

Plate 132

Muehlenbeckia complexa in Royal Botanic Gardens, Kew, England

Muehlenbeckia complexa f. *trilobata* in Royal Botanic Gardens, Kew, England

Plate 133

Morus nigra in the Royal Botanic Garden, Edinburgh, Scotland

Morus alba 'Multicaulis' in the Moreira Da Silva Nursery, Porto, Portugal

Plate 134

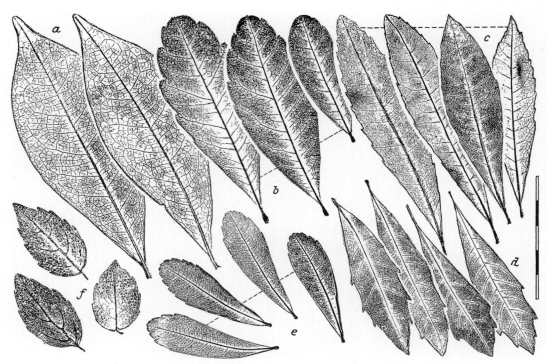

a. *Michelia figo.* — **Myrica.** b. *M. pensylvanica;* c. *M. californica;* d. *M. cerifera;* e. *M. gale* —
f. *Mitraria coccinea* (c, d, e. from wild plants)

Myrica gale var. *tomentosa* in its native habitat in Japan
Photo: Dr. Watari, Tokyo

Plate 135

Myrtus cheken
in Royal Botanic Gardens, Kew, England

Myrtus communis var. *tarentina*
in Royal Botanic Gardens, Kew, England

Myrica californica
in Royal Botanic Gardens, Kew, England

Myrtus luma
in Dereen, Ireland, 6 m high

Plate 136

Oldenburgia arbuscula in the Kirstenbosch Botanic Garden, South Africa

Myricaria germanica in its native habitat in the Ochotnica Riverbed in the Gorze Mts., Poland
Photo: K. Browicz, Poland

Plate 137

Nyssa. a. *N. aquatica;* b. *N. ogeche;* c. *N. silvatica;* d. *N. sinensis;* e. *N. silvatica* var. *biflora*
(material collected from wild plants)

Nothofagus antarctica in spring
in Dawyck, Scotland

Nothofagus procera
in Crarae, Scotland

Plate 138

Nolina longifolia
Isola Bella, Lake Maggiore, Italy

Nothofagus dombeyi
in Birr Castle Garden, Ireland

Notelaea excelsa
in the Hillier Arboretum, England

Notobuxus macowanii
in Kirstenbosch Botanical Garden, South Africa

Plate 139

Olea europaea, an old olive oil tree at the Sao Jorge Castle in Lisbon, Portugal;
twig above somewhat reduced

Plate 140

Olearia nummulariifolia var. *cymbifolia* in the Royal Botanic Garden, Edinburgh, Scotland

Olearia nummulariifolia in the Royal Botanic Garden, Edinburgh

Plate 141

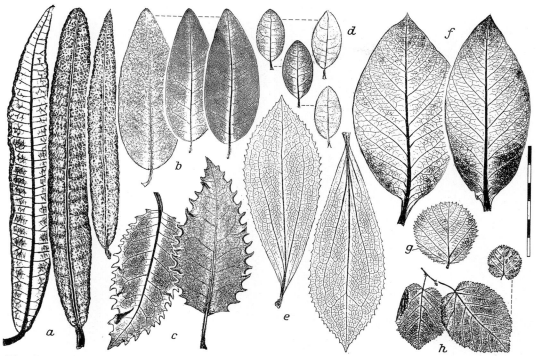

Olearia. a. *O. lacunosa;* b. *O. albida;* c. *O. macrodonta;* d. *O. haastii;* e. *O. chathamica.* — f. *Osmaronia cerasiformis.* — **Ostryopsis.** g. *O. nobilis;* h. *O. davidiana* (the *Ostryopsis* leaves are materials from China, others from Kew Gardens, England)

Olearia arborescens
in the Glasnevin Botanic Garden, Dublin, Ireland

Olearia traversii
in Royal Botanic Gardens, Kew, England

Plate 142

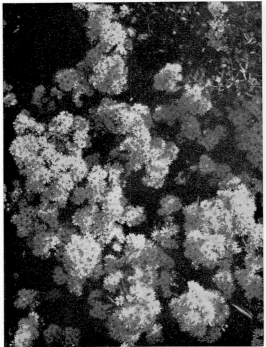

Olearia phlogopappa
in the Royal Botanic Garden, Edinburgh, Scotland

Olearia pinifolia
in the Knighthayes Park Court, Devon, England

Olearia moschata
in Malahide, Ireland

Osmanthus serrulatus
in the Hillier Arboretum, England

Plate 143

Oplopanax horridus in its native habitat in North America

Orixa japonica in the Dortmund Botanic Garden, W. Germany

Plate 144

Osmanthus heterophyllus 'Variegatus'
on Madre Island, Lake Maggiore, Italy

Osmanthus heterophyllus 'Rotundifolius'
in the Hillier Arboretum, England

Osmanthus heterophyllus 'Myrtifolius' in Westfalenpark, Dortmund

narrow to broadly obovate, 2–5 cm long, 0.5–2 cm wide, very thin, eventually totally glabrous above, thin silvery tomentose beneath; flowers small, white, in clusters of 2 to 5. New Zealand. z9 #

O. ilicifolia Hook. f. Medium-sized shrub, broad growing, to 2.5 m in its habitat, densely branched, all parts musky scented, leaves linear-oblong, stiff and leathery, 5–10 cm long, gray-green above, scabrous and coarsely dentate, white tomentose beneath; flowers white, 12 mm wide, fragrant, in 5–10 cm wide corymbs, June. New Zealand. One of the hardiest species, similar to *O. macrodonta*, but the venation departing the midrib at right angles (those of *O. macrodonta* at acute angles). z9 # ∅

O. insignis see: **Pachystegia insignis**

O. lacunosa Hook. f. Stiffly branched shrub, about 1.5 m high (in its habitat also 4–5 m high), young shoots, petioles and leaf undersides gray or brown tomentose; leaves very stiff, leathery, linear-oblong, 8–16 cm long, 8–25 mm wide, entire or with indistinct teeth, rugose and glabrous above, tomentose beneath with an involuted margin, petiole short; flowers in terminal, 10 cm wide panicles, heads small, 5 mm wide, white (= *O. alpina* Buch.-Ham.). New Zealand. 1864. Very easily recognizable by the long, narrow leaves. z9 Plate 141. # ∅

O. lineata see: **O. virgata** var. **lineata**

O. macrodonta Baker. Shrub, 1.5–5 m high, bark of older plants exfoliating in strips, young shoots angular and pubescent, these and the leaves scented like musk when crushed; leaves alternate, 5–10 cm long, ovate, oval-oblong to narrowly oblong, acute, base round, totally undulate and with long, sharp teeth, dark green and glossy above, silvery-white tomentose beneath; flowers in 7–15 cm wide, branched clusters at the ends of the current year's shoots, heads 12 mm wide, white, disk reddish, usully flowers abundantly, July. BM 7065; JRHS 90: 95. New Zealand. 1886. z9 Plate 141. # ∅ ✧

O. moschata Hook. f. Compact shrub, about 1–1.5 m high, with a musky scent, densely branched, somewhat glutinous, shoots, leaf undersides and calyx white tomentose; leaves elliptic to obovate, 1–2 cm long, 0.5–1 cm wide, entire, gray-green above, white tomentose beneath; flowers white, yellow in the center, in axillary corymbs, July. New Zealand. z9 Plate 142. #

O. nitida see: **O. arborescens**

O. nummulariifolia Hook. f. Abundantly and densely branched shrub, twigs stiff, yellowish tomentose when young, often also glutinous, later often quite glabrous; leaves alternate, very densely arranged, upright or more appressed, rounded to broadly oblong or obovate, 5 to 12 mm long, apex rounded, margins reflexed, very thick and leathery tough, glossy green above with reticulate venation, very densely yellow tomentose beneath; flower heads 8–12 mm wide, solitary in the leaf axils, yellowish-white, July. New Zealand, both islands, in the mountains. z9 Plate 140. # ∅

var. **cymbifolia** Hook. f. Like the species, but with outspread or curving leaves, oblong, obtuse, convex above, margins very involuted, therefore nearly navicular, more or less white stellate tomentose; flowers like those of the species, but the scales of the involucre are more tomentose. New Zealand; in the mountains of the South Island. Plate 140. #

O. odorata Petrie. Upright, abundantly branched shrub, 1.5–2.5 m high in its habitat, twigs divaricately spreading, stiff, cylindrical, furrowed; leaves opposite, mostly in clusters, linear-spathulate, 2–3 cm long, rounded at the apex, entire, margins flatly tapering to the petiole, glossy green and glabrous above, silvery-white glossy beneath from the appressed hairs; flower heads 6 mm wide, dull gray-brown, fragrant, in groups of 2–5 together and in opposite fascicles, June; involucre brown, glandular glutinous. New Zealand, S. Island. Of only slight garden merit, for collectors. z9 #

O. paniculata Cheesem. Abundantly branched shrub, to 2 m high, to 5 m high in its habitat, young shoots furrowed and angular, with a dark brown scabby tomentum, as are the leaves and flower stalks; leaves alternate, oval-oblong to broad ovate, 3–7 cm long, mostly obtuse, entire, but somewhat undulate, base round to somewhat cordate, glossy green and glabrous above, gray-white tomentose beneath; flower heads dull white, in small, conical, axillary, about 5 cm long panicles, October–December, fragrant, but otherwise of little ornamental value. KF 137 (= *O. forsteri* Hook. f.). New Zealand, both islands. 1866. z9 # ∅

O. phlogopappa DC. Abundantly branched, upright shrub, 1–2.5 m high, young shoots densely white tomentose; leaves alternate, oblong to narrow obovate, 1.2–3 cm long, apex rounded, margins very shallowly dentate or slightly sinuate, base cuneate, dull green above, white tomentose beneath, very short petioled; flowers white, disk yellow, heads 2.5 cm wide, many in loose, terminal, large cymes, June. BS 2: 364; LAu 226; BM 4638 (as *O. stellulata*) (= *O. gunniana* Hook. f.). Tasmania. Around 1848. Also discovered in 1930 by Comber in Tasmania with lilac, purple and blue flowers. z9 Plate 142. # ✧

O. pinifolia Hook. f.) Benth. Steeply erect shrub with thick shoots, occasionally tomentose when young, often branched only at the shoot tips like a small tree, 1–3 m high; leaves alternate, clustered, sessile, erect or outspread, narrowly linear, stiff and with prickly tips, 1.5–4.5 cm long, margins tightly involuted, leaves therefore appearing cylindrical and only 4 mm wide, young leaves more or less pubescent, with T-form silky hairs beneath; flowers in solitary heads, axillary near the shoot tips, often very numerous, white. Tasmania. Rather unusual in cultivation. z9 Plate 142. #

O. ramulosa Benth. An attractive, abundantly branched shrub, about 1 m high, shoots very thin, nodding, scabby or pubescent when young, leaves linear to more obovate or rounded, only 8 mm long, margins involuted, deep green above and somewhat warty, persistently woolly beneath; flowers in small heads, 1.5 cm wide, erect on the bowed twigs, very numerous, August. BM 8205.

Tasmania, Australia. 1822. One of the most tender species but a prolific bloomer. z10 # ⊕

O. × scilloniensis Dorrien-Smith (*O. lyrata* × *O. phlogopappa*). Shrub, nearly globose, very densely branched, 1–2 m high or more; leaves gray; flowers white, flowers abundantly in May. Developed before 1951 at Tresco Abbey on the Scilly Isles, SW. England, but very tender. z9 # ⊕

O. semidentata Decne. Erect, sparsely branched shrub, 0.5–3 m high, young shoots thin white woolly tomentose; leaves numerous, densely arranged, acute, 3–5 cm long, gradually tapering to a short petiole at the base, margins sparsely serrate to the apex, dark green and rugose above, silvery-white woolly beneath; flower heads solitary at the shoot tips, about 5 cm wide, purple-pink, disk violet, July. BM 8550; DRHS 1410. Chatham Islands; in moist areas. By far the most attractive *Olearia*, but quite tender. z9 # ⊘ ⊕

O. solandri Hook. f. Upright abundantly branched shrub, 1–4 m high in its habitat, twigs broom-like, young shoots angular and yellowish pubescent; leaves opposite on young plants, linear-obovate to more spathulate, very thin, tapering to a short petiole, 8–12 mm long, flat, white beneath, leaves of older plants in opposing clusters, mostly narrow linear to linear-spathulate, 5–8 mm long, leathery tough, glabrous above, yellow tomentose beneath, margins involuted, flower heads solitary and sessile in the leaf clusters, with 5–14 short, yellowish ray florets, August–October (= *O. fasciculifolia* Col.). New Zealand, both islands. z9 #

O. stellulata see: **O. phlogopappa**

O. traversii Hook. f. A small a tree in its habitat, 4–9 m high, young shoots 4 sided, with tightly appressed silky pubescence, as are the leaf undersides and inflorescence axis; leaves opposite, oval-oblong to broadly ovate, 3–6

cm long, acute, entire, short petioled, flat and glabrous above; flowers in 2.5–5 cm long, axillary panicles, ray florets absent, June. KF 34. Chatham Island, in the forest. 1840. z9 Plate 141. # ⊘

O. virgata Hook. f. Shrub, upright, abundantly branched, 4 × 5 m in size, developing thickets in its habitat, twigs outspread, thick or thin, angular to nearly cylindrical, smooth or furrowed, glabrous or pubescent, red when young; leaves linear-obovate, 6–18 mm long, opposite or in opposing clusters, usually glabrous and dark green above, white tomentose beneath, nearly sessile; flower heads about 4 mm wide and with 3–6 yellowish-white ray flowers, but blooming only sparsely, June. New Zealand, both islands. Slight garden merit. z7 #

var. **lineata** Kirk. Shoots thinner, more outspread or pendulous, silky pubescent when young; leaves in opposite clusters, 12–30 mm long, very narrow linear, about 1.5 mm wide, glabrous or lightly silky pubescent above, white tomentose beneath, margins involuted; flowers yellowish-white, heads 6 mm wide, with 8–14 ray florets, flowering very abundantly, but the flowers are very small (= *O. lineata* [Kirk] Cockayne). New Zealand. z7 #

O. 'Zennorensis' (*O. ilicifolia* × *O. lacunosa*). Attractively branched shrub, to 2 m high, young shoots and leaf petioles thick brownish tomentose; leaves linear, to 10 cm long, 12 mm wide, margins with small, sharp teeth, dark olive-green above, white beneath; flowers white. Developed in the garden of Arnold Foster, Zennor, Cornwall. z9 # ⊕

The *Olearia* species generally need a frost free climate and high humidity. Most are cultivated for their attractive foliage, a few for their flowers. Thrives in a sandy humus, fertile garden soil and easily propagated.

Lit. Talbot de Malahide, M.: The genus *Olearia*; (I) Jour. RHS **90**, 207–217; (II) 245–250, 1965.

ONONIS L. — Restharrow — LEGUMINOSAE

Shrubs, subshrubs or herbaceous plants, some with thorny twigs; leaves alternate, trifoliate, stipules adnate to the petiole; flowers 2–3 in axillary racemes clustered near the shoot tips, purple-pink or yellow; calyx campanulate, occasionally tubular, 5 parted; stamens connate; ovaries mostly short stalked, styles filamentous; fruit an oblong or rounded, bivalvate pod, longer than the persistent calyx; seeds without a hilum bulge. — About 75 species in Europe, mostly in the Mediterranean region, but only a few are cultivated.

Ononis aragonensis Asso. Deciduous shrubs, 15–50 cm high, twigs spirally twisted, gray when young, flowering shoots glandular-glutinous; leaves fascicled, petiolate, trifoliate, leaflets rounded, about 1 cm long, irregularly dentate, somewhat leathery, the center leaflet petiolate, the lateral leaflets sessile; flowers yellow, solitary or to 12 pairs in terminal, 10 cm long racemes, calyx glandular

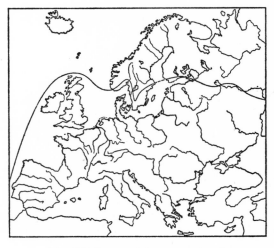

Fig. 233. Range of the genus *Ononis*

pubescent, persistent, lobes longer than the tube, June–July; fruit to 7 mm long, glandular. BS 2: 374. Spain, French Pyrenees, Algeria. 1816. z9

O. fruticosa L. Upright, abundantly branched shrub, 0.3–1 m high, young shoots finely pubescent, later glabrous and gray; leaves in sessile clusters, usually trifoliate, leaflets oblanceolate, 2–3 cm long, gray-green, glabrous, tough, entire or also irregularly dentate; flowers 2–3 axillary, grouped into terminal, glandular pubescent racemes, whitish-pink or pink with dark red venation, June to August; pods 2–2.5 cm long, pubescent. BM 317. S. Europe, N. Africa. 1680. z8

O. speciosa Lag. Thin branched shrub, 0.7–1 m high in its habitat, twigs glandular-glutinous, pubescent; leaves petiolate, trifoliate, densely glandular pubescent, middle leaflet petiolate, the lateral ones sessile, oval-elliptic; flowers 1–3 axillary in the bract axils, but grouped into terminal racemes, standard golden-yellow with red stripes, calyx persistent, glandular, lobes linear, twice as long as the calyx, May–June; pod nearly triangular, glandular. S. Spain, common. Most attractive species. z9 ✿

For sunny areas in a gravelly clay soil, best in a rock garden. Not difficult to cultivate in warmer zones.

OPLOPANAX Miq. — Devil's Club — ARALIACEAE

Deciduous shrubs with thorned shoots and leaves; leaves alternate, long petioled, palmately 5–7 lobed, without stipules; flowers in umbels, grouped into terminal panicles, greenish-white; calyx teeth indistinct; petals 5, valvate; stamens 5; styles 2, distinct; fruit a compressed drupe (= *Echinopanax* Decne. & Planch.). — 3 species in E. Asia and western N. America.

Oplopanax elatus (Nakai) Nakai. Distinguished from the following (more common) species in the larger leaves, lobes not further lobed or only with 1 small lobe on either side at the base of the middle lobe, margins double serrate, venation bristly pubescent on the underside. NK 16: 11. Korea. 1917. z5 ⊘

O. horridus (Sm.) Miq. Erect shrub, to 4 m high in its habitat, shoots, leaf petioles and inflorescence rachis densely covered with thin erect prickles; leaves nearly circular, 15–25 cm wide, with 5–7 incised lobes, sharply serrate, glabrous above and light green, somewhat shaggy pubescent beneath, both sides prickly on the venation; flowers white in 8–15 cm long, pubescent panicles, July–August; fruits 8 mm long, scarlet-red. BM 8572; Gs 2: 195; RFW 245 (= *Echinopanax horridum* Decne & Planch.; *Fatsia horrida* Benth. & Hook.). North America, Alaska to California. 1828. z6 Plate 143. ⊘

O. japonicus (Nakai) Nakai. Leaves often umbrella-shaped, deeply incised, distinctly ciliate, lobes and

secondary lobes long acuminate, flower stalk thinner; fruit stalk to 1 cm long. MJ 857 (= *Echinopanax japonicus* Nakai). Japan. 1915. z6 ⊘

All species winter hardy, but rarely cultivated. All prefer a cool, moist site and humus soil in semishade.

OREODOXA

Oreodoxa oleracea see: **Roystonea oleracea**

O. regia see: **Roystonea regia**

OREOHERZOGIA W. Vent — Rhamnaceae

The section *Eurhamnus* Boiss. of the genus *Rhamnus* was elevated to genus rank by W. Vent. The properties of the vegatative and generic organs were, until Vent's work, not considered distinctive. Therefore, the following nomenclatural changes:

Oreoherzogia alpina (L.) W. Vent
        = Rhamnus alpinus L.
O. cornifolia (Boiss. & Hohen) W. Vent
        = R. cornifolius Boiss. & Hohen.
O. depressa (Grub.) W. Vent
        = R. depressus Grub.
O. fallax (Boiss.) W. Vent
        = R. Fallax Boiss.
O. glaucophylla (Sommier) W. Vent
        = R. glaucophyllus Sommier
O. guiccardii (Heldr. & Sart.) W. Vent
        = R. guiccadii Heldr. & Sart.
O. imeretina (Booth) W. Vent
        = R. imeretinus Booth
O. legionensis (Rothm.) W. Vent
        = R. pumilus ssp. legionensis Rothm.
O. libanotica (Boiss.) W. Vent
        = R. libanoticus Boiss.
O. microcarpa (Boiss.) W. Vent
        = R. microcarpus Boiss.
O. pubescens (Sibth. & Smith) W. Vent
        = R. pubescens Sibth. & Smith
O. pumila (Turra) W. Vent
        = R. pumilus Turra
O. taurica W. Vent
        = R. libanoticus Boiss. pr. p.

The name × *Rhamzogia* W. Vent has been suggested for hybrids between *Rhamnus* and *Oreoherzogia*.

Lit. Vent, W.: Monographie der Gattung *Oreoherzogia* W. Vent gen. nov.; Feddes Rep. **65**, 3–132, 1962.

ORIXA Thunb. — RUTACEAE

Monotypic genus; aromatic, deciduous, erect shrub; leaves alternate, entire, flowers inconspicuous, unisexual, dioecious, greenish, in racemes, originating outside the leaf axils; calyx 4 parted; petals 4; male flowers with a 4 sided disk and 4 stamens; female flowers with a 4 lobed disk and 4 ovaries; fruit an aggregate composed of 4 fleshy fruitlets. — Japan.

Orixa japonica Thunb. An open, divaricate shrub, to 3 m high in its habitat, twigs light green, later gray; leaves broad lanceolate to obovate, 5 to 12 cm long, obtusely acuminate, glossy dark green above, occasionally somewhat bullate, lighter beneath, translucent punctate, crushed leaves aromatic; flowers on the older wood, May; fruitlets very small. KIF 4: 17; Gfl 35: 1232; MJ 1170. Japan, in subalpine mountain forests. 1856. z6 Plate 143. ⌀

'Variegata'. Leaves silver-gray, margin white. ⌀

Hardy and easily cultivated, but without particular garden merit. Not always in cultivation.

ORPHANIDESIA Boiss. — ERICACEAE

Monotypic genus; evergreen, procumbent shrub; leaves alternate, bristly pubescent; flowers cupulate, 5 toothed, sessile, but in small racemes, stamens 10, anthers dehiscing by a longitudinal slit; fruit a 5 chambered capsule. — Transcaucasia, NE. Asia Minor.

Orphanidesia gaultherioides Boiss. Shrub, procumbent, 15–30 cm high, young shoots glandular pubescent; leaves ovate to oblong, 8–12 cm long, abruptly short acuminate, base round, margin indistinctly undulate and irregularly bristly ciliate, otherwise bristly pubescent only on the venation of both sides, bristles larger on the underside; flowers 1–3 in small racemes in the apical leaf axils, corolla flat campanulate, to 5 cm wide, soft pink, pubescent on the interior of the basal half, calyx teeth 5, elliptic and acuminate, glabrous, March–April. BMns 14; GC 139: 133 + 357; JRHS 97: 94. E. Lazistan; (now NE. Turkey) Transcaucasia. 1934. Closely related to *Epigaea* and equally hardy, but rarely found in cultivation. z3 Fig. 234. # ✣

Cultivated in a humus soil with a top dressing of leaf mold in semishade. Flowers very attractive.

Fig. 234. *Orphanidesia gaultherioides* (from B. M.)

OSMANTHUS Lour. — OLEACEAE

Evergreen shrubs or trees (to 20 m high in habitat), young shoots glabrous or pubescent; leaves opposite, glabrous, occasionally with the petiole and midrib pubescent, usually leathery tough, lanceolate or ovate to oblanceolate or obovate, entire to scabrous dentate, occasionally crenate; flowers in axillary clusters or small panicles; flowers white or yellowish, occasionally yellow, usually fragrant, calyx 4 toothed, corolla more or less campanulate to tubular, limb with 4 lobes; stamens 2; fruit a single seeded, hard shelled drupe. — About 15 species in E. and S. Asia and Polynesia, 2 in N. America.

Osmanthus americanus (L.) Gray. Shrub or tree, to 15 m high in its habitat; leaves lanceolate-oblong to obovate, 5–15 cm long, acute or obtuse, entire (!), margins somewhat involuted, petiole 1–2 cm long; flowers nearly sessile, in small panicles, white, fragrant, April; fruit ovate, 1–1.4 cm long, dark blue. WT 186; SS 279–280; SM 857. SE. USA. 1785. Without ornamental merit; only occasionally found in botanic gardens. z9 #

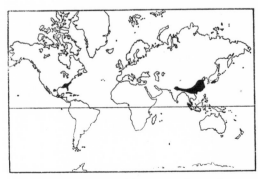

Fig. 235. Range of the genus *Osmanthus*

O. aquifolium see: **O. heterophyllus**

O. armatus. Diels. Shrub, 2–4 m high, young shoots densely pubescent; leaves oblong-lanceolate, 7–14 cm long, sparsely and coarsely dentate, base rounded to nearly cordate, fragrant above, dull green, petiole reddish; flowers cream-white, fragrant, in axillary fascicles, September; fruits dark violet, nearly 2 cm long. DRHS 1452; BM 9232; Gs 20: 53; ICS 4664. W. China. 1902. z9 Plate 145. # 〵

O. delavayi Franch. Broad, rounded shrub, 1.5–3 m high and wide, young shoots somewht pubescent; leaves ovate to elliptic, 1.5 to 3 cm long, finely and sharply dentate, deep green and glossy above, petiole short; flowers grouped 4–8 in small axillary and terminal clusters, pure white, fragrant, corolla tubular, 12 mm long, limb 12 mm wide, April–May; fruits oval, blue-black. BS 2: 380; BM 8459; 9176; DRHS 1452 (= *Siphonosmanthus delavayi* [Franch.] Stapf.). China. 1890. z8 to 9 # ✛

O. forrestii see: **O. yunnanensis**

O. × fortunei Carr. (*O. fragrans × O. heterophyllus*). Broad, rounded, glabrous shrub, 1.5–3 m high, but also twice as high in very mild areas; leaves elliptic-ovate, slender and thorny acuminate, 6–10 cm long, with 6–10 large, *Ilex*-like, triangular, marginal thorns on either side, also often entire at the twig base, similar to those of *O. heterophyllus*, but larger, less glossy above and with more marginal thorns; flowers white, fragrant, 8–10 in axillary fascicles, September. RH 1864: 69; MJ 3217; DL 1: 88 pp. Japan, but known only in cultivation. 1856. z8–9 Plate 145. # ⌀ ✛

O. fragrans Lour. Shrub or a small tree in its habitat, young shoots glabrous; leaves oblong-lanceolate to elliptic, 6–10 cm long, long acuminate, finely dentate, leathery tough, distinctly veined beneath; flowers white, solitary or few in stalked clusters, very fragrant, June–August; fruits bluish, 12 mm long. MJ 654; HKS 74; ICS 4661. Himalayas, Japan, China. z9 Plate 145. # ⌀ ✛

'Aurantiacus'. Like the above, but the leaves usually entire; flowers orange colored. BM 9211; MJ 655. China; only known in cultivation. z9 ✛

O. heterophyllus (G. Don) P. S. Green. Round, upright, dense shrub, 2.5–5 m high; leaves elliptic, ovate or elliptic-oblong, 2–6 cm long, very leathery tough, with 2–4 large, thorny teeth on either side, flat, deep green and glossy above, yellow-green with reticulate venation beneath; flowers white, fragrant, in axillary clusters, September–October; fruits oval-oblong, about 12 mm long, blue. MJ 656; KIF 2: 78 (= *O. aquifolium* S. & Z.; *O. ilicifolius* [Hassk.] Mouillef.). Japan. 1856. Hardy. z7 # ⌀

var. *argenteo-marginatus* see: **'Variegatus'**

'Aureomarginatus'. Leaves yellow bordered (= var. *aureovariegatus* Hort.). Before 1877.

var. *aureovariegatus* see: **'Aureomarginatus'**

'Gulftide'. Dense growing; leaves somewhat lobed or twisted, particularly green, stout thorned. Attractive and commendable. ⌀

var. **heterophyllus.** The type; description above. Rationale for the name change in accordance with the ICBN by Green, l. c. 512.

'Myrtifolius'. Growth broad and low; leaves elliptic to elliptic-oblong, 2.5–4.5 cm long, blade totally flat, margins thornless, except for the thorny tip. Before 1894. Plate 144.

'Purpureus'. New growth blackish-red; fully developed leaves deep green with a reddish trace. Gs 20: 52. Before 1896. More winter hardy than the species. ⌀

'Rotundifolius'. Slow growing; leaves more or less rounded obovate, 2.5–4 cm long, entire, margins undulate (!), base cuneate. Japanese cultivar, known since 1866. Plate 144.

'Variegatus'. Leaf margins cream-white, otherwise like the species (= var. *argenteo-marginatus* Hort.). Before 1861. Plate 144 ⌀

O. ilicifolius see: **O. heterophylla**

O. rehderianus see: **O. yunnanensis**

O. serrulatus Rehd. Shrub, 2–3 m high, young shoots finely pubescent; leaves ovate-lanceolate, 5–10 cm long, entire to scabrous to sharply and finely serrate, with 29–35 teeth about 1 mm long on either edge, glossy green above, lighter beneath and finely punctate; flowers in clusters of 4–9, white, very fragrant; March–April. China; Szechwan, Kwangsi Provinces. 1912. z9 Plate 142, 145. # ⌀ ✛

O. suavis King. Shrub, to 3 m high, or also a small tree, young shoots gray and finely pubescent; leaves oblong-lanceolate, 3–6 cm long, acuminate, cuneately tapering to the base, finely crenate; flowers white, 3–5 in axillary and terminal clusters, fragrant, the 4 corolla lobes rounded, calyx ciliate; fruits ovate, 1 cm long, blue-black. SNp 114; BM 9176 (= *Siphonosmanthus suavis* [Clarke] Stapf). India to SE. Tibet. z9 Plate 145. # ⌀ ✛

O. yunnanensis (Franch.) P. S. Green. Tall shrub, 3–5 m high, 6–8 m high in its habitat, young shoots glabrous; leaves more ovate-lanceolate to oblong, 7–20 cm long, with slender tips, base cuneate to nearly cordate, very tough and hardy, glabrous, margins with 25–30 short, sharp teeth on either edge or entire; flowers waxy, cream-white to light yellow, in axillary clusters of 5 to 7, very fragrant; fruits ovate, 12–15 mm long, dark blue. ICS 4662 (= *O. forrestii* Rehd.; *O. rehderianus* Hand.-Mazz.). China; Yunnan, Szechwan Provinces. 1923. z9 Plate 145. # ⌀

O. heterophyllus is the hardiest species. Cultivated like the evergreen *Ligustrum* species. The more tender species only suited for warmer climates. Very attractive.

Lit. Green, P. S.: A monographic revision of *Osmanthus* in Asia and America; in Not. Bot. Gard. Edinburgh **22**, 439–543, 1958 ● Green, P. S., & J. Keenan: *Osmanthus heterophyllus* and the application of the term Cultivar. Baileya **7**, 73–77, 1959 (with good ills.)

× OSMAREA (OSMANTHUS × PHILLYREA) Burkw. & Skipwith — OLEACEAE

Evergreen shrub, intermediate between the parents in appearance; differing from *Phillyrea* in the corolla tube which is as long as the corolla lobes; differing from *Osmanthus* in the thin flower stalks and the shorter corolla tube. — Only one hybrid known to date.

× **Osmarea burkwoodii** Burkw. & Skipwith (= *Osmanthus delavayi* × *Phillyrea vilmoriniana*). Smaller, evergreen shrub, to about 2 m high and wide, open habit, branches finely pubescent; leaves ovate-elliptic, 2–4 cm long, acute, more or less serrate, glossy green and glabrous above, short petioled; flowers 5–7 in axillary fascicles, milky-white, fragrant, corolla tube 4–5 mm long, anthers slightly exserted, style as long as the 2 mm long, greenish-white calyx, April–May. Developed before 1919 by Burkwood & Skipwith, Kingston-on-Thames, England. z7 Plate 145, 147. #

Culture and uses like *Ilex* or *Osmanthus;* generally quite hardy.

OSMARONIA Greene — Oregon Plum — ROSACEAE

Monotypic genus; deciduous shrub; leaves alternate, simple, entire, stipules small, abscising; flowers polygamous, in short racemes; sepals and petals 5 each; stamens 15, in 3 rings; carpels 5, distinct, style short; 1–3 drupes develop from each flower with 1 seed, a leathery inner shell and with only a thin fruit pulp (= *Oemleria* Reichb.).

Osmaronia cerasiformis (Torr. & Gray) Greene. Multistemmed shrub, growth narrowly upright, about 2 m high in cultivation (to 5 m high in its habitat), twigs with a chambered pith, stoloniferous, new growth appears very early, March; leaves oblong-lanceolate, 7–10 cm long, deep green above, gray-green and pubescent beneath; flowers white, in short, nodding racemes, about 1 cm wide, fragrant, flower stalks with 2 pale green bracts, May; fruit an aggregate composed of 1–5 blue-black, pruinose, to 1.5 cm long, plum-like fruitlets. BS 2: 357 (= *Nuttallia cerasiformis* Torr. & Gray). Western N. America. 1848. z7 Plate 141, 146.

Prefers a wooded site, in a humus, moist soil, protected from direct sunlight. Only suitable for collectors.

OSTEOMELES Lindl. — ROSACEAE

Deciduous to semi-evergreen shrubs; leaves alternate, small and finely pinnate, stipules linear-lanceolate, leaflets small, entire; flowers in small terminal corymbs; white, calyx with 5 acute teeth, petals 5, oval-oblong; stamens 15–20; styles 5, distinct; fruit a small pome with a persistent calyx and 5 seeds. — 3 species in E. Asia and Polynesia.

Osteomeles anthyllidifolia Lindl. Shrub, semi-evergreen, to 1.8 m high, young shoots densely pubescent; leaflets 13–19, obovate-oblong, obtuse or with small tips, 8–12 mm long, both sides densely pubescent when young, later glossy dark green above, silky pubescent beneath; inflorescences 2.5–3 cm wide, gray pubescent, styles shaggy; fruits globose, 8 mm

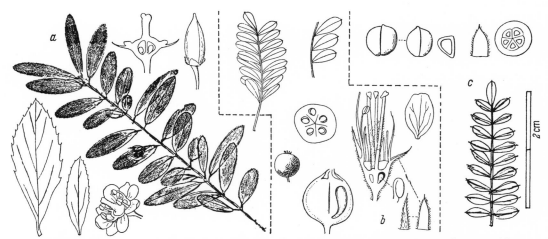

Fig. 236. **Pachistima**. a. *P. myrsinites,* twig of a plant from British Columbia, Canada; leaves, flower and fruit. — **Osteomeles**. b. *O. anthyllidifolia,* leaves, flower and flower parts; fruit shown in longitudinal and cross section. c. *O. schweriniae,* leaf, sepals; fruit in cross section, seed in cross section. (partly from Schneider)

thick, blue-black, sweet, pubescent at first, later nearly glabrous. Hawaii, Polynesia. Usually incorrectly labeled in cultivation. z9 Plate 145; Fig. 236. ⊘

O. schweriniae Schneid. Deciduous or semi-evergreen shrubs, to 3 m high and wide in mild areas, twigs gracefully pendulous, thin, petiole and leaves gray pubescent; leaves 3–7 cm long, rachis slightly winged, leaflets 15–31, elliptic to obovate-oblong, 4–12 mm long; flowers in 3–6 cm wide, loose cymes, individual flowers to 1.5 cm wide, white, styles pubescent, calyx teeth eventually becoming glabrous, May–June; fruits oval-rounded, blue-black, 6 to 8 mm long, glabrous. BM 7354 (as *O. anthyllidifolia*); BS 2: 382. W. China. 1888 z9 Fig. 236. ⊘

OSTEOSPERMUM L. — COMPOSITAE

Shrubs or subshrubs, occasionally herbaceous; leaves alternate or also opposite; ray florets yellow, lilac or violet; marginal flowers ligulate, the females fertile, disk florets tubular, 5 toothed, male, receptacle naked, fruitlets nut-like, with a pappus, thick and hard, straight, or slightly curved, not beaked. — About 70 species in S. Africa and on St. Helena.

Earlier included with *Dimorphotheca*, but differing in the drupaceous fruits.

Osteospermum ecklonis (DC.) Norl. Shrub or subshrub, 1 m high or more in its habitat, branched; leaves obovate to oblanceolate, 5–10 cm long, 0.5–2 cm wide, acute, margins sparsely glandular dentate, base drawn out to the petiole, soft glandular pubescent; flowers with a long stalk, the heads solitary or in loose corymbs at the branch tips, 5–7 cm wide, ray florets white above, blue or white beneath, disk florets violet-blue, July–September. BM 7535 (= *Dimorphotheca ecklonis* DC.). S. Africa. 1897. z9 Fig. 237. ✤

Often cultivated as an annual plant but shrubby in its habitat.

OSTRYA Scop. — CORYLACEAE

Deciduous trees, very similar to *Carpinus*, but easily distinguished in the normally biserrate and pubescent leaves and the *Betula*-like male flower catkins developed in the fall; male catkins lateral, with 12 or more stamens on short and branched stalks at the base of the large scales; female catkins terminal, paired by a surrounding, abscising scale; 2 filamentous stigmas; fruits surrounded by a sack-like hull. — 7 species in Europe, Asia and America.

Outline (from Rehder, altered)

● Leaves acuminate, 4–12 cm long;
  + Leaves slightly pubescent to nearly glabrous, with 11–15 vein pairs;

   1. Leaves usually rounded at the base, stalk 5–10 mm long, without glands; nutlets ovoid:
      O. carpinifolia

var. **microphylla** Rehd. & Wils. Leaflets less numerous, only 3–5 mm long, elliptic to obovate, becoming glabrous; inflorescences smaller and denser. W. China.

O. subrotunda K. Koch. Smaller, slower growing shrub with twisted branches, pubescent when young; leaflets 9–17, rounded to obovate, 4–8 mm long, ciliate, with a thin appressed pubescence beneath; flowers white, 1 cm wide, in loose, 2.5–3 cm wide corymbs in June, style glabrous. MJ 1389. E. China. 1894. Distinguished by the slow, rigid growth habit. z9? ⊘

Collector's plant; best grown in pots or tubs and over-wintered in a cool greenhouse or conservatory. Mature shrubs are occasionally seen in southern botanic gardens.

Fig. 237. *Osteospermum ecklonis*
(from Gard. Chron.)

   2. Leaves usually slightly cordate at the base, petiole 3–6 mm long, often with stalked glands; nutlets spindle-form:
      O. virginiana

  ++ Leaves usually soft pubescent beneath, with 9–12 vein pairs; nutlets narrow ovate:
      O. japonica

● ● Leaves acute to obtuse, 3–5 cm long, with 5–8 vein pairs:
      O. knowltonii

Ostrya carpinifolia Scop. Hop Hornbeam. Round crowned tree, bark gray, twigs pubescent at first, later glossy and olive-brown; leaves acutely oval-oblong, base rounded, 4–10 cm long, sharply double serrate, dark green above and sparsely pubescent, pubescent on the 11–15 vein pairs beneath; fruit cluster 4–6 cm long, nutlets ovoid, 6–10 mm long, with hair fascicles at the apex. HW 2: 35; HM Pl. 84 (= *O. vulgaris* Willd.; *O. italica* Spach.). S. Europe, Asia Minor. 1724. Ⓕ Italy, Yugoslavia. z6 Fig. 238. ⚭

O. italica see: **O. carpinifolia**

O. japonica Sarg. Tree, 10–15(25) m high, twigs densely pubescent; leaves ovate, long acuminate, coarse and scabrous, thereby irregularly serrate, with 9–12 vein pairs, soft pubescent on both sides; fruit clusters 3–4 cm long, nutlets ovate, 5–6 mm long. LF 77; MJ 1998; GF 6: 384; KIF 1: 40. Japan, China, NE. Asia. 1897. z6 Plate 147. ⚭

O. knowltonii Coville. Small tree, 6–9 m high, shoots soft pubescent, buds cylindrical, very pubescent; leaves ovate to elliptic, obtuse or acute, 3–5 cm long, pubescent on both sides, but more densely so beneath, with 5–8 vein pairs; fruit clusters about 3 cm long, nutlets oval-oblong, 6 mm long, pubescent at the apex. GF 7: 115; SS

446. North America; N. Arizona, Utah. 1914. z5 Fig. 238. ⚭

O. virginiana (Mill.) K. Koch. Tree, 15–20 m high, young twigs pubescent, often with glandular hairs, later glabrous and dark brown, bark gray-brown, thin; leaves oval-lanceolate, 6 to 12 cm long, long acuminate, base normally more or less cordate, with 11–15 vein pairs, pubescence like that of *O. carpinifolia*; fruit clusters 3–6 cm long, nutlets spindle-shaped (!!), 6–8 mm long, glabrous at the apex. BB 1208; GTP 148; RMi 98; SS 545. Eastern N. America. 1690. z5 Plate 147; Fig. 238. ⚭

var. **glandulosa** (Spach) Sarg. Young shoots, leaves and inflorescences covered with stalked glands, otherwise like the species. In the northern range of the species.

O. vulgaris see: **O. carpinifolia**

Suitable for large park areas; not particular as to soil and location, about like *Carpinus*. All species very winter hardy.

Lit. Schrafetter, R.: Die Hopfenbuche, *Ostrya carpinifolia* Scop., in den Ostalpen; in Mitt. DDG 1928, 11–18.

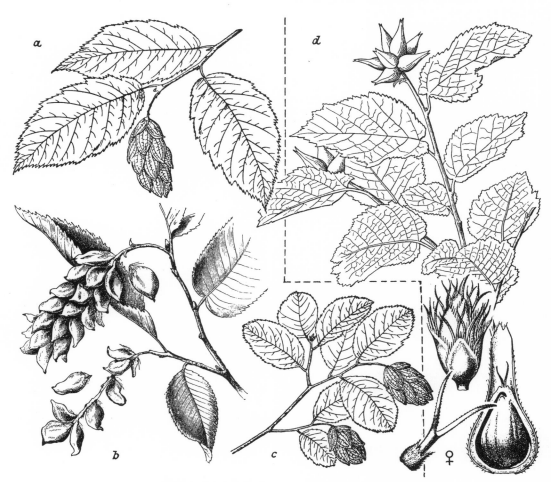

Fig. 238. **Ostrya.** a. *O. virginiana*; b. *O. carpinifolia*; c. *O. knowltonii*.
—**Ostryopis.** d. *O. davidiana* (from Sargent, Kerner, Decaisne and Lavallé)

OSTRYOPSIS Decne. — CORYLACEAE

Deciduous shrubs, intermediate between *Ostrya* and *Corylus;* leaves ovate, double serrate, folded in the bud stage; flowers appearing with the leaves, male flowers developed in early fall, naked over winter, in pendulous, cylindrical catkins, filaments split at the apex, anthers pubescent at the apex; female flowers in short, terminal, dense spikes, each individual flower surrounded by a 3 part involucral bract; fruits surrounded by a tubular involucre with a short 3 part apex. — 2 species in China.

Ostryopsis davidiana (Baill.) Decne. Stoloniferous shrub, resembling a small leaved *Corylus,* to 3 m high, young twigs pubescent; leaves broadly cordate-ovate, 3–7 cm long, double serrate and finely lobed, densely pubescent and glandular beneath, somewhat the same above, red punctate, petiole 5 mm long; fruits 6–12 together in long stalked clusters at the branch tips, nutlets ovate, about 8 mm long. LF 76; ICS 796. China;

Kansu, Shansi, Hopei Provinces. 1865. Hardy. z6 Plate 141; Fig. 238. ⚥

var. **cinerascens** Franch. Leaves smaller, rounded or ovate, both sides more brownish pubescent; catkins 2–3 cm long, purple; fruits slender, sessile, gray pubescent. China; Yunnan, Szechwan Provinces.

O. nobilis Balf. & Sm. Shrub, 1–1.5 m high, twigs dark gray and somewhat rough; leaves round to oval, leathery, dull green above, yellowish-brown and pubescent beneath, 2–4 cm long, double serrate, rounded at the base and apex; bracts bluish-green, pubescent, catkins 2 cm long, red-brown. China; Yunnan, Szechwan Provinces. z6 Plate 141. ∅

O. davidiana lacks any particular garden merit, it is however an interesting, attractive shrub, with no specific requirements, cultivated somewhat like *Corylus.*

OSYRIS L. — SANTALACEAE

Small, evergreen, glabrous shrubs; leaves alternate, entire, lanceolate to ovate; flowers small, stalked, not very conspicuous, in groups of a few in small, axillary racemes, the female often also solitary, otherwise like Fig. 240; fruit a single seeded drupe, with a persistent perianth. — 6–7 species in the Mediterranean region and India.

Osyris alba L. Shrub, about 1 m high, twigs striped, cylindrical; leaves linear-lanceolate, 2.5 cm long; flowers white, July; fruits pea-sized, red, rather attractive. Mediterranean region. 1793. Rare in cultivation. z10 Fig. 240.

OVIDIA Meissn. — THYMELAEACEAE

Deciduous shrubs, similar to *Daphne,* but the flowers are 4 parted with 8 stamens (the *Daphne* corolla is tubular with 2 stamens), style long and thin (in *Daphne* short or absent); flowers dioecious; twigs very flexible. — 4 species in Chile.

Ovidia andina Meissn. Deciduous shrub, about 1–2 m high, young shoots pubescent and very flexible; leaves narrowly ovate to oblanceolate, sessile, 5–12 cm long, obtuse or rounded at the apex, tapering to the base, dull green and glabrous above, bluish and pubescent beneath; flowers grouped 30–40 together in a dense, terminal umbel, 2.5–3 cm wide on a 2.5 cm long, thick stalk, corolla 6 mm wide, perianth short funnelform,

OTHONNOPSIS See: **HERTIA**

Fig. 239. *Ovidia andina* (from Dimitri)

Fig. 240. *Osyris alba* (from Hieronymus)

limb with 4 obovate lobes, cream-white, pubescent, anthers red, the stalks of the individual flowers very thin and 6–12 mm long, July; fruits ovate, pure white, 6 mm long; with a persistent stigma (= *Daphne andina* Poepp.). Chile, in the mountains at 1500 m. 1926. z9 Fig. 239.

O. pillopillo Meissn. Similar to the previous species, but becoming 3–9 m high in its habitat; leaves oblanceolate,

2.5–7 cm long, 6–12 mm wide, totally glabrous, dull green above, bluish beneath; flowers with a densely pubescent exterior, white, 12 mm wide; fruits purple-red. Chile. 1927. z9

Cultivated somewhat like *Daphne mezereum;* likes a wooded site, humus soil, rather moist; but of only slight ornamental merit.

OXYCOCCOS

Oxycoccos quadripetala see: **Vaccinium oxycoccos**

OXYDENDRUM DC. — Sourwood — ERICACEAE

Monotypic genus; deciduous tree (or shrub in cultivation), leaves alternate; flowers in one-sided, terminal panicles; calyx 5 parted to the base, corolla small, tubular; stamens 10, dehiscing by a longitudinal slit at the apex; fruit a 5 part, many seeded capsule.

Oxydendrum arboreum (L.) DC. Tree, 15–25 m high in its habitat, bark deeply furrowed, twigs glabrous; leaves oblong-lanceolate, 8–20 cm long, finely serrate, tough, glossy and glabrous above, venation slightly pubescent

beneath, bronze-green on the new growth, light green in summer, gray-white beneath until leaf drop, fall color a gorgeous scarlet-red; flowers white, in 15–20 cm long, nodding panicles, June–August. BM 905; NBB 3: 19. Eastern USA, from Pennsylvania to Louisiana. 1747. z6 Plate 146; Fig. 242. ∅ ✧

Small plants are slow growing and somewhat difficult to transplant. Prefer an acid soil, preferably somewhat moist (but also occurring in dry forests in its habitat).

OXYPETALUM R. Br. — ASCLEPIADACEAE

Erect or twining perennials or subshrubs; leaves opposite; flowers in cymes or also in axillary, stalked heads; calyx 5 parted, corolla campanulate to nearly globose, often with 5 narrow limb tips, corolla tube short, blue, white or purple; pollen agglutinated into a waxy mass. — About 150 species in Mexico, West Indies and Brazil.

Oxypetalum caeruleum (D. Don) Decne. Twining

subshrub, 30–90 cm high, shoots limp, sparsely branched, tomentose; leaves oblong-cordate or more lanceolate, with small tips, both sides soft pubescent, sage-green; flowers in few flowered cymes, the individual flowers about 2–2.5 cm wide, opening azure-blue with a greenish trace, then somewhat purple, eventually lilac, June–September. BM 3630 (as *Tweedia versicolor*). S. Brazil, Uruguay. 1832. z10

OZOTHAMNUS R. Br. — COMPOSITAE

Evergreen shrubs, shoots occasionally thorned; leaves alternate, occasionally decurrent, simple, entire; flowers in corymbs, occasionally solitary, the heads surrounded by an involucre, the inner bracts on some species have a small, erect, white blade. — All species from Australia and Tasmania. #

This genus of Australian evergreen shrubs, usually classified as a subgenus of *Helichrysum,* is maintained here as in Bean, Trees and Shrubs, III (1977).

Ozothamnus antennaria Hook. f. Upright, densely branched shrub, 1–1.5 m high, shoots angular, ridges pronounced at first by lines of short hairs, but soon becoming glabrous; leaves oblanceolate to obovate, rounded, 1–3 cm long, 4–10 mm wide, base tapered to a short, thin petiole, leathery, usually flat, grass-green

above and somewhat glutinous, lighter beneath with resinous scales; flower heads in terminal umbellate panicles on the main and lateral shoots, involucre top-shaped, with 20–28 ray flowers, flowers white, June (= *Helichrysum antennaria* [DC.] F. Muell. ex Benth.). Tasmania. Before 1880. z9 # ✧

O. coralloides see: **Helichrysum coralloides**

O. ledifolius Hook. f. Shrub, to 1 m, sweet smelling, globose growth habit, shoots thick, stiff, very densely crowded, young shoots tomentose; leaves alternate, densely clustered (abscising after a few years), oblong-linear, outspread, obtuse, 7 to 15 mm long, leathery, somewhat glutinous, margins involuted, smooth above, white tomentose beneath, with a yellow resin; flower heads in dense, terminal corymbs, involucre ovate, ray

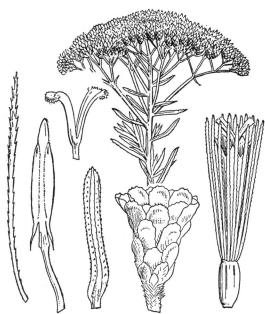

Fig. 241. *Ozothamnus rosmarinifolius*, flowers, flower parts and fruit, right a hermaphroditic flower (from Hooker)

Fig. 242. *Oxydendrum arboreum*, inflorescence (from Illick)

flowers 8–15, white, reddish in bud, June (= *Helichrysum ledifolium* [DC.] Benth.). Tasmania. z9 Plate 146. # ⊕

O. purpurascens DC. Upright shrub, 1–2 m high, sharply aromatic, young shoots tomentose; leaves densely crowded, outspread, usually reflexed with age, somewhat variable in size and pubescence, linear-oblong, obtuse, margins involuted, (3) 7–18 mm long, 0.8–1.6 mm wide, with a scattered pubescence above, tomentose beneath; flowers in dense umbellate panicles, terminally arranged on the main shoots and the numerous short side shoots, ray flowers 8 to 10, purple-pink, particularly before fully opening (= *Helichrysum purpurascens* [DC.] W. M. Curtis). Tasmania. z9 Plate 146. # ⊕

O. rosmarinifolius DC. Shrub, 2 m high in its habitat, abundantly branched, shoots white woolly; leaves linear, finely warty, 5–30 mm long, dull green above, whitish to rust-brown tomentose beneath; flower heads small, white, ray flowers 6–14, buds red, very attractive

(= *Helichrysum rosmarinifolium* [Labill.] Steud. ex Benth.). Tasmania, distributed throughout moist, marshy sites. z9 Fig. 241. # ⊕

O. selago see: **Helichrysum selago**

O. thyrsoideus DC. Shrub, to 3 m high, twigs thin and outspread, shoots somewhat angular, pubescent, particularly on the decurrent lines to the leaves; these narrowly linear, obtuse or acute, 2–5 cm long, 1.5–2 mm wide, margins finely involuted all around, deep green and usually glabrous above, densely white tomentose beneath, but the midrib and margins distinctly green; flowers in dense, nearly globose, small terminal panicles on the short shoots along the branch, ray florets 10–16, white, summer. CTa 77 (= *Helichrysum thyrsoideum* [DC.] Willis & Morris). Tasmania. z9 # ⊕

All species prefer a sunny area, dry soil and a frost free or very mild climate. The species described have been in cultivation on the British Isles for many years.

PACHISTIMA Raf. — CELASTRACEAE

Dwarf evergreen shrubs with angular, finely warty shoots; leaves opposite, densely arranged, small, entire or finely serrate, short petioled, stipules small and abscising early; flowers inconspicuous, green or reddish, few in axillary clusters; calyx with a short tube and 4 part limb, petals 4, stamens 4; fruit a 2 or unilocular, leathery capsule, seeds surrounded by a membranous, white, many segmented seed coat (=

Paxistima Raf.; *Pachystima* Raf.; *Pachystigma* Meissn.). — 5 species in N. America.

Pachistima canbyi Gray. To 25 cm high, twigs procumbent, rooting; leaves narrow oblong, 1–2 cm long, finely serrate, margins involuted, tough, light green; flowers brownish-red, 1–3 axillary, long stalked, April–May; fruits whitish, 4 mm thick. BB 2369; GSP

296. USA, occurring in the mountains of Virginia on dry calcareous slopes. 1880. z3 #

P. myrsinites (Pursh) Raf. Broad growing, to 50 cm high, twigs thick; leaves broad ovate to elliptic or obovate, 1–3 cm long, finely serrate to entire, leathery, light green; flowers short stalked, red-brown, in few flowered cymes, April; fruits whitish, about 8 mm long. DL 2: 237. Western N. America. 1879. z6 Fig. 236, 243. #

A beautiful ground cover; *P. canbyi* in full sun, the other species in semishade; both like a well drained, fertile, garden soil.

Fig. 243. *Pachistima myrsinites*, individual flower and fruit at right.

PACHYSANDRA Michx. — BUXACEAE

Evergreen or semi-evergreen, creeping shrubs with fleshy stems, branches ascending, becoming slightly woody; leaves alternate, coarsely dentate, clustered at the branch tips; flowers white, small, unisexual, in erect spikes; pistillate flowers at the base of the spike, the staminate flowers above much thickened filaments; fruit a 3-horned capsule. — About 5 species in E. Asia and eastern N. America.

Outline

P. axillaris:	Inflorescences in the leaf axils.
P. procumbens:	Inflorescences on the leafless stem base.
P. terminalis:	Inflorescences at the shoot tips.

Pachysandra axillaris Franch. Evergreen subshrub, 15–30 cm high, shoots white pubescent at first; leaves 3–6 at the shoot tips, ovate, 5–10 cm long, coarsely dentate on the apical half, base broadly cuneate to round; flowers white, in erect, 2–2.5 cm long spikes in the leaf axils, April; fruits pea-sized, with the persistent, involuted styles on the apex. China; Yunnan Province. 1901. z6 Plate 147. #

P. procumbens Michx. Semi-evergreen, more herbaceous, shoots unbranched and originating from the rootstock, 15–25 cm high, finely pubescent; leaves clustered at the shoot tips, broadly ovate to obovate or somewhat rhombic, 5–7 cm long, often nearly as wide, very coarsely dentate on the apical half, basal half entire and tapering to the 1.5–3 cm long petiole, basal leaves largest, finely scattered pubescent; flowers on the unbranched stems in erect, 5–10 cm long spikes, white, whitish-pink, March–April, very fragrant. BB 2345; NH 1955: 212. SE. USA. 1800 z6 Plate 147. #

P. terminalis S. & Z. Evergreen subshrub, spreading underground, to about 20 cm high, twigs glabrous; leaves obovate, 5 to 10 cm long, clustered at the branch tips, coarsely dentate at the apex, base cuneate, dark green, glossy above; flowers in 3–5 cm long, terminal spikes, white, April; fruits ovate, glass-like, white, 12 mm long, with 2–3 horns. MJ 1117. Japan. z5 Plate 147. # ∅

'**Green Carpet**' (Legendre). Much shorter than the species, more upright, shoots hardly procumbent; leaves smaller, more finely toothed, darker green. DB 1967: 11. Cultivated for many years in the Nantes Botanic Garden (France) and in Scotland under no particular name. Distributed under this name since around 1970 by Legendre, USA, and Hinrich Kordes, W. Germany. # ∅

'**Variegata**'. Slower growing; leaves white variegated. Introduced into Holland from Japan by Siebold in 1859. ∅

P. terminalis is a particularly valuable ground cover; thriving best in a wooded area, in semishade under an open canopy of trees; prefers a humus soil, somewhat moist. Develops a dense carpet quickly under these conditions.

PACHYSTEGIA Cheesem. — COMPOSITAE

Monotypic genus, earlier included in *Olearia* and differing primarily in the large, solitary flower heads on long stalks with a large, ovate involucre composed of numerous scales in many rings and always of even length, thickened at the apex, pappus hairs in only one ring (*Olearia* has smaller involucral scales of variable length).

Pachystegia insignis (Hook. f.) Cheesem. Low, rather stiff shrub, broad growing, 0.5–1(2) m high, twigs thick, densely tomentose; leaves in rosettes at the shoot tips, very thick and leathery, entire or occasionally somewhat emarginate, oblong to ovate or obovate, 6–12 cm long, 3–7 cm wide, eventually glabrous and deep green above, with a thick white to brownish tomentum beneath, petiole thick, to 5 cm long; flower stalks 10–30 cm long, densely tomentose, flowers *Aster*-like, 5–7 cm wide, with numerous white ray flowers, disk flowers yellow, involucre 1–2 cm wide, exterior tomentose, interior glabrous, pappus hairs 1 cm long, July–August. BM 7034 (= *Olearia insignis* Hook. f.). New Zealand, S. Island. 1850. z9 Plate 152; Fig. 244. # ∅ ✧

var. **minor** Cheesem. Much smaller and more slender; leaves 6–10 cm long, 2–4 cm wide (including the petiole); flower stalks 10 cm long, flower heads not over 3.5 cm wide. New Zealand. Recently introduced. z9 # ∅ ✧

Cultural requirements and uses similar to those of *Olearia*. Planted in many botanic gardens

Fig. 244. *Pachystegia insignis* (from Poole/Adams)

PACHYSTIGMA See: **PACHISTIMA**

PACHYSTIMA See: **PACHISTIMA**

PAEDERIA L. — RUBIACEAE

Twining shrubs with thin shoots, all parts unpleasant smelling when crushed; leaves opposite, occasionally in whorls of 3, petiolate, entire; flowers small, whitish or reddish, in axillary cymes, these grouped into large, terminal panicles; hermaphroditic or polygamous-dioecious, calyx 4 toothed, corolla tubular or funnelform, with 4–5 small, valvate limb lobes with undulate margins; fruits globose, with a brittle, glossy epidermis, later releasing the two seeds. — About 25 species in tropical to temperate Asia and America.

Paederia scandens (Lour.) Merr. Deciduous twining shrub, to 5 m high; leaves opposite, oval-lanceolate, 5–12 cm long, long acuminate, base round to slightly cordate, entire, dark green above, lighter and somewhat pubescent beneath; flowers tubular, whitish with a purple throat, about 1.5 cm long, in 20–40 cm long, terminal panicles, July–August; fruits pea-sized, orange. MJ 347; RH 1919: 298 (= *P. tomentosa* Maxim. non Bl.; *P. wilsonii* Hesse). Japan, China, Korea. 1907. Hardy. z6 Plate 147.

P. tomentosa see: **P. scandens**

P. wilsonii see: **P. scandens**

Of only slight ornamental merit; the large inflorescences are not very noticeable; no preference as to soil or planting site.

PAEONIA L. — PEONY — PAEONIACEAE

Herbaceous plants or deciduous shrubs with thick, stiff twigs and large winter buds; leaves very large, alternate, pinnatisect; flowers solitary, terminal, occasionally in groups, red, pink, white or yellow; sepals 5, persistent; petals 5–10, large; stamens numerous; carpels 2–5, these later developing into dehiscent follicles each with a few seeds. — The shrubby species are from China, the others come from Europe, Asia and western N. America.

Of the 33 species in this genus, only the woody types will be covered here. They comprise collectively the section *Moutan* DC. (Stem woody; petals much longer than the sepals, disk a fleshy plate surrounding the base of the carpel, ring- or sheath-like).

Outline of the cultivated species
(from Stern)

1. Leaflets deeply incised, tips of the lobes and teeth sharply acuminate:
    P. suffruticosa

2. Leaflets more or less unevenly 3 lobed with rounded sinuses; tips of the lobes obtuse:
    P. suffruticosa var. *spontanea*

3. Flowers with a distinct involucre of 8–12 green involucral bracts, these situated very close to the calyx of 5 leathery, greenish sepals; flowers red, 9–10 cm wide:
    P. delavayi

★ Flowers without an involucre; involucral bracts and sepals 5–7 in total, of these the 1–4 outermost more or less leaf-like, the innermost rounded and sepaloid;

4. Segments and lobes of the leaves usually 1.7–3 cm wide; flowers yellow or yellow with a red basal spot, 5–6 cm wide (to 8 cm wide in cultivation):
    P. lutea

> Segments and lobes of the leaves usually 5–10 mm wide, the lower portion of the less incised apical segments 1.5–2 cm wide; flowers to 6 cm wide;

5. Flowers red or white, widely opened:
    P. potaninii

6. Flowers yellow, slightly opening:
    P. potaninii var. *trollioides*

Paeonia arborea see: **P. suffruticosa**

P. delavayi Franch. Glabrous shrub, to 1.6 m high; leaves doubly trifoliate, 20–27 cm long, petiole 7–10 cm long, lobes oval-lanceolate, 5–10 cm long, entire or slightly dentate, dark green above, blue-green beneath; flowers dark red, about 9 cm wide, with about 8–12 involucral bracts immediately beneath the 5 sepals, petals obovate, rounded, 3 cm long, filaments dark carmine, anthers yellow, May; fruits about 2.5 cm long. StP 44, 95; DRHS 1470. China: Yunnan, Likiang Provinces. 1892. Native habitat is mountain forests and thickets between 2600–3600 m. z6 Plate 148; Fig. 245. ✤

var. *angustifolia* see: **P. potaninii**

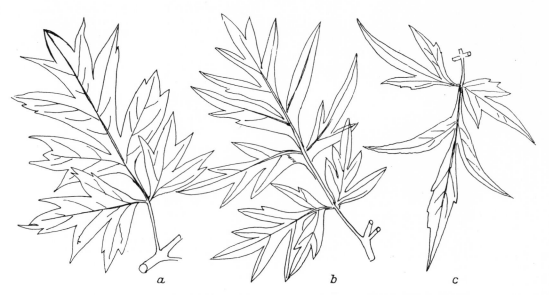

Fig. 245. **Paeonia**. Leaf parts. a. *P. lutea*; b. *P. delavayi*; c. *P. potaninii* (Original)

P. × lemoinei Rehd. (= *P. lutea × P. suffruticosa*). Includes the yellow flowering cultivars, resembling *P. suffruticosa* in habit and foliage. For the cultivars, refer to the List of Cultivars following the species descriptions. The oldest hybrid is 'L'Espérance' (Lemoine). ✣

P. lutea Delavay ex Franch. Erect shrub, about 1 m high, glabrous; leaves doubly trifoliate, long stalked, the individual leaflet groups with 2.5–5 cm long stalks, leaflets lobed and dentate, obtusely acuminate, dark green above, blue-green beneath, 5–10 cm long; flowers yellow, about 7 cm wide, with a distinct disk, involucral bracts and sepals 5–8, persistent, the outermost leaflike, petals obovate, rounded 2.5–3 cm long, somewhat hollow, June, stamens yellow; carpels glabrous, conical. BM 7788. China; Yunnan, Province, in thickets in the mountains at 2400–3400 m. 1900. z6 Fig. 245. ✣

var. **ludlowii** Stern & Taylor. Taller than the species, to about 2.5 m high, stem leafless to the current year's growth; flowers usually 4 on each shoot, on erect, 16 cm long stalks, 7–20(12) cm wide, involucral leaves and sepals 6 to 10, petals about 12, deep yellow, nearly circular; developed carpels only 1–2. BMns 209. SE. Tibet, in the mountains at 3000–3700 m. 1936. ✣

'**Superba**' (Lemoine 1886). Like the species, but with bronze-brown new growth, later greening; flowers to 12 cm wide, petals 9–11, yellow, later more or less pink at the base, filaments deep red, anthers orange-brown. ✣

P. moutan see: **P. suffruticosa**

P. potaninii Komar. Shrub, to 1.5 m high, but lower in cultivation, stoloniferous; leaves nearly trifoliately incised, the main lobe stalked and pinnatisect, segments outspread, further pinnately partite or lobed, oblong, drawn out to a long, slender apex, 5–10 mm wide, the basal lobes to 12 mm wide; flowers dark chestnut-brown, 5–6 cm wide, with a distinct, fleshy disk, May,

involucral bracts and sepals 5–7, partly acuminate, partly hemispherical, often reddish, petals obovate, apexes rounded or incised, irregularly limbed, filaments red; carpels 2–3, glabrous, fruits about 2.5 cm long. StP 48 (= *P. delavayi* var. *angustifolia* Rehd.). China; SW. Szechwan, NW. Yunnan Provinces. 1904. z6 Fig. 245.

'**Alba**'. Flowers white, filaments green, stigmas white. StP Pl. 9. ✣

var. **trollioides** (Stapf ex F. C. Stern) F. C. Stern. Segments of the leaves more oblong, short acuminate; flowers yellow, never fully opening, petals remaining curved inward like *Trollius* flowers; fruits larger than those of the species. StP Pl. 10. W. China; Tibet. ✣

P. suffruticosa Andr. Erect, 2 m high shrub, twigs thick, glabrous; leaves bipinnate, leaflets 5–10 cm long, ovate, with 3–5 large teeth or lobes, sessile or short petioled, terminal leaflets deeply 3 lobed, the lateral lobes to 5 cm, the terminal lobes 7–9 cm long and coarsely 3 toothed, lobes acute, light green above, bluish beneath, with a few scattered hairs on the venation; flowers pink to white, with a dark violet-red, and red bordered basal spot on each petal, the entire flower about 15 cm wide (much larger on the cultivars), petals numerous, 8 cm wide, concave, margins finely crenate, incised at the apex, filaments violet-red, becoming white at the apex; carpels 5, green, stigma white, May, BM 1154; StP 42 (= *P. arborea* Donn, *P. moutan* Sims). China, Tibet, Bhutan, but cultivated in Japan as early as the 6th century and further developed there. Cultivated in Chinese gardens since the Han Dynasty, more than 2000 years ago. "Moutan" is the Chinese word for Peony. z6 Plate 148; Fig. 246. ✣

var. **spontanea** Rehd. Lower growth habit, glabrous; leaves bipinnate, only 13.5 cm long, lateral leaflets nearly sessile, nearly rounded in outline, base rounded, with 3 large, round

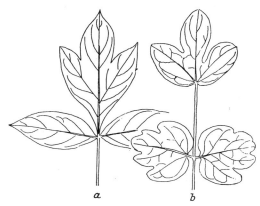

Fig. 246. Leaves from *Paeonia suffruticosa* (a)
and var. *spontanea* (b) (from Stern)

teeth, dark green and glabrous above, bluish and somewhat
pubescent beneath; flowers 11 cm wide. StP 42. China; found
only in Shensi Province to date. Fig. 246. ⊕

P. szechuanica Fang. Shrub, 1–1.5 m high, shoots 4 m
thick; very closely related to *P. suffruticosa*, but with
smaller, finer leaves, glabrous on both sides and not blue
beneath, carpels glabrous; leaves 9–12 cm long, 2–3
pinnate, with 3 to 4 primary leaflet pairs, leaflets ovate to
obovate or more oblong, 3 lobed, the lobes further 3
lobed, dark green above, light green beneath, totally
glabrous on both sides; flowers 12–13 cm wide, solitary,
terminal, involucral bracts 3–5, green, linear, 3 to 5 cm
long, 3 mm wide, sepals 4–5, often 3, petals 9–12, purple-
pink, obovate; carpels purple, glabrous, 12 mm long,
May. Ann. Soc. Hort. France 1958: 437. China; NE.
Szechwan Province, in the mountains at 2650 m.
Discovered in 1957. z5 ⊕

P. yunnanensis Fang. Shrub, to 0.8 m high, stem green,
glabrous; leaves doubly trifoliate, leaflets ovate to
oblong ovate, 8 to 12 cm long, 2–5 lobed, the lobes
obtuse, dark green above, lighter beneath, petiole 10–16
cm long, leaflet petiolules 2–4 cm long; flowers solitary,
terminal, to 16 cm wide, involucral bracts green, linear,
sepals obovate, petals white or reddish, oblong-obovate,
6 cm long, obtuse at the apex and slightly emarginate,
stamens numerous, April to May; carpels conical, with a
dense yellow tomentum (!), fruits not observed. Ann.
Soc. Hort. France 1958: 437. China; Yunnan Province, in
the mountains at 2500 m, discovered in 1937. z6 ⊕

List of Cultivars

All are forms of *P. suffruticosa* unless otherwise noted. A large
number of the cultivars planted in gardens today were
introduced in 1844 by Siebold from Japan (42 cultivars). The
Japanese cultivar names were changed to the French names
retained in the trade. These cultivars were crossed by Modeste
Guérin, Victor Lemoine (1823–1911) and Etienne Méchin
(1815–1895), all of France. Other hybridizers include Louis
Henry (1853–1903, of the Paris Museum of Natural History);
Donkelaer (1788–1858) of Belgium and Louis Van Houtte in
Ghent, Belgium. Eugene Verdier of Paris, in his catalogue of
1857/58, offered no less than 169 cultivars, Louis Van Houtte

offered 76. After 1900, further hybridizing was done by Goos &
Koenemann of W. Germany and G. Frutiger of Switzerland.
While all of these hybridizers developed relatively few
cultivars, hybridizers in the USA have more recently
developed many hybrids. Of the American breeders, A. P.
Saunders of Clinton, N.Y. deserves particular recognition
(1869–1953).

A complete list of the breeders may be found in the Manual of
the American Peony Society, 1928, including the supplement.
A shorter list appears in Nat. Hort. Mag. 1955, 52–53. See also
Bosse (1861).

A detailed history by Wister of these introductions may be
found in the same manual. See also notes by A. Harding (l. c.
187–205).

A "check list" of all the cultivars found in American collections
in 1954, including 250 names without descriptions, may be
found in Nat. Hort. Mag. 1955, 56–60.

Main list
(Japanese cultivar names omitted)

Only the more prominent species in cultivation today are
included.

'Age of Gold' (Saunders 1950). Flat rosette form, double, an
intense yellow. Excellent.

'Alice Harding' (Lemoine 1935) (*P. lemoinei*; *P. lutea* × 'Yaso-
okina'). Pure yellow, ball form, densely double.

'Amber Moon' (Saunders 1948). Huge flowers, double or only
semidouble, dark amber-yellow, with a carmine margin and
blush, early.

'Apricot' (Saunders 1948). Foliage gray and nearly fern-like,
shoots yellow; flowers erect, double, yellow and red.

'Archduc Ludovico' (before 1857). Fully double, pure pink
with a lilac shading.

'Argosy' (Saunders 1928) (*P. lemoinei*). Yellow, very large,
single. Plate 126.

'Arlésienne' (Dessert 1909). Carmine-pink, densely double.

'Athlète' (Mouchelet 1867). Strong growing; light pink, darker
interior, very large flowers.

'Aurore' (Lemoine 1935) (*P. lemoinei*). Coppery-red, single,
flowers abundantly, darker than 'Mme Louis Henry'.

'Baronne d'Alès' (Gombauld, before 1886). Compact habit;
flowers a bright, light pink, dark pink in the center.

'Beatrix' (from Japan, around 1905). White, single.

'Beauté de Twickel'. Strong growing; flowers carmine, with a
darker center, large flowers.

'Belle d'Orléans'. Pink, darker center, slightly double.

'Belle Japonaise'. Lilac-purple, with a blackish claw, single,
very early.

'Bijou de Chusan' (from China, before 1846). Pure white,
densely double. One of the best white forms.

'Blanche de Chateau Futu' (Mouchelet 1967). Lilac-white,
semidouble.

'Blanche Noisette' (Noisette, before 1864). White with salmon-
pink.

'Black Douglas' (Saunders 1948). Mahogany-red, loosely
double, anthers golden yellow, flowers abundantly.

'Black Pirate' (Saunders 1941). Glossy red-bronze with a blackish trace. JRHS 84: 118.

'Carnea Plena' (from Japan, before 1857). Ivory-white, a good double. Very attractive and long lasting.

'Caroline d'Italie' (before 1846). Very large flowers, loosely double, an attractive flesh-pink.

'Chromatella' (Lemoine 1928) (*P. lemoinei*). Pure sulfur-yellow, densely double, globose. A sport of 'Souvenir de Maxime Cornu'.

'Colonel Malcolm' (before 1846 from China). Bluish-purple.

'Comtesse de Tuder' (Gombault 1856). Salmon-pink, a good double, large flowers.

'De Bugny' (Amand, before 1846). White, violet striped in the center, flowers small but abundant.

'Demetra' (N. Daphnis, around 1960) (*P. lemoinei*). Loosely double, well developed, petals undulate, golden-yellow with a brown basal spot, very attractive. JRHS 1970: 67.

'Duchess de Morny'. Flowers lilac-pink.

'Eldorado' (Lemoine 1949) (*P. lemoinei*; *P. lutea* × 'Yaso-okina'). Pure yellow, limb lighter, very large flowers, distinctly double.

'Flambeau' (Lemoine 1930) (*P. lemoinei*). Salmon-red, purple veined, large flowers.

'Fragrans Maxima' (Koenig, before 1858). Dense shrub; flowers a soft salmon-pink, attractive habit.

'Grossherzog von Baden' (Siebold, 1844 from Japan). Very low; flowers crimson with pink, slightly double.

'Hybride de Vaumarcus' (Frutiger). Pure pink, double.

'Impératrice Josephine' (Hiss, before 1858). Dark red, densely double.

'Jeanne d'Arc'. Light pink, with a darker center.

'Josephine Sénéclause' (Sénéclause 1889). Light pink with salmon, flatly convex.

'Jules Pirlot' (Makoy, before 1867). Light pink, center a dark lilac-red. Very valuable, large flowering form.

'Kochs Weisse' (Koch, before 1889). Compact; flowers white with a soft lilac trace and claw, slightly double.

'Lactea' (David, 1839). White, claw violet, anthers yellow, stigma bright red.

'La Lorraine' (Lemoine 1909) (*P. lemoinei*). Soft yellow, a good double form, globose, very large, fragrant. MG 1912: 554.

'L'Espérance' (Lemoine 1909) (*P. lemoinei*). Pure primrose-yellow with a brownish trace, a carmine-pink basal spot, filaments red, anther golden-yellow, single, to 20 cm wide, petals 8–10, crispate. MG 1912: 554.

'Louise Mouchelet' (Mouchelet, before 1860). Light pink, darker at the base, large flowers.

'Ma bien Aimée' (Yvoire). Large flowers, very soft pink, late.

'Marquis de Clapiers' (Sénéclause, before 1899). Flesh-pink, a good double.

'Maxima Plena' (?, before 1858). Lilac, darker interior, densely double, flowers fade quickly.

'Mine d'or' (Lemoine 1956) (*P. lemoinei*; *P. lutea* × 'Yaso-okina'). Strong grower; foliage light green; flowers sulfur-yellow, single, to 8 cm wide, several at the shoot tips.

'Mme de Vatry' (Guérin, before 1858). Lilac-pink, darker center.

'Mme Henriette Caillot' (Paillet 1889). Pure pink, double.

'Mme Laffay' (before 1858). Limb pink, carmine at the base, densely double with very large flowers.

'Mme Louis Henry' (L. Henry 1919) (*P. lemoinei*). Yellowish-pink with a salmon overtone, cup-shaped, fragrant. Gs 1926: 148.

'Mme Marie Ratier'. Soft salmon-pink, large flowers, late.

'Mme Stuart Low' (Makoy 1863). Salmon-pink, semidouble, with a broad silvery limb, anthers yellow.

"Monsieur Edouard Sénéclause" = 'Reine Elisabeth'.

'Mont Vésuve' (Sénéclause, before 1889). Very densely double, medium red. Very popular in the USA.

"Moutan" is a synonym for *P. suffruticosa*, contrived from the Chinese word "Muh tang"; the Japanese word for this plant, "Botan", has the same origin.

'New York' (Goos & Koenemann, around 1930). White with a trace of violet, semidouble. Plate 148.

'Omar Pacha' (Paillet 1889). Dark lilac-pink, semidouble.

"Onyx" = 'Reine Elisabeth'.

'Osiris' (before 1858 from China). Very densely double, velvety black-purple, flowers very abundantly.

'Regina Belgica' (Siebold, Japan, 1844). Dark pink.

'Reine Amalie' (Paillet 1889). Bright, light pink, double, flowers carried just above the foliage.

'Reine des Saumons'. Salmon color, very large flowers.

'Reine des Violettes' (introduced by Fortune, before 1858 from China). Dark violet.

'Reine Elisabeth' (Casaretto 1846). Salmon-pink to a strong pink, very large flowers. One of the most attractive cultivars.

'Robert Fortune' (introduced from China before 1858). Light pink, darker inside, very large flowers.

'Rock's Var.'. Shrub, 2.5 m high, very vigorous; flowers single, very large, soft pink at first, gradually turning white, with a large chestnut-brown spot at the base of the petals, a good bloomer. JRHS 84: 104. Introduced in 1936 by J. Rock from China, SW. Kansu Province. Plate 126.

'Roman Gold' (Saunders 1941) (*P. lemoinei*). Golden-yellow with a darker blush, flowers abundantly.

'Sang Lorrain' (Lemoine 1939) (*P. delavayi* hybrids). Turbid red, claw black, semiduble, fragrant.

'Satin Rouge' (Lemoine 1926) (*P. lemoinei*). Ruby-red with some brick-red and salmon tones, large, densely double.

'Silver Sails' (Saunders 1940) (*P. lemoinei*). Totally soft yellow, with a darker trace. Very attractive.

'Souvenir de Ducher' (Ducher, before 1889). Deep violet. Prefers a semishady location.

'Souvenir d'Etienne Méchin' (Dessert, before 1899). Dark pink to salmon-red, densely double, middle-early.

'Souvenir de Maxime Cornu' (Henry 1919) (*P. lemoinei*). Deep yellow, often somewhat reddish, fragrant, very large, densely double, flowers abundantly.

'Souvenir de Mme Frutiger' (Frutiger). Lilac, huge flowers.

'Souvenir de Mme Knorr' (Van Houtte 1853). Ivory-pink, large flowers.

'Souvenir de Monsieur Frutiger' (Frutiger). Dark violet.

'Surprise' (Lemoine 1920) (*P. lemoinei*). Straw-yellow with salmon and purple, double, fragrant, very large flowers.

'Sybil Stern'. Dark cherry-red, 12–15 petals, obovate, anthers golden-yellow. Developed and introduced by Sir Frederick Stern, 1958. JRHS 87: 115.

'Tria' (N. Daphnis, around 1960) (*P. lemoinei*). Golden-yellow, anthers orange, petals 10–12, flowers very early, usually with 3 flowers on each shoot. JRHS 1970: 64 and 65.

"Triomphe de Flandres" = 'Triomphe de Vandermaelen'.

'Triomphe de Gand'. Ivory-white, with a darker center.

'Triomphe de Vandermaelen' (Vandermaelen 1849). Purple-pink, with a darker center. Very attractive.

'Velours Rouge' (Lemoine 1956). Velvety carmine-red, very dark toned, 21 cm wide.

'Victoire d'Alma' (Paillet 1889). Satiny pink, double.

'Ville de St. Denis' (Mouchelet 1854). Pale lilac, darker inside, densely double. Flowers fade quickly.

'Wilhelm Tell'. Compact; flowers carmine-pink, double.

'Zenobia' (imported from China by Fortune). Good growth habit; flowers purple-violet, semidouble.

Japanese Cultivars

Once again, because of the large number of cultivars in existence, only the more common will be listed; for further information, please refer to the *Picture Book of Peonies, Niigata Prefecture, Japan* by Miyazawa, which first appeared in 1920 and presents 54 cultivars in full size color plates.

'Fuso-no-tsukasa' (imported before 1931). White, densely double, petals irregularly arranged.

'Gessekai'. White, densely double, to 30 cm wide, petals wrinkled.

'Godaishu'. Strong grower; flowers white, with a yellow center, fully or semi double, globose, petals fimbriate and twisted.

'Hakugan'. Pure white, petal limb fimbriate, semidouble, flat.

'Hana-daigin' (imported before 1910). Violet-purple, semidouble, large.

'Hana-kisoi' (imported before 1929). Deep cherry-red, very large flowers, petals undulate. One of the best cultivars.

'Hinode-sekai'. Scarlet-red, fading lighter.

'Hodei' (imported before 1929). Carmine-red, very densely double.

'Ima-shojo' (imported before 1932). Glossy scarlet-red, globose, double, very large.

'Kenreimon' (imported before 1932). Deep purple with carmine, petals long and narrow, curved inward.

'Koka-mon' (imported before 1932). Chestnut-brown with white stripes, petals fimbriate, large flowers.

'Momo-yama' (imported before 1931). Soft pink, semidouble, large.

'Nissho' (imported before 1931). Glossy scarlet-red, densely double, very large flowers.

'Renkaku' (before 1893). Pure white, densely double.

'Rimpo' (before 1926). Strong grower; flowers luminescent purple, densely double, very large flowers, very late.

'Ruriban' (before 1893). Dark red with purple, densely double, large.

'Sakura-jishi'. Pink, fading darker, large, semidouble.

'Shintenchi' (before 1931). Cherry-pink, densely double, large.

'Tama-sudare' (imported before 1931). White, densely double, large.

'Yachiyo-tsubaki' (before 1931). Deep pink, semidouble, very symmetrically developed.

'Yae-zakura' (before 1931). Soft cherry-pink, double, very large.

'Yaso-okina' (before 1893). White, base soft pink, densely double.

'Yo-meimon' (before 1929). Scarlet-red.

Transplanting of the woody peonies should be done with care in March or April, just before the new growth appears; prefer a fertile soil, rich in humus with good drainage. Under these conditions the plants will thrive for 50 years or more, becoming more beautiful each year. Winter mulch advisable. Spent flowers should be removed.

Lit. Fang, Wen-Pei: Notes sur les Pivoines de Chine; in Ann Soc. Nat. Hort. France 1958, 434 to 441 ● Harding, A.: The Book of the Peony; Philadelphia 1917 ● Miyazawa: The Picture Book of Peonies, Niigata Prefecture, Japan; around 1920 (with 54 color plates) ● Stern, F. C.: A study of the genus *Paeonia*: London 1946 (with 15 color plates and 28 illustrations) ● Wister, J. C., & H. E. Wolfe: The Tree Peony; in Nat. Hort. Magazine 1955, 1–60 ● Haworth-Booth, M.: The Moutan or Tree Peony; London 1963 (169 pp., 11 Pl., with 30 pp. of cultivar descriptions). Armatys, L. J.: The new hybrids of Moutan; in Jour. RHS 96, 107–110, 1970.

PALIURUS Mill. — Christ's Thorn — RHAMNACEAE

Thorny shrubs or trees; twigs protected by 2 sharp stipular thorns at each node; leaves alternate, usually distichous, 3 veined, cordate to ovate, entire to serrate; flowers unattractive, small, yellowish, in small, axillary cymes, 5 parted; fruit dry, flat, with a large, horizontal winged limb. — 6 species, from S. Europe, Asia Minor, E. Asia, China, Japan, Korea.

Paliurus aculeatus see: **P. spina-christi**

P. aubletii see: **P. ramosissimus**

P. australis see: **P. spina-christi**

P. hemsleyanus Rehd. Tree, to 10(15) m high, twigs thin, glabrous, thorns black but occasionally thornless (!); leaves obovate to oval-lanceolate, 6–10 cm long, acute, base round and three veined, simple crenate, glossy green above, lighter beneath, somewhat leathery; fruits short stalked, 3 cm wide, surrounding wing reddish-brown. ICS 3234 (= *P. orientalis* Hemsl.). China; Yunnan, Szechwan, Hupeh and other provinces in the forest. z9 Plate 147. ⊘ ⚬

P. orientalis see: **P. hemsleyanus**

P. ramosissimus (Lour.) Poir. Small shrub, 1.5–3 m high, shoots flexuose, sharp thorned, densely silky pubescent; leaves obovate, 4 cm long, obtuse, crenate, base round and 3 veined, glossy green above, lighter and silky brown pubescent beneath; fruits cupulate, woody, 1 cm wide. LF 223; ICS 3235 (= *P. aubletii* Benth.). China; Szechwan, Hupeh Provinces among others. 1819. z9 ⚬

P. spina-christi Mill. Broad, thorny shrub, to 5 m high in its habitat, young twigs finely pubescent, thorny; leaves ovate to rounded, 2–4 cm long, slightly dentate to entire, dark green above, venation pubescent beneath; flowers yellowish-green, June–July; fruits flat, woody, 2–3 cm wide. HM 1886 (= *P. aculeatus* Lam.; *P. australis* Gaertn.). S. Europe, Orient. Ⓕ Yugoslavia (Croatia & Kras for reforestation). z8 Plate 147; Fig. 248. ⚬

Only for mild climates; likes a calcareous clay soil. *P. spina-christi* is the hardiest, but will freeze back in cold winters.

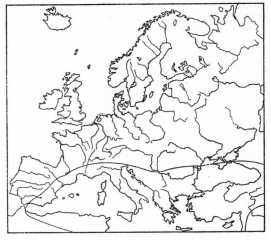

Fig. 247. Range of *Paliurus spina-christi* (from Rubtzov 1964)

Fig. 248. *Paliurus spina-christi*, twig, inflorescence, flower, fruit (from Weberbauer, Lauche)

PALMAE — PALMEN

In this work only a few species are included. The northern limit for the more common palm species will be found in the following list, based on Koch, Fischer and Anliker. Only those plants listed in bold print are described.

Explanation of the Table:

The more tender species are listed in Number 2.; these palms may be observed in parks and gardens of zone 10. The most tender, suitable only for the tropics or conservatories, are listed in Number 1. Number 3. contains the slightly hardier species found hardy in zone 9. Number 4. is the only palm which can be cultivated in zone 8.

Species	Origin	Northern limit
4. **Trachycarpus fortunei** Wendl.	China, S. Japan	z7 England; Atlanta, Ga.
3. **Erythea armata** Wats	S. California, USA	z8 Locarno, Italy; Raleigh, N. C.
Jubaea chilensis	Chile	z8 Locarno, Italy; Raleigh, N. C.
Brahea dulcis Mart.	S. Texas, USA	z8 Isola Bella, Italy; Houston, Texas
Butia yatay Becc.	Argentina	z9 Northern Italian seacoast; New Orleans, La.
Chamaerops humilis L.	Mediterranean region	z9 Northern Italian seacoast; New Orleans, La.
Phoenix canariensis Chab.	Canary Islands	z8 Brissago, Switzerland; Atlanta, Ga.
Erythrea edulis Wats.	Guadeloupe	z8 Garda Lake, Italy; Dallas, Texas
Phoenix silvestris Roxb.	E. India	z8 Garda Lake, Italy; Dallas, Texas
Washingtonia filifera Wendl.	S. California	z8 Garda Lake, Italy; Dallas, Texas
2. *Butia eriospatha* Becc.	S. Brazil	z10 Riviera; Miami, Fla.
— **capitata** (Kart.) Becc.	E. Brazil	z10 Riviera; Miami, Fla.
Cocos romanzoffiana Cham.	Brazil, Paraguay	z10 Riviera; Miami, Fla.
Corypha elata Roxb.	Bengal, Burma	z10 Riviera; Miami, Fla.
Howea belmoreana Becc.	Lord Howe Island, Australia	z10 Riviera; Miami, Fla.
— *forsteriana* Becc.	Lord Howe Island, Australia	z10 Riviera; Miami, Fla.
Livistona australia Mart.	SE. Australia	z10 Riviera; Miami, Fla.
— *chinensis* Mart.	S. China	z10 Riviera; Miami, Fla.
Phoenix dactylifera L.	N. Africa	z10 Riviera; Miami, Fla.
— **reclinata** Jacq.	W. Africa	z10 Riviera; Miami, Fla.
Sabal umbraculifera Mart.	Cuba	z10 Riviera; Miami, Fla.
1. *Nannorhops ritchieana* Wendl. & Hook.	India to S. Iran	NE. Spain (Blanes)

A worldwide list of the major palm collections may be found in "Cultivated Palms" (Living Palm Collections: 184).

Lit. Fischer, B.: Vorkommen und klimatische Bedeutung wärmeliebender fremdländischer Gehölze in der Schweiz; in Schweiz. Beitr. Dendrol. 1950, 7–32 (including 26–28 of the *Palmae* with the precise location and cultural requirements) ●

Anliker, J.: Die im Freien aushaltenden Palmen der Südschweiz; in Schweiz. Beitr. Dendrol. 1950. 33–51 ● Koch, F.: Palmen in Europa; in Mitt. DDG 1926, 92–97 ● For Palms in general, refer to the special issue of American Hort. Magazine of January 1961 (Cultivated Palms; 189 pp. with many photographs).

PANAX

Panax aculeatum see: **Acanthopanax trifoliatus**

P. arboreum see: **Neopanax arboreum**

P. davidii see: **Nothopanax davidii**

P. delavayi see: **Nothopanax delavayi**

P. discolor see: **Pseudopanax discolor**

P. lessonii see: **Pseudopanax lessonii**

P. sessiliflorus see: **Acanthopanax sessiliflorus**

PANDANUS L. f. — Screw Pine — PANDANACEAE

Upright, evergreen trees or shrubs of the tropics, stem simple or branched, rootstock occasionally prostrate and rooting, with many, strong aerial roots, these with "root caps"; leaves very long, linear or sword-like, usually sharply toothed at the margin and on the midrib beneath, spirally arranged around the stem; flowers dioecious, male inflorescences usually compound; female terminal, spike-like or racemose or solitary; fruit a drupe. — About 600 species, most in the Malaysian Archipelago, some on the Pacific Islands, Madagascar and in tropical Africa and Australia. Only the following 3 species are in cultivation in frost free areas of zone 10.

Pandanus candelabrum see: **P. utilis**

P. pygmaeus Thouars. Densely branched shrub, very broad, seldom over 60 cm high, quickly developing prop roots; leaves subulate-linear, 30–40 cm long, 7 mm wide, leaf margins and keel with fine brownish prickles; fruits short stalked. BM 4736. Madagascar. 1830. z10 # ∅

P. utilis Bory. Tree, 18–20 m high in its habitat where it also branches; leaves rather stiff and erect at first, later more pendulous, to 1.5 cm long, 10 cm wide, dark green, prickles red on the leaf margin, fruit clusters rather globose, hanging on long stalks, about 15 cm wide, with about 100 drupes. BM 5014 (as *P. candelabrum*); EKW 299. Madagascar. z10 Plate 150. # ∅

P. veitchii Dallière. Leaves about 50–60 cm long, drawn out to a long tip, more or less nodding, dark green, trimmed with a pure white or ivory-white marginal stripe, commonly also with a fine green stripe, margins thorny. PBl 1: 102. Polynesia. 1868. Very important cultivated plant, commonly used as a house plant. z10 # ∅

PANDOREA Spach. — BIGNONIACEAE

Evergreen shrubs, climbing without tendrils; leaves alternate, odd pinnate, leaflets entire or dentate; flowers many in terminal and axillary panicles, mostly white or pink; corolla campanulate or funnelform, with a narrow or broad limb, the limb lobes overlapping, calyx small; fruit an oblong pod with thick, non-keeled valves; seeds broadly elliptic, wingless. — 8 species in E. Malaysia and Central Australia.

Pandorea jasminoides (Lindl.) K. Schum. High climbing shrub, glabrous; leaflets 5–7(9), nearly sessile, acutely ovate, base rounded, terminal leaflet tapering to the petiole, 3–6 cm long, glossy above; flowers in few flowered panicles, corolla tube 4–5 cm long, white, exterior often turning pink, interior dark pink, limb much widened, 5 lobed, outspread, light pink, June–August. BM 4004; BR 2002; DRHS 1477 (= *Bignonia jasminoides* Hort.; *Tecoma jasminoides* Lindl.). E. Australia. z10 # ✜

'Alba'. Flowers totally white, somewhat larger.

'Rosea'. Flowers totally pink.

P. pandorana (Andr.) Van Steenis. Wonga-Wonga. Very fast growing, twining shrub; leaves quite variable in size, from 2–20 cm long, with 1–7(usually 5) leaflets, these ovate to more lanceolate, long acuminate, to linear, 3–6 cm long, entire to coarsely crenate, glossy above; flowers many, in panicles, corolla funnel-shaped with a narrow, 5 lobed limb, only 1.5–2 cm long, cream-white with purple spots on the throat, May–June. HV 17; DRHS 1477 (= *Bignonia australis* Ait.; *Bignonia pandorana* Andr.; *P. australis* [R. Br.] Spach). Australia to New Guinea. z10 Plate 137. # ✜

P. ricasoliana see: **Podranea ricasoliana**

PARAHEBE W. R. B. Oliver — SCROPHULARIACEAE

Very closely related to *Hebe* and usually included in that genus but differing in the fruits. Subshrubs, usually prostrate, shoots rooting on the basal portion; leaves opposite, sessile to short petioled, crenate or dentate; flowers in long or short, axillary panicles; calyx deep and evenly 4 lobed; corolla variable in form and size, with 4 or 5 uneven lobes, stamens 2; fruit capsule laterally compressed, with a dividing wall across the narrowest part (!; this wall on *Hebe* is across the widest part of the capsule). — 15 species, all in New Zealand.

Parahebe × bidwillii (Hook.) W. R. B. Oliver (*P. lyallii × P. decora*). Procumbent, rooting shrublet, mat-like growth habit or to 15 cm high; leaves oblong to obovate, 2–6 mm long, thick, leathery, entire or with one incision on either side; flowers in narrow, axillary, 7–20 cm long, erect racemes, corolla 8 mm wide, white with lilac lines, June. HI 814 (= *Veronica bidwillii* Hook.). New Zealand. 1850. z9 # ✜

P. catarractae (Forst. f.) W. R. B. Oliv. Subshrub, branches woody at the base, twigs procumbent, shoots ascending, thin, herbaceous, purple, to 60 cm high; leaves widely spaced, lanceolate to oval-lanceolate, coarsely dentate, usually 1.2–3 cm long, acute; flowers in erect, slender racemes, 7–20 cm long, white with pink-purple lines, center carmine, 8 to 12 mm wide, stalk 1–2.5 cm long, July–September, (= *Veronica catarractae* Forst. f.). New Zealand. z9 # ✜

P. lyallii (Hook. f.) W. R. B. Oliv. Subshrub, branches woody at the base, the procumbent twigs about 40 cm long; leaves ovate to rounded, 6–12 mm long, coarsely and obtusely dentate, glabrous, thick and leathery; flowers in erect, thin stalked, 7–20 cm long racemes, corolla white, pink veined, 8 mm wide, anthers blue, July–September (= *Veronica lyallii* Hook. f.). New Zealand. 1876. z9 # ✜

All species well suited for the rock garden, in sunny sites without a preference as to soil type.

PARABENZOIN See: LINDERA

PARASYRINGA W. W. Smith — OLEACEAE

Monotypic genus; similar to *Ligustrum* in appearance and flower, but the fleshy fruits eventually become dry, dehiscent capsules, with 1–2 seeds. — 1 species in China.

Parasyringa sempervirens (Franch.) W. W. Smith. Evergreen shrub, to 2.5 m high, densely branched, young shoots red-brown and pubescent at first; leaves lanceolate to cordate-rounded, 1.5–6 cm long, leathery, deep green and glossy above, lighter and black punctate beneath, petiolate; flowers cream-white, in dense, terminal, 5–10 cm long, branched panicles at the shoot tips, corolla 6 mm long, August–September; fruits fleshy at first, later dry and capsule-like, dehiscent, with 1–2 seeds. BM 9295; FN 6: 720; DRHS 1483 (= *Ligustrum sempervirens* [Franch.] Lingelsh.; *Syringa sempervirens* Franch.). China; Yunnan, Szechwan Provinces. 1916. z8 # ∅ ✜

A very attractive flowering shrub without a particular preference as to soil type and location.

PARKINSONIA L. — LEGUMINOSAE

Tropical trees; twigs thorned with simple to 3 part thorns; leaves evergreen, but usually abscising in the subtropics, alternate or clustered, bipinnate with a very short rachis, leaflets usually very small; flowers yellow or whitish, in loose, terminal or axillary racemes; calyx 5 parted, petals 5, clawed; fruit a leathery, many seeded, linear pod. — 5 species in S. America, N. America (S. California), S. Africa, W. Indies. Planted in the tropics and subtropics over a large range, primarily *P. aculeata.*

Parkinsonia aculeata L. Jerusalem Thorn. Small, evergreen tree, scarcely over 3 m high, twigs flexuose, often wide arching; leaves bipinnate (not easily recognizable since the rachis is very small!), 20–40 cm long, axes flat, leaflets very numerous, small, widely spaced, uneven sided linear-lanceolate, about 1 cm long, 2 mm wide; flowers yellow, fragrant, in loose pendulous racemes; pods 5–10 cm long. SS 131. Tropical America. Ⓕ In the Near East for erosion control and for hedges. z9 Plate 149; Fig. 249 # ∅ ✧

Fig. 249. *Parkinsonia aculeata* (from Sudworth)

PARROTIA C. A. Mey. — Persian Ironwood — HAMAMELIDACEAE

Monotypic species. Deciduous shrub or tree with exfoliating bark (plate-like); leaves alternate, simple, entire; flowers appear before the leaves, without petals; the 10–15 red stamens surrounded by a 5–7 lobed calyx, enveloped by brown bracts; ovaries 2 locular; fruit a woody, horned capsule dehiscing by 2 valves. — N. Iran.

Parrotia persica (DC.) C. A. Mey. Tall, broad sprawling shrub or (rarely) tree, to 10 m high, with exfoliating bark, twigs erect, olive-brown, somewhat clustered, buds dark brown, with a thick tomentum; leaves usually obovate-oblong, obtuse, coarsely crenate on the apical half, 6–10 cm long, dark green above, lighter beneath, both sides with a light stellate pubescence, often red margined on new growth, gorgeous golden-yellow and scarlet-red in fall (unfortunately there are some inferior types without particularly nice fall color in cultivation); flowers yellowish, with bright red anthers and deep brown bracts, in small, numerous globose heads, March. BM 5744; Gs 1921: 224. N. Persia. 1841. Ⓕ USSR; reforestation in Azerbashan. z5 Plate 149; Fig. 250. ∅ ✧

No particular requirements, but prefers a good, sufficiently moist garden soil.

Fig. 250. Inflorescence, individual flowers, stamens and fruits of *Parrotia persica* (upper row) and *Parrotiopsis jacquemontiana* (lower row) (from Harms)

PARROTIOPSIS (Niedenzu) Schneid. — HAMAMELIDACEAE

Monotypic genus. Deciduous shrub or tree, with all parts stellate pubescent; buds stalked; leaves petiolate, rounded, sharply dentate, stipules ovate; flowers hermaphroditic, without petals, in many flowered heads, surrounded by large, white bracts; sepals 5–7, small; stamens about 15, anthers yellow, opening by 2 valves; styles distinct, with linear stigmas; fruit a dehiscent capsule. — Himalayas.

Parrotiopsis jacquemontiana (Decne.) Rehd. A 7 m high tree in its habitat, often only a 2–3 m high shrub in cultivation, bark smooth, twigs with gray-yellow stellate pubescence at first; leaves rounded, 3–5(8) cm long, short and sharply toothed (somewhat resembling those of *Alnus glutinosa*), stellate pubescent above at first, light green beneath, venation densely pubescent, fall color golden-yellow; flowers in 3–5 cm wide heads, surrounded by white bracts, these brown lepidote beneath, appearing with the leaves, May. BM 7501; MD 1932: 4. NW. Himalayas, in the mountains from 900 to 3000 m. 1879. z6 Plate 147; Fig. 250. ✥

For a fertile, humus soil and a sunny, protected site.

PARTHENOCISSUS Planch. — Virginia Creeper — VITACEAE

Deciduous, occasionally evergreen climbers with smooth, lenticelate bark; white pith, never constricted above the nodes; tendrils usually with holdfasts; leaves pinnate or partly 3 lobed; flowers hermaphroditic, rarely polygamous, in compound cymes opposite the leaves; petals 5, occasionally 4, style short and thick; fruit a 1–4 seeded, dark blue or blue-black berry. — About 15 species in N. America, E. Asia, and the subtropical and tropical regions of SE. Asia and Mexico.

Parthenocissus henryana (Hemsl.) Diels & Gilg. Deciduous climbing shrub, to 5 m high in ideal conditions, young shoots angular, tendrils with 5–7 thin branches and holdfasts; leaves always 5 parted, leaflets 3–6 cm long, attractively red on the new growth, later dull green above with broad white venation, more or less purple beneath, toothed on the apical half; flowers in narrow, 7–15 cm long panicles; fruits dark blue, usually with 3 seeds. RH 1907: 71; HV 18. Central China. 1895. z7 Plate 150. ⊘

var. *glaucescens* see: **P. thomsonii**

P. heptaphylla (Buckl.) Brit. & Small. Deciduous climber, to 9 m high in its habitat, young shoots red-brown, slightly angular to rounded, tendrils long and forked, with holdfasts; leaves usually 7 parted, leaflets oblong-obovate, 3–6 cm long, coarsely serrate, base cuneate, dark green above, lighter beneath and glabrous; flowers in loose, 3–7 cm wide cymes, greenish, spring; fruits blue-black, 8–12 mm thick, globose, inedible. VT 712. USA; Texas; on rocky-gravelly sites. 1900. z9 Plate 150. ⊘

P. himalayana (Royle) Planch. Strong and tall growing, climbing shrub, shoots glabrous; leaves trifoliate, leaflets petiolate, 5 to 12 cm long, middle leaflets ovate to obovate, the lateral leaflets very assymetrically ovate, all short acuminate, dark green and glabrous above, midrib slightly pubescent beneath, coarsely serrate, base rounded to lightly cordate, fall color a beautiful red; inflorescences as long as the leaves. ICS 3281. Himalaya; China. 1894. z9 ⊘

var. **rubrifolia** (Lév. & Vaniot) Gagnep. Leaflets smaller and wider, purple when young; inflorescences smaller (= *Vitis rubrifolia* Lév. & Vaniot). W. China. 1907. z9 ⊘

P. inserta see: **P. vitacea**

P. laetevirens Rehd. Closely related to *P. quinquefolia*, tendrils with 5–8 thin brances, obovate to elliptic, coarsely serrate, both sides light yellow-green, 5–10 cm long, glabrous on both sides or pubescent on the venation beneath; flowers in large, terminal panicles, 15–25 cm long; fruits globose, 8 mm thick, dark blue, with 2–5 seeds. China; W. Hupeh Province, in the mountains at 600–1000 m. 1907. z9 ⊘

P. hirsuta see: **P. vitacea** var. **dubia**

P. quinquefolia (L.) Planch. High climbing shrub, young shoots reddish (!), tendrils with 5–8 arms, these terminating in disk-like holdfasts (!!); leaves 5 parted, leaflets petiolate, elliptic to oblong obovate, 4–10 cm long, long acuminate, base usually cuneate, coarsely serrate or also more crenate, dull green above, bluish beneath (!), bright red and carmine in fall; flowers normally in large terminal panicles (!), July–August; fruits blue-black, only slightly pruinose, 6 mm thick, usually with 2–3 seeds. VT 711; ST 88 (*Vitis hederacea* Ehrh.; *Ampelopsis quinquefolia* Michx; *Ampelopsis virginiana* Hort.). Eastern N. America. 1929. See the notes at *P. vitacea!* z3 Plate 150; Fig. 251. ⊘

var. **engelmannii** (Koehne & Graebn.) Rehd. Twigs thin, young shoots rounded, glabrous, tendrils with 4–6(7) branches and rounded to oblong holdfasts; leaf petiole glabrous, reddish to 15 cm long, leaflets to 12 cm long, 3 cm wide, broadly lanceolate to ovate, very sharp and regularly serrate, dark green above, distinctly bluish pruinose beneath, dark red in fall; inflorescence to 4 cm long. Eastern USA. Primarily differing from the species in the smaller leaflets. Fig. 252. ⊘

var. **hirsuta** (Pursh) Planch. Tendrils with 2–3 long and an equal number of short branches, shoots and leaves with erect, white pubescence, young shoots and leaves attractively red; leaflets 12–18 cm long, 5–8 cm wide, abruptly tapering to the nearly 1 cm long petiole, apex short, margins obtusely serrate, fall color bright red. Gfl 1462 (= *Ampelopsis graebneri* Bolle; *Ampelopsis hirsuta* Don; *Ampelopsis pubescens* Schlechtd.). Eastern N. America to Mexico. ⊘

Fig. 251. *Parthenocissus quinquefolia*
(from Sargent)

Fig. 252. *Parthenocissus quinquefolia*
var. *engelmannii*
(Original)

var. *latifolia* see: var. **murorum**

'Minor'. Presumably only a cultivar of var. *murorum*; tendrils with 10–12 short branchlets, leaflets broader, smaller, oval to elliptic, rounded on the base or nearly rounded, petiole 0.5–1 cm long (= *Ampelopsis hederacea minor* Graebn.).

var. **murorum** (Focke) Rehd. Young shoots glabrous, rounded, tendrils with 6–12 regularly distichous branchlets, clinging very tightly, holdfast disks often much longer than wide; leaflets shorter and broader than those of the species, thicker, the outermost leaflets rounded at the base, with a scattered shaggy pubescence beneath at first, later becoming glabrous, fall foliage a bright red; inflorescences about 10 cm long and wide, August–September. HM 1913; MG 21: 25 (= var. *latifolia* Rehd.; *Ampelopsis muralis* Hort.; *Ampelopsis radicantissima* Schelle). Plate 150. ⌀

var. **saint-paulii** (Koehne & Graebn.) Rehd. Resembling var. *murorum*, grows as vigorously, twigs often with small aerial roots, young twigs, leaf petioles and leaflets totally white haired beneath and above only on the midrib, tendrils with 8–12

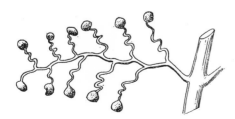

Fig. 253. *Parthenocissus quinquefolia* var. *saint-paulii*, typical tendril at actual size (from Engler-Prantl)

regularly distichous branchlets; leaflets elliptic to oblong-elliptic, 12–15(20) cm long, sharp and deeply serrate, gradually tapering to a short stalk, leaves persisting well into fall with a deep red fall color; inflorescences elongated. ST 88 (= *P. saint-paulii* Koehne & Graebn.). USA; Iowa to Texas. Fig. 253. ⌀

P. saint-paulii see: **P. quinquefolia** var. **saint-paulii**

P. semicordata (Wall.) Planch. Very similar to *P. himalayana* or possibly only a form thereof from a higher altitude; leaflets smaller, rough pubescent beneath as are the young shoots (= *Vitis semicordata* Wall.). Himalayas. Before 1914. The plant depicted in Plate 152 from Kew Gardens was drawn from seed sent from the Botanic Garden in Calcutta, India and is hardier than *P. himalayana*. z9 ⌀

P. spaethii see: **P. vitacea**

P. thomsonii (M. A. Laws.) Planch. Very similar to *P. henryana*; young shoots reddish, finely pubescent, angular, tedrils with 3–5 branchlets; leaves 5 parted, leaflets red when young, elliptic to obovate, 5–11 cm long, long acuminate, sharply dentate toward the apex, dull reddish above, glabrous beneath, midrib slightly pubescent, fall color purple-red; flowers in long stalked cymes; fruits black. ICS 3283 (= *P. henryi* var. *glaucescens* Diels & Gilg). Himalayas, Central China. 1900. z9 ⌀

P. tricuspidata (S. & Z.) Planch. High twining climber with tendrils, short but very well branched, with disk-like holdfasts; leaves long petioled, partly simple,

Fig. 254. *Parthenocissus tricuspidata* 'Lowii' (Original)

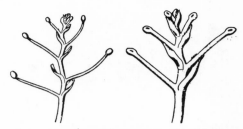

Fig. 255. *Parthenocissus tricuspidata* 'Veitchii', young tendrils without fully developed holdfasts, from above and beneath (from Troll)

broadly ovate, 10–20 cm wide, with 3 acuminate, coarsely serrate lobes, base cordate, partly trifoliate with ovate leaflets, much smaller on the shoot tips and then often lobed or trifoliate, glossy green on both sides and both sides glabrous or with the venation pubescent beneath, fall color orange-yellow and scarlet-red; flowers yellow-green, June–July; BM 8287; BC 2767; MJ 1023 (= *Ampelopsis veitchii* var. *robusta* Hort.). Japan, China, Korea. 1862. z5 Plate 150. ⊘

'Atropurpurea'. Strong growing; leaves mostly simple, large, dull bluish-green above and turning purple, red in spring and fall, blade bullate between the venation. Mutation of 'Purpurea' which is not very consistent and often reverts back to the original form. Plate 151. ⊘

'Aurata' (Schwerin). Leaves nearly golden yellow, slightly greenish marbled, with a scabrous, blood red margin. 1908.

'Beverley Brook' (Jackman, before 1952). Particularly small foliage, a climbing form with good red fall color.

'Gloire de Boskoop' see: **'Robusta'**

'Green Spring'. Leaves simple, to 25 cm long, smooth above, bright green, very glossy, dull beneath, fine reticulate venation, young leaves and shoots somewhat reddish toned, but quickly turning green (= *Ampelopsis veitchii* green leaved, Hort. Holl.). 1903. ⊘

'Lowii' (Low). Leaves only 2–3 cm long, simple or trifoliate, often wider than long, incised-serrate, bright green, reddish new growth, deep red in fall. RH 1917: 272; MG 1908: 261. A seedling from Low & Co. of Enfield, England. 1907. Fig. 254. ⊘

'Minutifolia' (Turbat). Like 'Lowii', but with larger leaves, green, very glossy on older shoots, purple in fall, eventually totally carmine, petioles 7–15 cm long. Plate 151. ⊘

'Purpurea'. Like 'Veitchii', but the leaves are always red, never becoming green; young shoots red; fall color appears 3 weeks before the other forms (= *Ampelopsis veitchii* var. *purpurea* Hort.). 1889. ⊘

'Robusta' (Pannebakker, around 1908). Growth vigorous; leaves often trifoliate, 10–20 cm wide, long petioled, orange and scarlet in fall, more glossy than the common 'Veitchii'. Presumably identical to **'Gloire de Boskoop'**, from Pannebakker, before 1920.

'Veitchii' (Record). Only a juvenile form; leaves smaller than those of the species, ovate, unlobed to trifoliate, leaflets with 1–

3 large teeth only on the outer margin, inner margins of the leaflets normally entire. RH 1877: 176; HM 1910 (= *Vitis veitchii* Hort.; *Ampelopsis veitchii* Record). Presumably introduced from Japan; known in England since 1861. Plate 150; Fig. 255. ⊘

P. vitacea Hitchcock. Shrub, usually low, occasionally high climbing, otherwise climbing on other shrubs, young shoots green (!), cylindrical, tendrils with 3–5 twining branchlets, only occasionally with holdfasts; leaves 5 parted, occasionally only trifoliate, leaflets elliptic-oblong, 5–12 cm long, long acuminate, coarse and sharply serrate, base cuneate, petiolules 8–12 mm long, dark green and glossy above, lighter beneath, glabrous and glazed, inflorescence forked (!), stalk 3–7 cm long, June–July; fruits blue-black, usually pruinose, 8 mm thick, with 3–4 seeds. HM 1912; VT 710; RM 305 (= *P. inserta* [Kern.] Fritsch; *P. spaethii* Koehne & Graebn.). Eastern N. America. Before 1800. This species without holdfasts on the tendrils is often mislabeled *P. quinquefolia*. (The author agrees with Suessenguth that the name *P. inserta* [Kerner] Fritsch, as used by Rehder, is doubtful because the illustration used by Kerner is not clearly distinguishable from the similar *Vitis inserta* and lacks a precise description.) z5 Plate 150. ⊘

var. **dubia** (Rehd.) Rehd. Stalk more or less densely rough haired, with bristly hairs at the nodes; petioles bristly at the base, leaflets 5 to 8 cm wide, more coarsely serrate than the species, both sides darker green, less glossy, rough pubescent on the venation beneath, leaves sometimes without a good fall color in cultivation (!); inflorescences denser than the species, rough pubescent; fruits ripening in late June (= *P. hirsuta* Graebn. non Small).

var. **laciniata** (Planch.) Rehd. Leaflets smaller, narrower, more deeply incised-serrate, usually yellow-green. USA; Wyoming to New Mexico. 1898.

'Macrophylla'. Young plants occasionally with scattered rough pubescence; leaflets large, elliptic, occasionally to 18 cm long and 10 cm wide, leaves hardly coloring in fall.

All species thrive in any good garden soil; the tender species, particularly *P. henryana* and *thomsonii*, need a very protected area. The beautiful fall foliage is most effective in full sun.

Lit. Graebner, P.: Die *Parthenocissus* species; in Mitt. DDG 1928, 1–10 ● Rehder, A.: Die amerikanischen Arten der Gattung *Parthenocissus*; in Mitt. DDG 1905, 469–476.

PASANIA

Pasania densiflora see: **Lithocarpus densiflorus**

PASSIFLORA L. — Passion Flower — PASSIFLORACEAE

Herbaceous or woody plants, usually climbing by tendrils, occasionally also shrubs or small trees, shoots often angular; leaves alternate, quite variable, usually 3–5 lobed, but also often 2 lobed or unlobed; petioles variable in length, usually with 2–8 glandular outgrowths; flowers usually solitary, axillary, calyx tubular, with 5 sepals, petals 5 (or absent), with a ring of filaments in several rows (corona) between the petals and stamens; ovaries and the normally 5 stamens on a gynophore; stigma capitate; fruit usually a juicy berry, quite variable in size– from pea-sized to the size of a small melon.—About 500 species in the tropical Americas, Asia, Australia and Polynesia.

The species with a tubular calyx were once combined into their own genus, *Tacsonia* Juss., but are included today in the *Passiflora*.

Key to the species described
(from G. H. M. Lawrence, simplified)

● Flowers usually less than 4 cm wide, greenish or whitish; corona with 1 or 2 rows; operculum usually plaited and without a row of filaments between this and the corona:

> (included here, among others, *P. coriacea, trifasciata, warmingii* [not covered in this work but cultivated in some botanic gardens])

●● Flowers usually not over 5 cm wide, conspicuously colored; corona with more than 2 rows; with a row of filaments between the operculum and the corona;
> + Calyx tube cylindrical, longer than the sepals;
>> 1. Petals 2.5–3.5 cm long, exterior greenish, interior pink; leaves deeply lobed:
>>> *P. mollissima*

>> 2. Petals 4–6.6 cm long, exterior brick-red, interior pink; leaves lobed nearly to the base:
>>> *P. exoniensis*

> ++ Calyx tube usually urceolate-campanulate (but cylindrical on *P. antioquiensis*!), shorter than the sepals;
>> 3. Flowers in 30 cm long racemes:
>>> *P. racemosa*

>> 4. Flowers solitary or paired in the leaf axils;
>>> X Flowers vermillion-red to scarlet, sepals and petals nearly equal in color;
>>>> a. Flower stalks much shorter than 8 cm:
>>>>> *P. manicata*

>>>> b. Flower stalks longer than 10 cm:
>>>>> *P. antioquiensis*
>>> XX Flowers white or green or pink toned or bluish-purple or, if the petals are scarlet or vermillion-red, then the sepals are another color;

v Shoots 4 sided, the edges more or less winged;
> § Leaves three lobed, sepals white inside, petals pink:
>> *P. alato-coerulea*

> §§ Leaves unlobed, sepals white inside, petals whitish pink:
>> *P. quadrangularis*

vv Shoots cylindrical or angular, the edges not winged;
> i Leaves unlobed:
>> *P. ligularis*

> ii Leaves 5–7 lobed:
>> *P. edulis*

> iii Leaves 5–9 lobed:
>> *P. caerulea*

Passiflora × alato-caerulea Lindl. (*P. alata × P. caerulea*). Climbing shrub, shoots winged; leaved 3 lobed, entire; flowers 7–10 cm wide, sepals white, petals pink, filaments blue-violet, in 3 rows, the outermost white on the tips, does not flower as abundantly as *P. caerulea*, but with larger flowers. BR 848 (= *P. pfordtii* Hort.). In cultivation before 1824. z9 # ⊕

'Empress Eugenie' see: **'Impératrice Eugénie'**

'Impératrice Eugénie' is by today's classification a selection with more violet, large flowers (= 'Kaiserin Eugenie'; 'Empress Eugenie'). Often culitivated as a pot plant since it blooms as a very young plant. # ⊕

'Kaiserin Eugenie' see: **'Impératrice Eugénie'**

P. × allardii Alland (*P. caerulea* 'Constance Elliott' × *P. racemosa*). Strong grower; leaves 3 lobed; flowers 9–11.5 cm wide, white with pink, corona white and dark purple-blue, flowering continuously from summer to fall. Developed in 1907 by E. J. Allard in the University of Cambridge Botanic Garden. z9 # ⊕

P. antioquiensis Karst. Climbing shrub, shoots and leaf undersides brown pubescent, as are the petioles and flower stalks; leaves in 2 forms, either lanceolate and unlobed or deeply 3 lobed, the lobes lanceolate, middle lobes much longer, long acuminate, margins dentate, flowers pendulous, an intensive pink-red, 10–13 cm wide, calyx tube 2.5–4 cm long, cylindrical, corona small, violet, solitary on long stalks, late summer. BM 5571 (= *Tacsonia van-volxemii* Lem.). Columbia, in the mountains at 2000–3000 m. 1858. z9 Fig. 256. # ⊕

P. caerulea L. Blue Passion Flower. Strong growing climber, more or less evergreen in mild climates, totally glabrous; leaves broad cordate, 5–7 lobed, 10–15 cm

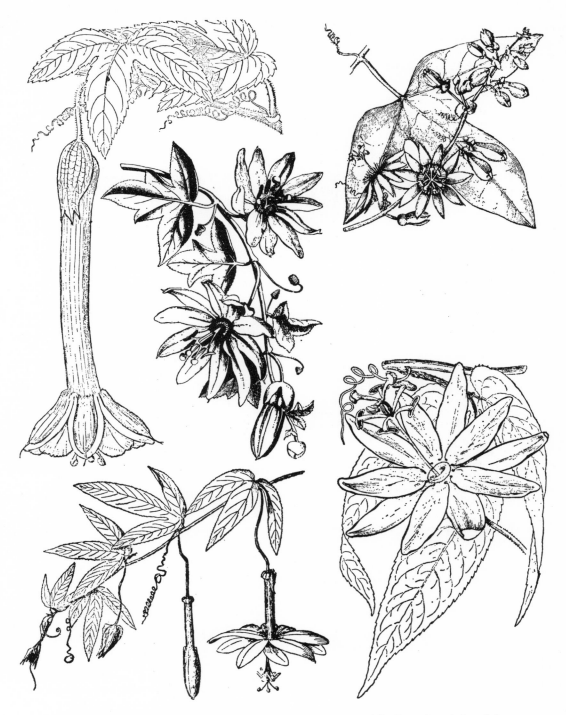

Fig. 256. **Passiflora.** Above, left *P. mollisima*, right *P. racemosa*; middle, *P. manicata*; below, left
P. exoniensis, right *P. antioquiensis* (from Marshall)

wide, lobes oblong-lanceolate; flowers solitary, on long, thin stalks in the leaf axils of the younger shoots, somewhat fragrant, 7–10 cm wide, petals white to slightly pink, the corona 5 cm wide, the filaments blue on the apical half, white in the middle, red on the base, styles purple, June–fall; fruits, egg-sized, orange. BM 28. Central America, western S. America. 1699. The hardiest species. z7 # ✧ ⚭

'Constance Elliott'. (Lucombe). Flowers ivory-white. Lucombe, Pince & Co., Exeter, England. 1884. Hardier than the species. Requires a moist, fertile soil or frequent watering. The fruits are as ornamental as the flowers. # ✧ ⚭

P. × caerulea-racemosa Sabine. Strong climber; leaves deeply 5 lobed, glabrous; flowers solitary, petals with a trace of dark violet, corona deep purple-violet. BC 573. Cultivated since 1821. z9 # ✧

P. edulis Sims. Purple Granadilla. Strong grower, shoots angular, somewhat pubescent or glabrous; leaves 10–20 cm wide, deeply 3 lobed, the lobes ovate, glandular dentate; flowers 6 cm wide, sepals oblong, inerior white, exterior green, petals equal in length, but narrower, corona with ruffled, white filaments, purple banded, flowering throughout the summer; fruits egg-sized, yellow to dark purple and speckled, edible. BM 1989. Tropical S. America. 1810. Cultivated throughout the tropics for its fruits. z9 # ✧ ✗

P. × exoniensis L. H. Bailey (*P. antioquiensis* × *P. mollissima*). Climbing shrub with soft haired shoots; leaves deeply 3 lobed, 10 cm long, 12 cm wide, base cordate, totally pubescent, petiole 2.5 cm long, with 2 glands; flowers pendulous, 10–13 cm wide, the calyx tube 6 cm long, glabrous, petals small, pink, corona small, whitish, sepals and petals oblong-lanceolate, throat violet (= *Tacsonia exoniensis*). Developed in 1870 by Veitch in Exeter. z10 Fig. 256. # ✧

P. ligularis Juss. Granadilla. Strong growing, glabrous climber; leaves oval-rounded with a deeply cordate base, 10 to nearly 20 cm long, short acuminate, blue-green beneath, petiole with unique, filamentous glands; flowers 7–10 cm wide, sepals greenish-white, 1 cm wide, petals smaller, filaments of the corona in 5–7 rings, of these the 2 outermost as long as the petals, white with purple, September–October; fruits ovate, 7 cm long, with a white, edible pith. BM 2967. Peru. 1819. z10 # ✧ ✗

P. manicata (Juss.) Pers. Strong grower, shoots angular and finely pubescent; leaves 3 lobed, 7–10 cm wide, leathery tough, the lobes ovate, 2.5–3 cm wide, finely serrate, lobed to the middle of the blade, lighter green underside, petiole with 3–4 glands; flowers on 10 cm long stalks, scarlet-red, 10 cm wide, calyx tube only 1.5–2 cm long, 10 ribbed at the base, sepals and petals oblong, obtuse, nearly equally formed, but the sepals are awl-shaped at the apex, outer corona composed of many short blue filaments, inner corona an inward curving membrane, long lasting flowers. HV 19; BM 6129 (= *Tacsonia manicata* Juss.). Columbia, Ecuador, Peru, Venezuela, in the mountains at 1500–2500 m. z10 Fig. 256. # ✧

P. mollissima (H. B. K.) L. H. Bailey. High climbing shrub, shoots cylindrical, soft pubescent; leaves cordate-ovate, 3 lobed nearly to the base, 7–12 cm long and about as wide, the lobes oval-lanceolate, serrate, densely white pubescent beneath, petiole 3 cm long, with 8–12 glands; flowers 7.5 cm wide, the calyx tube 7 cm long, sepals and petals oblong, pink, corona composed only of a nearly wart-like, raised ridge; fruit ovate, 7 cm long, yellowish, soft pubescent. FS 78; BM 4817; BR 32: 11 (= *Tacsonia mollissima* H. B. K.). Tropical S. America. 1843. Flowers abundantly when mature, all flowers pendulous. z10 Fig. 256. # ✧

P. princeps see: **P. racemosa**

P. pfordtii see: **P. × alato-caerulea**

P. quadrangularis L. Giant Granadilla. Shrub, very vigorous grower, shoots glabrous, thick, 4 sided and winged; leaves broadly ovate, with a cordate base, 10–20 cm long, simple, entire, acuminate above, petiole with 6 glands, 2.5–5 cm long; flowers about 11 cm wide, sepals ovate, 4 cm long, 2.5 cm wide, exterior greenish, interior white or soft pink, petals pink-white, corona 10 cm wide, composed of 5 rings, filaments white and purple banded; fruits ovate, 15–20 cm long, bittersweet, edible and often cultivated for its fruit in the tropics. BR 14; JRHS 85: 47. Tropical America. 1763. z10 # ✧ ✗

P. racemosa Brot. Red Passion Flower. Shrub, high climbing, glabrous; leaves normally 3 lobed or with only one lobe on either side or unlobed, 7.5–11 cm wide and equally long, leathery; flowers in 30 cm long or longer, pendulous racemes with 8–13 flowers, each 10–12 cm wide, scarlet-red, sepals with keel-like longitudinal wings, corona dark purple with a white apex, the inner filaments red and very short. BM 2001; BR 285 (= *P. princeps* Lodd.). Brazil. 1815. One of the most attractive species and cultivated in many botanic gardens. z10 Fig. 256. # ✧

Generally cultivated only in frost free regions in the open landscape.

Lit. Knock, F.: Passifloras for your Garden; Kansas City, Missouri 1965 ● Marshall, E. D.: The Passifloras; in Calif. Hort. Journ. **34**, 146–153, 1973 ● Lawrence, G. H. M.: Names of *Passiflora* hybrids. Baileya **8**, 118–120, 1960 ● Lawrence, G. H. M.: Identification of cultivated Passion Flowers; Baileya **8**, 121–132, 1960 (with ills.).

PAULOWNIA S. & Z.— SCROPHULARIACEAE*)

Tall deciduous trees without a taproot and with very thick, hollow shoots when young; twigs without terminal buds; leaves opposite, large to very large, cordate-ovate, entire or 3–5 lobed, also serrate on young seedlings; flowers at the ends of the previous year's shoots, panicles inconspicuous, but actually composed of numerous small corymbs, appearing before or with bud break of the leaves; calyx deeply 5 parted, tips thick and obtuse; corolla tube elongated, widened at the apex, violet to white, limb with 5 obliquely spreading lobes; stamens 4, styles perforated at the apex; fruit a leathery, ovate capsule with numerous, very small, winged seeds.— 17 species in E. Asia.

Outline of the genus (from Hu)

● Inflorescences pyramidal or cylindrical with simple or forked cymes; cymes looser, with 3–5 flowers, petioles nearly as long as the flower stalks; fruit capsules variable in size and form;

 * Capsule globose-ovate; pericarp tough or crusty; corolla *Digitalis*-like (except *P. elongata*):

Section 1. Paulownia Hu

1. Leaf undersides with long stalked dendroid hairs with very slender branches; crown with straight, segmented hairs; leaves of mature branches oval to cordate, angular or lobed on long shoots (Fig. 258):

P. tomentosa

2. Dendroid hairs sessile on the leaf undersides or nearly sessile; hair branches short and thick; crown with glandular hairs or with dendroid or glandular hairs;

  + Mature leaves cordate to ovate, underside nearly glabrous or with some scattered, rather short stalked dendroid hairs; calyx deeply lobed, tube as long or shorter than the lobes; corolla tube *Digitalis*-like, glandular pubescent:

P. glabrata

  ++ Mature leaves elongated to cordate, long acuminate; calyx obconical, shallowly lobed, tube twice as long as the lobes; corolla tube nearly funnelform, with dendroid and glandular hairs:

P. elongata

 ** Capsules oblong-ellipsoid, occasionally nearly ovate, constricted at the base; corolla nearly funnelform, 8–10 cm long, gradually narrowing to the base; calyx becoming partly glabrous at flowering time, tube obconical, distinctly narrowed at the base (Fig. 258):

*) The classical botanical literature (Endlicher, Engler-Prantl, Rehder, Bailey, Fernald, etc.) place *Paulownia* in the Scrophulariaceae, while Hallier, Campbell, Li and Lawrence consider the Bignoniaceae to be more correct. Nakai, 1949, disagreed with everyone and gave this genus its own family name, Paulowniaceae. In this book, because of an excellent monograph by Sh.-Y. Hu on the subject, the classical interpretation will be maintained.

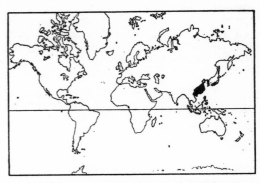

Fig. 257. Range map of the genus *Paulownia*

Section 2. Fortuneana Dode

(*P. fortunei*)

●● Inflorescences with thick lateral branches, similar to the rachis; cymes nearly sessile or umbellate; stalk of the cymes very short or absent; capsules rounded-ovate (Fig. 258):

Section 3. Kawakamia Hu

  + Corolla nearly campanulate, 3–4 cm long, tube shaggy pubescent; leaves pubescent beneath at flowering time:

P. kawakamii

  ++ Corolla *Digitalis*-like, 5.5 cm long; leaves woolly pubescent at flowering time, the hairs dendroid and long stalked:

P. fargesii

Paulownia coreana see: **P. tomentosa 'Coreana'**

P. duclouxii see: **P. fortunei**

P. elongata Sh.-Y. Hu. Large tree, annual shoots thick, glabrous, red-brown; leaves ovate and acuminate, 18–38 cm long, 8–22 cm wide, base cordate, nearly glabrous above, stellate tomentose beneath; flower rachis to 35 cm long, brownish woolly haired, glabrous at flowering time, calyx obconical, 16–19 mm long, corolla 6.5 to 7.5 cm long, exterior stellate pubescent and glandular, violet; capsule ovoid, 3.3 cm long, 2 cm thick, seeds 4 mm long. Hu, *Paulownia*: Pl. 3. China; Hupeh, Honan, Shantung, Hopei Provinces. Described by J. Hers as a very valuable lumber tree, widely cultivated in Honan Province. z6 Ø ✢

P. fargesii Franch. Tree, 9–12 m high, young shoots stellate pubescent, soon becoming glabrous; mature leaves elongated-ovate, 15–21 cm long, 12–14 cm wide, short acuminate, base cordate, entire and somewhat undulate, tomentose to densely woolly beneath; flowering twigs loosely branched, the main shoot 15–35 cm long, the lateral branches (1–2 pairs) 8–16 cm long, stalk yellow woolly tomentose, corolla *Digitalis*-like, 5–7 cm long, violet or lilac to white; capsule ovate, to 3.5 cm long, seeds 5–6 mm long. Hu, *Paulownia* Pl. 4. China; Hupeh, Yunnan, Szechwan Provinces. 1900. z7 Ø ✢

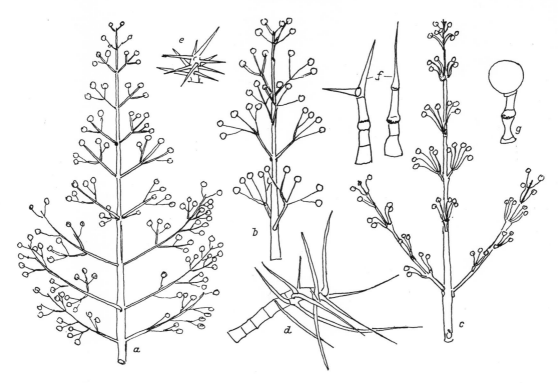

Fig. 258. **Paulownia.** Schematic of the inflorescences, d–g. hair forms. a. *tomentosa* type; b. *fortuneana* type; c. *kawakamia* type; d. dendroid hair of *P. tomentosa* (leaf undersides); e. nearly sessile, dendroid hair of *P. elongata*, stellate form as seen from above; f. straight hairs from the same twig as in d.; g. glandular hair from the same twig as in d. (from Sh.-Y. Hu)

P. fortunei (Seem.) Hemsl. Tree, to 27 m high, bark gray-brown; mature leaves elongated-ovate, 14–21 cm long, 7–12 cm wide, long acuminate, base cordate, paper thin, usually glabrous and glossy above, underside densely brownish-green woolly; flower shoots nearly cylindrical, the individual cymes usually 3 flowered, corolla tube nearly funnelform, gradually tapering to the base, 9–10 cm long, 4–6 cm wide at the base, 3 cm wide at the mouth, white to light purple; capsules woody, oblong-ellipsoid, 5–8 cm long, seeds 6–10 mm long. Hu, *Paulownia:* Pl. 5; JRHS 89: 301 (= *P. duclouxii* DIde; *P. mikado* Ito; *P. meridionalis* Dode; *P. longifolia* Hand.-Mazz.). China; Anhwei, Taiwan, Kwangtung, Yunnan. Ⓕ E. and Central China; Japan. z6 ⌀ ⌾

var. *tsinlingensis* see: **P. glabrata**

P. glabrata Rehd. Tree, to 15 m high, annual shoots glabrous, new growth glandular papillate, glabrous; mature leaves ovate to long ovate, 16 to 24 cm long, 10–17 cm wide, acute, entire, base round, to slightly cordate, with some dendroid hairs above; flowering shoots pubescent and somewhat glandular, base later glabrous and nearly cylindrical, all branches usually of equal size, cymes usually with 3–5 flowers, flower stalks brown woolly stellate pubescent, corolla tube funnelform-campanulate, lilac, 5.5 cm long, 1.2 cm wide, 5.5 cm wide at the limb; capsules globose-ovate, 2.5–3 cm long, 2–2.5

cm wide, glandular, seeds 5 mm long (= *P. fortunei* var. *tsinlingensis* Pai; *P. shensiensis* Pai). China; Szechwan, Honan Provinces in the Tsingling Mts. z6 ⌀ ⌾

P. imperialis see: **P. tomentosa**

P. japonica see: **P. tomentosa**

P. kawakamii Ito. Tree, 8–12 m high, bark gray, annual shoots eventually glabrous; leaves broadly cordate-ovate, 3–5 lobed, 11–30 cm long, 8 to 27 cm wide, abruptly short acuminate, base cordate, glandular pubescent above, glandular woolly beneath; flower shoots loosely branched, main shoot 17–30 cm long, lateral shoots 6–30 cm long. the individual cymes 3 flowered, sessile, calyx woolly, corolla nearly campanulate, soft lilac, 3–4.5 cm long, base 5 mm wide, throat with limb 3 cm wide; capsule ovate, 2–2.5 cm long. Kanehira, Formos. Trees: 612; LWT 337 (= *P. thyrsoides* Rehd.; *P. viscosa* Hand.-Mazz.; *P. rehderiana* Hand.-Mazz.). Taiwan; China; Fukien, Kwangsi, Kiangsi, Hupeh Provinces. z6 ⌀ ⌾

P. lilacina see: **P. tomentosa 'Lilacina'**

P. longifolia see: **P. fortunei**

P. meridionalis see: **P. fortunei**

P. mikado see: **P. fortunei**

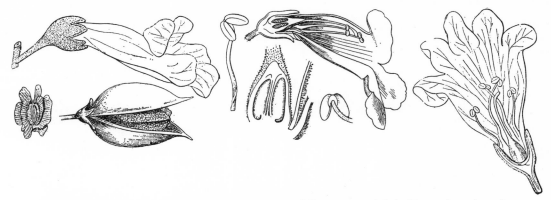

Fig. 259. *Paulownia tomentosa*, individual flowers and flower parts; left fruit capsule and seeds
(from Wettstein, Lauche)

P. recurva see: **P. tomentosa**

P. rehderiana see: **P. kawakamii**

P. thyrsoides see: **P. kawakamii**

P. tomentosa (Thunb.) Steud, Tree, 15–20 m high, broad crowned, shoots very thick when young, later much thinner, densely and softly pubescent at first; mature leaves broadly ovate, 17–30 cm long, 12–27 cm wide, short acuminate, base cordate, entire, bright green and pubescent above, densely gray woolly beneath, leaves often 3–5 lobed on the strong shoots and to 50 cm long and wide, irregularly serrate, both sides evenly pubescent, petiole to 40 cm long; flower shoots 25–40 cm long, pyramidal, main branch 20–38 cm long, lateral branches 3–15 cm long, stalks densely woolly, corolla *Digitalis*-like, 5–6 cm long, base 10 mm wide, 4.5 cm wide at the mouth of the limb, violet, yellow striped inside, fragrant, appearing before the leaves, May; capsule ovate, 3 cm long. BM 4666; DRHS 1495; JRHS 68: 78 (= *P. imperialis* S. & Z.; *P. tomentosa* var. *lanata* [Dode] Schneid.; *P. japonica* Réveil; *P. recurva* Rehd.). China; Hupeh, Kiangsi, Honan Provinces. Ⓕ Japan. z6 Plate 152; Fig. 259. ⌀ ✧

'Coreana'. Leaves cordate to ovate, yellowish woolly beneath; flowers violet, throat yellow speckled inside; seeds small, 3 mm long. MD 1932: 35 (= *P. coreana* Uyeki). 1925. A clone from Korean cultivation, not an independent species (from Hu). ⌀ ✧

var. *lanata* see: **P. tomentosa**

'Lilacina'. Young shoots green, glandular pubescent, brown in the second year; leaves broadly ovate, not lobed, long acuminate, 12–30 cm long, dull green above, tomentose beneath; flower clusters pyramidal, 25–30 cm long, corolla nodding, 7 cm long, pale lilac, limb outspread, pubescent outside, interior yellow; fruits ovate, 3–4 cm long. BM 8926 (= *P. lilacina* Sprague; *P. fargesii* Osborn non Franch.). Obtained as a cultivar and introduced in 1908 from W. China by Farges. ⌀ ✧

"var. *pallida* (Dode) Schneid." This plant's existence is only hearsay, never validated!

P. viscosa see: **P. kawakamii**

Attractive flowering trees for warmer climates. The flower buds develop in late fall and therefore freeze back in cooler climates. The tree can be made hardier by covering young trees in winter until the tree develops a 1 cm thick, solid stem instead of the 3–5 cm hollow stems which do not survive cold winters above ground. The tree is very fast growing in a moist fertile soil. Paulownias were once cultivated for the foliage which becomes nearly 50 cm wide on 3–4 m high plants which are cut back to the ground each year.

Lit. Hu, Shiu-Ying; A monograph of the genus *Paulownia*; in Quart. Jour. Taiwan. Mus. **12**, 1–54, 1959 (Pl. 1–7) ● Dode, L. A.: Notes Dendrologiques. VII. Sur les Paulownias; in Bull. Soc. Dendr. France, 1908, 159–163 ● Dening, K.: Zur Kenntnis von *Paulownia tomentosa* (Thunb.) Steud.; in Mitt. DDG 1937, 127–129.

PAVIA

Pavia californica see: **Aesculus californica**

PAXISTIMA see: **PACHISTIMA**

PENSTEMON Schmidel — Beardtongue — SCROPHULARIACEAE

Herbs or subshrubs; leaves opposite, occasionally whorled, the basal ones petiolate, the apical usually sessile, entire or serrate; flowers attractive, in panicles or racemes; calyx 5 parted; corolla more or less bilabiate, the tube elongated, more or less widened at the throat, upper lip 2 lobed, lower lip 3 lobed; stamens 5, of which one is sterile and often with a tufted pubescence; styles filamentous, with a capitate stigma; fruit a sectioned capsule — About 250 species in N. America and the cooler regions of Mexico, one species in Central America and one in E. Asia (*P. frutescens*).

Penstemon barrettiae Gray. Dense, erect, multistemmed shrub, 15–30 cm high, the inflorescences 10 cm higher,

older shoots densely foliate; leaves thick and fleshy, elliptic to oval-lanceolate, 3–5 cm long, 1.2–2 cm wide, the sessile leaves only half as large, obtuse, margins shallowly dentate, pale green and somewhat reddish, with a silvery shimmer, totally glabrous, flowers lilac-purple, in short panicles (! not racemes), sepals only 4–7 mm long, acute, anthers densely white woolly, May. GC 1920: 63. Western N. America, on basalt outcroppings in the Cascade Mts. z6 Fig. 260. # ✧

P. berryi see: **P. newberryi** var. **berryi**

P. cardwellii Howell. Open habit, flat, irregular, multistemmed, shoots erect to ascending, never mat-

Fig. 260. **Penstemon.** From left to right: upper row *P. fruiticosus, P. cardwellii, P. newberryi;* lower row *P. menziesii, P. barrettiae, P. rupicola* (from Abrams)

form, usually 7–15 cm high, the flowering shoots 5–10 cm higher, glabrous to the inflorescence, densely foliate; leaves dark green (either bluish or yellowish), oval-elliptic, 12–18 mm long, rounded, margins sparsely dentate, teeth round, margins often involuted, much smaller on flowering shoots; corolla 2.5–4 cm long, purple, often more bluish or reddish, sepals 7(12) mm long, margins distinctly glandular pubescent, May–June (often flowers for 6 weeks!!). USA; SW. Washington to SW. Oregon, on dry slopes in the higher mountains. z6 Fig. 260. # ⊕

P. davidsonii Greene. Mat-form habit, densely covering the ground, stoloniferous, flowering shoots hardly 5 cm higher (!); nearly all leaves entire, tiny, seldom longer than 5 mm; sepals elliptic, 6–8 mm long, acute, corolla 1.8–3.5 cm long, purple-lilac, anthers woolly. USA; Washington to California; in the higher mountains. z6 Plate 127. # ⊕

P. × edithae English (= *P. barrettae* × *P. rupicola*). Very much resembling *P. rupicola* in habit, height, form of the inflorescence, but totally glabrous; leaves elliptic, about 1 cm long, 6 mm wide, gray-green, glabrous. Hybridized by Carl English of Seattle, Washington, USA. # ⊕

P. fruticosus (Pursh) Greene. Flat, uneven habit, 10–40 cm high, with many erect or ascending shoots, never mat-like or hemispherical, shoots glabrous to the base, flower rachis somewhat glandular pubescent; leaves spathulate, 2–3(5) cm long, 5 to 15 mm wide, apical sessile leaves smaller, usually bluish-green, finely serrate, base tapered; calyx 1 cm long, sepals lanceolate, densely glandular pubescent, corolla 3–5 cm long, purple-blue (more blue), base often pink, anthers densely white woolly, May–June. Northwestern N. America. 1828. z6 Plate 151; Fig. 260. # ⊕

P. laetus Gray. Dwarf shrub, woody at the base, to 25 cm high, fine gray or yellow-green short haired (not blue-green!); leaves linear or oblong to oblanceolate, 1.5–7.5 cm long, sessile, entire; stalk with 1–4 flowers, these blue or bluish-purple, 2 to 3 cm long, in loose, glandular pubescent, racemose panicles, June–July. Calfornia, USA. z7

This species includes a few selected varieties of which only the following is found in cultivation:

var. **roezlii** (Regel) Jepson. Shrub, 5–20 cm high, differing from the species in the more linear, non-involuted leaves; flowers smaller, only 1.5–2 cm long, light to dark blue or violet, but not ruby-red (as occasionally described in the gardening literature!), July. GF 1872: 239 (= *P. roezlii* Regel). Western USA. 1872. Plate 127. # ⊕

P. menziesii Hook. Dense mat-form habit, twigs ascending, 10–15 cm high, finely pubescent; leaves obovate to broadly spathulate, thick and leathery, about 15 mm long, more or less finely dentate, never entire, short petioled; sepals 7–11 mm long, oval-elliptic, acute to acuminate, slightly glutinose, corolla purple, 2.5 cm long, rarely to 3 cm long, anthers densely white woolly, June–August. In the higher mountains of British Columbia to Alberta, Canada. 1902. z4 Fig. 260. # ⊕

P. newberryi Gray. Open habit, flat, uneven, about 15–25 cm high, the inflorescences 5–10 cm higher, shoots distinctly reddish (!) and finely pubescent to the inflorescence; leaves elliptic to ovate, 12–20 mm long, occasionally to 25 mm, yellow-green, glossy, usually with a fine red border, nearly totally finely crenate, glabrous; sepals long acuminate, always glandular pubescent, inflorescences quite variable, flowers either in terminal clusters or in long, narrow racemes, corolla tubular (!), 3–3.5 cm long, only 5–8 mm wide, pink to ruby-red, June–July. BC 2859; JRHS 89: 64. USA; Washington to California, in the higher mountains. 1872. z6 Fig. 260. # ⊕

var. **berryi** (Eastw.) Keck. Shoots glabrous; leaves very small, glossy green, rounded at both ends, those at the base of the inflorescence much smaller than the others, margins crenate, glabrous; flowers in small, dense racemes, appearing nearly like terminal clusters, corolla distinctly campanulate (!), throat 8–12 mm wide, pink to red (= *P. berryi* Eastw.). USA; SW Oregon to NW California, in the coastal mountains. # ⊕

var. *rupicola* see: **P. rupicola**

P. roezlii see: **P. laetus** var. *roezlii*

P. rupicola (Piper) Howell. Dense mat-like habit, to 50 cm wide, only 3 cm high, to 5 cm high when in flower; leaves normally conspicuously blue-gray, thick and stiff, rough to the touch, limb pubescent, elliptic, about 1 cm long, apical half sparsely dentate; flowers 2–4 in short racemes, corolla carmine-pink to coral-red, exterior glabrous, lower lip with a somewhat pubescent interior, anthers densely white woolly, June–July. BM 8660; RFW 335 (= *P. newberryi* var. *rupicola* Piper). Northwestern N. America, in the Cascade Mts. Quite variable. z6 Fig. 260. # ⊕

'Alba'. Like the species, but with apple-green leaves and a silvery shimmer; flowers pure white, flowers very abundantly. Known in various forms. # ⊕

'Six Hills Giant' (Elliott). To 15 cm high; leaves nearly entire, dark green, bronze-red beneath; flowers lilac-pink, flowers very abundantly. Six Hills Nursery, Cl. Elliott, Stevenage, England. (Probably *P. newberryi* × *P. rupicola*). # ⊕

P. scouleri Lindl. Bushy shrub, 10–40 cm high; leaves linear-lanceolate, usually about 2.5 cm long, only 2–5 (!) mm wide, entire or distinctly scabrous dentate; sepals 10–15 mm long, corolla violet-purple. BR 1277; NF 2: 110; BM 6834. Western N. America, in the mountains. 1828. z6 # ⊕

All species are well suited to full sun in the rock garden; thrive with little care (top dressing with gravel, hard pruning after flowering and covering in winter).

Lit. Bennett, R. W.: Studies in Penstemon; No. 2 — Subgenus *Dasanthera*, the "Shrubby Penstemons"; Arlington 1954 (86 pp.); Private publication of the American Penstemon Society.

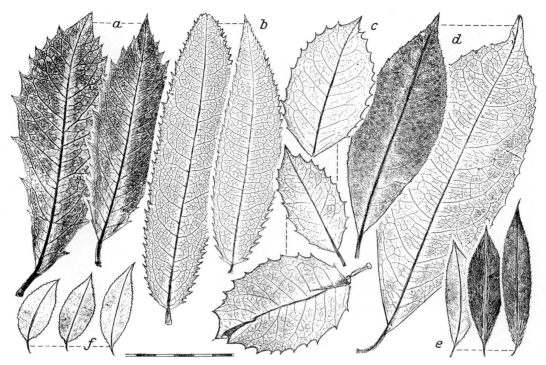

Osmanthus. a. *O. armatus*; b. *O. yunnanensis*; c. *O. fortunei*; d. *O. fragrans*; e. *O. serrulatus*; f. *O. suavis*
(mostly from material collected in the wild)

Osteomeles anthyllidifolia
in the Bonn Botanic Garden, W. Germany

Osmarea burkwoodii
in the Dortmund Bot. Garden, W. Germany

Plate 145

Plate 146

Osmaronia cerasiformis
in the Göteborg Botanic Garden, Sweden

Oxydendrum arboreum
in De Belder Arboretum, Kalmthout, Belgium

Ozothamnus ledifolius
in the Hillier Arboretum, England

Ozothamnus purpurascens
in Malahide, Ireland

Plate 147

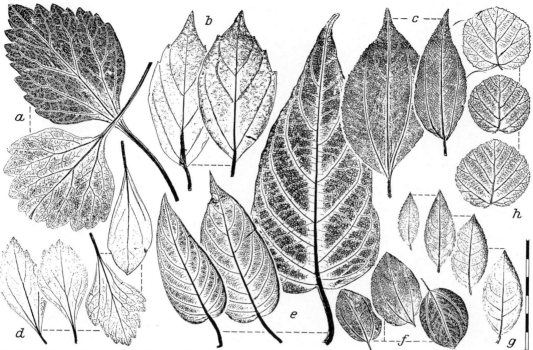

Pachysandra. a. *P. procumbens;* b. *P. axillaris;* c. *P. terminalis.* — d. *Paederia scandens.* — **Paliurus.** e. *P. hemsleyanus;* f. *P. spina-christi.* — g. *Osmarea burkwoodii.* — h. *Parrotiopsis jacquemonitiana* (most material collected from plants in the wild)

Ostrya japonica in its habitat
Photo: Dr. Watari, Tokyo

Ostrya virginiana
in the Morton Arboretum, USA

Plate 148

Paeonia suffruticosa
in the Botanic Garden of Dublin, Ireland

Paeonia delavayi
in the Royal Botanic Garden, Edinburgh, Scotland

Paeonia 'New York'
Photo: Dr. H. Goos

Plate 149

Parkinsonia aculeata in the Coimbra Botanic Garden, Portugal

Parrotia persica in the Pisa Botanic Garden, Italy (the oldest plant in Europe)

Plate 150

Parthenocissus. a. *P. heptaphylla*; b. *P. tricuspidata*; c. *P. vitacea*; d. *P. quinquefolia* var. *murorum*; e. *P. quinquefolia*; f. *P. tricuspidata* 'Veitchii'. — g. *Petteria ramentacea* (from herbarium material, but most collected from plants in the wild)

Pandanus utilis with fruit
in a park in Malaga, Spain

Parthenocissus henryana
in Malahide, Ireland

Plate 151

Parthenocissus tricuspidata 'Atropurpurea'
in the Minier Nursery, Angers, France

Parthenocissus tricuspidata 'Minutifolia'
in the Minier Nursery, Angers, France

Penstemon fruticosus
Photo: Dr. H. Goos, Niederwalluf

Plate 152

Parthenocissus semicordata

Pachystegia insignis
in the Bonn Botanic Garden, W. Germany

Paulownia tomentosa in the Geisenheim Arboretum, W. Germany
Photo Dr. P. Kiermeier

Plate 153

Perovskia atriplicifolia
Photo: Dr. H. Goos, Niederwalluf

Gaultheria thymifolia in the Dublin Botanic Garden, Ireland

Plate 154

Philadelphus inodorus var. *carolinus*
(Botanical Magazine, Plate 147)

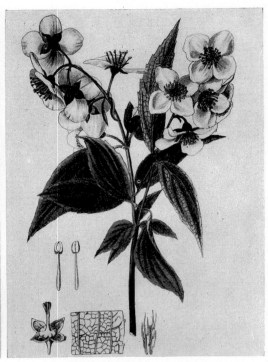

Philadelphus delavayi var. *melanocalyx*
(Botanical Magazine, Plate 9022)

Philadelphus sericanthus
(Botanical Magazine, Plate 8941)

Philadelphus purpurascens
(Botanical Magazine, Plate 8324)

Plate 155

Periploca sepium
in the Göteborg Botanic Garden, Sweden

Petteria ramentacea
in its native habitat in Yugoslavia

Philadelphus × lemoinei 'Fimbriatus'
in the Dortmund Botanic Garden, W. Germany

Philadelphus × falconeri
in a park in Prague, Czechoslovakia
Photo: Dr. Pilat, Prague

Plate 156

Philadelphus 'Virginal'
in the Dortmund Botanic Garden, W. Germany

Philadelphus 'Girandole'
in the Dortmund Botanic Garden, W. Germany

Philadelphus coronarius 'Duples'
in the Aarhus Botanic Garden, Denmark

Philadelphus pubescens
in the L. Späth Arboretum, Berlin, W. Germany

Plate 157

Philadelphus × *pendulifolius*
in the Dortmund Botanic Garden, W. Germany

Philadelphus carcasicus 'Aureus'
in a park in The Hague, Holland

Photinia glomerulata
in Malahide, Ireland

Phoenix roebelenii
in the Cape Town Botanic Garden, S. Africa

Plate 158

Phoenix canariensis
in Aveiro near Coimbra, Portugal

Phoenix loureirii
in the Kirstenbosch Garden, S. Africa

Phoenix dactylifera in Europe's only palm forest at Elche, Spain
Photo: Hinrich Kordes

Plate 159

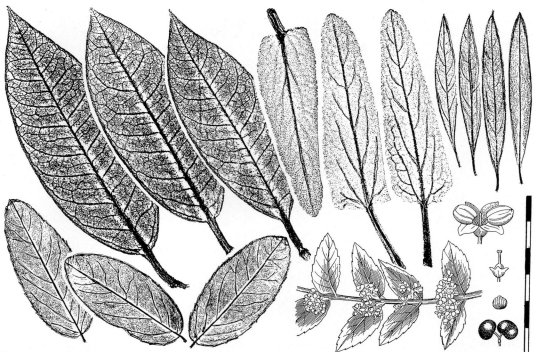

Upper row: *Phillyrea vilmoriniana* (3); *Phlomis fruitcosa* (3); *Phillyrea angustifolia* (4). Lower row: *Phillyrea latifolia* (3); *Phillyrea latifolia* var. *media* (from Sibth. & Sm.) (most material from wild plants)

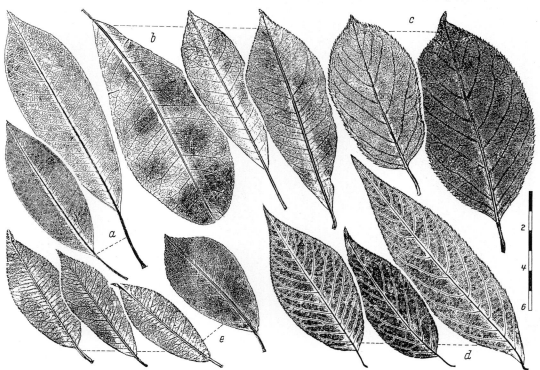

Photinia. a. *P. serrulata*; b. *P. davidsoniae*; c. *P. beauverdiana* var. *notabilis*; d. *P. beauverdiana*; e. *P. glabra* (material from wild plants)

Plate 160

Phyllodoce coerulea
Photo: C. R. Jelitto

Phyllodoce aleutica
in the Royal Botanic Garden, Edinburgh, Scotland

Phylica plumosa in its native habitat
in the Kirstenbosch Botanic Garden, S. Africa

Phormium tenax 'Veitchii'
on Ibiza, San Antonio, Balaeric Islands

PENTACTINA Nakai — ROSACEAE

Monotypic genus, closely related to *Spirea*. Deciduous shrub; leaves alternate, nearly sessile, dentate, without stipules; flowers hermaphroditic, small, in small, pendulous panicles; sepals 5; petals 5, linear, stamens 20; pistils 5, all distinct, then with 2 ovules; follicle glabrous, dehiscing on both seams; seeds spindle-shaped. — Korea.

Pentactina rupicola Nakai. Procumbent shrub, graceful habit, to 70 cm high, young shoots glabrous, angular, reddish; leaves elliptic to oblanceolate, 2–3 cm long, acute at both ends, with a few, large teeth, otherwise entire, underside bluish and somewhat pubescent; flowers in 6–8 cm long, terminal, pendulous panicles, white, June. GC 73: 158. Korea, Diamond Mts. 1918.

Very hardy but difficult to maintain in cultivation; attractive in the rock garden. Not particular as to soil type. z5

PENTAPERA Klotzsch — ERICACEAE

Monotypic genus very closely related to *Erica*, differing primarily in the 5 parted flowers. An evergreen shrublet, leaves needle-like; flowers in terminal, umbellate clusters; corolla urceolate, calyx 5 parted, anthers 2 parted, stamens 10, fruit a dry, loculicidal capsule. —One species in S. Europe.

Pentapera sicula Klotzsch. Evergreen shrub, 25–50 cm high, shoots thin, more or less pubescent; leaves in whorls of 4, erect to outspread, linear-filamentous, 12 mm long, obtuse, entire; flowers 3–4(7) in terminal clusters, corolla white to soft pink, urceolate, 8 mm long, with 5 small, reflexed lobes, calyx 5 parted, sepals lanceolate, pink-white, May–June. BM 7030; GC 131: 15; 134: 35 (= *Erica sicula* Guss.). Sicily, on limestone slopes along the West Coast, also in Malta and Syria. 1849. z7 # A ✛ ○

Differing from most of the *Ericaceae* in its excellent tolerance of soil alkalinity! Best cultivated as a pot plant in clay soil with many large limestone pieces, covered at the top with limestone gravel, and good drainage. Requires similar treatment if planted in the rock garden. Very long lived when left untouched.

PENTAPTERYGIUM

Pentapterygium rugosum see: **Agapetes rugosum**

P. serpens see: **Agapetes serpens**

PERAPHYLLUM Nutt. — ROSACEAE

Monotypic genus. Deciduous shrub, very closely related to *Amelanchier*, but with entire leaves, ovaries totally inferior, the 2 or 3 styles distinct; leaves partly clustered, nearly sessile to short petioled, entire or sparsely serrate; flowers 2–5 in terminal, clustered racemes; petals 5, outspread; stamens 20; styles 2–3, distinct; fruit a small, 4–6 locular drupe with more or less false partitions; calyx lobes persisting on the fruit.

Peraphyllum ramosissimum Nutt. Erect, about 1.5–2 m high, divaricate shrub, twigs gray tomentose at first; leaves spirally arranged, clustered on short shoots, lanceolate, light green, 2–5 cm long, tapered to both ends, silky pubescent at first, eventually bright glossy green; flowers 2–5 on short lateral shoots, white to pink-white, about 2 cm wide, stamens in 2 rings, May; fruits globose, about 1 cm thick, pendulous, yellow with brown-red cheeks, not edible. BS 2: 401; BM 7420; MS 237. Northwest N. America, Oregon to California and Colorado. 1870. z5 d ○

A collector's shrub. Totally hardy; for warm sites in sandy soil; only slight ornamental merit.

PERIPLOCA L. — Silk Vine — PERIPLOCACEAE

Deciduous or evergreen, twining shrubs; leaves opposite, simple, entire, secreting a milky sap (poisonous!); flowers few in cymes, corolla rotate, with 5 obtuse corolla sections; flowers dirty yellow or brown inside, greenish outside, terminal or axillary; fruit composed of 2 fused, cylindrical follicles. — 10 species in S. Europe, Asia and the African tropics and subtropics.

Periploca graeca L. Deciduous shrub, twining to 15 m high, glabrous; leaves normally elliptic to oblong-lanceolate, to 5 cm wide, dark green and glossy above, remaining green until leaf drop in late fall; flowers 8–12 in long stalked, open cymes, interior dull violet-brown, exterior yellowish-green, 2.5 cm wide, July–August; fruits 10–12 cm long. HM 3041; BM 2289; BR 803. S. Europe, Asia Minor. 1597. z7 Fig. 262.

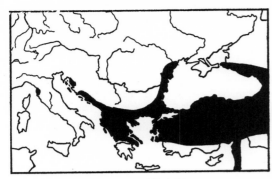

Fig. 261. Range of *Periploca graeca*
(from Browicz)

P. sepium Bge. Deciduous shrub, twining to 10 m high; leaves normally oblong-lanceolate, 4–10 cm long, but only 2.5 cm wide at the most, coloring yellow in fall; flowers like the previous species, but in only few flowered cymes, only 2 cm wide, corolla lobes reflexed (spreading outward on *P. graeca*), June–July; fruits 10–15 cm long, narrowed at both ends, in pairs. BC 2870. N. China. z7 1905. Plate 155.

Both species very strong growing and quite hardy; for sunny areas and a fertile soil.

Fig. 262. *Periploca graeca* (from Baillon and B. M.)

Lit. Browicz, K.: The Periplocaceae in Turkey and Cyprus; Feddes Rep. **72**, 124–131, 1966 ● Browicz, K.: History of the genera *Periploca* L. and *Cionura* Griseb.; Rocznik Dendrol. **20**, 53–73, 1966 (in Polish).

PERNETTYA Gaud. — ERICACEAE

Evergreen shrubs, dwarf or to 2 m high, twigs usually glabrous, occasionally short and finely pubescent and bristly; leaves alternate, entire or crenate, usually elliptic; flowers normally solitary and axillary, seldom racemose (*P. furens*), calyx 5 parted, sepals oblong-triangular, corolla usually campanulate, seldom urceolate, not narrowed at the apex, with 5 short limb lobes; stamens 10, anthers with 2 apicals horns each; fruit a globose berry with normally small, and dry, membranous sepals (but the New Zealand and Tasmanian species have thicker, fleshy sepals more or less [as with *Gaultheria*] enveloping the basal half of the fruit). — About 20 species from Mexico to Antarctic S. America, in New Zealand and Tasmania.

Pernettya andina see: **P. leucocarpa**

P. buxifolia Mart. & Gal. Very closely related to *P. ciliata*, but the young shoots with densely erect pubescence, later glabrous, to 30 cm high; leaves linear, about 15 mm long and 6 mm wide on wild plants, cultivated plants to 25 mm long; flowers solitary, pendulous, stalks 6 mm long, corolla urceolate, 7 mm long, white, often with pink limb lobes, exterior glabrous, interior partly pubescent, flowers abundantly, May; fruits to 1 cm thick, globose, red at first, then purple, eventually black. BMns 66. Mexico. z9 # ⚘

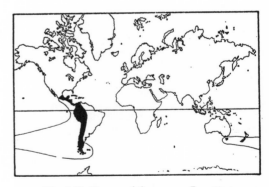

Fig. 263. Range of the genus *Pernettya*

P. ciliata (Cham. & Schlechtd.) Small. Divaricate shrub, 30–60 cm high, young shoots coarsely pubescent; leaves narrowly oval-oblong to lanceolate, 10–25 mm long, 3–9 mm wide, somewhat curved, glabrous above, somewhat bristly, leathery tough, acute, finely serrate; flowers solitary or in few flowered, axillary racemes, corolla oval to urceolate, white to somewhat pink, 6 mm long, calyx white, membranous, persistent, May; fruit deep red-brown, flat globose, about 6 mm thick. BM 3177 (as *Arbutus pilosa*) (= *P. pilosa* [Graham] G. Don). Mexico,

Guatemala; in the higher mountains. 1938. z9(?) # ⚭

P. empetrifolia see: **P. pumila**

P. furens (Hook. & Arn.) Klotzsch. Broader, about 0.6–1.2 m tall shrub, twigs outspread, young shoots pubescent; leaves oval-elliptic, 2.5–5 cm long, tapered to both ends, somewhat curved, leathery, finely and flatly serrate, both sides sparsely bristly, bluish beneath; flowers in axillary, 3–4 cm long, conspicuous racemes at the shoot tips, corolla campanulate-urceolate, white, 9 mm long, calyx somewhat enlarged and slightly fleshy, May; fruits globose, obtuse red-brown, 4–5 mm thick. BM 4920; DRHS 1530; PFC 86 (= *P. melanocarpa* Phil.). Central Chile. 1856. The fruits can cause intoxication, delirium or even death. The only species with attractive flowers. z9 # ⚭ ☉

P. gayana see: **P. leucocarpa**

P. lanceolata (Hook. f.) Burtt & Hill. A small, erect shrub; leaves oblong-elliptic, about 17 mm long, 4–5 mm wide; flowers solitary, axillary, corolla short campanulate; sepals becoming fleshy on the ripe berries, more or less enveloping the berries. Tasmania. z7 # ⚭

P. leucocarpa DC. Compact, rapidly spreading shrub, about 20–30 cm high, young twigs with short fine pubescence; leaves loosely arranged, usually elliptic, occasionally lanceolate-obovate, acute, 9–15 mm long, indistinctly serrate, leathery tough; flowers solitary in the apical leaf axils, corolla ovate-urceolate, white, stalks never longer than the petals, calyx always dry, membranous, sepals outspread, May; fruits flat globose, white to pink, about 6 mm thick (= *P. gayana* Decne.; *P. andina* Meigen). The cordillera of south central Chile and Argentina. Near the snow line. z9 # ⚭

'Harold Comber'. Differing from the species in the 8 mm thick, deep pink fruits. A selection from the species habitat in the Andes Mts., collected by Harold Comber in 1926. ⚭

P. macrostigma Col. Procumbent, wide spreading shrub, 50–70 cm high; leaves linear, narrowly elliptic to oblong, 12–14 mm long, 2–3 mm wide, with downward curving prickly tips, bristly serrate, sparsely pubescent above, glabrous beneath; flowers solitary in the apical leaf axils, corolla campanulate, white, 3 mm long, with short, reflexed limb tips, calyx irregularly enlarged and fleshy at fruiting, ovate and acute when in flower, May; fruits pink, flat globose, 6 mm thick (= *Gaultheria perplexa* Kirk). New Zealand. z9 # ⚭

P. melanocarpa see: **P. furens**

P. mucronata (L. f.) Gaud. Densely branched, broad, 0.5–1.5 m high shrub, spreading rapidly by stolons, young shoots thin, glabrous, or nearly so; leaves oval-elliptic to oblong-elliptic, 1–1.7(2) cm long, 4 to 6 mm wide, always distinctly acuminate, with 4–5 teeth on either side, base round or tapered, leathery tough, glossy green; flowers solitary in the leaf axils at the ends of the previous year's shoots, corolla ovate-urceolate, white, 6 mm long, with 5 short limb lobes, calyx persistent, dry membranous, sepals short and outspread, May–June; fruits flat globose, 8–12 mm thick, white to red or lilac. DL 1: 228;

Fig. 264. *Pernettya mucronata* (from Dimitri)

BM 3093 + 8023; BR 1695. S. America, Tierra del Fuego. 1828. z7 Fig. 264, 265. # ⚭

Includes many selections, but some may be known by more than one name in cultivation. Also listed are 2 varieties. All # ⚭

var. **angustifolia** (Lindl.) Reiche. Leaves larger, oblong-elliptic, narrow, 12–18 mm long, 3 mm wide, thinner, tapered to both ends, more bowed; flowers smaller. BM 3889. Central Chile. Fig. 285.

var. **rupicola** (Phil.) Reiche. Twigs finely pubescent; leaves thinner than those of the type, 8–10 mm long, 4 mm wide, narrow elliptic, always with a distinct prickly tip and 2–5 small teeth on either side, glossy with scattered pubescence; flowers oval-rounded, white with a pink trace (= *P. rupicola* Phil.). Central China.

'Alba'. Fruits white. Gs 1938:16.

'Atrococcinea'. Fruits large, dark ruby-red, glossy.

'Bell's Seedling' (Bell). Hermaphroditic form, young shoots reddish; leaves deep green, very glossy; fruits large, deep red, very persistent. Cultivated since about 1928. One of the best forms.

'Cherry Ripe'. Very similar to 'Bell's Seedling', but fruits somewhat smaller, cherry-red.

'Coccinea' (T. Davis). Fruits bright red. Named and cultivated before 1879.

'Crimsonia' (Gebr. Ellerbroek, Boskoop). Growth broad and strong; fruits carmine-red, 14–16 mm thick.

"Davis' Hybrids". Not a clone, rather a mix of large fruited seedlings.

'Edward Balls'. Shoots erect, thick, stiff, reddish, short bristly pubescent; leaves broadly oval. A clone selected in its habitat by E. K. Balls.

'Lilacina' (T. Davis). Fruits lilac. Before 1878.

'Lilian' (Gebr. Ellerbroek, Boskoop). Broad growing; fruits lilac-red, 14–16 mm thick.

Fig. 265. *Pernettya mucronata,* fruiting branch
(reduced; from M. G.)

'Mulberry Wine'. Young shoots green; large fruited, magenta at first, dark purple when ripe.

'Parelmoer' (Gebr. Ellerbroek). Growth rather broad; fruits light pink, 13–15 mm thick.

'Pink Pearl'. Berries medium-sized, lilac-pink.

'Purpurea' (T. Davis). Purple-violet. Before 1879.

'Rosalind' (W. Windhorst, Boskoop). Broad growing; fruits carmine-pink, 12–14 mm thick.

'Rosea' (T. Davis). Fruits pink. Before 1879.

'Rosie'. Young shoots red; leaves deep sea-green; fruits pink with a darker trace.

'Sea Shell'. Fruits large, totally light pink at first, later pure pink.

Fig. 266. *Pernettya mucronata* var. *angustifolia,*
flowering shoot (from B. R.)

'Signaal' (W. Windhorst, Boskoop). Broad growing; fruits deep pink, 12–14 mm thick.

'Sneeuwwitje' (Gebr. Ellerbroek). Broad growing, with a few protruding shoots; fruits pure white, with scattered red spots, 10–12 mm thick, in long fruit clusters.

'Thymifolia'. Leaves smaller than those of the species; male. An attractive form; interesting for its flowers in late May to early June.

'White Pearl'. A selection of 'Alba' with medium-sized or large, pure white fruits.

'Wintertime' (W. Windhorst, Boskoop). Rather compact; fruits pure white, 10–12 mm thick, ripening very late, but also persisting very long.

P. nana Col. Dwarf shrub, more or less prostrate on the ground or somewhat directed upward; leaves ovate, thick and leathery, 3–6 mm long, 1–2 mm wide, with numerous single celled hairs, finely glandular beneath, margins bristly serrate; flowers solitary, axillary at the end of the previous year's shoots, corolla campanulate, white, about 4 mm long and wide, filaments glabrous, scarcely widened at the base, anthers with small, but distinct horns, calyx irregularly enlarged at fruiting, then fleshy, May; fruits flat globose, reddish, 4 to 15 mm thick. New Zealand, S. Island. z9 # ⚥

P. pentlandii see: **P. prostrata** ssp. **pentlandii**

P. pilosa see: **P. ciliata**

P. prostrata (Cav.) Sleumer. Procumbent or creeping shrubs, about 15 cm high or more; leaves densely crowded, elliptic to oblong-elliptic, 4–7 mm long, 1.8–2.5 mm wide, finely bristly serrate, acute, midrib sparsely bristled beneath; flowers solitary in the apical leaf axils, white, urceolate, 5–6 mm wide, calyx enlarged and fleshy at fruiting, sepals whitish, triangular-ovate, May–June; fruits blue-black, globose, to 1.5 cm thick, poisonous. BM 6204 (= *Andromeda prostrata* Cav.). Costa Rica to Central Chile. 1870. z9 # ⚥

ssp. **pentlandii** (DC.) Burtt. Growth taller, to about 50 cm high, shoots pubescent, somewhat bristly; leaves oblong, elliptic, 2–3 cm long, 5–9 mm wide, glossy, sparsely and shallowly crenate; flowers solitary, axillary, pendulous, corolla white, ureolate, 3–5 mm long, filaments widened at the base, calyx only scarcely fleshy; fruits flat globose, about 1.5 cm wide. BMns 127 (= *P. pentlandii* DC.; *P. prostrata* var. *pentlandii* [DC.] Sleumer). Mountains from Costa Rica to N. Chile. # ⚥

P. pumila (L. f.) Hook. Procumbent, rapidly spreading dwarf shrub; leaves usually very dense and more or less distinctly distichous, usually glossy on both sides, oval-elliptic to lanceolate, about 6 mm long, 1.5 mm wide, obtuse, leathery thick; flowers solitary in the apical leaf axils, campanulate, white, petiole to 2 cm long, calyx dry, membranous, persistent, April–May; fruits globose, white to reddish, 1.2–2 cm thick (= *P. empetrifolia* Gaud.). Patagonia and Tierra del Fuego, the Falkland Islands. z7 # ⚥

P. rupicola see: **P. mucronata** var. **rupicola**

P. tasmanica Hook. f. Creeping dwarf shrub, 5–15 cm high, spreading cushion-like, young shoots somewhat pubescent; leaves narrowly oval-lanceolate, 4–8 mm long, 1.8–2.5 mm wide, indistinctly and finely bristly serrate; flowers solitary in the apical leaf axils, corolla campanulate, white, about 3 mm long, May; sepals fleshy and variable in size on the ripe fruits, berries half enclosed within the calyx, fruits flat-globose and somewhat lobed, light red, 1 cm thick. GC 130: 7; CFTa 108, Tasmania. 1930. z8–9 # &

This genus requires a humus, acid soil in light shade with sufficient soil moisture. *P. mucronata* and its colored fruit forms are more commonly found in the nursery trade.

Lit. Burtt, B. L., & A. W. Hill: The genera *Gaultheria* and *Pernettya* in New Zealand, Tasmania and Australia; in Jour. Linn. Soc. Bot. **49**, 611–644, 1935 ● Sleumer, H.: Revision der Gattung *Pernettya;* in Notizbl. Bot. Gart. Berlin **12**, 626–655, 1935 ● Stoker, F.: Hardy Pernettyas in cultivation; in New Flora and Silva 11, 165–174, 1939.

PEROVSKIA Karel. — LABIATAE

Deciduous subshrubs or herbs, resembling *Salvia*, aromatic; leaves opposite, serrate to pinnatisect; flowers whorled in false terminal spikes, calyx tubular-campanulate, bilabiate, upper lip entire or 3 toothed, lower lip 2 toothed; corolla bilabiate, upper lip unevenly 4 lobed, lower lip entire, rounded, 4 stamens, 2 of these fertile and spreading, 2 infertile, very small, hidden under the upper lip; fruit composed of 4 ovate nutlets. — 7 species from W. Asia to Himalaya and Tibet.

Perovskia abrotanoides Karel. Shrub, erect, later procumbent-ascending, to 50 cm high; leaves pinnatisect to bipinnate, 3–6 cm long, the lobes linear, soft pubescent; flowers lilac, August–September. Transcaspian to the W. Himalayas. 1935. Hardy. z5 Fig. 267.**d** ○

P. atriplicifolia Benth. Upright shrub, to 1.5 m high, but often with procumbent-ascending shoots, twigs white tomentose, rodlike, hardly branched; leaves oval-lanceolate, 3–6 cm long, unevenly and coarsely serrate, gray tomentose on both sides, later often glabrous, petiole 3–8 mm long; flowers 2–6 in widely spaced whorls, these in slender panicles grouped into 30–50 cm long, terminal panicles, corolla blue, August–September, calyx densely pubescent. BM 8481; RH

1905: 344; Gs 1926: 225. Afghanistan to W. Himalayas and W. Tibet; covering large areas in the Chitral Valley. 1904. Often freezing back in cooler climates, but resprouting well in spring. z6 Plate 153; Fig. 267. **d** ∅ ⊕ ○

'Blue Spire'. Leaves more deeply incised than those of the species, silver-gray; flowers very abundantly (Minier, 1975).

P. × 'Hybrida' (= *P. abrotanoides × P. atriplicifolia?*). Leaves ovate in outline, 4.7 cm long, simple to doubly pinnatisect, midrib broadly winged, gray-green; flowers blue-violet, flowering shoots 60–80 cm high. Cultivated by Hillier (England) since 1937 and distributed from the Geisenheim Research Institute since 1955 as *P.* 'Superba'. Fig. 267.

P. scrophulariifolia Bge. Shrub, erect, twigs 4 sided, gray tomentose; leaves ovate-oblong to elliptic-oblong, 2–4 cm long, obtuse, base cuneate, double crenate, very rugose above, venation slightly downy pubescent beneath when young; flowers lilac-pink, August–September. Turkestan. 1935. Hardy. z6 Fig. 267 **d** ○

All species need a sunny, dry site, preferring a lean, gravelly soil; very well suited for the rock garden, but requiring some space. A hard annual pruning is recommended.

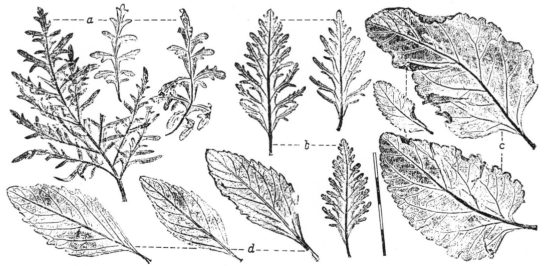

Fig. 267. **Perovskia**. a. *P. abrotanoides;* b. *P. × 'Superba'* (from Geisenheim, W. Germ.); c. *P. scrophulariifolia;* d. *P. atriplicifolia* (a., b. and d. from wild material collected by the Botanical Museum of Berlin-Dahlem)

PERSEA Mill. — LAURACEAE

Evergreen trees and shrubs; leaves alternate and scattered, simple, leathery tough, pinnately veined and 3 veined; flowers in panicles, corolla tube short, with 6 lobes, all either equal in size or the 3 outermost being smaller, perianth persistent, stamens usually 12; fruits globose or more ovate, commonly very large, fleshy, with a large seed. — 150 species in the tropics, some in SE. Asia and 1 on the Canary Islands.

Persea americana Mill. Avocado. Evergreen tree, 10 m high or higher, with a round crown, but multistemmed and lower in cultivation; leaves elliptic, ovate or oblong, 8–20 cm long, glabrous, dark glossy green above, blue-green beneath, 1.5 cm wide, fruits pear-shaped, dark green or brown according to variety, fruit pulp yellowish or greenish, nearly butter-like texture, tasty, seed acutely ovate, 3 to 4 cm thick. BM 4580 (= *P. gratissima* Gaertn.). Tropical America. 1739. Cultivated in the USA and many tropical countries. z10 # ⚬ ✄

P. borbonia Spreng. Evergreen tree, to 10 m high, twigs and shoots smooth; leaves oblong or more lanceolate, 5–7 cm long, glabrous, dark green and glossy above, blue-green beneath; fruit a small, dark blue drupe, about 1 cm long, with the orange colored remnants of the perianth at the base, stalk red. KTF 82 (= *P. carolinensis* Nees.). Eastern USA. z9 #

P. carolinensis see: **P. borbonia**

P. gratissima see: **P. americana**

Illustrations of other species:
P. indica BFCa 16
P. humilis KTF 83
P. palustris KTF 84

Lit. Kopp, L. E.: A taxonomic revision of the genus *Persea* in the Western Hemisphere; in Mem. New York, Bot. Gard. **14**, 1964 ● For culture, cultivars of fruiting varieties and additional literature see Hodgson, R. W.: The California Avocado Industry; Calif. Agric. Ext. Serv. Bull. Circ. **43**, 1–93, 1947.

PERTYA Schultz-Bip — COMPOSITAE

Herbaceous plants or subshrubs; leaves alternate, dentate, also compound on some species, sessile or petiolate; flower heads solitary or in racemes or panicles, all flowers tubular, fertile, involucre campanulate, scales imbricate, in many rows; corolla tubular, deeply 5 lobed; fruits much ribbed, with many pappus bristles. — 16 species from Afghanistan to Japan.

Pertya sinensis Oliv. Shrub, to 1.5 m high, deciduous; leaves on the current year's shoots much smaller, usually

several to a node, oval-lanceolate to elliptic, 5–8 cm long, acute at both ends, 2–2.5 cm wide, entire or with a few small teeth; flower heads solitary in the middle of the leaf clusters, on about 5 cm long, thin stalks, heads with 10–12 upward directed, lilac-pink ray flowers, June. ICS 6728. China. 1901. z6

Totally winter hardy, but without particular ornamental merit. Botanically interesting at most.

PETREA L. — Purple Wreath — VERBENACEAE

Evergreen shrubs, usually twining, occasionally becoming small trees; leaves opposite, simple, entire, leathery, flowers in long racemes in the axils of the apical leaves, the individual flowers short stalked, calyx teeth 5, large, colorful when in flower, corolla with a short tube, limb 5 lobed and oblique, usually very conspicuously colored. — 30 species in tropical America and the W. Indies.

Petrea volubilis L. Purple Wreath. Evergreen or

deciduous, high twining shrub, 6 m high or more (may be maintained to a free standing shrub by pruning, like *Wisteria*); leaves elliptic, rough, 3–20 cm long, 1–7 cm wide, short petioled; flower racemes very numerous, narrow, 7–30 cm long, violet, blue or lilac, calyx tube densely erect pubescent. BM 628. Mexico, Central America, W. Indies. 1731. Extraordinarily attractive flowering plant in frost free regions, nearly as intensive as *Forsythia*, but blue. z10 # ⊕

PETROPHYTUM (Nutt. ex Torr. & Gray) Rydb. — ROSACEAE

Evergreen, dwarf shrubs, shoots lying on the ground, very short; leaves crowded on the short shoots, oblanceolate to spathulate, leathery, entire; flowers in short, dense racemes elevated above the foliage; sepals 5, valvate, petals 5, imbricate, white, calyx cup hemispherical, stamens 20, filaments about twice as long as the sepals; fruit a leathery follicle, dehiscing along both seams. — 4 species in N. America; found in rocky crags high in the mountains.

Petrophytum caespitosum (Nutt.) Rydb. Annual growth very short; leaves blue-green, spathulate, 5–12 mm long, 2–4 mm wide, in rosettes, stem leaves linear, densely silky pubescent; sepals oval-lanceolate, 1.5 mm long, inflorescences 1–4 cm long, never branched, summer flowering. NF 11: 291; VT 431 (= *Spirea caespitosa* Nutt.). USA; Rocky Mountains. z3 Fig. 268. #

P. cinerascens (Piper) Rydb. Densely grasslike sub-

Fig. 268. **Petrophytum.** Left *P. cinerascens*; middle *P. caespitosum*; right *P. hendersonii*
(from Abrams)

shrub, shoots short, thick; leaves oblanceolate, 3 veined, 1–2.5 cm long, somewhat ash-gray, sparsely pubescent; inflorescence 5–15 cm long, sepals acute, June–September (= *Spiraea cinerascens* Piper). NW. North America, in cliffs along the Columbia River. z5 Fig. 268. #

P. elatius (S. Wats.). Heller. Annual growth often 2 to 3 cm long, erect; leaves oblanceolate, acute, silky pubescent, 1.5–2 cm long, 3–4 mm wide, usually distinctly petiolate; sepals lanceolate, petals oblanceolate, 2–3 mm long, inflorescences 4–10 cm long, often also branched,

July–September (= *Spiraea caespitosa elatior* S. Wats.). USA, Utah to Arizona, in the mountains. z3 #

P. hendersonii (Canby) Rydb. Dense grasslike habit, shoots short, thick; leaves spathulate, 1–2 cm long, thick, usually 3 veined; flower stalks 4–8 cm long, racemes 2.5–5 cm long, sepals obtuse, June–September (= *Spiraea hendersonii* Canby). N. America; on cliffs near the peaks in the Olympic Mts., Washington, USA. z4 Fig. 268. #

Only suitable for cracks in the rock garden; completely winter hardy, but generally cultivated only by collectors.

PETTERIA Presl — LEGUMINOSAE

Monotypic genus. Deciduous, erect shrub, closely related to *Laburnum*, but differing in the short, erect flower racemes; upper calyx lobes not connate.

Petteria ramentacea (Sieber). Presl. Narrowly upright shrub, to 2 m high, twigs glabrous, green, foliar buds covered after leaf drop by black-brown, persistent stipules; leaves trifoliate, leaflets elliptic-oblong, 2–5 cm long, ciliate, dark green above, lighter beneath, glabrous

on both sides; flowers yellow, fragrant, in 4–7 cm long, dense, erect racemes, May–June; pods flat, 4–5 cm long. BS 2: 407; HM 1304; HF 2307 (= *Cytisus ramentaceus* Sieber; *Cytisus weldenii* Visiani). Dalmatia, Istria to Albania (Yugoslavia, Caucasus Region). 1838. z6 Plate 150, 155. **d** ⊕ ○

Cultivated like *Cytisus*.

PEUMUS Molina emend. Pers. — MONIMIACEAE

Monotypic genus. Characteristics as follows.
Peumus boldus Molina. Boldo. Evergreen tree, 6 m high or higher in its habitat; leaves usually opposite, leathery, acutely ovate-elliptic, 2–4 cm long, 1.5–2.5 cm wide, obtuse, base round to broadly cuneate, short petioled, glabrous; flowers dioecious, few in small, terminal

cymes, not very conspicuous, white, male flowers somewhat larger than the female, sepals oval, twice as large as the narrowly elliptic petals; fruit small, pea-sized, sweet, eaten in Chile. PFC 135. Chile. 1844. Fruits have medicinal uses, the bark is used for tanning and dyeing. z10 # ∅

PHAEDRANTHUS Miers — BIGNONIACEAE

Monotypic genus, very closely realted to *Bignonia,* but differing in the protruding stamens, pubescent calyx and ovaries as well as the filamentous tendrils; otherwise as described below.

Phaedranthus buccinatorius (DC.) Miers. Evergreen climber, strong growing and high climbing; leaves opposite, composed of 2 leaflets and a 3 branched tendril, leaflets oblong-elliptic, or more oval-oblong, 6–7 cm long, obtuse, glabrous and glossy above, pubescent beneath when young, persisting only on the venation; flowers in terminal racemes, corolla tubular, bowed, 5 cm long, limb outspread, blood-red, yellow at the base, pendulous, June–August. BM 7516; NV 21; MCL 118 (as *Distictis buccinatoria*) (= *Bignonia buccinatoria* DC.). Mexico 1824. z10 Plate 127. # ⊕

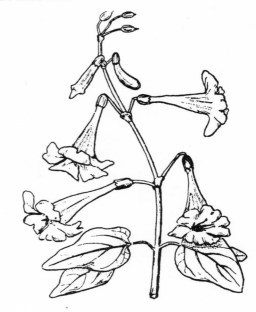

Fig. 269. *Phaedranthus buccinatorius*

PHELLODENDRON Rupr. — Corktree — RUTACEAE

Tall, deciduous, aromatic trees; bark often thick and corky; winter buds always hidden beneath the leaf petiole (important distinction from *Evodia*); leaves opposite, pinnate, leaflets translucent punctate; flowers dioecious, small, yellowish-green, in small, terminal panicles; petals 5–8, often as long as the 5–8 sepals; fruit a globose, pea-sized, black drupe, with 5 single seed pits. — About 10 species in subtropical and temperate E. Asia.

Phellodendron amurense Rupr. Tree, to 15 m high, broad crowned, bark thick and corky, deeply grooved, light gray, young twigs gray-yellow to orange-yellow; leaves to 35 cm long, rachis tomentose, leaflets 9–13, broadly ovate to oval-lanceolate, 5–10 cm long, long acuminate, dark green and glossy above (!), bluish beneath (!) and pubescent along the midrib, golden-yellow in fall, leaves dropping early, in September; flower panicles pubescent, about 6–8 cm long, June; fruits 1 cm thick, black with a strong turpentine scent. ST 93; KD 348; MJ 1166. N. China, Manchuria. 1856. Ⓕ NE. China, USSR (Ukraine and the Far East), Czechoslovakia, Romania. z3 Fig. 270. ⌀

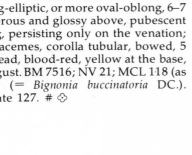

Outline of Species Characteristics

Species	Bark	Shoots	Leaflets	
			Above	Below
amurense	thick and corky	orange-yellow	glossy	bluish, glabrous
sachalinense	thin, with plates	red-brown	dull green	bluish
lavallei	thick, with cork	reddish-brown	dull yellow green	light green, pubescent
japonicum	thin, with plates, deep brown	red-brown	dull green	light green, tomentose
chinense	thin, gray-brown	red-brown	dark yellow-green, dull	light green, tomentose

Fig. 270. **Phellodendron.** a. *P. chinense;* b. *P. amurense;* c. *P. japonicum* (leaves from wild plants)

P. chinense Schneid. Tree, to 10 m high, bark thin, dark gray-brown, young shoots reddish-brown; leaflets 7–13, oblong-lanceolate, 7–14 cm long, the margins nearly parallel, long acuminate, base broadly cuneate, dark yellow-green above (!), underside with a long, soft pubescence, light green (!); flowers in dense, conical, 5 to 6 cm long panicles; ovaries pubescent, fruits black, 9 mm thick. SH 2: 79c. Central China. 1907. z5 Fig. 270 ⌀

P. japonicum Maxim. Small, erect tree, 5(10) m high, trunk bark thin, plate-like, finely channeled, deep brown, young shoots red-brown; leaves 25–35 cm long, rachis hairy-tomentose, leaflets 9–13, ovate to oval-oblong, 6–10 cm long, base very oblique, slightly cordate or truncate, dull green above, light green beneath and very tomentose, particularly on the venation; ovaries glabrous, fruits black, 1 cm thick, in erect, tomentose panicles, September. ST 95. Central Japan. 1863. z6 Fig. 270. ⌀

var. *lavallei* see: **P. lavallei**

P. lavallei Dode. Tree, 7–10(15) m high, bark thick and corky (like *P. amurense*, but the shoots are reddish-brown in winter!); leaves 20–35 cm long, rachis soft pubescent, leaflets 5–13, oval-elliptic to oblong-lanceolate, 5–10 cm long, acuminate, base cuneate, dull yellowish-green above, light green beneath and pubescent (at least when young), petiole 3–4 mm long; flowers in about 6–8 cm wide, loose, pubescent inflorescences, June; ovaries glabrous, fruits very numerous, black. BM 8945 (= *P. amurense* var. *lavallei* Sprague). Central Japan. Often confused in cultivation with *P. japonicum*. z6 ⌀

P. sachalinense (F. Schmidt) Sarg. Tree, 7(15) m high, with a straight trunk and broad crown, trunk bark dark brown, thin, not corky, finely channeled and eventualy plate-like, twigs red-brown; leaves 22–30 cm long, leaflets 7–11(15), ovate to oval-oblong, 6–12 cm long, acuminate, base round, dull green above (!), bluish and glabrous or nearly so beneath, margins slightly ciliate or not so; flowers in nearly glabrous (!), 6–8 cm long panicles; fruits 1 cm thick, black. LF 192; ST 94; MF 3484. Sachalin, N. Japan, Korea, W. China. 1877. The hardiest species. z3 ⌀

An attractive, medium-sized park tree, very disease resistant, with virtually no pest problems; requires a deep, moist, fertile soil and an open site.

Lit. Sprague, T.A.: *Phellodendron,* a revision; in Kew Bull. Misc. Inf. 1920, 231–235 ● Kern, E. E.: The Velvet Tree; in Bull. Appl. Bot. **27,** 3, 1–30, 1931 (fig. 1–16; Russian with an English summary) ● Wolf, E.: Die Korkbäume, *Phellodendron,* im Arboretum des Leningrader Forstinstitutes; in Mitt. DDG 1925, 215–218.

PHILADELPHUS L. — Mock Orange — PHILADELPHACEAE

Deciduous (also evergreen in Central America) shrubs, erect or broad growing, occasionally somewhat climbing and thorny, short shoots opposite, winter buds prominent or hidden under the leaf petiole; shoots with a solid pith, with persistent or exfoliating bark; leaves opposite, serrate or entire, 3 or 5 veined; flowers solitary or in 3's or in terminal racemes or few flowered panicles, usually fragrant, white or also reddish at the base of the petals, sepals 4, occasionally 5, petals 4, occasionally 5, stamens 13–90; ovaries inferior or subinferior, usually 4 locular, styles 4, occasionally 3 or 5; fruit a 4 valved capsule with numerous (on many species appendaged) seeds. — About 75 species from S. Europe to the Caucasus, E. Asia to Himalayas, N. and Central America.

A. Key to the subgenera, sections and important series
(from SH.-Y. Hu, enlarged)

● Lateral buds prominent:
> Seeds long caudate; stigmas distinct, widened, usually cristate; flowers solitary or in 3's or in few flowered panicles; fruits obovoid or nearly globose, with an encircling, persistent calyx:
Subgenus I. **Gemmatus** (Koehne) Sh.-Y. Hu

1. Leaves long acuminate, occasionally acute; flowers in few flowered panicles, one to many flowered, each flower on an acuminate stalk; stigmas elongated, but generally not cristate; fruits obovoid-ellipsoid:
Section 1. **Poecilostigma** Koehne
P. karwinskyanus, mexicanus

2. Leaves acute or obtuse; flowers solitary or in 3's; stigmas elongated, but generally not cristate; fruits nearly globose:
Section 2. **Coulterianus** Sh.-Y. Hu
P. coulteri

>> Seeds short caudate; stigmas distinct, clavate; flowers in racemose panicles; fruits ellipsoid with a subapical, persistent calyx:
Subgenus III. **Macrothyrsus** Sh.-Y. Hu
Section 6. **Californicus** Rydb.
P. californicus, cordifolius, insignis

>>> Seeds not caudate; stigmas connate, columnar or nearly capitate; flowers solitary or in 3's; fruits top-shaped to nearly globose, with an apical, persistent calyx:
Subgenus IV. **Deutzioides** Sh.-Y. Hu
1. Leaves rough haired or strigose, all hairs straight;
+ Styles 3–4 mm long; shrubs slightly climbing; leaves serrate:
Section 7. **Hirsutus** (Rydb.) Sh.-Y. Hu
P. hirsutus

++ Styles 1 mm long; shrubs somewhat thorny; leaves entire:
Section 8. **Pseudoserpyllifolius** Sh.-Y. Hu
— not covered in this text —

2. Leaves strigose pubescent and woolly beneath; styles 1–2 mm long; somewhat thorny or small, slender shrubs:
Section 9. **Serpyllifolius** Sh.-Y. Hu
P. serpyllifolius

●● Lateral buds enclosed:
Subgenus II. **Euphiladelphus** Sh.-Y. Hu

Flowers solitary, in 3's or in dichasial cymes;
× Large, wide arching shrubs; stamens 60–90; stigmas distinct, spatulate; leaves 4–10 cm long:
Section 3. **Pauciflorus** (Koehne) Sh.-Y. Hu
P. floridus, inodorus

×× Erect shrubs, habit low or compact or thorned and xerophytic; stamens 35–50; stigmas all distinct or partly connate, linear; leaves 0.5 to 2.5 cm, occasionally to 3 cm long:
Section **Microphyllus** (Koehne) Sh.-Y. Hu
P. argyrocalyx, microphyllus.

Flowers in distinct racemes:
Section 4. **Stenostigma** Koehne (see key B)

B. Key to the series of the section Stenostigma
(from Sh.-Y. Hu)

● Hypanthium (calyx cup) glabrous or occasionally with solitary hairs at the base;
  ★ The side of the stigma toward the axis is larger than the dorsal side;
    > Seeds long caudate, the embryonic portion only half as long as the appendage (except *P. triflorus*);
      + Leaves glabrous or becoming so; American species:

    Series 1. **Gordoniani** (Koehne) Rehd.
     P. intectus, lewisii

  ++ Leaves tomentose beneath; species from the Himalayas:
    Series 2. **Tomentosi** Sh.-Y. Hu
     P. tomentosus, triflorus

 >> Seeds medium to short caudate, the embryonic portion as long or longer than the appendage;
  + Seeds short caudate, embryonic portion longer than the appendage (except *P. brachybotrys*); North and East Chinese species:
    Series 3. **Pekinenses**
     P. brachybotrys, laxiflorus, pekinensis

  ++ Seeds medium long caudate, embryonic portion as long as the appendage; European and Caucasian species:
    Series 4. **Coronarii** (Koehne) Rehd.
     P. caucasicus, coronarius, salicifolius

★★ The side of the stigma toward the axis smaller than the dorsal side:
    Series 5. **Delavayani** Sh.-Y. Hu
     P. delavayi, purpurascens

●● Hypanthium regularly formed and generally densely pubescent;
  ★ Large shrubs, becoming 4–5 m high; leaf undersides, hypanthium and sepals equally long pubescent; species from the SE. USA:
    Series 6. **Pubescentes** Sh.-Y. Hu
     P. pubescens

  ★★ Medium high or low shrubs, becoming 1–3 m high; leaf undersides, hypanthium and sepals strigose or rough;
    ✕ Seeds short caudate; hypanthium pubescence strigose, rough or shaggy; Chinese species:
    Series 7. **Sericanthi** Rehd.
     P. dasycalyx, incanus, kansuensis, schrenkii, sericanthus, subcanus, tenuifolius

 ✕✕ Seeds long or medium caudate; hypanthium pubescence sparse, short, golden, crispate, sparse along the edges or over the entire surface; Japanese species:
    Series 8. **Satsumani** (Koehne) Sh.-Y. Hu
     P. satsumanus, satsumi, shikokianus

Philadelphus amurensis see: **P. schrenkii** var. **manshuricus**

P. argyrocalyx Woot. Erect, delicate shrub, about 1–2 m high, young shoots rust-brown pubescent, later bark brown and eventually exfoliating; leaves ovate, oval-oblong to elliptic, 1–3.5 cm long, 4–15 mm wide, usually obtuse at the ends, glabrous above, densely white haired and green beneath, distinctly 3 veined; flowers solitary, stalk 1–2 mm long, pubescent, sepals ovate and densely silvery-white woolly, as is the ovary (!), corolla cruciform, 3.5 cm wide, petals oval-oblong, apex rounded and emarginate, white, slightly fragrant, June–mid-August. JA 35: Pl. 14. USA; New Mexico. 1918. z7 Fig. 272. ⊕

P. billardii see: **P. insignis**

P. brachybotrys Koehne. Shrub, to 3 m high, twigs with a tight, gray-brown bark in the 2nd year (!), young shoots shaggy pubescent, later glabrous; leaves ovate, 2–6 cm long, 1–3 cm wide, short acuminate, base round, finely serrate to nearly entire, sparsely strigose above, strigose pubescence on the venation beneath; flowers 5–7, axis 2–4 cm long, sparsely pubescent, stalk 3–5 mm, calyx and calyx cup glabrous, sepals ovate, 4–6 mm long; corolla disc-shaped, cream-white, 2.5–3.5 cm wide, petals nearly circular, stamens 31–40, disk and style glabrous late June–July, (= *P. pekinensis* var. *brachybotrys* Koehne). China; Kiangsu Province. 1901. z7 ⊕

P. ✕ burkwoodii see: List of Hybrids and Cultivars

P. californicus Benth. Erect shrub, to 3 m high, twigs long and gracefully nodding, young shoots thin and softly pubescent, bark black-brown in the second year and somewhat exfoliating; leaves ovate-elliptic, 4.5–8 cm long and 3–5 cm wide on long shoots, 3–5 cm long and to 4.5 cm wide on flowering shoots, both sides glabrous, acute, base obtuse or acute, entire, ciliate or slightly serrate; flowers in panicles of lateral, usually 3(5) flowered cymes, fragrant, hypanthium campanulate, glabrous, sepals ovate, glabrous, corolla cruciform, 2.5 cm wide, petals oblong, 11 mm long, glabrous on both sides, stamens 25–37, disk and styles glabrous, only slightly incised or not so. SH 1: 234a to a2, 236a; MS 135 (as *P. lewisii* var. *californicus*); DL 1: 181 (as *P. lewisii*); JA

Fig. 271. **Philadelphus.** a. *P. salicifolius*; b. *P. coronarius* 'Dianthiflorus';
c. *P. coronarius* 'Zeyheri'; d. *P.* × *splendens* (drawn for this work by Sh.-Y. Hu)

35: Pl. 2 (= *P. lewisii* sensu Dipp.; *P. lewisii* var. *californicus*
[Benth.] Gray). California, USA; in moist sites in its
habitat, together with *Calycanthus* and *Toxicodendron*.
1858. z7 Fig. 272.

P. caucasicus Koehne. Erect shrub, some shoots
nodding, young twigs brown, slightly pubescent, second
year bark light to dark brown and exfoliating; leaves
elliptic to oval-elliptic, 5–9 cm long and 2–3(4) cm wide,
abruptly short acuminate, base obtuse to acute 3(5)
veined, with 0–8 small teeth on either edge, glabrous or
eventually so above, underside soft pubescent; flowers
5–9 in racemes, stalks with loose fine pubescence, 3–10
mm long, likewise the calyx and hypanthium, hairs
somewhat frizzy, corolla nearly disk-like, 3 cm wide,
petals obovate, stamens about 22, the longest about 6
mm long, disk and style dense yellowish pubescent (!).
JA 35: Pl. 3 + 4. Abchasian Republic (USSR), along the
Kuban River and the Black Sea region. Not yet in cultivation. Fig. 274.

'Aureus'. Shrub, to 1.5 m high; leaves ovate, more or less lobed,
3–5 cm long, 1.5–3 cm wide, short acuminate, base round,
serrate, glabrous above, major veins beneath pubescent;
flowers 3–7, sepals glabrous, hypanthium somewhat
pubescent at the base, corolla plate-like, 2.5 cm wide, petals
oblong-round, disk pubescent (!), style glabrous (= *P.*
coronarius 'Aureus'). Not uncommon in Dutch gardens. Before
1901. Plate 157. ⊘

var. **glabratus** Sh.-Y. Hu. Shrub, 2–2.5 m high, young shoots
sparsely pubescent, bark brown and exfoliating in the second
year, older shoots gray; leaves elliptic to oval-elliptic, 3–7 cm
long and 1.5–3 cm wide, short acuminate, base usually acute to

obtuse, glabrous above, sparsely strigose beneath; flowers 7–
11 in racemes, stalk, calyx and hypanthium glabrous, disk
pubescent, style glabrous, June (= *P. caucasicus* sensu Rehd.
non Koehne). Caucasus. 1888 (as *P. caucasicus*).

P. columbianus see: **P. lewisii** var. **gordonianus**

P. × **congestus** see: List of Hybrids and Cultivars

P. cordifolius Lange. Erect, 2 m high shrub, young shoots
glabrous, bark reddish-brown to gray in the second year,
exfoliating; leaves ovate, 4–10 cm long, 3–6 cm wide,
glabrous on both sides, the venation rarely with axillary
tufts of pubescence beneath, entire, base round to
slightly cordate (!); flowers in large, terminal panicles of
25–40 together, stalk 1–2 cm long, calyx campanulate,
glabrous, sepals ovate, corolla cruciform, 3–3.5 cm wide,
petals obovate-oblong, rounded at the apex, glabrous on
both sides, stamens about 27, disk and style glabrous,
June–July. SH 1: 234 l–p, 236f (= *P. lewisii* var. *cordifolius*
[Lange] Dipp.). California. According to Hu a dubious
species which is never true in cultivation! z7

P. coronarius L. Shrub, to 3 m high, rather stiff-upright,
usually with a few nodding shoots, young shoots slightly
pubescent to glabrous, bark dark brown and somewhat
exfoliating in the second year; leaves ovate, 4.5–9 cm
long and 2–4.5 cm wide, acuminate base obtuse to acute,
either edge with 6–11 small, widely spaced teeth,
glabrous except for the axillary tufts of pubescence in the
vein axils; flowers 5–9 (or more on strong shoots) in
racemes, cream-white, very fragrant, stalk slightly
pubescent, calyx, hypanthium, disk and style glabrous,
sepals ovate, 4–5 mm long, glabrous, corolla nearly disk-

form, 2.5–3 cm wide, petals oblong-obovate, apex rounded, late May to June; fruit capsule top-shaped. HM 978; SH 1: 237 x–z, 238 h–k (as *P. pallidus*); JA 35: Pl, 1, 3, 4; 37: Pl. 5 (= *P. pallidus* Hayek; *P. zeyheri* var. *kochianus* [Koehne] Rehd.). Habitat supposedly from Italy to the Caucasus but this has not been proven! 1569. The illustration BM 391 is unclear and more similar to *P. lewisii* (according to Sh.-Y. Hu). z5 Fig. 274. ✥

Key to the cultivars
(from Sh.-Y. Hu, revised)

● Leaves white variegated:
  'Variegatus'

●● Leaves all green;
  ★ Compact, to 1 m high shrubs;
    + Leaves opposite, always 2 at a node;
      # Flower stalk pubescent:
        'Duplex'

      ## Flower stalk glabrous;
        1. Flowers very densely double, petal tips acuminate:
          'Deutziiflorus'

        2. Flowers loosely double, petal tips rounded:
          'Dianthiflorus'

    ++ Leaves in whorls of 3 at each node:
      'Cochleatus'

  ★★ Shrubs over 2 m high, shoots stiff and open, rarely nodding; leaves ovate, oval-elliptic or elliptic;
    + Flowers single:
      'Zeyheri'
    ++ Flowers double:
      'Primuliflorus'

'Aureus' see: **P. caucasicus 'Aureus'**

'Cochleatus'. Low, glabrous shrub; leaves in whorls of 3, broadly elliptic, 2–2.5 cm long, 1 to 2 cm wide, obtuse at both ends, with 2–3 tiny teeth on either edge; flowers not observed. Cultivated in France since 1903.

'Deutziiflorus' (Lemoine 1874). Shrub, low, compact; leaves ovate, 2–5 cm long and 1.5–2.5 cm wide, glabrous on both sides, nearly entire or finely serrate on the apical half, to 9 cm long and 6 cm wide on long shoots; flowers usually in 3's or small racemes of 5, corolla 3 cm wide, petal tips more or less acuminate (= *P. coronarius* 'Multiflorus Plenus' Lemoine 1879).

'Duplex'. Small shrub, to 1 m high, young shoots with appressed pubescence, second year bark brown and exfoliating; leaves oblong-elliptic, 3–5 cm long and 1–2.5 cm wide, 6–8 cm long and 4–4.5 cm wide on long shoots, abruptly acuminate, coarsely dentate, both sides with a loose, erect pubescence; flowers in 2's or 3–5 together, double, corolla 2.5 cm wide, petals oblong, rounded at the apex. JA 35: Pl. 4 (= var. *nanus* Ait.; var. *pumilus* West.). First cultivated in England before 1770. Flowers very sparse and only on very old plants. Plate 156.

'Multiflorus Plenus' see: **'Deutziiflorus'**

var. *nanus* see: **'Duplex'**

'Primuliflorus' (Lemoine 1876). Compact shrub to 2 m high, shoots drooping, bark in the second year brown and exfoliating, young shoots warty pubescent (trichomes with thick bases); leaves broadly ovate, 3–7 cm long and 2–4 cm wide, to 9 cm long on long shoots and 7 cm wide, abruptly short acuminate, base round, strigose tomentose beneath; flowers grouped 2–3 or solitary, stalks 4–8 mm long, covered with straight hairs, calyx and hypanthium usually totally glabrous, corolla 3–3.5 cm wide, double. RH 1870: 47; JA 35: Pl. 3 (= var. *rosiflorus plenus* Hort.). Allegedly developed by Lemoine; presumably *P. coronarius* × *inodorus*, according to Hu.

var. *pumilus* see: **'Duplex'**

var. *rosiflorus plenus* see: **'Primuliflorus'**

var. *salicifolius* see: **P. salicifolius**

var. *tenuifolius* see: **P. tenuifolius**

var. *tomentosus* see: **P. tomentosus**

'Variegatus'. Sickly appearing shrub, bark gray in the second year, exfoliating late; leaves ovate, 2–3 cm long, usually somewhat deformed, margins white, base round; flowers grouped 3 or 5, corolla about 2.5 cm wide, slightly fragrant, June. Developed before 1770 in England, perhaps a mutation of *P. coronarius*.

'Zeyheri'. Shrub, erect, to 2 m high, young shoots slightly pubescent, bark chestnut-brown in the second year, exfoliating; leaves ovate, 6–10 cm long on long shoots, 3–6 cm wide, 4–5 cm long and 2.5–3.5 cm wide on flowering shoots, acuminate, serrate, base round, glabrous above, venation beneath somewhat pubescent; flowers 3–5 together, hypanthium glabrous, sepals ovate, 5 mm long, corolla disk-form, 2.5–3 cm wide, petals nearly circular, white, stamen filaments about 32, disk and style glabrous; fruits top-shaped. JA 35: Pl. 3 (= *P. zeyheri* Schrad. ex DC.). Developed around 1820 by Zeyher of W. Germany; not known in the wild. Fig. 271.

P. coulteri Wats. Weakly climbing shrub, to 1 m high, second year shoots only 2–3 mm thick, bark rough, young shoots rough and pubescent, trichomes thickened at the base; leaves ovate to more elliptic, 1.5–3 cm long, 1–1.5 cm wide, obtuse to acute, base round, nearly entire, strigose pubescent on both sides; flowers usually solitary, occasionally in 3's, corolla disk-form, 2.5–3 cm wide, petals round-obovate, pure white, without purple basal spots (often erroneously described!!), very fragrant, style and disk center pubescent. JA 37: Pl. 5 (= *P. purpusii* Brandegee). Mexico. Apparently never found to be true in cultivation and therefore not used in hybridizing. z9 Fig. 273.

P. × cymosus see: List of Hybrids and Cultivars

P. dasycalyx (Rehd.) Sh.-Y. Hu. Slightly climbing shrub, young shoots glabrous or soon glabrous, bark brown and exfoliating in the second year; leaves ovate or oval-elliptic, 3–6 cm long and 1.2–3.4 cm wide, short acuminate, base obtuse to round, nearly entire to slightly dentate or serrate, leaves on the flowering shoots quickly becoming glabrous on both sides except for a few individual hairs on the venation beneath; flowers grouped 5–7, floral axis 1.5–4.5 cm long and pubescent, hypanthium and calyx with sparse and outspread (!) pubescence, sepals ovate, 5–6 mm long, corolla nearly

Fig. 272. **Philadelphus.** a. *P. californicus;* b. *P. pubescens;* c. *P. argyrocalyx;* d. *P. microphyllus*
(drawn for this book by Sh.-Y. Hu)

plate-like, 2.5 cm wide, petals obovate, disk and styles glabrous, late May to early June; fruits obovoid, 6 mm long. (= *P. pekinensis* var. *dasycalyx* Rehd.). China; Honan, Shansi Province. z6 1926.

P. delavayi L. Henry. Shrub, 2–4 m high, young shoots glabrous, pruinose (!), usually gray-brown in the second year, bark smooth with horizontal grooves; leaves oval-lanceolate to oval-oblong, 5–14 cm long on long shoots, 2–7 cm wide, on flowering shoots 2–8 cm long and 2–5 cm wide, serrate, occasionally nearly entire, acuminate, base usually rounded, bristly pubescent above, densely appressed tomentose beneath; flowers grouped 5–9 (occasionally to 21) together, stalk 5–10 mm long, glabrous, hypanthium and calyx glabrous, pruinose (!), green with a trace of reddish-brown (!), sepals ovate, 6 mm long, corolla disk-form, 3 cm wide or more, petals rounded, stamens about 35, disk and style glabrous, styles columnar, stigma linear, June; fruits obovoid. RH 1903: 3; JA 35: Pl. 1,3,4. China, Tibet, Upper Burma. 1902. Inconsistent in cultivation with many poorly flowering types given this name. z6 Fig. 276. See also *P. purpurascens.*

var. *calvescens* see: **P. purpurascens**

var. **melanocalyx** Lemoine. Current year's shoots glabrous and striped; leaves oval-lanceolate, acuminate, base round, nearly entire or slightly dentate, bristly above, venation pubescent beneath, 5 veined; hypanthium and calyx glabrous, pruinose, purple (!), corolla cruciform, to 3 cm wide, petals oblong, about

13 mm long and 9 mm wide, stamens about 30, styles incised on the apical ⅓. BM 9022 (= *P. delavayi* sensu Stapf). China; Yunnan Province. 1903. Plate 154.

P. × falconeri see: List of Hybrids and Cultivars

P. floridus Beadle. Shrub, to 3 m high, young shoots brown, deep brown in the second year; leaves oval-elliptic to broadly elliptic, 4–10 cm long, 2–6 cm wide, abruptly acuminate, base round or obtuse, nearly entire or indistinctly and finely serrate, very sparsely, short, rough haired above, underside densely strigose; flowers usually in 3's, occasionally solitary or in racemes, stalks 3–7 mm long, usually glabrous, calyx and hypanthium sparsely shaggy pubescent, sepals ovate, 7–9 mm long, corolla plate-like, 4–5 cm wide, petals nearly circular, 2–2.5 cm wide, pure white, without fragrance, June. BR 2003 (as *P. speciosus* sensu Lindl.; *P. magnificus* sensu Rehd.). USA; Georgia. 1849. z6 Fig. 277.

'Faxonii'. Graceful shrub; leaves elliptic, 2.5–5 cm long, 1–2 cm wide, nearly entire or indistinctly and finely serrate, becoming glabrous above, strigose beneath; flowers grouped 1–3, calyx and hypanthium shaggy strigose, corolla cruciform (!), petals oblong, about 1.8 cm long. Known only in cultivation. 1907.

P. gloriosus see: **P. inodorus** var. **grandiflorus**

P. godohokeri see: **P. hirsutus**

P. gordonianus see: **P. lewisii** var. **gordonianus**

P. gordonianus var. *parviflorus* see: **P. satsumanus**

P. grandiflorus see: **P. inodorus** var. **grandiflorus** and **P. pubescens**

P. grandiflorus var. *laxus* see: **P. inodorus** var. **laxus**

P. helleri see: **P. lewisii** var. **helleri**

P. hirsutus Nutt. Shrub, to 2.5 m high, twigs divaricate and nodding, densely rough haired on the new growth, bark later red-brown and exfoliating; leaves oval-elliptic to oval-lanceolate, those at the tips of long shoots 2.5–7.5 cm long and 1–5 cm wide, acuminate, base round, sharply serrate, short rough haired above, but eventually nearly glabrous, underside with a persistent, dense, rough pubescence, gray, 3 veined at the base; flowers on very short shoots with only 1–2 leaf pairs, usually in 3's, occasionally solitary or in 5's, peduncle tomentose like the hypanthium, sepals triangular, 4 mm long, corolla plate-like, 2.5 cm wide, petals ovate, cream-white, without fragrance, disk glabrous, styles columnar, totally connate, June. BM 5334; BR 24: 14; VT 311; SH 1: 234 c–d2, 235 a–e; JA 35: Pl. 1,2,4 (= *P. villosus* Muhl.; *P. godohokeri* Hort.). SE. USA in the Appalachian Mts. 1820. z6 Fig. 277.

P. incanus Koehne. Shrub, stiffly erect, 1.5–3.5 m high, young shoots pubescent, bark gray and smooth in the second year or brown and eventually exfoliating; leaves oval-elliptic, occasionally elliptic-lanceolate, 4–8.5 cm long and 2–4 cm wide, to 10 cm long and 6 cm wide on long shoots, sparsely bristly above, underside with dense appressed pubescence, margins serrate; flowers grouped 7–11, axes pubescent, likewise the 5–10 mm long stalks, hypanthium and calyx with dense and appressed white pubescence (!), sepals ovate, 4–5 mm long, corolla broadly campanulate to plate-like, 2.5–3 cm wide, slightly fragrant, petals obovate-rounded, exterior somewhat strigose pubescent at the base, stamens about 34, disk and style glabrous (!), July (!). JA 35: Pl. 3. China; Shensi, W. Hupeh Provinces. 1896. z6 ⊕

P. inodorus L. A broad spreading, pendulous shrub, 1–3 m high, current year's shoots glabrous, second year shoots brown, bark exfoliating; leaves ovate to elliptic, occasionally broadly elliptic, 5–9 cm long, short acuminate, base acute, occasionally obtuse, entire or somewhat dentate, glabrous above, dark green and glossy, underside glabrous except for the small axillary pubescence in the basal vein axils, petioles glabrous; flowers usually solitary or in 3's at the end of short lateral shoots, shell-form, 4–5 cm wide, without fragrance, calyx totally glabrous (!), sepals ovate, short acuminate, petals oblong, 2 cm long, stamens 60–90, styles as long as the longest stamens, June. BB 2189; NBB 2: 273; JA 35: Pl. 4. Southeast USA. Before 1837. Often incorrectly described in the literature! z6 Fig. 273. ⊕

var. **carolinus** Sh.-Y. Hu. Low shrub, twigs nodding, 2–4 mm thick in the second year, bark exfoliating; current year's shoots glabrous; leaves tough, ovate, entire, 3–8 cm long, 2–4 cm wide, short acuminate, base usually rounded; flowers solitary or in 2's, occasionally 3–9, about 4–6 cm wide, petals obovate-rounded, stamens 60–90. BM 1748 (as *P. inodorus*) (= *P. inodorus* sensu Sims). USA, N. Carolina, Tennessee. 1810. Plate 154.

Fig. 273. **Philadelphus.** a. *P. mexicanus*; b. *P. coulteri*; c. *P. inodorus*; d. *P. inodorus* var. *laxus* (drawn for this book by Sh.-Y. Hu)

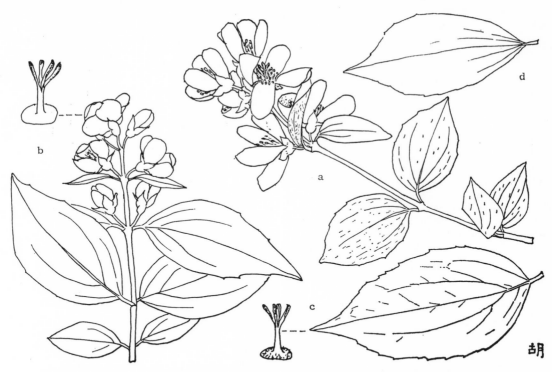

Fig. 274. **Philadelphus.** a. *P. lewisii;* b. *P. coronarius;* c. *P. caucasicus;* d. *P. lewisii* var. *gordonianus*
(drawn for this book by Sh.-Y. Hu)

var. **grandiflorus** (Willd.) Gray. Rounded shrub, shoots slender and nodding, bark of the second year shrubs brown and exfoliating; leaves ovate to oval-elliptic, 5–11 cm long, 2.5–5(6.5) cm wide, glabrous on both sides or very sparsely strigose, long acuminate, base acute, occasionally rounded, margins sparsely and coarsely serrate; flowers in 3's or solitary or grouped 7–9, stalks 3–15 mm long, sepals ovate, 8–12 mm long, long acuminate, corolla campanulate at first (!), later spreading plate-like, 4–5 cm wide, petals nearly circular at first, later oblong, pure white. BB 1862; SH 1: 234 g–j, 236e; JA 35: Pl. 3 (= *P. grandiflorus* Willd.; *P. speciosus* Koehne; *P. gloriosus* Beadle). Southeastern USA. 1811.

var. **laxus** Sh.-Y. Hu. Rounded shrub, twigs thin, nodding to the ground, bark exfoliating in the second year; leaves lanceolate or elliptic-lanceolate, 4–6 cm long, 1.5–2.5 cm wide, long acuminate, base cuneate to acute, sparsely appressed pubescent above, soft pubescent beneath primarily along the major veins, margins sparsely and finely dentate; flowers 4.5–5.5 cm wide, conspicuously cruciform(!), petals oblong, 1.5 cm wide. BR 25: 39; JA 35: Pl. 3 (= *P. laxus* Hort. ex Schrad.; *P. grandiflorus* var. *laxus* [Schrad.] Torr. & Gray). USA; N. Carolina. Around 1825. z5 Fig. 273. ✣

var. **strigose** Beadle. Shrub, to 2 m high, bark brown, exfoliating; leaves ovate to oval-elliptic, 3–7 cm long, 2–4 cm wide, entire or somewhat dentate, long acuminate, base obtuse or rounded, glabrous or somewhat strigose above, denser beneath, vein axils with tufts of pubescence; flowers solitary or in 3's, stalks and hypanthium glabrous, sepals ovate, exterior somewhat shaggy pubescent (!), corolla disk-form, 4–5 cm wide, petals nearly circular, 2 cm wide, stamens about 90 (= *P. strigosus* [Beadle] Rehd.). Southeastern USA. Around 1847.

P. insignis Carr. Shrub, erect, to 4 m high and wide, young shoots glabrous or pubescent, bark gray in the second year, smooth (occasionally brown and exfoliating); leaves ovate-elliptic, 3.5–8 cm long, 1.5–6 cm wide, short acuminate, base obtuse or acute or rounded, nearly entire to slightly dentate, evenly rough haired beneath or only on the venation; flowers in panicles, 3 flowered on lateral shoots, stalks 3–5 mm long, rough pubescent like the hypanthium and calyx, the latter occasionally also nearly glabrous, sepals ovate, corolla cruciform, 2.5–3.5 cm wide, petals oblong-obovate, stamens about 30, disk and styles glabrous, June–July. SH 1: 236g; JA 35: Pl. 3 (= *P. billardii* Koehne; *P.* 'Souvenir de Billard'). California. 1865. Long considered a garden hybrid. z7 Fig. 277.

P. intectus Beadle. Upright shrub, bark gray, smooth, second year shoots 3–4 mm thick, young shoots glabrous, not ciliate at the nodes (!); leaves ovate to oblong-elliptic, glabrous, 6–10 cm long on long shoots and 4–6 cm wide, 3–6 cm long and 1.5–3.5 cm wide on flowering shoots, acuminate, base rounded to obtuse, apical half sparsely and finely serrate, petiole 4–7 mm long; flowers grouped 5–9, stalks glabrous, hypanthium campanulate, glabrous, sepals ovate and caudate, 6 mm long, corolla plate-like, 3 cm wide, petals oblong-obovate, rounded at the apex and emarginate, disk and style glabrous, stigmas spathulate, June (= *P. pubescens* var. *intectus* [Beadle] A. H. Moore; *P. sanguineus* Hort.). USA; Kentucky, Tennessee, Oklahoma. 1894. z5

P. kansuensis (Rehd.) Sh.-Y. Hu. Shrub, 2 m high (in its habitat to 7 m), young shoots gray pubescent or quickly becoming glabrous, bark gray-brown in the second year, exfoliating; leaves ovate to oval-lanceolate, to 11 cm long and 6.5 cm wide on long shoots, on flowering shoots usually 3–5 cm long, 1–2 cm wide, nearly entire or finely serrate, acuminate, base obtuse to round, evenly bristly pubescent above, venation beneath densely strigose; flowers 5–7(11), fragrant, axis 2–8 cm long, somewhat frizzy pubescent, stalk 6 to 8 mm long, with gray outspread pubescence, as are the calyx and hypanthium, hairs thickened at the base, sepals triangular-ovate, 4 mm long, corolla nearly disk-form, 2.5 cm wide, petals oblong-rounded, slightly pubescent on the base of the dorsal side, disk only pubescent on the margin, otherwise glabrous as is the style, July. JA 35: Pl. 4 (= *P. pekinensis* var. *kansuensis* Rehd.). China; Kansu Province. 1926. z7

P. karwinskyanus Koehne. Evergreen shrub, to 4 m high in its habitat, slightly climbing, shoots long and nodding, the flowering shoots over 50 cm long, abundantly branched; leaves ovate, 4–7.5 cm long, 2–3.5 cm wide, long acuminate, base round and 5 veined, with 5–8 sharp, widely spaced teeth on either edge, both sides sparsely strigose-woolly; corolla 2.5–3 cm wide, petals obovate-rounded, glabrous on the inside, exterior somewhat pubescent on the midline, pure white, faintly scented, stamens about 45. JA 35: Pl. 2,4. Mexico; Oaxaca, in the mountains at 2000 m; cultivated in California and flowering there from August to May. z9 #

P. lasiogynus see: **P. schrenkii** var. **jackii**

P. latifolius see: **P. pubescens**

P. laxiflorus Rehd. Shrub, 2–3 m high, young shoots glabrous, yellow-brown, gray-brown to dark brown in the second year, bark exfoliating; leaves oval-elliptic to elliptic, 3–8 cm long, 1.5–3 cm wide, occasionally 2 cm long on weak shoots, acuminate, base obtuse (on long shoots also round), finely serrate, evenly bristly above and appressed pubescent (!), only the major veins pubescent beneath and often with axillary tufts; flowers 7 to 11, axis 6–12 cm long, glabrous, calyx and calyx cup glabrous, sepals ovate, 6 mm long, corolla disk-form, 2.5–3 cm wide, petals nearly circular, stamens 30–35, disk and style glabrous, June. China; Shansi, Shensi, Kansu, Hupeh Provinces. 1916. z6

P. lemoinei densiflorus see: List of Hybrids and Cultivars, **'Avalanche'**

P. × *lemoinei* see: List of Hybrids and Cultivars

P. lewisii Pursh. Erect, stiffly branched shrub, to 3 m high, young shoots long and glabrous, bark brown in the second year, occasionally yellow-brown, with horizontal grooves, eventually exfoliating; leaves ovate, 4–5.5 cm long, 2–3.5 cm wide, acute to short acuminate, base round, nearly entire or indistinctly dentate, more coarsely dentate on long shoots, bright green and somewhat strigose on the venation above, underside

more densely strigose with axillary tufts, margins ciliate; flowers 7–9–11, with fewer flowers on weak shoots, sepals ovate, 5–6 mm long, corolla cruciform, 3–4.5 cm wide, petals oblong, rounded at the apex and emarginate, pure white, without a fragrance, stamens 28–35, anthers glabrous (!), disk glabrous (!), styles shorter than the longer stamens, somewhat parted only at the apex, June–July. DL 1: 181; SH 1: 237 s–u; JA 35: Pl. 2. W. Canada, western USA. 1823. Fig. 274. z5 See also *P. californicus.*

var. *californicus* see: **P. californicus**

var. *cordifolius* see: **P. cordifolius**

var. **gordonianus** (Lindl.) Koehne. Upright shrub, to 4 m high, broad spreading habit, twigs slightly pubescent, bark gray-yellow, not exfoliating; leaves ovate, 5–6 cm long on long shoots, 2.5–4 cm wide, evenly sparsely pubescent beneath, with 4–5 coarse teeth with outward pointing tips on either edge, short acuminate, base round, leaves ovate on the flowering shoots, 3–5 cm long, 1.5–3.5 cm wide, nearly entire or totally finely serrate, both sides nearly totally glabrous; flowers in groups 3–9(11) in dense racemes, pure white, slightly scented, sepals and calyx cup glabrous, corolla disk-form, 3.5–4.5 cm wide, petals obovate, rounded at the apex, June–July. MS 136; BR 25: 32; SH 1: 236k (as *P. columbianus*) (= *P. gordonianus* Lindl.; *P. columbianus* Koehne). Western USA, from Washington to California. Around 1838. Fig. 274, 275.

var. **helleri** (Rydb.) Sh.-Y. Hu. Bark of the second year shoots brown to ash-gray, with horizontal grooves, otherwise smooth, young shoots somewhat pubescent, later glabrous; leaves oblong-ovate, 3–6 cm long, 1.5–2.8 cm wide, acute to obtuse, base round, nearly entire to indistinctly serrate, venation somewhat pubescent; flowers grouped 7–9 (occasionally 5), hypanthium and calyx somewhat pubescent, corolla cruciform, 3–4 cm wide, petals oblong, round to obtuse at the apex, disk and style glabrous (= *P. helleri* Rydb.). West coast of North America; British Columbia, Washington to California. Around 1800.

Fig. 275. *Philadelphus lewisii* var. *gordonianus* (from Dippel)

Fig. 276. **Philadelphus.** a. *P. sericanthus*; b. *P. pekinensis*; c. *P. satsumi*; d. *P. delavayi*
(drawn for this text by Sh.-Y. Hu)

P. magdalenae see: **P. sericanthus** and **P. subcanus** var. **magdalenae**

P. magnificus see: **P. floridus**

P. mandshuricus see: **P. schrenkii** var. **manshuricus**

P. matsumuranus see: **P. satsumanus**

P. × maximus see: List of Hybrids and Cultivars

P. mexicanus Schlechtd. Evergreen climbing shrub, to 5 m high, twigs long arching, loose bristly pubescent to nearly glabrous, second year twigs chestnut-brown, 3–4 mm thick; leaves ovate, very large, 5–11.5 cm long, 2–5 cm wide, both sides sparsely strigose, long acuminate, base round to slightly cordate, usually 5 veined, nearly entire or with 1–5 tiny teeth on either side, petiole 8–10 mm long; flowers solitary, yellowish-white, very fragrant, sepals leaflike, 1–2 cm long, corolla 3–4 cm wide, petals rounded, pubescent on both sides, disk and style pubescent, June. BM 7600; SH 1: 234 b to b2; JA 35: Pl. 4. Mexico, Guatemala; cultivated in California. z9 Fig. 273. #

P. microphyllus Gray. Slow growing, graceful, upright shrub, scarcely over 1 m high, young shoots sparse and appressed pubescent, second year bark dark brown, glossy and exfoliating; leaves oval-elliptic or occasionally nearly lanceolate, 1–1.5 cm long, 5–7 mm wide, entire and ciliate, deep green and glossy above, bluish and strigose beneath, usually obtuse at both ends,

petiole 2 mm long; flowering shoots 1.5–3(4) cm long with 3–4 leaf pairs, flowers solitary, occasionally in 2's, pure white, very fragrant, hypanthium campanulate, base and margins somewhat pubescent, sepals lanceolate, corolla cruciform, 3 cm wide, petals obovate-oblong, round and emarginate on the apex, stamens about 32, June. SH 1: 234 e to f2; VT 310; JA 35: Pl. 1,2,4. Southwestern USA. 1883. This species is often used in hybridizing; the origin of most of the slow growing cultivars. z6 Fig. 272. ⊕

P. × monstrosus see: List of Hybrids and Cultivars

P. nepalensis see: **P. tomentosus** and **P. triflorus**

P. × nivalis see: List of Hybrids and Cultivars

P. pallidus see: **P. coronarius**

P. pekinensis Rupr. A dense shrub to 2 m high, divaricate habit, twigs totally glabrous, often reddish, the second year bark is chestnut-brown and exfoliating; leaves ovate, 6–9 cm long and 2.5–4.6 cm wide on long shoots, long acuminate, base round, serrate, both sides usually totally glabrous, on flowering shoots 3–4 (occasionally to 7) cm long, 1.5–2.5 cm wide, glabrous on both sides or with with some axillary tufts on the venation beneath, serrate, sometimes nearly entire, petiole reddish-brown; flowers 5–9 in dense, glabrous racemes, occasionally only 3, stalk 3–6 mm long, glabrous calyx and calyx cup glabrous, sepals ovate, 4 mm long, corolla

Fig. 277. **Philadelphus.** a. *P. insignis*; b. *P. tomentosus*; c. *P. hirsutus*; d. *P. floridus*
(drawn for this book by Sh.-Y. Hu)

plate-like, 2–3 cm wide, very fragrant, petals obovate, about 1 cm long, yellowish-white, purple striped on the dorsal side before fully opening (!), disk and style glabrous, the latter somewhat incised only at the apex, late May–early June. SH 1: 237 p–r, 238 e–f; JA 35: plate 3, 4; 37: plate 5 (= *P. rubricaulis* Carr.). China; Hopei, Shansi, Szechwan Provinces. 1865. z6 Fig. 276.

var. *branchybotrys* see: **P. brachybotrys**

var. *dasycalyx* see: **P. dasycalyx**

var. *kansuensis* see: **P. kansuensis**

P. × pendulifolius see: List of Hybrids and Cultivars

P. × polyanthus see: List of Hybrids and Cultivars

P. pubescens Loisel. A tall, narrowly upright shrub, reaching 5 m high, twigs light gray, not exfoliating, young shoots green and glabrous; leaves ovate to broadly ovate, 4–8 on long shoots (occasionally to 15) cm long, 3.5–6.5 (occasionally to 10) cm wide, 4–8 cm long on flowering shoots and 3–5.5 cm wide, rounded at the base (!), abruptly short acuminate, sparsely dentate to nearly entire, glabrous above except for the rough pubescent venation, underside densely gray strigose; flowers grouped 7–9 (occasionally 5 or 11), flower stalks, calyx and hypanthium densely pubescent, sepals ovate, 6–7 mm long, corolla cruciform, 3.5 cm wide, petals obovate-oblong, about 15 mm long, cream-white, without fragrance, stamens about 37, disk and style

glabrous, June–July. VT 311; SH 1: 235 f–m; BR 570 (as *P. grandiflorus*); SH 1: 237 i–l (as *P. latifolius*); JA 35: Pl. 2 (= *P. grandiflorus* sensu Ker, non Willd.; *P. latifolius* Schrad.). Southeastern USA. Around 1820. This species may be offered by various names; flowers very abundantly. z6 Plate 156; Fig. 272. ⊕

var. *intectus* see: **P. intectus**

var. **verrucosus** (Schrad.) Sh.-Y. Hu. Tall shrub, branches gray, young shoots glabrous; leaves elliptic, oval-elliptic or oblong-elliptic, 6.5–15 cm long and 4–9 cm wide, 3.5–10 cm long and 1.5–5 cm wide on flowering shoots, abruptly short acuminate, base acute or obtuse (!), nearly entire or sparsely serrate or dentate, glabrous above, densely gray strigose beneath; flowers grouped 7–9 (or only 5) together (= *P. verrucosus* Schrad. ex DC.; *P. rhombifolius* Rehd.). USA; Illinois, Missouri, Tennessee, Arkansas, Oklahoma. 1828.

P. purpurascens (Koehne) Rehd. Shrub, 2–4 m high, young shoots glabrous, second year bark brown or gray, smooth; leaves ovate to oval-lanceolate, on long shoots 3.5–8 cm long and 1.5–4 cm wide, 1.5–6 cm long and 0.5–3 cm wide on flowering shoots, strigose pubescent above (often only on the apical half), pubescent only on the venation beneath, serrate; flowers usually 5–9, hypanthium and calyx glabrous, green with a trace of red (!), somewhat bluish pruinose, sepals ovate, 4–6 mm long, corolla campanulate (!), 3(4) cm wide, petals oblong, 1.5–2 cm long, very fragrant, stamens 25–30, style and disk glabrous, style 5 mm long, stigma linear,

Fig. 278. **Philadelphus.** Right *P. schrenkii*; left *P. schrenkii* var. *jackii* (from Nakai)

June; fruit obovoid. BM 8324 (as *P. delavayi* sensu Hutchins., non L. Henry); JA 35: Pl. 3 (= *P. delavayi* var. *calvescens* Rehd.). China; Sikang, Yunnan Provinces. 1911. z6 Plate 154. ✧

var. **venustus** (Koehne) Sh.-Y. Hu. Shrub, 2–3 m high, young shoots densely pubescent, bark gray or brown in the second year; leaves ovate, 1.5–4.5 cm long and 0.5–2 cm wide on flowering shoots, acuminate, base acute to round, appressed strigose above, likewise on the venation beneath; flowers grouped 5–7, flower stalks 3 to 10 mm long, glabrous, hypanthium and calyx glabrous, corolla somewhat campanulate, petals rounded-oblong, about 1.3 cm long, 1 cm wide, style and disk glabrous (= *P. venustus* Koehne). China; SW. Szechwan, Yunnan Provinces. 1906.

P. rhombifolius see: **P. pubescans** var. **verrucosus**

P. robustus see: **P. tenuifolius**

P. rubricaulis see: **P. pekinensis**

P. × purpureo-maculatus see: List of Hybrids and Cultivars

P. salicifolius Koch. Shrub, dense and compact, about 1 m high; leaves lanceolate, on flowering shoots 3.5–7 cm long and 1–1.5 cm wide, acuminate at both ends, both sides usually totally glabrous, sparsely and indistinctly finely serrate, 8–13 cm long on long shoots, 5–6 cm wide, long acuminate, base acute to round; flowers in 3's or solitary or in few flowered racemes, cream-white, slightly fragrant, flower stalk glabrous to somewhat pubescent, calyx and hypanthium glabrous, sepals oval-lanceolate, 5–6 mm long, corolla stellate, 2–3 cm wide,

petals oblong-lanceolate, to 14 mm long and 5 mm wide, petal tips acute and pubescent, stamens about 21, disk pubescent, style glabrous, incised nearly to the base, June (= *P. coronarius* var. *salicifolius* [Koch] Jäger). Origin unknown, possibly developed by L. Späth of Berlin, W. Germany, at least disseminated from there. 1869. z5 Fig. 271.

P. sanguineus see: **P. intectus**

P. satsumanus Sieb. ex Miq. Shrub, upright, 1.5–2.5 m high, young shoots pubescent to glabrous, bark brown in the second year, smooth, later exfoliating; leaves regularly formed above and sparsely pubescent, tomentose beneath, ovate on long shoots, 5–10 cm long, 2.5–6 cm wide, apex cordate (!), base usually round, ovate on flowering shoots, occasionally elliptic, 3–7 cm long, 1–3.5 cm wide, acuminate, sparsely and finely serrate, base usually obtuse; flowers grouped 5–7 in racemes, stalks 4–6 mm long, hypanthium tomentose, sepals ovate, 4 mm long, often tomentose, corolla 2.5–3 cm wide, petals oval-oblong, stamens 25–30, styles glabrous, June. JA 35: Pl. 3 (= *P. matsumuranus* Koehne; *P. satsumanus* var. *nikoensis* Rehd.; *P. gordonianus* var. *parviflorus* Dipp.). Japan. 1890. z6 See also *P. satsumi*.

P. satsumi Sieb. ex Lindl. & Paxt. Similar to *P. satsumanus*, but with densely tomentose leaf undersides. Upright shrub, 1.5–3 m high, young shoots thin, quickly becoming glabrous, second year bark brown and later exfoliating; leaves with sparse bristly pubescence above to glabrous, rough pubescence on the venation beneath, pubescent tufts in the vein axils, ovate to broadly elliptic on long shoots, 6–9 cm long, 3–5 cm wide, sparsely and coarsely serrate, caudate tipped, base obtuse to round, ovate to more lanceolate on flowering shoots, 4.5–7 cm long, 1.5–4.5 cm wide, acuminate to caudate, base acute to obtuse; flowers slightly fragrant, grouped in 5–7 in racemes, occasionally only 2–3, terminal, flower stalks 6–8 mm long, pubescent, hypanthium pubescent only on the margins and the venation, otherwise glabrous, hairs short, crispate, golden, corolla 3 cm wide, petals oblong-obovate, stamens 30, style glabrous, June; fruit ellipsoid. SH 1: 237 f–h; MJ 1461 (= *P. satsumanus* sensu Koch, sensu Rehd., non Sieb.). Japan. 1882. z6 Fig 276.

P. schrenkii Rupr. Upright shrub, 2–4 m high, densely branched, young shoots rough haired, often becoming glabrous, second year bark gray, smooth; leaves ovate, occasionally oval-elliptic, 7–13 cm long on long shoots, 4–7 cm wide, on flowering shoots 4.5–7.5 cm long, 1.5–4 cm wide, acuminate, base obtuse to acute, sparsely dentate to nearly entire, glabrous or occasionally somewhat pubescent above; flowers very fragrant, 5–7 in racemes, stalks 6–13 mm long, densely tomentose, likewise the hypanthium, sepals ovate, 4–7 mm long, glabrous or soon glabrous, corolla 2.5–3.5 cm wide, petals oblong-obovate, stamens 25–30, disk glabrous, style pubescent, inscised halfway or deeper, late June–early July; fruit ellipsoid. SH 1: 237 w–w2. E. Siberia, Manchuria, Korea. 1881. z5 Fig. 278. ☺

var. **jackii** Koehne. Leaves of the flowering shoots oval-elliptic, 3.5–7 cm long, 1.5–3.5 cm wide, sparsely pubescent on both sides, particularly on the venation beneath, short acuminate, base acute to obtuse, sparsely serrate; flowers very fragrant, grouped 5–7 together, corolla 2.5–3.5 cm wide, petals oblong-obovate, stamens 25–35, style and disk pubescent, mid-June (= *P. lasiogynus* Nakai). Manchuria, Korea. 1905. Fig. 278. ☺

var. **manshuricus** (Maxim.) Kitagawa. Leaves broadly-ovate, to 8 cm long and 5 cm wide, short acuminate, finely and sparsely serrate, base round or obtuse; flowers 5–7 together, corolla 3–3.5 cm wide, petals rounded, calyx cup pubescent, as are the style and disk (= *P. mandshuricus* [Maxim.] Nakai; *P. amurensis*; *P. schrenkii* sensu Rehd.). Manchuria.

P. sericanthus Koehne. Shrub, 1–3 m high, young shoots glabrous or soon glabrous, bark in the second year gray or gray-brown, tight, exfoliating in very small pieces; leaves oval-elliptic or elliptic-lanceolate, 4–11 cm long, 1.5–5 cm wide, acuminate, base obtuse to round, serrate, sparsely strigose above, only on the venation beneath; flowers grouped 7–15, only 5 or 3 on weak shoots, rachis glabrous or soon so, flower stalks 6–12 mm long, strigose, calyx and hypanthium densely white pubescent (!), sepals ovate, 6–7 mm long, corolla flat campanulate-disk form, 2.5 cm wide, petals obovate, exterior strigose pubescent at the base, stamens about 35, anthers glabrous (!), without fragrance, disk and style glabrous, style incised to the midpoint, June. BM 8941; SH 1: 236p, 237 m–o; JA 35: Pl. 3, 4 (= *P. magdalenae* sensu Rehd. non Koehne; *P. sericanthus* var. *rosthornii* Koehne). Central China. 1897. z6 Plate 154; Fig. 276.

var. *rehderianus* see: **P. subcanus**

var. *rosthornii* see: **P. sericanthus**

P. shikokianus Nakai. Upright, thin branched shrub, young shoots 1–2 mm thick, pubescent, bark brown in the second year, smooth or later exfoliating; leaves glabrous or sparsely rough haired above, glabrous beneath except for the venation and vein axils, ovate to broadly elliptic on long shoots, 8–10 cm long and 3.5–6 cm wide, caudate, base acute to round, sparsely serrate, lanceolate to ovate on flowering shoots, 4–8 cm long, 1.5–3.5 cm wide, acuminate, base acute to round, sparsely and finely serrate; flowers grouped 5–7, occasionally 9, stalks 8–12 mm long, pubescent, likewise the hypanthium and sepals, style and disk, corolla 2–2.5 cm wide, petals oblong, June (= *P. yokohamae* Sieb.). Japan; Shikoku. Around 1881. z6

P. speciosus see: **P. floridus**

P. × splendens see: List of Hybrids and Cultivars

P. strigosus see: **P. inodorus** var. **strigosus**

P. subcanus Koehne. Steeply erect shrub, 3–6 m high, young shoots brown, glabrous or quickly becoming glabrous, brown to gray-brown in the second year, bark smooth and later exfoliating; leaves ovate to oval-lanceolate, 4–14 cm long, 1.5–7 cm wide, acuminate, base round or obtuse, finely dentate, finely serrate on long shoots, sparsely rough haired above with erect pubescence, venation beneath densely pubescent; flowers grouped 5–29, usually 9–11 in racemens, rachis 2.5–22 cm long, pubescent, but later becoming glabrous, stalk 5–10 mm long, densely curly pubescent, sepals ovate, 6–7 mm long, corolla disk-form, 2.5–3 cm wide, pure white, slightly fragrant, petals circular to obovate, usually bristly pubescent at the base on the dorsal side, disk pubescent, also the basal half of the style, June (4 weeks earlier than the similar *P. incanus*). NF 6: 237 (= *P. wilsonii* Koehne; *P. sericanthus* var. *rehderianus* Koehne; *P. subcanus* var. *wilsonii* [Koehne] Rehd.). China; Szechwan, Yunnan Provinces. z6 1908.

var. **magdalenae** Sh.-Y. Hu. Shrub, to 4 m high, young shoots sparsely pubescent or soon glabrous, gray-brown in the second year, later exfoliating; leaves ovate, 3–6 cm long, 1–3 cm wide, short acuminate, base round, pubescent on both sides, rough above with short, erect hairs, pubescence beneath more appressed; flowers grouped 5–11 in racemes, occasionally panicles, pure white, slightly fragrant, stalks, hypanthium and the calyx slightly pubescent with curly hairs (!), corolla nearly disk-form, 2 cm wide, petals not pubescent at the base on the dorsal side (!), stamens about 28, style pubescent, June. JA 35: Pl. 3 (= *P. magdalenae* Koehne). China; Szechwan, Yunnan Provinces. 1906.

var. *wilsonii* see: **P. subcanus**

P. tenuifolius Rupr. & Maxim. Shrub, erect, 1–3 m high, twigs nodding, young shoots with appressed pubescence, second year bark usually gray-brown and exfoliating; leaves ovate on long shoots, to 10.5 cm long and 6 cm wide, sparsely denate, glabrous except for the loosely pubescent major veins beneath, base obtuse, ovate to elliptic on the flowering shoots, 4–6.5 cm long, 2–3 cm wide, acute, base obtuse to rounded, sparsely serrate to nearly entire, somewhat pubescent above, underside with only axillary pubescence; flowers 5–7 in racemes, stalks densely pubescent, hypanthium only pubescent at the base, sepals ovate, 5 mm long, sparsely pubescent, corola 2.5–3.4 cm wide, petals obovate-oblong, slightly fragrant, stamens 25–30, style and disk glabrous, style half split, June. JA 35: Pl. 3, 4 (= *P. coronarius* var. *tenuifolius* Maxim.; *P. viksnei* Zamelis; *P. robustus* Nakai). E. Siberia, Korea. 1890. z5

P. tomentosus Wall. Shrub, 2–3 m high, similar to *P. coronarius*, the previous year's shoots cinnamon-brown, bark later exfoliating, young shoots glabrous or quickly becoming so; leaves ovate, occasionally lanceolate, 4–10 cm long, 2–5 cm wide, caudately acuminate, base round 5–7 veined, soon glabrous above with regular appressed tomentum beneath; flowers cream-white, fragrant, grouped 5–7, occasionally 3, in false racemes, stalk 6–11 mm long, calyx and hypanthium more or less glabrous or often with a few slight hairs, sepals ovate, 5 mm long, corolla cruciform, 1.5–2.5 cm wide, petals obovate-oblong, 0.5–1 cm long, disk glabrous, styles glabrous, incised only on the apical third, stigmas clavate, June. DL 3: 177; SH 1: 237 c–e; JA 35: Pl. 3, 4; 37: Pl. 5 (= *P. nepalensis* Wall.; *P. coronarius* var. *tomentosus* Hook. f. & Thoms.). India; Himalayas. 1822. z6 Fig. 277.

P. triflorus Wall. Very closely related to *P. tomentosus*, but with the leaf undersides glabrous. Upright shrub, previous year's shoots glabrous; leaves ovate, 6–9 cm long, 3–5 cm wide, caudate-acuminate, base round, glabrous on both sides or somewhat pubescent on the venation beneath; flowers usually in 5's in racemes, occasionally grouped 3 or 7 together, stalks 10–15 mm long, glabrous, calyx and hypanthium glabrous, sepals ovate, 5 mm long, corolla nearly disk-form, 2–3 cm wide, petals ovate, disk and style glabrous, June; seeds quite short caudate. SH 1: 238 g (= *P. nepalensis* Koehne). India; Himalayas. Before 1893. z6

P. venustus see: **P. purpurascens** var. **venustus**

P. verrucosus see: **P. pubescens** var. **verrucosus**

P. viksnei see: **P. tenuifolius**

P. villosus see: **P. hirsutis**

P. × virginalis see: List of Hybrids and Cultivars

P. wilsonii see: **P. subcanus**

P. yokohamae see: **P. shikokianus**

P. zeyheri see: **P. coronarius**

Key to the hybrids and cultivars (from Sh.-Y. Hu)

● Flowers always white, without a dark spot in the middle;
  ★ Hypanthium glabrous;
    > Tall shrubs, 2–2.5 m high; leaves large, 3–11 cm long on flowering shoots, occasionally smaller;
      + Inflorescence compact, stalks of the lower flowers not elongated; flowers single; corolla disk-form; petals nearly circular:
           P. × splendens

      ++ Inflorescence loose, stalks of the lower flowers elongated, often branched; flowers in cymes;
        × Flowers single, corolla stellate, petals elliptic, acute at the apex:
           P. × falconeri

        ×× Flowers semidouble, double or single, corolla disk-form or cruciform, petals oblong or obovate, rounded at the apex:
           P. × cymosus

    >> Low shrubs, 1–1.5 m high; leaves small, usually under 3 cm long on the flowering shoots;
      + Flowers numerous; corolla cruciform or double; petals over 1 cm long:
           P. × lemoinei

      ++ Flowers few, corolla campanulate; petals shorter than 1 cm;
        × Leaves ovate to oblong, 1–2 cm wide, obtuse at both ends; bark smooth:
           'Patricia'

        ×× Leaves oval-lanceolate, 0.5–1 cm wide, base acute to rounded, apexes short acuminate; bark exfoliating:
           'Thelma'

  ★★ Hypanthium distinctly pubescent;
    > Petals oblong, obovate or rounded, 5 mm wide or more;
      + Shoots knotty, leaves glossy, pendulous:
           P. × pendulifolius

      ++ Shoots normal, not knotty; leaves not pendulous;
        × Flowers in 3's or 5's, in cymes or umbellate panicles, stalks of the basal flowers elongated or branched;
          § Lateral buds enclosed:
           P. × polyanthus

§§ Lateral buds more or less exserted:
   P. × burkwoodii

×× Flowers grouped in 9's or 7's, occasionally 5 or 3, in racemes, the stalks of the lowers flowers not elongated;
   § Flowers double or semidouble; plants 1–1.5 m high, twigs stiff:
      P. × virginalis

   §§ flowers single; shrubs tall for the genus; twigs usually nodding;
      > Particularly tall shrubs, 4–5 m high; stamens all sterile; fruits not observed:
         P. × maximus

      >> Medium high shrubs, 2–3 m high; stamens all fertile; always fruiting;
         * Hypanthium and sepals rough haired, hairs long and dense, some erect; corolla disk-form, petals nearly circular:
            P. × nivalis

         ** Hypanthium and sepals strigose, hairs short and appressed; corolla nearly disk-form; petals obovate:
            P. × congestus

         *** Hypanthium sparsely pubescent, hairs usually erect; corolla cruciform, petals oblong;
            ☆ Leaves ovate; flowers grouped 7–9 in loosely segmented racemes, the lower flower pairs in the axils of normal leaves; bark of the previous year's shoots gray, smooth:
               P. × monstrosus

            ☆☆ Leaves elliptic; flowers grouped in 5's or 3's or solitary, compact; bark of the previous year's shoots chestnut brown, exfoliating:
               'Slavinii'

      >> Petals oblanceolate, 3 mm wide:
         'Stenopetalus'

●● Flowers white, center pink to purple;
   > Flowers grouped 7 in racemes, hypanthium and sepals glabrous:
      'Maculiflorus'

   >> Flowers in 3's or solitary, occasionally 5's; hypanthium sparsely or only partly pubescent;
      + Flowers disk- or cruciform; chromosome count diploid:
         P. × purpureo-maculatus

      ++ Flowers nearly campanulate; chromosome count triploid;
         § Petals pubescent on the inside; anthers fertile:
            'Bicolore'

         §§ Petals glabrous, anthers sterile:
            'Belle Etoile'

List of the hybrids and cultivars
(for forms of P. coronarius, see the species description)

'Albâtre' (Lemoine 1912) (P. virginalis). More graceful habit than 'Virginal', broader and more compact, about 1.2 m high; leaves smaller; flowers grouped 5–7 together, double, pure white, moderate fragrance. NF 6: 42; Gs 1940: 104; AB 25: Pl. 7. According to Sh.-Y. Hu, hardly differs from 'Bouquet Blanc'. ⊕

'Amalthée' (Lemoine 1923) (P. cymosus). Upright habit; leaves oval-elliptic, 3–5 cm long, 11–18 mm wide, entire or with 1–2 tiny teeth on either side, glossy green above, loosely strigose pubescent beneath; flowers grouped 1–3, corolla disk-form, 3 cm wide, petals of spent flowers coloring reddish on the base, slightly fragrant.

'Argentine' (Lemoine 1913) (P. virginalis). Shrub, about 1 m high, young shoots glabrous, bark later becoming gray; leaves broadly ovate, 3–4.5 cm long, 1.5–3 cm wide, base round, with 2–4 teeth on either side or entire; flowers usually in 3's, corolla densely double, usually with about 30 petals, these round or obtuse on the apex, very fragrant, flowers abundantly, usually with 6 sepals. JA 35: Pl. 3. ⊕

'Atlas' (Lemoine 1923) (P. polyanthus). Strong growing, young shoots glabrous, bark brown in the second year; leaves ovate, 6–9 cm long, 3–6 cm wide, short acuminate, base obtuse, smooth above, strigose pubescent beneath; flowers 5–7 in racemes, calyx and sepals pubescent, corolla cruciform, 3.5 cm wide, petals oblong, 1.5 cm long, 1 cm wide, stamens fertile, disk and style glabrous, slightly fragrant.

'Avalanche' (Lemoine 1896) (P. lemoinei). Upright habit, 1–2 m high, twigs thin, nodding, bark gray; leaves elliptic, 2–2.5 cm long, 5–8 mm wide, entire, acute at both ends, glabrous on both sides; flowers usually in 7's, milk-white, single, corolla cruciform, 2–2.5 cm wide, fragrant, petals obovate, round at the apex, flowers abundantly. GC 21: 89; MG 1907: 379; AB 25: Pl. 7 (= P. lemoinei densiflorus Lemoine). ⊕

'Bannière' (Lemoine 1906) (P. cymosus). Upright shrub, to 2 m high, twigs long arching; leaves ovate, 5–7 cm long, 2.5–4 cm wide, short acuminate, base round, with 4–7 teeth on either edge, light green, glabrous beneath, vein axils pubescent; flowers grouped 1–5, slightly double, corolla 4–4.5 cm wide, outer petals obovate, 18 mm long, rounded at the apex, anthers sterile, partly petaloid, slightly fragrant, flowers abundantly. NF 6: 41. ⊕

'**Beauclerc**' (Palmer). Strong grower; flowers abundantly, flowers grouped 5 or 7 together, single to semidouble, about 5 cm wide, petals rounded, limb crispate, white, base reddish, moderate fragrance; aneuploid (diploid × triploid). GC 130: 21. Developed from 'Burfordiensis' × 'Sybille' by Lewis Palmer of Hedbourne, Worthy Grange, Winchester, England; brought into the trade in 1942. ⊕

'**Belle Etoile**' (Lemoine 1930). Compact upright habit, to 1.5 m high, twigs rather stiff, young shoots pubescent, winter buds exserted, second year bark brown, later exfoliating; leaves ovate, 1–4–5 cm long, 1–2 cm wide, short acuminate, base round, somewhat rough haired above, strigose beneath; flowers in 3's or solitary, calyx and sepals sparsely pubescent, sepals 1 cm long, corolla somewhat campanulate, 4.5–5 cm wide, nearly square in outline, petals white with a purple basal spot, ovate, 2.2 cm long, glabrous on both sides, slightly fringed, somewhat fragrant, stamens about 27, sterile, disk and style glabrous. JRHS 1951: 144, 272; NF 6; 239; JA 35: Pl. 3. ⊕

'**Bicolore**' (Lemoine 1918). Shrub, to 1 m high, young shoots somewhat pubescent, winter buds somewhat exserted; leaves oval-oblong, 2.5–4 cm long, 1 to 2 cm wide, acuminate, base obtuse, with 1–3 teeth on either edge in about the middle, somewhat rough above, strigose beneath; flowers solitary, occasionally in 3's, calyx and sepals pubescent, sepals leaflike, 11 mm long, corolla somewhat campanulate, 4.5–5 cm wide, petals obovate-rounded, white on the apical half, center yellowish and finely pubescent, base purple, stamens about 29, fertile (!), slightly fragrant. JA 35: Pl. 3, 4. Very attractive hybrid. ⊕

'**Boule d'Argent**' (Lemoine 1893) (*P. polyanthus*). Low, bushy growing, about 1.2 m high, twigs thin, sparsely pubescent when young, soon glabrous, bark brown in the second year, later exfoliating; leaves ovate, 3–4.5 cm long, 1.5–2 cm wide, acute, base rounded, only the venation slightly pubescent beneath; flowers grouped 5–7, loosely to densely double, in cymes, calyx somewhat pubescent, sepals glabrous, ovate, 5–6 mm long, corolla 3.5 cm wide, pure white, slightly fragrant, anthers usually sterile. GC 18: 18; 23: 331. ⊕

'**Bouquet Blanc**' (Lemoine 1903) (*P. cymosus*). Upright shrub, 1.5–2 m high, twigs nodding, young shoots scattered pubescent, bark gray-brown in the second year, smooth; leaves small, ovate, 2–5 cm long, 1–3 cm wide, base round, nearly entire, venation pubescent beneath; flowers grouped 7–9 together, double, about 2.5 cm wide, milk-white, stamens partly fertile, moderately fragrant, flowers very abundantly. Gw 17: 10; AB 25: Pl. 8. ⊕

'**Burfordensis**' (Lawrence) (*P. virginalis*). To 3 m high; flowers grouped 5–7 together, corolla shell-form, about 5 cm wide, pure white, fragrant, flowers very abundantly. Originated around 1920 as a sport on 'Virginal' at Lawrence Nursery in Burford, England. ⊕

'**Burkwoodii**' (Burkwood & Skipwith) ('Etoile Rose' × 'Virginal'). Small shrub, winter buds exserted, bark blackish in the second year, smooth, slowly exfoliating, young shoots strigose; leaves oval-elliptic, 3.5–6.6 cm long and 1.5–3 cm wide on long shoots, glabrous above, sparsely pubescent beneath; flowers grouped 1–3 or 5 together, sepals and hypanthium strigose, sepals ovate, caudate, to 1.5 cm long, corolla cruciform, to 4.5 cm wide, almost like a *Clematis montana* in appearance, petals narrow, acute, white with a violet-pink basal spot, interior sparsely pubescent, disk and style glabrous, incised to the base, moderately fragrant. Developed in England in 1929.

'**Candelabre**' (Lemoine 1894) (*P. lemoine*). Slow growing, low shrub, resembling *P. microphyllus* in habit, young shoots pubescent, bark chestnut-brown in the second year, later exfoliating; leaves ovate, 1.5–2.5 cm long, 6–11 mm wide, acuminate, base rounded, entire, venation beneath long haired; flowers solitary, occasionally in 3's, loosely arranged (candelabra-like), single, corolla nearly cup-form, 3.5 cm wide, later spreading cruciform, petals ovate, somewhat undulate, obtuse and somewhat incised at the apex, slightly fragrant. Very attractive form.

"Coles Glorious" has proven to be **P. intectus**! (JA 35: Pl. 3.)

P. × congestus Rehd. (= *P. inodorus* var. *laxus* × *P. pubescens* var. *verrucosus* ?). Shrub, drooping, young shoots pubescent, bark chestnut-brown in the second year, exfoliating; leaves elliptic, 3–3.5 cm long and 1–2.5 cm wide, acute at both ends, strigose pubescent beneath; flowers usually 5 in short racemes, occasionally in 3's or solitary, corolla nearly disk-form, 3–4 cm wide, petals obovate, hypanthium slightly pubescent, style shorter than the stamens, stigmas as wide as the anthers, June. Developed by Späth of Berlin, W. Germany and disseminated under the name *P. inodorus speciosus grandiflorus*. Before 1912.

'**Conquête**' (Lemoine 1903) (*P. cymosus*). Upright, outspread shrub, scarcely over 1 m high; leaves oval-lanceolate, 5.5–6.5 cm long (8–10 cm long on long shoots) and 2–3 cm wide, acuminate, base obtuse, entire, or with a few fine teeth (long shoots more coarsely toothed), light green, somewhat pubescent beneath; flowers in 3's or 5's, single to somewhat double, corolla cruciform, petals obovate, rounded at the apex and somewhat emarginate, pure white, fragrant, style parted to the base, stigmas small, calyx totally glabrous. Gw 17: 102. ⊕

'**Coupe d'Argent**' (Lemoine 1915) (*P. lemoinei*). Shrub, upright, nodding, bark brown, exfoliating; leaves ovate, short acuminate, base rounded, with 2–4 tiny teeth on either side, major veins sparsely pubescent beneath; flowers in 3's, occasionally in 5's, corolla disk-form, 3 cm wide, single, pure white, petals circular, 15 mm wide, slightly fragrant.

P. × cymosus Rehd. (parentage unknown). Strong growing to compact forms, bark exfoliating; leaves ovate, sparsely dentate, pubescent beneath, particularly on the venation; flowers in 3's or solitary or to 5 in cymose racemes, stalks of the lower raceme pairs elongated, often also branched, hypanthium glabrous (!), sepals leaflike, oval-lanceolate, style usually incised to the base (= *P. floribundus* Schrad.). 1852.

Includes the following cultivars (the descriptions may be found in alphabetical order):
'Amalthée', 'Bannière', 'Bouquet Blanc', 'Conquête', 'Dresden', 'Mer de Glace', 'Nué Blanche', 'Perle Blanche', 'Rosace', 'Velleda', 'Umbellatus'.

'**Dame Blanche**' (Lemoine 1910) (*P. lemoinei*). Low growing, compact, graceful, bark blackish, later exfoliating; leaves ovate, 1.5–2 cm long, 7–8 mm wide, acute, base obtuse to round, major veins sparsely pubescent beneath; flowers in 5's, occasionally in 2's, corolla disk-form, 14–18 mm wide, petals obovate, 8 mm long, very fragrant, flowers abundantly.

'**Dresden**' (Löbner) (*P. cymosus*). Low shrub, bushy habit, bark brown; leaves oval-elliptic, 2.5–5.5 cm long, 1.5–2.5 cm wide, acuminate, base obtuse, with 6–10 sharp teeth on either edge, glabrous except for the venation beneath; flowers in 3's or solitary, single, corolla disk-form, about 3 cm wide, petals nearly circular. Developed about 1900 by Max Löbner of Dresden, E. Germany. ✥

'Enchantement' (Lemoine 1923) (*P. virginalis*). Shrub, erect, long branched, young shoots densely pubescent, bark later becoming blackish; leaves ovate, 3–5 cm long, 1.2–3 cm wide, acute, base obtuse to round, serrate; flowers 7–9 in dense, erect racemes, corolla densely double, 2.5 cm wide, outer petals rounded, inner ones acute, flowering very late, slightly fragrant. ✥

'Erectus' (Lemoine 1890) (*P. lemoinei*). Very dense, upright, 1 m high shrub, twigs nodding with age, bark gray to blackish, later exfoliating; leaves long ovate to lanceolate, dark green, 1.5–4.5 cm long, 8–18 mm wide, gradually elongated and acuminate, base obtuse to round, either edge with 1–3 teeth at the apex, somewhat pubescent; flowers in 3's, occasionally 5's, sepals ovate, long acuminate, corolla cruciform, 3 cm wide, single, very fragrant, petals oblong. SpB 255; MG 1902: 383. ✥

'Etoile Rose' (Lemoine 1908) (*P. purpureo-maculatus*). Shrub, erect, only about 80 cm high, later somewhat nodding; leaves small; flowers solitary or in 3's, corolla slightly campanulate, to 4 cm wide, petals oblong, 2 cm long, carmine-pink at the base, gradually paling toward the apex, moderately fragrant. Developed from 'Fantaisie' × *P. purpureo-maculatus*.

P. × falconeri Sarg. (parentage unknown). Shrub, 2.5–3 m high, twigs thin, gracefully nodding, young shoots glabrous, second year bark chestnut-brown, exfoliating; leaves ovate to more elliptic, 3–6.5 cm long and 1–2.5 cm wide, acuminate, base obtuse to round, with 2–12 small teeth on either edge, nearly entire on flowering shoots, venation beneath sparsely strigose; flowers grouped 3–5 in racemes or also in compound cymes, corolla stellate, 3 cm wide, pure white, slightly fragrant, flowers abundantly, petals elliptic, acute (!), stamens not numerous, sterile, styles much longer than the stamens, deeply split, plaited in the bud stage, June. BC 2904; MG 14: 231; SpB 235; JA 35: Pl. 2,3. Developed in the USA. Before 1881. Plate 155.

'Fantaisie' (Lemoine 1900) (*P. purpureo-maculatus*). Shrub, to 80 cm high, shoots thin; leaves small, nearly sessile; flowers disk-form, large, white, basal spot soft pink, margin ciliate. Developed from *P.* "coulteri" Hort. × *P. lemoinei*.

'Favorite' (Lemoine 1916) (*P. polyanthus*). Young shoots sparsely pubescent, older bark gray-brown, later exfoliating; leaves ovate, 1.5–3 cm long, 0.5–2 cm wide, short acuminate, base round to obtuse, smooth above, strigose pubescent beneath; flowers grouped 3–5, compact, calyx and sepals pubescent, corolla cruciform, 3.5 cm wide, simple, petals obovate-oblong, pure white, stamens all fertile, style and disk glabrous. JA 35: Pl. 3. Origin unknown, but the *P. pubescens* influence is recognizable.

'Fimbriatus' (Lemoine 1900) (*P. lemoinei*). Densely erect shrub, only 80 cm high, twigs thin, flexible; leaves ovate, acuminate, base round, with 1–4 small teeth on the apical half, pubescent beneath; flowers usually in 3's, calyx and sepals glabrous, the latter ovate, long acuminate, 6–7 mm long, corolla disk-form, about 3 cm wide, loosely double, with 8–10 petals, these rounded to elliptic, ciliate or somewhat deformed, disk and style glabrous, split to the midpoint. Plate 155. ✥

'Fleur de Neige' (Lemoine 1915) (*P. virginalis*). Shrub, to 1 m high, resembling *P. inodorus* var. *grandiflorus*; flowers grouped

5–7, semidouble, to nearly 4 cm wide, flowers abundantly, good fragrance. Lemoine Cat. 1915: Pl. 2. ✥

'Fraicheur' (Lemoine 1933) (*P. virginalis*). Flowers very large, densely double, milk-white.

'Frosty Morn' (Guy D. Bush 1953). Growth broad and low; flowers grouped 3–5 together, double, about 3 cm wide, pure white, moderately fragrant. Patented. Recognized as exceptionally frost tolerant. ✥

'Galathée' (Lemoine 1915) (*P. purpureo-maculatus*). Shrub, medium sized, branches long and nodding; leaves medium-sized; flowers solitary, corolla cupulate, white, petals somewhat pink toned inside at the base.

'Gerbe de Neige' (Lemoine 1893) is the type of *P. polyanthus*!

'Girandole' (Lemoine 1915) (*P. virginalis*). Shrub, to 1.5 m high, weak grower, young shoots pubescent, bark brown and exfoliating in the second year; leaves ovate, 2.5–4.5 cm long and 1.3–2.3 cm wide, short acuminate, base round, nearly entire or with 1–6 teeth on either edge, pubescent on both sides; flowers in racemes of 5–7, calyx and sepals sparsely pubescent, sepals about 10 mm long, long caudate, corolla double, about 3.5 cm wide, milk-white, petals rounded at the apex, stamens all sterile, moderately fragrant. Lemoine Cat. 1915: Pl. 3. Attractive flowering on young plants. Plate 156. ✥

'Glacier'. (Lemoine 1913) (*P. virginalis*). Shrub, to 1.5 m high, stiff, young shoots glabrous, bark later becoming gray, smooth; leaves small, 2.5–3.5 cm long, 1.5–2.5 cm wide, acute to obtuse, base round, serrate; flowers 7–9 or only 5 together, in dense, compact racemes, corolla 2.5 cm wide, densely double, cream-white, very fragrant, blooms late. AB 25: Pl. 8. ✥

'Gracieux' (Lemoine 1909). Leaves rather large, like *P. coronarius*; flowers in racemes, double, milk-white, petals long, often toothed.

'Innocence' (Lemoine 1927) (*P. lemoinei*). Shrub, to 2 m high, shoots somewhat nodding; leaves yellow variegated (!); flowers single to semidouble, 3 to 3.5 cm wide, very fragrant.

P. × lemoinei Lemoine (*P. coronarius* × *P. microphyllus*). Low, compact shrubs, as wide as high, bark exfoliating; leaves ovate, 1.5–2.5 cm long, 7–12 mm wide, glabrous above, somewhat strigose beneath, short acuminate, round or obtuse at the base, with 2–3 teeth on either edge; flowers in 3's or solitary, occasionally in 5's, hypanthium glabrous, sepals ovate, corolla cruciform, 2.5–3 cm wide, petals ovate, emarginate at the apex, styles 3 mm long, split nearly to the base. GF 2: 616; JRHS 76: 272; JA 35: Pl. 3; SpB 255. Developed in 1892 by Lemoine; not often grown today. Fig. 279.

Includes the following cultivars (the descriptions may be found in alphabetical order):
'Avalanche', 'Candelabre', 'Coupe d'Argent', 'Dame Blanche', 'Erectus', 'Innocence' and 'Manteau d'Hermine'.

P. lemoinei densiflorus = **'Avalanche'**

P. lemoinei erectus = **'Erectus'**

P. lemoinei erectus grandiflorus = **'Mont Blanc'**

'Le Roy' (*P. virginalis*) Similar to *P. virginalis*, only about 1 m high, bark blackish, later exfoliating; leaves ovate-elliptic, 2–5 cm long, sparsely pubescent on both sides; flowers in racemes of 5–7, double, cream-white, very fragrant. Origin unknown;

Fig. 279. *Philadelphus lemoinei* (from G.F.)

disseminated from the Cole Nursery, Ohio, USA around 1954.

'Maculiflorus' (Koehne). Young shoots glabrous, bark brown in the second year; leaves ovate, 5–8 cm long, with 5–9 teeth on either edge, base round; flowers, about 7 in loose racemes, hypanthium and sepals glabrous, corolla nearly disk-form, 3 cm wide, petals obovate, white, reddish at the base (!), stamens about 35, all fertile, style and disk glabrous. Known only from the Botanic Gardens of Dresden and Leipzig, E. Germany, probably not in cultivation today. 1904. Considered a bud mutation of *P. coronarius* by Hu.

'Manteau d'Hermine' (Lemoine 1899) (*P. lemoinei*). Shrub, about 1 m high, dense, rounded habit, twigs thin, nodding, bark gray-brown; leaves yellowish-green, oval-elliptic, seldom ovate, very small, 7–25 mm long, 4–12 mm wide, acute at both ends, venation pubescent beneath; flowers grouped 3–5 together, loose to densely double, 2.5–3 cm wide, outer petals oblong, fragrance slight or absent, buds slightly reddish (influenced by *P. pekinensis*). ✣

P. × maximus Rehd. (probably *P. pubescens* × *P. tomentosus*). Shrub, to 5 m high, young shoots quickly becoming glabrous, bark light gray in the second year, smooth; leaves ovate to oval-elliptic, 8–12 cm long on long shoots and 4 to 6.5 cm wide, on flowering shoots 6–8 cm long and 3–4 cm wide, acuminate, base round, occasionally somewhat cordate, nearly entire to slightly dentate, sparsely pubescent above, underside tomentose; flowers grouped 7–9 together, hypanthium and sepals somewhat rough pubescent, sepals ovate, 5 mm long, corolla cruciform, 3–3.5 cm wide, petals obovate, stamens 26, anthers sterile, disk and style glabrous. JA 35: Pl. 3. Developed by Kew Gardens, England, in 1885 (as *P. tomentosus*).

'Mer de Glace' (Lemoine 1907) (*P. cymosus*). Small shrub, erect,

twigs short and thin; leaves light green, 4.5–7 cm long, 2.8–4 cm wide, short acuminate, base obtuse, with 6–15 sharp teeth on either edge, sparsely pubescent beneath; flowers grouped 3–5, occasionally solitary, semidouble to single, corolla 4.5 cm wide, outer petals obovate, rounded at the apex, slightly fragrant, style split to the base. Gw 17: 102.

'Minnesota Snowflake' (Guy D. Bush 1935) (*P. virginalis*). Shrub, to 1.5 m high, seedling of *P. virginalis;* flowers grouped 3–7, pure white, densely double, slightly fragrant, corolla to 4 cm wide. Patented. Very popular in North America. ✣

P. × monstrosus (Späth) Schelle. Shrub, medium-sized to 2 m, young shoots glabrous, bark gray in the second year, smooth; leaves ovate, 3.5–9 cm long and 2.5–5.5 cm wide, partly strigose above, rough tomentose beneath; flowers in racemes of 7 or 9, hypanthium with a sparse, soft pubescence, hairs erect, corolla cruciform, 3 to 5 cm wide, petals oblong, stamens all fertile, disk glabrous, as is the style or with a few hairs at the base and parted to the midpoint, June. Introduced into the trade in 1897 by Späth as *P. gordonianus monstrosus.*

'Mont Blanc' (Lemoine 1896) (*P. polyanthus*). Shrub, about 1 m high, narrowly upright, hard to distinguish from *P. lemoinei* in habit, but the flowers are totally different, young shoots pubescent; leaves ovate, 1.8–3 cm long, 8–18 mm wide, entire or with 1–3 teeth on either edge near the middle, very thin pubescence on the underside; flowers in 3's or 5's or solitary, calyx pubescent, sepals glabrous, corolla cruciform, 2.5 cm wide, pure white, petals obovate-oblong, stamens fertile, disk and style glabrous, flowers abundantly. Gs 1920: 65 (= *P. lemoinei erectus grandiflorus* Lemoine). ✣

P. × nivalis Jacq. (*P. coronarius* × *P. pubescens*). Shrub, to 2.5 m high, twigs nodding, young shoots glabrous, bark chestnut-brown in the second year, exfoliating; leaves oval-elliptic, 5–10 cm long, 2.5–6 cm wide, acuminate, base round or obtuse, nearly entire to slightly dentate, glabrous above, appressed tomentose beneath; flowers grouped 5–7 in broken racemes, hypanthium and sepals densely rough haired, the hairs partially erect, sepals to 7 mm long, corolla disk-form, 2.5–3.5 cm wide, petals rounded, stamens all fertile, style and disk glabrous, June (= *P. verrucosus* var. *nivalis* [Jacq.] Rehd.). Before 1841.

'Norma' (Lemoine 1910) (*P. polyanthus*). Shrub, to 1.5 m high, twigs long and rodlike, young shoots pubescent, bark later becoming gray-brown; leaves ovate, 3–5 cm long, 1.2–3 cm wide, acuminate, base round to obtuse, light green, serrate, partly pubescent beneath; flowers in 5's, occasionally 3's, calyx and sepals sparsely pubescent, sepals caudate, to 9 mm long, corolla disk-form, slightly double or often single, 4–4.5 cm wide, petals nearly circular, somewhat involuted on the margin, pure white, only slightly fragrant, stamens fertile, disk and style glabrous, flowers very abundantly. NF 6: 39; JA 35: Pl. 3. ✣

'Nuage Rose' (Lemoine 1916) (*P. purpureo-maculatus*). Low growing shrub; leaves small; flowers to 5 cm wide, single, white, soft pink in the center, petals somewhat fringed, fragrant.

'Nuée Blanche' (Lemoine 1903) (*P. cymosus*). Small shrub; leaves ovate, 2.5–5 cm long, 1.5 to 2.5 cm wide, short acuminate, base obtuse to acute, entire or with 1–3 teeth on either edge, appressed pubescent beneath; flowers solitary or in 3's, single to slightly double, corolla disk-form, 2.5–3 cm wide, cream-white, moderately fragrant, calyx glabrous, flowers very abundantly.

'Oeil de Pourpre' (Lemoine 1910) (*P. purpureo-maculatus*). Narrowly upright; leaves small, similar to *P. microphyllus*; flowers grouped 1–3 together, single, corolla cupulate, 3 cm wide, cream-white, blackish-red in the center, fragrant.

'Ophelie' (Lemoine 1913) (*P. purpureo-maculatus*). Flowers grouped 1–3 together, very abundantly flowering, corolla single to slightly double, white with a reddish center, to 4 cm wide, moderately fragrant.

'Patricia' (F. L. Skinner) (Origin unknown, but somewhat resembles *P. lewisii* var. *intermedius*). Shrub, erect, bushy, about 1 m high, young shoots rough, bark gray in the second year, smooth, winter buds enclosed; leaves oval-oblong, 2–4 cm long, 1–2 cm wide, obtuse at both ends, entire to slightly dentate with 1–2 teeth on either edge; flowers grouped 3–5, corolla cupulate, petals obovate, 10 mm long, 6–7 mm wide, stamens about 15, some fertile, the others stunted and sterile, hypanthium, disk and style glabrous, very fragrant, flowers sparsely. Developed in Canada in 1939.

'Pavillon Blanc' (Lemoine 1895) (*P. polyanthus*). Low, rounded shrub, narrowly upright, twigs long arching, young shoots pubescent, bark dark brown, smooth, later exfoliating; leaves ovate, 1.7–3.5 cm long, 1–1.8 cm wide, acute, base obtuse, entire, densely pubescent beneath; flowers in 5's, 7's or 3's, calyx and sepals pubescent; corolla cruciform, 3.5–4 cm wide, petals oblong, emarginate at the apex, cream-white, pleasantly fragrant, disk and style glabrous, June. NF 6: 40; Gs 1920: 64.

P. × pendulifolius Carr. Erect, densely branched shrub, twigs knotty, young shoots glabrous, second year bark chestnut-brown and smooth; leaves pendulous (!), glossy, ovate to broadly elliptic, 5–11 cm long, 3 to 7 cm wide, acuminate, base round, entire to slightly serrate, glabrous above, appressed pubescent beneath; flowers grouped 5–7, calyx and hypanthium densely tomentose, 2.5 cm wide, petals obovate, stamens sterile. Developed from seed before 1875 by Billard, but the true origin is unknown; earlier interpretations of this plant's origin have not been substantiated. Plate 157.

'Perle Blanche' (Lemoine 1900) (*P. cymosus*). Shrub, about 1.5 m high, compact habit, twigs short and thin, bark brown, smooth, slowly exfoliating; leaves oval-elliptic, 4–8 cm long, 1.5–3.2 cm wide, acuminate, base obtuse to acute, entire, light green, also serrate on long shoots; flowers grouped 5–7, corolla single to semidouble, fragrant, calyx pubescent, petals obovate-oblong. ⊕

P. × polyanthus Rehd. Shrub, erect, young shoots sparsely pubescent, bark chestnut-brown in the second year, smooth, later exfoliating; leaves ovate, 3–3.5 cm long, 1.5–2.5 cm wide, acute, base round to obtuse, entire or with 1–3 sharp teeth on either edge, glabrous above, sparsely pubescent beneath; flowers 3 or 5 in cymes, stalks of the lower flowers elongated and often branched, hypanthium and sepals pubescent, sepals ovate-caudate, 5–8 mm long, corolla cruciform, 3 cm wide, petals oblong, 15 mm long, 10 mm wide. Introduced into the trade in 1893 by Lemoine as 'Gerbe de Neige', confirmed as the type of this cross by Rehder in 1920 (? *P. insignis* × *P. lemoinei*).

Included here are the following cultivars (descriptions are in alphabetical order):
'Atlas', 'Boule d'Argent', 'Favorite', 'Mont Blanc', 'Norma', Pavillon Blanc'.

'Purity' (L. F. Skinner) (*P. virginalis*). Low shrub, flowers rather numerous, 3–5 together, single, to 5 cm wide, very fragrant. Seedling of *P. virginalis*.

P. × purpureo-maculatus Lemoine. Shrub, to 1.5 m high, bushy, twigs thin, later glabrous, second year bark blackish, slowly exfoliating; leaves broadly ovate, 1–3.5 cm long, 0.6–2.5 cm wide, acute, base round, entire or with 1–2 teeth on either edge, only very slightly pubescent beneath; flowers solitary, calyx and sepals bases partly pubescent, sepals ovate, 7 mm long, corolla disk-form, 2.5–3 cm wide, petals ovate, purple-red at the base, fragrant, June–July. BM 8193; GC 113: 55(= *P. × phantasia* Moore). Developed from seed of 'Fantaisie' in 1902 by Lemoine. z6

Includes the following cultivars:
'Etoile Rose', 'Fantaisie', 'Galathée', 'Nuage Rose', 'Oeil de Pourpre', 'Roméo', 'Sirène', 'Surprise', 'Sylviane', all $2x = 26$ (the triploid hybrids 'Belle Etoile', 'Bicolore' and 'Sybille' have been excluded here).

'Pyramidal' (Lemoine 1916) (*P. virginalis*) Shrub, strong growing, flowers abundantly, young shoots quickly becoming glabrous, bark gray-brown, smooth; leaves broadly ovate, 3–6 cm long, 2–4 cm wide, short acuminate, base round to somewhat cordate, dentate; flowers in 7's or 5's, corolla 3 cm wide, semidouble, fragrant, outer petals nearly circular.

'Romeo' (Lemoine 1913) (*P. purpureo-maculatus*). Leaves rather small; flowers grouped 1–3 together, single, milk-white, purple in the center, flowers abundantly, medium-sized.

'Rosace' (Lemoine 1904) (*P. cymosus*). Erect habit, dense and thinly branched, bark dark brown, slowly exfoliating; leaves oval-lanceolate, 6–10 cm long, 2.5–5 cm wide, acuminate, base obtuse, entire, dentate on long shoots, light green, sparsely pubescent beneath; flowers large, in 5's in true cymes, or in 3's, corolla flat, plate-like, usually single, but with a few staminoid petals in the center, cream-white, petals obovate, 2.5 cm long, moderately fragrant, styles split to the base, calyx glabrous on the exterior. Influence of *P. inodorus* distinctly recognizable.

'Savilos' (*P. virginalis*). Flowers single to semidouble, cream-white, to 5 cm wide, very fragrant, American hybrid?

'Schneesturm' (Krotz 1949) (*P. virginalis*?). Shrub, medium high, quick growing, twigs slightly nodding; flowers medium-sized, a good double, pure white.

'Silberregen' (Krotz 1949) (*P. lemoinei*?). Dense, rounded-erect shrub, scarcely 1 m high; leaves small, acute, base round; flowers solitary, but very numerous, pure white, corolla cruciform, about 4 cm wide, petals occasionally 5, oblong-obovate, flowering early, strawberry scent.

'Silver Showers'. Plants cultivated under this name in the USA appear identical to 'Silberregen'.

'Silvia' (*P. virginalis*). Flowers very abundantly, flowers double, pure white, about 3.5–4 cm wide, moderately fragrant. Hybridized by Leslie, Morden, Manitoba, Canada. Not to be confused with 'Sylviane'!

'Sirène' (Lemoine 1909) (*P. purpureo-maculatus*). Spreading habit, young shoots glabrous, bark later becoming blackish, eventually exfoliating, winter buds enclosed; leaves oval-elliptic, 5–9 cm long, 2.5–4 cm wide, acuminate, base obtuse, glabrous above, loosely pubescent on the venation beneath, sparsely dentate; flowers 3–5, occasionally solitary, to 5 cm wide, white with a soft pink basal spot. Distinguished from the other forms in this group by the larger leaves.

'Slavinii' (origin unknown). Shrub, to 3 m high, habit rounded, twigs nodding and pendulous, young shoots glabrous, bark brown in the second year, exfoliating; leaves elliptic-ovate, 6.5–9.5 cm long and 2–3.5 cm wide, acuminate, base acute, glabrous above, pubescent beneath, with 2–5 teeth on either edge; flowers in 5's, occasionally 3's or solitary, calyx and sepals pubescent, sepals ovate, 7 mm long, corolla cruciform, 4–6 cm wide (!), petals oblong, rounded at the apex, pure white, slightly fragrant, disk and style glabrous; capsule ellipsoid, seeds long caudate. USA. Before 1936. Flowers abundantly and attractively, 3 weeks later than *P. inodorus*. ⊕

'Souvenir de Billard' see: **P. insignis** .

P. × splendens Rehd. (origin unknown, perhaps *P. grandiflorus × P. gordonianus*). Erect shrub, young shoots glabrous, bark chestnut-brown in the second year, exfoliating; leaves of the flowering shoots oblong-elliptic to oval-oblong, 6–11.5 cm long and 2.5–5 cm wide, short acuminate, base obtuse to round, indistinctly and sparsely dentate to nearly entire, totally glabrous on both sides or with strigose pubescence on the venation beneath; flowers 5–7–9, crowded on a 1.5–3 cm long rachis, calyx and sepals glabrous, corolla disk-form, 3.5–4 cm wide, petals rounded, pure white, slightly fragrant, stamens 30, disk and style glabrous. Gs 1929: 214; JA 35: Pl. 3. Origin unknown. Disseminated by the Arnold Arboretum in 1912. Fig. 271.

'Stenopetalus' (Carr.) (perhaps a form of *P. pubescens?*). Very similar to *P. × falconeri*, but with inflorescences distinctly racemose, calyx pubescent, petals glabrous; leaves ovate, 4–7.5 cm long, 1.8–3 cm wide. acuminate, base obtuse, slightly serrate, pubescent beneath; flowers in 5's or 7's, corolla nearly campanulate, 1.5 cm wide, petals oblanceolate, 8 mm long, rounded at the apex, interior glabrous. Before 1870.

'Surprise' (Lemoine 1912) (*P. purpureo-maculatus*). Flowers rather large, white with purple-carmine basal spots, fragrant.

'Sybille' (Lemoine 1913) (*P. purpureo-maculatus*). Flowers grouped 1–3, corolla about 4–5 cm wide, single, shell-form, white with a pink basal spot, very fragrant, flowers rather abundantly. JRHS 76: 272. A very good cultivar.

'Sylviane' (Lemoine 1917) (*P. purpureo-maculatus*). Flowers grouped 5–9 together, shell-form, pure white, base of the brown petals pale pink toned, flowers abundantly. Not to be confused with 'Silvia'.

'Thelma' (F. L. Skinner) (origin unknown, but showing some influence from *P. purpurascens*). Low, graceful shrub, bark brown, exfoliating; leaves oval-lanceolate, 2–2.5 cm long, 5–10 mm wide, acuminate, base acute to round, entire or with 2–3 small, sharp teeth on either edge, sparsely strigose pubescent beneath; flowers in 3's, occasionally 5's, corolla campanulate, 1.5 cm wide, petals obovate, 9 mm long, June. Introduced into the trade in Canada in 1939.

'Velléda' (Lemoine 1922) (*P. cymosus*). Low shrub, bark exfoliating; leaves ovate, small, 1.5–3 cm long, 8–15 mm wide, nearly entire or with 2–4 large teeth on either edge; flowers usually solitary, corolla disk-form, to 2.5 cm wide, petals nearly circular, somewhat undulate, slightly fragrant, pure white. Influence of *P. inodorus* easily recognizable.

P. × virginalis Rehd. (*P. lemoinei* × ? *P. nivalis* 'Plenus'). Stiffly upright shrub, 1–2.5 m high, young shoots sparsely pubescent, thin glabrous, bark brown in the second year, smooth, later exfoliating; leaves ovate, 4–7 cm long and 2.5–4.5 cm wide, short acuminate, base round, appressed tomentose beneath; flowers in racemes, double, hypanthium and sepals tomentose, corolla pure white, very fragrant, 4–5 cm wide. Gs 1920: 65; 1934: 104. Developed by Lemoine in Nancy, France in 1909 and introduced into the trade under the name 'Virginal'. Elevated to the type of the group by Rehder in 1920. Plate 156. ⊕

Includes the following cultivars:
    'Argentine', 'Boule d'Argent', 'Enchantment', 'Fleur de Neige', 'Glacier', 'Girandole', 'Le Roi', 'Pyramidal' (the descriptions may be found in alphabetical order).

'Voie Lactée' (Lemoine 1905) (*P. cymosus*). Vigorous, upright habit, nodding; leaves ovate, 8–12 cm long on long shoots, coarsely dentate, bright green, acuminate; flowers 3–5 or solitary, single, corolla flat shell-form, 5–6 cm wide, milk-white, petals somewhat involuted on the margin, slightly fragrant, flowers very abundantly. SpB 253; Gw 17: 103; Gs 1920: 64. ⊕

'Waterton' (Grootendorst) (*P. lewisii*). Young shoots red-brown; flowers small, single, stellate, cream-white.

All species and cultivars are excellent flowering shrubs with few soil requirements; all prefer a fertile clay soil in full sun. The species will thrive in shade but with fewer flowers. Many "wild" *Philadelphus* species found in cultivation today are actually abundantly flowering selections with few of the characteristics of a true wild plant.

New selections of *Philadelphus* should be registered with the Arnold Arboretum in Cambridge, Mass., USA before their introduction.

Lit. Eitel, B. T.: A Study of the horticultural species, varieties and clones of the genus *Philadelphus*; in Ohio Nursery Notes **23** (10), 1–9, 1954 ● Hillier, H. G.: *Syringa, Philadelphus* etc., etc.; in Ornamental Flowering Trees and Shrubs; Report of the Conference held by the RHS 1938, 96–118 ● Hu, Shiu-Ying: A monograph of the genus *Philadelphus*; in Jour. Arnold Arboretum 1954, 275–333; 1955, 52–109, 325–368; 1956, 15–90 ● Koehne, E.: *Philadelphus*; in Gartenflora 1896, 450–451, 486–488, 500–508, 541–542, 561–563, 596–597, 618–619, 651–652 ● Koehne, E.: Zur Kenntnis der Gattung *Philadelphus*; in Mitt. DDG 1904, 200–209 ● Moore, A. H.: *Philadelphus*; in L. A. Bailey, Standard Cyclopedia of Horticulture III, 2579–2582, 1950 ● Pochlaks, H., & A. Zamelis; Selbstbestäubungs- und Kreuzungsversuche mit *Philadelphus*; in Act. Hort. Bot. Univ. Latvi. **11–12**, 229–232, 1939 ● Schneider, C.: Mock Oranges; in New Flora and Silva **6**, 113–117, 1934 ● Wyman, D.: *Philadelphus* popular for Spring White; in American Nurseryman 1961 (3/1), 14–15, 66–73.

× PHILAGERIA Mast. (LAPAGERIA × PHILESIA) — PHILESIACEAE

To date only one hybrid has been observed between these two genera:

× **Philageria veitchii** Mast. (*Lapageria rosea* × *Philesia buxifolia*). A small twining shrub, similar to *Lapageria rosea* in appearance, but with thinner shoots; leaves oblong, about 3 cm long, acute, leathery tough, quite finely dentate, with 3 veins, these running together at the apex, petiole 1.2 cm; flowers pendulous, 4–5 cm long, solitary, calyx composed of 3 fleshy, 2.5 cm long, acute sepals, corolla bright pink, consisting of 3, somewhat uneven, ovate, nearly 5 cm long petals, overlapping, acute, filaments pink punctate, anthers yellow, August–September. BMns 92. Developed in the Veitch Nursery in Chelsea, England, 1872; rather rare in England today. z9 # ✧

Fig. 280. *Philesia magellanica* (from Dimitri)

PHILESIA Comm. — PHILESIACEAE

Monotypic genus. Evergreen shrub, erect, shoots angular; leaves alternate; flowers hermaphroditic, solitary or few in groups, perianth consisting of 6 inward leaning segments, the 3 outermost segments oval-lanceolate, only about a third as large as the inner ones; stamens 6, fused into a row at the base, style elongated, stigmas capitate; fruit an ovate, many seeded berry.

Philesia magellanica Gmel. Dense, low shrub, 0.3–1 m high, very densely foliate, glabrous; leaves narrowly oblong, very stiff, single veined, 2.5–4 cm long, margins involuted, deep green above, bluish beneath; flowers usually solitary, pendulous, 6–7 cm long, short petioled, campanulate-tubular, sepals pink, fleshy-waxy, petals orange-green, (July) September–October. BM 4738 (= *P. buxifolia* Lam.). S. Chile, Tierra del Fuego, Strait of Magellan, in moist forests. 1850. z9 Fig. 280. # ✧

PHILLYREA L. — Mock Privet — OLEACEAE

Evergreen shrubs; leaves opposite, short petioled, entire or dentate; flowers small, white, fragrant, in axillary clusters on the previous year's shoots; calyx short, broad, 4 lobed, corolla with a short tube and 4 wide, short lobes; stamens 2, filaments short; fruit an ovate to rounded, single seeded drupe. — 4 species from the Mediterranean region to Asia Minor.

Phillyrea angustifolia L. Evergreen shrub, to 3 m high in its habitat, divaricate, twigs glabrous, gray-yellow; leaves linear, tough, 3 to 6 cm long, 0.5–1 cm wide, yellowish-green, usually entire, with 5–6 vein pairs; flowers greenish-white, fragrant, in axillary clusters, May–June; fruits oval-rounded, blue-black, 6 mm long. HM 2907. S. Europe, N. Africa. Before 1597. z9 Plate 159. #

var. **rosmarinifolia** Ait. Leaves usually smaller, more linear, 3–4 cm long, occasionally 6–8 cm long, 4–6 cm wide, more gray-green to blue-green (= *P. rosmarinifolia* Mill.). #

P. decora see: **P. vilmoriniana**

P. latifolia L. Evergreen, 5(9) m high, divaricate shrub, twigs with a short and fine pubescence when young; leaves ovate-elliptic, 2 to 6 cm long, glossy dark green above, lighter beneath, with 5–12 vein pairs, the large leaves more sharply dentate, the smaller ones indistinctly dentate; flowers yellowish, in short, axillary clusters, May; fruits blue-black, 6 mm thick. HM 2907. Mediterranean region; on sunny gravel slopes, on cliffs in the "Maquis" (dry brush similar to "chaparral" of western USA), usually on calcareous soils. 1597. z9 Plate 159. #

'Buxifolia'. Shrub, low, dense; leaves oval-oblong, rounded at the apex, small, entire. Kew Gardens. #

var. **media** (L.) Schneid. Usually shrubby, only occasionally to 5–6 m high; leaves ovate-elliptic to more lanceolate, 2–5 cm long, entire to slightly dentate, the teeth more obtuse (= *P. media* L.). Predominantly in the eastern Mediterranean region to Syria and Palestine. Plate 159. #

P. media see: **P. latifolia** var. **media**

P. rosmarinifolia see: **P. angustifolia** var. **rosmarinifolia**

P. vilmoriniana Boiss. & Bal. Evergreen, broad bushy shrub, 2–3 m high, twigs glabrous and stiff; leaves oblong-lanceolate, 8–12 cm long, 3–4 cm wide, acuminate, entire, glossy dark green above with reticulate venation, yellowish-green beneath, usually entire; flowers white, in small clusters, flowers abundantly, May; fruits ovate, to 1.5 cm long, blue-black. BM 6800 (= *P. decora* Boiss. & Bal.). Caucasus, Lazistan. 1868. z7 Plate 159. #

'Angustifolia'. Narrow leaved form, leaves 8–12 cm long, but only 1.5–2 cm wide. z6 #

Only the last named cultivar is generally hardy in a temperate climate. Soil requirements somewhat like those of Laurel and *Buxus*, with which it occurs in the Mediterranean "Maquis" thickets.

Lit. Sennen: *Phillyrea;* in Bull. Soc. Dendro. France 1935, 45–67 (with many ills.) ● Sébastian, C.: Etude du genre *Phillyrea*, 1–102; Rabat (Morocco) 1956 (16 plates).

PHLOMIS L. — LABIATAE

Shrubs, subshrubs or (usually) perennials; entire plant woolly-floccose; leaves opposite, very rugose, densely pubescent; flowers in dense mock whorls; calyx tubular, mucronulate; corolla with a short, bilabiate tube, ringed inside and hardly exserted past the calyx, yellow, white or purple, upper lip helmet-like, keeled, compressed; lower lip fitted to the outline of the upper lip, 3 parted; filaments arising from beneath the "helmet", style parted at the apex; nutlets 3 sided. — About 100 species from the Mediterranean region to China.

Phlomis chrysophylla Boiss. Small, evergreen shrub, totally covered with appressed, golden-yellow pubescence; leaves sage-like, broadly cordate, obtuse, rough beneath, petiolate; flowers in widely spaced whorls, golden-yellow, sometimes also in axillary pairs over stiff, awl-shaped, thorny tipped prophylls, calyx with tiny thorny teeth, June. Lebanon. z9 # ✿ ∅

P. fruticosa L. Evergreen subshrub, horizontally branched, 0.5–1 m high, stiff, dense gray-yellow woolly; leaves nearly sessile on the apical portion of the shoot, long petioled toward the base, oval-oblong, 5–10 cm long, entire to slightly crenate, very rugose, pale gray-green above, short soft pubescent, gray tomentose beneath; flowers many in cymes, dark yellow, June–July. BM 1843. Mediterranean region. 1596. z8–9 Plate 159; Fig. 281. # ✿

P. italica L. Subshrub, about 30 cm high, entire plant stellate pubescent, shoots cylindrical; leaves oblong to more lanceolate, obtuse, 3–5 cm long, tough, margins crenate, dark green above and loosely stellate pubescent, whitish beneath, woolly, petiolate; flowers light to dark pink, 2 cm long, in terminal spikes of 6 whorls. BM 9270. Balearic Islands. Around 1800. z9 # ✿ ∅

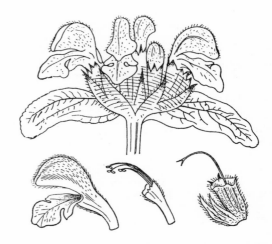

Fig. 281. *Phlomis fruticosa,* inflorescence and flower parts (from Sibt. & Sm.)

P. purpurea L. Subshrub, about 50 to 60 cm in cultivation, in its habitat to 2 m high, shoots woolly, but not glandular; basal leaves 4–9 × 2–4 cm, lanceolate to more oblong, leathery tough, crenate, stellate pubescent above, white woolly beneath and stellate pubescent, petiole 5 cm; prophylls very numerous, overlapping, petiolate, lanceolate, acute, with 10–12 flowers in a whorl, corolla purple or pink, occasionally white, 2.5 cm long, stellate tomentose, June. BMns 518. Spain, Portugal. 1661. z10 ✿

Only for warm protected areas and very lean, calcareous soils.

PHOEBE Nees — LAURACEAE

Evergreen trees and shrubs with aromatic wood; leaves alternate simple, pinnately veined; flowers hermaphroditic or polygamous, fragrant; perianth 6 parted, all parts rather regular, with 12 stamens, 9 fertile stamens in 3 rings, the 3rd ring with 2 glands at the base of each stamen, an innermost 4th ring with 3 sterile stamens having cordate or triangular apices; fruit a fleshy berry with a persistent 6 part perianth. — About 30 species in the Asiatic tropics, 7 in China.

Phoebe sheareri Gamble. Tree, 6–15 m high, bark brown, smooth, shoots dark brown, densely tomentose; leaves elliptic-oblong to more lanceolate, to 12 cm long and 3 cm wide, acute, leathery, base narrowly cuneate, glossy dark green above, dense brown woolly with raised reticulate venation beneath, petiole 2 cm; flowers in small, axillary panicles, white, sepals silvery pubescent on both sides, stalk 8 cm; fruits ellipsoid, 1 cm long, 0.6 cm wide, blue-black, perianth remnants brown. ICS 1657. China; Yunnan, Szechwan, Hupeh and other provinces, at 1000–2000 m. ℗ China. z10 #

PHOENIX L. — PALMAE

Short or tall palms, with or without a stem, often multistemmed; leaves terminal, bowed outward, odd pinnate, pinnae at the leaf apex nearly clustered or rather evenly wide spaced, long lanceolate, sword-like acuminate, stiff, entire, folded inward; petiole flat-convex, usually thorny; leaf sheaths short, separating into fibers; stem apex covered (under the green leaves) with leaf sheaths and petiole bases; the stem base smooth and with leaf scars; flowers dioecious, the inflorescences appearing among the petioles, small, yellow; fruit a drupe with a longitudinal furrow. — 17 species in tropical and subtropical Africa and Asia.

Phoenix canariensis Chabaud. Stem relatively thin and short; leaves very narrow, to 6 cm long, with numerous, slightly folded pinnae, deep green (= *P. jubae* Webb.). Canary Islands. 1888. z9 Plate 158. # ⌀

Often treated as a conservatory or cool greenhouse plant.

P. dactylifera L. Date Palm. Tree, 20–30 m high in its habitat; leaves gray-green, bowed, leaflets in 2 circles, irregularly spaced, narrowly linear, stiff, 2 parted at the apex, 20–40 cm long; inflorescences occasionally very extensive; fruits the well known, 3 to 5 cm long, brown or yellow dates. N. Africa, Saudi Arabia. 1597. Most prominant tree of the oases. Occasionally planted throughout the tropics. z10 Plate 158. # ⌀

P. jubae see: **P. canariensis**

P. loureirii Kunth. Dwarf species, to only 2 m high, single stemmed or bushy, stem somewhat bulbous at the base; young leaves with whitish fibers, younger shoots more pulverulent, pinnae in 2 rows, narrow, often sickle-form, soft, dark green, not prickly at the apex (= *P. roebelenii* O'Brien). Assam, Vietnam. 1889. Often used as a container plant. z10 Plate 157. # ⌀

P. reclinata Jacq. Trunk thin, scarcely over 8 m high, usually a multistemmed shrub; leaves gracefully nodding on the apical portion, leaflets in 2 rows, lanceolate, about 30 cm long, stiff, sharply acuminate-prickly; fruits brown. Tropical and S. Africa. 1972. z10 Plate 158. # ⌀

P. roebelenii see: **P. loureirii**

P. silvestris (L.) Roxb. Wild Date. Stem upright, thick, to 12 m high in its habitat; leaves 3–4.5 m long, glabrous, gray-green, leaflets numerous, in 2 or 4 rows, often also somewhat clustered, stiff, 15–45 cm long, petioles thorny; fruits 2.5 cm long, reddish-yellow when ripe. India. 1863. Very similar to *P. dactylifera* and possibly its wild form. z10 # ⌀

PHORMIUM J. R. & G. Forst. — New Zealand Flax — AGAVACEAE

Evergreen plants with a nest-like habit, to 3 m long, distichous, with hard, swordlike leaves; roots fleshy, fibrous; flowers in erect panicles, on a leafless shaft, perianth connate at the base, otherwise distinct, the 3 outermost segments parallel margined upright to the apex, the 3 innermost segments longer and outspread at the apex, red or yellow; fruit a leathery capsule. — 2 species in New Zealand, but commonly planted in Mediterranean climates.

Phormium colensoi Hook. f. Much smaller than the following species; leaves 60–150 cm long, 2.5–5 cm wide, long acuminate, light green to yellowish-green, red lines absent from the margins and midrib, usually rather limp; shaft 0.6–2 m high at the most, flowers 2.5–3 cm long, the outermost segments yellow to reddish-brown, the innermost green to greenish-yellow; fruit capsule 10–20 cm long, cylindrical, pendulous, spirally twisted. BM 6973 (as *P. hookeri*). New Zealand, from the mountains to the sea. 1868. Hardier than *P. tenax*. z9 # ⌀

'Tricolor'. Leaves very limp, pendulous, with several cream-white stripes on the margins, central part of the blade green, margins red; flowers yellowish, the outer segments golden-yellow to golden-brown, inner ones cream-yellow with green middle stripes. MNZ 52. Found in New Zealand in the late 1880's. z10 # ⌀

'Variegatum'. Leaves with 1 or 2 yellow-green or cream-white marginal stripes, without the red margins. z9 # ⌀

P. hookeri see: **P. colensoi**

P. tenax J. R. & G. Forst. Leaves 1–3 m long, 5–12 cm wide, linear, swordlike, acute to acuminate, apex eventually usually somewhat split or shredded, flat above, keeled beneath, yellowish-green to dark green above, often bluish beneath, midrib and margins normally with a red or orange line, very stiff and leathery; flower shaft 1.5–4.5 m, reddish, flowers numerous, usually dull red to red-brown, 2.5–5 cm long; fruit capsule 5–10 cm long, not twisted. BM 3199. New Zealand. 1789. z9 # ⌀

'Williamsii Variegata' (Duncan & Davies 1926). Very strong grower; leaves over 2 m long, bright yellow in the center, with some narrow yellow lines along the sides, partly nodding. JRHS 1977: 111.

7 cultivars grown in New Zealand (described in MNZ), but these are generally unknown in the rest of the world.

'Purpureum'. Leaves bronze-purple. Commonly found in English gardens. z9 # ⌀

'Variegatum'. Leaves yellow and green striped. z9 # ⌀

'Veitchii'. With a broad, sulfur-yellow band over the middle of the leaf. Plate 160. # ⌀

These plants prefer a very fertile, moist soil; they will thrive, however in other soils if the climate is mild enough. The huge leaves make these plants extraordinarily impressive.

PHOTINIA Lindl. — Christmas Berry — ROSACEAE

Evergreen or deciduous shrubs or trees; leaves alternate, simple, short petioled, usually finely serrate or entire, stipules sometimes nearly leaflike; flowers in terminal umbellate panicles, usually white; petals 5, circular; stamens about 20, styles 2, occasionally 3–5, carpels connate, sepals persisting on the fruit, seed core convex and rounded at the fruit apex, not acutely conical. — About 60 species in S. and E. Asia.

Deciduous species:
  P. beauverdiana, glomerata, parvifolia, villosa

Evergreen species:
  P. davidsoniae, × *fraseri, glabra, serrulata*

Photinia beauverdiana Schneid. Deciduous shrub or also a slender tree, to 9 m high, totally glabrous, otherwise similar to *P. villosa*; leaves narrow obovate to lanceolate, slender acuminate, 5–13 cm long, finely serrate, with 8–14 vein pairs, glabrous, petiole to 1 cm long; inflorescences fine warty, to 2.5 cm long and 5 cm wide, May; fruits ovate, 6 mm long, purple. ICS 2153. China. Around 1900. z6 Plate 159; Fig. 282. ∅ ⚭

var. **notabilis** (Schneid.) Rehd. & Wils. Leaves broad oblong-elliptic, 7–12 cm long, with about 12 vein pairs; inflorescences 8–10 cm wide; fruits ellipsoid, 7–8 mm long, purple (= *P. notabilis* Schneid.). China, W. Hupeh Province. 1908. Plate 159. ∅ ⚭

P. davidsoniae Rehd. & Wils. Evergreen shrub, also a tree in its habitat, 6–10 m high, young shoots reddish and pubescent; leaves oblanceolate to narrow elliptic, tapering to both ends, 8–14 cm long, base cuneate, leathery, dark green and glossy above, venation pubescent beneath at first; inflorescence pubescent, 7–10 cm wide, flowers 1 cm wide, white, May; fruits globose, orange-red, 8 mm long. GC 71: 102; ICS 2146. Central China. 1901. z9 Plate 159. # ∅ ⊕ ⚭

P. × **fraseri** (Dress) (*P. glabra* × *P. serrulata*). Large, evergreen shrub, intermediate between the parents in appearance; leaves elliptic to more obovate, 7–9 cm long, finely serrate, abruptly short acuminate, base broad cuneate, glossy bright green above, lighter beneath, petiole 1.2–2.3 cm long, pubescent above when young; flowers in umbellate panicles, 10–12 cm wide, white, petals pubescent on the inside at the base. z9 # ⊕ ∅

'Birmingham'. The type of this cross. Leaves usually obovate, bright coppery-red on the new growth, later dark green, thick, leathery. Originated as a chance seedling in the Fraser Nurseries of Birmingham, Alabama, USA about 1940. z9 # ⊕ ∅

'Red Robin'. Leaves sharply serrate, young leaves bright red, later glossy green (= *P. glabra* 'Red Robin'). Developed in New Zealand. z9 # ⊕ ∅

'Robusta'. Shrub, high or medium-sized; leaves more similar to those of *P. serrulata*, thick, leathery, oblong to obovate, young leaves coppery-red. JRHS 101: 441 (= *P. serrulata* 'Robusta'). Originated in the Hazlewood Nursery, Sydney, Australia. z9 Plate 128. # ⊕ ∅

'Rubens' (from Japan). Very attractive red shoots in spring, but less winter hardy than *P. serrulata*. z9 # ⊕ ∅

P. glabra (Thunb.) Maxim. Evergreen shrub, 3(6) m high, new growth red, totally glabrous; leaves elliptic to oblong-obovate, 5 to 8 cm long, cuneate beneath, petiole 1–1.5 cm, always totally glabrous; inflorescences 5–10 cm wide, petals with tufts of pubescence, May (often in March on good sites in mild climates); fruits globose, 5 mm thick, red at first, then black. MJ 1405; ICS 2147 (= *Crataegus glabra* Thunb.). Japan. 1903. z9 Plate 159; Fig. 282. # ∅ ⊕ ⚭

P. glabra 'Red Robin' see: **P.** × **fraseri** 'Red Robin'

P. glomerata Rehd. & Wils. 6–10 m high in its habitat, young shoots shaggy pubescent, red; leaves thin, leathery, deciduous, narrow oblong to oblanceolate, 12–18 cm long, 5–6 cm wide, short acuminate, often cuneate, often oblique, margins finely glandular serrate and slightly involuted, yellowish-green above, lighter beneath and with 6–9 vein pairs, these shaggy pubescent at first, eventually glabrous; inflorescence 6–10 cm wide, densely shaggy-tomentose, white, fragrant, the individual flowers in nearly sessile clusters; fruits ovate, red, 5 to 7 mm long. ICS 2148. China; Yunnan Province, 1500 to 1600 m. z9 Plate 157. # ⊕ ⚭

P. notabilis see: **P. beauverdiana** var. **notabilis**

P. parvifolia (Pritz.) Schneid. Deciduous shrub, 2–3 m high, young shoots dark red, glabrous; leaves ovate to somewhat obovate, 3–6 cm long, acuminate, base broad cuneate, sharply serrate, deep green above, lighter beneath and quickly becoming glabrous; flowers grouped 5–6 in about 3 cm wide false umbels, white, May–June; fruits ellipsoid, 1 cm long, scarlet-red. ICS 2158 (= *P. subumbellata* Rehd. & Wils.). China, Hupeh Province. 1908. Hardy. z6 Fig. 282. ⚭

P. serrulata Lindl. Evergreen shrub, in its habitat a tree 5–12 m high, totally glabrous or nearly so; leaves oblong to more oblanceolate, acuminate, base usually round, 10–18 cm long, serrate, dark green and glossy above, yellowish beneath, new growth reddish, petioles 2–4 cm long, pubescent; flowers in 10–18 cm wide panicles, the rachis rather thick and somewhat angular, white, petals totally glabrous, May–June; fruits pea-sized, red. BR 1956; BM 2105; ICS 2145. China. 1804. ⊕ Australia, as windbreaks and hedges. z9 Plate 159; Fig. 282. # ∅ ⊕ ⚭

'Robusta' see: **P.** × **fraseri** 'Robusta'

'Rotundifolia'. Lower growing; leaves smaller, rounded; corolla nearly globose. MA Pl. 47. Before 1892.

P. subumbellata see: **P. parvifolia**

P. variabilis see: **P. villosa**

P. villosa (Thunb.) DC. Deciduous shrub or small tree, to 5 m high, young shoots pubescent, thin, later glabrous; leaves obovate to oval-lanceolate, short acuminate, 3–8 cm long, dark green, very tough (!), light yellow-green and shaggy pubescent beneath, a glowing orange-scarlet in fall; flowers in loose, 5 cm wide, pubescent corymbs, stalk warty, June; fruits ellipsoid, 8 mm long, bright red.

Plate 161

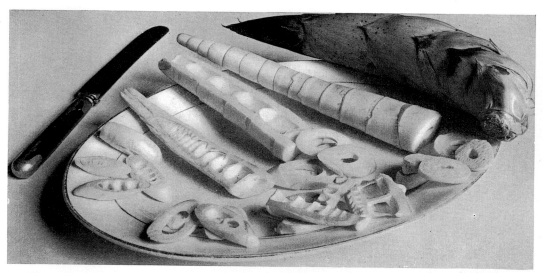

The young sprouts of all *Phyllostachys* species are edible if cooked. Pictured here *P. bambusoides*
Photo: U.S.D.A.

Phyllostachys viridis   *Phyllostachys bambusoides*
Photos: U.S. Plant Introduction Station, Savannah, Georgia

Plate 162

Phylostachys flexuosa in Georgia, USA.
Photo: U.S.D.A.

Phyllostachys nigra in Japan
Photo: Dr. Watari, Tokyo

Phyllostachys pubescens on a forest's edge in Japan
Photo: Dr. Watari, Tokyo

Plate 163

Phyllostachys bambusoides 'Allgold'
in the Prafrance Bambouseraie, S. France

Phyllostachys pubescens
in the Prafrance Bambouseraie, France

Pittosporum heterophyllum as a hedge
in the Suchumi Botanic Garden, USSR

Pittosporum anomalum
(¼ actual size) in Malahide, Ireland

Plate 164

Physocarpus. a. *P. amurensis;* b. *P. bracteatus;* c. *P. intermedia;* d. *P. glabratus;* e. *P. capitatus;* f. *P. monogynus;* g. *P. malvaceus;* h. *P. stellatus* (material collected from wild plants)

Physocarpus opulifolius in the Dortmund Botanic Garden, W. Germany

Plate 165

Phytolacca dioica in Lisbon, Portugal at Castelo Sao Jorge; flowering shoot below

Plate 166

Pieris japonica in a hoarfrost, in the Dortmund Botanic Garden, W. Germany

Pieris japonica 'Variegata' in the Dortmund Botanic Garden

Plate 167

Pieris formosa in the Brissago Botanic Garden, Tessin, Switzerland

Pieris floribunda in the Dortmund Botanic Garden

Plate 168

Pittosporum eugenioides 'Variegatum'
on Valentia Island, W. Ireland

Pittosporum dallii
in the Hillier Arboretum, England

Pittosporum tobira
Photo: Dr. Watari, Tokyo

Plate 169

Piptanthus laburnifolius
in a garden near Kenmare, W. Ireland

Pittosporum undulatum
in the Kirstenbosch Botanic Garden, S. Africa

Plantago cynops in the Dortmund Botanic Garden, W. Germany

Plate 170

Populus maximowiczii on a valley floor near Yamabe, Japan
Photo: T. Nitzelius, Göteborg, Sweden

Plate 171

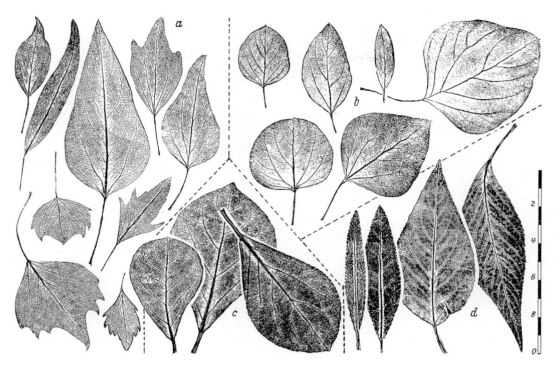

Populus. a. *P. euphratica*; b. *P. pruinosa*; c. *P. simonii* 'Fastigiata'; d. *P. angustifolia*
(material collected from wild plants, except c.)

Populus × berolinensis
in the Dortmund Botanic Garden, W. Germany

Populus tremuloides in its native habitat
in the Cascade Mts., Oregon, USA

Plate 172

Populus nigra 'Italica'
in Saalfeld, Thüringen, E. Germany

Populus nigra var. *thevestina*
near Pec, Yugoslavia, close to the Albanian border

Podranea ricasoliana
in Sintra, Portugal

Populus deltoides
in the Morton Arboretum, Lisle, Illinois, USA

Plate 173

Populus. a. *P. alba* 'Pyramidalis'; b. *P. alba* 'Richardii'; c. *P. koreana*; d. *P. yunnanensis*; e. *P. cathayana*; f. *P. lasiocarpa* (c–e. from wild plants)

Populus wilsonii
in the Aarhus Botanic Garden, Denmark

Populus canadensis 'Eugenei'
in Royal Botanic Gardens, Kew, England

Plate 174

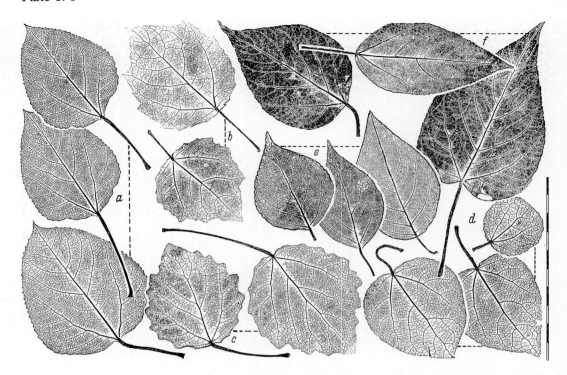

Populus I. a. *P. tremuloides*; b. *P. canescens*; c. *P. tremula*; d. *P. sieboldii*; e. *P. suaveolens*; f. *P. balsamifera* (material collected from wild plants)

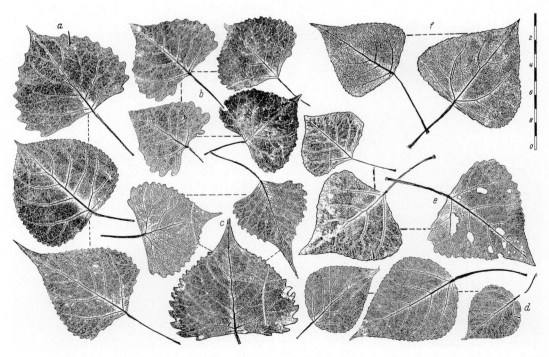

Populus II. a. *P. deltoides*; b. *P. fremontii*; c. *P. sargentii*; d. *P. jackii*; e. *P. fremontii* var. *arizonica*; f. *P. canadensis* 'Eugenei' (material from wild plants, except f.)

Plate 175

Potentilla fruticosa 'Friedrichsenii'

Potentilla fruticosa 'Parvifolia'

Prinsepia uniflora
in the Bergianska Arboretum, Stockholm, Sweden

Potentilla fruticosa 'Walton Park'
All photos: Dortmund Botanic Garden

Plate 176

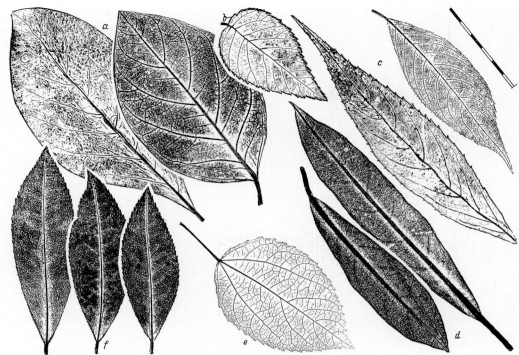

a. *Pinckneya pubens*; b. *Planera aquatica*; c. *Platycrater arguta*; d. *Pileostegia viburnoides*;
e. *Poliothyrsis sinensis*; f. *Pieris formosa* var. *forrestii* (material from wild plants)

Potentilla fruticosa f. *rigida*
in the Uppsala Botanic Garden, Sweden

Potentilla fruticosa 'Farreri'

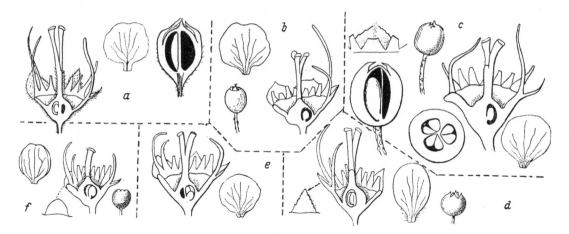

Fig. 282. **Photinia.** Flower parts. a. *P. villosa;* b. *P. beauverdiana;* c. *P. villosa* var. *laevis;* d. *P. glabra;* e. *P. parvifolia;* f. *P. serrulata* (from Schneider, altered)

GF 1: 67; ICS 2159 (= *P. variabilis* Hemsl.; *Pourthiaea villosa* Decne.). Japan, Korea, China. 1865. z5 Fig. 282. ⌀ ✛ ⚭

var. **laevis** (Thunb.) Dipp. Twigs very quickly becoming totally glabrous, likewise the leaves and inflorescences, leaves smaller and narrower, longer acuminate; fruits to 12 mm long. BM 9275. Fig. 282. ⌀ ✛ ⚭

f. **maximowicziana** (Lévl.) Rehd. Leaves obovate, rounded at the apex and short acuminate, cuneate at the base, distinctly indented venation above, petiole 1–2 mm long. Korea. 1876.

var. **sinica** Rehd. & Wils. Very similar to the species, but often tree-like, to 10 m high; leaves not tough (!), elliptic to oblong-elliptic, quickly becoming glabrous, fall foliage bright red; flowers grouped 5–8 together, occasionally to 15, loosely pubescent; fruits more ovate, 12 mm long, orange-red, stalk distinctly warty. Central and W. China. 1901. ⌀ ⚭

Illustrations of further species in ICS:

P. lasiogyna 2149	*P. impressivena* 2155
P. prunifolia 2150	*P. benthamiana* 2156
P. integrifolia 2151	*P. schneideriana* 2157
P. crassifolia 2152	*P. hirsuta* 2160
P. lucida 2154	

Attractive plants for a fertile, not too heavy soil; the deciduous types are good winter hardy species for sunny sites, the evergreen species are limited to mild areas but very ornamental.

Lit. Dress, W. J.: *Photinia × fraseri,* a new hybrid; Baileya **9,** 101–103, 1961.

PHYGELIUS E. Mey — SCROPHULARIACEAE

Evergreen shrubs or subshrubs, shoots angular; leaves opposite, the uppermost leaves smaller and also alternate, simple, margins crenate; flowers attractive, scarlet-red, calyx with 5 overlapping lobes, corolla long tubular, somewhat curved, limb with 5 short, erect lobes, grouped 3–7 together in small cymes, these united into one sided, terminal panicles; stamens 4, inserted in the throat; fruit a capsule. — 2 species in S. Africa.

Phygelius capensis E. Mey. Shrub, to 1 m high, woody only on the basal portion, shoots rather thick and 4 sided; leaves ovate to oval-lanceolate, 5–7 cm long, dentate, obtuse, base slightly cordate or rounded, petiole 2–5 cm long; flowers scarlet-red, pendulous, tube somewhat bowed, limb oblique, 5 lobed, inside yellowish, narrowed at the base, August–October. BM 4881. Cape Province, S. Africa. 1855. z9 Fig. 283. # ✛

Best treated as a tender perennial in cooler climates, easily rooted from cuttings and flowering in the same year; evergreen in milder climates.

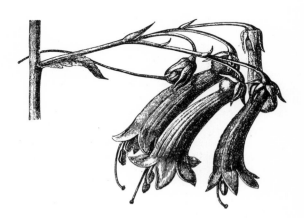

Fig. 283. *Phygelius capensis,* part of the inflorescence (nearly actual size) (from Kerner)

PHYLICA L. — Cape Myrtle — RHAMNACEAE

Low, evergreen shrubs, resembling *Erica* and *Myrtus* when not in bloom (however, not related to either); leaves alternate, densely pubescent, margins revolute; flowers usually in terminal heads, spikes or racemes; sepals 5, petals usually absent, when present, bristly or filamentous and pubescent; stamens 5, style undivided. — About 150 species, most in S. Africa; also in Madagascar, Tristan da Cunha. Only a few species in cultivation.

Phylica aethiopica see: **P. ericoides**

P. ericoides L. Shrub, 30–90 cm high, compact, all shoots erect and rather equal in length, finely floccose; leaves very densely arranged, heather-like, 6–12 mm long, linear, totally involuted; flower heads only 6 mm wide, globose, axillary and terminal, solitary or in clusters, surrounded by a dense, white woolly hull. BM 224 (= *P. microcephala* Willd.; *P. aethiopica* Hill). S. Africa; Cape Province. 1731. z9 # ✥

P. microcephala see: **P. ericoides**

P. plumosa L. Shrub, to 2 m high, entire plant long haired, all shoots tightly upright, not of equal length, densely foliate for their entire length; leaves linear-lanceolate, 2–3 cm long, glabrous above, woolly tomentose beneath, involuted, finely punctate; flowers in 2–3 cm long and equally wide spikes, surrounded by a thick hull of feathery, brownish-white bracts, very decorative (= *P. pubescens* Lodd.). S. Africa. z10 Plate 160. # ✥

P. pubescens see: **P. plumosa**

× PHYLLIOPSIS Cullen & Lancaster — ERICACEAE

Presumably hybrids between *Phyllodoce* and *Kalmiopsis*; differing from *Phyllodoce* in the nearly flat, wide leaves, with large yellowish glands on the underside, corolla campanulate with reflexed tips at the apex, interior loosely pubescent; differing from *Kalmiopsis* in the only slightly involuted leaves, raised midvein on the underside, corolla campanulate, tips reflexed, lightly pubescent interior, filaments glabrous.

× **Phylliopsis hillieri** Cullen & Lancaster (*Phyllodoce breweriana* × *Kalmiopsis leachiana*). Evergreen shrublets, presumably not reaching over 30 cm high; leaves alternate, compact, oblong-obovate, 15–20 × 6–8 mm in size, rounded apex, base cuneate, margins slightly involuted, dark green and glossy above, light green with brown-yellow glands beneath, petiole 1 mm; flowers many, in elongated racemes, stalks red, with stalked glands and crispate hairs, sepals 5, ciliate, obtuse, corolla campanulate, reddish-purple, about 1 cm wide, the lobes reaching to about the middle, base with a ring of short hairs, stamens 6–8, quickly abscising, filaments threadlike, glabrous, anthers light brown, ovaries globose, densely glandular, short crispate pubescent on the base, style longer than the stamens, April–May. z6

'Pinocchio'. The type of cross. JRHS 1977: 381. Chance seedling originating at the Hillier Nursery of Winchester, England in 1960. Supposedly winter hardy.

PHYLLODOCE Salisb. — Mountain Heath — ERICACEAE

Low, evergreen, heather-like shrubs; leaves alternate, linear, curved, finely serrate or bristly or glandular, soft pubescent beneath; flowers of all species (except *P. breweri*) in terminal heads; corolla urceolate or campanulate, 5 parted, calyx with 5 short teeth; stamens 10, inserted into the receptacle, anthers without appendages, dehiscing by a large opening; ovaries 5 chambered, the chambers with many ovules, the 5 valves separated by dividing walls. — 7 species in the arctic and alpine regions of the Northern Hemisphere.

Phyllodoce aleutica (Spreng.) A. Heller. Mat-like habit, to 20 cm high, shoots rather erect; leaves linear, obtuse, tapering to the base, 8–10 mm long, finely dentate, bright green above, yellowish-green beneath and with white lines; flowers grouped 6–12 in terminal heads, corolla globose-urceolate, yellowish-white, nodding, stalk glandular, sepals oval-lanceolate, 2/3 as long as the corolla, anthers pink, styles glabrous, April–May. TAP 117; NF 12: 34; MJ 744; BMns 496. NE. Asia to Alaska. z2 Plate 160. # ✥

P. alpina Koidz. Shrub, procumbent to erect, only 5–12 cm high, leaves linear, 3 to 10 mm long; flowers in terminal heads, corolla lilac-blue, ovate-urceolate, 6 mm long, glabrous (!), sepals ovate, elliptic, acute, glabrous, reddish-brown, 3 mm long, April–May. TAP 118. Japan. z3 # ✥

P. amabilis see: **P. nipponica** var. **amabilis**

P. breweri (Gray) A. Heller. Shrub, loose and bushy or somewhat prostrate, young shoots erect, densely foliate, 20–30 cm high; leaves linear, obtuse, 8–10 mm long, slightly glandular above, glossy dark green, margins involuted; flowers in erect, terminal, 5–10 cm long racemes (!), flower stalk 12 mm long, glandular, corolla shell-form, deeply incised, purple-pink, stamens widely exserted, sepals ovate, acute, ciliate, otherwise glabrous, green with a reddish trace, May–June. NF 12: 42; BM 8146; DRHS 1560 (= *Bryanthus breweri* A. Heller). USA, California; Rocky Mts. 1896. Often flowering again in the fall. z3 # ✥

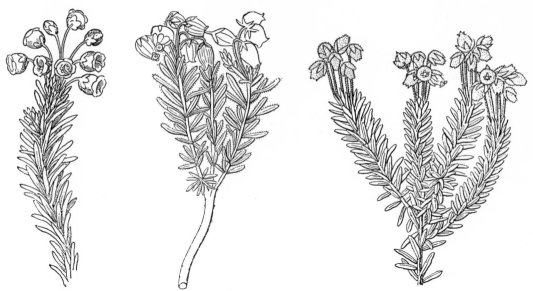

Fig. 284. **Phyllodoce.** Left *P. empetriformis;* center *P. coerulea;* right *P. glanduliflora*
(from B. M. & Hooker)

P. coerulea (L.) Bab. Bushy shrub, densely branched, about 10–15 cm high; leaves densely packed, linear, obtuse, 4–9 mm long, margins finely dentate, not involuted, glossy dark green, somewhat glandular at first; flowers in loose, terminal heads of 3–5, stalk to 3 cm long, glandular, corolla urceolate, 7–8 mm long, purple, becoming blue when dry, nodding, finely glandular pubescent, sepals oval-lanceolate, red-brown, glandular, May–June. BB 2760; TAP 119; NF 12: 44 (= *P. taxifolia* Salisb.). Arctic alpine region of Europe including Scotland and the Pyrenees. 1800. z2 Plate 160; Fig. 284. # ⊕

P. empetriformis (Sm.) D. Don. Small, broad growing shrub, 15–20 cm high, twigs very densely foliate; leaves linear, rounded at the apex, 6–15 mm long, margins finely dentate, glossy green above; flowers several at the shoot tips, nodding, stalks thin and glandular pubescent, 12–28 mm long, corolla campanulate, purple-pink, 7–9 mm long, with 5 rounded limb lobes, glabrous (!), sepals oval-lanceolate, glabrous (!), style exserted, April–June. NF 12: 38; RWF 266; BS 2: 432; BM 3176 (= *Bryanthus empetriformis* Gray). Western N. America, British Columbia to California. Around 1800. z3 Fig. 284. # ⊕

P. erecta see: × **Phyllothamnus erectus**

P. glanduliflora (Hook.) Cov. Upright shrublet, 10–20 cm high, very densely foliate; leaves linear, rounded, 6–12 mm long, finely dentate, deep green, with some white lines beneath, glandular above; flowers several at the shoot tips, fragrant, stalks not glandular pubescent, corolla urceolate-ovate, yellow-green, 6 mm long, exterior pubescent, narrow at the mouth, with 5 small, reflexed limb tips, stamens with purple anthers and pubescent filaments (!), style glabrous, sepals lanceolate,

acuminate, very glandular, April–May. NF 12: 75 (= *Bryanthus glanduliflora* Gray). Western N. America, Oregon to Alaska, to the snow line in the mountains. 1885. z3 Fig. 284. # ⊕

P. hybrida see: **P.** × **intermedia**

P. × **intermedia** (Hook.) Rydb. (*P. empetriformis* × *P. glanduliflora*). This is a group of naturally occurring hybrids, not particularly consistent and occasionally displaying the characteristics of one or the other of the parents. Bushy shrublet, 15–20 cm high, compact, strong growing; leaves linear, 6–12 mm long, obtuse or somewhat acute; flowers like those of *P. empetriformis,* but sepals more acute (!), long, ovate-campanulate to more urceolate, glabrous, sepals acutely lanceolate, red-brown, glandular pubescent only at the base, April–May. NF 12: 74; 211 (= *P. hybrida* Rydb.). British Columbia, Canada; occurring in the wild among the parents. # ⊕

'Drummondii'. Selection with dark purple flowers. ⊕

'Fred Stoker'. Selection with light purple flowers. ⊕

P. nipponica Mak. Somewhat similar to *P. empetriformis,* but with white flowers. Shrub, compact, more or less erect, 10–15 cm high, loosely foliate; leaves 6–12 mm long, linear, obtuse or somewhat acute, finely dentate, deep green and glabrous above, with white, finely pubescent lines beneath; flowers grouped 3–7 at the shoot tips, stalk glandular pubescent, corolla campanulate, open, 6 mm long and wide, limb tips round, white, lobes slightly reddish, sepals acute, 3 mm long, ciliate, green, stamens and style enclosed, May. JRHS 98: 30; TAP 120; NF 12: 35, MJ 743. N. Japan. 1915. z3 # ⊕

var. **amabilis** (Stapf) Stoker. Differing in the more reddish limb tips, red sepals and shorter anthers. BM 8405 (= *P. amabilis* Stapf). # ⊕

var. *tsugifolia* see: **P. tsugifolia**

P. taxifolia see: **P. coerulea**

P. tsugifolia Nakai. Very closely related to *P. nipponica*, but the flower stalks only glandular pubescent on the apical third (not totally!); leaves linear, obtuse or somewhat acute; flowers white, ovate, urceolate, sepals ovate, acute, glabrous, shorter than ⅓ of the corolla length. GC 119: 115; TAP 121 (= *P. nipponica* var.

tsugifolia Osborn). N. Japan. Existence in cultivation uncertain. z3 # ☺

Very attractive dwarf shrubs for experienced gardeners; best suited for semi-shady sites in the rock garden; requires a lime free, fertile, loose soil, humus; some winter protection advisable, but otherwise totally hardy. Gorgeous in flower!

Lit. Stoker, F.: The genus *Phyllodoce*; in New Flora and Silva **12**, 30–42 ● Camp, W. H.: *Phyllodoce* Hybrids; in the same volume, 207–211.

PHYLLOSTACHYS S. & Z. — Bamboo — GRAMINEAE

Normally a tall growing shrub with a creeping rootstock, occasionally spreading grasslike; stems with short, hollow internodes, flattened or furrowed on one side; lateral shoots usually 2–3 at a node, these further branched; stem sheaths quickly abscising; leaves petiolate, small or medium-sized, segmented with tessellate venation; leaf sheaths with small, rough bristles; flowers in terminal, foliate panicles; styles long, with 3 stigmas (flower characteristics not considered here). — About 40 species in E. Asia and Himalaya.

Key to the more prominent species (from Rehder)

● Stem sheath glabrous on the margins and as long or longer than the internodes; scales at the base of the twig not parted or at most parted only to the middle, persistent;

   > Stem dark colored when mature; leaf sheaths without bristles:
      P. flexuosa

   >> Stem green or yellow; leaf sheaths with bristles;

     + Stems more widely spaced apart; stem sheaths purple striped:
       P. viridi-glaucescens

     ++ Stems arranged in clusters;
       ★ Stem sheaths without or with only a few small spots; stems yellow; leaves 1–2 cm wide:
        P. aurea

      ★★ Stem sheaths with large spots;

       * Nodes clustered at the stem base; sheaths brownish punctate:
        P. pubescens

       ** Nodes not clustered; sheaths purple marbled:
        P. bambusoides

●● Stem sheaths with ciliate margins, shorter than the internodes when mature; scales deeply parted at the base of the shoots, quickly closing:
     P. nigra

Phyllostachys aurea Riv. Stems 3–5 m high (much taller in warm climates), 2–2.5 cm thick, light green when young, later yellow-green, eventually straw-yellow (like many bamboos, hence the name "aurea" can be somewhat misleading!), internodes about 15 cm long, with a ring-like bulge under each node (!!), not punctate; leaves linear, usually small, about 5 (10) cm long, 1–2 cm

wide, finely dentate only on one edge, dark green above, blue-green beneath, with 4–5 vein pairs, glabrous. MJ 2640; AH 114: 8–9 (= *Bambusa aurea* Hort. non Sieb.; *P. bambusoides* var. *aurea* [Riv.] Mak.). China; cultivated in Japan. Around 1870. z6 # Ø ✄

P. aureosulcata McClure. Medium height, but to 7 m or more in warm climates, stems erect, often zigzagging back and forth(!), young shoots green in the first year on the rounded part of the internodes, but golden-yellow on the flat portion, this more or less paling in the 2nd year(!), stem sheaths white striped; leaves usually 5–12 cm long, narrow, 3–5 on a twig. NH 39: 126; AH 114: 10–11. China; Chekiang Province. 1907. Widely grown in the southern USA as *"P. nevinii"*. z7 # Ø ✄

P. bambusoides S. & Z. Stems about 5–6 m high in milder areas (15–20 m high in its habitat and to 12 cm thick), internodes green (also yellow on some cultivars), glossy, without whitish pruinose, leaf scars from the fallen stem sheaths, these greenish to yellowish-brown, more or less dark brown speckled for the entire length; leaves 8 to 15 cm long, 1–3 cm wide, abruptly long acuminate, base tapered to a short petiole, finely serrate on one side, light green above, blue-green beneath and pubescent at the base. CBa 27; NH 25: 60; AH 114: 12–13 (= *P. reticulata* sensu Koch; *P. quilioi* Riv.). China. 1866. One of the more widely cultivated species in Japan and China for its many uses. z9 Plate 161. # Ø ✄

'Allgold' (McClure). Only 2–3 m high, stems light yellow at abscission of the sheaths, turning somewhat wine-red on the basal part of the stem, later becoming golden-yellow, now and then with some basal internodes having green stripes, stem sheaths punctate; leaves like those of the species, occasionally with light yellow to white stripes (= *P. sulphurea* A. & C. Riv.; *P. castillonis* var. *holochrysa* Pfitzer). China. 1865. Plate 163. # Ø

var. *aurea* see: **P. aurea**

'Castillon' (McClure). About 2.5–4 m high, stems slender, primary color of the twigs and internodes light yellow, basal internodes occasionally reddish, always green on the flat side (!); leaves 7–12 cm long, occasionally with light yellow stripes (!). NH 45: 291 (= *P. castillonis* [Marl.] Mitford; *P. nigra* var. *castillonis* [Mitford] Bean). China; brought to Japan centuries ago; introduced into Europe in 1886. # Ø

P. boryana see: **P. nigra 'Bory'**

P. castillonis see: **P. bambusoides 'Castillon'**

P. castillonis var. *holochrysa* see: **P. bambusoides 'Allgold'**

P. edulis see: **P. pubescens**

P. fastuosa see: **Semiarundinaria fastuosa**

P. flexuosa A. & C. Riv. 6–9 m high in its habitat, scarcely over 3 m high in cultivation (in warm areas), stems slender, delicately arranged, green at first, later nearly black (!); leaves 5–10 cm long, 1–1.5 cm wide, long acuminate, base abruptly tapering, dark green above, blue-green beneath, finely dentate on only one edge. CBa 31A; AH 114: 24–25 (= *Bambusa flexuosa* Carr. non Munro). China. 1864. z7 Plate 162. # ⌀ ✄

P. henonsis see: **P. nigra 'Henon'**

P. mitis see: **P. viridis**

"*P. nevinii*" see: **P. aureosulcata**

P. nidularia Munro. Stems to 10 m high in its habitat, erect, straight, occasionally bowed from the weight of the foliage, internodes green, more or less mealy pulverulent at the time of abscission of the stem sheaths (!), rings at the nodes very broad and prominent (!), stem sheath scars very thick, distinctly ciliate at first with long, brown hairs, stem sheaths olive-green to pale green and white, broadly spear-shaped (!), both sides with a large auricle at the base; leaves usually 7–10 cm long, 2 cm wide, usually totally glabrous on the underside. CBa 36A; NH 25: 43; AH 114: 32–33. China. z7 # ⌀ ✄

P. nigra (Lodd.) Munro. Stems 5–7 m high and to 3 cm thick in its habitat, internodes green at first, gradually becoming spotted, eventually nearly black from the purple-black or brown patches, often more or less mealy pulverulent at the time of abscission of the stem sheaths, stem sheaths greenish-brown, totally without spots, blade triangular, navicular, often undulate; leaves 5–10 cm long, 0.5–1.5 cm wide, very thin, dark green above, blue-green and usually glabrous beneath, occasionally pubescent, finely dentate. NH 24: 280; MJ 2643; AH 114: 34–35 (= *Bambusa nigra* Lodd.). China, Japan. 1823. z7 Plate 162. # ⌀ ✄

'Bory' (McClure). Taller growing, more narrowly upright, but with green stems at first, later light yellow and only with a few brown patches. CBa 28D (= *P. boryana* Mitford). Origin uncertain, but probably China. Before 1895. # ⌀

var. *castillonis* see: **P. bambusoides 'Castillon'**

'Henon'. (McClure). Taller than *P. nigra* in its habitat, but in cultivation seldom over 4 m high and very densely foliate, stems greenish at first, later yellowish, totally without spots; leaves green on both sides; otherwise hardly differing from the species. CBa 28A; NH 25: 33 (= *P. puberula* [Miq.] Munro; *P. henonis* Mitford; *P. nigra* var. *henonis* [Mitford] Stapf). S. China; cultivated in Japan. 1890. # ⌀

P. puberula see: **P. nigra 'Henon'**

P. pubescens Mazel. Stems 20 m high or more in its habitat, 5–7 cm thick or more at the base, straight and upright or also more or less bowed, internodes pale green, densely velvety pubescent when young, later gradually becoming glabrous and often with a waxy pulverulence, stem sheaths greenish-brown, with dark brown patches and brown pubescence, margins ciliate, lanceolate to triangular; leaves 6–12 cm long, 1–2 cm wide, long acuminate, base round, finely dentate on one side, dark green above, blue-green beneath. CBa 26A, 30B; MJ 2641; AH 114: 40–41 (= *P. edulis* [Carr.] Houzeau; *P. mitis* sensu Bean non Riv.; *Bambusa edulis* Carr.). China; cultivated in Japan. 1877. z7 Plate 162, 163. # ⌀ ✄

P. quilioi see: **P. bambusoides**

P. reticulata see: **P. bambusoides**

P. sulphurea see: **P. bambusoides 'Allgold'** and **P. viridis 'Robert Young'**

P. ruscifolia see: **Shibataea kumasca**

P. sulphurea var. *viridis* see: **P. viridis**

P. viridi-glaucescens (Carr.) Riv. Stems to 6 m high or more, internodes green, often unevenly and loosely mealy pulverulent at the time of abscission of the sheaths, totally glabrous and not furrowed, stem sheaths light brown, green toned, with small brown spots and patches, loosely mealy pulverulent, rough, often somewhat bristly; leaves 5–10 cm long, 1–2 cm wide, long acuminate, bright green above, blue-green beneath and pubescent at the base, with 4–7 faintly visible vein pairs. NH 24: 278; AH 114: 48–49 (= *Bambusa viridi-glaucescens* Carr.). China. 1846. z7 # ⌀ ✄

P. viridis (Young) McClure. Stems to 14 m high in its habitat, rather erect, often somewhat wavy, but not zigzagged, internodes pale-green, often bluish beneath the nodes, never striated, stem sheaths glabrous, often somewhat bluish, otherwise brownish-pink, the venation green, with brown spots and patches, auricles on the ligules totally absent (!); leaves usually 7–12 cm long, about 1.5–2 cm wide, often pubescent on one side, the underside occasionally pubescent only at the base. AH 114: 50–51 (= *P. mitis* Riv.; *P. sulphurea* var. *viridis* Young). China. 1928. z7 Plate 161. # ⌀ ✄

'Robert Young' (McClure). Low growing, internodes yellow-green with dark green stripes and a dark green band under the stem sheath scar, the primarily yellow-green internodes becoming golden-yellow with increasing age, but the green longitudinal stripes remain, stem sheaths somewhat lighter than those of the species (= *P. sulphurea* Riv.). China. 1865. # ⌀

All species thrive in a good garden soil, but for development of strong stems they require fertilization and good soil moisture. The species spread quickly by stolons; division should be undertaken in April and May. The new plants should be kept in a greenhouse until the new roots are developed before lining out. The plants will die after flowering, which rarely occurs in cultivation; in their habitat large areas become denuded by this means.

Lit. Please refer to the detailed listing in Vol. I, p. 188.

× PHYLLOTHAMNUS Schneid. — ERICACEAE

Generic hybrid between *Phyllodoce* and *Rhodothamnus;* only one hybrid is known to date. Flowers similar to those of *Rhodothamnus,* but the broadly campanulate corolla is surpassed by the stamens (!); leaves ciliate-serrate, margins involuted; differing from *Phyllodoce* in the form of the corolla and the narrowly involuted leaves with a raised, glabrous midrib beneath. — Originated in cultivation; as yet not observed in the wild.

× **Phyllothamnus erectus** (Lindl.) Schneid. (= *Phyllodoce empetriformis* × *Rhodothamnus chamaecistus*). Evergreen shrub to 30 cm high, narrowly upright, twigs densely foliate; leaves alternate, linear, 12 mm long, slightly serrate, glabrous; flowers 2–10 in terminal umbels, corolla broad campanulate, pink-red, 12 mm wide, flower stalks pubescent, sepals triangular, about ⅓ as long as the corolla, June–August. FS 659 (= *Bryanthus erectus* Lindl.; *Phyllodoce erecta* Drude). Developed about 1845 in the Cunningham & Fraser Nursery in Edinburgh, Scotland. Fig. 285. # ☙

Cultivated like *Phyllodoce.* According to Chittenden it is short lived, lasting hardly more than 8 years in the garden.

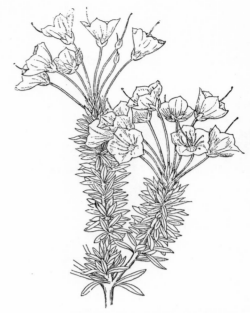

Fig. 285. × *Phyllothamnus erectus* (from Andr.)

PHYSOCARPUS (Cambess.) Maxim. — Ninebark — ROSACEAE

Tall, deciduous shrubs; leaves alternate, petiolate, large, usually 3 lobed, stipules quickly abscising, conspicuous (distinction from *Spirea*); flowers in corymbs, calyx broadly campanulate with a 5 lobed limb, petals 5, outspread, white; stamens very numerous, carpels usually 5, occasionally 1–4; fruit a follicle, bursting open at both seams (distinction from *Neillia*!); seeds wingless, stone hard, shell glossy. — 10 species in N. America, one in N. Asia.

Key to the more prominent species

● Carpels 4–5, occasionally 3;

  × Capsules glabrous;
  > Calyx glabrous to sparsely pubescent
      P. opulifolius

  >> Calyx densely stellate tomentose:
      P. capitatus

  ×× Capsules pubescent;
    § Shoots glabrous to slightly pubescent; lobes of the leaves acute to acuminate:
      P. amurensis

    §§ Shoots glabrous; lobes obtuse:
      P. intermedius

    §§§ Shoots stellate pubescent; leaves acuminate:
      P. stellatus

● ● Carpels usually 2, connate for half their length;

    × Styles outspread:
          P. monogynus

    ×× Styles erect:
          P. malvaceus

Physocarpus alabamensis see: **P. stellatus**

P. alternans (Jones) J. T. Howell. Shrub, 0.3–1 m high, densely branched, shoots stellate pubescent and often glandular; leaves rounded to rhombic, 8–20 mm long, double crenate or 3 lobed, more or less densely pubescent on both sides; flowers 3–6 in terminal corymbs, June–July; fruits solitary, about 6 mm long, densely pubescent. MS 207–209. USA; California to Nevada, in the mountains. z5

P. amurensis (Maxim.) Maxim. Shrub, 1–3 m high, very similar to *P. opulifolius*, twigs glabrous to slightly gray pubescent; leaves ovate, 5–10 cm long and nearly as wide, 3–5 lobed, double finely serrate-incised, with acute teeth, green and nearly glabrous above, whitish-green and scattered pubescent beneath; flowers about 1.5 cm wide (!), in loose corymbs, calyx densely stellate tomentose, anthers purple, June–July; fruits grouped 3–4 together. SH 1: 284 a–c, 285f, 286 d–e; Gf1 14: 489 (= *spiraea amurensis* Maxim.). Manchuria, Korea, Amur region. Around 1856. z5 Plate 164; Fig. 286. ☙

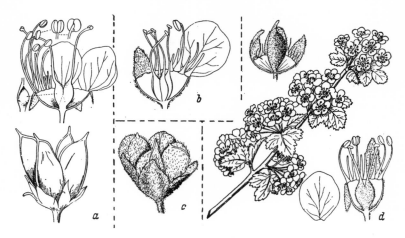

Fig. 286. **Physocarpus.** Flowers, fruits and twig. a. *P. opulifolius;* b. *P. capitatus;*
c. *P. amurensis;* d. *P. monogynus* (from Sargent, Schneider)

P. bracteatus (Rydb.) Rehd. Similar to *P. monogynus,* to about 1.8 m high or more, old bark papery, exfoliating, young shoots yellowish, glabrous; leaves broad ovate, 2–7 cm long, usually 3 lobed, the lobes double crenate; flowers about 12 mm wide, several in 5 cm wide, hemispherical cymes, surrounded by spathulate to obovate, leaflike bracts, stalk and calyx densely stellate pubescent, June; fruits in 2's, half connate. BS 2: 21. USA; Colorado. 1909. Plate 164.

P. capitatus (Pursh) Ktze. Erect shrub, 1–3 m high, often wide arching; leaves broad ovate on the flowering shoots, 3–7 cm long, 2–4 cm wide, 3–5 lobed, irregularly double serrate, glabrous above, loose or densely stellate tomentose beneath, larger on long shoots, more deeply lobed and sharper toothed; flowers many in dense, 5–7 cm wide, hemispherical corymbs, white to light pink, calyx and stalk pubescent, April–June; fruits grouped 3–5 together, inflated, reddish, glabrous. MS 206 (= *Spiraea capitata* Pursh). Western N. America, near water and on moist slopes. 1827. z5 Plate 164; Fig. 286. ✥

P. glabratus (Rydb.) Rehd. Similar to *P. monogynus,* but the flower stalks and calyx glabrous or nearly so, flowers somewhat larger, leaves not so deeply lobed, the lobes double crenate. USA; Colorado. 1908. z5 Plate 164.

P. intermedius (Rydb.) Schneid. Shrub, to 1.5 m high, young shoots glabrous or nearly so; leaves oval-rounded, 2–6 cm long, shallowly 3 lobed, the lobes obtuse double crenate, somewhat stellate pubescent beneath to nearly glabrous; flowers 12 mm wide, in dense cymes, flower stalks and calyx stellate pubescent to nearly glabrous; fruits stellate pubescent, inflated, twice as long as the sepals. BB 2: 144 (= *P. ramaleyi* A. Nel.; *P. missouriensis* Daniels). USA. 1908. z3 Plate 164.

var. **parvifolius** Rehd. Lower, denser habit, shoots erect; leaves more ovate, acute, only 1.5–2 cm long, base round to broadly cuneate; flowers smaller, calyx and stalks glabrous; fruits usually in 3's, finely pubescent. Before 1918.

P. malvaceus (Greene) Ktze. Shrub, 1–2 m high, erect habit, shoots densely stellate tomentose; leaves rounded to broadly ovate, 2–6 cm long, base round to cordate, 3 lobed (also 5 lobed on long shoots), as wide as long, the lobes broad-rounded, double crenate, stellate pubescent on both sides, but often glabrous above; flowers 1 cm wide, white, few in 3 cm wide corymbs, calyx and stalk pubescent, June; fruits in 2's, somewhat keeled and flat, with erect (!) beaks. BM 7758 (as *Neillia torreyi*) (= *P. pauciflorus* Piper; *Neillia malvacea* Greene). Western N. America. 1897. z5 Plate 164.

P. missouriensis see: **P. intermedius**

P. monogynus (Torr.) Coult. Shrub, somewhat resembling *Ribes alpinum,* scarcely 1 m high, shoots glabrous or somewhat stellate pubescent; leaves broad ovate to kidney-shaped, 2–4 cm long, deeply 3–5 lobed, the lobes incised-serrate, glabrous or nearly so on both sides; flowers 1 cm wide, white (often soft pink), few in small, flat-convex corymbs, late May to June; fruits in 2's, connate to the middle, densely stellate tomentose. VT 426; GF 2: 5; DRHS 1563 (= *P. torreyi* S. Wats.; *Spiraea monogyna* Torr.; *Neillia torreyi* S. Wats.). Central USA. 1889. z5 Plate 164; Fig. 286. z5 ✥

P. opulifolius (L.) Maxim. Shrub, 2–3 m high and wide, branches with a brown, shredded, exfoliating bark, twigs glabrous; leaves oval-rounded, base usually cordate, usually 5 lobed, the lobes crenate or also acute; flowers to 1 cm wide white (often pale pink), in many flowered, about 5 cm wide corymbs, June; fruits grouped 3–5, totally glabrous, connate at the base, reddish, inflated. BB 1882; GSP 167; HM 1014; VT 427 (*Spiraea opulifolia* L.). Central and Eastern N. America. 1687. z2 Plate 164; Fig. 286. ∅ ✥

'Dart's Gold'. Like 'Luteus', but remaining lower. Introduced in 1969 by the Darthuizer Nursery of Leersum, Holland, but the origin is unknown. Quite meritorious! ∅

'**Luteus**'. Leaves yellow on new growth, later more or less yellowish-green or bronze-yellow. BMns 459. Before 1864. ⌀

'**Nanus**'. Low growing; leaves smaller, deep green, not so deeply lobed. Before 1864.

P. pauciflorus see: **P. malvaceus**

P. ramaleyi see: **P. intermedius**

P. stellatus (Rydb.) Rehd. Closely related to *P.*

intermedius, but young shoots stellate pubescent, as is the leaf underside; leaves on long shoots often acute to acuminate; inflorescences rather small and loose, stalks and calyx densely stellate pubescent; fruits usually in 4's, acute (= *P. alabamensis* Rydb.). USA; S. Carolina to Georgia. 1906. z6 Plate 164.

P. torreyi see: **P. monogynus**

Not particular to soil type, about like *Spiraea;* likes full sun.

PHYTOLACCA L. — Pokeberry — PHYTOLACCACEAE

Perennials, shrubs or trees, usually totally glabrous; leaves alternate, petiolate to nearly sessile, simple, ovate, elliptic to lanceolate; flowers small, in terminal, erect or nodding racemes, often opposite the leaves (the shoots continue growing); calyx 4–5 parted, persistent; petals absent; stamens 6–30, in 2 rings; ovaries nearly globose, with 5–16 distinct or connate styles; fruit a fleshy berry. — About 35 species in the tropics, most from America, some from Africa and Asia.

Phytolacca dioica L. The "Bella Sombra" of South America ("Beautiful Shade"). Evergreen tree with a short thick trunk, very fast growing, to 18 m high and wide, corolla very dense; leaves petiolate, elliptic to ovate, 6–10 cm long; flowers unisexual, white, in pendulous racemes, about as long as the leaves, staminate flowers with 20–30 stamens, widely exserted past the calyx, flowering in fall, styles 7–10; fruits black. S. America, commonly cultivated in California. z9 Plate 165. # ⌀ ⊕

Not suited for temperate climates.

PICRASMA Bl. — Bitterwood — SIMAROUBACEAE

Deciduous trees, similar to *Ailanthus,* but with flowers axillary and fruits not winged; leaves alternate, odd pinnate; flowers polygamous, small, in loose, axillary umbellate panicles; petals 4–5, sepals persistent, stamens 4–5, inserted into the base of the 4–5 lobed disk, longer than the petals, carpels 2–5, styles thin, apex split into 2–5 stigmas; fruits in clusters of 1–5 rather globose drupes; bark very bitter. — 6 species in the tropics and subtropics.

Picrasma ailanthoides see: **P. quassioides**

P. quassioides (D. Don) Bennet. To 10 m high, shoots reddish-brown, densely covered with yellowish lenticels; leaves 25–35 cm long, leaflets 9–15, nearly sessile, oval-oblong, 4–10 cm long, acuminate, finely crenate, base broadly cuneate to round and oblique, glossy above, light green beneath, orange-red in fall; flowers greenish, in 8–15 cm wide umbellate panicles, May–June; fruits oval-rounded, 6–7 mm long, red. LF 193; BMns 279; MJ 1152; LWT 143 (= *P. ailanthoides* Planch.). N. China and Korea to the Himalayas and Japan. 1880. z5 Fig. 287. ⌀

Very hardy (to 6 m high in Poland!); without particular cultural requirements. Best planted in the open to obtain the full effect of the fall foliage.

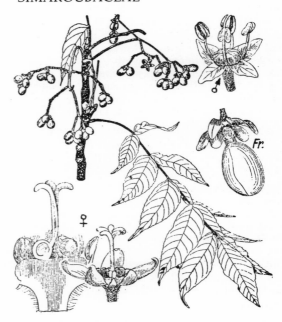

Fig. 287. *Picrasma quassioides,* fruiting twig, individual flowers and fruit (from Miyabe & Kudo, Engler)

PIERIS D. Don — ERICACEAE

Evergreen shrubs, occasionally a small tree, similar to *Andromeda*; leaves alternate, serrate to crenate; flowers in terminal panicles, usually leafless; corolla urceolate, short lobed, calyx lobes separate; stamens 10, anthers with 2 recurved awn-like appendages; fruit a 5 chambered, rounded capsule. — 10 species in N. America, E. Asia and the Himalayas.

Outline of the species characteristics

Species	Young Shoots	Inflorescences	Sepals	Corolla (length)
floribunda	stiffly pubescent	erect, 5–10 cm long	whitish	5–6 mm
formosa	red! finely pubescent	nodding, to 15 cm	green	6–9 mm
—var. *forrestii*	carmine (!), glabrous	pendulous	white (!)	9 mm
japonica	glabrous, often brown	pendulous, 6–12 cm	exterior finely pubescent	6–8 mm
phillyreifolia	glabrous	axillary (!), 5–6 cm	greenish	7–8 mm
taiwanensis	glabrous, yellow-green	erect, 8–15 cm	exterior glabrous	7–8 mm

Pieris bodinieri see: **P. formosa**

P. cavaleriei see: **Leucothoe griffithiana**

P. floribunda (Pursh) Benth. & Hook. f. Erect shrub, to 2 m high, but usually lower, twigs strigose; leaves elliptic to oblong-lanceolate, 3–8 cm long, crenate and ciliate, dull green above, with a scattered bristly pubescence on both sides; flowers in erect, dense, terminal panicles, 5–10 cm long, corolla white, urceolate, 5–6 mm long, angular, sepals whitish, April–May. BB 2768; BM 1566; DB 1951: 159; NBB 3: 17 (= *Andromeda floribunda* D. Don). Southeastern N. America, in mountain forests. 1800. z5 Plate 167; Fig. 288. # ⊕

P. forrestii see: **P. formosa var. forrestii**

P. formosa (Wall.) D. Don. To 6 m high in its habitat, with stolons, very similar to *P. japonica*, but larger in all respects, new growth red, young shoots finely pubescent; leaves oblong-lanceolate, 6–15 cm long, finely serrate, fine reticulate venation above (!); flowers very numerous, in somewhat erect, but generally nodding, 15 mm long panicles, corolla white, 6–9 mm long, sepals green (!), April–May. BM 8283 (= *Andromeda formosa* Wall.; *P. bodinieri* Lév.). E. Himalayas. 1858. z7 Plate 167; Fig. 288. # ⊘ ⊕

var. **forrestii** Airy-Shaw. To 3 m high in its habitat, new growth bright carmine-red, glabrous; leaves elliptic-lanceolate, acuminate, 6–10 cm long, finely serrate; flowers in terminal, nodding panicles, sepals white (!), lanceolate, corolla dull white, fragrant, 9 mm long, April–May. NF 3: 29; BM 9371 (= *P. forrestii* Harrow). W. China; Yunnan Province; Upper Burma. 1910. Since this plant is easily grown from seed those in cultivation are not very consistent. z8 Plate 176. # ⊘ ⊕

P. japonica (Thunb.) D. Don. Shrub, to 3 m high, bark brown, shredding in long strips, twigs glabrous, new growth often brownish to reddish; leaves clustered at the branch tips, obovate to lanceolate, 3–8 cm long, crenate, dark green above, lighter and glabrous beneath, very glossy on both sides; flowers in pendulous, 6–12 cm long panicles, corolla ovate, white, 6–8 mm long, sepals finely pubescent on the exterior, calyx often red-brown, February–May. DB 1951: 152; BS 2: 167 (= *Andromeda japonica* Thunb.). Japan. z6 Plate 166; Fig. 288. # ⊘ ⊕

P. japonica × **formosa** var. **forrestii**. To date including only 'Forest Flame'.

P. lucida see: **Lyonia lucida**

P. macrocalyx see: **Lyonia macrocalyx**

P. mariana see: **Lyonia mariana**

P. nana see: **Arcterica nana**

P. ovalifolia see: **Lyonia ovalifolia**

P. phillyreifolia (Hook.) DC. Evergreen shrub, 0.5–1 m high in cultivation, but in its habitat also epiphytic and climbing high (!) on the bark of *Taxodium distichum*; leaves oblong, elliptic-oblong, 3–5 cm long, with a few small teeth only at the apex, glossy above, margins somewhat involuted; flowers in numerous, axillary (!) racemes of 4–12 flowers, corolla ovate, narrowed at the apex, 7–8 mm long, white, occasionally turning somewhat reddish, stamens with S-form, bowed filaments, flowers appear January–February. HI 122; BR 30: 36; Bai 11: 42 (= *Andromeda phillyreifolia* Hook.). W. Florida, USA, in the *Taxodium* swamps. z7 #

P. taiwanensis Hayata. Erect, rounded, 1–2.5 m high shrub, shoots glabrous, yellow-green; leaves similar to those of *P. japonica*, 3–7 cm long, oblanceolate to obovate or narrow elliptic, glabrous, serrate on the apical half; flowers usually in erect panicles, the lateral branches, however, horizontal to pendulous, corolla urceolate, white, sepals green, exterior glabrous, March–April. BS 2: Pl. 36; BM 9016; LWT 287. Taiwan. 1918. Flowers occur so early that they are often damaged by frost. z7 # ⊘ ⊕

Outline of the Cultivars

'Bert Chandler' (*P. japonica*) Young leaves an attractive salmon-pink at first, turning glossy cream-yellow, then white, and finally deep green (= 'Chandleri'). Developed about 1954 in the Como Nurseries, The Basin, Vict., Australia. # ⊘ ⊕

'Blush' (*P. japonica*). Flowers a pretty pink in bud, opening white with a trace of soft pink. # ⊕

'Chandleri' see: **'Bert Chandler'**

'Charles Michael' (*P. formosa* var. *forrestii*). Flowers in large panicles, individual flowers pure white, surpassing all other forms in size. Grown at Caerhays Castle, Cornwall, England from seed collected by Forrest (F.27765). # ⊘ ⊕

'Christmas Cheer' (*P. japonica*). Flowers dark pink, flowers abundantly, beginning in March. Chance seedling originated in the K. Wada Hakoneya Nursery, Yokohama, Japan. 1967. Flowers noted for their frost tolerance. Good potential as a winter flowering pot plant. # ⊕

'Compacta' (*P. japonica*). Compact habit, but growing to 1.8 m high, very densely branched; leaves only half as large as those of the species. A selection by Westbury Rose Co., Long Island, N.Y., USA. 1949. #·

'Crispa' (*P. japonica*). Slow growing; leaves very undulate and crispate margined. Before 1946. # ⊘

'Crispa' (*P. taiwanensis*). Slow growing, low; young leaves coppery-red, very undulate and crispate, dull; flowers in large, loose racemes. # ⊘ ⊕

'Daisen' (*P. japonica*). Good grower, abundant foliage; buds dark pink, red when fully opened, gradually fading to pink (the color is more intense in the shade, paler in sun). Found on Mt. Daisen, introduced by K. Wada. Before 1967. # ⊕

'Dorothy Wyckhoff' (Hohman) (*P. japonica*). Compact habit; leaves deep green in summer and glossy, reddish-green in winter; flower buds dark red in winter, later dark carmine-red, opening pure pink (not pale pink or white!). Henry J. Hohman, Kingsville, Maryland, USA. 1960. # ⊘ ⊕

'Elongata' (*P. floribunda*). Inflorescences much longer than those of the species, the individual flowers also somewhat larger, flowering time later (= 'Grandiflora'). Originated around 1935. # ⊕

'Flame of the Forest' see: **'Forest Flame'**

'Flamingo' (Lambert) (*P. japonica*). New growth bronze-red; panicles about 11 cm long, flowers dark carmine-pink, not fading, individual flowers about 9 mm long, 7 mm wide. Lambert Gardens, Portland, Oregon. 1961. Plate 128. # ⊘ ⊕

'Forest Flame' (Sunningdale) (*P. japonica* × *formosa* var. *forrestii*). Combines the winter hardiness of *P. japonica* with the properties of *P. formosa* var. *forrestii* 'Wakehurst' (= 'Flame of the Forest'). Sunningdale Nurseries, Windlesham, Surrey, England. 1952. # ⊘ ⊕

'Grandiflora' see: **'Elongata'**

'Jermyns' (Hillier) (*P. formosa*). Leaves about 8 cm long and 2.5 cm wide, distinctly reticulate venation; flower stalk and sepals more or less reddish (!), developed by Hillier & Son, Winchester, England. Before 1959. # ⊘ ⊕

'Millstream' (*P. japonica*). Low growing and compact, very broad, small leaves. Selection of H. Lincoln Foster, Falk Village, Connecticut, USA. 1947. # ⊕

'Nana Compacta' see: **'Pygmaea'**

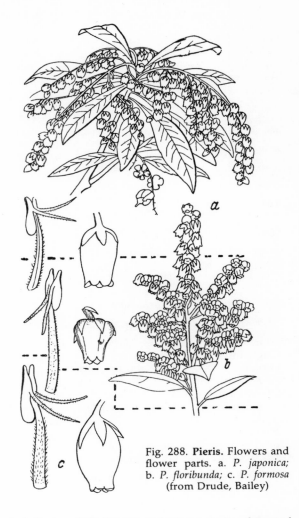

Fig. 288. **Pieris.** Flowers and flower parts. a. *P. japonica*; b. *P. floribunda*; c. *P. formosa* (from Drude, Bailey)

'Purity' (*P. japonica*). Flowers very large, pure white and nodding from their weight. Origin unknown, presumably a mutation. Introduced in 1967 by K. Wada, Hakoneya Nursery, Yokohama, Japan. # ⊕

'Pygmaea' (*P. japonica*). Much lower growing; leaves much smaller, linear-lanceolate, 1.5–2.5 cm long, margins crenate; flowers like the species, but usually in simple racemes (= 'Nana Compacta' H. A. Hesse). From Japan. 1873. #

'Splendens' (*P. japonica*). New growth an intense red-brown. Selected by J. Hachmann, Barmstedt, W. Germany. # ⊘ ⊕

'Valley Rose' (*P. japonica*). Slow growing (70–80 cm high in 10 years), winter buds dark brown, new foliage reddish-green; flowers pink. Developed by crossing 'Flamingo' with another red flowering type by R. L. Ticknor, North Willamette Experiment Station, Aurora, Oregon, USA. # ⊘ ⊕

'Variegata' (*P. japonica*). Often develops normal green shoots, these shorter; leaves usually smaller than those of the species, narrowly white bordered; few flowers. From Japan. 1873. Plate 166. # ⊘

'Wakehurst' (Price) (*P. formosa*). Leaves broader and shorter, mahogany-brown on new growth (not red!); panicles about 12 cm long, more erect than those of the species, nodding, corolla somewhat larger. JRHS 1957: 88. Developed at Wakehurst Place, Ardingly, Sussex, England from seed collected by George Forrest, in China, 1912. Plate 128. # ⌀ ✧

'Whitecaps' (*P. japonica*). Selection with particularly long inflorescences, pure white, flowering for about 6 weeks. From P. Vermeulen, Neshanic Station, N.J., USA. 1961. # ✧

'White Cascade' (*P. japonica*). Inflorescences much denser than those of the species, also much more densely foliate at the base, flowers abundantly every year, over a 5 week period, pure white. Selected by Raymond P. Korbobo, Middlesex, N.J., USA. 1961. # ✧

All species should be treated somewhat like *Rhododendron*, preferring a semi-shady site, humus and lime free soil; while the other species thrive in a fertile, clay soil, *P. floribunda* requires a sterile, sandy, but moist soil. They all grow more slowly in full sun, but flower more abundantly. *P. formosa* is gorgeous in flower but unfortunately limited to warmer climates.

Lit. Wagenknecht, B. L.: Registration Lists of Cultivar names in the genus *Pieris*; Arnoldia **21**, 47–50, 1961 ● Ingram, J.: Studies in the cultivated Ericaceae; 4. *Pieris*; Baileya **11**, 38–46, 1963.

PILEOSTEGIA Hook. & Thoms. — HYDRANGEACEAE

Evergreen shrubs, climbing in trees by tendrils; leaves opposite, entire or dentate; flowers small, whitish, similar to *Hydrangea* or *Schizophragma*, but without marginal flowers, calyx cup-form, with 4–5 sepals; petals 4 or 5; stamens 8–10, inserted with the petals into the calyx; style thick, short, 4–5 lobed at the apex; fruit a small, dehiscent capsule with many seeds. — 3 species in SE. Asia.

Pileostegia viburnoides Hook. & Thoms. Glabrous shrub, climbing 5–10 m high or more, young shoots and leaves lepidote at first, but soon totally glabrous; leaves narrow oblong, entire, 5–18 cm long, leathery tough, deep green and glossy above, venation very dense beneath; flowers white, about 1 cm wide, the filaments particularly conspicuous, in dense, terminal, numerous panicles about 10–15 cm long and wide, September–October. LWT 92; JRHS 97: 249; BM 9262 (= *Schizophragma viburnoides* Stapf). Khasi Mts. of India, China, Taiwan. 1908. z9 Plate 176. # ⌀ ✧

For mild regions; requires a moist, fertile soil; will climb on walls, trees and other supports. Good for late flowers.

PINCKNEYA Michx. — RUBIACEAE

Evergreen shrubs; leaves opposite, entire, petiolate; flowers in terminal and axillary cymes; corolla tubular, white, red punctate, limb with 5 reflexed lobes; sepals 5, narrow lanceolate, but 1 or 2 sepals on many flowers will be quite deviant, leaflike in form (ovate to elliptic, to 5 cm long and 3 cm wide, light pink to red); stamens 5, exserted, styles of about equal length; fruit a capsule with several, flat seeds. — One species each in N. and S. America.

Pinckneya pubens Michx. An evergreen shrub or small tree in its habitat, to 10 m high; leaves ovate to oblong, 10 cm long, acute at both ends, pubescent beneath; inflorescences 15–20 cm wide, flowers tubular to funnel-shaped, about 2 cm long, white with red spots, the large, leaflike sepals to 5 cm long and pink or red, May–June; fruits 2 cm long. FS 1937; KTF 183; JRHS 1977: 222; SS 227–228. SE. USA. 1786. z9 Plate 176; Fig. 289. ⌀ ✧

Fig. 289. *Pinckneya pubens*. Branch with flowers and fruit (from Sargent)

Thrives in very swampy sites and full sun in its habitat; requires very fertile soil.

Lit. Uphof, J. C. Th.: *Pinckneya pubens*; in Mitt. DDG 1937, 1–4.

PIPTANTHUS D. Don — LEGUMINOSAE

Deciduous or evergreen shrubs, also small trees in the habitat; young shoots with very wide pith; leaves alternate, 3 parted, the leaflets not petiolate, entire; flowers in erect racemes, yellow; calyx campanulate, with 5 teeth, standard petal large, long clawed, wings obovate, keel obovate, stamens distinct; ovaries linear; fruit a many seeded pod. — 8 species in Himalayas, Mongolia and SE. China.

Piptanthus bicolor see: **P. concolor** var. **yunnanensis**

P. concolor Harrow ex Craib. Deciduous shrub, to 2 m high, young shoots white pubescent at first, later glabrous and light brown; leaflets narrow ovate to oblanceolate, 5–10 cm long, finely pubescent only when young, green on both sides; flower racemes 5–10 cm long, flowers about 2.5 cm long, yellow, with somewhat brown markings, calyx silky pubescent; pod 8–10 cm long, glabrous, May. BM 9234 (= *P. concolor* var. *harrowii* Stapf). W. China. 1908. The hardiest species, but still only z7. ✛

var. *harrowii* see: **P. concolor**

var. **yunnanensis** Stapf. Differing from the species in the less pubescent, more quickly glabrous shoots, blue-green leaf undersides, larger flowers and narrower pods. BM 9234 (= *P. bicolor* Craib). SE. China. 1919. z9 ✛

P. forrestii Craib. Shrub, 3 m high, young shoots angular, pubescent; leaflets lanceolate to oblanceolate, acuminate, cm long, base cuneate, both sides with thin and appressed pubescence; flowers golden-yellow, 3 cm long, standard rounded, reflexed, about 2.5 cm wide, May to June. China. 1915. z7 ✛

P. laburnifolius (D. Don) Stapf. Evergreen Goldenrain Tree. Deciduous to semi-evergreen shrub, evergreen only in the mildest climates, young shoots very pithy, shaggy tomentose at first, later glabrous and a glossy olive-brown; leaflets lanceolate, finely pubescent at first, later glabrous, deep green above, blue-green beneath, 7–15 cm long; flowers light yellow, 3 cm long, in steeply erect, 5–7 cm long racemes, calyx campanulate, deeply incised, pubescent, May; pods 7–12 cm long, pubescent at first, later glabrous. BM 9234 (= *P. nepalensis* Sweet). Himalaya. 1821. z9 Plate 169. # ✛ ∅

P. nepalensis see: **P. laburnifolius**

P. tomentosus Franch. Similar to *P. forrestii*, but with denser, more velvety pubescence on both sides of the leaves; leaflets ovate to oval-lanceolate, 5–10 cm long; calyx funnelform, deeply incised, soft pubescent, flowers yellow, 2–2.5 cm long; pod 5–8 cm long, densely pubescent. China. 1887. Easily distinguished by its dense, persistent pubescence. z9 ✛

For any good garden soil in a sunny, dry site; the plants are often short lived, but readily seed themselves. Only for very mild climates.

PIRUS

Pirus delavayi see: **Docynia delavaya**

P. malus see: **Malus silvestris**

PISTACIA L. — Pistachio — ANACARDIACEAE

Deciduous or evergreen trees or shrubs; leaves alternate, simple or trifoliate or pinnate; the pinnate leaves either paired or unpaired pinnate; flowers inconspicuous, without petals, dioecious, in lateral panicles; male flowers with 3–5 short stamens and large anthers, female flowers with a globose ovary, styles short, 3-incised; fruit a drupe. — 10 species in the Mediterranean region; Asia Minor; E. Asia; North America, Texas, Mexico. In its native areas this plant is often of great economic importance.

Pistacia chinensis Bge. Deciduous, rounded tree to over 20 m high; leaves even pinnate (occasionally odd), about 20 cm long, leaflets 10–12, oval-lanceolate, long acuminate, asymmetrical, 5–8 cm long, nearly sessile, glabrous, fall foliage a beautiful carmine; flowers in clustered panicles near the shoot tips, the male panicles about 7 cm long, the female 2 to 3 times as large, looser; fruits peppercorn-sized, red at first, then blue. MD 1912: 25; LF 205; LWT 175 (= *P. formosana* Matsum.). Central China. 1890. z9 ∅

P. formosana see: **P. chinensis**

P. lentiscus L. Shrub, evergreen, 4–6 m high, densely branched, twigs unpleasant smelling, warty, not

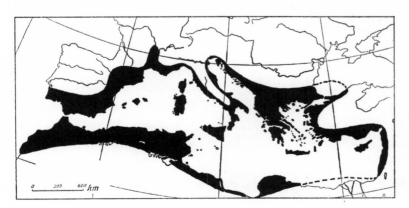

Fig. 290. Range of *Pistacia lentiscus* (from Rickli)

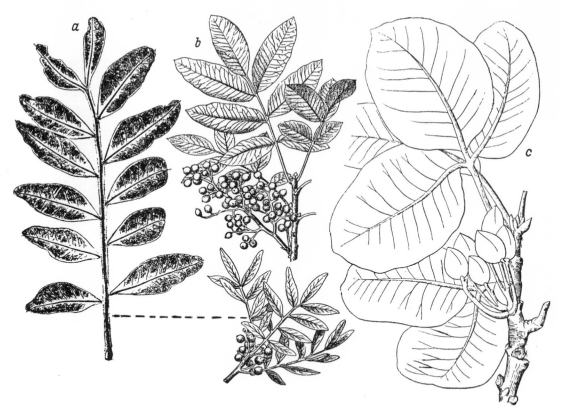

Fig. 291. **Pistacia**. a. *P. lentiscus;* b. *P. terebinthus;* c. *P. vera*
(from Hayne, Hempel & Wilhelm; left Original)

pubescent; leaves even pinnate, leaflets 8–10 narrow oblong to obovate, quite short acuminate, 2–4 cm long, rachis winged; fruits red at first, then black, peppercorn-sized. BM 1967; HM 1823. Mediterranean region. 1664. z9 Fig. 291. # ⌀

P. terebinthus L. Deciduous shrub, often also a tree in its habitat, to 9 m, twigs glabrous, pleasant smelling; leaves glabrous, odd pinnate, usually with 7–9 leaflets, these lanceolate-oblong, 3–6 cm long, entire, very glossy, rachis not winged; flowers greenish, in 5–15 cm long panicles; fruits oval-rounded, 8 mm long, red at first, then brownish. HM 1832. Mediterranean region. 1656. z9 Fig. 291. ⌀

P. vera L. True Pistachio. Small, deciduous tree, scarcely over 5 m high; leaves long petioled, odd pinnate, leaflets usually 3–5, ovate to obovate, sessile, pubescent on both sides, 3–6 cm long; flowers in 7–10 cm long, erect, but otherwise inconspicuous panicles; fruits hazelnut-sized (pistachio nut). DL 2: 385. Asia Minor, Syria, Mesopotamia. 1770. Ⓕ Soviet Union. z9 Fig. 291. ⌀ ⌀⌀

Pistachios are cultivated for their nuts in the Orient and SE. Europe as well as Sicily where the bright green fruited cultivars 'Minnulara' and 'Fimminedda' are grown. The trees bear only every other year. There are also cultivars with pink nuts. The hardiest species, but still tender, is *P. chinensis*, which is far too little known. The other species suitable only for very mild climates. They have no particular cultural requirements.

PITHECOCTENIUM Mart. ex Meissn. — BIGNONIACEAE

Evergreen climbers with tendrils; young shoots striped; leaves opposite, trifoliate or with the middle leaflet replaced by a 3 pronged tendril; flowers in terminal racemes or panicles; calyx campanulate, truncate or with small teeth; corolla campanulate, tubular at the base, bowed, tough, white or yellowish; stamens enclosed, disk large, ovaries warty; fruit a capsule with many

warts. — 7 species in Mexico, tropical S. America and the West Indies.

Pithecoctenium cynanchoides DC. Shoots striped, angular, pubescent when young; leaflets ovate, long acuminate, 3–5 cm long, slightly cordate to broad cuneate beneath, glabrous above; flowers few in terminal racemes, the basal pair long stalked, corolla tubular-campanulate, 4–5 cm long, limb outspread, white, pubescent exterior, flowers over a long period; fruit 6–7 cm long, covered with yellow thorns. BM 8556 (= *Bignonia alba* Hort. non Auth.). Argentina, Uruguay. 1884. z10 # ⬡

P. muricatum DC. Shrub, glabrous; leaflets ovate, acute or long acuminate, base round to slightly cordate; flowers many in terminal racemes, corolla 2.5 cm long, white, throat yellow; fruits oblong, 5–12 cm long, densely covered with prickles (= *Bignonia echinata* Jacq.). Mexico. z10 # ⬡

Cultural requirements like *Bignonia*, but for very mild climates.

PITTOSPORUM Banks — Australian Laurel — PITTOSPORACEAE

Evergreen trees or shrubs, glabrous or tomentose; leaves alternate to nearly whorled, usually entire, occasionally sinuately toothed or lobed; flowers axillary or terminal, solitary or in clusters or umbels or corymbs; sepals distinct or connate at the base; petals 5, more or less cupulate at the base or reflexed, often pleasantly scented; stamens 5; ovaries 2–4 locular, style short; fruit a globose or ovate, 2–4 valved, woody capsule with resinous, glutinous seeds. — About 150 species in New Zealand, S. Africa, subtropical Asia and Australia, as well as the Pacific Islands.

Outline of the species described

● Flowers white, yellow or greenish;

§ Leaves thick and leathery, obtuse at the apex:
      P. coriaceum, dallii, tobira, viridiflorum

§§ Leaves thin, acute at the apex;
   + Flowers in terminal clusters; leaves lanceolate or wider;

      × Young leaves glabrous or nearly so:
         P. eugenioides, heterophyllum, undulatum

      ×× Young leaves very pubescent, as is the fruit:
         P. revolutum

   ++ Flowers in axillary clusters; leaves linear:
         P. phillyreoides

●● Flowers pink to black-red;

§ Leaves densely tomentose to woolly beneath:
         P. bicolor, crassifolium, ralphii,

§§ Leaves glabrous beneath:
         P. colensoi, divaricatum, patulum, tenuifolium, turneri

Pittosporum anomalum Laing & Gourlay. Dwarf shrub, to 1 m high with age, very densely branched, twigs horizontally layered, thin, somewhat tomentose, very stoutly branched; juvenile foliage linear-oblong, irregularly serrate to pinnatisect, 1 cm long, 2 mm wide, mature foliage aromatic, alternate or clustered, linear to oblong, serrate, lobed or entire; flowers terminal and axillary, petals 3 mm long, yellow with a reddish limb; fruit a 5 mm long capsule, honey scented. New Zealand. Collector's plant; seldom found in cultivation. z9 Plate 163. # ⬡

P. bicolor Hook. f. Evergreen shrub or a small tree, 5–10 m high, young shoots with a dense, light brown tomentum; leaves linear, leathery, entire, 3–6 cm long, 4–8 mm wide, margins slightly involuted, dark green above, glabrous, white tomentose beneath at first, later brown; flowers solitary or in axillary clusters, dark brown-red, fragrant, 1 cm wide, sepals smaller, stamens yellow, spring. Tasmania, SE. Australia. z9 Fig. 292. #

P. colensoi Hook. f. Evergreen shrub or small tree, 3–9 m high, young shoots thick, loosely silky pubescent when young; leaves elliptic or more oblong, leathery, 3.5–10 cm long, 1.5–3 cm wide, acute to obtuse, entire, usually tapered to the base, pubescent only when young, dark green above with a yellow midrib, glossy, lighter beneath with reticulate venation; flowers grouped 1–3 together, axillary near the shoot tips, dark red, petals oblong, reflexed, 1.5 cm long, sepals broad ovate, April. BM 8305. New Zealand. z9 #

P. crassifolium Banks & Soland. ex A. Cunn. Evergreen shrub or small tree, 5 m high, to 9 m in its habitat, very densely branched, crown very narrow, nearly columnar; leaves alternate, obovate to elliptic or oblong, 5–7 (10) cm long, 2–2.5 cm wide, very leathery-tough, dark green above, white tomentose beneath, later light brown, margins somewhat involuted; flowers unisexual, in terminal clusters, dark purple, petals strap-form, recurved, 1 cm long; fruits oval, 2–3 cm long, with many black seeds. BM 5978. New Zealand. Ⓕ Tristan da Cunha (South Atlantic Islands) z9 Fig. 292. #

P. dallii Cheesem. A round crowned tree in its habitat, 4–7 m high, shoots glabrous, reddish, older bark light gray; leaves oval-lanceolate, 5–11 cm long, 2.5 cm wide, tapered to both ends, coarsely dentate or entire, leathery, glabrous, dull green above, clustered at the shoot tips; flowers white, 12 mm wide, fragrant, in dense, terminal clusters, June–July. New Zealand. 1913. One of the relatively hardier species. z9 Plate 168. # ∅

P. daphniphylloides Hayata. Evergreen shrub, also a 10 m high tree in its habitat, young shoots somewhat pubescent; leaves narrow oblong to more obovate, tapered to both ends, 5–20 cm long, 3–7 cm wide, dark green above, somewhat pubescent beneath when young; flowers in a large, terminal panicle, composed of many small, globose, 2–3 cm wide umbels, cream-white, very fragrant, April–July; fruit globose, 1 cm thick, red, rugose, W. China. z9 Fig. 292. # ⬡

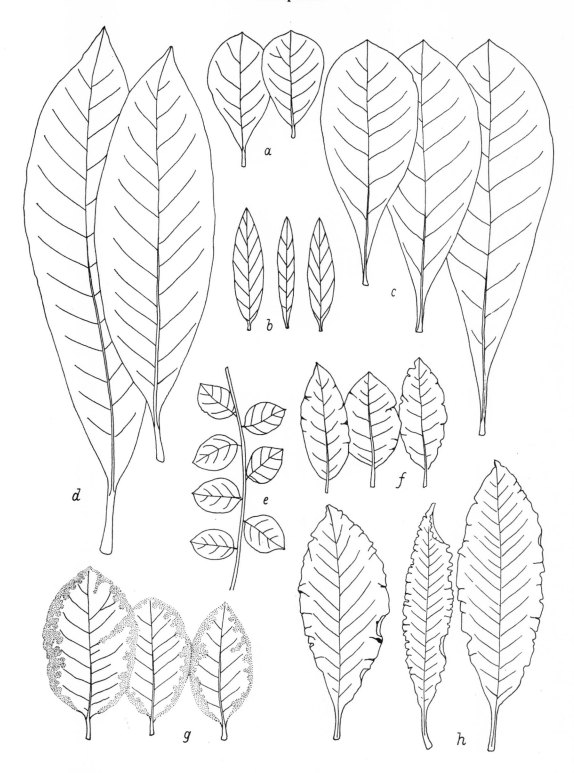

Fig. 292. **Pittosporum.** a. *P. crassifolium*; b. *P. bicolor*; c. *P. tobira*; d. *P. daphniphylloides*; e. *P. tenuifolium* 'James Stirling'; f. *P. tenuifolium* 'Purpureum'; g. *P. tenuifolium* 'Garnetti'; h. *P. eugenioides* (Original)

P. divaricatum Cockayne. Small to medium-sized shrub, to 4 m high in its habitat, normally only 1 m in cultivation, twigs divaricately spreading and tangled, stiff, twisted, pubescent when young; leaves quite variable, linear to obovate on young plants, 1.2–2 cm long, 3 mm wide, pinnatisect or dentate, 6–12 mm long on mature plants, linear-obovate to ovate, entire, lobed or also deeply toothed, leathery tough, deep green above; flowers solitary at the shoot tips, very small, 4 mm long, deep red-brown to nearly black, petals spoon-form, May; fruits ovate, 5–8 mm wide. New Zealand. A unique shrub, somewhat resembling *Corokia cotoneaster.* z9 #

P. eugenioides Cunn. Tree, 5–10 m high, very densely branched, twigs with a dark bark, glabrous; leaves narrow ovate to oblong, 5–10 cm long, margins undulate, dark green and glossy above, lighter beneath, totally glabrous on both sides, thin; flowers yellowish, many in terminal, short stalked corymbs (!), yellowish, very fragrant, summer. KF 49. New Zealand. z9 Fig. 292. # ⌀ ⊙

'Variegatum'. Leaf margins cream-yellow, otherwise like the species. A wonderful tree in very warm regions. Plate 168. # ⌀

P. flavum see: **Hymenosporum flavum**

P. fulvum see: **P. revolutum**

P. glabratum Lindl. Shrub, 1–1.5 m high, young shoots totally glabrous; leaves clustered at the shoot tips, narrow obovate, 5–12 cm long, long acuminate, base cuneate, entire, margins uneven, glossy green; flowers 6–10 in terminal clusters, dull yellow, fragrant, May; fruit capsule woody, 15 mm long. China; Ichang Province. 1845. One of the hardiest species, but of little garden interest. z9 # ⌀ ⊙

P. heterophyllum Franch. Evergreen shrub, very densely branched and foliate, 1–1.5 m high, occasionally to 3 m, shoots glabrous; leaves oblanceolate to obovate, or ovate, 3.8 cm long, 1–2.5 cm wide, obtuse, base cuneate; flowers in terminal and axillary clusters of up to 7 together, often grouped into large panicles, yellow, May–June. China. 1908. z9 Plate 163. #

P. mayi see: **P. tenuifolium**

P. patulum Hook. f. Large evergreen shrub or small, upright tree, conical habit, to about 5 m high, sparsely branched, young shoots and flower stalks pubescent, otherwise totally glabrous; leaves quite variable in form, very narrow on young plants, 3–5 cm long, 5 mm wide, distinctly lobed for the entire length, mature foliage more lanceolate, 4–5 cm long, 1–1.5 cm wide, entire or with a few teeth, gradually tapering to a short, thick petiole; flowers grouped 4–8 in terminal clusters, campanulate, dark red, very strongly and pleasantly scented, May; fruit capsule woody, 8 mm thick, globose. New Zealand. z9 # ⊙

P. phillyreoides DC. Shrub or small tree, glabrous except for the young shoots, twigs usually nodding; leaves linear-lanceolate, flat, 3–10 cm long, 3–10 mm wide, terminating in a small, hook-like mucro; flowers solitary or in axillary clusters, petals yellow, 10–15 mm long;

fruit an ovate, occasionally somewhat cordate, 1–2 cm long capsule. Australia. z9 #

P. ralphii Kirk. Shrub, 2–4 m high, bark dark brown, shoots, leaf undersides and petioles densely white or brown woolly; leaves oblong to more obovate, leathery tough, 7.5–12.5 cm long, 2.5–5.5 cm wide, petiole 1.5–2 cm long; flowers 3–10 in terminal groups, small, dark red, anthers yellow; fruit capsule to 1.5 cm long. New Zealand. Occasionally confused with the similar *P. crassifolium*, whose leaf blade, however, tapers gradually to the petiole and has an involuted margin, while *P. ralphii* has a leaf blade which abruptly narrows to the petiole and has a flat margin. z9 #

P. revolutum Ait. Evergreen shrub, 2–3 m high, young shoots brown tomentose-woolly; leaves lanceolate to narrow elliptic, 3–11 cm long, 0.8–2 cm wide, distinctly tapered on both ends, dark green above, dense brown woolly beneath, especially on the midrib; flowers solitary or in few flowered umbels at the shoot tips, 8–12 mm long, petals yellow, recurved, sepals awl-shaped, February–April. BR 186 (= *P. fulvum* Rudge). Australia; New South Wales. 1795. z9 #

P. tenuifolium Gaertn. Tree, to 9 m high, very densely branched, stem slender, young shoots nearly black; leaves obovate to oblong or elliptic, 3–7 cm long, entire, margins undulate, not dentate, glabrous, dull glazed; flowers usually in 2's (or more) in the leaf axils, fragrant, purple-brown. KF 46 (= *P. mayi* Hort.). New Zealand. z9 # ⌀

Includes the following cultivars:

'Garnettii'. Leaves 3–5 cm long, ovate-elliptic, margins with irregularly wide cream-white band, with pink or red spots in the white zone, margins flat. MNZ 220. Developed in Buxton's Nursery, Christchurch, New Zealand. Considered a hybrid of *P. tenuifolium* × *P. ralphii*. z9 Fig. 292. # ⌀

'Irene Paterson'. Leaves elliptic-obovate, 2.5–3.5 cm long, margins very undulate, young leaves often nearly totally white, green speckled and punctate, turning somewhat reddish in winter. MNZ 60. Found in the wild by G. Paterson in the vicinity of Christchurch, N.Z. 1970. z9 # ⌀

'James Stirling'. Shoots thin, blackish-red; leaves small, rounded-elliptic, a beautiful silvery-green. Developed by James Stirling, Wellington, N.Z. 1966. z9 Fig. 292. # ⌀

'Purpureum'. Leaves oblong to elliptic-obovate, margins very undulate, deep bronze-brown to nearly black (effective as a good purple beech!). Found in Melbourne, Australia. z9 Fig. 292. # ⌀

'Saundersii'. Differing from 'Garnettii' only in the more compact habit and somewhat more rounded leaves. Developed from seed of 'Garnettii' by F. J. Saunders, Invercargill, N. Z. z9 # ⌀

'Silver Queen'. Leaves silvery-gray. Found in a nursery in Ireland and distributed by Hillier. z9 # ⌀

'Variegatum'. Shoots gray (not black!!); leaves elliptic-lanceolate, 3–6 cm long, acute, margins flat, gray-green above, with a narrow white border. Not as attractive as the other white variegated forms. z9 # ⌀

'Warnham Gold'. Young leaves greenish-yellow, maturing to

golden-yellow, particular effective in fall and winter. Developed in Warnham Court, Sussex, England, 1959. z9 # ⌀

P. tobira Ait. A stiffly erect shrub, occasionally a 5 m high small tree; leaves obovate, rounded at the apex, 3–10 cm long, base cuneate, leathery tough, deep green above and very glossy, glabrous, midrib lighter; flowers cream-white, later becoming yellow, about 2.5 cm wide, very fragrant, summer, usually several together in terminal clusters, flowers abundantly, April–June. BM 1369. Japan, China. Often planted in Mediterranean regions. 1804. z9 Plate 168; Fig. 292. # ⌀ ✧

'Variegatum'. Leaves white variegated. # ⌀

P. turneri Petrie. A small conical tree, 4–9 m high in its habitat, twigs erect, glabrous; leaves obovate, 2.5–3 cm long, 1–1.2 cm wide, thin-leathery, base cuneate, sparsely pubescent at first, but quickly becoming totally glabrous; flowers 6–12 in terminal clusters, usually on the lateral shoots, pink to purple, petals narrow and recurved, May–June. New Zealand. z9 #

P. undulatum Vent. Tree, 9–12 m high in its habitat; leaves laurel-like, 7–15 cm long, acuminate at both ends, glabrous, glossy dark green above, lighter beneath, entire, limb undulate; flowers cream-white, 12–18 mm wide, pleasantly scented, in a terminal cluster of one or more umbels, May to July. BMns 234. Australia. 1789. z9 Plate 169. # ⌀ ✧

P. viridiflorum Sims. Shrub, to 3 m high, twice as high in its habitat, shoots somewhat pubescent; leaves quite variable, obovate, apex round, base cuneate, 2.5–10 cm long and 1.2–3 cm wide, glossy dark green above; flowers yellowish-green, jasmine-scented, several in terminal, 2.5–3 cm wide clusters. Petals acute, reflexed, May. BM 1684. S. Africa. z9 #

All species are very attractive as evergreen shrubs, the flowers usually are strongly honey-scented, especially in the evening. Soil and site requirements slight, prefers a clay-humus soil, but suited only to milder climates.

Lit. Gowda, M.: The genus *Pittosporum* in the Sino-Indian region; Jour. Arnold Arb. **32**, 263–301, 303–343 ● Laing, R. M., & H. W. Gourlay: The small leaved species of the genus *Pittosporum* occurring in New Zealand; Transact. R.S. N.Z. **65**, 44 to 62, 1935 ● Cooper, R. C.: The Australian and New Zealand species of *Pittosporum*; in Ann. Mo. Bot. Gard. **43**, 87–188, 1956.

PLAGIANTHUS J. R. & G. Forst. — MALVACEAE

Evergreen or deciduous trees or shrubs (also herbaceous in Australia); leaves alternate, simple or lobed or serrate; flowers white, small, in axillary clusters or panicles or solitary, hermaphroditic or unisexual; calyx 5 toothed or 5 incised; ovaries with one or 2–5 locules with an equal number of styles; fruit dehiscing into fruitlets; seeds solitary, pendulous — 15 species in Australia and New Zealand.

Plagianthus betulinus A. Cunn. Attractive, deciduous shrub, also a tree in mild climates to 10 m high or more, stem nearly 1 m thick in its habitat, young shoots and leaves pubescent; leaves ovate on young plants, 1.5–3 cm long, dentate and lobed, mature foliage ovate to oval-lanceolate, 3–7 cm long, long acuminate, less deeply lobed; flowers unisexual, small, but very numerous, in terminal and axillary panicles, 10–20 cm long, yellowish-

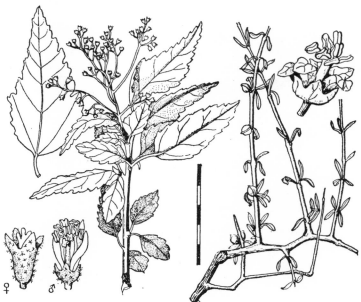

Fig. 293. **Plagianthus.**
Left *P. betulinus*;
right *P. divaricatus*
(from Metcalf)

white. New Zealand. 1870. z9 Fig. 293. ⊕

P. divaricatus J. R. & G. Forst. Evergreen shrub, 1–2 m high, glabrous, abundantly branched, twigs thin and divaricate; leaves alternate or clustered on short lateral shoots, linear-oblong and 2.5 cm long, entire or sinuate, narrow linear or obovate-linear on mature plants, 6–18 mm long, entire and leathery tough, single veined; flowers very small, solitary or in clusters, axillary, yellowish-white, June; fruits peppercorn-sized, pubescent. BM 3271. New Zealand. 1820. z9 Fig. 293. # ⊕

P. lyallii Hook. f. see: **Hoheria glabrata** and **H. lyallii**

P. pulchellus Gray. Tall, evergreen shrub, shoots and leaves stellate pubescent; leaves lanceolate, long acuminate, 5–11 cm long, base cordate, margins coarsely crenate; flowers small, in short, few flowered, axillary racemes near the shoot tips, white, 12 mm wide, July. BM 2753 (as *Sida pulchella*). Tasmania. 1820. z9 # ⊕

For use only in the mildest regions; prefers a sandy-clay soil; won't tolerate frost. Interesting for the variation between the juvenile and mature foliage.

PLANERA Gmel. — Water Elm — ULMACEAE

Deciduous trees, resembling *Ulmus,* but the flowers partly unisexual, the female axillary on young shoots, leaves simple serrate; fruits with a hard or crusty shell. — One species in N. America.

Planera aquatica (Walt.) Gmel. Broad crowned, deciduous tree, to 12 m high or only a tall shrub, shoots pubescent; leaves alternate, ovate, 3–7 cm long, acuminate, simple and somewhat unevenly serrate, base oblique, rough above, petiole 3–6 mm long; fruit an ovoid drupe in fleshy panicles (!). SS 316; BB 1254; KTF 64 (= *P. ulmifolia* Michx.). SE. USA, often in swampy areas. 1816. Plate 176; Fig. 294, 295.

P. davidii see: **Hemiptelea davidii**

Cultivated like *Ulmus;* totally hardy. Usually incorrectly labeled in cultivation! Often confused with *Ulmus,* but easily identified if carefully inspected.

Fig. 294. *Planera aquatica* (from Sargent)

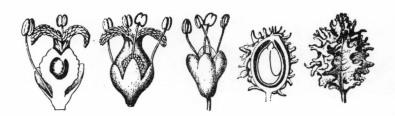

Fig. 295. *Planera aquatica.* Left, 2 female flowers; center, male; right, 2 fruits (from Sargent)

PLANTAGO L. — PLANTAGINACEAE

From the nearly 265 species for this genus, the following species is occasionally shrubby and therefore included here.

Plantago cynops L. Evergreen, about 30–40 cm high, abundantly branched shrub; leaves linear, opposite, 3–6 cm long, triangular in cross section, margins rough; flowers whitish, very small, inconspicuous, in ovoid, about 1 cm long heads on 3–10 cm long, axillary stalks, June–July. S. Europe. 1596. Plate 169. #

Totally hardy and without particular cultural requirements; interesting small shrub for the collector.

PLATANUS L.— Planetree — PLATANACEAE

Deciduous tree with plate-like, exfoliating bark; twigs and leaves stellate pubescent; leaves alternate, maple-like, with stipules, 3–5(7) lobed; flowers inconspicuous, monoecious, in globose heads, male and female flowers very similar to each other; fruits grouped into globose heads, 1–3 or more on a long, pendulous stalk.—1 species (*P. orientalis*) from SE. Europe to N. Persia; 1 (*P. kerrii*) in Indochina; 1 in eastern North America and 7 in southwest North America and Mexico; and the widely cultivated hybrid *P. acerifolia* (*P. orientalis* × *P. occidentalis*).

It is not yet unequivocally clear which is the correct nomenclature for the world's most popular Planetree, whether *P. hybrida* Brotero or *P. hispanica* Muenchh.; therefore the most frequently used name to date, *P. acerifolia*, will be maintained here.

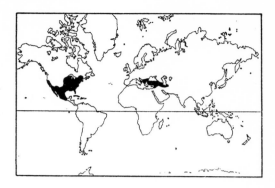

Fig. 296. Range of the genus *Platanus*

Outline of the more prominent species

	P. occidentalis	*P. acerifolia*	*P. orientalis*
Stem	not or rarely continuous to the peak	often continuous to the peak	
Bark	exfoliating in small scales	flaking in large plates	
Leaf form	usually trilobed	usually 5, occasionally 7 lobed; also trilobed on young shoots	5–7 lobed, occasionally trilobed on young shoots
Middle lobe	shorter than the basal width	slightly longer than its basal width	much longer than its basal width
Leaf margin	coarsely sinuate, also entire	shallowly sinuate to entire	usually shallowly sinuate
Leaf base	obtuse to cuneate	usually truncate, obtuse to slightly cordate	usually long cuneate, occasionally truncate
Leaf underside	vein axils persistently pubescent, often tomentose	becoming glabrous	becoming glabrous
Stipules	very large, often conical or tubular	medium-sized	small
Fruit clusters	solitary, occasionally in 2's	usually in 2's, occasionally 3's or solitary	grouped 3–4 or more, occasionally in 2's
Fruit	rounded at the apex	conical, style arising from the apex	

Platanus × acerifolia [Ait.] Willd. (*P. orientalis* × *P. occidentalis*). Trees, to 35 m high, stem often continuous through the apex, bark flaking in large plates, young shoots densely brown tomentose; leaves 3–5 lobed, 12–25 cm wide, the lobes triangular to broad triangular, middle lobes slightly longer than their basal width, margins sinuately toothed to entire, base usually truncate, obtuse to slightly cordate, underside eventually nearly or totally glabrous, stipules medium-sized, petiole 3–10 cm long; fruits about 2.5 cm thick, usually in groups of 2 together, occasionally in 3's or solitary, bristly, nutlets ovoid or globose at the apex with a remnant of the style. EH 204; HM 993; BD 7: 56 (= *P. hispanica* Muenchh.; *P. hybrida* Brotero). Origin unknown. Before 1700. An important street tree, thriving even in larger cities; withstands a dry atmosphere. ⓕ Yugoslavia; Argentina. z5 Fig. 297. ⦰

f. *argenteovariegata* see: 'Suttneri'

f. *aureovariegata* see: 'Kelseyana'

'Cantabrigiensis'. Leaves smaller, more deeply lobed, more slender, more delicate in all respects, base truncate with a small, cuneate central portion (= *P. cantabrigiensis* Henry). Fig. 297.

f. *fastigiata* see: 'Pyramidalis'

'Hispanica'. Leaves often to 30 cm wide, normally 5 lobed, the lobes toothed, base truncate to slightly cordate, venation beneath and leaf petiole with a persistent tomentum; fruits grouped 1–2 (= var. *hispanica* [Muenchh.] Bean). Fig. 297.

'Kelseyana'. Leaves yellow variegated (= f. *aureovariegata* Hort.). Before 1899. ⦰

'Mirkovec'. Leaves becoming reddish in summer, coloring purple-red in fall. Found by R. de Belder in the Mirkovec Nursery, Yugoslavia. 1965. ⦰

Fig. 297. **Platanus**. a. *P. acerifolia* 'Pyramidalis'; b. *P. acerifolia* 'Hispanica' (normal and
small leaf); c. *P. acerifolia* (typical!); d. *P. acerifolia* 'Cantabrigiensis' (from Hadfield)

'Pyramidalis'. Young plants form a tight, narrow upright cone, becoming broader with age, the lower branches not pendulous; leaves often also 3 lobed, wider than long, base rounded (= f. *fastigiata* Hort.). Found around 1850 in France. Fig. 287.

'Sutterni'. Leaves whitish punctate and speckled over the entire surface (= f. *argenteovariegata* Hort.). Before 1896. ⌀

'Tremonia'. Habit quite narrow conical, very fast growing. A chance seedling discovered in a shipment of plants from Italy around 1920, but going relatively unnoticed until 1951. The original tree still stands in the Dortmund Botanic Garden, W. Germany (= *Platanus acerifolia* 'Dortmund').

P. californica see: **P. racemosa**

P. cantabrigiensis see: **P. × acerifolia 'Cantabrigiensis'**

P. cuneata see: **P. orientalis** var. **cuneata**

P. densicoma see: **P. occidentalis** var. **glabrata**

P. digitata see: **P. orientalis** var. **digitata**

P. glabrata see: **P. occidentalis** var. **glabrata**

P. hispanica see: **P. × acerifolia**

P. hybrida see: **P. × acerifolia**

P. nepalensis see: **P. orientalis** f. **digitata**

P. occidentalis L. Tree, to 40 m high, trunk usually not continuous to the peak, branches directed more upward, bark exfoliating in small plates; leaves generally 3 lobed, middle lobe shorter than its basal width, margins coarsely sinuate, occasionally also entire, base obtuse to cuneate, venation with a persistent pubescence in the axils beneath, often tomentose, stipules very large, often conical to tubular; fruit clusters usually solitary, occasionally in 2's, nutlets rounded at the apex. BB 1881; GTP 218; VT 326; GF 2: 354. Southern N. America, along stream and river banks. 1640. Ⓕ USA. z5 Fig. 298.

var. **glabrata** (Fern.) Sarg. Leaves normally smaller, tougher, more deeply lobed, usually truncate at the base, sinuses about ⅓ into the leaf blade, the lobes long acuminate and often entire or also coarsely and sparsely dentate. VT 327 (= *P. glabrata* Fern.; *P. densicoma* Dode p.p.). N. America; Iowa to Mexico.

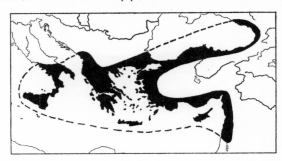

Fig. 299. Range of *Platanus orientalis*
(from Rickli)

Fig. 298. **Platanus.** Left and right *P. occidentalis,* with female flower and fruit;
middle *P. orientalis* f. *digitata* (Original)

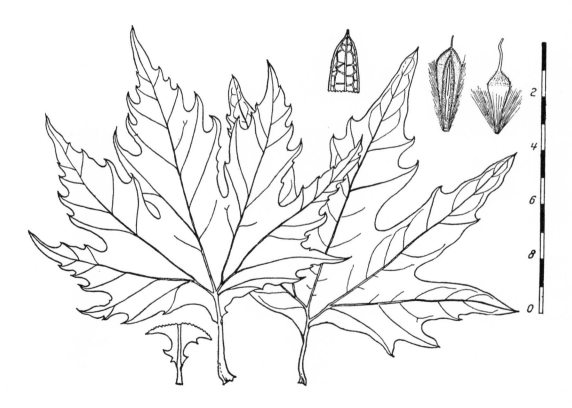

Fig. 300. *Platanus orientalis.* 2 leaves, leaf tips and base;
fruitlet, from the front and in cross section (from Hadfield, Gadeceau)

P. orientalis L. Tree, to 30 m high, trunk usually continuous to the peak, bark exfoliating in large plates, twigs usually more horizontally spreading; leaves deeply 5–7 lobed, occasionally 3 lobed on the younger shoots, 10–20 cm wide, lobes much longer than wide, the sinuses reaching nearly to the middle of the blade, eventually nearly totally glabrous, base usually long cuneate, occasionally truncate, stipules small; fruit clusters in groups of 3–4 or more, occasionally in 2's, nutlets conical, apex becoming the style. HM 994; GF 4: 91; EH 175. SE. Europe to Asia Minor. ⓕ Kashmir (erosion control). z7 Fig. 300. ∅

var. **cretica** (Dode). Leaves 5 lobed, the sinuses very deep, the lobes further lobed with ascending teeth, lobes very long acuminate at the apex, base more or less cordate; fruit heads only about 1.5 cm thick, usually 2–3 together, occasionally to 4, nutlets with acute, pubescent heads and a persisting, elongated style. BD 7: 60. Crete. Plants cultivated under this name are usually erroneously labeled.

var. **cuneata** (Willd.) Loud. Small tree, often shrubby; leaves 3–5 lobed, with ascending teeth, base cuneate to truncate, short stalked. BD 7: 55 (= *P. cuneata* Willd.). Greece, Asia Minor.

f. **digitata** (Gord.) Janko. Tree, conical when young, branches ascending; leaves 3–5 lobed, deeply incised, the lobes more distinct on larger leaves, only slightly dentate to entire on smaller leaves, base cuneate or truncate to slightly cordate; fruit heads grouped 2–4, occasionally to 5; nutlets obtuse at the apex, becoming glabrous, with a style. BD 7: 59 (= *P. digitata* Gord.; *P. nepalensis* Hort.; *P. orientalis laciniata* Hort.). Cypress, Caucasus. Fig. 298. ∅

P. orientalis laciniata see: **P. orientalis** f. **digitata**

P. racemosa Nutt. Tree 30(40) m high in its habitat, trunk often multistemmed; leaves usually cordate to truncate, 15–30 cm wide, deeply 3–5 lobed, thick and tough, deep green above, lighter and tomentose beneath, the lobes narrow and entire or sparsely dentate; fruit clusters grouped 2–7 together, nutlets tomentose when young, with a conical or globose apex. SPa 156; SS 328; BC 3061 (= *P. californica* Benth.). North America; Calfornia. 1897. z8 Fig. 301. ∅

P. wrightii S. Wats. Tree, to 25 m high; leaves deeply 3–5 lobed, tomentose beneath, often eventually nearly glabrous, the lobes lanceolate, entire to sparsely dentate;

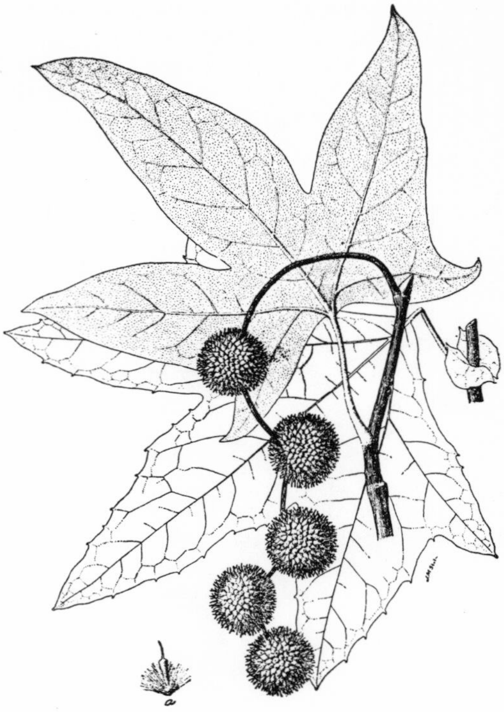

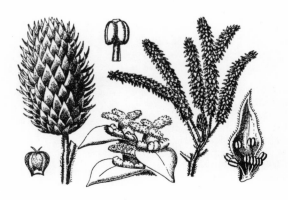

Fig. 301. *Platanus racemosa* (from Sudworth)

fruit clusters grouped 2–4, the individual heads usually stalked (!), normally glabrous, nutlets truncate to hemispherical, styles usually abscising. VT 327; SS 329. North America; Arizona to Mexico. z9 ∅

Excellent street trees with broad crowns, thriving in all situations, heat tolerant.

Lit. Jaennicke, F.: Studien über die Gattung *Platanus*; in Abh. Leop. Carl. Akad. Naturf. **77**, 113 to 226, 1899 ● Griggs, R. F.: Characters and relationships of Platanaceae; in Bull. Torrey Club **36**, 369–395, 1909 ● Brouwer, J.: Studies in Platanaceae; in Réc. Trav. Bot. Néerl. **21**, 369–382, 1924 ● Hadfield, M.: The mystery of the London Plane; in Gard. Chron. **148**, 422, 443, 462, 1960 ● Dode, L.-A.: Sur. les Platanes; in Bull. Soc. Dendr. France **7**, 27–68, 1908.

PLATYCARYA S. & Z. — JUGLANDACEAE

Deciduous trees; twigs with a solid pith; leaves alternate, odd pinnate; flowers in erect catkins at the tips of short, foliate shoots, the males in several catkins arranged around the female inflorescence, these cone-like when ripe, the individual fruitlets are 2 winged nutlets — 2 species in E. Asia.

Fig. 302. *Platycarpa strobilacea*. Inflorescence, individual flowers, fruit cluster and individual fruits (from Baillon, Shirasawa)

Platycarya strobilacea S. & Z. Deciduous tree, to 15 m high in its habitat, often a tall shrub or small tree in cultivation, young twigs pubescent; leaves 20–30 cm long, leaflets 7–15, sessile, ovate to oval-lanceolate, 4–10 cm long, long acuminate, biserrate, somewhat pubescent at first, eventually totally glabrous, yellow in fall; male catkins 5–8 cm long, June; fruit cones 3–4 cm long, brown, nutlets 5 mm long. LF 68–69; NK 20: 16; FS 331. China. 1845. z6 Plate in Vol. III; Fig. 302. ⊘ ⚭

Cultivated like *Juglans.*

PLATYCRATER S. & Z. — HYDRANGEACEAE

Monotypic genus. Procumbent shrub, deciduous, very closely related to *Hydrangea,* but the sepals on the sterile flowers connate, disk-like, occasionally also absent; stamens numerous; vegetative buds stalked; seeds numerous, linear, winged on both ends.

Platycrater arguta S. & Z. Procumbent shrub or to 1 m high, young shoots usually totally glabrous, eventually brown, bark exfoliating; leaves opposite, oblong to lanceolate, 10–20 cm long, long acuminate, thin and membranous, cuneate at the base, finely and sparsely serrate, scattered pubescence above, venation pubescent beneath; flowers in loose, terminal cymes of about 5–10 flowers, white, marginal flowers with a 3–4 lobed calyx, July; fruit a 2 locular capsule. SH 1: 245 a–l. Central and Southern Japan, along river banks. 1864. z8 Plate 176; Fig. 303. ⊕

'Hortensis'. The form in cultivation. Inflorescences with only 3–5 flowers, without marginal flowers. Gfl 516 (= var. *hortensis* Maxim.).

Of only slight ornamental value; cultivated about like *Hydrangea,* semishade with a moist, humus soil.

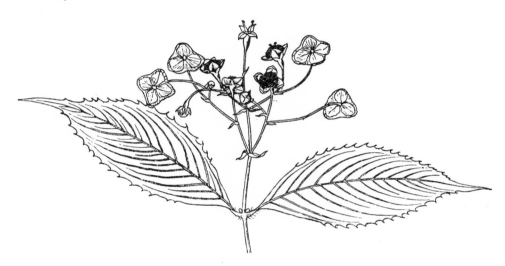

Fig. 303. *Platycrater arguta* (from Sieb. & Zucc.)

PLEIOBLASTUS See: **ARUNDINARIA** and **SASA**

PLEONOTOMA Miers — BIGNONIACEAE

A climbing shrub very closely related to *Bignonia;* twigs 4 sided; leaves pinnate, with branched leaf tendrils; flowers in terminal and axillary racemes; corolla campanulate-funnelform, usually distinctly bilabiate, stamens inserted, disk cushion form; calyx campanulate or tubular, truncate or with 5 short teeth, leathery tough; fruit an oblong, broad linear capsule with woody valves. — 12 species in tropical America and Trinidad.

Pleonotoma variabile (Jacq.) Miers. A glabrous climbing shrub; the basal leaves doubly trifoliate, the apical leaves pinnate; flowers in many flowered, short, terminal racemes, corolla tube greenish-yellow, the limb tips eventually white, June to August (= *Bignonia variabilis* Jacq.). Venezuela. 1819. z10 # ⊕

PLOCAMA Ait.— RUBIACEAE

Monotypic genus with the following characteristics.

Plocama pendula Ait. Deciduous to semi-evergreen, very strongly scented shrub, 2–4 m high, twigs limp and nodding in a short arch, strongly branched; leaves opposite or also in whorls of 4, linear-filamentous, 6–8(12) cm long, acute, dark green above, lighter beneath, limp; flowers tiny, terminal and axillary near the branch tips, calyx tube campanulate, 5 lobed, persistent, corolla funnelform-campanulate, tube short, limb with 5–7 oblong-lanceolate lobes, white, winter to spring; fruit a small, globose, black berry. KGC 48; BFC 48; BFCa 65, 215. Canary Islands. z10

Interesting, but only for collectors.

PLUMBAGO L.— Leadwort— PLUMBAGINACEAE

Shrubs or perennials, occasionally only annual plants; leaves alternate, often auricled at the base, entire; flowers in small racemes or spikes at the shoot tips; calyx tubular, 5 parted, occasionally somewhat angular and glandular; corolla with a longer, narrower tube and flat, 5 lobed limb; stamens 5; style with 5 stigmas. — 12 species in the warmer regions, but only the following shrubby species is generally found in cultivation.

Plumbago auriculata Lam. Climbing or procumbent shrub, 2–3 m high, but usually grown in cultivation as a small crowned tree, shoots angular, striped; leaves scattered, oval-oblong to somewhat spathulate, obtuse or with a small mucro, 5–6 cm long, 1–2 cm wide, light green above, with small white scales beneath, otherwise glabrous; flowers numerous in short, terminal spikes, light blue, corolla tube 3 times longer than the glandular pubescent calyx, May–October. BM 2110; BR 417; PBl 2: 337 (= *P. capensis* Thunb.). S. Africa. 1818. z10 # ⊕

P. capensis see: **P. auriculata**

Very popular as a tub plant as it flowers continuously; suitable for the landscape in frost free regions.

Lit. Wood, C. E.: *Plumbago auriculata* versus *P. capensis* (Plumbaginaceae); Baileya **16**, 137–139, 1968.

PLUMERIA L.— Frangipani— APOCYNACEAE

Deciduous, tropical trees with very thick shoots; leaves alternate, entire, pinnately veined, but with a connecting vein running parallel to the leaf margins; flowers in terminal cymes, large, corolla tube cylindrical, the 5 limb tips spirally twisted together before opening, large, white, yellowish or pink, very fragrant and waxy; fruit an inflated follicle, seeds winged. — 7 species in the warmer regions of America.

Plumeria acuminata see: **P. rubra** var. **acutifolia**

P. acutifolia see: **P. rubra** var. **acutifolia**

P. alba L. 5–6 m high in its habitat; leaves oblong to more lanceolate, to 30 cm long, 7 cm wide, acuminate at the apex or also rounded, without the marginal veins, margins somewhat involuted, tomentose beneath, distinctly petiolate; flowers in clusters, white, with a yellow center, very fragrant, July–August. W. Indies. z10 ⊕

P. arborescens see: **P. rubra**

P. rubra L. Frangipani, West Indian Jasmine. Shrub or small tree, 4–6 m high, shoots very thick with many, densely arranged leaf scars; leaves lanceolate to oblong-obovate, tough, acute or obtuse, to 30 cm long and 7 cm wide, with a distinct marginal vein; flowers in open spreading cymes, the broad-elliptic, spiraling (in the bud stage) limb tips longer than the corolla tube, usually pink, but also yellow or white, very fragrant, June–September. BM 279; DRHS 1614 (= *P. arborescens* R. Br.). Mexico to Guayana and Ecuador. z10 ⊕

var. **acutifolia** (Poir.) L. H. Bailey. Leaves to 30 cm long, long acuminate, broad lanceolate; limb tips elliptic, white with a yellow center. BM 3952; BR 114; BC 3082 (= *P. acutifolia* Poir.; *P. acuminata* Ait.). ⊕

PODALYRIA Lam. ex Willd.— LEGUMINOSAE

Evergreen shrubs; leaves alternate, simple, usually more or less silky pubescent, stipules subulate, small, normally abscising; flowers usually 1–2, seldom 3–4 on axillary stalks; calyx broad campanulate, dentate or unevenly lobed; petals of unequal length, standard nearly circular, emarginate at the apex, somewhat longer than the wings, keel shorter than the wings; fruit an ovate or oblong, swollen pod. — 25 species in S. Africa, but only a few known in cultivation.

Podalyria calyptrata Willd. Branched shrub, 1–2 m high, thin branched, shoots finely silky pubescent, as are the leaves, these elliptic to oblong, obtuse, 4–5 cm long, 2 to 3.5 cm wide; flowers usually solitary or also paired axillary on the shoot tips, standard deeply incised, bilabiate, 2.5 cm wide, reflexed, light pink, calyx broad campanulate, gray-green, silky pubescent, May–June. RCA 34. S. Africa. 1792. Flowers with a very strong fragrance! z10 # ⊕

PODRANEA Sprague — BIGNONIACEAE

Evergreen climber, closely related to *Pandorea*, but differing in the inflated calyx, the oblong ovaries, the oblong-linear fruit capsule with leathery, flexible valves; leaves opposite, odd pinnate; flowers in terminal panicles, pink to lilac, calyx regular, campanulate, 5 toothed; corolla campanulate, drawn out to a cylindrical tube; stamens 4, the 5th is a staminode. — 2 species in S. Africa.

Podranea ricasoliana (Baill.) Sprague. Evergreen climber; leaves opposite, leaflets 7–11, short stalked,

oval-elliptic, acute to acuminate, serrate, deep green above, lighter beneath, glabrous, about 2.5 cm long; flowers in loose, terminal panicles, corolla campanulate-funnelform, limb 5 lobed, outspread, pink with dark red stripes, 5 cm long, glabrous on both sides, August to September; fruits cylindrical, 25–30 cm long. Gw 2: 343 (= *Tecoma ricasoliana* Tanfani, *Pandorea ricasoliana* Baill.). S. Africa. z9 Plate 172. # ⊕

Used like *Campsis;* but requires a very warm site and is not very frost tolerant.

POINCIANA

Poinciana gilleisii see: **Caesalpinia gilliesii**

Poinciana regia see: **Delonix regia**

Poinsettia

Poinsettia pulcherrima see: **Euphorbia pulcherrima**

POLIOTHYRSIS Oliv. — FLACOURTIACEAE

Monotypic genus. Deciduous tree; leaves alternate, petiolate, simple, dentate, base 3–5 veined (!), stipules small and quickly abscising; flowers monoecious, in large, terminal panicles; sepals 5, valvate (!); male flowers with numerous, distinct stamens and crooked pistils; female flowers with staminodes at the base of the ovaries, styles 3, reflexed, incised at the apex; fruit a 3–4 valved capsule with many, broad winged seeds.

Poliothyrsis sinensis Oliv. Tree, to 15 m high in its habitat, only 3–4 m high in cultivation, young shoots finely pubescent; leaves oval-oblong, 8–16 cm long, acuminate, base usually rounded, dentate, deep green above, finely tomentose beneath to nearly glabrous, petioles 2–4 cm long; flowers greenish to yellowish, in 10–20 cm long, loose panicles, sepals whitish tomentose on the exterior, July; fruit capsules 2 cm long. LF 238; BMns 480. China. Introduced in 1908 by Späth of Berlin, W. Germany. A beautiful park tree. z7 Plate 176; Fig. 304. ∅ ⊕

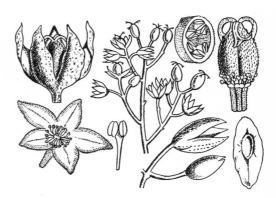

Fig. 304. *Poliothyrsis sinensis.* Inflorescence, individual flowers and flower parts, at lower right a fruit capsule and seed kernel (from Oliver)

POLYGALA L. — Milkwort — POLYGALACEAE

Perennials or shrubs, rarely trees; leaves alternate, opposite or in whorls, simple; flowers usually in racemes, hermaphroditic; sepals uneven, the 3 outermost are more "sepal-like", the 2 inner ones are much larger, petaloid ("wings"); petals 5, but often only 3, fused together and with the stamens; stamens usually 8, in 1 or 2 groups, connate, forming a deeply incised

tube; style straight or curved; fruit a loculicidal, 2 seeded capsule. — About 500–600 species, most herbaceous, the woody species are primarily found in the subtropics.

Polygala alpestris see: **P. chamaebuxus**

P. chamaebuxus L. Evergreen shrubs, procumbent and creeping, to 25 cm high, occasionally 30–50 cm wide, shoots 4 sided, yellow-green to reddish, glabrous; leaves elliptic to obovate, entire, 1–2 cm long, margins slightly recurved, leathery tough, somewhat glossy above, deep green, lighter to gray-green beneath, glabrous on both sides; flowers 1–3 axillary or terminal groups, about 1.2 cm long, wings cream-white, keel yellow, with a 4 incised crest, April–June; fruits flat-globose. HM 1723; BM 316 (= *P. alpestris* Spach). Central Europe to Italy. 1658. z7 # ✣

var. **grandiflora** Gaud. Flowers with red or pink colored wings and a yellow keel (= var. *purpurea* Neilr.; var. *rhodoptera* Brügg.). Switzerland. Plate 128. # ✣

var. *purpurea* see: **P. chamaebuxus** var. **grandiflora**

var. *rhodoptera* see: **P. chamaebuxus** var. **grandiflora**

P. grandiflora see: **P. myrtifolia**

P. myrtifolia L. Abundantly branched, evergreen shrub, 1–1.5 m high; leaves oblong to more obovate, with a small mucro, short petioled; flowers an intense purple, wings broad and oblique obovate, keel large, veined, stalk shorter than the flower (the flowers close up at night). HM 1721; NF 1: 117; BM 3616 (*P. grandiflora*); RCA 8. S. Africa, Cape to Natal Provinces; naturalized along the French and Italian Riviera. z9 # ✣

P. speciosa see: **P. virgata** var. **speciosa**

P. vayredae Costa y Cuxart. Normally only 5 cm high, more open growing than *P. chamaebuxus*; leaves linear and more or less recurved, 2–2.5 cm long; flowers purple-pink, wings light green to purple-brown, keel yellow on the tip. March–April. BM 9009. The Pyrenees, France. 1877. z7 # ✣

P. virgata Thunb. Nearly deciduous shrub (most of the leaves abscise in winter), 1.5–2 m high, shoots thin, reed-like, branched, narrowly erect; leaves linear-lanceolate, 2–2.5 cm long; flowers in terminal, leafless, many flowered racemes at the shoot tips, the individual flowers purple-pink, the lateral petals spathulate, slightly emarginate, wings oblong, obtuse, crest fringed. South Africa. z10 ✣

The following variety is an improvement on the species:

var. **speciosa** Harv. Totally glabrous; the basal leaves more obovate to cuneate, the apical leaves more linear, all obtuse; flowers in about 15 cm long racemes, purple-violet, flowers stalks outspread, bracts quickly abscising, April–June. BM 1780; BR 150 (= *P. speciosa* Sims). S. Africa. 1814. z10 ✣

The winter hardy species are best suited for a sandy-humus site in the rock garden, full sun to semishade, sufficiently moist; in a good site *P. chamaebuxus* will bloom all season long.

Lit. Blake, S. F.: A revision of the genus *Polygala* in Mexico, Central America, and the West Indies; in Contr. Gray Herb. **47**, 1–122, 1916 ● Chodat, R.: Monographia Polygalacearum; in Mem. Soc. Phys. Hist. Nat. Genève, 30–31 + Suppl., 1891–1893.

POLYGONUM L. — Fleece Vine — POLYGONACEAE

Procumbent or erect or twining herbs or shrubs; leaves alternate, stipules fused into a cone; flowers usually in axillary or terminal spiked racemes or panicles; perianth funnelform to campanulate, usually colored, the segments uneven in size; stamens usually 8, at the base of the perianth; ovaries compressed or triangular, with 2 or 3 styles; nutlets enclosed totally or nearly to the tip by the perianth, sharply triangular, shell glossy. — About 300 species worldwide, but most found in the temperate zones.

Characteristics of both the twining species

Flowers white; inflorescences finely rough haired:
    P. aubertii

Flowers pink; inflorescences nearly glabrous:
    P. baldschuanicum

Polygonum aubertii L. Henry. Shrub, very strong growing and twining, annual growth occasionally 8 m, shoots glabrous, new growth reddish; leaves ovate to oval-oblong, 4–9 cm long, bright green, base hastate, margins usually sinuate, densely foliate; floral axes somewhat rough haired, flowers white to greenish-white, never more reddish, in narrow, erect, axillary panicles, September–October. HV 22; RH 1907: 82 (= *Bilderdykia aubertii* [L. Henry] Moldenke). China; W. Szechwan Province; Tibet. 1899. Very rarely sets seed. z5 ∅ ✣

P. baldschuanicum Regel. Shrub, twining to 15 m high,

Fig. 305. *Polygonum baldschuanicum* (from Gfl.)

more woody than *P. aubertii*, twigs glabrous, green striped; leaves broad ovate, 4–10 cm long, base cordate to hastate, usually obtuse-acuminate, pale green, petiole 1–3.5 cm long; flowers small, white, gradually turning reddish, in large, terminal and axillary panicles, the axes nearly glabrous, July–October; fruits triangular, black, glossy. BS 2: 510; BM 7544; BC 3103 (= *Bilderdykia baldschuanica* [Reg.] D. A. Webb). Bukhara, USSR. z5 Fig. 305. #

P. vacciniifolium Wall. Subshrub, procumbent and creeping, shoots very thin and rooting easily, glabrous, usually not over 15 cm high; flowering shoots foliate, leaves small, elliptic to rounded, acute, entire, 1.5–2 cm long, somewhat bluish beneath, stipules about 1.2 cm long, brown, slitted; flowers small, pink, nearly sessile, in 5–7 cm long spikes on short, erect shoots, August–October. BM 4622; PS 925. Himalayas, in the higher mountains. z7 Ø ✧

Both climbing species require a good soil and will quickly cover a wall or other means of support; a hard pruning every other year is advisable. *P. vacciniifolium* is best suited to the rock garden; particularly with a north exposure, in a moist site, plants will quickly cover a large area.

Lit. Steward, A. N.: The polygoneae of Eastern Asia; in Contrib. Gray Herb. 80, 1930.

POLYLEPIS Ruiz & Pavon — ROSACEAE

From this genus (in the group Sanguisorbeae) of 35 South American species, only the following is found in cultivation.

Polylepis australis Bitt. Broad growing, deciduous shrub of medium height, to 3 m (?) high and wide, stem and branches bowed and twisted in zigzag fashion, with

Fig. 306.
Polylepis australis
(from Pizarro)

brown bark exfoliating in long strips, shoots with long internodes; leaves odd pinnate, the first leaf has only 1 leaflet, but no axillary shoots, the second leaf has 5 leaflets and the 3rd has 5–7 leaflets, both with an axillary shoot, leaflets elliptic, 2–4 cm long, 1 cm wide, emarginate at the apex, margin crenate, base cuneate, bluish-green, glabrous; flowers in pendulous, branched spikes, greenish-yellow, 1 cm wide with 4 petals; fruit 3 winged. Argentina. Probably first taken into cultivation at the Copenhagen Botanic Garden, Denmark; collected in 1949, first planted in the landscape in 1962 (the plant was 2.6 m high and 3 m wide in 1970). The climate of its habitat is cool and moist in summer and dry in winter. z6 Fig. 306.

Lit. Bitter, G.: Revision der Gattung, *Polylepis*; in Engler, Bot. Jahrb. **45**, 564–656, 1911 (7 plates, 9 maps).

POLYSCIAS

Polyscias guilfoylei see: **Nothopanax guilfoylei**

POMADERRIS Labill. — RHAMNACEAE

Evergreen shrubs, also small trees in their habitat; leaves alternate, all species more or less white or rust-brown stellate pubescent, quite variable in size, from needlelike to large and oblong; flowers in cymes, often grouped into large, terminal corymbs or panicles; calyx tube adnate to the ovary, with a 5 part limb; petals 5 or totally absent, stamens 5; fruit a capsule with 3 valves. — 45 species in New Zealand and Australia.

Pomaderris apetala Labill. Shrub or small tree, 1–4 m high, occasionally taller; leaves oval-oblong, obtuse or somewhat acute, 5–10 cm long, margins irregularly and finely crenate, glabrous and very rugose above; flowers very numerous in terminal and axillary panicles, 7–25 cm long, light yellow, petals absent, June. LHN 87. Australia, Tasmania. 1803. z10 # ✧

P. elliptica Labill. Shrub, 1–2 m high, occasionally a small tree in its habitat, shoots, leaves and calyx stellate pubescent; leaves ovate to elliptic-lanceolate, 5–10 cm long, 2–5 cm wide, totally glabrous above, densely white stellate pubescent beneath; flowers in flat clusters, 5–7 cm wide, light yellow, petals wide. LHN 86; BM 1510. Australia, New Zealand. 1805. z10 # ✧

P. phylicifolia Lodd. A shrub, heather-like in appearance, 0.5–1 m high, shoots densely woolly pubescent; leaves linear to more oblong, 4 to 12 mm long, 1 mm wide, but somewhat larger and wider on young plants, margins involuted, the white tomentose underside is hardly visible; flowers abundant in loose, axillary cymes, these grouped into large foliate panicles, light yellow, April, the individual flowers are very small. NF 12: 107. Australia; New Zealand. 1819. z10 # ✧

All species prefer a good soil, in an open frost free site, therefore suitable only for the conservatory in cooler climates.

PONCIRUS Raf. — RUTACEAE

Monotypic genus. Deciduous, usually thorny shrub to small tree; leaves trifoliate, translucent punctate, petiole winged; flowers white, axillary, appearing before the leaves, petals 5, much larger than the small, triangular sepals, stamens 8–10, distinct; ovary pubescent, style short and thick, fruit a small, pubescent lemon.

Poncirus trifoliata (L.) Raf. Small, deciduous tree or shrub, 1–7 m high, shoots dark green, with stout, stiff, green thorns; leaves trifoliate, leaflets obovate to elliptic, 3–6 cm long, crenate, somewhat leathery, the lateral ones somewhat smaller, base very oblique; flowers white, 3–5 cm wide, petals eventually flat spreading, April–May; fruit a dull yellow-green, finely pubescent, a 3–5 cm thick lemon, inedible. HM 1701; BM 6513; VT 588 (= *Citrus trifoliata* L.). N. China; Korea; Japan in cultivation only. 1850. z7–8 # ⚭

For a deep, fertile, clay soil and a warm, protected site.

POPULUS L.—Poplar—SALICACEAE

Deciduous, fast growing, normally very tall trees; winter buds often resinous; shoots angular to cylindrical; leaves alternate, usually ovate to lanceolate, entire or dentate, rolled up in the bud stage; flowers dioecious, in pendulous catkins before bud break of the leaves, each flower has an oblique, cupulate disk at the base, the axil contains an incised, occasionally nearly entire catkin scale; stamens 8–30, always distinct; ovary sessile, variable, usually glabrous, tapering to a short style or setting directly upon the stigma; fruit a 2–4 valved capsule, ripe before the leaves are fully developed; seeds numerous, small, with many long, silky hairs.—About 35 species in Europe, N. America, N. Africa and Asia.

Absolute distinction between the species is not simple. The nomenclature of the so-called "Canadian" poplars is still being contested. In this book, the nomenclature of the recent *Check List of Trees of the United States* will be followed.

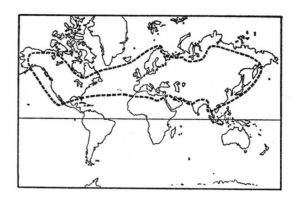

Fig. 307. Range of the genus *Populus*

Outline of the more prominent species, with Key
(from Rehder, altered)

Section 1. **Leuce** Duby—White and Quaking Aspen
   Bark smooth, rough only at the base of older trunks; winter buds tomentose, glabrous or glutinous; leaves tomentose or glabrous; petiole compressed to nearly cylindrical;

   a) White Poplars
      ★ Leaves on the long shoots with white or gray tomentose undersides, petiole round or compressed only on the basal portion; buds tomentose;
         § Leaves lobed on the long shoots:
            P. alba
      §§ Leaves not lobed:
            P. tomentosa
   b) Quaking Aspen
      ★★ Leaves glabrous or pubescent beneath or thin tomentose while developing; buds glabrous;

         ● Leaves without a translucent limb;
            > Petiole compressed; leaves often rounded;
               ✕ Leaves obtuse, serrate or short acuminate;
                  A) Leaves cuneate or rounded at the base, coarsely and irregularly dentate; young twigs lightly tomentose:
                     P. grandidentata

                  AA) Leaves rounded or slightly cordate at the base, crenate to open round toothed;
                     a) Glands at the leaf base on short shoots often absent;

                        / Leaves irregularly round toothed, often obtuse:
                           P. tremula

                        // Leaves regularly crenate, normally short acuminate:
                           P. tremuloides

                     aa) Glands at the leaf base normally well developed; buds and twigs slightly tomentose:
                        P. sieboldii

               ✕✕ Leaves long acuminate, glandular at the base:
                  P. adenopoda
            >> Petiole round;

               § Leaves light green beneath;

Section 2. **Leucoides** Spach—Large leaved Poplars
   Bark of older trunks rough, scaly; leaves of the long and short shoots hardly differing, petiole only somewhat compressed at the tip, otherwise cylindrical; winter buds conical, glabrous, somewhat glutinous:

   Leaves cordate at the base, white tomentose when young;

      a) Young shoots pubescent;
         / Young shoots floccose, dark green above, to 15 cm long:
            P. heterophylla

// Leaves pubescent beneath, glossy green above, with a red midrib, to 30 cm long:
P. lasiocarpa

aa) Young shoots glabrous, leaves dull bluish-green above:
P. wilsonii

Section 3. **Tacamahaca** Spach. — Balsam Poplars
Trunk bark furrowed; winter buds large, very glutinous, quite aromatic; leaves mostly whitish beneath, without a translucent margin; petiole cylindrical to 4 sided rectangular, furrowed above;

Leaves rounded at the base or broad cuneate, lanceolate to ovate-lanceolate, glabrous;
+ Petiole 1–4 cm long; leaves lanceolate to oval-lanceolate:
P. angustifolia

++ Petiole 3–7 cm long, leaves rhombic-lanceolate to oval-rhombic:
P. acuminata

§§ Leaves whitish beneath;
× Shoots brown;
/ Shoots glabrous (varieties of *P. balsamifera* slightly downy pubescent);
A) Leaves widest at or above the middle, 4–12 cm long, petiole 0.5–2.5 cm:
P. simonii

AA) Leaves widest beneath the middle;
a) Shoots more or less angular; capsule 3–4 valved;
+ Leaves broad cuneate at the base; petiole 0.5 to 2.5 cm:
P. yunnanensis

++ Leaves rounded at the base or slightly cordate, 10–20 cm long; petiole 2–4 cm:
P. szechuanica

aa) Shoots round, capsule bivalvate; petiole 3–5 cm:
P. balsamifera

// Shoots pubescent, occasionally glabrous on *P. trichocarpa*;
a) Leaves cordate at the base; capsule bivalvate;
P. balsamifera var. *subcordata*

aa) Leaves truncate to slightly cordate at the base, occasionally broad ovate;
+ Shoots angular, brown; petiole 3–6 cm; capsule pubescent, trivalvate:
P. trichocarpa

++ Shoots round, dark brown, petiole 1.5–4 cm:
P. tristis

×× Shoots yellowish-gray to orange-yellow;
/ Shoots round;
+ Young shoots glabrous or with a very fine, downy pubescence; petiole 1–4 cm;
a) Leaves glabrous, ovate or narrowly ovate, acuminate:
P. cathayana

aa) Leaves pubescent on the venation beneath, elliptic, abruptly tapering to a twisted apex:
P. maximowiczii

++ Young shoots glandular glutinous; petiole 0.5–2.5 cm long:
P. koreana

// Shoots sharply angular, pubescent when young; leaves sparsely pubescent beneath at least when young:
P. laurifolia

● ● Leaves with a distinctly translucent limb;

Section 4. **Aigeiros** Duby — Black Poplars
Bark of older trunks furrowed; winter buds glutinous; leaves with slitted openings on both sides and a translucent margin, usually triangular to oval-rhombic, more or less coarsely crenate;
+ Petiole cylindrical or nearly so; the translucent limb very narrow; young shoots only slightly angular:
P. berolinensis

++ Petiole compressed;
> Glands absent from the leaf margin; leaf blade not or only very short ciliate;
a) Leaves rhombic-ovate, cuneate at the base; capsule bivalvate:
P. nigra

aa) Leaves truncate at the base or broad cuneate;
× Leaves triangular, irregular and sparsely ciliate, occasionally with 1–2 glands at the base:
P. canadensis

XX Leaves broad triangular-ovate, sparsely ciliate, capsule 3–4 valved:
  P. fremontii

>> Glands present at the leaf base; leaves truncate or somewhat cordate, densely ciliate;
  § Shoots round, except on strong shoots; subtending leaves of the catkins slit;
    + Buds pubescent, occasionally glabrous; leaves very wide at the base and very coarsely dentate:
      P. sargentii

    ++ Buds glabrous; leaves triangular-ovate:
      P. deltoides

  §§ Shoots sharply angular; leaves ovate or oblong-ovate, to 28 cm long; subtending leaves of the catkins toothed:
      P. angulata

Populus × acuminata Rydb. (= *P. angustifolia* × *P. sargentii*). Tree, to 15 m high, crown rounded, shoots cylindrical, yellowish-brown, glabrous, buds brown, acuminate; leaves oval-rhombic (!), long acuminate, 5–10 cm long, crenate, base broadly cuneate, glossy dark green above, light green beneath (!) and glabrous, midrib very prominent, petiole 2–7 cm long; capsule 3 or 2-valved. BB 1167; MD 1912: 119; SS 731 (= *P. andrewsii* Sarg.). N. America, east side of the Rocky Mts.

P. adenopoda Maxim. Tree, to 25 m or more, shoots thin, pubescent when young, later gray-brown to brown, winter buds acutely conical, glabrous; leaves of the long shoots ovate and long acuminate (!), base truncate to cordate and with 2 protruding glands (!), crenate, the tips of the teeth curved inward and with glands, 7–10(15) cm long, usually glabrous above, lighter beneath and gray pubescent, at least when young, leaves smaller on the short shoots, more ovate to oval-rounded, 5–8 cm long, petiole 1.5–3 (occasionally to 6) cm long; catkins 6–10 cm long, subtending leaves deeply lobed and long ciliate; fruit catkins 12–16 cm long, fruit capsules short stalked. BD 1922: 23 (= *P. duclouxiana* Dode). Central and W. China. 1907. Ⓕ China. z5

P. alba L. Silver Poplar, White Poplar. Tree, to 30 m high, crown broad rounded, bark gray-white, rough only on older trunks, young twigs and buds white tomentose; leaves ovate, 3–5 lobed on long shoots, the lobes coarsely dentate, 6–12 cm long, base rounded-cordate, dark green above, white tomentose beneath, smaller on short shoots, oval-elliptic, sinuately dentate, more gray tomentose beneath, petiole pubescent; catkins 5–8 cm long, subtending leaves dentate, limb with long pubescence. BB 1164; HM Pl. 83; HW 2: 135; MD 1914: 277; HF 944. Central and Southern Europe to N. Africa and Central Asia, found in moist forests as well as dry sites. Ⓕ W. Germany, Italy, USSR, Greece, Iran, Syria, India. z3 Fig. 308. ⊘

f. *arembergiana* see: 'Nivea'

var. **bachofenii** Hartig. Trunk and branches smooth, chalk-white; leaves of the long shoots deeply lobed, otherwise more triangular, base truncate, 5–8 cm long, dense white tomentose beneath. Central Asia to Himalayas. Fig. 309.

'Globosa' (Späth). Tall shrub, habit broad rounded, new growth pink, gray-tomentose beneath. 1887.

'Intertexta' (Späth). Young leaves chalk-white, the older ones yellow speckled.

var. *macrophylla* see: **P. × canescens 'Macrophylla'**

'Nivea'. Only a juvenile form, young twigs, petiole and leaf undersides an intense chalky white; leaves deeply lobed; these characteristics become less distinctive with maturity (= f. *arembergiana* Hort.). First named in England in 1789.

'Pendula'. Twigs pendulous (= *P. alba salomonis* [Carr.] Wesm.). 1886.

'Pyramidalis'. Tall tree, habit narrow conical, bark gray-green; leaves more deeply lobed and larger on the long shoots, more rounded on the short shoots, often glabrous and becoming green beneath. MD 1916: Pl. 28; 1917: Pl. 38 (= *P. bolleana* Lauche). Place of origin not known for certain, but commonly cultivated in central Asia before 1841. Introduced in 1841 from the Karataw Mts., between Bokhara and Samarkand. Plate 173.

'Raket'. Extremely slender growth habit, all branches tightly upright, stem very straight. Introduced from the Bosbouwproefstation, Holland in 1956. Resistant to foliage diseases.

'Richardii' (Richard). Leaves golden-yellow above, white tomentose beneath. Discovered in Holland around 1910. Needs a semishady location. Plate 173. ⊘

var. *salomonis* see: 'Pendula'

P. andrewsii see: **P. × acuminata**

P. × andrewsii Sarg. (= *P. acuminata* × *P. sargentii*). Very similar to *P. sargentii*, but the current year's shoots are a light orange-brown; leaves oblong-ovate, 9–10 cm long, finely crenate, base rounded or cuneate, petiole cylindrical. BD 1921: 5. Colorado. 1913. z3

P. angulata Ait. Broad crowned tree, to 30 m high, young shoots thick, very angular, yellow-brown, glabrous, winter buds greenish, somewhat glutinous; leaves ovate to broad oval-oblong, abruptly short acuminate, to 18 cm long and 12 cm wide, crenate and ciliate, base cordate and with 2 or more glands, light green on both sides; catkin scales concave, margins crenate, not fimbriate (!). EH 384. SE. USA. Before 1789. Origin unclear. z6 ⊘

'Cordata' (Simon-Louis). Leaves larger, more conspicuously cordate at the base, 15–20 cm long and 12–15 cm wide; only the male is known. More winter hardy than the species.

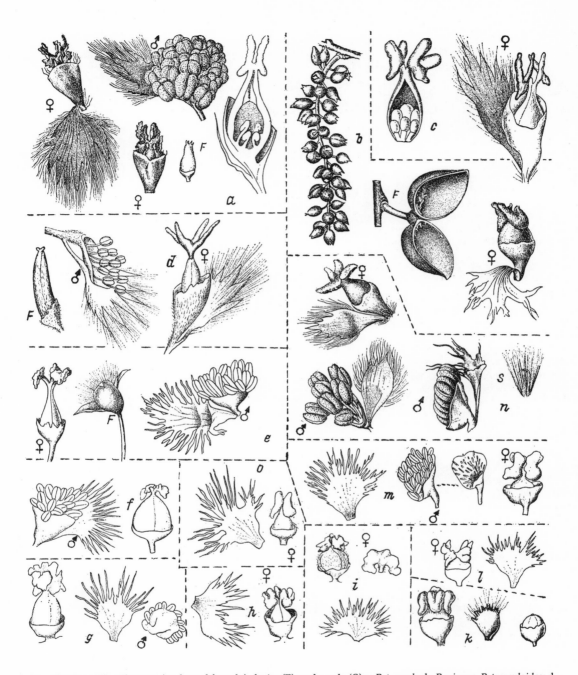

Fig. 308. **Populus.** Flowers (male and female), fruits (F) and seeds (S) a. *P. tremula;* b. *P. nigra* c. *P. tremuloides;* d. *P. grandidentata;* e. *P. heterophylla;* f. *P. angustifolia;* g. *P. suaveolens;* h. *P. laurifolia;* i. *P. trichocarpa;* k. *P. deltoides;* l. *P. balsamifera* var. *subcordata;* m. *P. balsamifera;* n. *P. alba;* o. *P. berolinensis* (from Hempel & Wilhelm, Schnitzlein, Sargent, Schneider, Shirasawa)

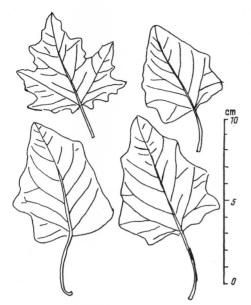

Fig. 309. *Populus alba* var. *bachofenii*
(from Bugala)

P. angustifolia James. Tree, to 20 m high, narrow crowned, conical, twigs thin, cylindrical, glabrous, eventually orange-brown, buds small, sharply acuminate, glabrous; leaves willow-like, lanceolate to oval-lanceolate, glandular serrate with involuted margins, 8–10 cm long, light green beneath; fruit capsule bivalvate. BB 1160; SS 492; VT 88; MD 1912: 119 (= *P. coloradensis* Dode). W. USA, in river valleys. 1893. Ⓕ France. Plate 171; Fig. 308.

P. ariana see: **P. euphratica**

P. 'Bachelieri' see: **P. × canadensis 'Robusta'**

P. balsamifera L. Balsam Poplar. Tree, to 30 m high, branches ascending, current year's shoots cylindrical, glabrous, winter buds large, to 2.5 cm long, with a yellowish, fragrant, glutinous pith; leaves ovate to oval-lanceolate, acuminate, rather tough, 7–12 cm long, crenate and quite finely ciliate, base usually round or somewhat broad cuneate, smooth above, whitish beneath, petiole cylindrical, 3–5 cm long; catkins 5–7 cm long, fruit catkins 12–14 cm long; capsules bivalvate. SS 490; EH 387 (= *P. tacamahaca* Mill.). Northern N. America. 1689. After 1919 the name *P. balsamifera* was used by Sargent and Farwell for *P. deltoides* Marsh. which created much confusion; therefore the former name was discarded by Davy, Houtzagers and Cansdale as "nomen ambiguum". Eventually Rouleau succeeded in proving that Linnaeus' original name for the Balsam Poplar should be retained. Ⓕ N. USA; USSR. z6 Plate 174; Fig. 308, 310. ⌀

var. *candicans* see: **P. balsamifera** var. **subcordata**

var. **subcordata** Hylander. Tree, to 30 m high, round crowned, annual shoots red-brown to olive-brown, somewhat angular on particularly strong shoots, otherwise cylindrical, winter buds large, acute, appressed, brown, fragrant and very glutinous; leaves broad cordate-ovate or triangular, acuminate, 12–16 cm long, leathery, glossy dark green above, underside whitish metallic and somewhat pubescent, particularly dense on the venation (!), margins dentate and ciliate, petiole cylindrical, 2–6 cm long, channeled above, reddish, very pubescent; fruit catkins to 16 cm long, capsule stalked, bivalvate. GTP 97; SS 491; SPa 107 (= *P. balsamifera* var. *candicans* [Ait.] A. Gray; *P. candicans* A. Gray). Northeastern N. America and Canada. Ⓕ E. France, Yugoslavia, Canada. Fig. 308, 311. ⌀

'Aurora'. Leaves yellowish-white, with light and dark green spots in the center of the plant, therefore regular shearing is necessary to maintain good foliage color (= *P. candicans* 'Variegata'). Brought into the trade before 1954 by Treseder's Nursery, Truro, Cornwall, England. ⌀

'Balm of Gilead' is nothing more than a sterile clone of var. *subcordata*, long planted over a considerable range in NE. USA and SE. Canada (= *P. gileadensis* Rouleau). The origin of this male clone is unknown. ⌀

var. *viminalis* see: **P. laurifolia**

P. × berolinensis Dipp. (*P. laurifolia* × *P. nigra* 'Italica'). Berlin Laurel Poplar. Tree, broad columnar, to 25 m high, branches spreading and ascending, shoots somewhat angular, gray-yellow, slightly pubescent, winter buds greenish, glutinous; leaves acutely ovate, 8–12 cm long, crenate, margins undulate and translucent,

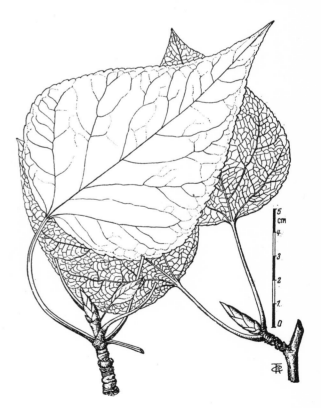

Fig. 310. *Populus balsamifera* (from Sudworth)

dark green above, underside whitish or more green, petiole round, scattered pubescent; only the male observed. EH 410 (= *P. certinensis* Hort. ex Schmidt). Developed in the Berlin Botanic Garden before 1870. One of the more prominent Poplars for park trees in Europe. Ⓕ W. Germany, Czechoslovakia, Yugoslavia, France, England, USA. z2 Plate 171; Fig. 308. ⊘

P. bolleana see: **P. alba 'Pyramidalis'**

P. brevifolia see: **P. simonii**

P. × canadensis Moench (*P. deltoides* × *P. nigra*). Canadian Poplar. Tall, usually broad crowned tree, young shoots cylindrical to somewhat angular, glabrous, occasionally somewhat pubescent, winter buds glutinous, lenticels round or linear; leaves delta-shaped, long acuminate, 7–10 cm long, margins crenate and ciliate at first, but quickly becoming glabrous or with remnants of ciliate hairs, the teeth at the truncate leaf base more widely spaced, petiole reddish, with or without 1(2) glands at the leaf base (!!); male catkins to 7 cm long, stamens 15–25 (!) (= *P. euramericana* Guiner). Originating spontaneously around 1750 in France. Ⓕ Europe, S. America, China. z5

Including a number of forms, most used in commercial forestry.

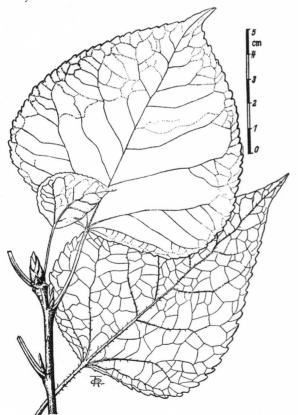

5
cm
#

3

2

1

0

Fig. 311. *Populus balsamifera* var. *subcordata* (from Sudworth)

Key to the more prominent hybrids
(from Hesmer)

● Annual shoots glabrous;
  ★ Leaves often somewhat cuneate at the base;

       > Thick, light green to yellow-green leaves, apex gradually drawn out in a long tip, base cuneate; petiole pure green; annual shoots gray and very flexible; female:
               'Marilandica'
     >> Smaller, light green leaves, less cuneate at the base, straighter; petiole predominantly green, but not as pure green as 'Marilandica'; annual shoots more brown, less flexible:
               'Eugenei'

  ★★ Leaves straight at the base, somewhat rounded or cordate;
    × Male plants;
      + Typically white bark, gray twigs, leaves bright green;
        § Tree densely covered with cankerous knots:
           'Brabantica'
        §§ Tree not covered with cankers; leaves somewhat lighter green:
           'Gelrica'

      ++ Leaves darker green, base wide and straight, apex short acuminate; petiole turning red; annual shoots brown:
           'Serotina'

    ×× Female plants; petiole often more or less turning red; annual shoots brown:
           'Regenerata'

●● Annual shoots more or less pubescent; leaves large, thick, dark green, leathery, triangular, drawn out to a short apex; non-woody shoots reddish; second year and older shoots with many, short, outspread branchlets; male:
           'Robusta'

'Aurea' (Van Geert). Somewhat like 'Serotina', but slower growing. Leaves golden-yellow in early summer, then gradually becoming yellow-green, petiole red. IH 232 (= *P. canadensis van geertii* Witte; *P. serotina aurea* Henry). Originated as a bud mutation around 1867 in Van Geert Nursery, Belgium. ⊘

'Brabantica'. Trunk usually crooked, suffering greatly from cankers and therefore easily recognizable (!), shoots glabrous, gray, somewhat angular, without short shoots on the second year and older branches, winter buds smooth, appressed, acute like 'Serotina', new growth appears in late April, brownish, but quickly turning green; leaves somewhat lighter green than 'Serotina' and 'Marilandica', larger and also more coarsely dentate, petiole green; male. Developed in Holland about 1850. Difficult to identify as a young plant. Fig. 313.

'Eugenei' (Simon-Louis). Fast growing, upright, crown rather narrow, twigs somewhat angular, glabrous, directed upward, buds small and appressed, bark in the crown area conspicuously light, new growth appears early, brownish; leaves triangular to rhombic, 5–8 cm long and wide (somewhat smaller than most of the other forms), short acuminate, coarsely crenate and ciliate, petiole green; flowers only male, only about 3.5–5 cm long. EH 409. Developed in 1832 by Simon Louis in France. Plate 174; Fig. 313.

'Gelrica'. Strong growing, trunk somewhat crooked, continuous through the crown, crown densely foliate, bark remaining whitish, with dark rings and spots, twigs less flexible, lead-gray, buds outspread, new growth red-brown, appearing with or somewhat before 'Regenerata'; leaves triangular, wider than long, glossy bright green, coarsely dentate, glabrous on both sides, petiole flat, normally green, turning somewhat reddish; male. Developed in Holland about 1850. Fig. 312.

'Marilandica'. May Poplar. Tall tree, usually with a short, thick trunk and a wide, well branched crown, major branches often very crooked, arising at obtuse angles from the stem, the basal branches often directed downward, new growth appears middle-early, brown, but very quickly turning green; leaves particularly light green, rhombic with long tips, these not dentate, base broad cuneate, to 10 cm long and 6–8 cm wide, petiole green; flowers only female. EH 409. Presumably developed in France about 1800. Fig. 312.

'Regenerata'. Old trees have a rather straight trunk, branches frequently somewhat whorled and spreading at steep angles, bark light gray, annual shoots brown, more gray at the base, buds glabrous, appressed, acute, rather short, new growth light brown, but becoming green after a few days; leaves triangular, about as long as wide, light green, petiole often somewhat reddish; only the male known. EH 409. Discovered in France about 1814. Fig. 312.

'Robusta' (Simon-Louis). Columnar habit, regularly upright, branches nearly whorled (!), but not arising at exactly the same height, young twigs finely pubescent, green, turning red, lenticels round and forming lines (usually round), buds appressed, acute, smooth, glabrous, rather long; leaves conspicuously red-brown on new growth, triangular, tough, glossy, glabrous, 10–12 cm long and 8–10 cm wide, apex serrate nearly to the tip, petiole turning red; flowers only male, catkins 7–9 cm long, with about 20 stamens per flowers (= *P. robusta* Schneid.; *P.* 'Bachelieri' [Solemacher]; *P. vernirubens* Henry). Developed by Simon-Louis in Plantières near Metz, France in 1895, now widely planted. Ⓕ W. Germany, Belgium, Holland, France, England. Fig. 312.

'Serotina'. Late Poplar. Tree, to 40 m high, crown broad conical, eventually rounded, open, annual shoots smooth, brown, glabrous, rather flexible, more gray at the base, buds appressed, acute, glabrous, glutinous, new growth appears very late, mid-May, red-brown; leaves nearly an equilateral triangle, 7–10 cm long and wide, dark green, not glossy, base straight, truncate, petiole usually reddish; only the male is known. EH 386. Discovered in France about 1700. Fig. 312.

'Serotina Erecta'. A selection with a straight stem and fast growth, lateral shoots more ascending, trunk gray-white, remaining smooth, bark developing late; leaves triangular, wider than long, short acuminate, young leaves green, petiole green, only slightly reddish; male. EH 385. Discovered in Belgium about 1818.

var. *van geertii* see: 'Aurea'

P. × **canescens** (Ait.) Smith (= *P. alba* × *P. tremula*). Gray Poplar. Large tree, similar to *P. alba*, often considered a separate species, new growth appears early, somewhat later than *P. alba*, young shoots gray (!, not white); leaves on the long shoots triangular-ovate, irregularly glandular dentate, margins slightly undulate and ciliate at first, dark green above, loose gray tomentose beneath, 6–12 cm long, base slightly cordate; leaves on the short

shoots oval-rounded, not ciliate, quickly becoming rather glabrous and light green beneath; catkins 6–10 cm long. HW 2: 137; EH 7: 382; HF 944 (= *P. hybrida* Reichenb.). Europe, Aisa Minor. In cultivation for centuries. Ⓕ W. Germany, France, Yugoslavia, USSR, Iran. z5 Plate 174; Fig. 314.

'Aureovariegata'. Poor grower; leaves smaller and yellow marbled, but often reverting back to the green form. ∅

'Macrophylla'. Very fast growing form; leaves on long shoots often to 15 cm long (= *P. alba macrophylla* Hort.; *P. picardii* Hort.).

'Pendula'. Shoots gracefully weeping, otherwise like the species.

'Pyramidalis'. Regularly broad-conical crown, trunk straight, continuing through the crown, bark gray-green, branches steeply ascending, the lower branches more outspread; leaves like those of the species. IDS Year Book 1971: 22. Discovered in 1934 in the vicinity of Warsaw, Poland by A. Wrobeswski.

P. candicans see: **P. balsamifera** var. **subcordata**

P. candicans 'Variegata' see: **P. balsamifera** 'Aurora'

P. cathayana Rehd. Tree, to 30 m high, annual shoots cylindrical, olive-green at first, later orange-yellow to gray-yellow, buds stretched; leaves on the short shoots ovate to narrow ovate, 6–10 cm long, 3.5–7 cm wide, acuminate, base rounded or (occasionally) somewhat cordate, glossy green above, whitish beneath, with 5–7 curving veins, otherwise with reticulate venation beneath; leaves of the long shoots 10–20 cm long, often slightly cordate, petiole glabrous, 2–6 cm long, male catkins 5–6 cm, female catkins 10–20 cm long, capsules glabrous, 7–9 mm long, nearly sessile, usually 3–4 valved. EH 410; LF 59; BD 1923: 16 (= *P. suaveolens* sensu Schneid.). NW. China, Manchuria, Korea. 1908. Ⓕ China, Yugoslavia. z5 Plate 173. ∅

P. cercidifolia see: **P. tremuloides** var. **aurea**

P. certinensis see: **P.** × **berolinensis**

P. × **charkowiensis** Schroed. (= *P. nigra* or *P. nigra* 'Italica' × ?). Columnar habit, but somewhat broader than 'Italica', more broadly conical, young bark conspicuously yellow (!), later more yellow-green, pubescent at first; leaves like those of *P. nigra*. Developed about 1900 by J. J. Gabeschtoff in Charkow, Russia. This poplar was once considered both the most rapidly growing poplar and the one with the lightest colored wood. Recent experience has brought this view into question.

P. coloradensis see: **P. angustifolia**

P. davidiana see: **P. tremula** var. **davidiana**

P. deltoides Marsh. American Poplar (Cottonwood). A broad crowned tree in its habitat, to 30 m high, twigs glabrous, only the long shoots angular ribbed, otherwise cylindrical, the non-woody shoots green, only slightly reddish, with long, linear, white lenticels, winter buds brownish, glutinous, long and sharply acuminate, appressed; leaves triangular-ovate to somewhat

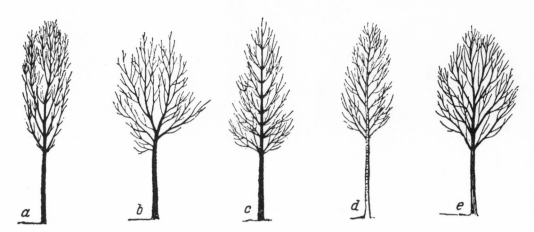

Fig. 312. **Populus.** Younger trees in the dormant state.
a. 'Regenerata'; b. 'Marilandica'; c. 'Robusta'; d. 'Gelrica'; e. 'Serotina' (from Houtzagers)

rhombic, 7–12 cm long, coarsely serrate with bowed teeth, margins dense and persistently ciliate (!!), otherwise light green, apex entire, base slightly cordate to truncate, with 2–3 glands (!!), petiole more or less reddish; flowers each with 40–60 stamens, male catkins 7–10 cm long; fruit clusters 20 cm long, the capsules 3–4 valved, short stalked. BB 1172; GTP 103; SM 137 (= *P. balsamifera* L. p.p.; *P. monilifera* Ait.). Central USA, Mississippi region, Virginia, Carolina. Around 1750. Ⓕ USA, W. Europe, Asia, N. and S. Africa, S. America. z2 Plate 172; Fig. 308, 313.

It is doubtful whether the common division of this species into 2 geographic varieties (var. *missouriensis* Henry for the northern types and var. *virginiana* for the southern) should be retained since many transitional forms are to be found between these two ranges. The division of this variable species is therefore not sustainable.

P. duclouxiana see: **P. adenopoda**

P. euphratica Oliv. Slender tree or shrub, shoots without terminal buds, winter buds pubescent, not glutinous; leaves quite variable, partly nearly circular or ovate or kidney-shaped to narrow lanceolate or linear, occasionally coarsely dentate, partly entire, with reticulate venation, blue-green or gray-green on both sides, petiole 1 to 4 cm long (= *P. ariana* Dode). N. Africa to Asia Minor and Central Asia. 1920. Often mentioned in the Bible as the "willow" of the children of Israel. Ⓕ N. and E. Africa to China and Mongolia. z9 Plate 171. ⊘

P. euramericana see: **P. × canadensis**

P. europaea see: **P. nigra**

P. fargesii see: **P. lasiocarpa**

P. fastigiata see: **P. nigra 'Italica'**

P. fremontii S. Wats. Tree, to 30 m high, somewhat similar to 'Marilandica', but with a straighter stem, crown more conical, slower growing, annual shoots round, thick, smooth, glabrous, brown-gray at the apex, more

gray toward the base, buds appressed, smooth, obtuse; leaves broad oval-rounded to rhombic, short acuminate, usually broader than long, 7–8 cm long, margins dentate, teeth coarse, often also curving upward, margins distinctly translucent, somewhat ciliate, later glabrous, base broad cuneate to slightly cordate; catkins 5–10 cm long. VT 86; MD 1922: 11; SPa 110; SS 496. N. America; California to Arizona. Ⓕ France. z7 Plate 174; Fig. 315.

var. *arizonica* (Sarg.) Jeps. (= *P. arizonica* Sarg.) Cannot be distinguished from the species. Plate 174.

var. *macdougalii* (Rose) Jeps. (= *P. macdougalii* Rose). Cannot be distinguished from the species.

var. **wislizenii** S. Wats. Differing from the species primarily in the more slender stalked flowers; leaves triangular to broad ovate, 5–10 cm long and wide, short acuminate, coarse and irregularly crenate, base truncate or slightly cuneate; fruit catkins very slender, 5–15 cm long, capsule above 1 cm long, with 3–4 valves, petiole about 1.5 cm long. SS 732 (= *P. wislizenii* [S. Wats.] Sarg.). W. Texas (Rio Grande Valley), New Mexico. 1894. Ⓕ France.

P. × generosa Henry (= *P. angulata* × *P. trichocarpa*). Tall, vigorous growing tree, similar to *P. trichocarpa*, annual shoots green-brown, younger shoots gray-green, slightly angular, winter buds brown, long acuminate, glutinous; leaves rather regularly oval, base truncate to slightly cordate, margins serrate to dentate, translucent, ciliate, somewhat leathery, glossy dark green, gray-green to whitish beneath, never angular, long drawn out, 7–11 cm long, small leaves very similar to those of *P. trichocarpa*; male catkins 5–8 cm long, 40–60 stamens in each flower; fruit catkins eventually 20–25 cm long, capsules with 3–4 valves. GC 56: 103. Developed by Royal Botanic Gardens, Kew, England in 1912. Ⓕ Belgium, France. z6

P. gileadensis see: **P. balsamifera 'Balm of Gilead'**

P. glaucicomans see: **P. pruinosa**

Fig. 313. **Populus.** Young trees in the dormant state. From left to right: *P. deltoides*; *P. canadensis* 'Eugenei'; *P. canadensis* 'Brabantica'; *P. nigra* (Original)

P. grandidentata Michx. Tree, to 20 m high, crown narrow and rounded, bark smooth, twigs gray tomentose at first, eventually glossy brown, winter buds gray tomentose (!); leaves on the long shoots acuminate-ovate, 7–10 cm long, coarsely sinuate, dark green above, gray tomentose beneath at first, later glabrous and blue-green, more elliptic and sharper toothed on the short shoots; male catkins 4–6 cm long, fruit catkins to 10 cm long. GTP 94; SS 488. Eastern N. America. 1722. Ⓕ USA. z3 Fig. 308, 316.

var. pendula see: **P. pseudograndidentata**

P. hastata see: **P. trichocarpa** var. **hastata**

P. heterophylla L. Broad, round crowned tree, 7–15 m high, to 30 m in its habitat, often only a shrub, irregularly branched, twigs thick, very tomentose at first, then dull brown and glabrous, pith orange (!); leaves broad ovate, 10–18 cm long, crenate, rounded at the apex, base cordate, gray-white tomentose on the new growth, quickly becoming glabrous and dark green above, tomentum persisting longer beneath, midrib yellow, petiole cylindrical (!), 6–8 cm long; catkins 3–6 cm long; ripe fruit clusters 6–12 cm long, capsules with 3–4 valves. SS 489; BB 1168; VT 90; KTF 15. E. USA, river banks and swamps. 1765. Rarely seen in cultivation and then not thriving. z5 Fig. 308, 316. ∅

P. hudsonica see: **P. nigra** var. **betulifolia**

P. hybrida see: **P. × canescens**

P. × jackii Sarg. (= *P. balsamifera* × *P. deltoides*). Broad crowned tree, shoots cylindrical; very closely related to *P. balsamifera*, but with more broadly ovate leaves, long acuminate, more coarsely serrate with a narrow, translucent limb, base cordate, petiole not compressed, rectangular in cross section. BC 3145. N. America. 1900. Plate 174.

P. koreana Rehd. Tree, to 25 m high, conical habit, bark light gray, annual shoots cylindrical, glandular glutinous at first, later light brown, new growth appears 2 weeks earlier than other poplars (!); leaves on the long shoots elliptic to oval-oblong, 7–15 cm long, glandular crenate, glutinous at first, glossy with aromatic young foliage,

dark green above and very rugose, whitish with a red midrib beneath, leaves on the short shoots only 4 to 10 cm long, often pubescent on the venation beneath and the petiole; fruit capsule 2–4 valved, sessile. PB 23; NK 18: Pl. 49. Korea. 1923. Not very tolerant of air pollution. Ⓕ Great Britain, Yugoslavia. z5 Plate 173. ∅

P. lasiocarpa Oliv. Round crowned tree, to 20 m high, twigs thick, densely tomentose at first, later yellow-brown and glabrous, angular, buds coarse, glutinous, basal scales pubescent; leaves cordate-ovate, to 30 cm long, acute, glandular crenate, glossy gray-green above, lighter and pubescent beneath; venation and petiole red (!); catkins 8–9 cm long. LF 60–61; MD 1921: 10; PB 10; BM 8625 (= *P. fargesii* Franch.). China. 1900. Slow growing, difficult to propagate; requires a protected site. Ⓕ SW. China. z6 Plate 173, 5 (Vol. III). ∅

P. laurifolia Ledeb. Laurel Poplar. Tree, 15 to 20 m high, wide-spreading crown, current year's shoots thin, sharp

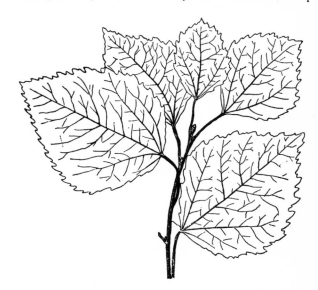

Fig. 314. *Populus × canescens* (from Dippel)

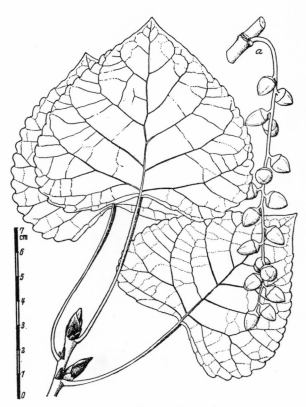

Fig. 315. *Populus fremontii.*
Fruit cluster at right (from Sudworth)

angular (!), gray-yellow, somewhat pubescent at the apex when young, winter buds outstretched, erect, not appressed, glutinous; leaves on the long shoots ovate-lanceolate to lanceolate, acuminate, 6–13 cm long, base rounded, finely glandular serrate, glossy green above, whitish and somewhat pubescent beneath, at least on the midrib, petiole short, pubescent, leaves on the short shoots elliptic-ovate, base rounded, petioles longer; capsule slightly pubescent, 2–3 valved. EH 410; BD 1922: 23; PB 24 (= *P. balsamifera* var. *viminalis* Loud.). NW. India to NE. Siberia and Japan. 1830. Ⓕ China, USSR. z5 Fig. 308.

f. **lindleyana** (Carr.) Rehd. Leaves narrower, usually lanceolate on long shoots and narrow elliptic on short shoots, base round to broad cuneate, limb usually undulate. BC 3144 (= *P. lindleyana* Carr.; *P. salicifolia* Hort.). Before 1876.

P. lindleyana see: **P. laurifolia** f. **lindleyana**

P. macdougalii see: **P. fremontii** var. *macdougalii*

P. maximowiczii Henry. Large, broad crowned tree, to 30 m high, bark gray, deeply furrowed, annual shoots reddish at first (!) and densely pubescent, later gray; leaves elliptic to oval-elliptic, abruptly short acuminate with a twisted apex (!), 6–12 cm long, glandular serrate and ciliate, dull dark green and rugose above, whitish beneath, leathery, venation finely pubescent on both

sides; catkins 5–10 cm long; capsules glabrous, nearly sessile, 3–4 valved. NK 18: 50; NF 1: 13; PB 22 (= *P. suaveolens* sensu Maxim. non Fisch.). NE. Asia to N. Japan. 1890. Ⓕ NE. China, Japan. z5 Plate 170. ⊘

P. monilifera see: **P. deltoides**

P. nigra L. Black Poplar. Tree, to 30 m high, broad crowned, trunk often with thick knots, branches stout, widespreading, old bark deeply furrowed, new growth cylindrical (!), gray-yellow, glabrous, buds long, outstretched, red-brown, glutinous, bowed outward at the apex; leaves rhombic, long acuminate, 5–10 cm long, the lateral edges rounded, finely crenate, not ciliate (!), lighter green beneath, glabrous, leaves smaller on the short shoots, broader, often truncate or rounded at the base, petioles thin; flowers with 20–30 stamens, catkins 4 to 10 cm long; capsules bivalvate (!). BB 1171; HW 139; HF 946; PB 25 (= *P. europaea* Dode). N. Africa to central Europe, reaching eastward to the Jenisse River in central Siberia. In cultivation for centuries. Ⓕ Europe; Turkey, Iran, Israel, Syria, Argentina, Chile. z2 Fig. 308, 313.

var. **betulifolia** (Pursh) Torr. Like the species, but with somewhat more obtuse buds, twigs pubescent at first, the 1–2 year shoots brown-orange (not gray), new growth starts earlier than the species; leaves somewhat smaller, gradually drawn out to a point, young leaves and petioles pubescent, the latter yellow-green (green on *P. nigra*); observed as male and female plants. PB 25c; EH 409; BM 8298 (= *P. hudsonica* Michx.). England, France. 1790.

'**Italica**'. Pyramidal Poplar, Italian Poplar, Lombardy Poplar. Columnar habit, branches ascending at a very sharp angle, young shoots cylindrical, more brown at the apex, later more gray, new growth about 3 weeks earlier than that of the species; leaves somewhat smaller, more rounded-rhombic, petiole turning red; nearly all male plants observed but occasionally female specimens may be found, these with a somewhat broader crown. HW 2: 140 (= *P. pyramidalis* Salisb.; *P. nigra pyramidalis* [Borkh.] Spach; *P. fastigiata* Desf.). Origin uncertain, but probably found in the latter half of the 18th century in Italy (Lombardy); widely planted since that time. Plate 172.

'**Plantierensis**' (Simon-Louis). Similar to 'Italica', but not quite as columnar, twigs pubescent (!); leaves somewhat smaller, petiole reddish and pubescent (!); both male and female plants (= *P. plantierensis* Simon-Louis; *P. nigra* var. *elegans* Bailey). Originated in Metz, France before 1855.

var. *elegans* see: '**Plantierensis**'

var. *pyramidalis* see: '**Italica**'

var. **thevestina** (Dode) Bean. Like 'Italica', but with the bark of older trunks nearly white, younger shoots light gray, pubescent at first; leaves triangular-ovate, base broad cuneate, petiole pubescent (= *P. thevestina* Dode). N. Africa, Algeria where it is planted over a considerable range. 1903. Plate 172.

P. occidentalis see: **P. sargentii**

P. pekinensis see: **P. tomentosa**

P. × petrowskiana (Regel) Schneid. (= *P. deltoides* × *P. laurifolia*). Very similar to *P. berolinensis* (perhaps only a geographical mutation), shoots and petiole pubescent and angular; leaves ovate, long acuminate, about 8–12 cm long, base rounded to slightly cordate, sharply and finely dentate. SH 1: 4 t–v. Before 1882.

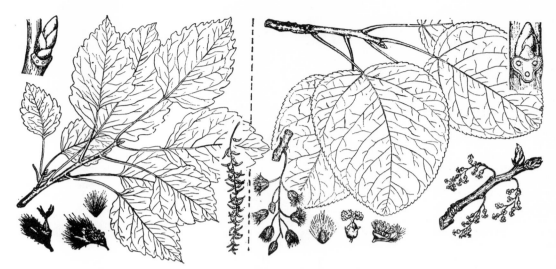

Fig. 316. *Populus grandidentata* (left) and *P. heterophylla* (right) (from Illick)

P. picardii see: **P. × canescens 'Macrophylla'**

P. plantierensis see: **P. nigra** var. **betulifolia 'Plantierensis'**

P. pruinosa Schrenk. Bark of older trunks furrowed like *Fraxinus* and *Ulmus;* closely related to *P. euphratica,* but the leaves usually larger, kidney-shaped to ovate-triangular, rounded at the apex and on the sides, normally entire or with a few small teeth at the apex, blue-green on both sides (= *P. glaucicomans* Dode). Turkestan to SW. Siberia. 1911. z9 Plate 171. ∅

P. przewalski see: **P. simonii**

P. pseudograndidentata Dode. Very closely related to *P. grandidentata,* but with pendulous twigs, tomentose at first; leaves very similar to *P. tremula,* but much tougher and 7–10 cm wide with a cartilaginous margin; stamens only 5, disk glabrous. BC 3128; EH 409; BB 1169 (= *P. tremula* var. *pseudograndidentata* Aschers. & Graebn.; *P. grandidentata pendula* Hort.). Origin unknown. Before 1838. z6

P. purdomii Rehd. Tall tree, very closely related to *P. cathayana;* leaves ovate to narrow ovate, 10–13 cm long, to 25 cm long on long shoots and then more oval-oblong, base round to somewhat cordate, more coarsely glandular serrate, pubescent beneath on the distinctly raised venation (even on the 2nd and 3rd order). NW. China. 1914. z5

P. pyramidalis see: **P. nigra 'Italica'**

P. × rasumowskiana (Regel) Dipp. (= *P. ? laurifolia* × *P. nigra*). Similar to *P. berolinensis,* but with twigs more angular and glabrous, buds acuminate, glutinous; leaves on the stout shoots oval-rounded, to 14 cm long, rounded to slightly cordate at the base, apex drawn out in a long acuminate, glandular dentate apex, nearly totally glabrous, margins glandular crenate, not ciliate,

petiole cylindrical, furrowed above. DL 2: 49. Before 1882. Weak grower and of only slight ornamental value.

P. robusta see: **P. × canadensis 'Robusta'**

P. × rogalinensis Wroblewski (*P. alba* × *P. tremula*). Actually only a form of the polymorphic hybrids of *P. canescens* Smith, but differing as an older plant; leaves partly circular, partly ovate, finely to coarsely dentate or with 1–2 small lobes on either side. PDR 9: 155. Discovered in Poland before 1930.

P. salicifolia see: **P. laurifolia** f. **lindleyana**

P. sargentii Dode. Tree, similar to *P. deltoides,* but smaller, young shoots lighter, yellow, glossy, buds pubescent; leaves often wider than long, with an abruptly tapered, long apex, broad triangular, usually with a few, large teeth, becoming larger toward the base, glabrous, yellow-green. SM 135; VT 89 (= *P. occidentalis* Rydb.) USA, Great Plains, commonly found along rivers. 1908. Ⓕ USA. z4 Plate 174.

P. serotina see: **P. × canadensis 'Aurea'**

P. sieboldii Miq. Small tree, 6–9 m high, to 20 m in its habitat, stoloniferous, twigs rather thick, tomentose until summer; leaves 5–8 cm long, with a short, triangular apex, base round to somewhat cuneate and with 2 glands at the point of attachment of the petiole, margins finely glandular serrate, dark green above, underside more or less white pubescent at first, eventually rather glabrous, petiole pubescent, 1–4 cm long. EH 408; PB 7; MJ 2025. Japan. 1881. z5 Plate 174.

P. simonii Carr. Narrow crowned tree, 12–15 m high, growth slender and regular, trunk with a gray-green bark, current year's shoots angular, glabrous, red-brown, winter buds acute, glutinous, new growth appears early; leaves acute, rhombic-elliptic, 6–12 cm long, crenate, bright green above, whitish to greenish beneath, petiole 1–2 cm long, red; catkins 2–3 cm long. LF 62; BD 1923: 16; PB 13 (= *P. przewalskii* Maxim.; *P. brevifolia* Carr.). N. China. 1862. Ⓕ Austria, Bulgaria, Yugoslavia, N. China. z2 ∅

'Fastigiata'. Growth nearly columnar, shoots thin, less angular, dark brown, ascending at a very acute angle; leaves smaller, obovate, base long, cuneate, petiole shorter. PB 13. Around 1915. Plate 171. ∅

P. suaveolens Fisch. A slow growing, upright tree, to 30 m high in its habitat, current year's shoots cylindrical, yellow-brown, slightly pubescent above the nodes (!), winter buds brown, acute, glutinous; leaves thick and tough, oblong-elliptic, 5–12 cm long, widest at about the middle, abruptly acuminate with a very short, often twisted apex, venation above somewhat indented, whitish and pubescent beneath, petiole 1–4 cm long; fruit catkins to 10 cm long, rather thick. SL 206. E. Siberia, Manchuria, Korea, N. Japan. 1834. Rarely found in cultivation. See also *P. cathayana* and *P. maximowiczii*. Ⓕ NW. China, Mongolia. z3 Plate 174; Fig. 308.

var. *woobstii* see **P. × woobstii**

P. szechuanica Schneid. Tall tree, to 40 m high in its habitat, young twigs angular, reddish, eventually nearly cylindrical and yellow-brown, winter buds purple, glabrous, glutinous; leaves reddish on the new growth (!), ovate-oblong on the long shoots, 7–20 cm long, acuminate, base round to slightly cordate, crenate, glandular dentate, bright green above, silvery-white beneath, red veined, petiole red, 2–4 cm long, broad ovate on the short shoots, petiole 3–7 cm long; fruit capsules nearly sessile, 3–4 valved. BD 1921: 25; PB 15. W. China. 1908. Ⓕ China. z4 ∅

P. tacamahaca see: **P. balsamifera**

P. thevestina see: **P. nigra** var. **thevestina**

P. tomentosa Carr. Similar to *P. alba*, but often growing taller, twigs gray tomentose, buds less tomentose; leaves on the long shoots triangular-ovate, acuminate, double dentate, to 15 cm long (!), dark green above, gray tomentose beneath, not lobed, smaller on short shoots and older trees, more sinuately toothed, underside eventually nearly glabrous. BD 1926: 16 (= *P. pekinensis* L. Henry). N. China. 1867. Ⓕ N. China. z4 ∅

P. tremula L. Quaking Aspen. Tree, to 30 m high, stoloniferous, round crowned, bark smooth, yellowish-gray, later channeled and black-gray, twigs totally glabrous, winter buds glutinous; leaves oval to nearly circular, undulate, crenate, 3–8 cm long and wide, new growth tomentose, but soon becoming totally glabrous, bluish beneath, petiole compressed, glabrous, as long at the blade, leaves of the root sprouts ovate, to 15 cm long, pubescent beneath; catkins 8–10 cm long. HM Pl. 83; HW 131; HF 945. Europe, N. Africa, Asia Minor, Siberia. Ⓕ W. Germany, Austria, Czechoslovakia, Poland, Hungary, Sweden, Norway. z2 Plate 174; Fig. 308.

var. **davidiana** (Dode) Schneid. Leaves smaller, often rounded at the apex and with a small, prominent tip, margins less deeply

serrate. LF 63; NK 18: 47–48 (= *P. davidiana* Dode). NE. Asia, China.

'Erecta'. Very tight and narrow upright habit, otherwise like the species. Discovered in 1911 in Sweden, but cultivated there since 1847. Seldom planted outside Sweden.

'Gigas' (Nilsson). Triploid, shoots thick; leaves much larger and tougher; only male. 1935. Supposedly grows twice as vigorously as the species. Propagated only by grafting onto hybrid poplars.

'Pendula'. Twigs very pendulous; catkins an attractive purple and appearing very early; male. First observed in England around 1787.

var. *pseudograndidentata* see: **P. pseudograndidentata**

'Purpurea'. Leaves turning reddish, but not very conspicuous; female. England.

'Tapiau'. Fast growing form, one season's growth to 2 m. Originating from East Prussia.

var. **villosa** (Lang) Wesm. Young shoots with pubescence persisting into the second year; leaves also remaining pubescent longer than the species (= *P. villosa* Lang).

P. tremuloides Michx. Tree, to 30 m high in its habitat, but rarely taller than 15 m in cultivation, trunk slender, lighter than that of *P. tremula*, young shoots glabrous, eventually red-brown, winter buds somewhat glutinous, stoloniferous; leaves oval-rounded (!), short acuminate, 3–7 cm long, finely serrate (!), truncate to broad cuneate at the base, blue-green and glabrous beneath, fall color pale yellow, petiole 3–9 cm long; catkins 5–8 cm long, more slender than those of *P. tremula*. BB 1170; SPa 103–105; SS 487; PB 6. N. America, Canada to Mexico. 1779. Ⓕ USA. z1 Plate 171, 174; Fig. 308.

var. **aurea** (Tidestr.) Daniels. Small tree, bark light gray; leaves rounded kidney-shaped, base slightly cordate, about 3.5 cm wide, entire or with the margins slightly undulate or also sparsely and irregularly serrate, dull green above, lighter beneath, fall color bright orange (!) (= *P. cercidifolia* Britt.). Rocky Mts. ∅

'Pendula'. Twigs longer, much more ornamental than the pendulous form of *P. tremula*; female. Discovered in 1865 in Troyes, France and distributed as "Parasol de St. Julien". Not as attractive in flower as *P. tremula* 'Pendula'.

var. **vancouveriana** (Trel.) Sarg. Young shoots finely soft pubescent; leaves tomentose on the new growth, later becoming glabrous, coarsely crenate. SM 122 (= *P. vancouveriana* Trel.). British Columbia to Oregon. 1922. z5

P. trichocarpa Torr. & Gray. Tree, to 30 m high, crown ascending broadly, open, young twigs slightly angular, olive-brown, glabrous to pubescent, buds elongated, glabrous; leaves ovate to rhombic-oblong, widest beneath the middle, 8–12 cm long, to 25 cm long on particularly strong shoots, acuminate, base truncate to rounded, finely crenate, leathery tough, dark green above and glabrous to very finely downy, whitish to brownish beneath, with reticulate venation, petiole 3–6 cm long; male catkins 3.5 to 6 cm long, female 6–8 cm long; ovaries tomentose, capsule pubescent, 3 valved, SPa 108–109; SS 493; MD 1927: 44; PB 18–19. Pacific North America. 1892. Ⓕ USA, Yugoslavia, Great Britain, W. Germany, Holland. z5 Fig. 308. ∅

var. **hastata** (Dode) Henry. Leaves narrower, usually oblong-ovate, long acuminate, base round, glabrous or nearly so; ovaries less pubescent, fruit capsule often totally glabrous (= *P. hastata* Dode). USA; N. California and northward. 1892.

P. tristis Fisch. Tree-like shrub, divaricate habit, twigs cylindrical, dark red-brown, pubescent, winter buds 3 cm long, glutinous; leaves oblong-ovate, 7–12 cm long, acuminate, crenate and ciliate, base round to slightly cordate, tough, blackish-green above, whitish and finely pubescent beneath. EH 410; PB 20. Central Asia. Before 1831. The dead leaves persist on the branches well into winter (!). Ⓕ USSR. z4 ∅

P. vancouveriana see: **P. tremuloides** var. **vancouveriana**

P. vernirubens see: **P. × canadensis 'Robusta'**

P. villosa see: **P. tremula** var. **villosa**

P. violascens Dode. Similar to *P. lasiocarpa,* but with somewhat thinner shoots, buds smaller; leaves smaller, oval-elliptic, ovate on the lateral shoots, 10–22 cm long, new growth an attractive violet-red (!), later dull green with red venation, long acuminate, cordate, venation pubescent beneath, petiole violet. BD 1921: 24. China. 1921. z6 ∅

P. wilsonii Schneid. Tree, to 25 m high, crown regularly conical, current year's shoots thick, stiff, reddish at first, soon becoming green or gray-brown and glabrous, cylindrical, buds large, glossy, somewhat glutinous, glabrous; leaves broad cordate, 8–18 cm long, crenate, truncate at the apex, base cordate, dull green above to more blue-green, gray-green beneath, both sides quickly becoming glabrous; fruit catkins to 15 cm long, pubescent at first, the capsules later nearly totally glabrous. Gw 31: 684; BD 1921: 25. Central and W. China. 1907. It is possible that all plants of this species in cultivation are of a single clone since they are uncommonly uniform in habit and all are female. Ⓕ SW. China. z5 Plate 173. ∅

P. wislizenii see: **P. fremontii** var. **wislizenii**

P. × woobstii (Regel) Dode (= *P. laurifolia × P. tristis*?). Tree, very similar to *P. laurifolia,* but the current year's shoots are slightly ribbed, glabrous; leaves lanceolate, widest in the center. DH 2: 101 (= *P. suaveolens* var.

woobstii Regel). Origin unknown.

P. yunnanensis Dode. Tree with angular, glabrous shoots, new growth very reddish, eventually brown, winter buds glabrous, glutinous; leaves reddish on new growth, acute-ovate on long shoots, 6–15 cm long, crenate, base broad cuneate, bright green above, whitish beneath, with a red midrib and petiole, leaves of the fruiting branches more ovate, to 15 cm long, base more cordate (PB 14); fruit catkins 10–15 cm long, capsules 3–4 valved. BD 1922: 81; MD 1938: 15. SW. China, Yunnan. 1905. Ⓕ China; Southern France. z5 Plate 173. ∅

All poplars prefer a fertile, moist soil and are fast growing. The white Poplars are, however, not particular and thrive even on dry soil.

The International Poplar Commission at Via delle Terme di Caracalla, 00100 Rome, Italy, keeps a register of all new poplar cultivars (primarily for forestry use). All new hybrids and cultivars should be registered with the Commission before their introduction.

Lit. Hesmer, H.: Das Pappelbuch; Bonn 1951 (304 pp., 184 fig. (the bibliography contains 291 titles) ● Houtzagers, G.; Handboek voor de Populieren-Teelt; Wageningen 1941 (231 pp., ill.) ● Schneider, C.: Die bisher bekannten Pappel-Bastarde; in Mitt. DDG 1932, 25–30 ● Cansdale, L. S.: The Black Poplars and their Hybrids, cultivated in Britain; Oxford 1938 ● Dode, L.-A.: Extraits d'une monographie inédite du genre *Populus;* in Bull. Soc. Hist. Nat. d'Autun 18, 161–231, 1905 ● Wesmael, A.: Monographie des Peupliers; in Mém. Soc. Sci. Hainaut, Sér. III, **3**, 183–253, 1869 (pl. 1–23) ● Houtzagers, G.: 2e beschrijvende Rassenlijst voor Populieren, Wilgen en Iepen; I.V.T. Wageningen 1947 (132 pp.) ● Boom, B. K.: *Populus canadensis* Moench versus *P. euramericana* Guinier; in Act. Bot. Neerl. **6**, 54–59, 1957 ● Hamaya, T., & I. Inokuma: Native Species of *Populus* in Japan; Tokyo 1957 (17 pp., 9 plates) ● Bugala, W.: The North American Poplars of the *Aigeiros* Section and their influence upon the cultivation of the Poplar in Europe; in Rocznik Dendrologiczny **11**, 225–261, 1956 (Polish with a summary in English) ● Müller, R.: Wirtschaftspappelsorten; Erfahrungen und Ergebnisse der Identifizierung von Nutzpappeln; in Holz-Zentr. Bl. Nr. 142, 1954, Stuttgart ● Müller, R.: Altstammsorten der Schwarzpappelbastarde für den Anbau in Deutschland; 1. Teil. Merkmale der Schwarzpappelbastarde; in Holz-Zentr. Bl. 1957, Stuttgart ● Rohmeder, E.: *Populus* 'Bachelieri' und *P.* 'Vernirubens' gibt es nicht! in Dtsch. Baumschl. 1965, 190 to 195.

POTENTILLA L. — Cinquefoil — ROSACEAE

Low, deciduous perennials or shrubs, leaves alternate, 3 to many parted, palmate or pinnate, uneven, stipules adnate to the stem clasping leaf blade base; flowers solitary or in few flowered, terminal panicles or cymes, yellow or white (also red), hermaphroditic, occasionally also dioecious; petals 5, sepals 5, stamens 20–30, ovaries numerous, with one ovule; fruits numerous, nut-like, small, hard, falling individually. — About 500 species in the colder and temperate zones; of these, only 2 are included here.

Potentilla arbuscula var. *unifoliolata* see: **P. fruticosa** var.

unifoliolata

P. davurica see: **P. fruticosa** var. **davurica**

P. farreri prostrata see: **P. fruticosa 'Pyrenaica'**

P. friedrichsenii leucantha see: **P. fruticosa 'Beanii'**

P. friedrichsenii ochroleuca see: **P. fruticosa 'Ochreleuca'**

P. fruticosa L. Deciduous shrub, erect, to 1.5 m high and wide, very densely branched; leaves normally trifoliate, leaflets sessile, elliptic to linear-oblong, 1 to 3 cm long, margins involuted, entire, more or less silky pubescent;

flowers pure yellow, 2–3 cm wide, solitary or several together, May to August. SH 1: 313 (= *Dasiphora fruticosa* Raf.). Northern Hemisphere. In cultivation for centuries with many cultivars. z2 Fig. 318.

Positive identification of the cultivars is best done with spring blooms since the important characteristics are often somewhat altered on the autumn flowers.

Outline of the more prominent P. fruticosa cultivars
(from F. Schneider, expanded)

Leaf Color	Flower Color	Habit			
		tall, erect	medium height	wide	wide, flat
green	white	Snowflake Mount Everest Veitchii	Sandved		
	light yellow	Purdomii Friedrichsenii Goldfinger			
	yellow	*fruticosa* Jackman Friesengold	Klondike Farreri Hachmanns Gigant Sommerflor	Goldstar	Donard Gold Goldteppich
	orange			Tangerine Daydawn Sunset	
	coppery			Red Ace	
blue-green	white light yellow yellow	Maanelys	Abbottswood	Katherine Dykes	*mandshurica* Longacre Elizabeth (Sutters Gold)
gray	white yellow	Vilmoriniana		Primrose Beauty	

'Abbottswood' (Tustin). Half as high, broad growing; leaves blue-green; flowers large, pure white. The best white flowering form to date. ⊕

var. **albicans** Rehd. & Wils. Leaves 5 parted, leaflets elliptic-oblong, dull gray-green and pubescent above, white tomentose beneath; flowers medium yellow, 2 cm wide. China; W. Szechwan Province, high in the mountains. 1909. Resembling 'Vilmoriniana' which has, however, cream-white flowers.

'Arbuscula' see: **'Elizabeth'**

var. **arbuscula** D. Don. Only about 30 cm high, but to 1 m wide, twigs arching; leaflets usually 5, about 15 mm long, blue-green, white pubescent beneath; flowers golden-yellow, solitary, to 4 cm wide, calyx wide. Himalayas. This wild form should not be confused with the 'Arbuscula' of cultivation, which should more correctly be called 'Elizabeth'!

'Beanii'. Strong growing, erect, like 'Friedrichsenii', but with white flowers, 25 mm wide (= *P. friedrichsenii leucantha* Späth). Developed by Späth before 1910, but the name was changed in England.

'Beesii' (Bees). Growth wider than high; leaves gray and silky pubescent on both sides; flowers in conical inflorescences, corolla 2–2.5 cm wide, yellow (= 'Nana Argentea' in England). Developed in 1925 in England from seed brought from Tibet by Kingdon Ward. Fig. 317.

'Berlin Beauty' see: **'Friedrichsenii'** (name changed in the USA)

'Buttercup'. Lower habit; with 7 leaflets; flowers dark yellow, 2.7 cm wide, long flowering. Probably a seedling of var. *parvifolia*.

'Coronation Triumph'. Medium high; leaflets 5–7; flowers medium yellow, 2.5 cm wide, long flowering period. Introduced in 1950 by J. Walker, Indian Head Forestry Station, Indian Head, Saskatchewan, Canada.

var. **davurica** (Nestler) Ser. Shrub, scarcely over 50 cm high; leaflets 5, elliptic, about 2 cm long, dark green above, bluish beneath; flowers usually solitary, about 2–3 cm wide, white to slightly yellowish; diploid. Bai 2: 27; BM 3676 (= *P. davurica* Nestler; *P. glabrata* Willd.; *P. glabra* Lodd.). China to Siberia. 1822. Fig. 318.

'Daydawn'. Flowers salmon-pink. Mutation of 'Tangerine'.

'Donard Gold' (Donard). Quite low, creeping, to 50 cm high at most, but much wider; foliage green; flowers large, golden-yellow, but presumably sometimes orange. Slieve Donard Nursery, N. Ireland. 1956.

'Elizabeth'. Broadly hemispherical habit, about 1 × 1 m; flowers golden-yellow, 3.5 cm wide, flowering from early summer to fall. Before 1965. This is the correct name for the

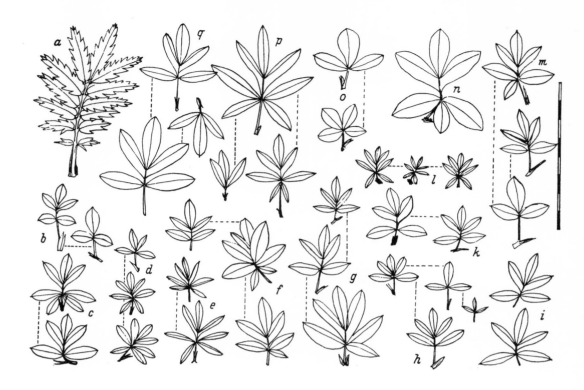

Fig. 317. **Potentilla.** Leaves. a. *P. salessowiana*; b.–q. forms of *P. fruticosa*: b. *'Beesii'*; c. *'Elizabeth'*; d. *'Farreri'*; e. *'Klondike'*; f. *'Friedrichsenii'*; g. *'Mount Everest'*; h. *'Walton Park'*; i. *'Maanely's*; k. *'Lady Daresborough'*; l. *'Parvifolia'*; m. *'Veitchii'*; n. *'Vilmoriniana'*; o. f. *rigida*; p. *'Jackman'*; q. *'Grandiflora'* (Original)

often incorrectly listed 'Arbuscula'. Developed by Hillier as a cross between var. *arbuscula* and var. *mandshurica*. (Known in the USA as 'Sutters Gold'.) Fig. 317. ⊕

'Farreri'. About 60 cm high, to 90 cm wide, finely branched; leaves very small, leaflets 7, only 5–8 mm long; flowers bright golden-yellow, 2–3 cm wide, dorsal side similarly dark yellow (!). GC 120: 109 (= f. *farreri* Besant). Collected by R. Farrer in 1920 in Tibet at 2400 m. Surpassed today by 'Klondike' with larger, darker flowers; grown in the USA as 'Gold Drop'. Plate 176; Fig. 317.

"farreri prostrata" see: **'Pyrenaica'**

"Farrer's Red Form". Farrer sent seed from China under the name *Potentilla fruticosa* Red Form. When germinated in England, only yellow flowers were produced at first. The best color was by the Slieve Donard Nursery and distributed as 'Donard Gold'. Eventually a red spotted flower was obtained and introduced into the trade as 'Tangerine', which see.

'Farrer's White'. To 75 cm high and wide; foliage yellowish-green, otherwise like 'Farreri', but with white flowers. Introduced by Jackman in 1955.

'Francis Lady Daresbury' see: **'Lady Daresbury'**

'Friedrichsenii' (Späth). Upright shrub, to about 1.5 m high and 1.8 m wide; leaves rather large, leaflets 5–7, to 3 cm long, light green, whitish beneath; flowers light yellow, whitish on the dorsal side, 2.5–3 cm wide (= 'Berlin Beauty'). Introduced by Späth, Berlin, W. Germany in 1895. Plate 175; Fig. 317.

'Friesengold' (Herm. A. Hesse, before 1950). Bushy upright habit, about 1 m high; leaves 5, to 4 cm long, elliptic-oblong, dark green above, gray-green beneath; flowers golden-yellow, 2.5 cm wide, sepals acutely triangular, very long.

'Gold Drop' see: **'Farreri'**

'Goldfinger'. Erect habit, 80–100 cm high; foliage green; flowers bright yellow, 3.5–4 cm wide, flowers very abundantly and over a long period. The best of the bright yellow, tall growing types to date. Developed by H. Knol, Gorssel, Holland. ⊕

'Goldkissen' (J. J. Sörensen, Marzahne, E. Germany). Low, tight globose form; flowers medium-sized, yellow. Not generally cultivated outside of E. Germany.

'Goldstar' ('Goldteppich' × 'Hachmanns Gigant'). Semi-erect habit, to 80 cm, spreading; flowers golden-yellow, 4–5 cm wide, long flowering period. J. Hachmann, Barmstedt, W. Germany. 1976.

'Goldteppich' ('Elizabeth' × 'Jackman'). Flatter and lower growing than 'Elizabeth'; leaves green; flowers also somewhat smaller, 3–4 cm wide, golden-yellow, flowers very abundantly,

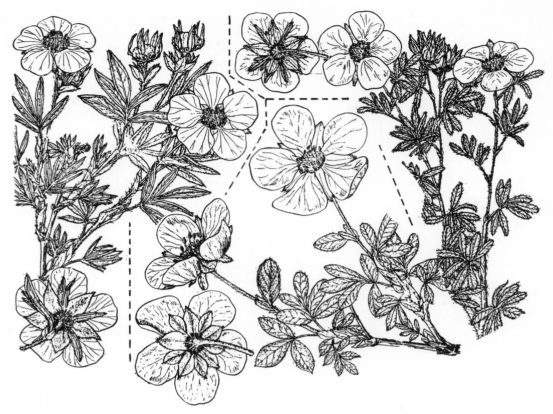

Fig. 318. **Potentilla.** Left, *P. fruticosa*; center *P. fruticosa* var. *davurica*; right, *P. fruticosa* 'Parvifolia'. Each twig with flowers and a single bloom as seen from below (from Rhodes)

particularly in June and July (like 'Jackman'), then only with scattered flowers (like 'Elizabeth'). J. Hachmann 1969. ✧

'Grandiflora'. Very strong growing, presumably to 1.8 m high; leaves to 5 cm long, leaflets 5–7, to 1 cm wide, blue-green beneath and pubescent, green above; flowers pure yellow, about 3 cm wide, but flowers sparsely. A very old cultivar, originating from England around 1929. Fig. 317.

'Hachmanns Gigant' ('Elizabeth' × 'Klondike'). Like 'Elizabeth', leaves green; flowers deep golden-yellow, larger, 4–5 cm wide, particularly large at the beginning of its flowering period, also flowers abundantly. J. Hachmann 1967. ✧

f. *hersii* see: **'Snowflake'**

f. *humilis* see: **'Parvifolia'**

'Jackman' (Jackman). Erect habit, coarse, 1–1.2 m high, shoots very pubescent; leaves deep green, to 6 cm long, leaflets usually 7, lanceolate to elliptic, to 1 cm wide, blue-green and pubescent beneath; flowers numerous, in clusters, corolla golden-yellow, 3.5–4 cm wide, the dry calyces persistent. Developed from seed of 'Grandiflora' by Jackman in 1940. Fig. 317.

'Katherine Dykes' (Dykes). To 1.5 m high, wider than tall, twigs nodding; leaves about 3 cm long, with 5 leaflets, gray-green, rather densely pubescent beneath; flowers light yellow, about 3 cm wide, flowers well. JRHS 71: 39. Developed about 1925 in England by M. W. R. Dykes. ✧

'Klondyke' (Kruyt). Medium high, erect, habit and foliage similar to that of 'Farreri', but leaflets somewhat larger, 10–18 mm long; flowers darker yellow, 3.5–4 cm wide, petals somewhat reflexed. Dfl 4:65. Developed in Holland about 1950. Has more garden merit than 'Farreri'. Fig. 317. ✧

'Lady Daresbury'. Seedling of 'Purdomii', but somewhat lower in habit, to 90 cm, broad bushy, shoots outspread; leaves distinctly blue-green, with 5 leaflets, more blue-gray beneath with densely pubescent venation; flowers medium yellow, 3.5 cm wide, long flowering. Developed in England before 1955 and known at first as 'Francis Lady Daresbury'. Fig. 317.

'Longacre' (Slieve Donard). Low, broad cushion form habit; foliage blue-green, leaflets 5; flowers large, 3 cm wide, medium yellow, flowers abundantly. Introduced in 1956 by the Slieve Donard Nursery. Presumably a selection of var. *arbuscula* and differing from 'Elizabeth' in the more erect habit and smaller flowers. ✧

'Maanelys' (Olsen). Shrub, to 1.2 m high, erect; leaves blue-green, about 4 cm long, with 5 leaflets; flowers medium yellow (somewhat darker than 'Katherine Dykes'), 2.5–3 cm wide, petals light yellow on the dorsal side, flowers abundantly over a long period. Developed about 1950 by Axel Olsen in Kolding, Denmark. The best erect form to date, better than 'Jackman', since the calyces remain green long after the flower drops rather than turning brown as does 'Jackman'. Fig. 317. ✧

var. **mandshurica** Maxim. Shrub, low, only 30–40 cm high,

shoots reddish; leaves densely gray silky pubescent on both sides; flowers pure white, 2.5 cm wide, but few in number. Manchuria. 1911. Easily damaged by red spider mites.

'Micrandra'. Grows to scarcely 50 cm high, but twice as wide; leaves dull green, to 3.5 cm long, leaflets usually 7, narrow oblong, blue-green beneath and pubescent; flowers bright yellow, 2.5–3 cm wide, stamens very short. SH 1: 313 i–l (= *P. micrandra* Koehne). Developed about 1890.

'Moonlight'. US name for 'Maanelys', which see.

'Mount Everest' (Kruyt). Strong upright grower, about 1 m high, open; leaves unattractively yellowish-green, leaflets 5, narrow; inflorescences conical, corolla 3–3.5 cm wide, white, sepals abscising very early (!). Developed about 1950 in Holland, but surpassed in habit and foliage by 'Snowflake'. Fig. 317.

'Nana Argentea' see: 'Beesii'

'Ochroleuca' (Späth). Erect, to about 1.2 m high and 1.8 m wide, shoots outspread; leaves 3 cm long, leaflets 5–7, oblong-elliptic, about 5 mm wide, green above, bluish beneath; flowers light yellow, 3 cm wide, petals whitish on the dorsal side, sepals ovate (= *P. friedrichsenii ochroleuca* Späth). Before 1902.

'Parvifolia'. Slow growing, somewhat conical form, to 80 cm high; leaflets usually 7, very small, green above, blue-green beneath, pubescent; flowers about 2 cm wide, somewhat lighter yellow than 'Farreri', but whitish-yellow (!!) on the dorsal side of the petals, styles clavate formed (not filamentous!). Bai 2: 28 (= f. *humilis* Regel). Before 1924. Plate 175; Fig. 317, 318.

'Primrose Beauty' (Cannegieter). An attractive, elegant, broad growing shrub, shoots more or less nodding; leaflets to 15 mm long, gray pubescent; flowers dark cream-yellow with a somewhat darker center. Holland. ⊕

'Purdomii'. Erect habit, to about 1 m high, shoots pubescent; leaves about 3 cm long, normally with 7 leaflets, these oblong-elliptic, light green above, light gray-green beneath; flowers light yellow, whitish on the dorsal side, 2–2.5 cm wide, long flowering period. Collected by Purdom in S. China in 1911.

'Pyrenaica'. Procumbent shrub, scarcely over 20 cm high; leaves about 2 cm long, leaflets usually in 3's, oblong-elliptic, green above, bluish-green beneath; flowers golden-yellow, about 2.5 cm wide, petals dark yellow on the dorsal side (!), long flowering period (= *P. prostrata* Lap.; *P. farreri prostrata* Hort.). 1809. A clone originating in the Pyrenees Mts. and therefore not a botanical variety as in var. *pyrenaica* Willd. ●

'Red Ace' (Barker 1973). Grows about 50–60 cm high and 1 m wide; flowers light red on the interior, exterior yellow. JRHS 1976: 253. Originated as a chance seedling at Hopleys Plants, Much Hadham, Herts., England. Selected by Dr. Barker and introduced to the trade in 1976 by A. Bloom of Bressingham Gardens, Diss, England. The first true red cultivar. ⊕

'Rhodocalyx'. Low, broad growing shrub; leaflets 5; flowers white, 2.5 cm wide, cup form, calyx conspicuously reddish.

f. **rigida** (Lehm.) Wolf. About 60 cm high, but becoming wider, shoots short and thick, greenish pubescent; leaves 3 cm long, with 3 leaflets, elliptic, about 2.5 cm long, 1 cm wide, green above, blue-green beneath; flowers golden-yellow, 3 cm wide, petals yellow on the underside, stamens more orange, few and late flowering. Lehmann, *Potentilla*, pl. 1. Himalayas. 1906. Plate 176; Fig. 317.

'Ruth'. A selection of 'Rhodocalyx' with cream-white flowers and a more reddish calyx.

'Sandved'. Shrub, 1 m high, bushy, loosely branched; foliage somewhat yellowish-green; flowers cream-white, 3.5 cm wide. Introduced by Bloch Sandved, Sandnes, Norway; distributed by Rud. Schmidt, Rellingen, Norway, 1958.

'Snowflake'. Medium size; foliage large, deep green, leaflets 5; flowers white, single to double (!), flowering period not continuous, dry sepals persistent; triploid (= f. *hersii* Hort.). Quite meritorious.

'Sommerflor' ('Goldfinger' × 'Goldteppich'). Compact and upright habit, like 'Klondike'; flowers golden-yellow, 3.5 cm wide, flowers abundantly over a long period, therefore superior to 'Klondike', 'Farreri' and others. J. Hachmann 1976. ⊕

'Sommerfreude' (J. J. Sörensen, Marzahne, E. Germany). Medium height, evenly bushy and tight; flowers large, golden-yellow. Virtually unknown outside E. Germany.

'Sonnenglut' (J. J. Sörensen). Medium height, loose habit; flowers large, yellow. Not cultivated outside E. Germany.

'Sternwiese' (J. J. Sörensen). Low, bushy habit; flowers light yellow. Not cultivated outside E. Germany.

'Sunset'. Low growing; flowers quite variable between orange-yellow and brick-red. Mutation of 'Tangerine'. Surpassed by 'Red Ace'.

'Sutter's Gold' see: 'Elizabeth'

'Tangerine' (Slieve Donard). Low growing, loose; foliage small, green, leaflets 7; flowers about 3 cm wide, coppery-yellow to golden-yellow, orange only in cooler weather. Developed from "Farrer's Red Form", 1955. Surpassed by 'Red Ace'.

var. **tenuiloba** Ser. Very low habit (to only 45 cm high and 90 cm wide in 40 years at the Arnold Arboretum); leaves very small, linear, leaflets 5; flowers golden-yellow, 2.5 cm wide. Western N. America.

var. **unifoliolata** Ludl. Shrub, erect, 1 m high, densely foliate; leaves simple (one leaflet!), occasionally with 2 leaflets, 7.5–15 mm long; flowers solitary, 2–3 cm wide, golden-yellow. Bull. Brit. Mus. (Nat. Hist.) Bot. 2: 67, Pl. 1 (as *P. arbuscula* var. *unifoliolata* Ludl.). Bhutan. 1937.

'Veitchii'. Broad to upright growing, to 1.5 m high, dense, twigs nodding; leaves 2.5 cm long, leaflets usually 5, about 1 cm long, light green; flowers white, 2.5 cm wide, stamens reddish. BM 8637; BS 2: 529 (= *P. veitchii* Wils.). Central and W. China, in the mountains at 1800 m. 1902. Fig. 317.

'Vilmoriniana'. Shrub, strong growing, about 0.9 to 1.2 m high and to 1.5 m wide; leaves gray pubescent above, white tomentose beneath; flowers cream-white to pale yellow. RH 1910: 57. W. China. 1905. Somewhat difficult to cultivate. Fig. 317. ∅

'Walton Park'. Shrub, broad growing, about 0.6 m high, rounded; leaves deep green, to 4 cm long, blue-green beneath; flowers golden-yellow, grouped 1–5 at the ends of short lateral shoots, corolla 3.5–4 cm wide, long flowering period (= 'Waltoniensis'). England. Before 1955. Very susceptible to powdery mildew! Plate 175. Fig. 317. ⊕

'Waltoniensis' see: 'Walton Park'

P. glabra see: **P. fruticosa** var. **davurica**

P. glabrata see: **P. fruticosa** var. **davurica**

P. micrandra see: **P. fruticosa 'Micrandra'**

P. prostrata see: **P. fruticosa 'Pyrenaica'**

P. salessowiana Steph. Shrub, narrow upright habit, 0.5–1 m high, sparsely branched; leaflets 7 to 13, sessile, the terminal leaflet short stalked, narrow oblong, 2–4 cm long, coarsely serrate, dark green above, white tomentose beneath; flowers grouped 3–7 in cymes, white with a reddish trace, 3 cm wide, of little garden merit, June–August. BM 7258. Turkestan, SE. Siberia. N. China, Himalayas. 1823. z4 Fig. 317.

P. veitchii see: **P. fruticosa 'veitchii'**

All species prefer full sun and a well drained soil; they are long lived, requiring little care, thriving for 50 years in some cases; very deeply rooted and therefore tolerant of dry periods.

Bowden's cytotaxonomic research suggests that *P. fruticosa* is a collective species. The forms from northern Europe are tetraploid (2n = 28), the North American forms diploid (2n = 14) as well as 'Parvifolia' and var. *davurica* from E. Asia. However, var. *arbuscula*, also from E. Asia is a polyploid complex composed of diploids, a triploid (2n = 21), hexaploid (2n = 42) and an octaploid (2n = 56).

Lit. Lehmann, J. G. C.: Revisio Potentillarum iconibus illustrata; Bonn 1856 (only pp. 1–23 with plates 1–5) ● Wold, T.: Monographie der Gattung *Potentilla*; in Bibl. Bot. **16**, 1–714, 1908 ● Rhodes, H. L. J.: The cultivated shrubby Potentillas; in Baileya **2**, 89–96, 1954 ● Wyman, D.: *Potentilla fruticosa*; in Arnoldia 1955, 45–49 ● Bowden, W. M.: Cytotaxonomy of *Potentilla fruticosa*, allied species and cultivars; in Jour. Arnold Arb. 1957, 381–388 ● Wyman, D.: *Potentilla fruticosa* varieties in the Arnold Arboretum; Arnoldia **28**, 125–131, 1968 ● Schneider, F.: *Potentilla fruticosa*; in Dendroflora **4**, 42–50, 1967.

POTERIUM L. — ROSACEAE

Monotypic genus. Deciduous thorny shrub, leaves pinnate; flowers monoecious, unisexual, inconspicuous, in spikes, the basal flowers male, the uppermost female; sepals 4, petals absent, stamens numerous, ovaries 2, stigmas brush-like; fruit a fleshy nutlet.

Poterium spinosum L. Shrub, 30–60 cm high, erect, twigs divaricate and very thorny, thorns forked; leaves alternate, odd pinnate, 3–5 cm long, leaflets 9–15, ovate, sharply dentate, 6–8 mm long, densely pubescent beneath; inflorescences 3–5 cm long, greenish, April–August. DL 3: 240 (= *Sarcopoterium spinosum* [L.] Spach). Eastern Mediterranean region, Asia Minor. z8

POURTHIAEA

Pourthiaea villosa see: **Photinia villosa**

PRINSEPIA Oliv. — ROSACEAE

Thorny, deciduous shrubs; branches with a chambered pith; leaves alternate, usually in clusters, simple, entire or serrate; stipules small, persistent; flowers axillary, grouped 1–4 on the previous year's shoots (except on *P. utilis*); petals 5, clawed, sepals 5, ovaries with a laterally attached stigma (!); fruit drupe-like, compressed, a portion of the fruit wall is leathery. — 3–4 species in E. Asia to NW. China and Himalayas.

Prinsepia sinensis (Oliv.) Oliv. Shrub, to 2 m high, twigs light gray-brown, thorns 6–10 mm long, new growth appears very early; leaves oval-lanceolate to obovate-lanceolate, 5–8 cm long, entire to slightly serrate, finely ciliate, glossy green above, lighter beneath; flowers yellow (!), 1.5 cm wide, fragrant, clustered 1–4, April; fruit cherry-like, red, juicy, good flavor, 1.5 cm long. MD 1903: 1; NK 5: 27; BS 2: 531; ICS 2335. Manchuria. 1866. z3 Fig. 319. ✕

P. uniflora Batal. Shrub, to 1.5 m high, twigs light gray, thorns 6–10 mm long; leaves linear-oblong, 3–6 cm long, entire or sparsely finely serrate, dark green above, lighter beneath; flowers white (!), 1.5 cm wide, fragrant, grouped 1–3, April–May; fruits globose, 1–1.5 cm thick,

Fig. 319. *Prinsepia sinensis*. Branch with flower, flower parts, fruit, seed (from Oliver; Komarov)

black, purple pruinose. ICS 2336. N. China, Shensi Province. 1911. z3 Plate 175.

var. **serrata** Rehd. Leaves wider, serrate; internodes larger, flowers more abundantly and fruits well. S. Kansu Province. Supposedly hardier than the species.

P. utilis Royle. Shrub, very thorny, to 3 m high in its habitat, twigs and thorns green, thorns thick, 3–5 cm long, occasionally bearing some leaves (modified short shoots); leaves lanceolate to 10 cm long, serrate, long acuminate, base cuneate; flowers cream-white, fragrant,

in 3–5 cm long axillary racemes (!), or solitary to a few in groups, appearing in fall and on the current year's wood (!), stamens numerous; fruits oblong, purple, 1 cm long, ripening the following year. BMns 194. W. China. z7

Cultivated somewhat like *Prunus;* full sun and a moist soil. Of only slight ornamental merit; more suitable for collectors. The fruits of *P. sinensis* are eaten in E. Asia.

Lit. Woeikoff: *Prinsepia sinenesis* und seine biologischen Eigenshaften; in Mitt. DDG 1932, 46–52.

PRINUS

Prinus glabra see: **Ilex glabra**

PRITCHARDIA

Pritchardia filifera see: **Washingtonia filifera**

PROSOPIS L. — LEGUMINOSAE

Deciduous trees or shrubs, with or without thorns, the thorns axillary, solitary or paired or with only the stipules thorned; leaves bipinnate, with 1 or 2 pair of pinnae, leaflets usually numerous, small, entire; flowers in cylindrical or globose axillary spikes, small, greenish. — About 40 species in the subtropical and tropical regions of the world.

The sweet fruits of many species are a form of livestock feed.

Prosopis juliflora (Sw.) DC. Mesquite. Broad crowned, deciduous tree, 12–15 m high, but often much lower and shrubby, with many, 1.2–5 cm long thorns; leaves normally with 2 pinnae, very rarely with 4, each of these 7–15 cm long and with 8–15 pairs of leaflets, these linear-oblong, 1.5–5 cm long, 3–6 mm wide, terminal leaflet absent; flowers in 5 to 10 cm long, cylindrical spikes, 1.5 cm wide, the individual flowers small, densely clustered, fragrant, golden-yellow (from the anthers!), May–July; the fruits look like green beans, 10–20 cm long, sometimes eaten by the Indians and Mexicans. SM 453; LWT 166. Western N. America, southward to Chile and Argentina. 1827. z10 Fig. 320.

Quite a valuable tree for hot climates with little rainfall, very deeply rooted.

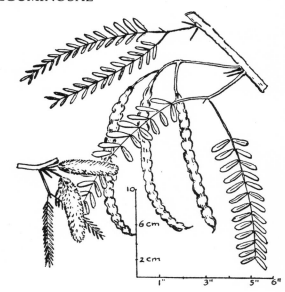

Fig. 320. *Prosopis juliflora.*
Branch, left with flowers, right with fruits

PROSTANTHERA Labill. — LABIATAE

Evergreen shrubs or subshrubs, occasionally small trees, with many, scattered resin glands, strongly scented; leaves opposite; flowers axillary, occasionally developing foliate racemes, or also grouped into panicles; calyx and corolla bilabiate, corolla tube short,

throat broad campanulate, stamens 4, in 2 pairs; upper lip 2 lobed, lower lip 3 lobed. — 50 species in Australia.

Attractive plants, flowering very abundantly.

Prostanthera lasianthos Labill. Medium-sized shrub,

also larger in its habitat, upright growing, shoots rectangular; leaves lanceolate to more oblong, rather large for the genus, 5–10 cm long, 1.2–2.5 cm wide, dark green above, lighter and punctate beneath, margins coarsely dentate; flowers white with a trace of purple, in racemes, these grouped into 10–15 cm long and half as wide, terminal panicles, corolla 2–2.5 cm wide, pubescent, June to July. BM 2434. Australia, Tasmania. 1807. z9 Fig. 321. # ✣

P. melissifolia R. Br. Slender, upright shrub, 3–4 m high in its habitat, short pubescent; leaves ovate, 2.5–5 cm long; flowers in leafless racemes, lilac. Australia; Victoria, along river banks. A good park shrub. z9 Fig. 321.

P. ovalifolia R. Br. Shrub, to 2 m high; leaves small, olive-green, soft, long petioled, flat; flowers purple-red, in large numbers on nodding shoots in spring. AP 5: 188. Australia. z9 # ✣

P. rotundifolia R. Br. Peppermint Shrub. Tall shrub, glandular, aromatic, twigs pubescent; leaves occasionally clustered, nearly circular, 4–7 mm wide, glabrous, entire or slightly crenate, tapering to a small petiole; flowers short stalked, usually in very short axillary racemes or clusters, grouped 5 or more together, purple-blue or lilac, corolla 1 cm wide, twice as long as the 10 ribbed calyx and with a pubescent exterior, April to May. BM 9061. Tasmania. 1824. Gorgeous shrub, when in bloom a cloud of heliotrope-colored flowers. z9 Fig. 321. # ✣

All species well suited and beautiful in the open landscape in milder climates.

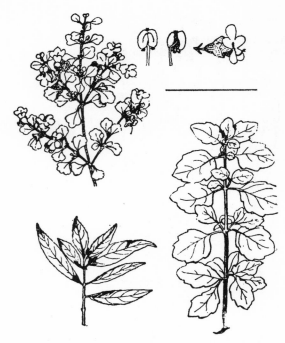

Fig. 321. **Prostanthera.** *P. rotundifolia* above; lower left *P. lasianthos*; right *P. melissifolia* (from Blake, Audas)

PROTEA L. — PROTEACEAE

Evergreen shrubs; leaves alternate or scattered in arrangement, leathery, quite variable in form and size, often silky glossy or blue-green pubescent, sessile to long petioled; flowers hermaphroditic, in dense heads, surrounded at the base by a dense ring of bracts, these usually conspicuously colored; corolla tube long, narrow, so densely arranged that only the outer tips are visible; 2 lobes of the perianth are fused into a lip; ovary pubescent, style straight or bowed; fruit single seeded, small, densely pubescent. — About 130 species in the tropics and S. Africa.

Protea barbigera Meisner. Queen Protea. About 1 m high and wide; leaves elliptic-oblong, blue-green, leathery, to 15 cm long and 2.5 cm wide, margins ciliate, midrib yellow, raised; flowers 15(20) cm wide, globose (resembling giant powder puffs), usually soft pink, but also all tones between white and dark red depending on the variety, densely white pubescent in the center, hair tips black, involucral leaves with white pubescent tufts at the tips. RPr 19 to 25 (various colors). S. Africa, Cape Province. The second largest flowers after *P. cynaroides*. z10 # ✣

P. cynaroides (L.) L. King Protea, Giant Protea. Shrub, to 1 m high and wide, shoots glabrous, reddish; leaves leathery tough, from rounded to oval, 5–12 cm long, 5–7 cm wide, base cuneate, long petioled; flowers 15(30!) cm long and wide, involucral leaves in 12–13 rings, mostly pink, but also all tones from pure white to deep red, silky pubescent, May to June. RPr 3; NF 11: 246. S. Africa, Cape Province. 1774. The national flower of South Africa. z10 # ✣

P. longifolia Andrews. (not to be confused with *P. longiflora*!). Shrub, 1–2 m high, shoots finely pubescent at first; leaves linear, 7–15 cm long, 6–12 mm wide, obtuse, glabrous, gradually tapered to the base; flowers 10–15 cm long, 10 cm wide, obconical, involucral leaves in 9 or 10 rings, the outermost short, ovate, green with black tips, the innermost slender lanceolate, purple to black, February to April. RPr 10. S. Africa, Cape Province. 1798. z10 # ✣

P. neriifolia R. Br. Shrub, to 3 m high in its habitat and nearly as wide; leaves narrow lanceolate, 10–11 cm long, 1.5–2 cm wide, acute to obtuse, margins finely

Fig. 322.
Protea neriifolia (from Palmer and Pitman)

pubescent; flowers extraordinarily numerous, to 15 cm long, 7 cm wide, involucral leaves with a silky shimmer, usually pink, but also seen in all tones from white to deep red, dark red or black bearded at the apex. RPr 12. S. Africa, Cape Province. z10 Fig. 322. # ✧

P. speciosa L. Brown Bearded Protea. Shrub, upright, 1–1.5 m high, stiff; leaves oblong-obovate, linear-oblanceolate, curved inward at the apex, stiff, 7–14 cm long, 2.5–4 cm wide, margins undulate, ciliate, pubescent at the base; flowers to 12 cm long and wide, involucral leaves in 9–11 rings, yellowish-white, the outermost silky pubescent, shorter than the inner ones, apexes brown bearded, April. RPr 14. S. Africa, Cape Province. 1786. z10 # ✧

Nearly all species require a well drained, acid soil with sufficient moisture; tolerate light frost occasionally without harm in its habitat. Grown in frost free regions as a cut flower crop.

Lit. Rousseau, F.: The Protaceae of South Africa; Cape Town 1970 (a very beautiful picture book with many close-ups, all in color) ● Eliovson, S.: Proteas for Pleasure; 2nd edition, Cape Town 1967 (many plates in black and white and color).